HANDBOOK OF OPTICS

Volume III
Classical Optics, Vision Optics, X-Ray Optics

Second Edition

**Sponsored by the
OPTICAL SOCIETY OF AMERICA**

Michael Bass   Editor in Chief

*School of Optics / The Center for Research and Education in Optics and Lasers (CREOL),
University of Central Florida
Orlando, Florida*

Jay M. Enoch   Associate Editor

*School of Optometry, University of California at Berkeley
Berkeley, California
and
Department of Ophthalmology
University of California at San Francisco
San Francisco, California*

Eric W. Van Stryland   Associate Editor

*School of Optics / The Center for Research and Education in Optics and Lasers (CREOL),
University of Central Florida
Orlando, Florida*

William L. Wolfe   Associate Editor

*Optical Sciences Center, University of Arizona
Tucson, Arizona*

McGRAW-HILL

New York   San Francisco   Washington, D.C.   Auckland   Bogotá
Caracas   Lisbon   London   Madrid   Mexico City   Milan
Montreal   New Delhi   San Juan   Singapore
Sydney   Tokyo   Toronto

McGraw-Hill

A Division of The **McGraw·Hill** *Companies*

Copyright © 2001 by The McGraw-Hill Companies, Inc. All rights reserved. Printed in the United States of America. Except as permitted under the United States Copyright Act of 1976, no part of this publication may be reproduced or distributed in any form or by any means, or stored in a data base or retrieval system, without the prior written permission of the publisher.

1 2 3 4 5 6 7 8 9 0   DOC/DOC   0 6 5 4 3 2 1 0

ISBN 0-07-135408-5

The sponsoring editor for this book was Stephen S. Chapman and the production supervisor was Sherri Souffrance. It was set in Times Roman by North Market Street Graphics.

Printed and bound by R. R. Donnelley & Sons Company.

 This book was printed on recycled, acid-free paper containing a minimum of 50% recycled, de-inked fiber.

McGraw-Hill books are available at special quantity discounts to use as premiums and sales promotions, or for use in corporate training programs. For more information, please write to the Director of Special Sales, Professional Publishing, McGraw-Hill, Two Penn Plaza, New York, NY 10121-2298. Or contact your local bookstore.

CONTENTS

Chapter 5. Xerographic Systems   *Howard Stark*   5.1

Chapter 6. Photographic Materials   *John D. Baloga*   6.1

Chapter 7. Radiometry and Photometry: Units and Conversions
James M. Palmer   7.1

Part 2.   Vision Optics

Chapter 8. Update to Part 7 ("Vision") of Volume I of the *Handbook of Optics*
Theodore E. Cohn   8.1

Chapter 13. Optics and Vision of the Aging Eye   *John S. Werner and Brooke E. Schefrin*   13.1

Chapter 14. Radiometry and Photometry Review for Vision Optics   *Yoshi Ohno*   14.1

Chapter 15. Ocular Radiation Hazards   *David H. Sliney*   15.1

Chapter 16. Vision Problems at Computers   *James E. Sheedy*   16.1

Part 3. X-Ray and Neutron Optics

SUBPART 3.1. INTRODUCTION

Chapter 19. An Introduction to X-Ray Optics and Neutron Optics

SUBPART 3.2. REFRACTIVE OPTICS

Chapter 20. Refractive X-Ray Optics *B. Lengeler, C. Schroer, J. Tümmler,*

SUBPART 3.3. DIFFRACTIVE AND INTERFERENCE OPTICS

Chapter 21. Gratings and Monochromators in the VUV and Soft X-Ray Spectral Region *Malcolm R. Howells*

SUBPART 3.7. SUMMARY AND APPENDIX

Chapter 37. Summary of X-Ray and Neutron Optics *Walter M. Gibson
and Carolyn A. MacDonald*   **37.3**

Appendix. X-Ray Properties of Materials *E. M. Gullikson*   **A.1**

Cumulative Index, Volumes I through IV, follows Appendix

CONTRIBUTORS

Jan P. Allebach   *Electronic Imaging Systems Laboratory, Purdue University, School of Electrical and Computer Engineering, West Lafayette, Indiana* (CHAP. 17)

John D. Baloga   *Eastman Kodak Company, Imaging Materials and Media, R&D, Rochester, New York* (CHAP. 6)

B. Benner   *Zweites Physikalisches Institut, RWTH Aachen, Aachen, Germany* (CHAP. 20)

Donald H. Bilderback   *Cornell High Energy Synchrotron Source, Cornell University, Ithaca, New York* (CHAP. 29)

William Cassarly   *Optical Research Associates, Pasadena, California* (CHAP. 2)

Franco Cerrina   *Department of Electrical and ComputerEngineering, University of Wisconsin—Madison, Madison, Wisconsin* (CHAP. 27)

Theodore E. Cohn   *University of California—Berkeley, School of Optometry, Berkeley, California* (CHAP. 8)

Aristide Dogariu   *School of Optics/The Center for Research and Education in Optics and Lasers (CREOL), University of Central Florida, Orlando, Florida* (CHAP. 3)

Jay M. Enoch   *University of California—Berkeley, Berkeley, School of Optometry, California* (CHAP. 9)

Edward D. Franco   *Advanced Research and Applications Corporation, Sunnyvale, California* (CHAP. 29)

Andreas Freund   *European Synchrotron Radiation Facility (ESRF), Grenoble, France* (CHAP. 26)

Robert Q. Fugate   *Starfire Optical Range, Directed Energy Directorate, Air Force Research Laboratory, Kirkland Air Force Base, New Mexico* (CHAP. 1)

Ralph Garzia   *University of Missouri—St. Louis, St. Louis, Missouri* (CHAP. 11)

Walter M. Gibson   *University at Albany, State University of New York, Albany, New York* (CHAPS. 19, 30, 34, 35, 37)

E. M. Gullikson   *Lawrence Berkeley National Laboratory, Berkeley, California* (APPENDIX)

Gerald C. Holst   *JCD Publishing, Winter Park, Florida* (CHAP. 4)

Malcolm R. Howells   *Advanced Light Source, Lawrence Berkeley National Laboratory, Berkeley, California* (CHAP. 21)

S. L. Hulbert   *National Synchrotron Light Source, Brookhaven National Laboratory, Upton, New York* (CHAP. 32)

Marshall K. Joy   *Department of Space Sciences, NASA/Marshall Space Flight Center, Huntsville, Alabama* (CHAP. 28)

Vasudevan Lakshminarayanan   *School of Optometry and Department of Physics and Astronomy, University of Missouri—St. Louis, St. Louis, Missouri* (CHAP. 9)

B. Lengeler   *Zweites Physikalisches Institut, RWTH Aachen, Aachen, Germany* (CHAP. 20)

William F. Long   *University of Missouri St. Louis, St. Louis, Missouri* (CHAP. 11)

Carolyn A. MacDonald   *University at Albany, State University of New York, Albany, New York* (CHAPS. 19, 30, 31, 35, 37)

Alan Michette   *King's College London, Physics Department, Strand, London, United Kingdom* (CHAPS. 23, 33)

David Mildner *Nuclear Methods Group, National Institute of Standards and Technology, Gaithersburg, Maryland* (CHAP. 36)

Donald T. Miller *Graduate Program in Visual Science, School of Optometry, Indiana University, Bloomington, Indiana* (CHAP. 10)

Yoshi Ohno *Optical Technology Division, National Institute of Standards and Technology, Gaithersburg, Maryland* (CHAP. 14)

James M. Palmer *Research Professor, Optical Sciences Center, University of Arizona, Tucson, Arizona* (CHAP. 7)

Thrasyvoulos N. Pappas *Electrical and Computer Engineering, Northwestern University, Evanston, Illinois* (CHAP. 17)

Bernice E. Rogowitz *IBM T.J. Watson Research Center, Hawthorne, New York* (CHAP. 17)

Brooke E. Schefrin *University of California, Davis Medical Center, Department of Ophthalmology, Sacramento, California* (CHAP. 13)

Clifton Schor *University of California, School of Optometry, Berkeley, California* (CHAP. 12)

C. Schroer *Zweites Physikalisches Institut, RWTH Aachen, Aachen, Germany* (CHAP. 20)

James E. Sheedy *University of California—Berkeley, School of Optometry, Berkeley, California* (CHAP. 16)

Qun Shen *Cornell High Energy Synchrotron Source and Department of Materials Science and Engineering, Cornell University, Ithaca, New York* (CHAP. 25)

Martin Shenker *Martin Shenker Optical Designs, Inc., White Plains, New York* (CHAP. 18)

Peter Siddons *National Synchrotron Light Source, Brookhaven National Laboratory, Upton, New York* (CHAP. 22)

David H. Sliney *U.S. Army Center for Health Promotion and Preventive Medicine, Laser/Optical Radiation Program, Aberdeen Proving Ground, Maryland* (CHAP. 15)

A. Snigirev *European Synchrotron Radiation Facility (ESRF), Grenoble, France* (CHAP. 20)

I. Snigireva *European Synchrotron Radiation Facility (ESRF), Grenoble, France* (CHAP. 20)

Eberhard Spiller *Lawrence Livermore National Laboratories, Livermore, California* (CHAP. 24)

Howard Stark *Xerox Corporation, Corporate Research and Technology—Retired* (CHAP. 5)

Brian H. Tsou *Air Force Research Laboratory, Wright Patterson Air Force Base, Ohio* (CHAP. 18)

J. Tümmler *Zweites Physikalisches Institut, RWTH Aachen, Aachen, Germany* (CHAP. 20)

Johannes Ullrich *University at Albany, State University of New York, Albany, New York* (CHAP. 31)

Jeffrey L. Weaver *American Optometric Association, St. Louis, Missouri* (CHAP. 11)

John S. Werner *University of California, Davis Medical Center, Department of Ophthalmology, Sacramento, California* (CHAP. 13)

G. P. Williams *National Synchrotron Light Source, Brookhaven National Laboratory, Upton, New York* (CHAP. 32)

PREFACE

In the preface to Vols. I and II of the *Handbook of Optics,* Second Edition, we indicated that the limitations of keeping within the confines of two books caused us to restrict the contents to optics that was clearly archival. There was much more that could have been included. We hope to have remedied some of that in this current volume with expanded parts on Classical optics, vision optics, and X-ray optics. References are provided to chapters in Vols. I, II, and IV, as well as to other chapters in Vol. III.

Optics has been developing so rapidly during the past 20 years that some of the material that has been included, although clearly of archival value, has not yet had the benefit of developed secondary references such as texts and long review articles. Thus, in the tutorial chapters presented in Vol. III, some of the references are made to primary publications.

Part 1, entitled "Classical Optics" (Chaps. 1–7), contains treatments of adaptive optics and nonimaging techniques. Chapters include solid-state cameras and electrophotography (e.g., how your laser printer or copying machine works) because they are ubiquitous and very important aspects of optics.

Part 2, "Vision Optics"(Chaps. 8–18), represents a major expansion over the similar part in Vol. I. The new chapters specifically address the factors that impact the design of optical instruments and devices. For example, because we are faced with a rapidly aging population, the designer must consider both the associated reduction in accommodation and physiological changes that accompany the aging process. Also discussed are visual functions not affected by aging and the many forms of visual corrections found in spectacles, multifocal and mono-vision designs, diffractive formats, contact lenses, intraocular implant lenses, and refractive surgery. The use and properties of computer/VDT displays, heads-up displays, and advanced display formats are discussed. Chapters are also presented concerning the unique features of biological fiber-optic elements and on laser eye-safety issues.

In Vol. I, there was a single chapter on X-ray optics. In Vol. III, we have expanded the coverage of X-ray optics in Part 3, "X-Ray and Neutron Optics" (Chaps. 19–37, plus Appendix). This part has articles covering such subjects as optical components, sources, and detectors for the X-ray region of the spectrum. We thank Professors Carolyn A. MacDonald and Walter M. Gibson of SUNY—Albany for their work in designing and developing this special section.

The *Handbook of Optics,* Second Edition, is only possible through the support of the staff of the Optical Society of America and, in particular, Alan N. Tourtlotte, Kimberly Street, and Laura Lee. We also thank Steve Chapman of McGraw-Hill Publishers for his leadership in the production of this volume.

Michael Bass, Editor in Chief
Jay M. Enoch, Associate Editor
Eric W. Van Stryland, Associate Editor
William L. Wolfe, Associate Editor

P · A · R · T · 1

CLASSICAL OPTICS

William L. Wolfe Part Editor

Optical Sciences Center, University of Arizona
Tucson, Arizona

CHAPTER 1
ADAPTIVE OPTICS*

Robert Q. Fugate
Starfire Optical Range
Directed Energy Directorate
Air Force Research Laboratory
Kirtland Air Force Base, New Mexico

1.1 GLOSSARY

C_n^2	refractive index structure parameter
d_0	section size for laser beacon adaptive optics
D_n	refractive index structure function
D_ϕ	phase structure function
E_f^2	mean square phase error due to fitting error
E_{FA}^2	mean square phase error due to focus anisoplanatism
E_n^2	mean square phase error due to sensor read noise
E_s^2	mean square phase error due to servo lag
f_G	Greenwood frequency
f_{T_G}	Tyler G-tilt tracking frequency
$F_\phi(f)$	phase power spectrum
$F_{\phi G}(f)$	G-tilt power spectrum
$H(f, f_c)$	servo response function
N_{pde}	number of photo-detected electrons per wavefront subaperture
r_0	Fried's coherence length
SR_{HO}	Strehl ratio resulting from higher-order phase errors
SR_{tilt}	Strehl ratio due to full-aperture tilt
$Z_j(\rho, \theta)$	Zernike polynomials

* *Acknowledgments:* I would like to thank David L. Fried, Earl Spillar, Jeff Barchers, John Anderson, Bill Lowrey, and Greg Peisert for reading the manuscript and making constructive suggestions that improved the content and style of this chapter. I would especially like to acknowledge the efforts of my editor, Bill Wolfe, who relentlessly kept me on course from the first pitiful draft.

σ_θ rms full-aperture tracking error

$\sigma_{\theta G}$ rms full-aperture G-tilt induced by the atmosphere

θ_0 isoplanatic angle

1.2 INTRODUCTION

An Enabling Technology

Adaptive optics (AO) is *the* enabling technology for many applications requiring the real-time control of light propagating through an aberrating medium. Today, significant advances in AO system performance for scientific, commercial, and military applications are being made because of improvements in component technology, control algorithms, and signal processing. For example, the highest-resolution images of living human retinas ever made have been obtained using AO.[1-3] Increased resolution of retinal features could provide ophthalmologists with a powerful tool for the early detection and treatment of eye diseases. Adaptive optics has brought new capabilities to the laser materials processing industry, including precision machining of microscopic holes and parts.[4, 5] Beaming power to space to raise satellites from low earth orbit to geosynchronous orbit, to maneuver satellites that are already in orbit, and to provide electricity by illuminating solar panels remains a topic of commercial interest.[6,7] Recent work has shown that high-energy pulsed lasers, corrected for atmospheric distortions with AO, could be used to alter the orbits of space debris objects,[8] causing them to reenter the atmosphere much sooner than natural atmospheric drag—creating an opportunity for the environmental restoration of space. High-speed optical communication between the ground, aircraft, and spacecraft (including very deep space probes) is a topic of continuing interest that would benefit from the use of AO.[9] Adaptive optics offers significant benefit to a large variety of military applications, including the U.S. Air Force Airborne Laser program, in which a high-powered chemical oxygen-iodine laser will be mounted in an aircraft to engage boosting missiles at long ranges.[10] Adaptive optics will be used to predistort the laser beam as it leaves the aircraft to compensate for the atmospheric distortions that the beam encounters as it propagates to the missile target.

Perhaps the most widely known and discussed application for AO is imaging through the atmosphere. Military sites are now using AO on medium-sized ground-based telescopes [of the order of 4 meters (m)] to inspect low-earth-orbiting satellites. However, the most dramatic use of AO is in astronomy. Astronomers have been plagued by the atmosphere for centuries—since the invention of the telescope. Indeed, distortion caused by turbulence is the principal motivation for launching large telescopes into space. However, if ground-based telescopes were able to achieve diffraction-limited resolution and high Strehl ratios, a significant obstacle to new discoveries and more productive research would be overcome. Adaptive optics, originally proposed by the astronomer Horace Babcock in 1953,[11] may enable that to happen, creating a revolution in ground-based optical astronomy. Adaptive optics is especially important in light of the large new telescope mirrors. Incredible advances in mirror technology have made 8-m-diameter monolithic mirrors almost commonplace (four have seen first light and five more are under construction). These are in addition to the two 10-m Keck telescopes and the 11-m Hobby-Eberly telescope, all of which use segmented primaries. The atmosphere limits the resolution of ground-based telescopes to an equivalent diameter that is equal to Fried's coherence length, r_0 (a few tens of centimeters at visible wavelengths at the best sites).[12] Adaptive optics offers the potential for telescopes to achieve diffraction-limited imaging and high throughput to spectroscopic instruments. Furthermore, interferometric arrays of large telescopes will enable unprecedented imaging resolution of faint objects, given that the unit telescopes can each produce nearly distortion-free wave fronts. Because imaging through the atmosphere is a topic of great interest, and because it embodies all aspects of AO technology, this discussion is oriented toward that application.

The chapter is organized as follows: Sec. 1.3 describes the basic concept of AO as applied to ground-based telescopes. Section 1.4 is a short summary of the classical description of atmospheric turbulence and parameters that are important for designing and evaluation of AO systems. Section 1.5 is a description of the hardware and the software that are needed for practical implementation of AO. This includes tracking, wavefront sensing, processors, and wavefront correctors. Section 1.6 is a discussion of design issues for top-level system performance trades using scaling laws and formulas.

1.3   THE ADAPTIVE OPTICS CONCEPT

This section provides a brief, qualitative overview of how AO imaging systems work. Figure 1 is a highly simplified diagram that illustrates the principles. In this figure, a conventional telescope is shown on the left and one that is equipped with AO is shown on the right. The science object (*a*) and a natural guide star (*b*) are at the top of the figure. The turbulent atmosphere (*c*) creates higher-order phase distortion and an overall tilt on wavefronts reaching the telescope (*d*). The telescope (*e*) forms an aberrated image (*f*) at the location of the camera or spectrograph. The natural guide star is used, either manually or automatically, to correct pointing errors in the telescope mount and the overall wavefront tilt that are induced by atmospheric turbulence.

The objectives of the AO system are to continuously remove the higher-order distortion and to stabilize the position of the image by removing the overall tilt. The components are shown on the right side of Fig. 1. One can think of the AO system as a dynamic optical system that simultaneously relays the original image to a new focal plane [the camera (*t*)] while removing both the higher-order aberrations (those distortions having spatial frequencies higher than tilt) with the deformable mirror (*k*) and the tilt aberrations with the fast-steering mirror (*i*), leaving only a minor amount of residual error in the wave front (*l*). The appropriate optical relay telescopes (*g* and *o*) are used as required by the particular implementation.

Figure 2 illustrates how a deformable mirror removes phase aberrations that are induced by atmospheric turbulence. The conjugate of the measured wavefront distortion is imposed on the surface of the deformable mirror so that on reflection, the distortions are removed. The AO system is never able to perfectly match the distortions in the wavefront because of a number of error sources that will be discussed in later sections.

To set the figure on the deformable mirror, the AO system must get information about the turbulent atmosphere by measuring its effect on the wavefronts from a beacon—a source of light such as a bright star (see Fig. 1, item *b*) located at or near the science object. The object itself may serve as the beacon if it is bright enough (either by self-emission, as in the case of a star or astronomical object, or by reflection of natural or man-made light, as in the case of artificial satellites). When a natural star does not exist or when the object itself is not bright enough, it may be possible to use an artificial beacon generated by a low-power laser (see Fig. 1, item *u*). A laser beacon can be created either by Rayleigh scattering up to a range of about 20 km, or by resonant scattering of atomic species such as sodium in the mesosphere, at an altitude of 90 km above Earth's surface. In the implementation that is shown in Fig. 1, a laser beam that is capable of exciting the D_2 line in atomic sodium in the mesosphere is projected from behind the secondary mirror of the telescope.

Control signals for the wavefront corrector mirrors are generated by a full-aperture tracking sensor (see Fig. 1, item *r*) and a wavefront sensor (*p*) by observing the residual error in the beacon wavefront (*l*). An aperture-sharing element (*m*) directs light from the beacon into the wavefront sensor, which is located at an image of the entrance pupil of the telescope. The wavefront sensor samples the slope of the wavefront over subaperture regions of the order of r_0 in size— for which the wavefront is essentially an undistorted, but tilted, plane wave—measuring the tilt and reporting that tilt as a wavefront phase gradient. The computer (*q*) combines the subaperture gradient measurements and "reconstructs" a best estimate of the residual phase error at

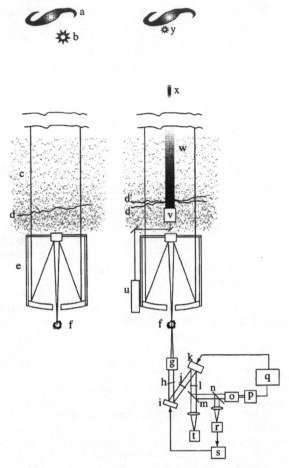

FIGURE 1 Conventional telescope with no AO (left) and additional components needed for AO (right). (*a*) Object of interest, (*b*) natural guide star, (*c*) atmospheric turbulence, (*d*) aberrated wavefront (including tilt and higher-order distortions) after passing through the turbulent atmosphere, (*d′*) aberrated wavefront from the laser beacon with full-aperture tilt removed, (*e*) telescope, (*f*) aberrated image, (*g*) relay optics, (*h*) demagnified aberrated wavefront including full-aperture tilt, (*i*) fast-steering mirror, (*j*) tilt-removed wavefront with only higher-order aberrations remaining, (*k*) deformable mirror, (*l*) corrected wavefront, (*m*) aperture-sharing element, (*n*) tilt sensor pickoff beam splitter, (*o*) relay optics, (*p*) higher-order wavefront sensor, (*q*) electronic processor to compute deformable mirror commands, (*r*) full-aperture tilt sensor, (*s*) tilt mirror processor and controller, (*t*) science camera, (*u*) laser for generating laser beacons, (*v*) launch telescope for the beacon laser, (*w*) Rayleigh laser beacon, (*x*) mesospheric sodium laser beacon, (*y*) faint natural guide star for full-aperture tilt sensing.

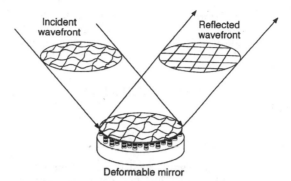

FIGURE 2   Deformable mirror concept in which phase aberrations on an incident wavefront are removed by setting the deformable mirror to the conjugate of the aberration.

specific points in the aperture, generally the locations of the actuators of the deformable mirror. Error signals derived from the reconstructed wavefront are sent to the deformable mirror to further reduce the residual error.

The tracking sensor measures the overall tilt of the beacon wavefront by computing the centroid of a focused image of the beacon that is formed by the full aperture of the telescope (or by other, more sophisticated means). The tracker processor (s) generates an error signal that controls the fast-steering mirror to keep the image of the beacon centered on the tracking sensor. It is also possible to derive full-aperture tilt information from the higher-order wavefront sensor, but optimum tracking performance usually requires a specialized sensor and processor. When a laser beacon is used to obtain higher-order wavefront information, it is still necessary to use a natural guide star to derive full-aperture tilt information: The laser beacon's position in the sky is not known with respect to an inertial reference (like a star), because its path on the upward propagation is random. To first order the beacon does not appear to move at all in the tracker due to near perfect reciprocity on the upward and downward paths, generating a return wavefront having no full aperture tilt as shown at (d') in Fig. 1. The requirement to use a natural guide star for tracking is not as serious as it first sounds, however, because it is much more likely that a natural guide star can be found near the science object that satisfies tracking requirements but is still too faint for higher-order wavefront measurements. This is possible because the full aperture of the telescope and a small number of specialized detectors can be used for tracking so that stars fainter by a factor of $\sim(D/r_0)^2$ than are available to the wavefront sensor become available to the tracking sensor.

Figure 1 is highly simplified. The details of the optics have been left out to emphasize the main principles. For instance, it is customary to form an image of the entrance pupil of the telescope onto the fast-steering mirror, the deformable mirror, and again on the wavefront sensor. At least two powered optical elements are required to generate each of these images. Even more sophisticated approaches have been proposed in which multiple deformable mirrors are reimaged to layers in the turbulence. The nature of the aperture-sharing element will vary significantly depending on the adopted philosophy for the system. For example, if obtaining enough light for the wavefront sensor is a problem (as is usually the case in astronomy), it is prudent to send the entire spectrum over which the detectors are responsive to the wavefront sensor and divert only the absolute minimum needed part of the spectrum to the imaging sensor. This approach also means that the optics must be designed to work over a wide spectral range (for example, from 450 to 1000 nm for a visible wavefront sensor) and also requires careful design of atmospheric dispersion correction. Furthermore, there are innovative approaches for implementing the wavefront corrector mirrors. The Steward Observatory group is developing a tilting, deformable secondary mirror for the new 6.5-m primary replacement at the Multiple Mirror Telescope (MMT) and 8.4-m Large Binocular Telescope (LBT) observatories.[13] This arrangement requires only two reflecting surfaces between the sky and the camera.

1.4   THE NATURE OF TURBULENCE AND ADAPTIVE OPTICS REQUIREMENTS

Turbulence Generation and the Kolmogorov Model

This section summarizes the classical description of atmospheric turbulence and its effect on the propagation of light. The literature on this topic is enormous. Our purpose here is to introduce the principles and define a few important parameters that are relevant to the operation of adaptive optical systems. The references cited throughout provide more detail.

Atmospheric turbulence is generated by solar heating of the Earth's surface. Air at the surface of the Earth is warmed in the day and cooled at night. Temperature gradients develop, creating a convective flow of large masses of air. Turbulence develops as large-scale masses of air break up into smaller spatial scales and dissipate their energy to the surrounding air. The effects of turbulence on electromagnetic wave propagation is governed by the nature of the spatial and temporal fluctuations of the index of refraction of air. The refractive index of air at optical wavelengths obeys the formula[14]

$$n = 1 + 77.6 \left[1 + \frac{7.52 \cdot 10^{-3}}{\lambda^2} \right] \frac{P}{T} \cdot 10^{-6} \tag{1}$$

where λ is the wavelength of light in micrometers, P is the atmospheric pressure in millibars, and T is the atmospheric temperature (K). The effect of atmospheric pressure changes on n are small and can, for problems of interest in this chapter, be neglected. The dominant influence on variations of n is the air temperature. At visible wavelengths, $dn/dT \sim 10^{-6}$ at standard pressure and temperature. This means that two 1-meter-long columns of air having a temperature difference of 1 K create roughly one wave of optical path difference for visible light, a very significant effect. We can model the index of refraction as the sum of two parts,

$$n(\mathbf{r}, t, \lambda) = n_0(\mathbf{r}, \lambda) + n_1(\mathbf{r}, t) \tag{2}$$

where n_0 represents the deterministic, slowly changing contribution (such as variation with height above the ground), and n_1 represents the random fluctuations arising from turbulence. The wavelength dependence of n_1 is ignored. Furthermore, typical values of n_1 are several orders of magnitude smaller than unity.

During the 1940s, Andrey Nikolayevich Kolmogorov (1903–1987) developed theories[15] describing how energy is dissipated in the atmosphere and modeled the spatial power spectrum of turbulent velocity fluctuations. V. I. Tatarskii postulated the applicability of the Kolmogorov spatial power spectrum to refractive index fluctuations and then solved the wave equation for the power spectrum of Kolmogorov's model and determined the effect on propagation through weak turbulence.[16] David L. Fried subsequently extended Tatarskii's results to describe phase distortions of turbulence in terms of Zernike polynomials[17] and derived his famous atmospheric coherence diameter parameter r_0, as a measure of optical resolution.[12] Nearly all subsequent work in statistical atmospheric optics is based on the contributions of Kolmogorov, Tatarskii, and Fried. The effects of turbulence on wave propagation are critical because the spatial and temporal frequency distribution and optical depth of the phase aberrations drive the design requirements for beacon brightness, wavefront sensing, tracking, electronic processing, and wavefront corrector mirror subsystems in a real-time phase compensation system.

For refractive index fluctuations, the Kolmogorov spatial power spectral density, $\Phi_n(\kappa)$, is given by

$$\Phi_n(\kappa) = 0.033 C_n^2 \kappa^{-11/3} \tag{3}$$

where κ is the spatial wavenumber and C_n^2 is the refractive index structure constant, a measure of the optical strength of turbulence. The wavenumber κ is inversely related to the size of the turbulent eddies in the atmosphere. Equation 3 is valid between two limits of κ called the inertial subrange. The scale size of eddies for the smallest values of κ in the inertial subrange is called the outer scale, denoted $L_0 = 2\pi/\kappa_0$, and is of the order of meters to kilometers. The inner scale, $l_0 = 2\pi/\kappa_m$, or smallest eddies in the inertial subrange, is of the order of millimeters. For values of κ outside the inertial subrange, the von Kármán spectrum is often used in place of Eq. 3 to avoid mathematical complications as κ approaches zero. The von Kármán spectrum has the form

$$\Phi_n(\kappa) \cong \frac{0.033 C_n^2}{(\kappa^2 + \kappa_0^2)^{11/6}} \exp\left(-\frac{\kappa^2}{\kappa_m^2}\right) \tag{4}$$

where κ_0 and κ_m are the spatial wavenumbers for the outer and inner scales.

The Variation of n and the C_n^2 Parameter

Tatarskii introduced structure functions to describe the effect of turbulence on wave propagation. Structure functions describe how a physical parameter is *different* between two points in space or time. The structure function of the index of refraction, $D_n(\mathbf{r})$, is defined as

$$D_n(\mathbf{r}) = \{[n_1(\mathbf{r}') - n_1(\mathbf{r}' - \mathbf{r})]^2\} \tag{5}$$

where $n_1(\mathbf{r})$ represents the fluctuating part of the index of refraction and $\{\ldots\}$ represents an ensemble average over the turbulent conditions. The autocorrelation function and power spectral density form a three-dimensional Fourier transform pair (a useful relation used many times in statistical optics), and using that relationship, it can be shown that[18]

$$D_n(r) = C_n^2 r^{2/3} \tag{6}$$

where we have also assumed that the fluctuations are isotropic and dropped the vector dependence on r. Equation 6 is the defining equation for the refractive index structure constant, C_n^2, in that the numerical coefficient on the right side of Eq. 6 is unity. Defined in this way, C_n^2 is a measure of the optical strength of turbulence.

A few meters above the ground, C_n^2 has an average value of the order of 10^{-14} m$^{-2/3}$, rapidly decreasing by three or four orders of magnitude at heights above 10 km. Several means have been developed to measure the C_n^2 profile, including in situ instruments on balloons as well as remote sensing optical and radar techniques.[19, 20] Several mathematical models of C_n^2 profiles have been developed based on experimental data. One of the most widely used was suggested by Hufnagel[21] and modified to include boundary layer effects by Valley.[22] The expression for the Hufnagel-Valley model for C_n^2 is

$$C_n^2(h) = 5.94 \cdot 10^{-23} h^{10} e^{-h} \left(\frac{W}{27}\right)^2 + 2.7 \cdot 10^{-16} e^{-2h/3} + A e^{-10h} \tag{7}$$

where h is the height above the site in kilometers, W is an adjustable wind correlating parameter, and A is a scaling constant, almost always taken to be $A = 1.7 \cdot 10^{-14}$. Winker[23] suggested $W = 21$ producing the HV$_{5/7}$ model named from the fact that the resulting profile yields a value of Fried's coherence diameter of $r_0 = 5$ cm and an isoplanatic angle of $\theta_0 = 7$ μrad for zenith propagation at 0.5 μm. (The parameters r_0 and θ_0 are defined and their relationships to C_n^2 are given in Sec. 1.4.) The HV$_{5/7}$ model has been widely used in the evaluation of AO system performance. Figure 3 shows how C_n^2, computed with this model, varies with altitude.

The Hufnagel-Valley model is useful for continental sites or sites with well-developed boundary layers. The C_n^2 profile, which is associated with mountaintop sites that are surrounded by water (e.g., Mauna Kea or La Palma) or mountaintop sites that are close to a coastline [so as to be above the marine boundary layer (e.g., Paranal, Chile)], often exhibits distinct layers of turbulence, but little or no boundary layer in the first few kilometers above the ground. A C_n^2 profile that is representative of average seeing for Mauna Kea is also shown in Fig. 3 and yields an r_0 of 18 cm and θ_0 of 15 μrad. Figure 4 shows a C_n^2 profile for Mauna Kea representative of the 90th-percentile best seeing (note the distinct layering), giving an r_0 of 45 cm, and an analytical model that is used for average seeing at Paranal,[24] giving an r_0 of 18 cm.

Wave Propagation in Turbulence

An electromagnetic wave propagating in the atmosphere must satisfy Maxwell's equations—in particular, the electric and magnetic fields must satisfy the time-independent Helmholtz equation:

$$\nabla^2 U(\mathbf{r}) + n^2 k^2 U(\mathbf{r}) = 0 \qquad (8)$$

where $k = 2\pi/\lambda$ and n is the refractive index of air. $U(\mathbf{r})$ is the complex phasor representation of a spatial wave:

$$U(\mathbf{r}) = A(\mathbf{r})e^{i\phi(\mathbf{r})} \qquad (9)$$

where the amplitude $A(\mathbf{r})$ and phase $\phi(\mathbf{r})$ are random variables governed by the statistics of the fluctuations of the refractive index, n. There is no closed-form solution for $A(\mathbf{r})$ and $\phi(\mathbf{r})$.

The surfaces in space that are defined by $\phi_0(\mathbf{r}) = $ constant are wavefronts. For starlight arriving at the top of the atmosphere, the wavefronts are planes of infinite extent. When a plane wave passes through the atmosphere, whose refractive index varies randomly along each propagation path, the plane wave becomes distorted. The total phase change from that in a vacuum along a path z is

$$\triangle\phi = \phi_0 - k \int_0^z n_1(z)dz \qquad (10)$$

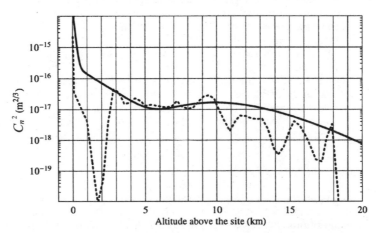

FIGURE 3 C_n^2 profiles for the HV$_{5/7}$ model (solid curve) and for average seeing conditions at Mauna Kea (dotted curve).

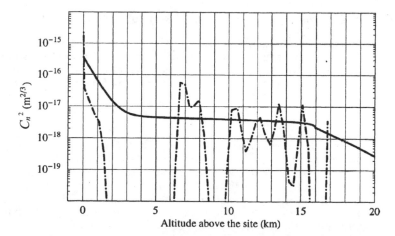

FIGURE 4 Average seeing C_n^2 profile analytical model for Paranal, Chile [site of European Southern Observatory's (ESO's) Very Large Telescope (VLT) (solid curve)], and a C_n^2 profile that is representative of best seeing conditions (occurring less than 10 percent of the time) at Mauna Kea (dotted curve) and represents 0.25 arcsec seeing. Note how the turbulence appears as layers and the lack of a boundary layer in the Mauna Kea profile. Similar conditions have also been observed at Paranal and La Palma.

where $k = 2\pi/\lambda$, and $n_1(z)$ is the fluctuating part of the index of refraction. If this equation could be evaluated along every direction, one could construct the three-dimensional shape of the wavefront at any position along its propagation path due to phase changes induced by atmospheric turbulence. This is, of course, an intractable problem in terms of a closed-form solution since the fluctuations of the index of refraction are a random process.

A phase deformation that propagates far enough becomes an amplitude fluctuation as different parts of the propagating wavefront combine to create regions of constructive and destructive interference. This effect is called scintillation—the twinkling of starlight is a famous example. Amplitude fluctuations are generally not severe for astronomical observing, and 70 to 80 percent or more of the distortions can be removed by phase-only compensation (by using a steering mirror and deformable mirror). However, both the amplitude and phase of the wave are random variables, and the most general representation of the wave is as given in Eq. 9.

The historically accepted approach to dealing with wave propagation in turbulence was introduced by Tatarskii by use of the Rytov transformation. The Rytov transformation defines a complex quantity ψ as the natural logarithm of U: $\psi = \ln U$. This substitution allows a solution to the wave equation in the form of a multiplicative perturbed version of the free space field:

$$U(\mathbf{r}) = A_0(\mathbf{r})e^{i\phi_0(\mathbf{r})}e^{\ln [A(\mathbf{r})/A_0(\mathbf{r})] - i[\phi_0(\mathbf{r}) - \phi(\mathbf{r})]} \tag{11}$$

Since the field at any point is the superposition of many independent contributions of propagating waves in the turbulent medium, we expect the fluctuating parts of the field to obey Gaussian statistics (invoking the central limit theorem). This means if $\ln(A/A_0)$ obeys Gaussian statistics, we expect the amplitude fluctuations to be log-normally distributed. Most experimental evidence supports this for weak fluctuations. Tatarskii's results make it possible to compute important characteristic functions and to use the properties of Gaussian random processes to develop practical descriptions of the effects of turbulence.

Fried's Coherence Diameter, r_0, and the Spatial Scale of Turbulence

Hufnagel and Stanley[25] extended Tatarskii's work and developed an expression for the modulation transfer function in terms of the mutual coherence function. Fried[26] developed an expression for *phase* structure function, D_ϕ, of a propagating electromagnetic plane wave showing that it is proportional to the 5/3 power of spatial separation, r, and is given by*

$$D_\phi(r) = \{[\phi(r') - \phi(r' - r)]^2\} = \left[2.91 \left(\frac{2\pi}{\lambda}\right)^2 \int C_n^2(z)dz\right] r^{5/3} \tag{12}$$

where λ is the wavelength of propagation, $C_n^2(z)$ is the position-dependent refractive index structure constant and integration is over the optical path to the source.

Fried[26] developed a very useful relationship between the phase structure function and a particular measure of optical resolution—the volume under the two-dimensional optical transfer function. Defined in this way, the seeing-limited resolving power asymptomatically approaches a limiting value as the aperture size increases. The limiting value is set by the strength of the turbulence. Fried defined a quantity called the coherence diameter, r_0, such that the limiting resolution that is obtained in the presence of atmospheric turbulence is the same as that that is obtained by a diffraction-limited lens of diameter r_0 in a vacuum. Its value can be computed from the definition,[17]

$$r_0 = 0.185\lambda^{6/5}(\cos \psi)^{3/5} \left[\int_0^\infty C_n^2(h)dh\right]^{-3/5} \tag{13}$$

where λ is the wavelength, $C_n^2(h)$ is the vertical profile of the index of refraction structure constant, h is the height above the ground, ψ is the angle between the propagation direction and the zenith, and the integral is evaluated from the ground to an altitude at which $C_n^2(h)$ no longer contributes significantly (typically 20 km). This equation explicitly shows r_0 scales with wavelength as $\lambda^{6/5}$ and with zenith angle as $(\cos \psi)^{3/5}$. By convention, the numerical values of r_0 are usually expressed at 0.5 μm for zenith viewing. At the best astronomical sites, the median value of r_0 is from 15 to 25 cm, corresponding to seeing conditions of $\lambda/r_0 = 0.7$ to 0.4 arcsec. The best intracontinental mountaintop sites exhibit median values of r_0 between 10 and 20 cm, and average intracontinental sites have r_0 values between 5 and 10 cm.[20]

Given the definition of r_0 in Eq. 13, the phase structure function becomes

$$D_\phi(r) = 6.88 \left(\frac{r}{r_0}\right)^{5/3} \tag{14}$$

Fried defined r_0 so that the knee in the curve showing resolving power as a function of diameter occurs at $D/r_0 = 1$, where D is the aperture diameter of the imaging telescope. Most scaling laws of interest to AO are expressed in terms of the ratio D/r_0.

Zernike Polynomial Description of Turbulence

R. J. Noll[27] developed a description of the average spatial content of turbulence-induced wavefront distortion in terms of Zernike polynomials. The Zernike polynomials are often used to describe the classical aberrations of an optical system—tilt, defocus, astigmatism, coma, and spherical aberration. Noll expresses the phase of the distorted wavefront over a circular aperture of radius, R, as

$$\phi(\rho, \theta) = \sum_j a_j Z_j(\rho, \theta) \tag{15}$$

---

* In the more general case where amplitude effects are significant, the *wave* structure function, $D(r) = D_x(r) + D_\phi(r)$, should be used, but since it has exactly the same value as given by Eq. 12, and since conventional AO does not correct intensity fluctuations, the amplitude term will be dropped for the material that is presented here.

where Z_j is a modified set of the orthogonal Zernike polynomials defined over points in the aperture at reduced radius $\rho = r/R$ and azimuthal angle θ. The Z_j is given by

$$\left. \begin{array}{l} Z_{\text{even}j} = \sqrt{n+1}\, R_n^m(r)\, \sqrt{2}\, \cos m\theta \\[2mm] Z_{\text{odd}j} = \sqrt{n+1}\, R_n^m(r)\, \sqrt{2}\, \sin m\theta \end{array} \right\} \quad m \neq 0 \tag{16}$$

$$Z_j = \sqrt{n+1}\, R_n^0(r), \qquad\qquad m = 0 \tag{17}$$

where

$$R_n^m(r) = \sum_{s=0}^{(n-m)/2} \frac{(-1)^s (n-s)!}{s!\,[(n+m)/2 - s]!\,[(n-m)/2 - s]!}\, r^{n-2s} \tag{18}$$

The indices n and m are always integers and satisfy $m \leq n$ and $n - |m| = $ even. The index, j, is a mode-ordering number and depends on n and m.

The a_j are coefficients having mean square values that accurately weight the particular Zernike mode to represent the Kolmogorov distribution of turbulence. Using this approach, Noll computed the values of a_j and showed that the piston-removed wavefront distortion for Kolmogorov turbulence averaged over an aperture of diameter, D, expressed as a mean square value in units of square radians (rad^2) of optical phase is

$$\langle \phi^2 \rangle = 1.0299 \left(\frac{D}{r_0} \right)^{5/3} \tag{19}$$

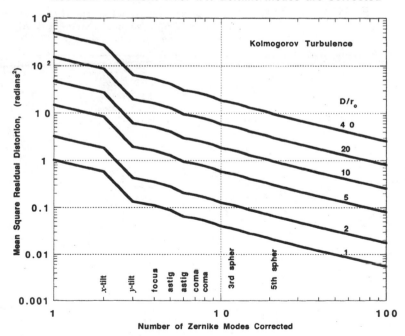

Residual wavefront error if n Zernike modes are corrected

FIGURE 5 Residual mean square phase distortion when n Zernike modes of atmospheric turbulence are corrected for various values of D/r_0.

Table 1 lists the strengths of the first 21 Zernike modes, and the residual mean square distortion as each component is removed. Data in this table show that the two components of tilt (X and Y axes) make up 87 percent of the total wavefront distortion.

Noll's results are useful for estimating the performance of an AO system if one can estimate how many Zernike modes the system will correct (a rule of thumb is approximately one mode is corrected per deformable mirror actuator for systems having large numbers of actuators). Figure 5 shows the mean square residual phase error for a range of values of D/r_0. Generally, the mean square residual error should be much less than 1 rad^2. Even if all 21 modes listed in Table 1 are corrected, the mean square residual error for conditions in which D/r_0 is 20, for example, is $0.0208(20)^{5/3} = 3.06$ rad^2, a very significant, generally unacceptable, error.

When more than 21 modes are corrected, the residual mean square wavefront error is given by[27]

$$\sigma_M^2 = 0.2944 M^{-(\sqrt{3}/2)}(D/r_0)^{5/3} \tag{20}$$

where M is the number of modes corrected. Figure 6 shows the required number of Zernike modes to be removed as a function of D/r_0 for three image quality conditions. As an example, if r_0 is 12 cm at 0.8 μm and the telescope aperture is 3.5 m, $D/r_0 = 29$, and more than 1000 modes must be removed to achieve λ/15 image quality, but only 100 modes need be removed if λ/5 quality is sufficient. If r_0 is 60 cm at 1.6 μm and $D = 10$ m, we must remove nearly 400 modes to achieve λ/15 image quality. For imaging applications, several techniques for postprocessing are well established[28] and allow significant enhancements of images that are obtained in real time with AO. Postprocessing may be justification to allow the relaxation of wavefront quality requirements in an AO system. However, for spectroscopy (or other applications requiring high Strehl ratios in real time), there may be little choice but to design the

TABLE 1   Mean Square Wavefront Distortion Contributions (in rad^2) from the First 21 Zernike Modes of Atmospheric Aberrations and Residual Error As Each Term Is Corrected

Aberration	Contribution	Distortion
All terms	$1.0299(D/r_0)^{5/3}$	$1.0299(D/r_0)^{5/3}$
Z_2 X-tilt	$0.4479(D/r_0)^{5/3}$	$0.582(D/r_0)^{5/3}$
Z_3 Y-tilt	$0.4479(D/r_0)^{5/3}$	$0.134(D/r_0)^{5/3}$
Z_4 Defocus	$0.0232(D/r_0)^{5/3}$	$0.111(D/r_0)^{5/3}$
Z_5 X-astigmatism	$0.0232(D/r_0)^{5/3}$	$0.0880(D/r_0)^{5/3}$
Z_6 Y-astigmatism	$0.0232(D/r_0)^{5/3}$	$0.0649(D/r_0)^{5/3}$
Z_7 3rd-order coma	$0.0061(D/r_0)^{5/3}$	$0.0587(D/r_0)^{5/3}$
Z_8 3rd-order coma	$0.0061(D/r_0)^{5/3}$	$0.0525(D/r_0)^{5/3}$
Z_9 Coma	$0.0062(D/r_0)^{5/3}$	$0.0463(D/r_0)^{5/3}$
Z_{10} Coma	$0.0062(D/r_0)^{5/3}$	$0.0401(D/r_0)^{5/3}$
Z_{11}	$0.0024(D/r_0)^{5/3}$	$0.0377(D/r_0)^{5/3}$
Z_{12}	$0.0025(D/r_0)^{5/3}$	$0.0352(D/r_0)^{5/3}$
Z_{13}	$0.0024(D/r_0)^{5/3}$	$0.0328(D/r_0)^{5/3}$
Z_{14}	$0.0024(D/r_0)^{5/3}$	$0.0304(D/r_0)^{5/3}$
Z_{15}	$0.0025(D/r_0)^{5/3}$	$0.0279(D/r_0)^{5/3}$
Z_{16}	$0.0012(D/r_0)^{5/3}$	$0.0267(D/r_0)^{5/3}$
Z_{17}	$0.0012(D/r_0)^{5/3}$	$0.0255(D/r_0)^{5/3}$
Z_{18}	$0.0012(D/r_0)^{5/3}$	$0.0243(D/r_0)^{5/3}$
Z_{19}	$0.0011(D/r_0)^{5/3}$	$0.0232(D/r_0)^{5/3}$
Z_{20}	$0.0012(D/r_0)^{5/3}$	$0.0220(D/r_0)^{5/3}$
Z_{21}	$0.0012(D/r_0)^{5/3}$	$0.0208(D/r_0)^{5/3}$

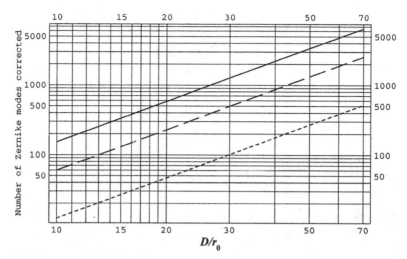

FIGURE 6   Required number of modes to be corrected (roughly the number of actuators or degrees of freedom in the deformable mirror) as a function of D/r_0. Solid curve: $\lambda/15$ image quality; dashed curve: $\lambda/10$; dotted curve: $\lambda/5$ image quality. In these curves, r_0 is evaluated at the imaging wavelength.

AO system with the required number of degrees of freedom needed to achieve desired Strehl ratios.

The results in Table 1 can be used to estimate the maximum stroke requirement for the actuators in the deformable mirror. Assuming that tilt will be corrected with a dedicated two-axis beam-steering mirror, the root-mean-square (rms) higher-order distortion is (from Table 1) $0.366(D/r_0)^{5/6}$ rad of phase. A practical rule of thumb is that the deformable mirror surface should be able to correct five times the rms wavefront distortion to get 99 percent of the peak values. Since a reflected wavefront has twice the distortion of the mirror's surface (reducing the stroke requirement by a factor of 2), the maximum actuator stroke requirement for operation becomes

$$\sigma_m(\mu m) = 0.075 \left(\frac{D}{r_0} \right)^{5/6} \tag{21}$$

where σ_m is the maximum actuator stroke in μm, and r_0 is the value of Fried's coherence diameter at a wavelength of 0.5 μm along the line of sight corresponding to the maximum zenith angle of interest. Figure 7 shows the maximum stroke that is required for several aperture diameters as a function of seeing conditions. This figure assumes the outerscale is much larger than the aperture diameter and Kolmogorov turbulence. If the outerscale is of the order of the aperture size or even smaller, the stroke requirements will be reduced.

Atmospheric Tilt and Its Effect on the Strehl Ratio

Tilt comprises 87 percent of the power in atmospheric turbulence-induced wavefront distortion (see the preceding section). The Strehl ratio of an image that is due to jitter alone is approximated to better than 10 percent by the equation[29, 30]

$$SR_{\text{tilt}} = \frac{1}{1 + \dfrac{\pi^2}{2} \left(\dfrac{\sigma_\theta}{\lambda/D} \right)^2} \tag{22}$$

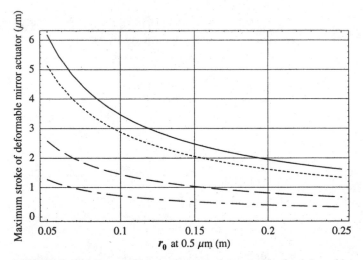

FIGURE 7 Maximum stroke requirement of actuators in the deformable mirror. The curves are for aperture diameters (from top to bottom) of 10, 8, 3.5, and 1.5 m. The value of r_0 is for an observing wavelength of 0.5 μm along the line of sight corresponding to the maximum zenith angle of interest, and the stroke is in micrometers. These curves are independent of wavelength since they do not include any dispersion effects.

where σ_θ is the one-axis rms angular jitter. Figure 8 shows how the one-axis rms jitter affects system Strehl ratio. Greenwood and Fried[31] derived bandwidth requirements for AO systems and developed expressions for the variance of full-aperture tilt that is induced by turbulence. The one-axis angular jitter variance (units of angular rad^2) is given by

$$\sigma_{\theta G}^2 = 0.170 \left(\frac{\lambda}{D} \right)^2 \left(\frac{D}{r_0} \right)^{5/3} \tag{23}$$

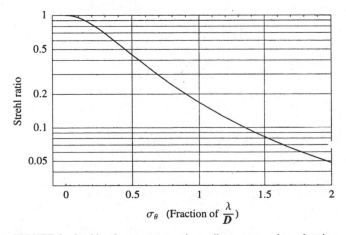

FIGURE 8 Strehl ratio versus one-axis rms jitter expressed as a fraction of λ/D.

where $\sigma_{\theta G}$ is the rms G-tilt. G-tilt is the average gradient over the wavefront and is well approximated by most centroid trackers. For Z-tilt (the direction of the normal of the best-fit plane to the wavefront distortion), the coefficient in Eq. 23 is 0.184. Consequently, a tracking sensor that measures the centroid of the focused image improperly estimates the Z-tilt; this has been called centroid anisoplanatism.[30]

Figure 9 shows the dependence of $\sigma_{\theta G}$ on aperture diameter for two seeing conditions, one representative of a continental site and one of a mountaintop island like Mauna Kea, Hawaii.

The effect of atmospheric tilt on the Strehl ratio can be seen by substituting Eq. 23 into Eq. 22. Figure 10 shows that the loss in the Strehl ratio due to jitter alone is significant. For $D/r_0 = 1$, the atmospherically induced jitter limits the Strehl ratio to 0.54. It is most important to effectively control image motion that is created by full-aperture atmospheric tilt.

Tracking: Bandwidth and Steering Mirror Stroke Requirements

Tyler[32] has considered the problem of required tracking bandwidth by computing the tilt power spectrum and determining the error rejection achieved by a steering mirror under control of an RC-type servo. He considers both G-tilt and Z-tilt. Temporal fluctuations are derived by translating "frozen-turbulence" realizations past the telescope's aperture. The hypothesis is that the internal structure of the turbulence changes more slowly than its mass transport by the wind (but this assumption may not suffice for today's giant-aperture telescopes). The results for power spectra are in integral form, requiring numerical evaluation. Sasiela[33] uses Mellin transforms to derive a solution in terms of a power series of ratios of gamma functions. The asymptotic limits for high and low frequencies are, however, tractable in a form that is easily evaluated. We consider here the results for only G-tilt since that is the most likely implementation in a real system (approximated by a centroid tracker). The low-frequency limit for G-tilt is

$$\lim_{f \to 0} F_{\phi G}(f) = 0.804 D^{-1/3} \sec \psi f^{-2/3} \int_0^\infty C_n^2(h)[v(h)/D]^{-1/3} dh \qquad (24)$$

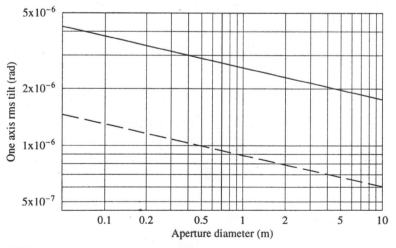

FIGURE 9   One-axis rms tilt for atmospheric turbulence. The upper line corresponds to an r_0 of 5 cm and the lower curve corresponds to an r_0 of 18 cm, both referenced to 0.5 μm and zenith. The value of full-aperture tilt is independent of wavelength for a given seeing condition.

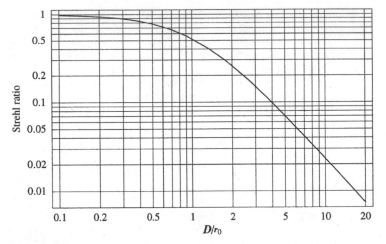

FIGURE 10 The effect of turbulence-induced jitter on system Strehl ratio.

and the high-frequency limit is

$$\lim_{f \to \infty} F_{\phi G}(f) = 0.0110 D^{-1/3} \sec \psi f^{-11/3} \int_0^\infty C_n^2(h)[v(h)/D]^{8/3} dh \tag{25}$$

where $F_{\phi G}(f)$ is the one-axis G-tilt power spectrum in rad^2/Hz, D is the aperture diameter, ψ is the zenith angle, and $v(h)$ is the transverse wind component along a vertical profile. Note that the tilt spectrum goes as $f^{-2/3}$ at low frequencies and as $f^{-11/3}$ at high frequencies.

The disturbances described in Eqs. 24 and 25 can be partially corrected with a fast-steering mirror having an RC filter response, $H(f, f_c)$, of the form

$$H(f, f_c) = \frac{1}{1 + i\dfrac{f}{f_c}} \tag{26}$$

where f_c represents a characteristic frequency [usually the 3 decibel (dB) response]. The uncorrected power in the residual errors can be computed from control theory using the relation

$$\sigma_{\theta G}^2 = \int_0^\infty |1 - H(f, f_c)|^2 \, F_{\phi G}(f) df \tag{27}$$

where $F_{\phi G}(f)$ is the power spectrum (rad^2/Hz) of G-tilt induced by optical turbulence.

For a steering mirror response function given by Eq. 26, the G-tilt variance is given by

$$\sigma_{\theta G}^2 = \int_0^\infty \frac{(f/f_{3\,dB})^2}{1 + (f/f_{3\,dB})^2} \, F_{\phi G}(f) df \tag{28}$$

In the limit of a small (zero) servo bandwidth, this equation represents the tilt that is induced by atmospheric turbulence. Expressing the variance of the jitter in terms of the diffraction angle (λ/D) results in the same expression as Eq. 23.

What does the servo bandwidth need to be? Tyler defines a G-tilt tracking frequency characteristic of the atmospheric turbulence profile and wind velocity as

$$f_{T_G} = 0.331 D^{-1/6} \lambda^{-1} \sec^{1/2} \psi \left[\int_0^\infty C_n^2(h) v^2(h) dh \right]^{1/2} \tag{29}$$

such that the variance of the tilt can be expressed as

$$\sigma_\theta^2 = \left(\frac{f_{T_G}}{f_{3\,dB}}\right)^2 \left(\frac{\lambda}{D}\right)^2 \tag{30}$$

Three wind velocity profiles are shown in Fig. 11. The analytical model of Bufton[34] is given by

$$v_B(h) = 5 + 37e^{-(h-12)^2/25} \tag{31}$$

where h is the height above ground in km and v_B the wind speed in m/s. The other two curves in Fig. 11 are based on measurements made with National Weather Service sounding balloons launched in Albuquerque, New Mexico. They show a marked contrast between winter and summer months and the effect of the jet stream.

When the 3-dB closed-loop tracking servo frequency equals the G-tilt tracking frequency (the Tyler frequency), the one-axis rms jitter is equal to λ/D. The track bandwidth needs to be four times the Tyler frequency in order for $\sigma_{\theta G}$ to be $1/4(\lambda/D)$, a condition that provides a Strehl ratio due to tilt of greater than 80 percent. Figure 12 shows the required 3-dB tracking bandwidths to meet this criterion for three combinations of wind and C_n^2 profiles as a function of aperture diameter. Other combinations can be computed easily by using Eq. 29 and the wind profiles that are shown in Fig. 11.

The one-axis rms tilt that is induced by the atmosphere is given by Eq. 23. When specifying the maximum excursion of a tilt mirror, the mirror should be able to move five times $\sigma_{\theta G}$ in order to accommodate over 99 percent of the tilt spectrum. We also need to account for the optical magnification between the tilt mirror and the primary mirror of the telescope. If, for instance, the tilt mirror has a diameter of 10 cm, the telescope aperture is 10 m, and r_0 is 18 cm (dashed curve in Fig. 9), the required maximum stroke of the tilt mirror is 0.6 μrad (from Fig. 9) times 50 ($5\sigma_{\theta G}$ × optical magnification of 10) or 30 μrad.

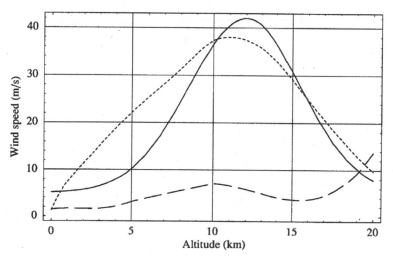

FIGURE 11   Wind profiles. The solid curve is an analytical model by Bufton. The dotted and dashed curves are based on two decades of rawinsonde measurements at Albuquerque, New Mexico, and are known as SOR winter and SOR summer models, respectively. These two models show clearly the absence of a jet stream in the summer months.

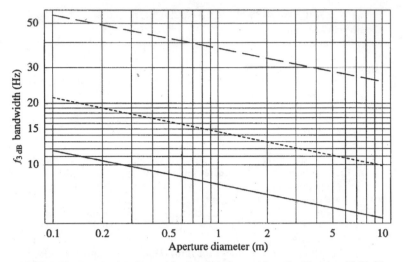

FIGURE 12   Required 3-dB tracking bandwidth to achieve Strehl ratio of 0.82. The dashed curve is for the SOR winter wind and $HV_{5/7}$ profiles; the dotted curve is for the Mauna Kea average C_n^2 profile and Bufton wind profile; the solid curve is for the SOR summer wind and $HV_{5/7}$ C_n^2 profiles. All three curves are for 0.5 μm wavelength.

Higher-Order Phase Fluctuations and Bandwidth Requirements

Time delays that are caused by detector readout and data processing create errors in the phase correction that is applied to the deformable mirror—the correction being applied is relevant to an error that is measured at an earlier time. The control system concepts discussed in the preceding two sections are also relevant to higher-order errors. The residual errors can be computed knowing the closed-loop servo response, $H(f, f_c)$, where f_c represents a characteristic frequency (usually the 3-dB response) and the disturbance. The uncorrected power is similar to Eq. 27 and is given by

$$\sigma_r^2 = \int_0^\infty |1 - H(f, f_c)|^2 \, F_\phi(f) df \tag{32}$$

where $F_\phi(f)$ is the power spectrum (rad²/Hz) of the phase distortions that are induced by optical turbulence. At high frequencies, Greenwood[35] showed that $F_\phi(f)$ is proportional to $f^{-8/3}$ and is given by

$$\lim_{f \to \infty} F_\phi(f) = 0.0326 k^2 f^{-8/3} \int_0^\infty C_n^2(z) v^{5/3}(z) dz \tag{33}$$

where $k = 2\pi/\lambda$, f is the frequency in Hz, and $v(z)$ is the transverse component of the wind profile. The closed-loop response of the AO servo can be modeled as a conventional resistor-capacitor (RC) filter making

$$H(f, f_c) = \frac{1}{1 + i\dfrac{f}{f_c}} \tag{34}$$

the characteristic frequency becomes

$$f_c = \left[0.102 \left(\frac{k}{\sigma_r} \right)^2 \int_0^\infty C_n^2(z) v^{5/3}(z) dz \right]^{3/5} \tag{35}$$

Figure 11 shows three wind profiles, one analytical and two based on measured data. For the $HV_{5/7}$ C_n^2 profile and the SOR summer and winter wind models and $\sigma_r = 0.2\pi$, $(\lambda/10)$, then $f_c = 23$ and 95 Hz, respectively.

The value of f_c for which $\sigma_r = 1$ is known as the Greenwood frequency and is given explicitly as

$$f_G = \frac{2.31}{\lambda^{6/5}} \left[\int_0^L C_n^2(z) v(z)^{5/3} dz \right]^{3/5} \tag{36}$$

Tyler[36] shows that the mean square residual phase in an AO control system having a -3-dB closed-loop bandwidth of $f_{3\,dB}$ Hz is

$$\sigma_{\phi_{servo}}^2 = (f_G/f_{3\,dB})^{5/3} \tag{37}$$

In a real system, this means that the closed-loop servo bandwidth should be several times the Greenwood frequency in order to "stay up with the turbulence" and keep the residual wavefront error to acceptable levels. Note that the Greenwood frequency scales as $\lambda^{-6/5}$ (the inverse of r_0 and θ_0 scaling) and that zenith angle scaling depends on the specific azimuthal wind direction.

Anisoplanatism

Consider two point sources in the sky (a binary star) and an instrument to measure the wavefront distortion from each. As the angular separation between the sources increases, the wavefront distortions from each source become decorrelated. The *isoplanatic angle* is that angle for which the difference in mean square wavefront distortion is 1 rad². Fried[37] defined the isoplanatic angle, θ_0, as

$$\theta_0 = 0.058 \lambda^{6/5} (\sec \psi)^{-8/5} \left[\int_0^\infty C_n^2(h) h^{5/3} dh \right]^{-3/5} \tag{38}$$

where the integral is along a vertical path through the atmospheric turbulence. Note that θ_0 scales as $\lambda^{6/5}$ (the same as r_0), but as $(\cos \psi)^{8/5}$ with zenith angle. Fried has shown that if a very large diameter ($D/r_0 \gg 1$) imaging system viewing along one path is corrected by an AO system, the optical transfer function of the system viewing along a different path separated by an angle θ is reduced by a factor $\exp[-(\theta/\theta_0)^{5/3}]$. (For smaller diameters, there is not so much reduction.) The numerical value of the isoplanatic angle for zenith viewing at a wavelength of 0.5 µm is of the order of a few arc seconds for most sites. The isoplanatic angle is strongly influenced by high-altitude turbulence (note the $h^{5/3}$ weighting in the preceding definition). The small value of the isoplanatic angle can have very significant consequences for AO by limiting the number of natural stars suitable for use as reference beacons and by limiting the corrected field of view to only a few arc seconds.

Impact of Turbulence on Imaging and Spectroscopy

The optical transfer function (OTF) is one of the most useful performance measures for the design and analysis of AO imaging systems. The OTF is the Fourier transform of the optical system's point spread function. For an aberration-free circular aperture, the OTF for a spatial frequency **f** is well known[38] to be

$$\mathbf{H}_0(\mathbf{f}) = \frac{2}{\pi} \left[\arccos\left(\frac{\mathbf{f}\bar{\lambda}F}{D} \right) - \frac{\mathbf{f}\bar{\lambda}F}{D} \sqrt{1 - \left(\frac{\mathbf{f}\bar{\lambda}F}{D} \right)^2} \right] \tag{39}$$

where D is the aperture diameter, F is the focal length of the system, and $\bar{\lambda}$ is the average imaging wavelength. Notice that the cutoff frequency (where the OTF of an aberration-free system reaches 0) is equal to $D/\bar{\lambda}F$.

Fried[26] showed that for a long-exposure image, the OTF is equal to $\mathbf{H}_0(\mathbf{f})$ times the long-exposure OTF due to turbulence, given by

$$\mathbf{H}_{\text{LE}}(\mathbf{f}) = \exp\left\{-\frac{1}{2}\, D(\bar{\lambda}F|\mathbf{f}|)\right\} = \exp\left\{-\frac{1}{2}\, 6.88 \left(\frac{\bar{\lambda}F|\mathbf{f}|}{r_0}\right)^{5/3}\right\} \tag{40}$$

where D is the wave structure function and where we have substituted the phase structure function that is given by Eq. 14. Figure 13 shows the OTF along a radial direction for a circular aperture degraded by atmospheric turbulence for values of D/r_0 ranging from 0.1 to 10. Notice the precipitous drop in the OTF for values of $D/r_0 > 2$. The objective of an AO system is, of course, to restore the high spatial frequencies that are lost due to turbulence.

Spectroscopy is a very important aspect of observational astronomy and is a major contributor to scientific results. The goals of AO for spectroscopy are somewhat different than for imaging. For imaging, it is important to stabilize the corrected point spread function in time and space so that postprocessing can be performed. For spectroscopy, high Strehl ratios are desired in real time. The goal is flux concentration and getting the largest percentage of the power collected by the telescope through the slit of the spectrometer. A 4-m telescope is typically limited to a resolving power of $R \sim 50,000$. Various schemes have been tried to improve resolution, but the instruments become large and complex.

However, by using AO, the corrected image size decreases linearly with aperture size, and very high resolution spectrographs are, in principle, possible without unreasonable-sized gratings. A resolution of 700,000 was demonstrated on a 1.5-m telescope corrected with AO.[39] Tyler and Ellerbroek[40] have estimated the sky coverage at the galactic pole for the Gemini North 8-m telescope at Mauna Kea as a function of the slit power coupling percentage for a 0.1-arcsec slit width at J, H, and K bands in the near IR. Their results are shown in Fig. 14 for laser guide star (top curves) and natural guide star (lower curves) operation.

The higher-order AO system that was analyzed[40] for the results in Fig. 14 employs a 12-by-12 subaperture Shack-Hartmann sensor for both natural guide star (NGS) and laser guide star (LGS) sensing. The spectral region for NGS is from 0.4 to 0.8 μm and the 0.589-μm LGS is produced by a 15-watt laser to excite mesospheric sodium at an altitude of 90 km. The wave-

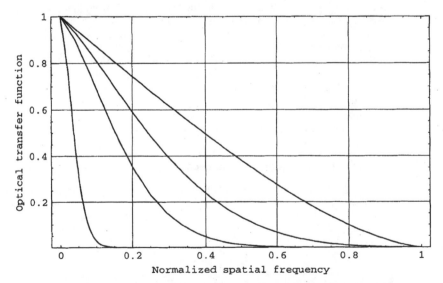

FIGURE 13 The OTF due to the atmosphere. Curves are shown for (top to bottom) $D/r_0 = 0.1$, 1, 2, and 10. The cutoff frequency is defined when the normalized spatial frequency is 1.0 and has the value $D/\lambda F$, where F is the focal length of the telescope and D is its diameter.

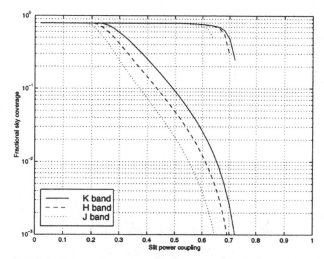

FIGURE 14   Spectrometer slit power sky coverage at Gemini North using NGS (lower curves) and LGS adaptive optics. See text for details.

front sensor charge-coupled device (CCD) array has 3 electrons of read noise per pixel per sample and the deformable mirror is optically conjugate to a 6.5-km range from the telescope. The tracking is done with a 2-by-2-pixel sensor operating at J + H bands with 8 electrons of read noise. Ge et al.[41] have reported similar results with between 40 and 60 percent coupling for LGSs and nearly 70 percent coupling for NGSs brighter than 13th magnitude.

1.5   *AO HARDWARE AND SOFTWARE IMPLEMENTATION*

Tracking

The wavefront tilt averaged over the full aperture of the telescope accounts for 87 percent of the power in turbulence-induced wavefront aberrations. Full-aperture tilt has the effect of blurring images and reducing the Strehl ratio of point sources. The Strehl ratio due to tilt alone is given by[29, 30]

$$SR_{\mathrm{tilt}} = \cfrac{1}{1 + \cfrac{\pi^2}{2}\left(\cfrac{\sigma_\theta}{\lambda/D}\right)^2} \tag{41}$$

where σ_θ is the one-axis rms full-aperture tilt error, λ is the imaging wavelength, and D is the aperture diameter. Figure 8 is a plot showing how the Strehl ratio drops as jitter increases. This figure shows that in order to maintain a Strehl ratio of 0.8 due to tilt alone, the image must be stabilized to better than $0.25\lambda/D$. For an 8-m telescope imaging at 1.2 μm, $0.25\lambda/D$ is 7.75 milliarcsec.

Fortunately, there are several factors that make tilt sensing more feasible than higher-order sensing for faint guide stars that are available in any field. First, we can use the entire aperture, a light gain over higher-order sensing of roughly $(D/r_0)^2$. Second, the image of the guide star will be compensated well enough that its central core will have a width of approximately λ/D rather than λ/r_0 (assuming that tracking and imaging are near the same wavelength). Third, we can track with only four discrete detectors, making it possible to use photon-counting avalanche photodiodes (or other photon-counting sensors), which have essentially no noise at the short integration times required (~10 ms). Fourth, Tyler[42] has shown that the fundamental frequency that determines the tracking bandwidth is considerably less (by as much as a factor of 9) than the Greenwood frequency, which is appropriate for setting the servo bandwidth of the deformable mirror control system. One must, however, include the vibrational disturbances that are induced into the line of sight by high-frequency jitter in the telescope mount, and it is dangerous to construct a simple rule of thumb comparing tracking bandwidth with higher-order bandwidth requirements.

The rms track error is given approximately by the expression

$$\sigma_\theta = 0.58 \frac{\text{angular image size}}{\text{SNR}_V} \tag{42}$$

where SNR_V is the voltage signal-to-noise ratio in the sensor. An error of $\sigma_\theta = \lambda/4D$ will provide a Strehl ratio of approximately 0.76. If the angular image size is λ/D, the SNR_V needs to be only 2. Since we can count on essentially shot-noise-limited performance, in theory we need only four detected photoelectrons per measurement under ideal conditions (in the real world, we should plan on needing twice this number). Further averaging will occur in the control system servo.

With these assumptions, it is straightforward to compute the required guide star brightness for tracking. Results are shown in Fig. 15 for 1.5, 3.5, 8, and 10-m telescope apertures, assuming a 500-Hz sample rate and twice-diffraction-limited AO compensation of higher-order errors of the guide star at the track wavelength. These results are not inconsistent with high-performance tracking systems already in the field, such as the HRCam system that has been so successful on the Canada-France-Hawaii Telescope at Mauna Kea.[43]

As mentioned previously, the track sensor can consist of just a few detectors implemented in a quadrant-cell algorithm or with a two-dimensional CCD or IR focal plane array. A popular approach for the quad-cell sensor is to use an optical pyramid that splits light in four directions to be detected by individual detectors. Avalanche photodiode modules equipped with photon counter electronics have been used very successfully for this application. These devices operate with such low dark current that they are essentially noise-free and performance is limited by shot noise in the signal and quantum efficiency.

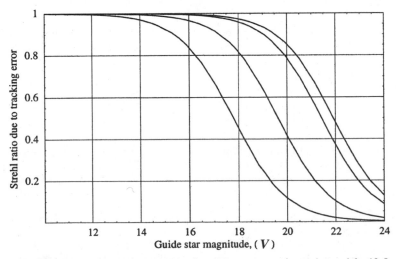

FIGURE 15   Strehl ratio due to tracking jitter. The curves are (top to bottom) for 10, 8, 3.5, and 1.5-m telescopes. The assumptions are photon shot-noise-limited sensors, 25 percent throughput to the track sensor, 200-nm optical bandwidth, and 500-Hz sample rate. The track wavelength is 0.9 µm.

For track sensors using focal plane arrays, different algorithms can be used depending on the tracked object. For unresolved objects, centroid algorithms are generally used. For resolved objects (e.g., planets, moons, asteroids, etc.), correlation algorithms may be more effective. In one instance at the Starfire Optical Range, the highest-quality images of Saturn were made by using an LGS to correct higher-order aberrations and a correlation tracker operating on the rings of the planet provided tilt correction to a small fraction of λ/D.[44]

Higher-Order Wavefront Sensing and Reconstruction: Shack-Hartmann Technique

Higher-order wavefront sensors determine the *gradient,* or slope, of the wavefront measured over subapertures of the entrance pupil, and a dedicated controller maps the slope measurements into deformable mirror actuator voltages. The traditional approach to AO has been to perform these functions in physically different pieces of hardware—the wavefront sensor, the wavefront reconstructor and deformable mirror controller, and the deformable mirror. Over the years, several optical techniques (Shack-Hartmann, various forms of interferometry, curvature sensing, phase diversity, and many others) have been invented for wavefront sensing. A large number of wavefront reconstruction techniques, geometries, and predetermined or even adaptive algorithms have also been developed. A description of all these techniques is beyond the scope of this document, but the interested reader should review work by Wallner,[45] Fried,[46] Wild,[47] and Ellerbroek and Rhoadarmer.[48]

A wavefront sensor configuration in wide use is the Shack-Hartmann sensor.[49] We will use it here to discuss wavefront sensing and reconstruction principles. Figure 16 illustrates the concept. An array of lenslets is positioned at a relayed image of the exit pupil of the telescope. Each lenslet represents a subaperture—in the ideal case, sized to be less than or equal to r_0 at the sensing wavelength. For subapertures that are roughly r_0 in size, the wavefront that is sampled by the subaperture is essentially flat but tilted, and the objective of the sensor is to measure the value of the subaperture tilt. Light collected by each lenslet is focused to a spot on a two-dimensional detector array. By tracking the position of the focused spot, we can determine the X- and Y-tilt of the wavefront averaged over the subaperture defined by the lenslet. By arranging the centers of the lenslets on a two-dimensional grid, we generate gradient values at points in the centers of the lenslets. Many other geometries are possible resulting in a large variety of patterns of gradient measurements.

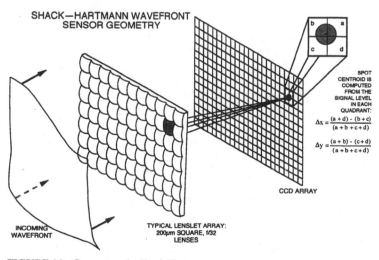

FIGURE 16   Geometry of a Shack-Hartmann sensor.

Figure 17 is a small-scale example from which we can illustrate the basic equations. In this example, we want to estimate the phase at 16 points on the corners of the subapertures (the Fried geometry of the Shack-Hartmann sensor), and to designate these phases as ϕ_1, ϕ_2, $\phi_3, \ldots, \phi_{16}$, using wavefront gradient measurements that are averaged in the centers of the subapertures, $S_{1x}, S_{1y}, S_{2x}, S_{2y}, \ldots, S_{9x}, S_{9y}$. It can be seen by inspection that

$$S_{1x} = \frac{(\phi_2 + \phi_6) - (\phi_1 + \phi_5)}{2d}$$

$$S_{1y} = \frac{(\phi_5 + \phi_6) - (\phi_1 + \phi_2)}{2d}$$

$$\ldots$$

$$S_{9x} = \frac{(\phi_{16} + \phi_{12}) - (\phi_{11} + \phi_{15})}{2d}$$

$$S_{9y} = \frac{(\phi_{15} + \phi_{16}) - (\phi_{11} + \phi_{12})}{2d}$$

This system of equations can be written in the form of a matrix as

$$
\begin{bmatrix} S_{1x} \\ S_{2x} \\ S_{3x} \\ S_{5x} \\ S_{6x} \\ S_{7x} \\ S_{8x} \\ S_{9x} \\ S_{1y} \\ S_{2y} \\ S_{3y} \\ S_{4y} \\ S_{5y} \\ S_{6y} \\ S_{7y} \\ S_{8y} \\ S_{9y} \end{bmatrix}
= \frac{1}{2d}
\begin{bmatrix}
-1 & 1 & 0 & 0 & -1 & 1 & 0 & 0 & 0 & 0 & 0 & 0 & 0 & 0 & 0 & 0 \\
0 & -1 & 1 & 0 & 0 & -1 & 1 & 0 & 0 & 0 & 0 & 0 & 0 & 0 & 0 & 0 \\
0 & 0 & -1 & 1 & 0 & 0 & -1 & 1 & 0 & 0 & 0 & 0 & 0 & 0 & 0 & 0 \\
0 & 0 & 0 & 0 & -1 & 1 & 0 & 0 & -1 & 1 & 0 & 0 & 0 & 0 & 0 & 0 \\
0 & 0 & 0 & 0 & 0 & -1 & 1 & 0 & 0 & -1 & 1 & 0 & 0 & 0 & 0 & 0 \\
0 & 0 & 0 & 0 & 0 & 0 & -1 & 1 & 0 & 0 & -1 & 1 & 0 & 0 & 0 & 0 \\
0 & 0 & 0 & 0 & 0 & 0 & 0 & 0 & -1 & 1 & 0 & 0 & -1 & 1 & 0 & 0 \\
0 & 0 & 0 & 0 & 0 & 0 & 0 & 0 & 0 & -1 & 1 & 0 & 0 & -1 & 1 & 0 \\
0 & 0 & 0 & 0 & 0 & 0 & 0 & 0 & 0 & 0 & -1 & 1 & 0 & 0 & -1 & 1 \\
-1 & -1 & 0 & 0 & 1 & 1 & 0 & 0 & 0 & 0 & 0 & 0 & 0 & 0 & 0 & 0 \\
0 & -1 & -1 & 0 & 0 & 1 & 1 & 0 & 0 & 0 & 0 & 0 & 0 & 0 & 0 & 0 \\
0 & 0 & -1 & -1 & 0 & 0 & 1 & 1 & 0 & 0 & 0 & 0 & 0 & 0 & 0 & 0 \\
0 & 0 & 0 & 0 & -1 & -1 & 0 & 0 & 1 & 1 & 0 & 0 & 0 & 0 & 0 & 0 \\
0 & 0 & 0 & 0 & 0 & -1 & -1 & 0 & 0 & 1 & 1 & 0 & 0 & 0 & 0 & 0 \\
0 & 0 & 0 & 0 & 0 & 0 & -1 & -1 & 0 & 0 & 1 & 1 & 0 & 0 & 0 & 0 \\
0 & 0 & 0 & 0 & 0 & 0 & 0 & 0 & -1 & -1 & 0 & 0 & 1 & 1 & 0 & 0 \\
0 & 0 & 0 & 0 & 0 & 0 & 0 & 0 & 0 & -1 & -1 & 0 & 0 & 1 & 1 & 0 \\
0 & 0 & 0 & 0 & 0 & 0 & 0 & 0 & 0 & 0 & -1 & -1 & 0 & 0 & 1 & 1
\end{bmatrix}
\begin{bmatrix} \phi_1 \\ \phi_2 \\ \phi_3 \\ \phi_4 \\ \phi_5 \\ \phi_6 \\ \phi_7 \\ \phi_8 \\ \phi_9 \\ \phi_{10} \\ \phi_{11} \\ \phi_{12} \\ \phi_{13} \\ \phi_{14} \\ \phi_{15} \\ \phi_{16} \end{bmatrix}
\tag{43}
$$

These equations express gradients in terms of phases,

$$\mathbf{S} = H\mathbf{\Phi} \tag{44}$$

where $\mathbf{\Phi}$ is a vector of the desired phases, H is the measurement matrix, and $\mathbf{S}$ is a vector of the measured slopes.

However, in order to control the actuators in the deformable mirror, we need a control matrix, M, that maps subaperture slope measurements into deformable mirror actuator control commands. In essence, we need to invert Eq. 44 to make it of the form

$$\mathbf{\Phi} = M\mathbf{S}, \tag{45}$$

where M is the desired control matrix. The most straightforward method to derive the control matrix, M, is to minimize the difference in the measured wavefront slopes and the actual slopes on the deformable mirror. We can do this by a maximum a posteriori method account-

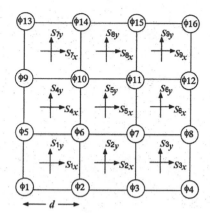

FIGURE 17 A simple Shack-Hartmann sensor in the Fried geometry.

ing for actuator influence functions in the deformable mirror, errors in the wavefront slope measurements due to noise, and statistics of the atmospheric phase distortions. If we do not account for any effects except the geometry of the actuators and the wavefront subapertures, the solution is a least-squares estimate (the most widely implemented to date). It has the form

$$\Phi = [H^T H]^{-1} H^T S \qquad (46)$$

and is the *pseudoinverse* of H. (For our simple geometry shown in Fig. 17, the pseudoinverse solution is shown in Fig. 18.) Even this simple form is often problematical since the matrix $[H^T H]$ is often singular or acts as a singular matrix from computational roundoff error and cannot be inverted.

However, in these instances, singular-value-decomposition (SVD) algorithms can be used to directly compute a solution for the inverse of H. Singular value decomposition decomposes an $m \times n$ matrix into the product of an $m \times n$ matrix (U), an $n \times n$ diagonal matrix (D), and an $n \times n$ square matrix (V). So that

$$H = UDV^T \qquad (47)$$

and H^{-1} is then

$$H^{-1} = VD^{-1}U^T \qquad (48)$$

If H is singular, some of the diagonal elements of D will be zero and D^{-1} cannot be defined. However, this method allows us to obtain the closest possible solution in a least-squares sense by zeroing the elements in the diagonal of D^{-1} that come from zero elements in the matrix D. We can arrive at a solution that discards only those equations that generated the problem in the first place. In addition to straightforward SVD, more general techniques have been proposed involving iterative solutions for the phases.[46, 50, 51]

In addition, several other "tricks" have been developed to alleviate the singularity problem. For instance, piston error is normally ignored, and this contributes to the singularity since there are then an infinite number of solutions that could give the same slope measurements. Adding a row of 1s to the measurement matrix H and setting a corresponding value of 0 in the slope vector for piston allows inversion.[52]

In the implementation of Shack-Hartmann sensors, a reference wavefront must be provided to calibrate imperfections in the lenslet array and distortions that are introduced by any relay optics to match the pitch of the lenslet array to the pitch of the pixels in the detector array. It is general practice to inject a "perfect" plane wave into the optical train just in front

$$
\begin{bmatrix}
\frac{7}{16} & \frac{9}{80} & -\frac{1}{20} & \frac{9}{80} & \frac{3}{20} & \frac{1}{80} & -\frac{1}{20} & \frac{1}{80} & -\frac{1}{16} & \frac{7}{16} & \frac{9}{80} & \frac{1}{20} & \frac{9}{80} & \frac{3}{20} & \frac{1}{80} & -\frac{1}{20} & \frac{1}{80} & -\frac{1}{16} \\[4pt]
\frac{11}{60} & -\frac{61}{240} & -\frac{1}{16} & \frac{29}{240} & \frac{1}{20} & -\frac{19}{240} & \frac{1}{16} & -\frac{11}{240} & \frac{1}{60} & \frac{11}{60} & -\frac{61}{240} & \frac{1}{16} & \frac{29}{240} & \frac{1}{20} & -\frac{19}{240} & \frac{1}{16} & \frac{11}{240} & \frac{1}{60} \\[4pt]
\frac{1}{16} & \frac{61}{240} & -\frac{11}{60} & \frac{19}{240} & -\frac{1}{20} & \frac{29}{240} & \frac{1}{60} & \frac{11}{240} & -\frac{1}{16} & \frac{1}{16} & \frac{61}{240} & \frac{11}{60} & \frac{19}{240} & -\frac{1}{20} & \frac{29}{240} & \frac{1}{60} & \frac{11}{240} & -\frac{1}{16} \\[4pt]
\frac{1}{20} & \frac{9}{80} & \frac{7}{16} & \frac{1}{80} & \frac{3}{20} & -\frac{9}{80} & \frac{1}{16} & -\frac{1}{80} & \frac{1}{20} & \frac{1}{20} & \frac{9}{80} & \frac{7}{16} & \frac{1}{80} & -\frac{3}{20} & -\frac{9}{80} & \frac{1}{16} & \frac{1}{80} & \frac{1}{20} \\[4pt]
-\frac{11}{60} & \frac{29}{240} & -\frac{1}{16} & \frac{61}{240} & \frac{1}{20} & \frac{11}{240} & \frac{1}{16} & -\frac{19}{240} & -\frac{1}{60} & \frac{11}{60} & \frac{29}{240} & \frac{1}{16} & \frac{61}{240} & \frac{1}{20} & -\frac{11}{240} & \frac{1}{16} & \frac{19}{240} & \frac{1}{60} \\[4pt]
\frac{1}{16} & \frac{9}{80} & -\frac{1}{20} & \frac{9}{80} & \frac{3}{20} & \frac{1}{80} & -\frac{1}{20} & \frac{1}{80} & -\frac{1}{16} & \frac{1}{16} & \frac{9}{80} & \frac{1}{20} & \frac{9}{80} & \frac{3}{20} & \frac{1}{80} & -\frac{1}{20} & \frac{1}{80} & -\frac{1}{16} \\[4pt]
\frac{1}{20} & \frac{9}{80} & -\frac{1}{16} & \frac{1}{80} & \frac{3}{20} & \frac{9}{80} & \frac{1}{16} & -\frac{1}{80} & \frac{1}{20} & \frac{1}{20} & \frac{9}{80} & \frac{1}{16} & \frac{1}{80} & \frac{3}{20} & \frac{9}{80} & \frac{1}{16} & \frac{1}{80} & \frac{1}{20} \\[4pt]
\frac{1}{16} & \frac{29}{240} & \frac{11}{60} & \frac{11}{240} & \frac{1}{20} & \frac{61}{240} & \frac{1}{60} & \frac{19}{240} & -\frac{1}{16} & \frac{1}{16} & \frac{29}{240} & \frac{11}{60} & \frac{11}{240} & \frac{1}{20} & \frac{61}{240} & \frac{1}{60} & \frac{19}{240} & \frac{1}{16} \\[4pt]
\frac{1}{16} & \frac{19}{240} & -\frac{1}{60} & \frac{61}{240} & -\frac{1}{20} & -\frac{11}{240} & \frac{29}{240} & \frac{1}{16} & \frac{1}{16} & \frac{19}{240} & \frac{1}{60} & \frac{61}{240} & -\frac{1}{20} & \frac{11}{240} & \frac{11}{60} & \frac{29}{240} & \frac{1}{16} & \frac{1}{16} \\[4pt]
\frac{1}{20} & \frac{1}{80} & \frac{1}{16} & \frac{9}{80} & \frac{3}{20} & \frac{1}{80} & \frac{1}{16} & \frac{9}{80} & \frac{1}{20} & \frac{1}{20} & \frac{1}{80} & \frac{1}{16} & \frac{9}{80} & \frac{3}{20} & \frac{1}{80} & \frac{1}{16} & \frac{9}{80} & \frac{1}{20} \\[4pt]
\frac{1}{16} & -\frac{1}{80} & \frac{1}{20} & \frac{1}{80} & \frac{3}{20} & -\frac{9}{80} & \frac{1}{20} & \frac{9}{80} & -\frac{1}{16} & \frac{1}{16} & \frac{1}{80} & \frac{1}{20} & \frac{1}{80} & \frac{3}{20} & \frac{9}{80} & \frac{1}{20} & \frac{9}{80} & -\frac{1}{16} \\[4pt]
\frac{1}{60} & \frac{19}{240} & \frac{1}{16} & \frac{11}{240} & \frac{1}{20} & \frac{61}{240} & \frac{1}{16} & \frac{29}{240} & \frac{11}{60} & -\frac{1}{60} & \frac{19}{240} & \frac{1}{16} & \frac{11}{240} & -\frac{1}{20} & \frac{61}{240} & \frac{1}{16} & \frac{29}{240} & \frac{11}{60} \\[4pt]
\frac{1}{20} & \frac{1}{80} & \frac{1}{16} & \frac{9}{80} & \frac{3}{20} & \frac{1}{80} & -\frac{7}{16} & \frac{9}{80} & \frac{1}{20} & \frac{1}{20} & \frac{1}{80} & \frac{1}{16} & \frac{9}{80} & \frac{3}{20} & \frac{1}{80} & \frac{7}{16} & -\frac{9}{80} & \frac{1}{20} \\[4pt]
\frac{1}{16} & \frac{11}{240} & \frac{1}{60} & \frac{29}{240} & \frac{1}{20} & \frac{19}{240} & \frac{11}{60} & \frac{61}{240} & \frac{1}{16} & \frac{1}{16} & \frac{11}{240} & \frac{1}{60} & \frac{29}{240} & \frac{1}{20} & \frac{19}{240} & \frac{11}{60} & \frac{61}{240} & \frac{1}{16} \\[4pt]
\frac{1}{60} & \frac{11}{240} & \frac{1}{16} & \frac{19}{240} & -\frac{1}{20} & \frac{29}{240} & \frac{1}{16} & \frac{61}{240} & \frac{11}{60} & \frac{1}{60} & \frac{11}{240} & \frac{1}{16} & \frac{19}{240} & \frac{1}{20} & \frac{29}{240} & \frac{1}{16} & \frac{61}{240} & \frac{11}{60} \\[4pt]
\frac{1}{16} & -\frac{1}{80} & \frac{1}{20} & \frac{1}{80} & \frac{3}{20} & -\frac{9}{80} & \frac{1}{20} & \frac{9}{80} & \frac{7}{16} & \frac{1}{16} & \frac{1}{80} & \frac{1}{20} & \frac{1}{80} & \frac{3}{20} & \frac{9}{80} & \frac{1}{20} & \frac{9}{80} & \frac{7}{16}
\end{bmatrix}
$$

FIGURE 18 Least-squares reconstructor matrix for the geometry shown in Fig. 17.

of the lenslet array and to record the positions of all of the Shack-Hartmann spots. During normal operation, the wavefront sensor gradient processor then computes the difference between the spot position of the residual error wavefront and the reference wavefront.

Laser Beacons (Laser Guide Stars)

Adaptive optical systems require a beacon or a source of light to sense turbulence-induced wave distortions. From an anisoplanatic point of view, the ideal beacon is the object being imaged. However, most objects of interest to astronomers are not bright enough to serve as beacons.

It is possible to create artificial beacons (also referred to as synthetic beacons) that are suitable for wavefront sensing with lasers as first demonstrated by Fugate et al.[53] and Primmerman et al.[54] Laser beacons can be created by Rayleigh scattering of focused beams at ranges of between 15 and 20 km or by resonant scattering from a layer of sodium atoms in the mesosphere at an altitude of between 90 and 100 km. Examples of Rayleigh beacon AO systems are described by Fugate et al.[55] for the SOR 1.5-m telescope and by Thompson et al.[56] for the Mt. Wilson 100-inch telescope; examples of sodium beacon AO systems are described by Olivier et al.[57] for the Lick 3-m Shane telescope and by Butler et al.[58] for the 3.5-m Calar Alto telescope. Researchers at the W. M. Keck Observatory at Mauna Kea are in the process of installing a sodium dye laser to augment their NGS AO system on Keck II.

The laser beacon concept was first conceived and demonstrated within the U.S. Department of Defense during the early 1980s (see Fugate[59] for a short summary of the history). The information developed under this program was not declassified until May 1991, but much of the early work was published subsequently.[60] The laser beacon concept was first published openly by Foy and Labeyrie[61] in 1985 and has been of interest in the astronomy community since.

Even though laser beacons solve a significant problem, they also introduce new problems and have two significant limitations compared with bright NGSs. The new problems include potential light contamination in science cameras and tracking sensors, cost of ownership and operation, and observing complexities that are associated with propagating lasers through the navigable airspace and near-earth orbital space, the home of thousands of space payloads. The technical limitations are that laser beacons provide no information on full-aperture tilt, and that a "cone effect" results from the finite altitude of the beacon, which contributes an additional error called focus (or focal) anisoplanatism. Focus anisoplanatism can be partially alleviated by using a higher-altitude beacon, such as a sodium guide star, as discussed in the following.

Focus Anisoplanatism   The mean square wavefront error due to the finite altitude of the laser beacon is given by

$$\sigma_{FA}^2 = (D/d_0)^{5/3} \tag{49}$$

where d_0 is an effective aperture size corrected by the laser beacon that depends on the height of the laser beacon, the C_n^2 profile, the zenith angle, and the imaging wavelength. This parameter was defined by Fried and Belsher.[62] Tyler[63] developed a method to rapidly evaluate d_0 for arbitrary C_n^2 profiles given by the expression

$$d_0 = \lambda^{6/5} \cos^{3/5}(\psi) \left[\int C_n^2(z) F(z/H) dz \right]^{-3/5} \tag{50}$$

where H is the vertical height of the laser beacon and the function $F(z/H)$ is given by

$$
\begin{aligned}
F(z/H) = {} & 16.71371210((1.032421640 - 0.8977579487u) \\
& \times [1 + (1 - z/H)^{5/3}] - 2.168285442 \\
& \times \left\{ \frac{6}{11} \, {}_2F_1\left[-\frac{11}{6}, -\frac{5}{6}; 2; \left(1 - \frac{z}{H}\right)^2 \right] \right. \\
& - \frac{6}{11} (z/H)^{5/3} - u \frac{10}{11}\left(1 - \frac{z}{H}\right) \\
& \left. \times {}_2F_1\left[-\frac{11}{6}, \frac{1}{6}; 3; \left(1 - \frac{z}{H}\right)^2 \right] \right\})
\end{aligned}
\tag{51}
$$

for $z < H$ and

$$F(z/H) = 16.71371210(1.032421640 - 0.8977579487u) \qquad (52)$$

for $z > H$. In these equations, z is the vertical height above the ground, H is the height of the laser beacon, u is a parameter that is equal to zero when only piston is removed and that is equal to unity when piston and tilt are removed, and $_2F_1[a, b; c; z]$ is the hypergeometric function.

Equations 50 and 51 are easily evaluated on a programmable calculator or a personal computer. They are very useful for quickly establishing the expected performance of laser beacons for a particular C_n^2 profile, imaging wavelength, and zenith angle view. Figures 19 and 20 are plots of d_0 representing the best and average seeing at Mauna Kea and a site described by the HV_{57} turbulence profile. Since d_0 scales as $\lambda^{6/5}$, values at 2.2 µm are 5.9 times larger, as shown in the plots.

It is straightforward to compute the Strehl ratio due only to focus anisoplanatism using the approximation $SR = e^{-\sigma_{FA}^2}$, where $\sigma_{FA}^2 = (D/d_0)^{5/3}$. There are many possible combinations of aperture sizes, imaging wavelength, and zenith angle, but to illustrate the effect for 3.5- and 10-m apertures, Figs. 21 and 22 show the focus anisoplanatism Strehl ratio as a function of wavelength for 15- and 90-km beacon altitudes and for three seeing conditions. As these plots show, the effectiveness of laser beacons is very sensitive to the aperture diameter, seeing conditions, and beacon altitude. A single Rayleigh beacon at an altitude of 15 km is essentially useless on a 10-m aperture in $HV_{5/7}$ seeing, but it is probably useful at a Mauna Kea–like site. One needs to keep in mind that the curves in Figs. 21 and 22 are the upper limits of performance since other effects will further reduce the Strehl ratio.

Generation of Rayleigh Laser Beacons   For Rayleigh scattering, the number of photodetected electrons (pdes) per subaperture in the wavefront sensor, N_{pde}, can be computed by using a lidar equation of the form

$$N_{pde} = \eta_{QE} T_t T_r T_{atm}^2 \, \frac{A_{sub}\beta_{BS}\Delta l}{R^2} \, \frac{E_p\lambda}{hc} \qquad (53)$$

where

η_{QE} = the quantum efficiency of the wavefront sensor
T_t = the laser transmitter optical transmission

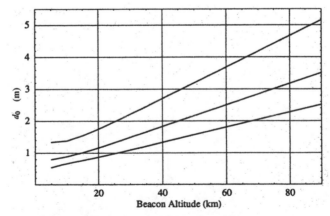

FIGURE 19   Values of d_0 versus laser beacon altitude for zenith imaging at 0.5 µm and (top to bottom) best Mauna Kea seeing, average Mauna Kea seeing, and the $HV_{5/7}$ turbulence profile.

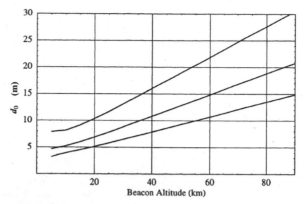

FIGURE 20 Values of d_0 versus laser beacon altitude for zenith imaging at 2.2 μm and (top to bottom) best Mauna Kea seeing, average Mauna Kea seeing, and the $HV_{5/7}$ turbulence profile.

T_r = the optical transmission of the wavefront sensor
T_{atm} = the one-way transmission of the atmosphere
A_{sub} = the area of a wavefront sensor subaperture
β_{BS} = the fraction of incident laser photons backscattered per meter of scattering volume [steradian $(sr)^{-1}\ m^{-1}$]
R = the range to the midpoint of the scattering volume
Δl = the length of the scattering volume—the range gate
E_p = the energy per pulse
λ = the laser wavelength
h = Planck's constant
c = the speed of light

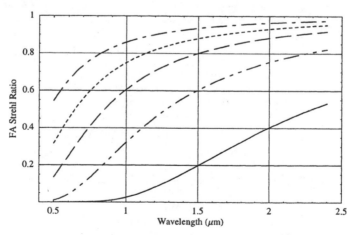

FIGURE 21 The telescope Strehl ratio due to focus anisoplanatism only. Conditions are for a 3.5-m telescope viewing at 30° zenith angle. Curves are (top to bottom): best seeing at Mauna Kea, 90-km beacon; average seeing at Mauna Kea, 90-km beacon; $HV_{5/7}$, 90-km beacon; best seeing at Mauna Kea, 15-km beacon; and $HV_{5/7}$, 15-km beacon.

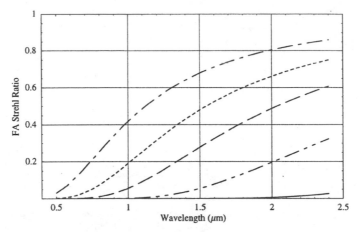

FIGURE 22 The telescope Strehl ratio due to focus anisoplanatism only. Conditions are for a 10.0-m telescope viewing at 30° zenith angle. Curves are (top to bottom): best seeing at Mauna Kea, 90-km beacon; average seeing at Mauna Kea, 90-km beacon; HV$_{5/7}$, 90-km beacon; best seeing at Mauna Kea, 15-km beacon; and HV$_{5/7}$, 15-km beacon.

The laser beam is focused at range R, and the wavefront sensor is gated on and off to exclude backscattered photons from the beam outside a range gate of length Δl centered on range R. The volume-scattering coefficient is proportional to the atmospheric pressure and is inversely proportional to the temperature and the fourth power of the wavelength. Penndorf[64] developed the details of this relationship, which can be reduced to

$$\beta_{BS} = 4.26 \times 10^{-7} \frac{P(h)}{T(h)} \text{ sr}^{-1} \text{ m}^{-1} \tag{54}$$

where values of number density, pressure, and temperature at sea level (needed in Penndorf's equations) have been obtained from the U.S. Standard Atmosphere.[65] At an altitude of 10 km, $\beta_{BS} = 5.1 \times 10^{-7}$ sr^{-1} m^{-1} and a 1-km-long volume of the laser beam scatters only 0.05 percent of the incident photons per steradian. Increasing the length of the range gate would increase the total signal that is received by the wavefront sensor; however, we should limit the range gate so that subapertures at the edge of the telescope are not able to resolve the projected length of the scattered light. Range gates that are longer than this criterion increase the size of the beacon's image in a subaperture and increase the rms value of the measurement error in each subaperture. A simple geometric analysis leads to an expression for the maximum range gate length:

$$\Delta l = 2 \frac{\lambda R_b^2}{D r_0} \tag{55}$$

where R_b is the range to the center of the beacon and D is the aperture diameter of the telescope. Figure 23 shows the computed signal in detected electrons per subaperture as a function of altitude for a 100-watt (W) average power laser operating at 1000 pulses per second at either 351 nm (upper curve) or 532 nm (lower curve). Other parameters used to compute these curves are listed in the figure caption. When all first-order effects are accounted for, notice that (for high-altitude beacons) there is little benefit to using an ultraviolet wavelength laser over a green laser—even though the scattering goes as λ^{-4}. With modern low-noise detector arrays, signal levels as low as 50 pdes are very usable; above 200 pdes, the sensor generally no longer dominates the performance.

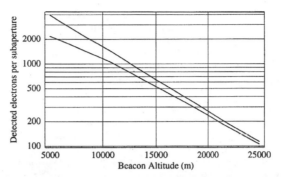

FIGURE 23 Rayleigh laser beacon signal versus altitude of the range gate for two laser wavelengths: 351 nm (top curve) and 532 nm (bottom curve). Assumptions for each curve: range gate length $= 2\lambda R_b^2/(Dr_0)$, $D = 1.5$ m, $r_0 = 5$ cm at 532 nm and 3.3 cm at 351 nm, $T_t = 0.30$, $T_r = 0.25$, $\eta_{OE} = 90$ percent at 532 nm and 70 percent at 351 nm, $E_p = 0.10$ J, subaperture size $= 10$ cm², one-way atmospheric transmission computed as a function of altitude and wavelength.

Equations 53 and 55 accurately predict what has been realized in practice. A pulsed copper-vapor laser, having an effective pulse energy of 180 millijoules (mJ) per wavefront sample, produced 190 pdes per subaperture on the Starfire Optical Range 1.5-m telescope at a backscatter range of 10 km, range gate of 2.4 km, and subapertures of 9.2 cm. The total round-trip optical and quantum efficiency was only 0.4 percent.[55]

There are many practical matters on how to design the optical system to project the laser beam. If the beam shares any part of the optical train of the telescope, then it is important to inject the beam as close to the output of the telescope as feasible. The best arrangement is to temporally share the aperture with a component that is designed to be out of the beam train when sensors are using the telescope (e.g., a rotating reflecting wheel as described by Thompson et al.[56]). If the laser is injected optically close to wavefront sensors or trackers, there is the additional potential of phosphorescence in mirror substrates and coatings that can emit photons over a continuum of wavelengths longer than the laser fundamental. The decay time of these processes can be longer than the interpulse interval of the laser, presenting a pseudo-continuous background signal that can interfere with faint objects of interest. This problem was fundamental in limiting the magnitude of natural guide star tracking with photon counting avalanche photo diodes at the SOR 1.5-m telescope during operations with the copper-vapor laser.[66]

Generation of Mesospheric Sodium Laser Beacons   As the curves in Figs. 21 and 22 show, performance is enhanced considerably when the beacon is at high altitude (90 versus 15 km). The signal from Rayleigh scattering is 50,000 times weaker for a beacon at 90 km compared with one at 15 km. Happer et al.[67] suggested laser excitation of mesospheric sodium in 1982; however, the generation of a beacon that is suitable for use with AO has turned out to be a very challenging problem. The physics of laser excitation is complex and the development of an optimum laser stresses modern materials science and engineering. This section only addresses how the temporal and spectral format of the laser affects the signal return. The engineering issues of building a laser are beyond the present scope.

The sodium layer in the mesosphere is believed to arise from meteor ablation. The average height of the layer is approximately 95 km above sea level and it is 10 km thick. The column density is only 2–$5 \cdot 10^9$ atoms/cm², or roughly 10^3 atoms/cm³. The temperature of the layer is approximately 200 K, resulting in a Doppler broadened absorption profile having a full width half maximum (FWHM) of about 3 gigahertz (GHz), which is split into two broad resonance peaks that are separated by 1772 MHz, caused by splitting of the $3S_{1/2}$ ground state. The natural lifetime of an excited state is only 16 ns. At high laser intensities (roughly 6 mW/cm^2 for lasers with a natural line width of 10 MHz or 5 W/cm^2 for lasers having spectral content that covers the entire Doppler broadened spectrum), saturation occurs and the return signal does not increase linearly with increasing laser power.

The three types of lasers that have been used to date for generating beacons are the continuous-wave (CW) dye laser, the pulsed-dye laser, and a solid-state sum-frequency laser. Continuous-wave dye lasers are available commercially and generally provide from 2 to 4 W of power. A specialized pulsed-dye laser installed at Lick Observatory's 3-m Shane Telescope, built at Lawrence Livermore National Laboratory by Friedman et al.[68], produces an average power of 20 W consisting of 100- to 150-ns-long pulses at an 11-kHz pulse rate. The sum-frequency laser concept relies on the sum-frequency mixing of the 1.064- and 1.319-µm lines of the Nd:YAG laser in a nonlinear crystal to produce the required 0.589-µm wavelength for the spectroscopic D_2 line. The first experimental devices were built by Jeys at MIT Lincoln Laboratory.[69, 70] The pulse format is one of an envelope of macropulses, lasting of the order of 100 µs, containing mode-locked pulses at roughly a 100-MHz repetition rate, and a duration of from 400 to 700 ps. The sum-frequency lasers built to date have had average powers of from 8 to 20 W and have been used in field experiments at SOR and Apache Point Observatory.[71, 72]

A comprehensive study of the physics governing the signal that is generated by laser excitation of mesospheric sodium has been presented by Milonni et al.[73, 74] for short, intermediate, and long pulses, and CW, corresponding to the lasers described previously. Results are obtained by numerical computations and involve a full-density matrix treatment of the sodium D_2 line. In some specific cases, it has been possible to approximate the results with analytical models that can be used to make rough estimates of signal strengths.

Figure 24 shows results of computations by Milonni et al. for short- and long-pulse formats and an analytical extension of numerical results for a high-power CW laser. Results are presented as the number of photons per square centimeter per millisecond received at the primary mirror of the telescope versus average power of the laser. The curves correspond to (top to bottom) the sum-frequency laser; a CW laser whose total power is spread over six narrow

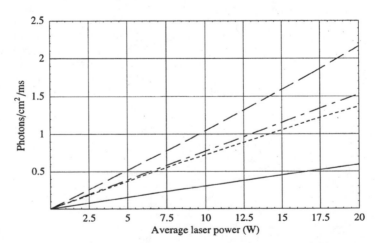

FIGURE 24   Sodium laser beacon signal versus average power of the pump laser. The lasers are (top to bottom) sum-frequency laser with a micropulse-macropulse format (150-µs macropulses filled with 700-ps micropulses at 100 MHz), a CW laser having six 10-MHz-linewidth lines, a CW laser having a single 10-MHz linewidth, and a pulsed-dye laser having 150-ns pulses at a 30-kHz repetition rate. The sodium layer is assumed to be 90 km from the telescope, the atmospheric transmission is 0.7, and the spot size in the mesosphere is 1.2 arcsec.

lines that are distributed across the Doppler profile; a CW laser having only one narrow line, and the pulsed-dye laser. The specifics are given in the caption for Fig. 24. Figure 25 extends these results to 200-W average power and shows significant saturation for the pulsed-dye laser format and the single-frequency CW laser as well as some nonlinear behavior for the sum-frequency laser. If the solid-state laser technology community can support high-power, narrow-line CW lasers, this chart says that saturation effects at high power can be ameliorated by spreading the power over different velocity classes of the Doppler profile. These curves can be used to do first-order estimates of signals that are available from laser beacons that are generated in mesospheric sodium with lasers having these temporal formats.

Real-Time Processors

The mathematical process described by Eq. 45 is usually implemented in a dedicated processor that is optimized to perform the matrix multiplication operation. Other wavefront reconstruction approaches that are not implemented by matrix multiplication routines are also possible, but they are not discussed here.

Matrix multiplication lends itself to parallel operations and the ultimate design to maximize speed is to dedicate a processor to each actuator in the deformable mirror. That is, each central processing unit (CPU) is responsible for multiplying and accumulating the sum of each element of a row in the M matrix with each element of the slope vector S. The values of the slope vector should be broadcast to all processors simultaneously to reap the greatest benefit. Data flow and throughput is generally a more difficult problem than raw computing power. It is possible to buy very powerful commercial off-the-shelf processing engines, but

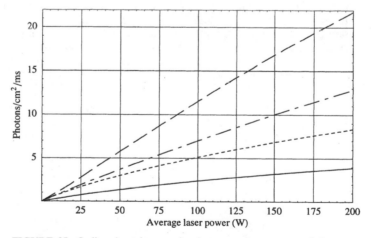

FIGURE 25 Sodium laser beacon signal versus average power of the pump laser. The lasers are (top to bottom) sum-frequency laser with a micropulse-macropulse format (150-μs macropulses filled with 700-ps micropulses at 100 MHz), a CW laser having six 10-MHz-linewidth lines, a CW laser having a single 10-MHz linewidth, and a pulsed-dye laser having 150-ns pulses at a 30-kHz repetition rate. The sodium layer is assumed to be 90 km from the telescope, the atmospheric transmission is 0.7, and the spot size in the mesosphere is 1.2 arcsec. Saturation of the return signal is very evident for the temporal format of the pulsed-dye laser and for the single-line CW laser. Note, however, that the saturation is mostly eliminated (the line is nearly straight) in the single-line laser by spreading the power over six lines. There appears to be very little saturation in the micropulse-macropulse format.

getting the data into and out of the engines usually reduces their ultimate performance. A custom design is required to take full advantage of component technology that is available today.

An example of a custom-designed system is the SOR 3.5-m telescope AO system[75] containing 1024 digital signal processors running at 20 MHz, making a 20-billion-operations-per-second system. This system can perform a $(2048 \times 1024) \times (2048)$ matrix multiply to 16-bit precision (40-bit accumulation), low-pass-filter the data, and provide diagnostic data collection in less than 24 μs. The system throughput exceeds 400 megabytes per second (MB/s). Most astronomy applications have less demanding latency and throughput requirements that can be met with commercial off-the-shelf hardware and software. The importance of latency is illustrated in Fig. 26. This figure shows results of analytical predictions showing the relationship between control loop bandwidth of the AO system and the wavefront sensor frame rate for different data latencies.[76] These curves show that having a high-frame-rate wavefront sensor camera is not a sufficient condition to achieve high control loop bandwidth. The age of the data is of utmost importance. As Fig. 26 shows, the optimum benefit in control bandwidth occurs when the latency is significantly less than a frame time (~½).

Ensemble-averaged atmospheric turbulence conditions are very dynamic and change on the scale of minutes. Optimum performance of an AO system cannot be achieved if the system is operating on phase estimation algorithms that are based on inaccurate atmospheric information. Optimum performance requires changing the modes of operation in near–real time. One of the first implementations of adaptive control was the system ADONIS, which was implemented on the ESO 3.6-m telescope at La Silla, Chile.[77] ADONIS employs an artificial intelligence control system that controls which spatial modes are applied to the deformable mirror, depending on the brightness of the AO beacon and the seeing conditions.

A more complex technique has been proposed[48] that would allow continuous updating of the wavefront estimation algorithm. The concept is to use a recursive least-squares adaptive algorithm to track the temporal and spatial correlations of the distorted wavefronts. The algorithm uses current and recent past information that is available to the servo system to predict the wavefront for a short time in the future and to make the appropriate adjustments to the deformable mirror. A sample scenario has been examined in a detailed simulation, and the system Strehl ratio achieved with the recursive least-squares adaptive algorithm is essentially

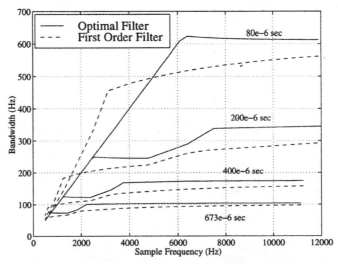

FIGURE 26   Effect of data latency (varying from 80 to 673 μs in this plot) on the control loop bandwidth of the AO system. In this plot, the latency includes only the sensor readout and wavefront processing time.

the same as an optimal reconstructor with *a priori* knowledge of the wind and turbulence profiles. Sample results of this simulation are shown in Fig. 27. The requirements for implementation of this algorithm in a real-time hardware processor have not been worked out in detail. However, it is clear that they are considerable, perhaps greater by an order of magnitude than the requirements for an ordinary AO system.

Other Higher-Order Wavefront Sensing Techniques

Various forms of shearing interferometers have been successfully used to implement wavefront sensors for atmospheric turbulence compensation.[78, 79] The basic principle is to split the wavefront into two copies, translate one laterally with respect to the other, and then interfere them. The bright and dark regions of the resulting fringe pattern are proportional to the slope of the wavefront. Furthermore, a lateral shearing interferometer is self-referencing—it does not need a plane wave reference like the Shack-Hartmann sensor. Shearing interferometers are not in widespread use today, but they have been implemented in real systems in the past.[78]

Roddier[80, 81] introduced a new concept for wavefront sensing based on measuring local wavefront curvature (the second derivative of the phase). The concept can be implemented by differencing the irradiance distributions from two locations on either side of the focal plane of a telescope. If the two locations are displaced a distance, *l*, from the focus of the telescope and the spatial irradiance distribution is given by $I_1(\mathbf{r})$ and $I_2(\mathbf{r})$ at the two locations, then the relationship between the irradiance and the phase is given by

$$\frac{I_1(\mathbf{r}) - I_2(-\mathbf{r})}{I_1(\mathbf{r}) + I_2(-\mathbf{r})} = \frac{\lambda F(F-l)}{2\pi l}\left[\frac{\partial\phi}{\partial n}\left(\frac{F}{l}\mathbf{r}\right)\delta_c - \nabla^2\phi\left(\frac{F}{l}\mathbf{r}\right)\right] \tag{56}$$

where F is the focal length of the telescope, and the Dirac delta δ_c represents the outward-pointing normal derivative on the edge of the phase pattern. Equation 56 is valid in the geometrical optics approximation. The distance, *l*, must be chosen such that the validity of the

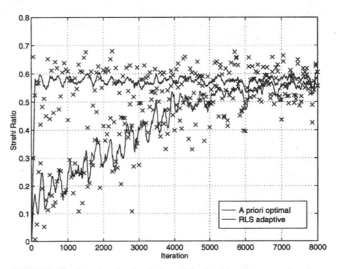

FIGURE 27 Results of a simulation of the Strehl ratio versus time for a recursive least-squares adaptive estimator (lower curve) compared with an optimal estimator having *a priori* knowledge of turbulence conditions (upper curve). The ×'s represent the instantaneous Strehl ratio computed by the simulation, and the lines represent the average values.

geometrical optics approximation is ensured, requiring that the blur at the position of the defocused pupil image is small compared with the size of the wavefront aberrations desired to be measured. These considerations lead to a condition on l that

$$l \geq \theta_b \, \frac{F^2}{d} \tag{57}$$

where θ_b is the blur angle of the object that is produced at the positions of the defocused pupil, and d is the size of the subaperture determined by the size of the detector. For point sources and when $d > r_0$, $\theta_b = \lambda/r_0$ and $l \geq \lambda F^2/r_0 d$. For extended sources of angular size $\theta > \lambda/r_0$, l must be chosen such that $l \geq \theta F^2/d$. Since increasing l decreases the sensitivity of the curvature sensor, in normal operations l is set to satisfy the condition $l \geq \lambda F^2/r_0 d$, but once the loop is closed and low-order aberrations are reduced, the sensitivity of the sensor can be increased by making l smaller. To perform wavefront reconstruction, an iterative procedure can be used to solve Poisson's equation. The appeal of this approach is that certain deformable mirrors such as piezoelectric bimorphs deform locally as nearly spherical shapes, and can be driven directly with no intermediate mathematical wavefront reconstruction step. This has not been completely realized in practice, but excellent results have been obtained with two systems deployed at Mauna Kea.[82] All implementations of these wavefront sensors to date have employed single-element avalanche photodiodes operating in the photon-counting mode for each subaperture. Since these devices remain quite expensive, it may be cost prohibitive to scale the curvature-sensing technique to very high density actuator systems. An additional consideration is that the noise gain goes up linearly with the number of actuators, not logarithmically as with the Shack-Hartmann or shearing interferometer.

The problem of deriving phase from intensity measurements has been studied extensively.[83–86] A particular implementation of multiple-intensity measurements to derive phase data has become known as phase diversity.[87] In this approach, one camera is placed at the focus of the telescope and another in a plane a known amount of defocus from focus. Intensity gathered simultaneously from both cameras can be processed to recover the phase in the pupil of the telescope. The algorithms needed to perform this operation are complex and require iteration.[88] Such a technique does not presently lend itself to real-time estimation of phase errors in an adaptive optical system, but it could in the future.

Artificial neural networks have been used to estimate phase from intensity as well. The concept is similar to other methods using multiple-intensity measurements. Two focal planes are set up, one at the focus of the telescope and one near focus. The pixels from each focal plane are fed into the nodes of an artificial neural network. The output of the network can be set up to provide most any desired information from Zernike decomposition elements to direct-drive signals to actuators of a zonal, deformable mirror. The network must be trained using known distortions. This can be accomplished by using a deterministic wavefront sensor to measure the aberrations and using that information to adjust the weights and coefficients in the neural network processor. This concept has been demonstrated on real telescopes in atmospheric turbulence.[89] The concept works for low-order distortions, but it appears to have limited usefulness for large, high-density actuator adaptive optical systems.

Wavefront Correctors

There are three major classes of wavefront correctors in use today: segmented mirrors, bimorph mirrors, and stacked-actuator continuous-facesheet mirrors. Figure 28 shows the concept for each of these mirrors.

The segmented mirror can have piston-only or piston-and-tilt actuators. Since the individual segments are completely independent, it is possible for significant phase errors to develop between the segments. It is therefore important that these errors be controlled by real-time interferometry or by strain gauges or other position-measuring devices on the individual actu-

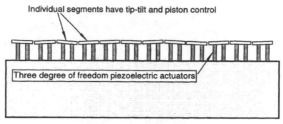

Individual segments have tip-tilt and piston control

Three degree of freedom piezoelectric actuators

SEGMENTED DEFORMABLE MIRROR

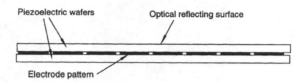

Piezoelectric wafers

Optical reflecting surface

Electrode pattern

BIMORPH DEFORMABLE MIRROR

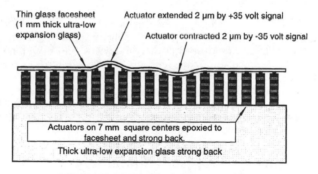

Thin glass facesheet
(1 mm thick ultra-low
expansion glass)

Actuator extended 2 µm by +35 volt signal

Actuator contracted 2 µm by -35 volt signal

Actuators on 7 mm square centers epoxied to
facesheet and strong back.

Thick ultra-low expansion glass strong back

STACKED ACTUATOR CONTINUOUS
FACESHEET DEFORMABLE MIRROR

FIGURE 28   Cross sections of three deformable mirror designs.

ators. Some large segmented mirrors have been built[90] for DOD applications and several are in use today for astronomy.[91, 92]

The stacked actuator deformable mirror is probably the most widely used wavefront corrector. The modern versions are made with lead-magnesium-niobate—a ceramic-like electrostrictive material producing 4 µm of stroke for 70 V of drive.[93] The facesheet of these mirrors is typically 1 mm thick. Mirrors with as many as 2200 actuators have been built. These devices are very stiff structures with first resonant frequencies as high as 25 kHz. They are typically built with actuators spaced as closely as 7 mm. One disadvantage with this design is that it becomes essentially impossible to repair individual actuators once the mirror structure is epoxied together. Actuator failures are becoming less likely with today's refined technology, but a very large mirror may have 1000 actuators, which increases its chances for failures over the small mirrors of times past. Figure 29 is a photograph of the 941-actuator deformable mirror in use at the 3.5-m telescope at the SOR.

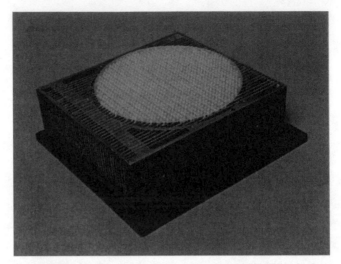

FIGURE 29 The 941-actuator deformable mirror built by Xinetics, Inc., for the SOR 3.5-m telescope.

New Wavefront Corrector Technologies

Our progress toward good performance at visible wavelengths will depend critically on the technology that is available for high-density actuator wavefront correctors. There is promise in the areas of liquid crystals, MEM devices, and nonlinear-optics processes. However, at least for the next few years, it seems that we will have to rely on conventional mirrors with piezoelectric-type actuators or bimorph mirrors made from sandwiched piezoelectric layers. Nevertheless, there is significant development in the conventional mirror area that has the potential for making mirrors with very large numbers of actuators that are smaller, more reliable, and much less expensive.

1.6 HOW TO DESIGN AN ADAPTIVE OPTICAL SYSTEM

Adaptive optical systems are complex and their performance is governed by many parameters, some controlled by the user and some controlled by nature. The system designer is faced with selecting parameter values to meet performance requirements. Where does he or she begin? One approach is presented here and consists of the following six steps:

1. Determine the average seeing conditions (the mean value of r_0 and f_G) for the site.
2. Determine the most important range of wavelengths of operation.
3. Decide on the minimum Strehl ratio that is acceptable to the users for these seeing conditions and operating wavelengths.
4. Determine the brightness of available beacons.
5. Given the above requirements, determine the optimum values of the subaperture size and the servo bandwidth to minimize the residual wavefront error at the most important wavelength and minimum beacon brightness. (The most difficult parameters to change after the system is built are the wavefront sensor subaperture and deformable mirror actuator geometries. These parameters need to be chosen carefully to address the highest-priority requirements in terms of seeing conditions, beacon brightness, operating wavelength, and required Strehl ratio.) Determine if the associated Strehl ratio is acceptable.

6. Evaluate the Strehl ratio for other values of the imaging wavelength, beacon brightness, and seeing conditions. If these are unsatisfactory, vary the wavefront sensor and track parameters until an acceptable compromise is reached.

Our objective in this section is to develop some practical formulas that can be implemented and evaluated quickly on desktop or laptop computers using programs like Mathematica™ or MathCad™ that will allow one to iterate the six aforementioned steps to investigate the top-level trade space and optimize system performance for the task at hand.

The most important parameters that the designer has some control over are the subaperture size, the wavefront sensor and track sensor integration times, the latency of data in the higher-order and tracker control loops, and the wavelength of operation. One can find optimum values for these parameters since changing them can either increase or decrease the system Strehl ratio. The parameters that the designer has little or no control over are those that are associated with atmospheric turbulence. On the other hand, there are a few parameters that always make performance better the larger (or smaller) we make the parameter: the quantum efficiency of the wavefront and track sensors, the brightness of the beacon(s), the optical system throughput, and read noise of sensors. We can never make a mistake by making the quantum efficiency as large as physics will allow and the read noise as low as physics will allow. We can never have a beacon too bright, nor can we have too much optical transmission (optical attenuators are easy to install). Unfortunately, it is not practical with current technology to optimize the AO system's performance for changing turbulence conditions, for different spectral regions of operation, for different elevation angle, and for variable target brightness by changing the mechanical and optical design from minute to minute. We must be prepared for some compromises based on our initial design choices.

We will use two system examples to illustrate the process that is outlined in the six aforementioned steps: (1) a 3.5-m telescope at an intracontinental site that is used for visible imaging of low-earth-orbiting artificial satellites, and (2) a 10-m telescope that operates at a site of excellent seeing for near-infrared astronomy.

Establish the Requirements

Table 2 lists a set of requirements that we shall try to meet by trading system design parameters. The values in Table 2 were chosen to highlight how requirements can result in significantly different system designs.

In the 3.5-m telescope example, the seeing is bad, the control bandwidths will be high, the imaging wavelength is short, and the required Strehl ratio is significant. Fortunately, the objects (artificial earth satellites) are bright. The 10-m telescope application is much more forgiving with respect to the seeing and imaging wavelengths, but the beacon brightness is four magnitudes fainter and a high Strehl ratio is still required at the imaging wavelength. We now investigate how these requirements determine an optimum choice for the subaperture size.

TABLE 2 Requirements for Two System Examples

Parameter	3.5-m Requirement	10-m Requirement
C_n^2 profile	Modified HV_{57}	Average Mauna Kea
Wind profile	Slew dominated	Bufton
Average r_0 (0.5 μm)	10 cm	18 cm
Average f_G (0.5 μm)	150 Hz	48 Hz
Imaging wavelength	0.85 μm	1.2–2.2 μm
Elevation angle	45°	45°
Minimum Strehl ratio	0.5	0.5
Beacon brightness	$m_v = 6$	$m_v = 10$

Selecting a Subaperture Size

As was mentioned previously, choosing the subaperture size should be done with care, because once a system is built it is not easily changed. The approach is to develop a mathematical expression for the Strehl ratio as a function of system design parameters and seeing conditions and then to maximize the Strehl ratio by varying the subaperture size for the required operating conditions that are listed in Table 2.

The total-system Strehl ratio is the product of the higher-order and full-aperture tilt Strehl ratios:

$$SR_{Sys} = SR_{HO} \cdot SR_{tilt} \tag{58}$$

The higher-order Strehl ratio can be estimated as

$$SR_{HO} = e^{-[E_n^2 + E_f^2 + E_s^2 + E_{FA}^2]} \tag{59}$$

where E_n^2, E_f^2, E_s^2, and E_{FA}^2 are the mean square phase errors due to wavefront measurement and reconstruction noise, fitting error, servo lag error, and focus anisoplanatism (if a laser beacon is used), respectively. Equation 59 does not properly account for interactions between these effects and could be too conservative in estimating performance. Ultimately, a system design should be evaluated with a detailed computer simulation, which should include wave-optics atmospheric propagations, diffraction in wavefront sensor optics, details of hardware, processing algorithms and time delays, and many other aspects of the system's engineering design. A high-fidelity simulation will properly account for the interaction of all processes and provide a realistic estimate of system performance. For our purposes here, however, we will treat the errors as independent in order to illustrate the processes that are involved in system design and to show when a particular parameter is no longer the dominant contributor to the total error. We will consider tracking effects later, but for now we need expressions for components of the higher-order errors so that we may determine an optimum subaperture size.

Wavefront Measurement Error, E_n^2   We will consider a Shack-Hartmann sensor for the discussion that follows. The wavefront measurement error contribution, E_n^2, can be computed from the equation[73]

$$E_n^2 = \alpha \left[0.09 \ln (n_{sa}) \sigma_{\theta_{sa}}^2 \, d_s^2 \left(\frac{2\pi}{\lambda_{img}} \right)^2 \right] \tag{60}$$

where α accounts for control loop averaging and is described below, n_{sa} is the number of subapertures in the wavefront sensor, $\sigma_{\theta_{sa}}^2$ is the angular measurement error over a subaperture, d_s is the length of a side of a square subaperture, and λ_{img} is the imaging wavelength. The angular measurement error in a subaperture is proportional to the angular size of the beacon [normally limited by the seeing for unresolved objects but, in some cases, by the beacon itself (e.g., laser beacons or large astronomical targets)] and is inversely proportional to the signal-to-noise ratio in the wavefront sensor. The value of $\sigma_{\theta_{sa}}$ is given by[94]

$$\sigma_{\theta_{sa}} = \pi \left(\frac{\lambda_b}{d_s} \right) \left[\left(\frac{3}{16} \right)^2 + \left(\frac{\theta_b/2}{4\lambda_b/d_s} \right)^2 \right]^{1/2} \left(\frac{1}{n_s} + \frac{4n_e^2}{n_s^2} \right)^{1/2} \tag{61}$$

where λ_b is the center wavelength of the beacon signal, θ_b is the angular size of the beacon, r_0 is the Fried seeing parameter at 0.5 μm at zenith, ψ is the zenith angle of the observing direction, n_s is the number of photodetected electrons per subaperture per sample, and n_e is the read noise in electrons per pixel. We have used the wavelength and zenith scaling laws for r_0 to account for the angular size of the beacon at the observing conditions.

The number of photodetected electrons per subaperture per millisecond, per square meter, per nanometer of spectral width is given by

$$n_s = \frac{(101500)^{\sec \psi}}{(2.51)^{m_v}} \, d_s^2 T_{ro} \eta_{QE} t_s \Delta\lambda \tag{62}$$

where m_v is the equivalent visual magnitude of the beacon at zenith, d_s is the subaperture size in meters, T_{ro} is the transmissivity of the telescope and wavefront sensor optics, n_{QE} is the quantum efficiency of the wavefront sensor, t_s is the integration time per frame of the wavefront sensor, and $\Delta\lambda$ is the wavefront sensor spectral bandwidth in nanometers.

The parameter α is a factor that comes from the filtering process in the control loop and can be considered the mean square gain of the loop. Ellerbroek[73, 95] has derived an expression (discussed by Milonni et al.[73] and by Ellerbroek[95]) for α using the Z-transform method and a simplified control system model. His result is $\alpha = g/(2 - g)$, where $g = 2\pi f_{3dB} t_s$ is the gain, f_{3dB} is the control bandwidth, and t_s is the sensor integration time. For a typical control system, we sample at a rate that is 10 times the control bandwidth, making $t_s = 1/(10 f_{3dB})$, $g = 0.628$, and $\alpha = 0.458$.

Fitting Error, $\mathbf{E}_f^2$ The wavefront sensor has finite-sized subapertures, and the deformable mirror has a finite number of actuators. There is a limit, therefore, to the spatial resolution to which the system can "fit" a distorted wavefront. The mean square phase error due to fitting is proportional to the $-5/6$th power of the number of actuators and $(D/r_0)^{5/3}$ and is given by

$$E_f^2 = C_f \left(\frac{D}{r_0 \left(\dfrac{\lambda_{img}}{0.5\mu m} \right)^{6/3} (\cos \psi)^{3/5}} \right)^{5/3} n_a^{-5/6} \tag{63}$$

where C_f is the fitting error coefficient that is determined by the influence function of the deformable mirror and has a value of approximately 0.28 for thin, continuous-facesheet mirrors.

Servo Lag, $\mathbf{E}_s^2$ The mean square wavefront error due to a finite control bandwidth has been described earlier by Eq. 37, which is recast here as

$$E_s^2 = \left(\frac{f_G \left(\dfrac{\lambda_{img}}{0.5\mu m} \right)^{-6/5} (\cos \psi)^{-3/5}}{f_{3dB}} \right)^{5/3} \tag{64}$$

where f_G is the Greenwood frequency scaled for the imaging wavelength and a worst-case wind direction for the zenith angle, and f_{3dB} is the control bandwidth -3-dB error rejection frequency. Barchers[76] has developed an expression from which one can determine f_{3dB} for a conventional proportional integral controller, given the sensor sample time (t_s), the additional latency from readout and data-processing time (δ_0), and the design gain margin [G_M (a number between 1 and ∞ that represents the minimum gain that will drive the loop unstable)].[96] The phase crossover frequency of a proportional integral controller with a fixed latency is $\omega_{cp} = \pi/(2\delta)$ rad/s, where $\delta = t_s + \delta_0$ is the total latency, which is made up of the sample period and the sensor readout and processing time, δ_0. The loop gain that achieves the design gain margin is $K = \omega_{cp}/G_M$ where G_M is the design gain margin. Barchers shows that f_{3dB} can be found by determining the frequency for which the modulus of the error rejection function is equal to 0.707 or when

$$|S(i\omega_{3dB})| = \left| \frac{i\omega_{3dB}}{i\omega_{3dB} + \dfrac{\pi/2(t_s + \delta_0)}{G_M} e^{-i(t_s + \delta_0)\omega_{3dB}}} \right| = 0.707 \tag{65}$$

where $f_{3dB} = \omega_{3dB}/2\pi$. This equation can be solved graphically or with dedicated, iterative routines in math packages that find roots. Figure 30 shows the error rejection curves for four combinations of t_s, δ_0, and G_M, which are detailed in the caption. Note in particular that decreasing the latency δ_0 from 2 ms to 0.5 ms increased f_{3dB} from 14 Hz to 30 Hz (compare two curves on the left), illustrating the sensitivity to readout and processing time.

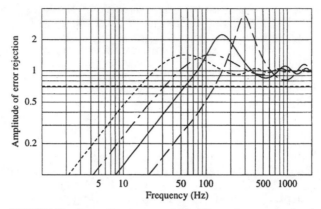

FIGURE 30 Control loop error rejection curves for a proportional integral controller. The curves (left to right) represent the following parameters: dotted (t_s = 1 ms, δ_0 = 2 ms, G_M = 4); dash-dot (t_s = 1 ms, δ_0 = 0.5 ms, G_M = 4); solid (t_s = 667 μs, δ_0 = 640 μs, G_M = 2); dashed (t_s = 400 μs, δ_0 = 360 μs, G_M = 1.5). f_{3dB} is determined by the intersection of the horizontal dotted line with each of the three curves and has values of 14, 30, 55, and 130 Hz, respectively.

Focus Anisoplanatism, $\mathbf{E}_{FA}^2$ If laser beacons are being used, the effects of focus anisoplanatism, which are discussed in Sec. 1.5, must be considered since this error often dominates the wavefront sensor error budget. Focus anisoplanatism error is given by

$$E_{FA}^2 = \left(\frac{D}{d_0 \left(\dfrac{\lambda_{img}}{0.5\mu m} \right)^{6/5} (\cos \psi)^{3/5}} \right)^{5/3} \tag{66}$$

where d_0 is the section size given by Eqs. 50 and 51, scaled for the imaging wavelength and the zenith angle of observation.

Tracking The tracking system is also very important to the performance of an AO system and should not be overlooked as an insignificant problem. In most instances, a system will contain tilt disturbances that are not easily modeled or analyzed arising most commonly from the telescope mount and its movement and base motions coupled into the telescope from the building and its machinery or other seismic disturbances. As described earlier in Eq. 22, the Strehl ratio due to full-aperture tilt variance, σ_θ^2, is

$$SR_{tilt} = \frac{1}{1 + \dfrac{\pi^2}{2} \left(\dfrac{\sigma_\theta}{\lambda/D} \right)^2} \tag{67}$$

As mentioned previously, the total system Strehl ratio is then the product of these and is given by

$$SR_{Sys} = SR_{HO} \cdot SR_{tilt} \tag{68}$$

Results for 3.5-m Telescope AO System

When the preceding equations are evaluated, we can determine the dependence of the system Strehl ratio on d_s for targets of different brightness. Figure 31 shows the Strehl ratio versus

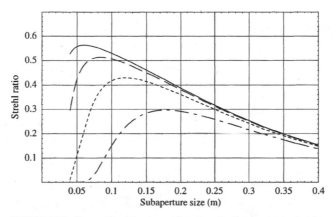

FIGURE 31   System Strehl ratio as a function of subaperture size for the
3.5-m telescope example. Values of m_V are (top to bottom) 5, 6, 7, and 8.
Other parameters are: $r_0 = 10$ cm, $f_G = 150$ Hz, $t_s = 400$ μs, $\delta_0 = 360$ μs,
$G_M = 1.5$, $f_{3dB} = 127$ Hz, $\lambda_{img} = 0.85$ μm, $\psi = 45$ degrees, $T_{ro} = 0.25$, $\eta_{QE} = 0.90$,
$\Delta\lambda = 400$ nm.

subaperture size for four values of target brightness corresponding to (top to bottom) $m_v = 5$,
6, 7, and 8 and for other system parameters as shown in the figure caption.

Figure 31 shows that we can achieve the required Strehl ratio of 0.5 (see Table 2) for an
$m_V - 6$ target by using a subaperture size of about 11 to 12 cm. We could get slightly more performance ($SR - 0.52$) by reducing the subaperture size to 8 cm, but at significant cost since we
would have nearly twice as many subapertures and deformable mirror actuators. Notice also
that for fainter objects, a larger subaperture (18 cm) would be optimum for an $m_V = 8$ target,
but the SR would be down to 0.4 for the $m_V = 6$ target and would not meet the requirement.

Figure 32 shows the performance of the system for sixth- and eighth-magnitude targets as
a function of wavelength. The top two curves in this figure are for 12-cm subapertures (solid
curve) and 18-cm subapertures (dashed curve) for the $m_V = 6$ target. Notice that the 12-cm

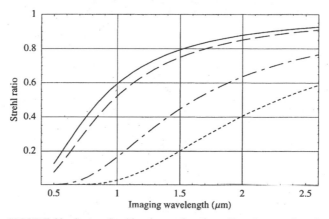

FIGURE 32   System Strehl ratio as a function of imaging wavelength
for the 3.5-m telescope example. Curves are for (top to bottom) $m_V = 6$,
$d_s = 12$ cm; $m_V = 6$, $d_s = 18$ cm; $m_V = 8$, $d_s = 18$ cm; $m_V = 8$, $d_s = 12$ cm. Other
parameters are: $r_0 = 10$ cm, $f_G = 150$ Hz, $t_s = 400$ μs, $\delta_0 = 360$ μs, $G_M = 1.5$,
$f_{3dB} = 127$ Hz, $\psi = 45°$, $T_{ro} = 0.25$, $\eta_{QE} = 0.90$, $\Delta\lambda = 400$ nm.

subapertures have better performance at all wavelengths with the biggest difference in the visible. The bottom two curves are for 18-cm subapertures (dash-dot curve) and 12-cm subapertures (dotted curve) for the $m_V = 8$ target. These curves show quantitatively the trade-off between two subaperture sizes and target brightness.

Figure 33 shows how the system will perform for different seeing conditions (values of r_0) and as a function of target brightness. These curves are for a subaperture choice of 12 cm. The curves in Figs. 31 through 33 give designers a good feel of what to expect and how to do top-level system trades for those conditions and parameters for which only they and the users can set the priorities. These curves are easily and quickly generated with modern mathematics packages.

Results for the 10-m AO Telescope System

Similar design considerations for the 10-m telescope example lead to Figs. 34 through 36. Figure 34 shows the Strehl ratio for an imaging wavelength of 1.2 μm versus the subaperture size for four values of beacon brightness corresponding to (top to bottom) $m_V = 8, 9, 10,$ and 11. Other system parameters are listed in the figure caption. These curves show that we can achieve the required Strehl ratio of 0.5 for the $m_v = 10$ beacon with subapertures in the range of 20 to 45 cm. Optimum performance at 1.2 μm produces a Strehl ratio of 0.56, with a subaperture size of 30 cm. Nearly 400 actuators are needed in the deformable mirror for 45-cm subapertures, and nearly 900 actuators are needed for 30-cm subapertures. Furthermore, Fig. 34 shows that 45 cm is a better choice than 30 cm in the sense that it provides near-optimal performance for the $m_V = 11$ beacon, whereas the Strehl ratio is down to 0.2 for the 30-cm subapertures.

Figure 35 shows system performance as a function of imaging wavelength. The top two curves are for the $m_V = 10$ beacon, with $d_s = 45$ and 65 cm, respectively. Note that the 45-cm subaperture gives excellent performance with a Strehl ratio of 0.8 at 2.2 μm (the upper end of the required spectral range) and a very useful Strehl ratio of 0.3 at 0.8 μm. A choice of $d_s = 65$ cm provides poorer performance for $m_V = 10$ (in fact, it does not satisfy the requirement) than does $d_s = 45$ cm due to fitting error, whereas a choice of $d_s = 65$ cm provides better performance for $m_V = 12$ due to improved wavefront sensor signal-to-noise ratio.

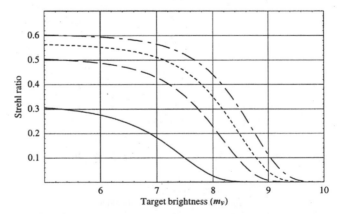

FIGURE 33   System Strehl ratio as a function of target brightness for the 3.5-m telescope example. Curves are for values of r_0 of (top to bottom) 25, 15, 10, and 5 cm. Other parameters are: $d_s = 12$ cm, $f_G = 150$ Hz, $t_s = 400$ μs, $\delta_0 = 360$ μs, $G_M = 1.5$, $f_{3dB} = 127$ Hz, $\lambda_{img} = 0.85$ μm, $\psi = 45°$, $T_{ro} = 0.25$, $\eta_{QE} = 0.90$, $\Delta\lambda = 400$ nm.

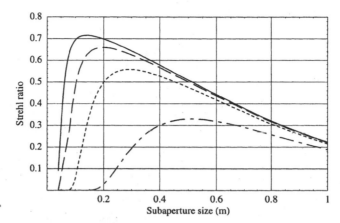

FIGURE 34 System Strehl ratio as a function of subaperture size for the 10-m telescope example. Values of m_V are (top to bottom) 8, 9, 10, and 11. Other parameters are: $r_0 = 18$ cm, $f_G = 48$ Hz, $t_s = 1000$ μs, $\delta_0 = 1000$ μs, $G_M = 2$, $f_{3dB} = 39$ Hz, $\lambda_{img} = 1.2$ μm, $\psi = 45°$, $T_{ro} = 0.25$, $\eta_{QE} = 0.90$, $\Delta\lambda = 400$ nm.

Figure 36 shows system performance in different seeing conditions as a function of beacon brightness for an intermediate imaging wavelength of 1.6 μm. These curves are for a subaperture size of 45 cm. This chart predicts that in exceptional seeing, a Strehl ratio of 0.5 μm can be achieved using an $m_V - 12.5$ beacon.

As in the 3.5-m telescope case, these curves and others like them with different parameters can be useful in performing top-level design trades and in selecting a short list of design candidates for further detailed analysis and simulation.

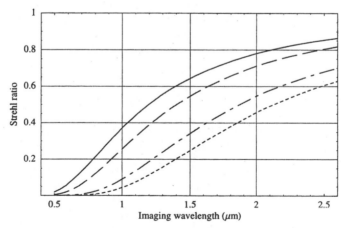

FIGURE 35 System Strehl ratio as a function of imaging wavelength for the 10-m telescope example. Curves are for (top to bottom) $m_V = 10$, $d_s = 45$ cm; $m_V = 10$, $d_s = 65$ cm; $m_V = 12$, $d_s = 65$ cm; $m_V = 12$, $d_s = 45$ cm. Other parameters are: $r_0 = 18$ cm, $f_G = 48$ Hz, $t_s = 1000$ μs, $\delta_0 = 1000$ μs, $G_M = 2$, $f_{3dB} = 39$ Hz, $\psi = 45°$, $T_{ro} = 0.25$, $\eta_{QE} = 0.90$, $\Delta\lambda = 400$ nm

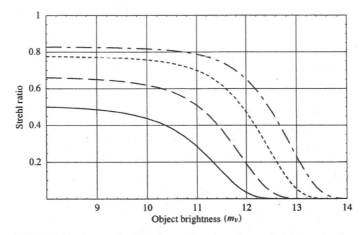

FIGURE 36 System Strehl ratio as a function of target brightness for the 10-m telescope example. Curves are for values of r_0 of (top to bottom) 40, 25, 15, and 10 cm. Other parameters are: $d_s = 45$ cm, $f_G = 48$ Hz, $t_s = 1000$ μs, $\delta_0 = 1000$ μs, $G_M = 2$, $f_{3dB} = 39$ Hz, $\lambda_{img} = 1.6$ μm, $\psi = 45$ degrees, $T_{ro} = 0.25$, $\eta_{QE} = 0.90$, $\Delta\lambda = 400$ nm.

1.7 REFERENCES

1. D. R. Williams, J. Liang, and D. T. Miller, "Adaptive Optics for the Human Eye," *OSA Technical Digest 1996* **13**:145–147 (1996).

2. J. Liang, D. Williams, and D. Miller, "Supernormal Vision and High-Resolution Retinal Imaging Through Adaptive Optics," *JOSA A* **14**:2884–2892 (1997).

3. A. Roorda and D. Williams, "The Arrangement of the Three Cone Classes in the Living Human Eye," *Nature* **397**:520–522 (1999).

4. K. Bar, B. Freisleben, C. Kozlik, and R. Schmiedl, "Adaptive Optics for Industrial CO2-Laser Systems," *Lasers in Engineering,* vol. 4, no. 3, unknown publisher, 1961.

5. M. Huonker, G. Waibel, A. Giesen, and H. Hugel, "Fast and Compact Adaptive Mirror for Laser Materials Processing," *Proc. SPIE* **3097**:310–319 (1997).

6. R. Q. Fugate, "Laser Beacon Adaptive Optics for Power Beaming Applications," *Proc SPIE* **2121**:68–76 (1994).

7. H. E. Bennett, J. D. G. Rather, and E. E. Montgomery, "Free-Electron Laser Power Beaming to Satellites at China Lake, California," *Proc SPIE* **2121**:182–202 (1994).

8. C. R. Phipps, G. Albrecht, H. Friedman, D. Gavel, E. V. George, J. Murray, C. Ho, W. Priedhorsky, M. M. Michaels, and J. P. Reilly, "ORION: Clearing Near-Earth Space Debris Using a 20-kw, 530-nm, Earth-Based Repetitively Pulsed Laser," *Laser and Particle Beams* **14**:1–44 (1996).

9. K. Wilson, J. Lesh, K. Araki, and Y. Arimoto, "Overview of the Ground to Orbit Lasercom Demonstration," *Space Communications* **15**:89–95 (1998).

10. R. Szeto and R. Butts, "Atmospheric Characterization in the Presence of Strong Additive Measurement Noise," *JOSA A* **15**:1698–1707 (1998).

11. H. Babcock, "The Possibility of Compensating Astronomical Seeing," *Publications of the Astronomical Society of the Pacific* **65**:229–236 (October 1953).

12. D. Fried, "Optical Resolution Through a Randomly Inhomogeneous Medium for Very Long and Very Short Exposures," *J. Opt. Soc. Am.* **56**:1372–1379 (October 1966).

13. R. P. Angel, "Development of a Deformable Secondary Mirror," *Proceedings of the SPIE* **1400**:341–351 (1997).

14. S. F. Clifford, "The Classical Theory of Wave Propagation in a Turbulent Medium," *Laser Beam Propagation in the Atmosphere,* Springer-Verlag, New York, 1978.

15. A. N. Kolmogorov, "The Local Structure of Turbulence in Incompressible Viscous Fluids for Very Large Reynolds' Numbers," *Turbulence, Classic Papers on Statistical Theory,* Wiley-Interscience, New York, 1961.

16. V. I. Tatarski, *Wave Propagation in a Turbulent Medium,* McGraw-Hill, New York, 1961.

17. D. Fried, "Statistics of a Geometrical Representation of Wavefront Distortion," *J. Opt. Soc. Am.* **55**:1427–1435 (November 1965).

18. J. W. Goodman, *Statistical Optics,* Wiley-Interscience, New York, 1985.

19. F. D. Eaton, W. A. Peterson, J. R. Hines, K. R. Peterman, R. E. Good, R. R. Beland, and J. H. Brown, "Comparisons of vhf Radar, Optical, and Temperature Fluctuation Measurements of c_n^2, r_0, and θ_0," *Theoretical and Applied Climatology* **39**:17–29 (1988).

20. D. L. Walters and L. Bradford, "Measurements of r_0 and θ_0: 2 Decades and 18 Sites," *Applied Optics* **36**:7876–7886 (1997).

21. R. E. Hufnagel, *Proc. Topical Mtg. on Optical Propagation through Turbulence, Boulder, CO* **1** (1974).

22. G. Valley, "Isoplanatic Degradation of Tilt Correction and Short-Term Imaging Systems," *Applied Optics* **19**:574–577 (February 1980).

23. D. Winker, "Unpublished Air Force Weapons Lab Memo, 1986," U.S. Air Force, 1986.

24. E. Marchetti and D. Bonaccini, "Does the Outer Scale Help Adaptive Optics or Is Kolmogorov Gentler?" *Proc. SPIE* **3353**:1100–1108 (1998).

25. R. E. Hufnagel and N. R. Stanley, "Modulation Transfer Function Associated with Image Transmission Through Turbulent Media," *J. Opt. Soc. Am.* **54**:52–61 (January 1964).

26. D. Fried, "Limiting Resolution Looking down Through the Atmosphere," *J. Opt. Soc. Am.* **56**:1380–1384 (October 1966).

27. R. Noll, "Zernike Polynomials and Atmospheric Turbulence," *J. Opt. Soc. Am.* **66**:207–211 (March 1976).

28. J. Christou, "Deconvolution of Adaptive Optics Images," *Proceedings, ESO/OSA Topical Meeting on Astronomy with Adaptive Optics: Present Results and Future Programs* **56**:99–108 (1998).

29. G. A. Tyler, *Reduction in Antenna Gain Due to Random Jitter,* The Optical Sciences Company, Anaheim, CA, 1983.

30. H. T. Yura and M. T. Tavis, "Centroid Anisoplanatism," *JOSA A* **2**:765–773 (1985).

31. D. P. Greenwood and D. L. Fried, "Power Spectra Requirements for Wave-Front-Compensative Systems," *J. Opt. Soc. Am.* **66**:193–206 (March 1976).

32. G. A. Tyler, "Bandwidth Considerations for Tracking Through Turbulence," *JOSA A* **11**:358–367 (1994).

33. R. J. Sasiela, *Electromagnetic Wave Propagation in Turbulence,* Springer-Verlag, New York, 1994.

34. J. Bufton, "Comparison of Vertical Profile Turbulence Structure with Stellar Observations," *Appl. Opt.* **12**:1785 (1973).

35. D. Greenwood, "Tracking Turbulence-Induced Tilt Errors with Shared and Adjacent Apertures," *J. Opt. Soc. Am.* **67**:282–290 (March 1977).

36. G. A. Tyler, "Turbulence-Induced Adaptive-Optics Performance Degradation: Evaluation in the Time Domain," *JOSA A* **1**:358 (1984).

37. D. Fried, "Anisoplanatism in Adaptive Optics," *J. Opt. Soc. Am.* **72**:52–61 (January 1982).

38. J. W. Goodman, *Introduction to Fourier Optics,* McGraw-Hill, San Francisco, 1968.

39. J. Ge, "Adaptive Optics," *OSA Technical Digest Series* **13**:122 (1996).

40. D. W. Tyler and B. L. Ellerbroek, "Sky Coverage Calculations for Spectrometer Slit Power Coupling with Adaptive Optics Compensation," *Proc. SPIE* **3353**:201–209 (1998).

41. J. Ge, R. Angel, D. Sandler, C. Shelton, D. McCarthy, and J. Burge, "Adaptive Optics Spectroscopy: Preliminary Theoretical Results," *Proc. SPIE* **3126**:343–354 (1997).

42. G. Tyler, "Rapid Evaluation of d0," Tech. Rep. TR-1159, The Optical Sciences Company, Placentia, CA, 1991.

43. R. Racine and R. McClure, "An Image Stabilization Experiment at the Canada-France-Hawaii Telescope," *Publications of the Astronomical Society of the Pacific* **101**:731–736 (August 1989).

44. R. Q. Fugate, J. F. Riker, J. T. Roark, S. Stogsdill, and B. D. O'Neil, "Laser Beacon Compensated Images of Saturn Using a High-Speed Near-Infrared Correlation Tracker," *Proc. Top. Mtg. on Adaptive Optics, ESO Conf. and Workshop Proc.* **56**:287 (1996).

45. E. Wallner, "Optimal Wave-Front Correction Using Slope Measurements," *J. Opt. Soc. Am.* **73**:1771–1776 (December 1983).

46. D. L. Fried, "Least-Squares Fitting a Wave-Front Distortion Estimate to an Array of Phase-Difference Measurements," *J. Opt. Soc. Am.* **67**:370–375 (1977).

47. W. J. Wild, "Innovative Wavefront Estimators for Zonal Adaptive Optics Systems, ii," *Proc. SPIE* **3353**:1164–1173 (1998).

48. B. L. Ellerbroek and T. A. Rhoadarmer, "Real-Time Adaptive Optimization of Wave-Front Reconstruction Algorithms for Closed Loop Adaptive-Optical Systems," *Proc. SPIE* **3353**:1174–1185 (1998).

49. R. B. Shack and B. C. Platt, *J. Opt. Soc. Am.* **61**:656 (1971).

50. B. R. Hunt, "Matrix Formulation of the Reconstruction of Phase Values from Phase Differences," *J. Opt. Soc. Am.* **69**:393 (1979).

51. R. H. Hudgin, "Optimal Wave-Front Estimation," *J. Opt. Soc. Am.* **67**:378–382 (1977).

52. R. J. Sasiela and J. G. Mooney, "An Optical Phase Reconstructor Based on Using a Multiplier-Accumulator Approach," *Proc SPIE* **551**:170 (1985).

53. R. Q. Fugate, D. L. Fried, G. A. Ameer, B. R. Boeke, S. L. Browne, P. H. Roberts, R. E. Ruane, G. A. Tyler, and L. M. Wopat, "Measurement of Atmospheric Wavefront Distortion Using Scattered Light from a Laser Guide-Star," *Nature* **353**:144–146 (September 1991).

54. C. A. Primmerman, D. V. Murphy, D. A. Page, B. G. Zollars, and H. T. Barclay, "Compensation of Atmospheric Optical Distortion Using a Synthetic Beacon," *Nature* **353**:140–141 (1991).

55. R. Q. Fugate, B. L. Ellerbroek, C. H. Higgins, M. P. Jelonek, W. J. Lange, A. C. Slavin, W. J. Wild, D. M. Winker, J. M. Wynia, J. M. Spinhirne, B. R. Boeke, R. E. Ruane, J. F. Moroney, M. D. Oliker, D. W. Swindle, and R. A. Cleis, "Two Generations of Laser-Guide-Star Adaptive Optics Experiments at the Starfire Optical Range," *JOSA A* **11**:310–324 (1994).

56. L. A. Thompson, R. M. Castle, S. W. Teare, P. R. McCullough, and S. Crawford, "Unisis: A Laser Guided Adaptive Optics System for the Mt. Wilson 2.5-m Telescope," *Proc. SPIE* **3353**:282–289 (1998).

57. S. S. Olivier, D. T. Gavel, H. W. Friedman, C. E. Max, J. R. An, K. Avicola, B. J. Bauman, J. M. Brase, E. W. Campbell, C. Carrano, J. B. Cooke, G. J. Freeze, E. L. Gates, V. K. Kanz, T. C. Kuklo, B. A. Macintosh, M. J. Newman, E. L. Pierce, K. E. Waltjen, and J. A. Watson, "Improved Performance of the Laser Guide Star Adaptive Optics System at Lick Observatory," *Proc. SPIE* **3762**:2–7 (1999).

58. D. J. Butler, R. I. Davies, H. Fews, W. Hackenburg, S. Rabien, T. Ott, A. Eckart, and M. Kasper, "Calar Alto Affa and the Sodium Laser Guide Star in Astronomy," *Proc. SPIE* **3762**: in press (1999).

59. R. Q. Fugate, "Laser Guide Star Adaptive Optics for Compensated Imaging," *The Infrared and Electro-Optical Systems Handbook,* S. R. Robinson, ed., vol. 8, 1993.

60. See the special edition of *JOSA A* on Atmospheric-Compensation Technology (January-February 1994).

61. R. Foy and A. Labeyrie, "Feasibility of Adaptive Telescope with Laser Probe," *Astronomy and Astrophysics* **152**:L29–L31 (1985).

62. D. L. Fried and J. F. Belsher, "Analysis of Fundamental Limits to Artificial-Guide-Star Adaptive-Optics-System Performance for Astronomical Imaging," *JOSA A* **11**:277–287 (1994).

63. G. A. Tyler, "Rapid Evaluation of d_0: The Effective Diameter of a Laser-Guide-Star Adaptive-Optics System," *JOSA A* **11**:325–338 (1994).

64. R. Penndorf, "Tables of the Refractive Index for Standard Air and the Rayleigh Scattering Coefficient for the Spectral Region Between 0.2 and 20.0 μm and Their Application to Atmospheric Optics," *J. Opt. Soc. Am.* **47**:176–182 (1957).

65. *U.S. Standard Atmosphere,* National Oceanic and Atmospheric Administration, Washington, D.C., 1976.

66. R. Q. Fugate, "Observations of Faint Objects with Laser Beacon Adaptive Optics," *Proceedings of the SPIE* **2201**:10–21 (1994).

67. W. Happer, G. J. MacDonald, C. E. Max, and F. J. Dyson, "Atmospheric Turbulence Compensation by Resonant Optical Backscattering from the Sodium Layer in the Upper Atmosphere," *JOSA A* **11**:263–276 (1994).

68. H. Friedman, G. Erbert, T. Kuklo, T. Salmon, D. Smauley, G. Thompson, J. Malik, N. Wong, K. Kanz, and K. Neeb, "Sodium Beacon Laser System for the Lick Observatory," *Proceedings of the SPIE* **2534**:150–160 (1995).

69. T. H. Jeys, "Development of a Mesospheric Sodium Laser Beacon for Atmospheric Adaptive Optics," *The Lincoln Laboratory Journal* **4**:133–150 (1991).

70. T. H. Jeys, A. A. Brailove, and A. Mooradian, "Sum Frequency Generation of Sodium Resonance Radiation," *Applied Optics* **28**:2588–2591 (1991).

71. M. P. Jelonek, R. Q. Fugate, W. J. Lange, A. C. Slavin, R. E. Ruane, and R. A. Cleis, "Characterization of artificial guide stars generated in the mesospheric sodium layer with a sum-frequency laser," *'OSA A* **11**:806–812 (1994).

72. E. J. Kibblewhite, R. Vuilleumier, B. Carter, W. J. Wild, and T. H. Jeys, "Implementation of CW and Pulsed Laser Beacons for Astronomical Adaptive Optics," *Proceedings of the SPIE* **2201**:272–283 (1994).

73. P. W. Milonni, R. Q. Fugate, and J. M. Telle, "Analysis of Measured Photon Returns from Sodium Beacons," *JOSA A* **15**:217–233 (1998).

74. P. W. Milonni, H. Fern, J. M. Telle, and R. Q. Fugate, "Theory of Continuous-Wave Excitation of the Sodium Beacon," *JOSA A* **16**:2555–2566 (1999).

75. R. J. Eager, "Application of a Massively Parallel DSP System Architecture to Perform Wavefront Reconstruction for a 941 Channel Adaptive Optics System," *Proceedings of the ICSPAT* **2**:1499–1503 (1977).

76. J. Barchers, Air Force Research Laboratory/DES, Starfire Optical Range, Kirtland AFB, NM, private communication, 1999.

77. E. Gendron and P. Lena, "Astronomical Adaptive Optics in Modal Control Optimization," *Astron. Astrophys.* **291**:337–347 (1994).

78. J. W. Hardy, J. E. Lefebvre, and C. L. Koliopoulos, "Real-Time Atmospheric Compensation," *J. Opt. Soc. Am.* **67**:360–367 (1977); and J. W. Hardy, *Adaptive Optics for Astronomical Telescopes,* Oxford University Press, Oxford, 1998.

79. J. Wyant, "Use of an AC Heterodyne Lateral Shear Interferometer with Real-Time Wavefront Correction Systems," *Applied Optics* **14**:2622–2626 (November 1975).

80. F. Roddier, "Curvature Sensing and Compensation: A New Concept in Adaptive Optics," *Applied Optics* **27**:1223–1225 (April 1988).

81. F. Roddier, *Adaptive Optics in Astronomy,* Cambridge University Press, Cambridge, England, 1999.

82. F. Roddier and F. Rigault, "The VH-CFHT Systems," *Adaptive Optics in Astronomy,* ch. 9, F. Roddier, ed., Cambridge University Press, Cambridge, England, 1999.

83. J. R. Fienup, "Phase Retrieval Algorithms: A Comparison," *Appl. Opt* **21**:2758 (1982).

84. R. A. Gonsalves, "Fundamentals of wavefront sensing by phase retrival," *Proc SPIE* **351,** p. 56, 1982.

85. J. T. Foley and M. A. A. Jalil, "Role of Diffraction in Phase Retrival from Intensity Measurements," *Proc SPIE* **351**:80 (1982).

86. S. R. Robinson, "On the Problem of Phase from Intensity Measurements," *J. Opt. Soc. Am.* **68**:87 (1978).

87. R. G. Paxman, T. J. Schultz, and J. R. Fineup, "Joint Estimation of Object and Aberrations by Using Phase Diversity," *JOSA A* **9**:1072–1085 (1992).

88. R. L. Kendrick, D. S. Acton, and A. L. Duncan, "Phase Diversity Wave-Front Sensor for Imaging Systems," *Appl. Opt.* **33**:6533–6546 (1994).

89. D. G. Sandler, T. Barrett, D. Palmer, R. Fugate, and W. Wild, "Use of a Neural Network to Control an Adaptive Optics System for an Astronomical Telescope," *Nature* **351**:300–302 (May 1991).

90. B. Hulburd and D. Sandler, "Segmented Mirrors for Atmospheric Compensation," *Optical Engineering* **29**:1186–1190 (1990).

91. D. S. Acton, "Status of the Lockheed 19-Segment Solar Adaptive Optics System," *Real Time and Post Facto Solar Image Correction,* Proc. Thirteenth National Solar Observatory, Sacramento Peak, Summer Shop Series 13, 1992.

92. D. F. Busher, A. P. Doel, N. Andrews, C. Dunlop, P. W. Morris, and R. M. Myers, "Novel Adaptive Optics with the Durham University Electra System," *Adaptive Optics,* Proc. OSA/ESO Conference Tech Digest, Series 23, 1995.

93. M. A. Ealey and P. A. Davis, "Standard Select Electrostrictive PMN Actuators for Active and Adaptive Components," *Optical Engineering* **29**:1373–1382 (1990).

94. G. A. Tyler and D. L. Fried, "Image-Position Error Associated with a Quadrant Detector," *J. Opt. Soc. Am.* **72**:804–808 (1982).

95. B. L. Ellerbroek, Gemini Telescopes Project, Hilo, Hawaii, private communication, 1998.

96. C. L. Phillips and H. T. Nagle, *Digital Control System: Analysis and Design,* Prentice Hall, Upper Saddle River, NJ, 1990.

CHAPTER 2
NONIMAGING OPTICS: CONCENTRATION AND ILLUMINATION*

William Cassarly
Optical Research Associates, Pasadena, California

2.1 INTRODUCTION

Nonimaging optics is primarily concerned with efficient and controlled transfer of radiation. *Nonimaging* refers to the fact that image formation is not a fundamental requirement for efficient radiation transfer; however, image formation is not excluded and is useful in many cases. The two main aspects of radiation transfer that nonimaging optics attempts to solve are maximizing radiation transfer and creating a controlled illuminance distribution. These two problem areas are often described as *concentration* and *illumination*. Many systems require a blending of the concentration and illumination aspects of nonimaging optics, with an important example being the creation of a uniform illuminance distribution with maximum collection efficiency.

Solar collection is an area in which one of the dominant requirements is to maximize the concentration of flux collected by a receiver of a given size (e.g., the irradiance). Any small patch of area on the receiver surface can collect radiation from a hemispherical solid angle. A related problem exists in the detection of low levels of radiation (e.g., Cherenkov counters,[1] infrared detection,[2] and scintillators[3]). Nonimaging optics provides techniques to simultaneously maximize the concentration and collection efficiency. The compound parabolic concentrator[4] is the most well known nonimaging device and has been extensively investigated over the past 30 years.

In addition, there are situations in which collected solid angle must be less than a hemisphere, but the flux density must still be maximized. For a given solid angle and receiver size, the problem is then to maximize the average radiance rather than the irradiance. For example, many fibers only propagate flux that enters the lightpipe within the fiber's numerical aperture (NA), and minimizing the size of the fiber is a key aspect to making practical fiber

* *Acknowledgments:* Special acknowledgement is given to Doug Goodman, whose list of uniformity references and OSA course notes provided an important starting point for portions of this chapter. The support of Optical Research Associates (ORA) during the manuscript preparation is also appreciated, as are the editorial and terminology discussions with many of the technical staff at ORA. Simulation and ray trace results were all generated using either LightTools® or CodeV®.

optic lighting systems.[5] Projector systems often require maximum flux density at the "film gate," but only the flux reimaged by the projection lens is of importance.[6,7]

The collected flux must often satisfy certain uniformity criteria. This can be especially important when the luminance of the source of radiation varies or if the flux is collected in a way that introduces nonuniformities. Nonimaging optics provides techniques to control the uniformity. Uniformity control can sometimes be built into the optics used to collect the radiation. In other cases, a separate component is added to the optical system to improve the uniformity.

Uniformity control tends to be created using two main approaches: tailoring and superposition. Tailoring transfers the flux in a controlled manner and uses knowledge of the source's radiance distribution.[8] Superposition takes subregions of a starting distribution and superimposes them to provide an averaging effect that is often less dependent on the details of the source than tailoring. In many systems, a combination of tailoring and superposition is used. Examples of nonimaging approaches to uniformity control include mixing rods, lens arrays, integrating spheres, faceted reflectors, and tailored surfaces.

The emphasis of this chapter is on the transfer of incoherent radiation. Many of the techniques can be applied to coherent radiation, but interference and diffraction are only mentioned in passing.

2.2   BASIC CALCULATIONS

Some of the terms used in nonimaging optics have yet to be standardized. This section is an effort to collate some of the most common terms.

Photometric and Radiometric Terminology

Understanding nonimaging optics can sometimes be confusing if one is not familiar with photometric and radiometric terminology. This is especially true when reading papers that cross disciplines because terminology usage has not been consistent over time. Reference 9 discusses some of these terminology issues.

This chapter is written using photometric terminology. Photometry is similar to radiometry except that photometry weights the power by the response of the human eye. For readers more familiar with radiometric terminology, a chart showing the relationship between some of the most common photometric terms and radiometric terms is shown in Table 1. Reference 10 describes radiometry and photometry terminology in more detail.

TABLE 1   Photometric and Radiometric Terminology

Quantity	Radiometric	Photometric
Power or flux	Watt (W)	Lumen (lm)
Power per unit area	Irradiance, W/m^2	Illuminance, $lm/m^2 = lux$ (lx)
Power per unit solid angle	Radiant intensity, W/sr	Luminous intensity, $lm/sr = candela$ (cd)
Power per unit solid angle per unit projected area or Power per unit projected solid angle per unit area	Radiance, W/m^2-sr	Luminance, cd/m^2

Exitance and *emittance* are similar terms to *irradiance;* however, they denote the case of flux at the surface of a source, whereas irradiance applies to any surface. There are numerous textbooks and handbooks on radiometry and photometry.[11–16]

Etendue

Etendue describes the integral of the area and the angular extents over which a radiation transfer problem is defined. Etendue is used to determine the trade-off between the required area and angular extents in nonimaging optic designs. Reference 17 provides a brief review of various etendue descriptions with copious references to other work.

One definition of etendue is

$$\text{etendue} = n^2 \int \int \cos{(\theta)} \, dA \, d\Omega \tag{1}$$

where n is the index of refraction and θ is the angle between the normal to the differential area dA and the centroid of the differential solid angle $d\Omega$.

In phase space nomenclature (e.g., Ref. 4, Appendix A), etendue is described by

$$\text{etendue} = \int \int dx \, dy \, ndL \, ndM = \int \int dx \, dy \, dp \, dq \tag{2}$$

where dL and dM are differential changes in the direction cosines (L, M, N), $dx \, dy$ is the differential area, and $dp \, dq$ is the differential projected solid angle within the material of index n. The term *phase space* is often used to describe the area and solid angle over which the etendue integral is performed.

Luminance

Luminance divided by the index of refraction squared is the ratio of the differential flux $d\Phi$ to the differential etendue:

$$L/n^2 = d\Phi/d \text{ etendue} = d\Phi/[n^2 \cos{(\theta)} \, dA \, d\Omega] \tag{3}$$

The L/n^2 over a small area and a small angular extent is constant for a blackbody source (e.g., Ref. 18, p. 189). A consequence of constant L/n^2 is that if optical elements are added to modify the apparent area of the small region, then the angular extent of this small area must also change.

If the source of radiation is not a blackbody source, then flux that passes back to the source can either pass through the source or be reflected by the source. In this case, L/n^2 can increase. One example occurs with discharge sources where a spherical mirror is used to reflect flux back through the emitting region. Another example is observed by evaluating the luminance of tungsten in a coiled filament. The luminance is higher at the filament interior surface than the exterior surface because the interior filament surface emits radiation and also reflects radiation that is emitted from other coils.

Lambertian

In many situations the luminance of a radiation source does not vary as a function of angle or position. Such a source is often called a *Lambertian radiator.* In some nonimaging literature, the term *isotropic* is used. The context in which *isotropic* is used should be evaluated because *isotropic* is sometimes used to describe isotropic intensity instead of isotropic luminance.

If constant L/n^2 can be assumed for a given system, then etendue is an important tool for understanding the trade-offs between angular and spatial distributions. Etendue has been used to understand the limits to concentration,[4] projection display illumination,[19] and backlit display illumination.[20] Reference 21 investigates the situation where there is a spectral frequency shift.

In the imaging community, etendue conservation arises in many situations and is often described using the Lagrange invariant. Because imaging systems can often make paraxial assumptions, the approximation $\tan(\theta) = \sin(\theta)$ is often used; however, $\sin(\theta)$ should be used when the collection angles become large.

Clipped Lambertian

An aperture with a clipped Lambertian distribution is one where the source of flux appears to be Lambertian, but only over a finite range of angles. Outside of that range of angles, there is no flux and the range of angles is assumed to be constant across the aperture. The most common example is when the source is at infinity and the flux at a planar aperture is considered. A clipped Lambertian distribution is also often found at an intermediate surface within an optical system. The terms *limited Lambertian* and *restricted Lambertian* are also used.

Generally, a clipped Lambertian distribution is defined across an aperture and the illuminance across the aperture is constant (e.g., a spatially uniform clipped Lambertian distribution.) Another common distribution is a spatially uniform apodized Lambertian. In this case, the angular distribution is spatially uniform but the luminance is not Lambertian. A clipped Lambertian is a special case of an apodized Lambertian. The output of a fiber optic cable is often assumed to have a spatially uniform apodized Lambertian distribution.

The etendue for clipped Lambertian situations is straightforward to compute. Consider the case of an infinite strip of width $2R$ with a clipped Lambertian distribution defined between $\pm\theta_{max}$, relative to the surface normal. The etendue per unit length for this 2D clipped Lambertian case is

$$\text{etendue}_{2D} = n(2R)(2\sin\theta_{max}) \tag{4}$$

2D refers to a trough or extruded geometry and typically assumes that the distribution is infinite in the third dimension. Mirrors can often be placed at the ends of a finite length source so that the source appears to be infinitely long.

For a Lambertian disk with a half cone angle of θ_{max}, the etendue is

$$\text{etendue}_{3D} = n^2\text{Area}\,\pi\sin^2\theta_{max} = n^2\pi R^2\pi\sin^2\theta_{max} \tag{5}$$

In a system where the etendue is preserved, these etendue relationships highlight that increasing either θ_{max} or R requires a reduction in the other, as depicted in Fig. 1.

Hottel Strings

The etendue relationships described by Eqs. (4) and (5) are primarily defined for the case of a clipped Lambertian source. Such situations arise when the source is located at infinity (e.g.,

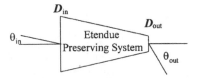

FIGURE 1 Graphical representation of etendue preservation.

solar collection) or when considering the output of a fiber optic cable. When the angular distributions vary spatially, Ref. 22 provides a method to compute the etendue that is very simple to use with 2D systems. The method is straightforward for the case of a symmetric system or an off-axis system, and even if there is an absorber positioned between the two apertures. The method is depicted in Fig. 2, where the etendue of the radiation that can be transferred between AB and CD is computed.

For a rotationally symmetric system, the etendue between two apertures (e.g., left side of Fig. 2) has been shown by Ref. 23 to be $(\pi^2/4)(AD-AC)^2$. Reference 24 has provided a generalized treatment of the 3D case.

Solid Angle and Projected Solid Angle

Solid angle is the ratio of a portion of the area of a sphere to the square of the sphere radius (see *OSA Handbook,* Chap. 24, and Refs. 24 and 25) and is especially important when working with sources that radiate into more than a hemisphere. Solid angle is directly applicable to point sources; however, when the size of the sphere is large enough, the solid angle is still used to characterize extended sources. For solid angle to be applied, the general rule of thumb is that the sphere radius should be greater than 10 times the largest dimension of the source, although factors of 30 or more are required in precision photometry (e.g., Ref. 26, pp. 4.29–4.31).

The etendue is sometimes called the *area–solid angle product;* however, this term can often cause confusion because etendue is actually the projected-area-solid-angle product [$\cos \theta \, dA \, d\Omega$] or area-projected-solid-angle product [$dA \cos \theta \, d\Omega$]. Reference 27 discusses some distinctions between projected solid angle (PSA) and solid angle. An important implication of the cosine factor is that the solid angle for a cone with half angle θ is

$$\text{Solid Angle}_{cone} = 2\pi[1 - \cos \theta] = 4\pi \sin^2 (\theta/2) \tag{6}$$

but the projected solid angle for the cone is

$$\text{Projected Solid Angle}_{cone} = \pi \sin^2 \theta \tag{7}$$

For a hemisphere, the solid angle is 2π and the PSA is π. This factor of 2 difference is often the source of confusion. PSA and solid angle are pictured in Fig. 3.

If the luminance at the receiver is constant, then PSA times luminance provides illuminance. Nonuniform luminance can be handled using weighted averages.[28]

In the 2D case, the projected solid angle analog is simply projected angle, and is $2 \sin \theta$ for angles between $\pm\theta$ or $|\sin \theta_1 - \sin \theta_2|$ for angles between θ_1 and θ_2.

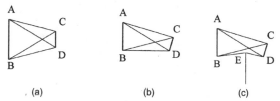

(a)   (b)   (c)

FIGURE 2 Hottel's crossed string relationship for computing etendue. This figure shows a symmetric case (*a*) and an asymmetric case (*b*) where the 2D etendue between aperture AB and CD is (AD-AC + BC-BD). In (*c*), radiation is blocked below point E and the etendue is (AD-AC + BC-BED).

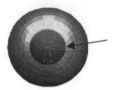 

Projected Area of
Intersected Surface Area of Sphere
max area = πR^2
max projected solid angle = π

Intersected Surface Area of Sphere
max area = $4\pi R^2$
max solid angle = 4π

FIGURE 3   PSA (*left*) vs. solid angle (*right*).

Concentration

Concentrators can be characterized by the ratio of the output area to the input area.[29] For an *ideal* system, the etendue at the input aperture and the output aperture are the same, which leads to the ideal concentration relationships

$$\text{Concentration}_{2D} = n_{out} \sin \theta_{out}/(n_{in} \sin \theta_{in}) \tag{8}$$

$$\text{Concentration}_{3D} = n^2_{out} \sin^2 \theta_{out}/(n^2_{in} \sin^2 \theta_{in}) \tag{9}$$

where the input and output distributions are clipped Lambertians, θ_{in} is the maximum input angle, and θ_{out} is the maximum output angle. Maximum concentration occurs when $\sin \theta_{out} = 1$, so that the maximum concentration ratios for 2D and 3D symmetric systems in air are $1/\sin \theta_{in}$ and $1/\sin^2 \theta_{in}$, respectively. Concentrating the flux within a dielectric medium further increases the concentration ratio by a factor of n^2_{out} in the 3D case and n_{out} in the 2D case.

In general, maximum concentration occurs when the phase space of the receiver is completely filled (i.e., all portions of the receiver are illuminated from all angles). Maximum concentration can be obtained if the etendue of the input is greater than the etendue of the output. In this case, however, the efficiency of the system is less than 1. One example of this is provided by a two-stage system where the second stage is an ideal maximum concentrator and the etendue of the first stage is greater than the etendue of the second stage.[30, 31] Such a system is not conventionally considered ideal because not all of the flux from the first stage is coupled into the entrance aperture of the second stage.

Dilution

Dilution is a term that is used to describe the situation in which the phase space of the receiver is unfilled. Consider the situation where a Lambertian disk of radius 1 is coupled directly to a receiver of radius 2. In this case, the area of the receiver is underfilled and represents *spatial dilution*. An example of *angular dilution* occurs when flux is transferred to a receiver using multiple discrete reflector elements. If there are gaps between the reflector elements, then the angular distribution of the flux incident on the receiver possesses gaps. The phrase *etendue loss* is also used to describe situations in which dilution is present.

2.3   SOFTWARE MODELING OF NONIMAGING SYSTEMS

The computer simulation of nonimaging systems is quite different from the computer simulation of imaging systems. Modeling of nonimaging systems typically requires three major items that imaging designers either do not need or tend to use infrequently. These are nonsequential ray tracing, spline surfaces, and modeling extended sources. Extended sources are often modeled using Monte Carlo techniques. Other differences include the need to model more complex surface properties including scattering surfaces; new methods to present simulation results including luminance, illuminance, and intensity distributions; and improved visualization because of the more complex geometries typically involved in nonimaging systems.

Nonsequential Ray Tracing

Nonsequential ray tracing means that the order of the surfaces with which the rays interact is not predetermined. In fact, rays can hit a single surface multiple times, which is especially common when skew rays are investigated. Nonsequential surfaces are especially important in the software modeling of lightpipes[32, 33] and prisms.

Spline Surfaces

Spline surfaces have been successfully applied to the design of nonimaging optical systems. Design work in this area includes genetic algorithm-based approaches,[34, 35] neural networks,[36] and variational principles.[37] Automotive headlamp designs also use spline surfaces routinely. Spline surfaces are also extremely important in computer-aided drafting (CAD) data exchange formats (e.g., IGES, STEP, and SAT) where Non-Uniform Rational B Splines (NURBS) have found tremendous utility.

Monte Carlo Techniques

Monte Carlo simulations are used to determine the intensity and/or illuminance distribution for optical systems. Typically, Monte Carlo techniques are used to model the source and/or surface properties of nonspecular surfaces. Reference 38 discusses some of the issues regarding Monte Carlo simulations. Some nonimaging systems that have been evaluated using Monte Carlo ray traces include an end-to-end simulation of a liquid crystal display (LCD) projector system,[39] light-emitting diodes,[40] integrating spheres,[41, 42] scintillating fibers,[43, 44] stray light,[45, 46] and automotive headlamps.[47]

Source Modeling

Monte Carlo modeling of sources typically falls into three categories: point sources, geometric building blocks, and multiple camera images. Point sources are the simplest approach and are often used in the early stages of a design because of simplicity. Measured data, sometimes called *apodization data,* can be applied to a point source so that the intensity distribution of the source matches reality. The projected area of the source as seen from a given view angle can often be used to estimate the apodization data[8, 48] when measurements are unavailable. Since the luminance of a point source is infinite, point source models are of limited utility when etendue limitations must be considered. Geometric models create the source through the superposition of simple emitters (e.g., disks, spheres, cylinders, cubes, ellipsoids, toroids). The simple emitters are typically either Lambertian surface emitters or isotropic volume emitters, and the emitters can have surface properties to model effects that occur when flux propagates back through the emitter. The simple emitters are then combined with models of the nonemissive geometry (e.g., bulb walls, electrodes, scattering surfaces) to create an accurate source model. Apodization of the spatial and/or angular distributions of sources can help to improve the accuracy of the source model. In the extreme, independent apodization files can be used for a large number of view angles. This often results in the use of multiple source images,[49, 50] which has seen renewed interest now that charge-coupled device (CCD) cameras have matured.[51-55]

The most typical Monte Carlo ray trace is one in which the rays traced from the source are independent of the optical system under investigation. When information about the portions of the source's spatial and angular luminance distribution that are contributing to a region of the intensity/illuminance distribution is available, importance sampling can be used to improve the efficiency of the ray trace. An example of such an approach occurs when an f/1 condenser is used to collect the flux from a source. Use of importance sampling means that rays outside of the f/1 range of useful angles are not traced.

Backward Trace

When a simulation traces from the source to the receiver, rays will often be traced that land outside of the region of the receiver that is under investigation. A backward ray trace can eliminate those rays by only tracing from the observation point of interest. The efficiency of the backward ray trace becomes dependent upon knowing which ray directions will hit the receiver. Depending upon the details of the system and prior knowledge of the system, backward tracing can often provide far more efficient use of traced rays when only a few observation points are to be investigated. Use of the backward ray trace and examples for a number of cases have been presented.[28] Backward ray tracing has also been called the *aperture flash mode.*[56]

Field Patch Trace

Another trace approach is the field patch, where rays are traced backward from a point within the optical system and those rays are used to determine which rays to trace forward. Reference 28 describes the prediction of a distribution by summing results from multiple field patches. This type of approach is especially useful if the effect of small changes in the system must be quantified because shifting the distributions and resuming can approximate small changes.

Software Products

There are many optical software vendors that have features that are primarily aimed toward the illumination market. A detailed comparison of the available products is difficult because of the speed at which software technology progresses. A survey of optical software products has been performed by Illuminating Engineering Society of North America (IESNA) and published in *Lighting Design and Application.*[57, 58]

2.4 BASIC BUILDING BLOCKS

This section highlights a number of the building blocks for designing systems to collect flux when the angles involved are large. Many of these building blocks are also important for imaging optics, which reflects the fact that image formation is not a fundamental requirement for nonimaging optics but that image formation is often useful.

Spherical Lenses

Spherical lenses are used in nonimaging systems although the aberrations can limit their use for systems with high collection angles. The aplanatic points of a spherical surface[59] are sometimes used to convert wide angles to lower angles and remain free of all orders of spherical aberration and low orders of coma and astigmatism. There are three cases[60] in which a spherical surface can be aplanatic:

1. The image is formed at the surface (e.g., a field lens).
2. Both object and image are located at the center of curvature.
3. The object and image are located on radii that are rn/n' and rn'/n away from the center of curvature of the spherical surface.

Case 3 is used to construct hyperhemispheric lenses that are often used in immersed high-power microscope objectives. Such lenses can suffer from curvature of field and chromatism (see Ref. 60, pp. 258–262, and Ref. 61). Some example aplanats are shown in Fig. 4.

Hyperhemispheric lenses can also be used in LED packages[62] and have been used in photographic-type objectives and immersed IR detectors. Aplanatic surfaces for use as concentrators have been investigated[63] and can be nearly ideal if the exit surface is nonplanar (Ref. 4, pp. 47–48).

Aspheric Lenses

Spherical lenses have been the workhorse of the imaging industry because of the simplicity and accuracy with which they can be manufactured. Design requirements, especially the speed of the system, often necessitate the use of aspheric surfaces in nonimaging optics. Fortunately, the accuracy with which the surface figure must be maintained is often less severe than that of an imaging system. Aspheric lenses are often used as condensers in projection systems including LCD projectors, automotive headlamps, and overhead projectors.

The classic plano aspheric examples are the conic lenses where the eccentricity is equal to $1/n$ with the convex surface facing the collimated space and eccentricity $= n$ with the plano surface facing the collimated space (see, for example, Ref. 64, pp. 112–113, and Ref. 65, pp. 100–103). Examples are shown in Fig. 5. The paraxial focal lengths are $Rn/(n-1)$ and $R/(n-1)$, respectively, where R is the radius of curvature. These examples provide correction for spherical aberration but still suffer from coma.

Refractive aspheric surfaces are often used in the classic Schmidt system.[60, 66] Reference 67 has described a procedure for determining the aspheric profile for systems with two aspheric surfaces that has been adapted to concentrators by Minano.[68]

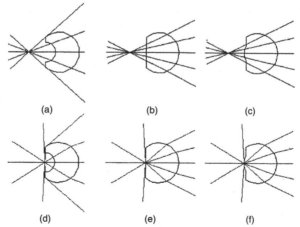

(a)   (b)   (c)

(d)   (e)   (f)

FIGURE 4 Several aplanats used to concentrate a virtual source. (*a–c*) The rays superimposed on top of the aplanats. (*d–f*) The ray trace. (*d*) A meniscus lens with hyperhemisphic outer surface and a spherical surface centered about the high concentration point. (*e*) A hyperhemisphere. (*f*) A hyperhemisphere with curved output surface.

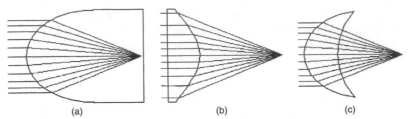

(a)   (b)   (c)

FIGURE 5 Conic collimators. (a) $e = 1/n$ and $f = Rn/(n-1)$. (b) $e = n$ and $f = R/(n-1)$. (c) $e = 1/n$ with other surface being spherical. e = eccentricity, n = index of refraction, $k = -e^2$.

Fresnel Lenses

The creation of a lens using a symmetric arrangement of curved prisms is typically attributed to A. J. Fresnel and commonly called a *Fresnel lens*. General descriptions of Fresnel lenses are available.[69, 70]

Fresnel lenses provide a means to collect wide angular distributions with a device that can be easily molded. Standard aspheric lenses can be molded, but the center thickness compared to the edge thickness adds bulk and fabrication difficulties.

In imaging applications, Fresnel lenses are used, but the system designer must be careful to consider the impact of the facets on the image quality. The facet structure is also important in nonimaging applications where the use of facets introduces losses either by blocking flux or by underfilling the aperture, which typically results in an etendue loss.

Fresnel lenses have two main variations: constant-depth facet structures and constant-width facet structures (see Fig. 6). As the width of the facets becomes smaller, the effects of diffraction should be considered.[71] Although Fresnel lenses are typically created on planar substrates, they can be created on curved surfaces.[72, 73]

The design of a Fresnel lens requires consideration of the riser angle (sometimes called the *draft angle*) between facets.[74, 75] A typical facet is depicted in Fig. 7. The riser angle should be designed to minimize its impact on the rays that are controlled by the facets that the riser connects.

Total Internal Reflection Fresnel Lenses. As the Fresnel lens collection angle increases, losses at the outer facets increase.[76] One solution is the use of total internal reflection (TIR) facets. In this case, the flux enters the substrate material, hits the TIR facet, and then hits the output surface. The entering, exiting, and TIR facets can also have power in more sophisticated designs. The basic TIR Fresnel lens idea has been used in beacon lights since at least the

EQUI-WIDTH FACETS   EQUI-DEPTH FACETS

FIGURE 6 Two main types of Fresnel lenses: equi-width and equi-depth.

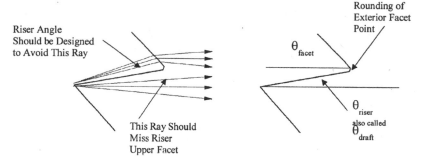

FIGURE 7   Facet terminology and design issues.

1960s.[77] TIR Fresnel lenses have received more recent design attention for applications including fiber illumination, LEDs, condensers, and concentrators.[76, 78–82] Vanderwerf[83] investigates achromatic issues with catadioptric Fresnel lenses.

In some applications, the TIR Fresnel lens degenerates into one refractive facet and two or more TIR facets. Combined TIR and refractive designs have been proposed for small sources with significant attention to LEDs.[84–86]

Conic Reflectors

Reflectors made from conic sections (parabolas, ellipses, and hyperbolas) are commonly used to transfer radiation from a source to a receiver. They provide spherical-aberration-free transfer of radiation from one focal point to the other. The problem is that the distance from a focal point to a point on the surface of the reflector is not constant except for the case of a sphere. This introduces a nonuniform magnification (coma) that can result in a loss of concentration.

Many textbooks describe conic surfaces using equations that have an origin at the vertex of the surface or at center of symmetry. Reference 17 (page 1.38) shows a number of different forms for a conic surface. In nonimaging systems, it is often convenient to define conic reflectors using an origin shifted away from the vertex. Reference 87 describes these surfaces in Cartesian coordinates as *apo-vertex surfaces*. They can also be described in polar coordinates where the focal point is typically the origin. The polar definition of a conic surface[8] is

$$r = \frac{f(1 + e)}{1 + e \cos \phi}$$

where r is the distance from the foci to the surface, f is the distance from one focus to its nearest vertex, ϕ is the angle from the foci to the vertex, e is $s/(s + 2f)$, and s is the distance between the two foci. In Cartesian coordinates, $z = r \sin \phi$ and $y = r \cos \phi$. An example is shown in Fig. 8.

Macrofocal Reflectors

One can consider the standard conic reflector to map the center of a sphere to a single point. What can sometimes be more valuable for nonimaging optics is to map the edge of the sphere to a single point. Reflectors to provide this mapping have been called macrofocal reflectors.[88] They have also been called extinction reflectors[8] because of the sharp edge in the illuminance distribution that they can produce. Reference 89 describes an illumination system that provides a sharp cutoff using tilted and offset parabolic curves.

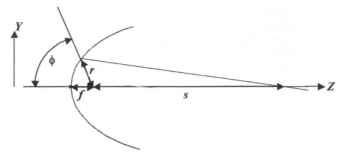

FIGURE 8   Conic reflector showing r, f, s, and ϕ.

Involute

A reflector that is used in many nonimaging systems is an *involute*. An involute reflector sends tangential rays from the source back onto themselves. One way to produce an involute for a circle is to unwind a string that has been wrapped about the circle. The locus of points formed by the string equals the involute. In Cartesian coordinates, the equation for the involute of a circle is[90]

$$x = r(\sin \theta - \theta \cos \theta) \tag{10}$$

$$y = -r(\cos \theta + \theta \sin \theta) \tag{11}$$

where r is the radius of the circle, $\theta = 0$ at the cusp point, and the origin is at the center of the circle. An example is shown in Fig. 9, where rays are drawn to highlight the fact that rays tangent to the circle reflect off of the involute and retrace their path.

The term *full involute* has been used to describe an involute that is continued until the output aperture is tangent to the surface of the source. A *partial involute* is an involute that is less than a full involute. An involute with gap is also shown in Fig. 9, and is sometimes called a *modified involute*.[91] Involute segments can also be applied to noncircular convex shapes and to disjoint shapes.[92]

2.5 CONCENTRATION

The transfer of radiation in an efficient manner is limited by the fact that the luminance of the radiation from a source in a lossless system cannot increase as the radiation propagates

Full Involute   Partial Involute   Partial Involute with Gap

FIGURE 9   Full involute, partial involute, and involute with gap. Rays tangent to the source reflect back toward the same location on the source.

through the system. A consequence is that the etendue cannot be reduced and maximizing the luminance means avoiding dilution. A less understood constraint that can impact the transfer of radiation is that rotationally symmetric optical systems conserve the skewness of the radiation about the optical axis (see for example Ref. 4, Chap. 2.8, and more recently Ref. 93). This section describes some of the basic elements of systems used to transfer radiation with an emphasis on those that preserve etendue.

Many papers on nonimaging optics can be found in the Society of Photooptical Instrumentation Engineers (SPIE) proceedings. SPIE conferences specifically focused on nonimaging optics include Refs. 94–99. Reference 100 also provides a large number of selected papers. Textbooks on nonimaging optics include Refs. 4 and 101. Topical articles include Refs. 102 and 103.

Discussions of biological nonimaging optic systems have also been written.[104–108]

Tapered Lightpipes

Lightpipes can be used to transport flux from one location to another. This can be an efficient method to transport flux, especially if total internal reflection (TIR) at the lightpipe surface is utilized to provide lossless reflections. Lightpipes can provide annular averaging of the flux. In addition, if the lightpipe changes in size from input to output over a given length (e.g., the lightpipe is tapered), then the change in area produces a corresponding change in angles. If the taper occurs over a long enough distance, then in most cases the etendue will be preserved (see, for example, Ref. 109). If the taper occurs over too short a length, the etendue will not be preserved, which can result in an increase in angles or possibly rays lost, such as when they are reflected back toward the input.

An interesting geometric approach to determining if meridional rays in a conical lightpipe can propagate from input to output end was shown by Williamson,[110] and is informally called the Williamson construction or a tunnel diagram. An example is shown in Fig. 10, where the duplicate copies of the primary lightpipe are all centered about a single point. Where the rays cross the copies shows the position and at what angle the rays will hit the primary lightpipe. As seen in Fig. 10, the rays that do not cross the output end of any of the copies of the cone are also the rays that are reflected back out the input end of the cone. Other analyses of tapered geometries have been performed by Welford[4] (pp. 74–76), and Meyer[111] provides numerous references.

Witte[112] has performed skew ray analysis using a variation of the Williamson construction. Witte also shows how a reference sphere centered at the point where the cone tapers to a point can be used to define a pupil when cones are used in conjunction with imaging systems analysis. Burton[113] simplified the expressions provided by Witte.

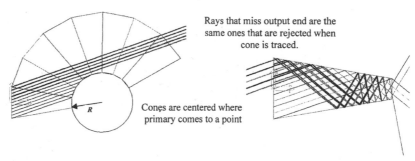

FIGURE 10 Williamson construction. Multiple copies of the primary cone are drawn. Where rays cross, there copies are the same as where they hit the actual cone. Rays that do not cross the output end are reflected back to the input end.

Vector flux investigations have shown that a cone is an ideal concentrator for a sphere (Ref. 114, p. 539). An important implication of this result, which is similar to the information presented by Witte,[112] is that a cone concentrator can be analyzed by tracing from the cone input aperture to the surface of a sphere. If the rays hit the sphere surface, then they will also hit the sphere after propagating through the lightpipe. This is true for both meridional and skew rays.

CPC

A compound parabolic collector (CPC) can be used to concentrate the radiation from a clipped Lambertian source in a nearly etendue-preserving manner. Welford[4] provides a thorough description of CPCs. An example CPC is shown in Fig. 11, where the upper and lower reflective surfaces are tilted parabolic surfaces. The optical axis of the parabolic curves is not the same for the upper and lower curves, hence the use of the term *compound*.

In 2D, the CPC has been shown to be an ideal concentrator.[4] In 3D, a small fraction of skew rays are rejected and bound the performance. For a CPC, this skew ray loss increases as the concentration ratio increases. Optimization procedures have produced designs that provide slightly better performance than the CPC in 3D.[37] Molledo[115] has investigated a crossed cylindrical CPC configuration.

The length of a CPC concentrator is

$$\text{Length} = (R_{\text{in}} + R_{\text{out}})/\tan (\theta_{\text{in}}) \tag{12}$$

A particularly important aspect of the CPC is that the length is minimized for the case where the input port can see the exit port. If the distance between the input and output ports is made smaller than the CPC length, then rays higher than the maximum design angle could propagate directly from input to output aperture. The reduction in concentration is often small, so many practical systems use truncated CPC-like designs.[90, 116–118] A cone that is the same length as an untruncated CPC will provide lower concentration, but as the cone is lengthened, the performance can exceed the performance of the CPC.

In the literature, CPC is sometimes used to denote any ideal concentrator. For nonplanar receivers, the reflector shape required to create an ideal concentrator is not parabolic. The convention is to use CPC-type.

An example of a CPC-type concentrator for a tubular receiver is shown in Fig. 12. Part of the reflector is an involute and the remainder reflects rays from the extreme input angle to the tangent of the receiver. The design of CPC-type concentrators for nonplanar sources has been described.[4, 116, 119, 120]. Designs with gaps between reflector and receiver[121–123] and with prisms attached to the receiver[124, 125] have also been described.

CPC-type geometries are also used in laser pump cavities.[126]

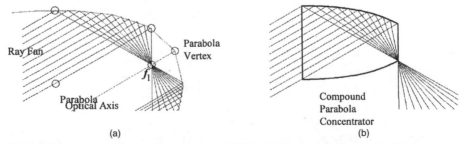

(a)   (b)

FIGURE 11   Compound parabolic concentrator (CPC). (*a*) Upper parabolic curve of CPC. Circles are drawn to identify the parabola vertex and the edges of the CPC input and output apertures.

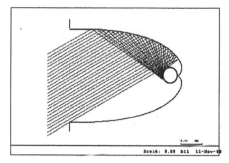

FIGURE 12   A CPC-type reflector for a tubular source.

Arrays of CPCs or CPC-like structures have also been applied to the liquid crystal displays,[127, 128] illumination,[89] and solar collection.[129]

CEC

When the source of radiation is located at a finite distance away from the input port of the concentrator, a construction similar to the CPC can be used; however, the reflector surface is now elliptical.[130] In a similar manner, if the source of radiation is virtual, then the reflector curvature becomes hyperbolic. These two constructions are called compound elliptical concentrators (CEC) and compound hyperbolic concentrators (CHC). A CEC is shown in Fig. 13, where the edge of the finite size source is reimaged onto the edge of the CEC output aperture. Hottel strings can be used to compute the etendue of the collected radiation (see Sec. 2.6). The CPC is similar to the CEC, except the edge of the source is located at infinity for the CPC.

CHC

A CHC is shown in Fig. 14 where the rays that pass through the finite-size aperture create a virtual source. The CHC concentrates the flux. Arrangements where the CHC is just a cone are possible.[131] Gush[132] also describes a hyperbolic cone-channel device.

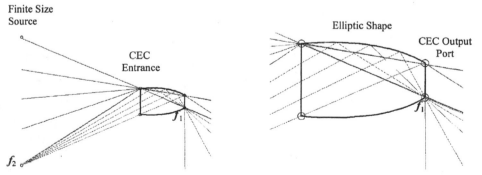

FIGURE 13   Compound elliptical concentrator (CEC). Rays originating at finite size source and collected at the CEC entrance are concentrated at the CEC output port. Edge of source is imaged onto edge of output aperture.

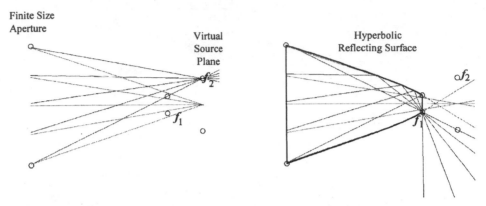

FIGURE 14 Compound hyperbolic concentrator for use with virtual source. Edge of virtual source is imaged onto the edge of the output port.

If the virtual source is located at the output aperture of the concentrator, then an ideal concentrator in both 2D and 3D can be created (see Fig. 15). Generically, this is a hyperboloid of revolution and is commonly called a *trumpet*.[133] The trumpet maps the edge of the virtual source to ±90 degrees. The trumpet can require an infinite number of bounces for rays that exit the trumpet at nearly 90°, which can limit the performance when finite-reflectivity mirrors are used.

DCPC

The CPC construction can also be applied when the exit port is immersed in a material with a high index of refraction. Such a device is called a *dielectric compound parabolic concentrator* (DCPC).[134] This allows concentrations higher than those found when the exit port is immersed in air. In this immersed exit port case, the standard CPC construction can still be used, but the exit port area is now n times smaller in 2D and n^2 smaller in 3D. Another motivation for a DCPC is that the reflectivity losses for a hollow CPC can be minimized because TIR can provide lossless reflections. A DCPC is shown in Fig. 16. A DCPC that uses frustrated TIR at the DCPC to receiver interface has been investigated.[135]

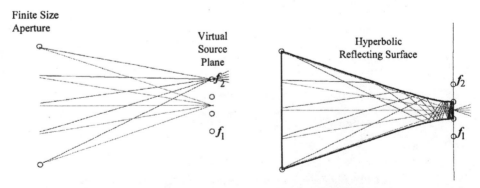

FIGURE 15 Trumpet with exit aperture located at virtual source plane. Rays "focused" at edge of virtual source exit the trumpet at ±90°.

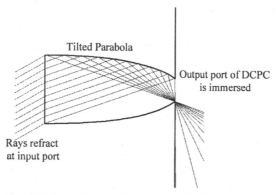

FIGURE 16   Dielectric CPC: same as CPC, except output port is immersed.

A hollow CPC can also be attached to a DCPC to minimize the size of the DCPC.[136] Extraction of flux from a high-index medium into a lower-index medium has been investigated[137, 138] using faceted structures.

Multiple Surface Concentrators

CPC-type concentrators can produce a long system when θ_{in} is small. To avoid excessively long systems, other optical surfaces are typically added. Often, a CPC-type concentrator is combined with a primary element such as a condensing lens or a parabolic reflector. There are also designs where the optical surface of the primary and the optical surface of the concentrator are designed together, which often blurs the distinction between primary and concentrator.

Lens + Concentrator.   There are numerous examples of lenses combined with a nonimaging device in the literature. One approach is to place a lens with the finite-size aperture shown in Fig. 13, 14, or 15 and illuminate the lens with a relatively small angular distribution. The lens produces higher angles that are then concentrated further using the conventional CPC, CEC, CHC, or trumpet. Such a combination typically produces a shorter package length than a CPC-type concentrator.

A lens/CHC can be attractive because the CHC is placed between the lens and lens focal plane, whereas the CEC is placed after the lens focal plane. Thus, the lens/CHC provides a shorter package than a lens/CEC. Both the lens curvature and the CHC reflector can be optimized to minimize the effects of aberrations, or the combination can be adjusted so that the CHC turns into a cone. Investigations of lens/cone combinations include Williamson,[10] Collares-Pereira,[75] Hildebrand,[139] Welford,[4] and Keene.[140] Use of a cone simplifies fabrication complexity.

After the theoretical implications of the trumpet were recognized, Winston[141] and O'Gallagher[133] investigated the case of a lens/trumpet.

When TIR is used to implement the mirrors, the mirror reflection losses can be eliminated, with the drawback that the package size grows slightly. A design procedure for a lens mirror combination with maximum concentration was presented by Ning,[142] who coined the term *dielectric total internal reflecting concentrator* (DTIRC) and showed significant package improvements over the DCPC.

Eichhorn[143] describes the use of CPC, CEC, and CHC devices in conventional optical systems, and uses explicit forms for the six coefficients of a generalized quadric surface.[144] A related "imaging" lens and reflector configuration is the Schmidt corrector system.[145]

Arrays of lenses with concentrators have also been investigated for use in illumination.[146]

Mirror + Concentrator.   There have been numerous investigations of systems that have a parabolic primary and a nonimaging secondary.[137, 147-150] Other two-stage systems have been investigated.[29-31, 151-153] Kritchman[154] has analyzed losses due to aberrations in a two-stage system.

Asymmetric concentrators have also been investigated. Winston[152] showed that an off-axis parabolic system can improve collection by tilting the input port of a secondary concentrator relative to the parabolic axis. Other asymmetric designs have been investigated.[155, 156]

Related two-mirror "imaging" systems (e.g., Ref. 60, Chap. 16) such as the Ritchey-Chretien can be used as the primary. Ries[157] also investigated a complementary Cassegrain used for concentration.

Simultaneous Multiple Surfaces.   Minano and coworkers have investigated a number of concentrator geometries where the edge of the source does not touch the edge of a reflector. The procedure has been called simultaneous multiple surfaces (SMS) and builds on the multisurface aspheric lens procedure described by Schultz.[67] Minano uses the nomenclature R for refractive, X for reflective (e.g., refleXive), and I when the main mode of reflection is TIR. The I surface is typically used for refraction the first time the ray intersects the surface and TIR for the second ray intersection. The I surface is sometimes mirrored over portions of the surface if reflection is desired but the angles do not satisfy the TIR condition. Illustrative citations include RR,[158] RX,[159, 160] and RXI.[161, 162] Example RX and RXI devices are shown in Fig. 17. SMS can be used to design all reflecting configurations.

Restricted Exit Angle Concentrators with Lenses

If the angular distribution of the flux is constant across an aperture and the centroid of the distribution is normal to the aperture, then the distribution is called *telecentric.* Many optical systems require the transfer of flux from one location to another. If the desired distribution of flux at the second location is also telecentric, then the system is called doubly telecentric. An afocal imaging system can be modified to provide a doubly telecentric imaging system when used with finite conjugates. An example is shown in Fig. 18.

In nonimaging systems, flux must often be transferred from one aperture to another and the angular distribution at the second aperture must be constant across the aperture. However, the point-by-point mapping is not required in nonimaging systems. A single lens can provide this transfer of radiation, as shown in Fig. 19. This type of lens is a collimator and is similar to a Fourier transform lens; it has been called a *beam transformer.*[163]

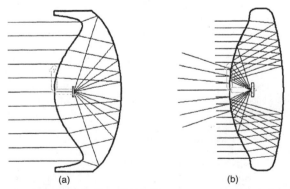

(a)   (b)

FIGURE 17   RX (*a*) and RXI (*b*) concentrators. The receiver is immersed in both cases. In the RXI case, the central portion of the refractive surface is mirrored.

FIGURE 18 Doubly telecentric system showing a twofold increase in size and a twofold decrease in angles.

The aberrations introduced by a lens can limit the etendue preservation of the aggregate flux collected by the lens. The increase in etendue tends to become more severe as range of angles collected by the lens increases. Somewhere around f/1, the change can become quite significant.

θ_1/θ_2

θ_1/θ_2 concentrator[29, 163, 164] maps a limited-angle Lambertian distribution with maximum angle θ_1 into another limited-angle Lambertian with maximum angle θ_2. One version of the θ_1/θ_2 is a compound parabolic construction with a cone replacing part of the small end of the CPC (Ref. 4, pp. 82–84). A picture is shown in Fig. 20. The length of a θ_1/θ_2 is given by the same equation as the CPC [e.g., Eq. (12)]. If the θ_1/θ_2 is hollow and $\theta_2 = 90°$, then the cone disappears and the θ_1/θ_2 becomes a CPC.

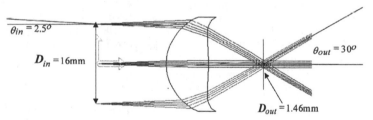

FIGURE 19 Condenser to provide an angle to area transformation and produce a distribution that is symmetric about the optical axis at both input and output planes. $D_{in} \sin \theta_{in} = D_{out} \sin \theta_{out}$.

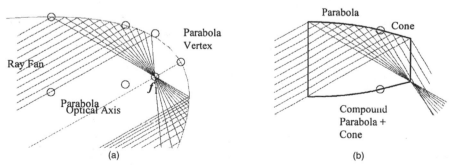

FIGURE 20 θ_1/θ_2 implemented using cone with compound parabolic. (*a*) Construction information. (*b*) θ_1/θ_2.

To reduce losses introduced by nonunity mirror reflectivities, a θ_1/θ_2 converter can also be created using a solid piece of transparent material. The design is similar to the hollow θ_1/θ_2 and must take into account the refraction of rays at the input and output port. A dielectric θ_1/θ_2 converter is shown in Fig. 21.

If the θ_1/θ_2 is an all-dielectric device, nearly maximum concentration for an air exit port can be obtained without the mirror losses of a hollow CPC; however, Fresnel surface reflections at the input and output ports should be considered.

Similar to the lens-mirror combination for maximal concentration, the θ_1/θ_2 can be implemented with a lens at the input port. Going one step further, the output port of the θ_1/θ_2 can also incorporate a lens. Similar to a Galilean telescope, the θ_1/θ_2 with lenses at both the input and output ports can provide a short package length compared to embodiments with a flat surface at either port.[164]

In many cases, a simple tapered cone can be used to implement a θ_1/θ_2 converter. One reason is that when the difference between θ_1 and θ_2 is small, the compound parabolic combined with a cone is nearly identical to a simple cone. By using an approximation to the ideal shape, some of the flux within θ_1 maps to angles outside of θ_2. In many practical systems, the input distribution does not have a sharp transition at θ_1, so a small mapping outside of θ_2 has minor impact. In general, the transition can be made sharper by increasing the cone length; however, there may be multiple local minima for a specific situation.[165]

Uniformity at the output port of a θ_1/θ_2 compared to a CPC has been investigated.[166] Tabor[118] describes hot spots that can occur with CPCs used in solar energy systems. Edmonds[125] uses a prism where hot spots that may occur are defocused after propagating through the prism. Emmons[167] discusses polarization uniformity with a dielectric 90/30 converter.

Two CPCs can be placed "throat-to-throat" to provide a θ_1/θ_2 converter that also removes angles higher than the desired input angle.[168] The removal of higher angles is a result of the CPC's ability to transfer rays with angles less than θ_1 and reject rays with angles greater than θ_1. Two nonimaging concentrators have also been investigated for use in fiber-optic coupling.[169]

Tight lightpipe bends are also made possible by inserting a bent lightpipe between the throats of the two concentrators.[33] In the special case of $\theta_1 = \theta_2 = 90°$, a bent lightpipe has also been used as an ideal means to transport the flux for improved packaging.[170]

2D vs. 3D Geometries

Skew Ray Limits.   One of the standard design methods for concentrators has been to design a system using a meridional slice and then simply rotate the design about the optical axis. In some cases, such as the CPC used with a disk receiver, this may only introduce a small loss; however, there are cases where the losses are large.

The standard CPC-type concentrator for use with a circular receiver[90, 171] is an example where this loss is large. To assess the loss, compare the 2D and 3D cases. In 2D a concentrator

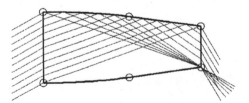

FIGURE 21   Dielectric θ_1/θ_2 implemented using cone with compound parabolic. The rays refract at the input port, total internally reflect (TIR) at the sidewalls, and then refract at the output port.

can be designed where the input port is $2R_{in}$ with a $\pm\theta_{in}$ angular distribution; the size of the tube-shaped receiver is $2\pi R_{tube} = 2R_{in} \sin \theta_{in}$. The 2D case provides maximum concentration. In the 3D case where the 2D design is simply spun about the optical axis, the ratio of the input etendue to the output etendue is $\pi R_{in}^2 \pi \sin^2 \theta_{in}/(4\pi R_{tube}^2) = \pi^2/4 \sim 2.5$. The dilution is quite large. Feuermann[172] provides a more detailed investigation of this type of 2D-to-3D etendue mismatch.

Ries[93] shows how skew preservation in a rotationally symmetric system can limit system performance even though the etendue of the source and receiver may be the same. The skewness distributions (detendue/dskewness as a function of skewness) for an example disk, sphere, and cylinder aligned parallel to the optical axis are shown in Fig. 22. The etendues of the disk, sphere, and cylinder are all the same. A second example where a cylinder is coupled to a disk is shown in Fig. 22b. This figure highlights losses and dilution. For skew values where the skewness distribution for the receiver is greater than the source, dilution occurs. Where the source is greater than the receiver, losses occur. If the size of the disk is reduced, then the dilution will also decrease, but the losses increase. If the disk is made bigger, then the losses decrease, but the dilution increases. Skew ray analysis of inhomogeneous sources and targets has also been considered.[173]

Star Concentrator.   One way of avoiding the skew ray limit in a system with a rotationally symmetric source and a rotationally symmetric receiver is to avoid the use of rotationally symmetric optics to couple the flux from source to target. A device called a *star concentrator* has been devised[174] and shown to improve the concentration. The star concentrator derives its name from the fact that the cross section of the concentrator looks like a star with numerous lobes. The star concentrator is designed using a global optimization procedure. Performance of the star concentrator, assuming a reflectivity of 1, provides performance that exceeds the performance possible using a rotationally symmetric system. The star concentrator approach has also been investigated for the case of coupling from a cylindrical source to a rectangular aperture.[7, 175]

Image Dissectors.   The problem of coupling flux from a rotationally symmetric source to a spectrometer slit motivated some earlier efforts to overcome the skewness limit. Benesch[176] shows a system with two sets of facets that "dissect" an X by Y aperture into N regions of size X/N by Y and relay those images to create an X/N by NY region. The dissection can be done while preserving the NA of the input (e.g., the etendue is preserved but the skew is changed). Benesch calls this system an *optical image transformer* and it is similar in function to Bowen's image slicer.[177] Refractive versions are also possible (see for example Ref. 178). One issue that

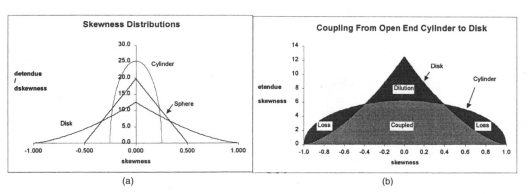

(a)   (b)

FIGURE 22 Skewness distributions (Ries[93]). (*a*) Skewness distribution for a disk with unit radius, a cylinder with length = 2 and radius = 0.25, and a sphere of radius 0.5. The etendue for all three cases is constant. (*b*) Comparison of the coupling of a disk to a cylinder where the coupled flux is distinguished from the losses and the dilution.

must be addressed for some systems, which use these dissectors, is that the adjoining regions of the X/N by Y areas can produce a nonuniform illuminance distribution. Laser arrays can also use dissectors to improve performance.[179, 180]

Tandem-lens-array approaches can be used that provide the image dissection and superimpose all the images rather than placing them side by side.[181, 182]

An example of a refractive image dissector compared to a lens array is shown in Fig. 23. The image dissector breaks an input square beam into four smaller square regions that are placed side by side at the output. The lens array breaks the input beam into four rectangles that are superimposed at the output. Both systems use field lenses at the output. The lens array does not have a gap between mapped regions at the output plane compared to the dissector.

Fiber bundles can also be used as image dissectors. For example, the fibers can be grouped in a round bundle at one end and a long skinny aperture at the other end. Feuermann[183] has provided a recent investigation of this idea. If the fibers in the bundle have a round cross section, tapering the ends of each of the fibers can help improve the uniformity at the output.[184] Special systems that use slowly twisted rectangular light guides have also been investigated for scintillators,[185–187] document scanning,[188, 189] and microlithography.[190]

Bassett[191] also shows that arrays of ideal fibers can be used to create ideal concentrators, and Forbes[192] uses the same approach to create a θ_1/θ_2 transformer.

Geometrical Vector Flux

The edge ray design method does not show how the jumble of rays from multiple reflections yields a Lambertian output. The geometric vector formalism was investigated in an attempt to understand the process. Summaries can be found in Bassett[101] and Welford.[4] The proof that a trumpet provides ideal performance has been performed using the flow line concept.[114]

The vector flux formalism provides the lines of flow and loci of constant geometric vector flux J. The vector J is the PSA for three orthogonal patches where the PSA on the front side of a patch can subtract from the PSA on the back side. The magnitude of J describes what the PSA would be for a surface aligned perpendicular to J. The flow line concept uses the idea that an ideal mirror can be placed along lines of flow without modifying the flow.

In the context of vector flux, the CPC, θ_1/θ_2, and light cone are discussed in Winston.[141] The light cone is demonstrated using an orange in Winston.[95] The 2D CEC and θ_1/θ_2 are discussed by Barnett.[193, 194] Gutierrez[195] uses Lorentz geometry to show that other rotationally symmetric ideal concentrators can be obtained using surfaces that are paraboloid, hyperboloid, and ellipsoid. Greenman[196] also describes vector flux for a number of ideal concentrators.

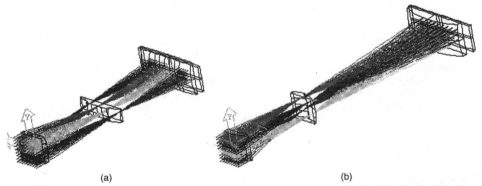

(a)   (b)

FIGURE 23   Square-to-rectangular mapping with an image dissector (*a*) and a lens array (*b*) where the output angular distribution is symmetric. The image dissector uses an array of discrete field lenses. The lens array uses a single field lens.

Edge Rays

The design of nonimaging optics has been facilitated by the use of the edge ray method[4, 68, 101, 197] and is sometimes called the string method.[4, 8] In the design of an ideal concentrator, the main principle is that extreme rays at the input aperture must also be extreme rays at the output aperture. Minano[198] describes the edge rays as those that bound the others in phase space. In many cases, the result is that the edge of the source is imaged to the edge of the receiver, and the reflector surface to perform the mapping is a compound conic. When the location of the source edge or the receiver edge is not constant (e.g., a circular source), the required reflectors are not simple conics (e.g., see the preceding section). Gordon[199] describes a complementary construction that highlights the fact that what is conventionally the "outside" of a concentrator can also be used in nonimaging optics.

Davies[200] further explores the continuity conditions required by the edge-ray principle. Ries[201] refines the definition of the edge ray principle using topology arguments. Rabl[202] shows examples where the analysis of specular reflectors and Lambertian sources can be performed using only the edge rays.

Tailoring a reflector using edge rays is also described in the next section.

Inhomogeneous Media

A medium with a spatially varying index of refraction can provide additional degrees of freedom to the design of nonimaging optics. In imaging optics, examples of inhomogeneous media include Maxwell's fisheye and the Luneburg lens.[4, 18] In concentrator design, 2D and 3D geometries have been investigated.[198, 203]

2.6   UNIFORMITY AND ILLUMINATION

In many optical systems, an object is illuminated and a lens is used to project or relay an image of the illuminated object to another location. Common examples include microscopes, slide projectors, overhead projectors, lithography, and machine vision systems. In a projection system the uniformity of the projected image depends upon the object being illuminated, the distribution of flux that illuminates the object, and how the projection optics transport the object modulated flux from object to image. The portion of the system that illuminates the object is often called the *illumination subsystem*.

Beam forming is another broad class of illumination systems. In beam forming, there is often no imaging system to relay an image of the illuminated object. Since the human eye is often used to view the illuminated object, the eye can be considered an imaging subsystem. Application examples include commercial display cases, museum lighting, room lighting, and automotive headlamps.

Approaches to provide uniform object illumination are discussed in this section. The discussed approaches include Kohler/Abbe illumination, integrating cavities, mixing rods, lens array, tailored optics, and faceted structures. Many of the approaches discussed in this section can be designed to preserve etendue, but the design emphasis is on uniformity so they have been placed in this section rather than the preceding one.

Classic Projection System Uniformity

In classical projection systems, there is typically a source, a transparency (or some other item to be illuminated), and a projection lens. There are two classic approaches to obtaining uniformity in those systems: Abbe and Kohler illumination (see Fig. 24). In both cases, a source illu-

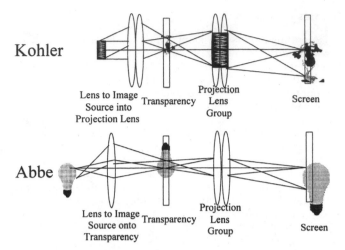

FIGURE 24   Abbe vs. Kohler layout comparison.

minates a transparency that is then often imaged onto a screen. *Transparency* is a general term that includes items such as spatial light modulators, film, slides, gobos, and microscope slides.

Some general references regarding Abbe/Kohler illumination include Jones,[204] Kingslake,[205] O'Shea,[206] Wallin,[207] Weiss,[208] Ray (pp. 455–461),[66] Bradbury,[209] Inoue,[210] and Born.[18] A description of Kohler with partial coherence is included in Lacombat.[211]

Abbe.   Abbe (also called Nelsonian and Critical illumination) images the source onto the transparency. When the source is spatially uniform, the Abbe approach works fine. A frosted incandescent bulb is a good example. Sometimes a diffuser is inserted in the system and the diffuser is imaged onto the transparency. Fluorescent tubes are another case where a light source provides excellent spatial uniformity. The luminance of sources that are spatially uniform is typically lower than that of sources that are not uniform.

Kohler.   Even though the source's illuminance distribution may be nonuniform, there are often uniform regions of the source's intensity distribution. This phenomenon is used in the design of Kohler-type systems where the source is imaged into the projection lens aperture stop and a mapping of the source intensity distribution illuminates the transparency. One reason this approach works well is that far from the source, the illuminance distribution over a small area is independent of where the flux originated. This tends to provide uniformity over portions of the source's intensity distribution.

A nearly spherical reflector is sometimes added behind the source to improve the collection efficiency of a Kohler illumination system. Typically, the spherical reflector produces an image of the source that is aligned beside the primary source. The direct radiation from the source and the flux from the image of the source then become the effective source for the Kohler illuminator. If the source is relatively transparent to its own radiation, then the image of the source can be aligned with the source itself. Blockage by electrodes and leads within the bulb geometry often limit the flux that can be added using the spherical reflector. Scattering and Fresnel surface reflections also limit the possible improvement. Gordon[212] provides a recent description of a back reflector used with filament sources.

In Kohler illuminators, tailoring the illuminance distribution at the transparency to provide a desired center-to-edge relative illumination can be performed using the edge ray approach described by Zochling.[213] Medvedev[82] also applied the edge ray principle to the design of a projection system. The other tailoring approaches described later in this section may also be relevant.

Integrating Cavities

Integrating cavities are an important example of nonimaging optics that are used in precision photometry,[16] backlighting systems, collection systems (Basset,[101] Sec. 7, and Steinfeld[214]), scanners,[41] and reflectometers.[215] They provide a robust means for creating a distribution at the output port that is independent of the angular and spatial luminance distributions at the input port, thereby removing "hot spots" that can be obtained using other methods. Integrating cavities are limited by a trade-off between maximizing the transmission efficiency and maximizing the luminance at the output port. The angular distribution exiting the sphere is nominally an unclipped Lambertian, but there are a number of techniques to modify the Lambertian distribution.[216]

Nonuniformities with Integrating Spheres. Integrating cavities, especially spherically shaped cavities, are often used in precision photometry where the flux entering the sphere is measured using the flux exiting the output port of the sphere. For precision photometry, the angular distribution of the exiting flux must be independent of the entering distribution. In general, the flux is scattered multiple times before exiting the sphere, which provides the required independence between input and output distributions. The two main situations where this independence is not obtained are when the output port views the input port and when the output port views the portion of the sphere wall in which the direct illumination from the input port illuminates the sphere wall.

Figure 25 shows an integrating sphere with five ports. A source is inserted into the upper port and the other four ports are labeled. Figure 26 shows the intensity distribution that exits output port 1 of the sphere shown in Fig. 25. The structure in the exiting intensity distribution is roughly Lambertian, but there are nonuniformities in the distribution that occur at angles that can "see" the source or the other output ports. To highlight this effect, rotated views of the sphere are included in Fig. 26. The bright peak in the intensity distribution occurs at the view angle where the output port can "see" direct radiation from the source. The dips in the intensity distribution occur at view angles where the output port can "see" the unused ports in the sphere.

If the source is replaced by a relatively collimated source that illuminates only a small region of the cavity interior (c.g., a collimated laser), then the output port intensity also shows a peak as shown in Fig. 27.

Three methods that are often used in precision photometry to create an output intensity distribution that is independent of the input involve the use of baffles, satellite spheres, and diffusers. Baffles block radiation from exiting the sphere directly, thereby forcing the exiting flux to scatter multiple times before exiting the sphere. A *satellite sphere* is a secondary sphere that is added to the primary sphere. Diffusers superimpose flux from multiple angles to provide an averaging effect that removes observable structure.

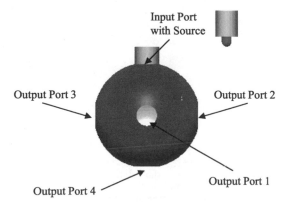

FIGURE 25   Spherical integrating cavity with one input port and four output ports.

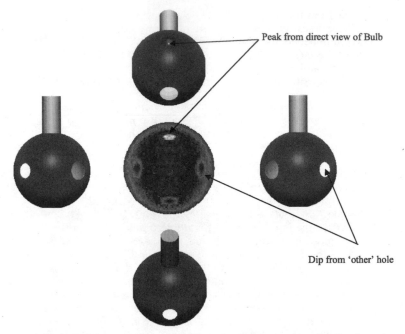

FIGURE 26 Relationship between the intensity distribution from the flux that exits an integrating sphere to what you see when looking back into the sphere.

Finite-thickness walls at the exit port will also alter the view angle of external ports and can improve the on-axis intensity because light that reflects off of the walls can go back into sphere and have a chance to add to the output.

Efficiency vs. Luminance. An important aspect of an integrating cavity is that structure in the angular and spatial distribution of the source's light flux are removed by multiple reflec-

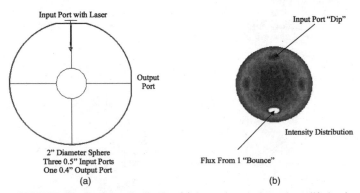

FIGURE 27 Intensity distribution (*a*) for an integrating sphere (*b*) that is illuminated with a highly collimated source. The flux that exits the sphere with only one bounce produces a higher intensity than the flux that rattles around and exits in a more uniform manner. The intensity distribution is shown normal to the orientation of the output port.

tions from the diffuse reflector. Under the assumption that the flux inside of an integrating sphere is scattered in a Lambertian manner, Goebel[217] has provided a description of the relationship between the fraction of flux that enters the sphere and the fraction of flux that exits the sphere (see also the Technical Information chapter in the LabSphere catalog). Goebel's results can be used to show that the ratio of the flux that exits the sphere via one port to the flux that exits out another port is

$$\text{Efficiency} = \eta = \frac{f_{\text{output}} R_{\text{sphere}}}{1 - f_{\text{input}} R_{\text{input}} - f_{\text{output}} R_{\text{output}} - f_{\text{sphere}} R_{\text{sphere}}} \qquad (13)$$

where

$f_{\text{sphere}} = (\text{total sphere surface area} - \text{port areas})/\text{total sphere surface area}$
$f_{\text{output}} = \text{output port area}/\text{total sphere surface area}$
$f_{\text{input}} = \text{input port area}/\text{total sphere surface area}$
$R_x = \text{reflectivity of } x$

When the input and output have zero reflectivity, the result is

$$\text{Efficiency} = \eta = \frac{f_{\text{output}} R_{\text{sphere}}}{1 - f_{\text{sphere}} R_{\text{sphere}}} \qquad (14)$$

Averaging the input flux over a hemisphere gives a ratio of the luminance at the output port to the luminance at the input port of

$$\frac{\text{Output Luminance}}{\text{Input Luminance}} = \frac{\eta/\text{Area}_{\text{output}}}{1/\text{Area}_{\text{input}}} = \frac{\eta f_{\text{input}}}{f_{\text{output}}} = \frac{R_{\text{sphere}} f_{\text{input}}}{1 - f_{\text{input}} R_{\text{input}} - f_{\text{output}} R_{\text{output}} - f_{\text{sphere}} R_{\text{sphere}}} \qquad (15)$$

When the input and output ports have zero reflectivity, then the result is

$$\frac{\text{Output Luminance}}{\text{Input Luminance}} = \frac{R_{\text{sphere}}/f_{\text{input}}}{1 - f_{\text{sphere}} R_{\text{sphere}}} \qquad (16)$$

The efficiency and the ratio of input to output luminances are shown in Fig. 28 for the case of $R_{\text{sphere}} = 98.5\%$.

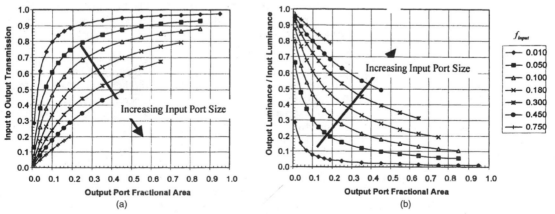

FIGURE 28 Input to output transmission (*a*) and relative output luminance (*b*) for the case of sphere reflectivity = 0.985. The different curves are for various input port fractional areas.

When the fractional input port size equals the fractional output port size, the sphere transmission is less than 50% and the output luminance is less than 50% of the input luminance. Increasing the output port size provides higher transmission, but the luminance of the output port is now less than 50% of that of the input port. Reducing the output port size can increase the output luminance to values greater than the input luminance; however, the transmission is now less than 50%. Thus, an integrating cavity offers the possibility of improving uniformity at the expense of either luminance or efficiency.

Integrating Cavity with Nonimaging Concentrator/Collectors.   The flux exiting an integrating cavity tends to cover a hemisphere. Adding a nonimaging collector, such as a CPC, to the output port can reduce the angular spread. The integrating cavity can also include a concentrator at the input port so as to minimize the size of the input port if the angular extent of the radiation at the input port does not fill a hemisphere.

A CEC can be added to the output port of an integrating cavity so that the portion of the sphere surface that contributes to the output distribution is restricted to a region that is known to have uniform luminance.[215] Flux outside of the CEC field of view (FOV) is rejected by the CEC. CPCs and lenses with CHCs have also been investigated.[215, 218, 219]

Modifying the Cavity Output Distribution.   Prismatic films can also be placed over the exit port of an integrating cavity, which is an approach that has been used by the liquid crystal display community (e.g., the BEFII films commercialized by 3M). The prismatic films increase the intensity at the angles that exit the sphere by reflecting the undesired angles back into the sphere so that they can be rescattered and have another chance to exit the sphere. For systems that require polarized light, nonabsorbing polarizing films can also be used to return the undesired polarization state back into the sphere.

Another approach to selectively controlling the exit angles is the addition of a hemispherical or hemiellipsoid reflector with a hole in the reflector. Webb[220] described this approach for use with lasers.

In a similar manner, reflecting flux back through the source has been used with discharge sources, and is especially effective if the source is transparent to its own radiation.

Mixing Rods (Lightpipes)

When flux enters one end of a long lightpipe, the illuminance at the other end can be quite uniform. The uniformity depends upon the spatial distribution at the lightpipe input, the intensity distribution of the input flux, and the shape of the lightpipe. Quite interestingly, straight round lightpipes often do not provide good illuminance uniformity, whereas some shapes like squares and hexagons do provide good illuminance uniformity. This is shown in Fig. 29.

Shapes (Round vs. Polygon).   Square lightpipes have been used very successfully to provide extremely uniform illuminance distributions. To understand how a lightpipe works, consider the illuminance distribution that would result at the output face of the lightpipe if the lightpipe's sidewalls were removed. Including the sidewalls means that various regions of the illuminance distribution with no lightpipe are superimposed. In the case of a square, the multiple regions add in a controlled manner such that there is improved uniformity by virtue of the superposition of multiple distributions. If the source distribution is symmetric about the optical axis, then distributions with a positive slope are superimposed onto distributions with a negative slope. This means that the uniformity from N distributions can be better than the addition of N uncorrelated distributions that would provide uniformity with a $1/\sqrt{N}$ standard deviation.

One way to view the operation of a rectangular mixing rod is to consider the illuminance distribution that would occur at the output face of the mixing rod if no sidewall reflections were present. The sidewalls force subregions of this unreflected illuminance distribution to superimpose in a well-controlled manner.[222] This is shown in Fig. 30, where a Williamson-type

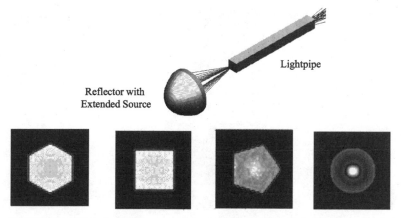

FIGURE 29   When flux from a source is coupled into a mixing rod, the illuminance at the output end of the mixing rod is significantly affected by the shape of the lightpipe. Square and hexagonal lightpipes work much better than round or many other regular polygons.

construction is shown on the left. The illuminance distribution at the lightpipe output with no sidewalls is also shown, as is the superposition of the subregions when sidewall reflection is included. Every other subregion is inverted because of the reflection at the sidewall. This inversion provides a difference between a lightpipe used for homogenization and a lens array.

Figure 31 shows the results of a Monte Carlo simulation where a source with a Gaussian illuminance distribution and a Gaussian intensity distribution are propagated through a

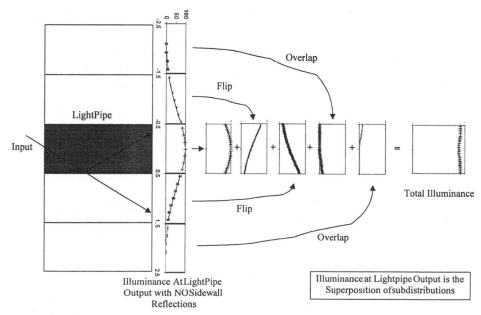

FIGURE 30   Flip-and-fold approach[222] for use with rectangular mixing rods. Subregions of the illuminance at the lightpipe output with no sidewall reflections are superimposed to create a uniform distribution. The lightpipe is shown shaded and a Williamson construction is included. The path of one ray is shown with and without sidewall reflection.

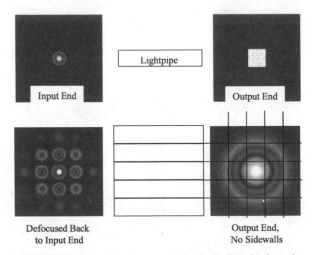

FIGURE 31 Example with a source that has a Gaussian intensity distribution. If the source is propagated to the end of the lightpipe without sidewall reflections, the illuminance distribution at the output can be sliced into distinct regions. These regions are superimposed to create the illuminance distribution that actually occurs with the sidewall reflections. If the flux at the lightpipe output is propagated virtually back to the input end, the multiple images can be observed.

square lightpipe. The illuminance at the lightpipe input and output faces is shown in the top of the figure. The illuminance distribution in the lower right shows the illuminance distribution if no sidewall reflections occurred. Lines are drawn through the no-sidewall reflection illuminance distribution to highlight the subregions that are superimposed to create the illuminance distribution when the lightpipe is present. The lower left illuminance distribution results from propagating the flux from the output end of the lightpipe back to the input end. Multiple images of the source are present, similar to the multiple images observed in a kaleidoscope. This kaleidoscope effect is one reason why lightpipe mixing is sometimes considered to be the superposition of multiple virtual sources.

In general, if the cross-sectional area is constant along the lightpipe length, and the area is covered completely by multiple reflections with respect to straight sidewalls of the lightpipe, then excellent uniformity can be obtained. The shapes that provide this *mirrored tiling* include squares, rectangles, hexagons, and equilateral triangles.[221] A square sliced along its diagonal and an equilateral triangle sliced from the center of its base to the apex will also work. Pictures of these shapes arranged in a mirror-tiled format are shown in Fig. 32. The lightpipe length must be selected so as to provide an adequate number of mirrored regions. In particular, the regions near the edge that do not fill a complete mirror-tiled subregion must be controlled so that they do not add a bias to the output distribution.

Shapes that do not mirror-tile may not provide ideal uniformity as the lightpipe length is increased, at least for a small clipped Lambertian source placed at the input to the lightpipe. One explanation is shown in Fig. 33.

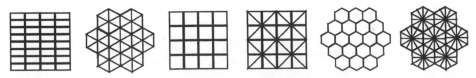

FIGURE 32 Mirror-tiled shapes that can ensure uniformity with adequate length.

Hexagon 'Tiles'   Pentagon does not 'Tile'

FIGURE 33   Diagram showing why pentagon does not tile like the
hexagon. Simulation results are shown in Fig. 29.

Round lightpipes can provide annular averaging of the angular distribution of the flux that
enters the input end. Because of this, the illuminance distribution along the length of the
lightpipe tends to be rotationally symmetric; however, the illuminance distribution tends to
provide a peak in the illuminance distribution at the center of the lightpipe. The magnitude of
this peak relative to the average varies along the length of the lightpipe. If the distribution
entering the lightpipe is uniform across the whole input face, and the angular distribution is a
clipped Lambertian, then the round lightpipe maintains the uniformity that was present at the
input end. However, if either the input spatial distribution or the input angular distributions
are not uniform, then nonuniform illuminance distributions can occur along the lightpipe
length. This presents the interesting effect that a source can be coupled into a square lightpipe
of adequate length to provide a uniform illuminance distribution at the lightpipe output face;
however, if the angular distribution is nonuniform and coupled into a round lightpipe, then
the illuminance at the output face of the round lightpipe may not be uniform.

Periodic Distributions.   If the unmirrored illuminance distribution has a sinusoidal distribution with certain periods that match the width of the lightpipe, then the peaks of the sinusoid
can overlap. A Fourier analysis of the lightpipe illuminance distribution can help to uncover
potential issues.[222]

Length.   The length of the lightpipe that is required to obtain good uniformity will depend
upon the details of the source intensity distribution. In general, overlapping a 9×9 array of
mirrored regions provides adequate uniformity for most rotationally symmetric distributions.
For an f/1 distribution with a hollow lightpipe, this means that the lightpipe aspect ratio
(length/width) should be greater than 6:1. For a lightpipe made from acrylic, the aspect ratio
needs to be more like 10:1.

Solid vs. Hollow.   In addition to a solid lightpipe requiring a longer length to obtain a given
uniformity, the design of the solid case should consider the following: Fresnel surface losses at
the input and output ends, material absorption, cleanliness of the sidewalls (heat shrink
Teflon can help here), and chipping of the corners. For the hollow case, some considerations
include dust on the inside, coatings with reflectivities of less than 100 and angular/color
dependence, and chips in mirrors where the sidewalls join together.[222]

With high-power lasers and high-power xenon lamps, a solid coupler may be able to handle the average power density, but the peak that occurs at the input end may introduce damage or simply cause the solid lightpipe to shatter.

Hollow lightpipes can also be created using TIR structures.[223, 224]

Angular Uniformity.   Although the illuminance at the output end of a lightpipe can be
extremely uniform, the angular distribution may not be, especially the "fine" structure in the
distribution. One method to smooth this fine structure issue is to add a low-angle diffuser at

the lightpipe output end.[225] A diffuser angle that is about the angular difference between two virtual images is often sufficient. Since multiple images of the source are usually required to obtain illuminance uniformity at the lightpipe output, the diffuser does not increase the etendue much. Making the lightpipe longer reduces the angular difference between neighboring virtual images and the required diffuser angle is reduced accordingly. Combinations of lightpipes and lens arrays have also been used.[226]

Tapered Lightpipes. If the lightpipe is *tapered,* meaning that the input and output ends have different sizes, excellent uniformity can still be obtained. Since the lightpipe tends to need a reasonable length to ensure uniformity, the tapered lightpipe can provide reasonable preservation of etendue. This etendue preservation agrees with Garwin's[3] adiabatic theorem. In general, for a given light pipe length, better mixing is obtained by starting with high angles at the light pipe input and tapering to lower angles than if the input distribution starts with the lower angles.[227] A Williamson construction can be used to explain why the tapered pipe has more virtual images than a straight tapered case. Tapered lightpipes have been used with multimode lasers[228] and solar furnaces.[221]

An example with a small Lambertian patch placed at the entrance to a tapered lightpipe and a straight lightpipe is shown in Fig. 34. The input NA for the untapered case is 0.5/3 for the 100-mm-long straight lightpipe and 0.5 for the 100-mm-long tapered lightpipe. The input size of the straight lightpipe is a square with a 30-mm diagonal. The input size of the tapered lightpipe is a 10-mm diagonal and the output size is a 30-mm diagonal. The Williamson construction shows that the extreme rays for the 0.5-NA meridian have almost three bounces for the tapered case, but only slightly more than one bounce for the extreme ray in the 0.5/3-NA straight lightpipe case. The illuminance distribution that would occur at the lightpipe output if the lightpipe sidewalls were removed is shown in Fig. 34. Gridlines are superimposed on the picture to highlight that the tapered case has numerous regions that will be superimposed, but the superposition will be small for the straight lightpipe case. The tapered case provides superior uniformity, although the output NA in the tapered case is higher than 0.5/3 because the

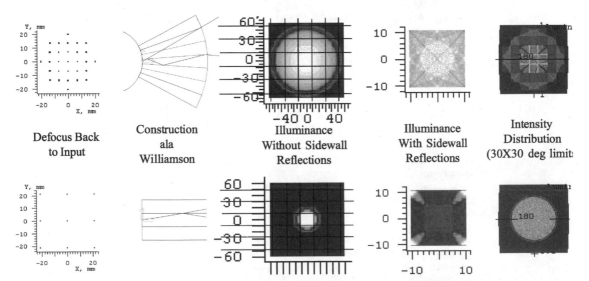

FIGURE 34 Tapered lightpipe example. The upper and lower rows are for the tapered and untapered cases, respectively. Moving left to right, the drawings show the source when defocused from the lightpipe output back to the lightpipe input, a Williamson construction, the no-sidewall illuminance at the lightpipe output, the illuminance at the lightpipe output, and the angular distribution of the flux exiting the lightpipe.

angular distribution has "tails." These results are obtained using a source that is a small Lambertian patch. The comparison is typically far less drastic when the input face of the lightpipe is better filled.

Applications of Mixing Rods.   Mixing rods have found use in a number of applications. Examples include: projection systems,[229] providing uniformity with a bundle of fibers using discharge lamps,[227] fiber-to-fiber couplers,[230] and solar simulators.[221]

Mixing rods have also been used with coherent sources. Some applications include semiconductor processing;[231] hyperpigmented skin lesions;[232, 233] lithography, laser etching, and laser scanning;[228] and high-power lasers.[234] If the coherence length of the laser is not small compared to the path length for overlapping distributions, then speckle effects must be considered.[231, 234] Speckle effects can be mitigated if the lightpipe width is made large compared to the size of the focused laser beam used to illuminate the input end of the lightpipe (e.g., Refs. 231 and 236). A negative lens for a laser illuminated lightpipe is described by Grojean;[235] however, this means that etendue of the source is less than the etendue of the lightpipe output.

Speckle can be controlled in lightpipes by adding a time-varying component to the input distribution. Some methods include placing a rotating diffuser at the lightpipe input end and moving the centroid of the distribution at the input end. In principle, the diffuser angle can be small so that the change in etendue produced by the diffuser can also be small.

Coherent interference can also be used advantageously, such as in the use of mixing rods of specific lengths that provide sub-Talbot plane reimaging. This was successfully applied to an external cavity laser.[237]

In addition to uniformity and the classic kaleidoscope, mixing rods can also be used in optical computing systems by reimaging back to the lightpipe input.[238, 239]

Lens Arrays

Lens arrays are used in applications including Hartmann sensors,[240] diode laser array collimators,[241] beam deflection,[242] motion detectors,[243] fiber-optic couplers,[244] commercial lighting, and automotive lighting. In commercial and automotive lighting, lens arrays are often used to improve uniformity by breaking the flux into numerous subregions and then superimposing those subregions together. This superposition effect is illustrated in Fig. 35, where fans of rays illuminate three lenslets and are superimposed at the focal plane of a condensing lens. The condensing lens is not used if the area to be illuminated is located far from the lens array.

If the incident flux is collimated, then the superposition of wavefronts provides improved uniformity by averaging the subregions. However, when the wavefront is not collimated, the off-axis beams do not completely overlap the on-axis beams (see Fig. 35b). Tandem lens arrays add additional degrees of design freedom and can be used to eliminate the nonoverlap problem (see Fig. 35c).

Literature Summary.   Some journal citations that describe the use of lens arrays with incoherent sources include Zhidkova's[245] description of some issues regarding the use of lens arrays in microscope illumination. Ohuchi[246] describes a liquid crystal projector system with an incoherent source and describes the use of two tandem lens arrays with nonuniform sizes and shapes for the second lens array. This adjustment to the second lens array provides some adjustment to the coma introduced by the conic reflector.

Rantsch[247] shows one of the earliest uses of tandem lens arrays. Miles[248] also uses a tandem lens array system and is concerned about the distribution in the pupil. Ohtu[249] uses two sets of tandem lens arrays in a lithography system where one array controls the spatial distribution of the image and the other controls the angular (pupil) distribution of the image. Kudo[226] combines a lens array and lightpipe to provide the same type of control over the illuminance and intensity distributions. Matsumoto[250] adds a second set of tandem lens arrays with a

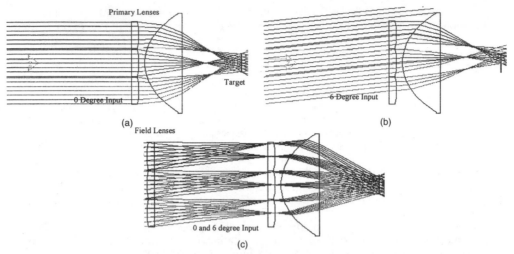

FIGURE 35   Lens arrays used for illumination. (*a*) Superposition of three subregions of the input wavefront. (*b*) Off-axis angles are displaced relative to the on-axis angles. (*c*) Tandem lens array case, where the addition of an array of field lenses allows the superposition to work with a wide range of angles.

sparse fill to correct the fact that an odd spatial frequency term in the illuminance distribution can result in a final nonuniformity, independent of the number of lenslets. Van den Brandt[251] shows numerous configurations using tandem lens arrays, including the use of nonuniform sizes and shapes in the second lens array. Watanabe[252] provides design equations for tandem lens arrays in a converging wavefront.

Journal publications that include lens arrays often involve the use of lens arrays with coherent sources rather than incoherent sources. Deng[253] shows a single lens array with a coherent source and shows that the distribution at the target is a result of diffraction from the finite size of each lenslet and interference between light that propagates through the different lenslets. Nishi[254] modifies the Deng system to include structure near the lenslet edges so as to reduce the diffraction effects. Ozaki[255] describes the use of a linear array of lenses with an excimer laser, including the use of lenslets where the power of the lenslets varies linearly with distance from the center of the lens array. Glockner[256] investigates the performance of lens arrays under coherent illumination as a function of statistical variation in the lenslets. To reduce interference effects with coherent sources, Bett[257] describes the use of a hexagonal array of Fresnel zone plates where each zone incorporates a random phase plate. Kato[258] describes random phase plates.

Single-Lens Arrays.   Single-lens-array designs for uniformity tend to have two extremes. One occurs when adding the lens array significantly changes the beam width. This is the beam-forming category where the shape of the beam distribution is essentially determined by the curvature and shape of the lenslets. The other extreme occurs when adding the lens array does not significantly change the beam width. This is the beam-smearing category and is useful for removing substructure in the beam, such as structure from filament coils or minor imperfections in the optical elements. There are many designs that lie between these two extremes.

Figure 36 depicts the beam-forming case where the source divergence is assumed to be small. Positive and/or negative lenses can be used to create the distribution. Hybrid systems where mixtures of positive and negative lenses are used provide the added benefit that the slope discontinuity between neighboring lenses can be eliminated. Slope discontinuities can degrade performance in many systems.

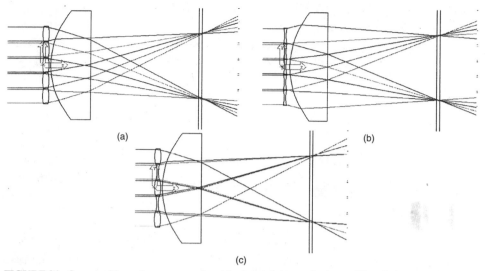

FIGURE 36 Superposition using an array of positive lenses (*a*), negative lenses (*b*), and also a hybrid system (*c*) that uses negative and positive lenses. A collimated wavefront passing through each lenslet is mapped over the target. The hybrid system offers the possibility of a continuous surface with no slope discontinuities.

Tandem-Lens Arrays. The use of a single-lens array has two issues that sometimes limit its use: First, the lens array increases the etendue because the angles increase but the area stays the same. Second, the finite size of the source introduces a smearing near the edges of the "uniform" distribution. Adding a second lens array in tandem with the first can minimize these limitations.

Figure 37*a* shows a 1:1 imager in an Abbe (critical) illumination system. Figure 37*b* shows the same imager with two-lens arrays added. Adding the lens arrays has changed the system from Abbe to Kohler (see the preceding text). In addition, as shown in Fig. 38, the lenslets provide channels that can be designed to be independent of one another. There are three main reasons why two tandem-lens arrays provide uniformity:

1. Each channel is a Kohler illuminator, which is known to provide good uniformity assuming a slowly varying source-intensity distribution.

2. Adding all the channels provides improved uniformity as long as the nonuniformities of one channel are independent of the nonuniformities of the other channels (e.g., a factor of $1/\sqrt{N}$ improvement by adding independent random variables).

3. For symmetric systems, the illuminance from channels on one side of the optical axis tends to have a slope that is equal but opposite to the slope on the other side of the optical axis. The positive and negative slopes add to provide a uniform distribution.

Reflector/Lens-Array Combinations. Using a reflector to collect the flux from the source can provide a significant increase in collection efficiency compared to the collection efficiency of a single lens. The drawback is that a single conic reflector does not generally provide uniform illuminance. The use of a lens array in this situation allows the illuminance distribution to be broken into multiple regions, which are then superimposed on top of each other at the output plane.

Examples showing the illuminance distribution at the first lens array, the second lens array, and the output plane are shown in Fig. 39 for the case of a lens collector, Fig. 40 for the case

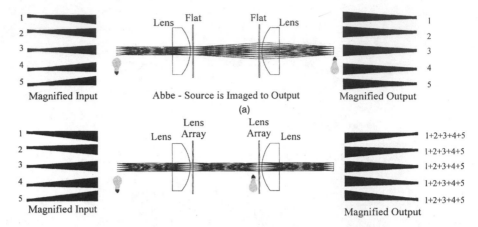

FIGURE 37   Critical illumination (Abbe, *a*) compared to Kohler (*b*). In Abbe, the source is imaged to the output. In Kohler, most of the source is "seen" from all output points.

of a parabola collector, and Fig. 41 for the case of an elliptical collector. These examples show that nonuniform illuminance distribution at the first lens array is broken into regions. Each region is focused onto the second set of lenses, which creates an array of images of the source. The lenslets in the second array image the illuminance distribution at the first lens array onto the output plane. Enhanced uniformity is obtained because the distributions from each of the

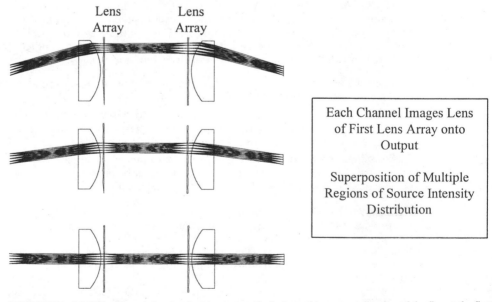

FIGURE 38   Multiple channels in a tandem-lens array. Each channel images a subregion of the flux at the first lens array. The flux from the different channels is superimposed.

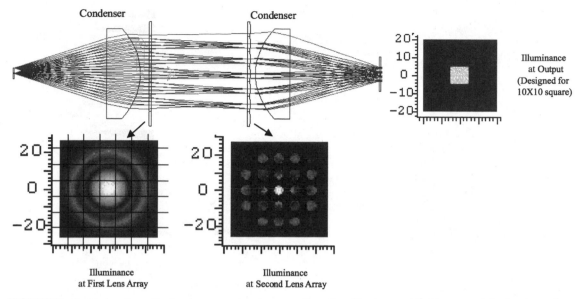

FIGURE 39 Source/condenser/tandem lens arrays/condenser configuration. Illuminance at first lens array is broken into subregions that are superimposed at the target by the second lens array/second condenser. All units are millimeters.

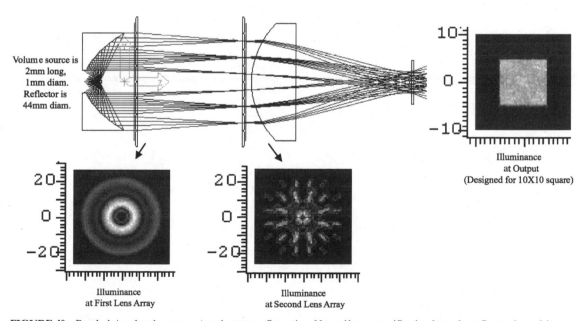

FIGURE 40 Parabola/tandem lens arrays/condenser configuration. Nonuniform magnification from the reflector (coma) is seen in the array of images at the second lens array. The illuminance at the target is the superposition of subregions at the first lens array. All units are millimeters.

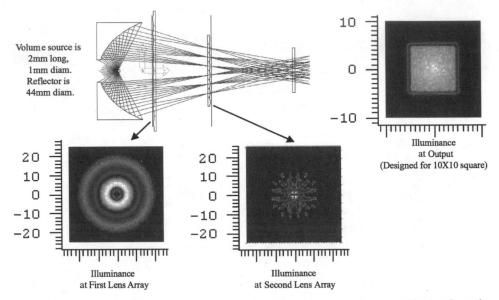

FIGURE 41   Ellipse/tandem lens arrays/condenser configuration. Nonuniform magnification from the reflector (coma) is seen in the array of images at the second lens array. The illuminance at the target is the superposition of subregions at the first lens array. The spacing between lenses is different for the two lens arrays. All units are millimeters.

channels are superimposed at the target. The discrete images of the source that are seen at the second lens array indicate that the pupil of the output distribution is discontinuous.

Improvements to this basic idea include the use of lens arrays where the lenslets in the second lens array do not all have the same size/shape.[251] The curvature of each lenslet in the first array is decentered to "aim" the flux toward the proper lenslet in the second array. When nonuniform lenslet size/shapes are used, the etendue loss that results from the use of a conic reflector can be minimized. Such a system can be used to create asymmetric output distributions without a correspondingly misshaped pupil distribution (see earlier section on image dissectors).

Tailored Optics

The tailoring of a reflector to provide a desired distribution with a point or line source has been explored.[8, 259–265] Some simple cases are available in closed form.[88, 266, 267] Refractive equivalents have also been investigated.[268, 269]

In a manner similar to the lens configurations shown in Fig. 36, tailored reflectors can produce converging or diverging wavefronts. Examples are shown in Fig. 42. The converging reflector creates a smeared image of the source between the reflector and the target. The diverging reflector creates a smeared image of the source outside of the region between the reflector and target. This terminology was developed assuming that the target is far from the reflector. The terms *compound hyperbolic* and *compound elliptic* are also used to describe these two types of reflector configurations.[270]

If the flux from the two sides of the reflector illuminates distinct sides of the target, then there are four main reflector classifications (e.g., Ref. 26, pp. 6.20–6.21; Refs. 91 and 260). Examples of the four cases are pictured in Fig. 43. They are the crossed and uncrossed versions of the converging and diverging reflectors, where *crossed* is relative to the center of the

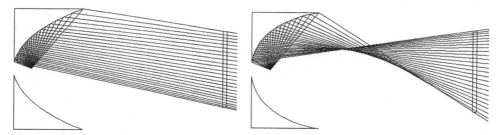

FIGURE 42 Two main classes of reflector types—diverging (*left*) and converging (*right*). The target is shown on the right of each of the two figures.

optical axis. Elmer[271] uses the terms *single* and *divided* for uncrossed and crossed, respectively. The uncrossed converging case tends to minimize issues with rays passing back through the source. The four cases shown provide one-to-one mappings because each point in the target receives flux from only one portion of the reflector.

Tailored Reflectors with Extended Sources. Elmer[8] describes some aspects of the analysis/design of illumination systems with extended sources using what are now called *edge rays* (see also Ref. 271). Winston[266] sparked renewed interest in the subject. The following is a brief summary of some of the edge-ray papers that have been published in recent years. Some concentration related citations are included. Edge rays were also discussed earlier.

Gordon[272] adds a gap between the source and reflector to offset the cosine cubed effect and obtain a sharp cutoff. Gordon[273] explores the case where the extreme direction is a linear function of polar angle. Friedman[147] uses a variable angle to design a secondary for a parabolic primary. Gordon[274] describes a tailored edge ray secondary for a Fresnel primary, in particular for a heliostat field. Winston[270] describes the tailoring of edge rays for a desired functional rather than maximal concentration. Ries[275] describes the tailoring based on a family of edge

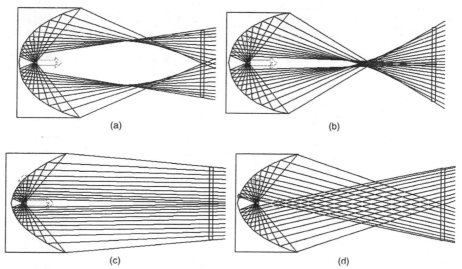

FIGURE 43 Four types of one-to-one mapping reflectors where the flux from each side of the reflector does not overlap at the target. (*a*) Uncrossed converging. (*b*) Crossed converging. (*c*) Uncrossed diverging. (*d*) Crossed diverging.

rays. Rabl[276] tailors the reflector by establishing a one-to-one correspondence between target points and edge rays for planar sources. Ong[277, 278] explores tailored designs using full and partial involutes. Jenkins[279] also describes a partial involute solution and generalizes the procedure as an integral design method.[280] Ong[91] explores the case of partial involutes with gaps. Gordon[281] relates the string construction to tailored edge ray designs for concentrators.

Faceted Structures

Nonuniform distributions can be made uniform through the use of reflectors where the reflector aims multiple portions of the flux from the source toward common locations at the target. In many cases, the resulting reflector has a faceted appearance. If the source etendue is significantly smaller than the etendue of the illumination distribution, then smearing near the edges of the uniform region can be small. However, if the source etendue is not negligible compared to the target, then faceted reflectors tend to experience smearing similar to the case of one lens array.

The term *faceted reflector* identifies reflectors composed of numerous distinct reflector regions. Elmer[270] uses the term *multiphase* and Ong[91] uses the term *hybrid* to describe these types of reflectors. If there are more than two regions, then the regions are often called *facets*. As with lens arrays, there are two extremes for faceted reflector designs. One is the beam-smearing category, where the facets remove substructure in the beam. The other extreme is the beam-forming category, where each facet creates the same distribution. If all facets create the same distribution, Elmer[271] calls the reflector *homogeneous*. If the distributions are different, then the term *inhomogeneous* is used.

For a given meridional slice, each facet can create an image of the source either in front of or behind the reflector. Using the convergent/divergent terminology to distinguish where the blurred image of the source occurs (see Fig. 42), a meridional slice of a faceted reflector can be composed of an array of convergent or divergent facets. These two cases are shown in Fig. 44 *a* and *b*. Flat facets are often desired because of the simplicity of fabrication and fall under the divergent category because the image of the source is behind the reflector.

The intersection between facets of the same type introduces a discontinuity in the slope of the reflector. Mixed convergent/divergent versions are possible, as shown in Fig. 44*c*. Mixed versions offer the possibility of minimizing discontinuities in the slope of the reflector curve.

If all the distributions are superimposed, then improved uniformity can arise through superposition if sufficient averaging exits. Consider a manufacturing defect where a subregion of the reflector that is about the size of a facet is distorted. If a superposition approach is used, then the effect of the defect is small because each facet provides only a small contribution to target distribution. However, if a one-to-one mapping approach is used, then the same defect distorts an entire subregion of the target distribution because there is no built-in redundancy.

The effects of dilution should be considered when faceted optics are used. A single-lens array provides insufficient degrees of freedom to provide homogeneous superposition without introducing dilution, which is why a second lens array is often added (e.g., the tandem-lens-array configuration). Schering[282] shows lenslets placed on a reflector combined with a second array of lenslets. A straightforward extension of lenslets on a reflector is the use of concave facets, as pictured in Fig. 45.

Some Literature on Faceted Structures.   If the source is highly collimated (e.g., a laser) and has a Gaussian distribution, then only a couple of regions of the wavefront need to be superimposed to provide excellent uniformity. Descriptions of optical implementations that emphasize nearly Gaussian wavefronts include a four-surface prism by Kawamura,[283] a wedge prism by Lacombat,[211] and a multipass wedge prism by Sameshima.[284] Austin[285] has described a prism system for an excimer laser. Brunsting[286] has described a system with nonplanar wedges. Bruno[287] and Latta[228] have discussed examples of mirrored systems with very few superimposed wavefronts.

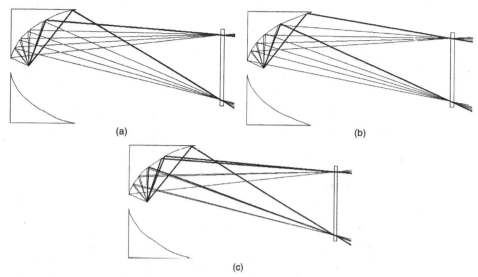

FIGURE 44   Some faceted reflector profiles. Convergent (*a*), divergent (*b*), and mixed convergent/divergent (*c*) profiles are all shown with five facets for the upper half of the reflector. Rays from the source that hit near the edges of the facets are shown. The rays for the upper facets are highlighted for all three cases. The rays for the second facet are highlighted in the mixed case.

For the more general case, uniformity can be obtained by superimposing a number of different regions of the beam. Dourte[287] describes a 2D array of facets for laser homogenization. Ream[289] discusses a system with a convex array of flat facets for a laser. Doherty[290] describes a segmented conical mirror that can be used to irradiate circular, annular, and radial surfaces with a laser source. Dagenais[291] describes the details of a system with nonplanar facets for use with a laser in paint stripping. Dickey[292] shows a laser-based system composed of two orthogonal linear arrays of facets and includes curvature in the facet structures to minimize the effects of diffraction. Experimental data for two orthogonal linear arrays of facets [called a *strip mirror integrator* (SMI)] are provided by Geary.[293] Lu[294] has described a refractive structure using a multiwedge prism structure. Henning[295] and Unnebrink[296] show curved facets for use with UV laser systems.

Laser-based systems provide the degree of freedom that the wavefront divergence can be small, although the effects of diffraction and coherence must be considered in those designs. The divergence of nonlaser sources is often an important issue.

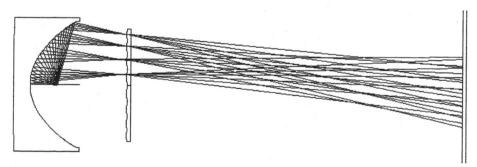

FIGURE 45   Reflector with array of concave facets and an array of lenslets. The lenslets image the facets.

Jolley[259] describes a faceted mirror design with disc and spherical sources. David[297] describes faceted reflectors like the ones used in commercial lighting products and highlights the fact that facets on a conic reflector base can improve uniformity with very little increase in "spot size" for long linear sources. Coaton[298] provides a brief description of a faceted reflector design with a point source. Donohue[299] describes faceted structures in headlamps.

There are many patents that describe the use of faceted structures for incoherent sources. Examples include Wiley,[300] who uses the facets to remove filament image structure; Laudenschlarger,[301] who describes design equations for a faceted reflector; Heimer,[302] who uses a faceted reflector in a lithography system; and Hamada,[303] who uses a faceted structure with a linear lamp.

Ronnelid[304] uses a corrugated reflector to improve the angular acceptance of a 2D CPC in solar applications. Receiver uniformity can also be improved in CPCs using irregular surfaces (Ref. 4, pp. 153–161).

Faceted structures have often been used in complex lightpipes such as those used in the illumination of backlit liquid-crystal displays and instrument panels. For instrument-panel application, more recent efforts have been directed toward using aspheric surfaces.[305]

2.7 REFERENCES

1. H. Hinterberger, L. Lavoie, B. Nelson, R. L. Sumner, J. M. Watson, R. Winston, and D. M. Wolfe, "The Design and Performance of a Gas Cerenkov Counter with Large Phase-Space Acceptance," *Rev. Sci. Instrum.* **41**(3):413–418 (1970).

2. D. A. Harper, R. H. Hildebrand, R. Stiening, and R. Winston, "Heat Trap: An Optimized Far Infrared Field Optics System," *Appl. Opt.* **15**:53–60 (1976) (Note: References are on page 144 of vol. 15).

3. R. L. Garwin, "The Design of Liquid Scintillation Cells," *Rev. Sci. Instrum.* **23**:755–757 (1952).

4. W. T. Welford and R. Winston, *High Collection Nonimaging Optics,* Academic Press, San Diego, CA, 1989.

5. W. J. Cassarly and J. M. Davenport, "Fiber Optic Lighting: The Transition from Specialty Applications to Mainstream Lighting," *SAE* 1999-01-0304 (1999).

6. A. E. Rosenbluth and R. N. Singh, "Projection Optics for Reflective Light Valves," *Proc. SPIE* **3634**:87–111 (1999).

7. J. Bortz, N. Shatz, and R. Winston, "Advanced Nonrotationally Symmetric Reflector for Uniform Illumination of Rectangular Apertures," *Proc. SPIE* **3781**:110–119 (1999).

8. W. Elmer, *The Optical Design of Reflectors,* 3d ed., 1989.

9. J. Palmer, Frequently Asked Questions, www.optics.arizona.edu/Palmer/rpfaq/rpfaq.htm., 1999.

10. J. M. Palmer, "Radiometry and Photometry: Units and Conversions," in *Handbook of Optics,* 2nd ed., vol. 3, chap. 7, McGraw-Hill, New York, 2000, pp. 7.1–7.20.

11. A. Stimson, *Photometry and Radiometry for Engineers,* John Wiley & Sons, 1974.

12. F. Grum and R. J. Becherer, *Optical Radiation Measurements, vol. 1,* Radiometry, Academic Press, 1979.

13. W. Budde, *Optical Radiation Measurements,* vol. 4, *Physical Detectors of Optical Radiation,* Academic Press, 1983.

14. A. D. Ryer, *Light Measurement Handbook,* International Light, Inc., 1997.

15. W. L. Wolfe, *Introduction to Radiometry,* Proc. SPIE Press, 1998.

16. W. R. McCluney, *Introduction to Radiometry and Photometry,* Artech, 1994.

17. D. Goodman, "Geometric Optics," in *OSA Handbook of Optics,* 2nd ed., vol. 1, chap. 1, McGraw-Hill, New York, 1995.

18. M. Born and E. Wolf, *Principles of Optics,* Cambridge Press, pp. 522–525, 1980.

19. M. S. Brennesholtz, "Light Collection Efficiency for Light Valve Projection Systems," *Proc. SPIE* **2650**:71–79 (1996).

20. B. A. Jacobson, R. Winston, and P. L. Gleckman, "Flat-Panel Fluorescent Backlights with Reduced Illumination Angle: Backlighting Optics at the Thermodynamic Limit," *SID 92 Digest,* 423–426 (1992).

21. H. Ries, "Thermodynamic Limitations of the Concentration of Electromagnetic Radiation," *J. Opt. Soc. Am.* **72**(3):380–385 (1982).

22. H. Hottel, "Radiant Heat Transmission," in *Heat Transmission,* W. H. McAdams, ed., 3d ed., McGraw Hill, New York, 1954.

23. R. Winston, "Cone-Collectors for Finite Sources," *Appl. Opt.* **17**(5):688–689 (1978).

24. G. H. Derrick, "A Three-Dimensional Analogue of the Hottel String Construction for Radiation Transfer," *Opt. Acta* **32**:39–60 (1985).

25. C. Weiner, *Lehrbuch der darstellenden Geometrie,* vol. 1, Leipzig, 1884 (see F. E. Nicodemus, *Self Study Manual on Optical Radiation Measurements,* part 1, *Concepts,* NBS, Washington, March 1976).

26. M. S. Rea (ed.), *Lighting Handbook,* Illuminating Engineering Society of North America, 1993.

27. F. O. Bartell, "Projected Solid Angle and Black Body Simulators," *Appl. Opt.* **28**(6):1055–1057 (March 1989).

28. D. G. Koch, "Simplified Irradiance/Illuminance Calculations in Optical Systems," *Proc. SPIE* **1780**:226–242, 1992.

29. A. Rabl and R. Winston, "Ideal Concentrators for Finite Sources and Restricted Exit Angles." *Appl. Optics* **15**:2880–2883 (1976).

30. E. Harting, D. R. Mills, and J. E. Giutronich, "Practical Concentrators Attaining Maximal Concentration," *Opt. Lett.* **5**(1):32–34 (1980).

31. A. Luque, "Quasi-Optimum Pseudo-Lambertian Reflecting Concentrators: An Analysis," *Appl. Opt.* **19**(14):2398–2402 (1980).

32. R. M. Saraiji, R. G. Mistrick, and M. F. Modest, "Modeling Light Transfer through Optical Fibers for Illumination Applications," *IES* 128–139 (Summer 1996).

33. T. L. Davenport, W. J. Cassarly, R. L. Hansler, T. E. Stenger, G. R. Allen, and R. F. Buelow, "Changes in Angular and Spatial Distribution Introduced into Fiber Optic Headlamp Systems by the Fiber Optic Cables," *SAE,* Paper No. 981197, 1998.

34. S. Doyle and D. Corcoran, "Automated Mirror Design using an Evolution Strategy," *Opt. Eng.* **38**(2):323–333 (1999).

35. I. Ashdown, "Non-imaging Optics Design Using Genetic Algorithms," *J. Illum. Eng. Soc.* **3**(1):12–21 (1994).

36. Y. Nakata, "Multi B-Spline Surface Reflector Optimized with Neural Network," *SAE* 940638: 81–92 (1994).

37. N. E. Shatz and J. C. Bortz, "Inverse Engineering Perspective on Nonimaging Optical Design," *Proc. Proc. SPIE* **2538**:136–156, 1995.

38. C. Gilray and I. Lewin, "Monte Carlo Techniques for the Design of Illumination Optics," IESNA Annual Conference Technical Papers, Paper #85, pp. 65–80, July 1996.

39. J. C. Schweyen, K. Garcia, and P. L. Gleckman, "Geometrical Optical Modeling Considerations for LCD Projector Display Systems," *Proc. SPIE* **3013**:126–140 (1997).

40. D. Z. Ting and T. C. McGill, "Monte Carlo Simulation of Light-Emitting Diode Light-Extraction Characteristics," *Opt. Eng.* **34**(12):3545–3553 (1995).

41. M. Kaplan, "Monte Carlo Calculation of Light Distribution in an Integrating Cavity Illuminator," *Proc. SPIE* **1448**:206–217 (1991).

42. B. G. Crowther, "Computer Modeling of Integrating Spheres," *Appl. Opt.* **35**(30):5880–5886 (1996).

43. R. C. Chaney, "Monte Carlo Simulation of Gamma Ray Detectors using Scintillation Fibers," EUV, X-ray, and Gamma-Ray Instrumentation for Astronomy and Atomic Physics; Proceedings of the Meeting, San Diego, California, Aug. 7–11, 1989 (A90-50251 23-35), SPIE 1989.

44. J. A. Bamberg, "Scintillation Detector Optimization Using GUERAP-3 Radiation Scattering in Optical Systems, Proceedings of the Seminar, Huntsville, Alabama, September 30–October 1, 1980 (A81-36878 16-74) p. 86–93, SPIE 1981.

45. B. K. Likeness, "Stray Light Simulation with Advanced Monte Carlo Techniques," Stray-Light Problems in Optical Systems; Proceedings of the Seminar, Reston, VA, April 18–21, 1977 (A78-40270 17-35) pp. 80–88, SPIE 1977.

46. E. R. Freniere, "Simulation of Stray Light in Optical Systems with the GUERAP III," Radiation Scattering in Optical Systems; Proceedings of the Seminar, Huntsville, AL, September 30–October 1, 1980 (A81-36878 16-74) pp. 78–85, SPIE 1981.

47. N. Shatz, J. Bortz, and M. Dassanayake, "Design Optimization of a Smooth Headlamp Reflector to SAE/DOT Beam-Shape Requirements," *SAE* 1999.

48. T. Hough, J. F. Van Derlofske, and L. W. Hillman, "Measuring and Modeling Intensity Distributions of Light Sources in Waveguide Illumination Systems," *Opt. Eng.* **34**(3):819–823 (1995).

49. R. E. Levin, "Photometric Characteristics of Light-Controlling Apparatus," *Illum. Eng.* **66**(4):202–215 (1971).

50. R. J. Donohue and B. W. Joseph, "Computer Synthesized Filament Images from Reflectors and Through Lens Elements for Lamp Design and Evaluation," *Appl. Opt.* **14**(10):2384–2390 (1975).

51. I. Ashdown, "Near Field Photometry: A New Approach," *J. IES:* 163–180 (Winter 1993).

52. R. D. Stock and M. W. Siegel, "Orientation Invariant Light Source Parameters," *Opt. Eng.* **35**(9):2651–2660 (1996).

53. M. W. Siegel and R. D. Stock, "Generalized Near-Zone Light Source Model and Its Application to Computer Automated Reflector Design," *Opt. Eng.* **35**(9):2661–2679 (1996).

54. P. V. Shmelev and B. M. Khrana, "Near-Field Modeling Versus Ray Modeling of Extended Halogen Light Source in Computer Design of a Reflector," *Proc. SPIE* (July 27–28, 1997).

55. R. Rykowski and C. B. Wooley, "Source Modeling for Illumination Design," 3130B-27, *Proc. SPIE* (July 27–28, 1997).

56. T. P. Vogl, L. C. Lintner, R. J. Pegis, W. M. Waldbauer, and H. A. Unvala, "Semiautomatic Design of Illuminating Systems," *Appl. Opt.* **11**(5):1087–1090 (1972).

57. "1998 IESNA Software Survey," *Lighting Design and Application* **28**(10):53–62 (1998).

58. "1999 IESNA Software Survey," *Lighting Design and Application* **29**(12):39–48 (1999).

59. W. T. Welford, *Aberrations of Optical Systems,* pp. 158–161, Adam Hilger, 1986.

60. W. J. Smith, *Modern Lens Design,* McGraw Hill, New York 1992.

61. Y. Shimizu and H. Takenaka, "Microscope Objective Design," in *Advances in Optical and Electron Microscopy,* vol. 14, Academic Press, San Diego, CA, 1994.

62. G. P. Smestad, "Nonimaging Optics of Light-Emitting Diodes: Theory and Practice," *Proc. SPIE* **1727**:264–268 (1992).

63. W. T. Welford and R. Winston, "Two-Dimensional Nonimaging Concentrators with Refracting Optics," *J. Opt. Soc. Am.* **69**(6):917–919 (1979).

64. R. Kingslake, *Lens Design Fundamentals,* Academic Press, San Diego, CA, 1978.

65. E. Hecht and A. Zajac, *Optics,* Addison-Wesley, 1979.

66. S. F. Ray, *Applied Photographic Optics,* Focal Press, 1988.

67. G. Schultz, "Achromatic and Sharp Real Image of a Point by a Single Aspheric Lens," *Appl. Opt.* **22**(20):3242–3248 (1983).

68. J. C. Minano and J. C. Gonzalez, "New Method of Design of Nonimaging Concentrators," *Appl. Opt.* **31**(16):3051–3060, 1992.

69. E. A. Boettner and N. E. Barnett, "Design and Construction of Fresnel Optics for Photoelectric Receivers," *J. Opt. Soc. Am.* **41**(11):849–857 (1951).

70. Fresnel Technologies, Fort Worth, TX. See www.fresneltech.com/html/FresnelLenses.pdf.

71. S. Sinzinger and M. Testorf, "Transition Between Diffractive and Refractive Microoptical Components," *Appl. Opt.* **34**(26):5670–5676 (1995).

72. F. Erismann, "Design of Plastic Aspheric Fresnel Lens with a Spherical Shape," *Opt. Eng.* **36**(4):988–991 (1997).

73. D. J. Lamb and L. W. Hillman, "Computer Modeling and Analysis of Veiling Glare and Stray Light in Fresnel Lens Optical Systems," *Proc. SPIE* **3779**:344–352 (1999).

74. J. F. Goldenberg and T. S. McKechnie, "Optimum Riser Angle for Fresnel Lenses in Projection Screens," U.S. Patent 4,824,227, 1989.

75. M. Collares-Pereira, "High Temperature Solar Collector with Optimal Concentration: Non-focusing Fresnel Lens with Secondary Concentrator," *Solar Energy* **23**:409–419 (1979).

76. W. A. Parkyn and D. G. Pelka, "Compact Nonimaging Lens with Totally Internally Reflecting Facets," *Proc. SPIE* **1528**:70–81 (1991).

77. R. I. Nagel, "Signal Lantern Lens," U.S. Patent 3,253,276, 1966.

78. W. A. Parkyn, P. L. Gleckman, and D. G. Pelka, "Converging TIR Lens for Nonimaging Concentration of Light from Compact Incoherent Sources," *Proc. SPIE* **2016**:78–86 (1993).

79. W. A. Parkyn and D. G. Pelka, "TIR Lenses for Fluorescent Lamps," *Proc. SPIE* **2538**:93–103 (1995).

80. W. A. Parkyn and D. G. Pelka, "New TIR Lens Applications for Light-Emitting Diodes," *Proc. SPIE* **3139**:135–140 (1997).

81. V. Medvedev, W. A. Parkyn, and D. G. Pelka, "Uniform High-Efficiency Condenser for Projection Systems," *Proc. SPIE* **3139**:122–134 (1997).

82. V. Medvedev, D. G. Pelka, and W. A. Parkyn, "Uniform LED Illuminator for Miniature Displays," *Proc. SPIE* **3428**:142–153 (1998).

83. D. F. Vanderwerf, "Achromatic Catadioptric Fresnel Lens," *Proc. SPIE* **2000**:174–183 (1993).

84. J. Spigulis, "Compact Illuminators, Collimators and Focusers with Half Spherical Input Aperture," *Proc. SPIE* **2065**:54–59 (1994).

85. D. Silvergate, "Collimating Compound Catadioptric Immersion Lens," U.S. Patent 4,770,514, 1988.

86. McDermott, "Angled Elliptical Axial Lighting Device," U.S. Patent 5,894,196, 1999.

87. D. Korsch, *Reflective Optics,* Academic Press, San Diego, CA, 1991.

88. D. E. Spencer, L. L. Montgomery, and J. F. Fitzgerald, "Macrofocal Conics as Reflector Contours," *J. Opt. Soc. Am.* **55**(1):5–11 (1965).

89. D. R. Philips, "Low Brightness Louver," U.S. Patent 2,971,083, 1961.

90. H. P. Baum and J. M. Gordon, "Geometric Characteristics of Ideal Nonimaging (CPC) Solar Collectors with Cylindrical Absorber," *Solar Energy* **33**(5):455–458 (1984).

91. P. T. Ong, J. M. Gordon, and A. Rabl, "Tailored Edge-Ray Designs for Illumination with Tubular Sources," *Appl. Opt.* **35**(22):4361–4371 (1996).

92. I. M. Bassett and G. H. Derrick, "The Collection of Diffuse Light onto an Extended Absorber," *Optical Quantum Electronics* **10**:61–82 (1978).

93. H. Ries, N. Shatz, J. Bortz, and W. Spirkl, "Performance Limitations of Rotationally Symmetric Nonimaging Devices," *J. Opt. Soc. Am. A* **14**(10):2855–2862 (1997).

94. M. Ruda (ed.), "International Conference on Nonimaging Concentrators," *Proc. SPIE* **441** (1983).

95. R. Winston and R. L. Holman (eds.), "Nonimaging Optics: Maximum Efficiency Light Transfer," *Proc. SPIE* **1528** (1991).

96. R. Winston and R. L. Holman (eds.), "Nonimaging Optics: Maximum Efficiency Light Transfer II," *Proc. SPIE* **2016** (1993).

97. R. Winston (ed.), "Nonimaging Optics: Maximum Efficiency Light Transfer III," *Proc. SPIE* **2538** (1995).

98. R. Winston (ed.), "Nonimaging Optics: Maximum Efficiency Light Transfer IV," *Proc. SPIE* **3139** (1997).

99. R. Winston (ed.), "Nonimaging Optics: Maximum Efficiency Light Transfer V," *Proc. SPIE* **3781** (1999).

100. R. Winston (ed.), "Selected Papers on Nonimaging Optics," *Proc. SPIE Milestone Series* **106** (1995).

101. I. M. Bassett, W. T. Welford, and R. Winston, "Nonimaging Optics for Flux Concentration," in *Progress in Optics,* E. Wolf (ed.), pp. 161–226, 1989.

102. R. Winston, "Nonimaging Optics," *Sci. Am.* 76–81 (March 1991).

103. P. Gleckman, J. O'Gallagher, and R. Winston, "Approaching the Irradiance of the Sun Through Nonimaging Optics," *Optics News:* 33–36 (May 1989).

104. M. F. Land, "The Optical Mechanism of the Eye of Limulus," *Nature* **280**:396–397 (1979).

105. M. F. Land, "Compound Eyes: Old and New Optical Mechanisms," *Nature* **287**:681–685 (1980).

106. R. Levi-Seti, D. A. Park, and R. Winston, "The Corneal Cones of Limulus as Optimized Light Concentrators," *Nature* **253**:115–116 (1975).

107. D. A. Baylor and R. Fettiplace, "Light Path and Photon Capture in Turtle Photoreceptors," *J. Physiol.* **248**(2):433–464 (1975).

108. R. Winston and J. Enoch, "Retinal Cone Receptor as Ideal Light Collector," *J. Opt. Soc. Am.* **61**(8):1120–1121 (1971).

109. R. L. Garwin, "The Design of Liquid Scintillation Cells," *Rev. Sci. Inst.* **23**:755–757 (1952).

110. D. E. Williamson, "Cone Channel Condensor," *J. Opt. Soc. Am.* **42**(10):712–715 (1952).

111. J. H. Myer, "Collimated Radiation in Conical Light Guides," *Appl. Opt.* **19**(18):3121–3123 (1980).

112. W. Witte, "Cone Channel Optics," *Infrared Phys.* **5**:179–185 (1965).

113. C. H. Burton, "Cone Channel Optics," *Infrared Phys.* **15**:157–159 (1975).

114. R. Winston and W. T. Welford, "Ideal Flux Concentrators as Shapes That Do Not Disturb the Geometrical Vector Flux Field: A New Derivation of the Compound Parabolic Concentrator," *J. Opt. Soc. Am.* **69**(4):536–539 (1979).

115. A. G. Molledo and A. Luque, "Analysis of Static and Quasi-Static Cross Compound Parabolic Concentrators," *Appl. Opt.* 2007–2020 (1984).

116. R. Winston and H. Hinterberger, "Principles of Cylindrical Concentrators for Solar Energy," *Solar Energy* **17**:255–258 (1975).

117. W. R. McIntire, "Truncation of Nonimaging Cusp Concentrators," *Solar Energy* **23**:351–355 (1979).

118. H. Tabor, "Comment—The CPC Concept—Theory and Practice," *Solar Energy* **33**(6):629–630 (1984).

119. A. Rabl and R. Winston, "Ideal Concentrators for Finite Sources and Restricted Exit Angles," *Appl. Opt.* **15**:2880–2883 (1976).

120. A. Rabl, N. B. Goodman, and R. Winston, "Practical Design Considerations for CPC Solar Collectors," *Solar Energy* **22**:373–381 (1979).

121. W. R. McIntire, "New Reflector Design Which Avoids Losses Through Gaps Between Tubular Absorber and Reflectors," *Solar Energy* **25**:215–220 (1980).

122. R. Winston, "Ideal Flux Concentrators with Reflector Gaps," *Appl. Opt.* **17**(11):1668–1669 (1978).

123. R. Winston, "Cavity Enhancement by Controlled Directional Scattering," *Appl. Opt.* **19**:195–197 (1980).

124. F. Bloisi, P. Cavaliere, S. De Nicola, S. Martellucci, J. Quartieri, and L. Vicari, "Ideal Nonfocusing Concentrator with Fin Absorbers in Dielectric Rhombuses," *Opt. Lett.* **12**(7):453–455 (1987).

125. I. R. Edmonds, "Prism-Coupled Compound Parabola: A New Look and Optimal Solar Concentrator," *Opt. Lett.* **11**(8):490–492 (1986).

126. J. D. Kuppenheimer, "Design of Multilamp Nonimaging Laser Pump Cavities," *Opt. Eng.* **27**(12):1067–1071 (1988).

127. D. Lowe, T. Chin, T. L. Credelle, O. Tezucar, N. Hariston, J. Wilson, and K. Bingaman, "SpectraVue: A New System to Enhance Viewing Angle of LCDs," *SID 96 Applications Digest* 39–42 (1996).

128. J. L. Henkes, "Light Source for Liquid Crystal Display Panels Utilizing Internally Reflecting Light Pipes and Integrating Sphere," U.S. Patent 4,735,495, 1988.

129. R. Winston, "Principles of Solar Concentrators of a Novel Design," *Solar Energy* **16**:89–94 (1974).

130. R. Winston, "Cone-Collectors for Finite Sources," *Appl. Opt.* **17**(5):688–689 (1978).

131. M. Collares-Pereira, A. Rabl, and R. Winston, "Lens-Mirror Combinations with Maximal Concentration," *Appl. Opt.* **16**(10):2677–2683 (October 1977).

132. H. P. Gush, "Hyberbolic Cone-Channel Condensor," *Opt. Lett.* **2**:22–24 (1978).

133. J. O'Gallagher, R. Winston, and W. T. Welford, "Axially Symmetric Nonimaging Flux Concentrators with the Maximum Theoretical Concentration Ratio," *J. Opt. Soc. Am. A* **4**(1):66–68 (1987).

134. R. Winston, "Dielectric Compound Parabolic Concentrators," *Appl. Opt.* **15**(2):291–292 (1976).

135. J. R. Hull, "Dielectric Compound Parabolic Concentrating Solar Collector with a Frustrated Total Internal Reflection Absorber," *Appl. Opt.* **28**(1):157–162 (1989).

136. R. Winston, "Light Collection Within the Framework of Geometrical Optics," *J. Opt. Soc. Am.* **60**(2):245–247 (1970).

137. D. Jenkins, R. Winston, R. Bliss, J. O'Gallagher, A. Lewandowski, and C. Bingham, "Solar Concentration of 50,000 Achieved with Output Power Approaching 1kW," *J. Sol. Eng.* **118**:141–144 (1996).

138. H. Ries, A. Segal, and J. Karni, "Extracting Concentrated Guided Light," *Appl. Opt.* **36**(13):2869–2874 (1997).

139. R. H. Hildebrand, "Focal Plane Optics in Far-Infrared and Submillimeter Astronomy," *Opt. Eng.* **25**(2):323–330 (1986).

140. J. Keene, R. H. Hildebrand, S. E. Whitcomb, and R. Winston, "Compact Infrared Heat Trap Field Optics," *Appl. Opt.* **17**(7):1107–1109 (1978).

141. R. Winston and W. T. Welford, "Geometrical Vector Flux and Some New Nonimaging Concentrators," *J. Opt. Soc. Am.* **69**(4):532–536 (1979).

142. X. Ning, R. Winston, and J. O'Gallagher, "Dielectric Totally Internally Reflecting Concentrators," *Appl. Opt.* **26**(2):300–305 (1987).

143. W. L. Eichhorn, "Designing Generalized Conic Concentrators for Conventional Optical Systems," *Appl. Opt.* **24**(8):1204–1205 (1985).

144. W. L. Eichhorn, "Generalized Conic Concentrators," *Appl. Opt.* **21**(21):3887–3890 (1982).

145. G. H. Smith, *Practical Computer-Aided Lens Design,* Willmann-Bell, VA, pp. 379–383, 1998.

146. K. W. Beeson, I. B. Steiner, and S. M. Zimmerman, "Illumination System Employing an Array of Microprisms," U.S. Patent 5,521,725, 1996.

147. R. P. Friedman, J. M. Gordon, and H. Ries, "New High-Flux Two-Stage Optical Designs for Parabolic Solar Concentrators," *Solar Energy* **51**:317–325 (1993).

148. P. Gleckman, J. O'Gallagher, and R. Winston, "Concentration of Sunlight to Solar-Surface Levels using Non-imaging Optics," *Nature* **339**:198–200 (May 1989).

149. H. Ries and W. Sprikl, "Nonimaging Secondary Concentrators for Large Rim Angle Parabolic Trough with Tubular Absorbers," *Appl. Opt.* **35**:2242–2245 (1996).

150. J. O'Gallagher and R. Winston, "Test of a 'Trumpet' Secondary Concentrator with a Paraboloidal Dish Primary," *Solar Energy* **36**(1):37–44 (1986).

151. A. Rabl, "Comparison of Solar Concentrators," *Solar Energy* **18**:93–111 (1976).

152. R. Winston and W. T. Welford, "Design of Nonimaging Concentrators as Second Stages in Tandem with Image-Forming First-Stage Concentrators," *Appl. Opt.* **19**(3):347–351 (1980).

153. X. Ning, R. Winston, and J. O'Gallagher, "Optics of Two-Stage Photovoltaic Concentrators with Dielectric Second Stages," *Appl. Opt.* **26**(7):1207–1212 (1987).

154. E. M. Kritchman, "Second-Stage Concentrators—A New Formalism," *J. Opt. Soc. Am. Lett.* **73**(4):508–511 (1983).

155. D. R. Mills and J. E. Giutronich, "New Ideal Concentrators for Distant Radiation Sources," *Solar Energy* **23**:85–87 (1979).

156. D. R. Mills and J. E. Giutronich, "Asymmetrical Non-imaging Cylindrical Solar Concentrators," *Solar Energy* **20**:45–55 (1978).

157. H. Ries and J. M. Gordon, "Double-Tailored Imaging Concentrators," *Proc. SPIE* **3781**:129–134 (1999).

158. J. C. Minano and J. C. Gonzalez, "New Method of Design of Nonimaging Concentrators," *Appl. Opt.* **31**(16):3051–3060 (1992).

159. J. C. Minano, P. Benitez, and J. C. Gonzalez, "RX: A Nonimaging Concentrator," *Appl. Opt.* **34**(13):2226–2235 (1995).

160. P. Benitez and J. C. Minano, "Ultrahigh-Numerical-Aperture Imaging Concentrator," *J. Opt. Soc. Am. A* **14**(8):1988–1997 (1997).

161. J. C. Minano, J. C. Gonzalez, and P. Benitez, "A High-Gain, Compact, Nonimaging Concentrator: RXI," *Appl. Opt.* **34**(34):7850–7856 (1995).

162. J. L. Alvarez, M. Hernandez, P. Benitez, and J. C. Minano, "RXI Concentrator for 1000× Photovoltaic Conversion," *Proc. SPIE* **3781**:30–37 (1999).

163. J. F. Forkner, "Aberration Effects in Illumination Beams Focused by Lens Systems," *Proc. SPIE* **3428**:73–89 (1998).

164. X. Ning, "Three-Dimensional Ideal Θ_1/Θ_2 Angular Transformer and Its Uses in Fiber Optics," *Appl. Opt.* **27**(19):4126–4130 (1988).

165. A. Timinger, A. Kribus, P. Doron, and H. Ries, "Optimized CPC-type Concentrators Built of Plane Facets," *Proc. SPIE* **3781**:60–67 (1999).

166. J. P. Rice, Y. Zong, and D. J. Dummer, "Spatial Uniformity of Two Nonimaging Concentrators," *Opt. Eng.* **36**(11):2943–2947 (1997).

167. R. M. Emmons, B. A. Jacobson, R. D. Gengelbach, and R. Winston, "Nonimaging Optics in Direct View Applications," *Proc. SPIE* **2538**:42–50 (1995).

168. D. B. Leviton and J. W. Leitch, "Experimental and Raytrace Results for Throat-to-Throat Compound Parabolic Concentrators," *Appl. Opt.* **25**(16):2821–2825 (1986).

169. B. Moslehi, J. Ng, I. Kasimoff, and T. Jannson, "Fiber-Optic Coupling Based on Nonimaging Expanded-Beam Optics," *Opt. Lett.* **14**(23):1327–1329 (1989).

170. M. Collares-Pereira, J. F. Mendes, A. Rabl, and H. Ries, "Redirecting Concentrated Radiation," *Proc. SPIE* **2538**:131–135 (1995).

171. A. Rabl, "Solar Concentrators with Maximal Concentration for Cylindrical Absorbers," *Appl. Opt.* **15**(7):1871–1873 (1976). See also an erratum, *Appl. Opt.* **16**(1):15 (1977).

172. D. Feuermann and J. M. Gordon, "Optical Performance of Axisymmetric Concentrators and Illuminators," *Appl. Opt.* **37**(10):1905–1912 (1998).

173. J. Bortz, N. Shatz, and H. Ries, "Consequences of Etendue and Skewness Conservation for Nonimaging Devices with Inhomogeneous Targets," *Proc. SPIE* **3139**:28 (1997).

174. N. E. Shatz, J. C. Bortz, H. Ries, and R. Winston, "Nonrotationally Symmetric Nonimaging Systems that Overcome the Flux-Transfer Performance Limit Imposed by Skewness Conservation," *Proc. Proc. SPIE* **3139**:76–85 (1997).

175. N. E. Shatz, J. C. Bortz, and R. Winston, "Nonrotationally Symmetric Reflectors for Efficient and Uniform Illumination of Rectangular Apertures," *Proc. Proc. SPIE* **3428**:176–183 (1998).

176. W. Benesch and J. Strong, "The Optical Image Transformer," *J. Opt. Soc. Am.* **41**(4):252–254 (1951).

177. I. S. Bowen, "The Image-Slicer, a Device for Reducing Loss of Light at Slit of Stellar Spectrograph," *Astrophysical J.* **88**(2):113–124 (1938).

178. W. D. Westwood, "Multiple Lens System for an Optical Imaging Device," U.S. Patent 4,114,037, 1978.

179. J. Bernges, L. Unnebrink, T. Henning, E. W. Kreutz, and R. Poprawe, "Novel Design Concepts for UV Laser Beam Shaping," *Proc. SPIE* **3779**:118–125 (1999).

180. J. Endriz, "Brightness Conserving Optical System for Modifying Beam Symmetry," U.S. Patent 5,168,401, 1992.

181. T. Mori and H. Komatsuda, "Optical Integrator and Projection Exposure Apparatus Using the Same," U.S. Patent 5,594,526, 1997.

182. J. A. Shimizu and P. J. Janssen, "Integrating Lens Array and Image Forming Method for Improved Optical Efficiency," U.S. Patent 5,662,401, 1997.

183. D. Feuermann, J. M. Gordon, and H. Ries, "Nonimaging Optical Designs for Maximum-Power-Density Remote Irradiation," *Appl. Opt.* **37**(10):1835–1844 (1998).

184. F. C. Genovese, "Fiber Optic Line Illuminator with Deformed End Fibers and Method of Making Same," U.S. Patent 4,952,022, 1990.

185. P. Gorenstein and D. Luckey, "Light Pipe for Large Area Scintillator," *Rev. Sci. Instrum.* **34**(2):196–197 (1963).

186. W. Gibson, "Curled Light Pipes for Thin Organic Scintillators," *Rev. Sci. Instrum.* **35**(8):1021–1023 (1964).

187. H. Hinterberger and R. Winston, "Efficient Design of Lucite Light Pipes Coupled to Photomultipliers," *Rev. Sci. Instrum.* **39**(3):419–420 (1968).

188. H. D. Wolpert, "A Light Pipe for Flux Collection: An Efficient Device for Document Scanning," *Lasers Appl.* 73–74 (April 1983).

189. H. Karasawa, "Light Guide Apparatus Formed from Strip Light Guides," U.S. Patent 4,824,194, 1989.

190. D. A. Markle, "Optical Transformer Using Curved Strip Waveguides to Achieve a Nearly Unchanged F/Number," U.S. Patent 4,530,565, 1985.

191. I. M. Bassett and G. W. Forbes, "A New Class of Ideal Non-imaging Transformers," *Opt. Acta* **29**(9):1271–1282 (1982).

192. G. W. Forbes and I. M. Bassett, "An Axially Symmetric Variable-Angle Nonimaging Transformer," *Opt. Acta* **29**(9):1283–1297 (1982).

193. M. E. Barnett, "The Geometric Vector Flux Field Within a Compound Elliptical Concentrator," *Optik* **54**(5):429–432 (1979).

194. M. E. Barnett, "Optical Flow in an Ideal Light Collector: The Θ_1/Θ_2 Concentrator," *Optik* **57**(3):391–400 (1980).

195. Gutierrez, J. C. Minano, C. Vega, and P. Benitez, "Application of Lorentz Geometry to Nonimaging Optics: New 3D Ideal Concentrators," *J. Opt. Soc. Am. A* **13**(3):532–540 (1996).

196. P. Greenman, "Geometrical Vector Flux Sinks and Ideal Flux Concentrators," *J. Opt. Soc. Am. Lett.* **71**(6):777–779 (1981).

197. W. T. Welford and R. Winston, "On the Problem of Ideal Flux Concentrators," *J. Opt. Soc. Am.* **68**(4):531–534 (1978). See also addendum, *J. Opt. Soc. Am.* **69**(2):367 (1979).

198. J. C. Minano, "Design of Three-Dimensional Nonimaging Concentrators with Inhomogeneous Media," *J. Opt. Soc. Am. A* **3**(9):1345–1353 (1986).

199. J. M. Gordon, "Complementary Construction of Ideal Nonimaging Concentrators and Its Applications," *Appl. Opt.* **35**(28):5677–5682 (1996).

200. Davies, "Edge-Ray Principle of Nonimaging Optics," *J. Opt. Soc. Am. A* **11**:1256–1259 (1994).

201. H. Ries and A. Rabl, "Edge-Ray Principle of Nonimaging Optics," *J. Opt. Soc. Am.* **10**(10):2627–2632 (1994).

202. A. Rabl, "Edge-Ray Method for Analysis of Radiation Transfer Among Specular Reflectors," *Appl. Opt.* **33**(7):1248–1259 (1994).

203. J. C. Minano, "Two-Dimensional Nonimaging Concentrators with Inhomogeneous Media: A New Look," *J. Opt. Soc. Am. A* **2**(11):1826–1831 (1985).

204. M. T. Jones, "Motion Picture Screen Light as a Function of Carbon-Arc-Crater Brightness Distribution," *J. Soc. Mot. Pic. Eng.* **49**:218–240 (1947).

205. R. Kingslake, *Applied Optics and Optical Engineering,* vol. II, Academic Press, pp. 225–226, 1965.

206. D. O'Shea, *Elements of Modern Optical Design,* pp. 111–114 and 384–390, Wiley, 1985.

207. W. Wallin, "Design of Special Projector Illuminating Systems," *J-SMPTE* **71**:769–771 (1962).

208. H. Weiss, "Wide-Angle Slide Projection," *Inf. Disp.* 8–15 (September/October 1964).

209. S. Bradbury, *An Introduction to the Optical Microscope,* pp. 23–27, Oxford University Press, 1989.

210. S. Inoue and R. Oldenbourg, "Microscopes," in *OSA Handbook of Optics,* McGraw-Hill, New York, vol. II, chap. 17, 1995.

211. M. Lacombat, G. M. Dubroeucq, J. Massin, and M. Brevignon, "Laser Projection Printing," *Solid State Technology* **23**(115): (1980).

212. J. M. Gordon and P. T. Ong, "Compact High-Efficiency Nonimaging Back Reflector for Filament Light Sources," *Opt. Eng.* **35**(6):1775–1778 (1996).

213. G. Zochling, "Design and Analysis of Illumination Systems," *Proc. SPIE* **1354**:617–626 (1990).

214. A. Steinfeld, "Apparent Absorptance for Diffusely and Specularly Reflecting Spherical Cavities," *Int. J. Heat Mass Trans.* **34**(7):1895–1897 (1991).

215. K. A. Snail and L. M. Hanssen, "Integrating Sphere Designs with Isotropic Throughput," *Appl. Opt.* **28**(10):1793–1799 (1989).

216. D. P. Ramer and J. C. Rains, "Lambertian Surface to Tailor the Distribution of Light in a Nonimaging Optical System," *Proc. SPIE* **3781** (1999).

217. D. G. Goebel, "Generalized Integrating Sphere Theory," *Appl. Opt.* **6**(1):125–128 (1967).

218. L. Hanssen and K. Snail, "Nonimaging Optics and the Measurement of Diffuse Reflectance," *Proc. SPIE* **1528**:142–150 (1991).

219. D. B. Chenault, K. A. Snail, and L. M. Hanssen, "Improved Integrating-Sphere Throughput with a Lens and Nonimaging Concentrator," *Appl. Opt.* **34**(34):7959–7964 (1995).

220. R. H. Webb, "Concentrator for Laser Light," *Appl. Opt.* **31**(28):5917–5918 (1992).

221. M. M. Chen, J. B. Berkowitz-Mattuck, and P. E. Glaser, "The Use of a Kaleidoscope to Obtain Uniform Flux over a Large Area in a Solar or Arc Imaging Furnace," *Appl. Opt.* **2**:265–271 (1963).

222. D. S. Goodman, "Producing Uniform Illumination," OSA Annual Meeting, Toronto, October 5, 1993.

223. L. A. Whitehead, R. A. Nodwell, and F. L. Curzon, "New Efficiency Light Guide for Interior Illumination," *Appl. Opt.* **21**(15):2755–2757 (1982).

224. S. G. Saxe, L. A. Whitehead, and S. Cobb, "Progress in the Development of Prism Light Guides," *Proc. SPIE* **692**:235–240 (1986).

225. K. Jain, "Illumination System to Produce Self-Luminous Light Beam of Selected Cross-Section, Uniform Intensity and Selected Numerical Aperture," U.S. Patent 5,059,013, 1991.

226. Y. Kudo and K. Matsumoto, "Illuminating Optical Device," U.S. Patent 4,918,583, 1990.

227. W. J. Cassarly, J. M. Davenport, and R. L. Hansler, "Uniform Lighting Systems: Uniform Light Delivery," *SAE 950904,* **SP-1081**:1–5 (1995).

228. M. R. Latta and K. Jain, "Beam Intensity Uniformization by Mirror Folding," *Opt. Comm.* **49**:27 (1984).

229. C. M. Chang, K. W. Lin, K. V. Chen, S. M. Chen, and H. D. Shieh, "A Uniform Rectangular Illuminating Optical System for Liquid Crystal Light Valve Projectors," *Euro Display '96:* 257–260 (1996).

230. L. J. Coyne, "Distributive Fiber Optic Couplers Using Rectangular Lightguides as Mixing Elements," *Proc. FOC '79:* 160–164 (1979).

231. M. Wagner, H. D. Geiler, and D. Wolff, "High-Performance Laser Beam Shaping and Homogenization System for Semiconductor Processing," *Meas. Sci. Technol. UK* **1**:1193–1201 (1990).

232. K. Iwasaki, Y. Ohyama, and Y. Nanaumi, "Flattening Laserbeam Intensity Distribution," *Lasers Appl.* 76 (April 1983).

233. K. Iwasaki, T. Hayashi, T. Goto, and S. Shimizu, "Square and Uniform Laser Output Device: Theory and Applications," *Appl. Opt.* **29**:1736–1744 (1990).

234. J. M. Geary, "Channel Integrator for Laser Beam Uniformity on Target," *Opt. Eng.* **27**:972–977 (1988).

235. R. E. Grojean, D. Feldman, and J. F. Roach, "Production of Flat Top Beam Profiles for High Energy Lasers," *Rev. Sci. Instrum.* **51**:375–376 (1980).

236. B. Fan, R. E. Tibbetts, J. S. Wilczynski, and D. F. Witman, "Laser Beam Homogenizer," U.S. Patent 4,744,615, 1988.

237. R. G. Waarts, D. W. Nam, D. F. Welch, D. R. Scrifes, J. C. Ehlert, W. Cassarly, J. M. Finlan, and K. Flood, "Phased 2-D Semiconductor Laser Array for High Output Power," *Proc. SPIE* **1850**:270–280 (1993).

238. J. R. Jenness Jr., "Computing Uses of the Optical Tunnel," *Appl. Opt.* **29**(20):2989–2991 (1990).

239. L. J. Krolak and D. J. Parker, "The Optical Tunnel—A Versatile Electrooptic Tool," *J. Soc. Motion Pict. Telev. Eng.* **72**:177–180 (1963).

240. T. D. Milster, "Miniature and Micro-Optics," in *OSA Handbook of Optics,* vol. II, McGraw-Hill, New York, p. 7.12, 1995.

241. R. A. Sprague and D. R. Scifres, "Multi-Beam Optical System Using Lens Array," U.S. Patent 4,428,647, 1984.

242. K. Flood, B. Cassarly, C. Sigg, and J. Finlan, "Continuous Wide Angle Beam Steering Using Translation of Binary Microlens Arrays and a Liquid Crystal Phased Array," *Proc. SPIE* **1211**:296–304 (1990).

243. W. Kuster and H. J. Keller, "Domed Segmented Lens System," U.S. Patent 4,930,864, 1990.

244. W. J. Cassarly, J. M. Davenport, and R. L. Hansler, "Uniform Light Delivery Systems," *SAE:* Paper No. 960490 (1996).

245. N. A. Zhidkova, O. D. Kalinina, A. A. Kuchin, S. N. Natarovskii, O. N. Nemkova, and N. B. Skobeleva, "Use of Lens Arrays in Illuminators for Reflected-Light Microscopes," *Opt. Mekh. Promst.* **55**:23–24 (September 1988); *Sov. J. Opt. Technol.* **55** 539–541 (1988).

246. S. Ohuchi, T. Kakuda, M. Yatsu, N. Ozawa, M. Deguchi, and T. Maruyama, "Compact LC Projector with High-Brightness Optical System," *IEEE Trans. Consumer Electronics* **43**(3):801–806 (1997).

247. K. Rantsch, L. Bertele, H. Sauer, and A. Merz, "Illuminating System," U.S. Patent 2,186,123, 1940.

248. J. R. Miles, "Lenticulated Collimating Condensing System," U.S. Patent 3,296,923, 1967.

249. M. Ohtu, "Illuminating Apparatus," U.S. Patent 4,497,013, 1985.

250. K. Matsumoto, M. Uehara, and T. Kikuchi, "Illumination Optical System," U.S. Patent 4,769,750, 1988.

251. A. H. J. Van den Brandt and W. Timmers, "Optical Illumination System and Projection Apparatus Comprising Such a System," U.S. Patent 5,098,184, 1992.

252. F. Watanabe, "Illumination Optical System," U.S. Patent 5,786,939, 1998.

253. X. Deng, X. Liang, Z. Chen, W. Yu, and R. Ma, "Uniform Illumination of Large Targets Using a Lens Array," *Appl. Opt.* **25**:377–381 (1986).

254. N. Nishi, T. Jitsuno, K. Tsubakimoto, M. Murakami, M. Nakatsuma, K. Nishihara, and S. Nakai, "Aspherical Multi Lens Array for Uniform Target Illumination," *Proc. SPIE* **1870**:105–111 (1993).

255. Y. Ozaki and K. Takamoto, "Cylindrical Fly's Eye Lens for Intensity Redistribution of an Excimer Laser Beam," *Appl. Opt.* **28**:106–110 (1989).

256. S. Glocker and R. Goring, "Investigation of Statistical Variations Between Lenslets of Microlens Arrays," *Appl. Opt.* **36**(19):4438–4445 (1997).

257. T. H. Bett, C. N. Danson, P. Jinks, D. A. Pepler, I. N. Ross, and R. M. Stevenson, "Binary Phase Zone-Plate Arrays for Laser-Beam Spatial-Intensity Distribution Conversion," *Appl. Opt.* **34**(20):4025–4036 (1995).

258. Y. Kato, K. Mima, N. Miyanaga, S. Arinaga, Y. Kitagawa, M. Nakatsuka, and C. Yamanaka, "Random Phasing of High-Power Lasers for Uniform Target Acceleration and Plasma-Instability Suppression," *Phys. Rev. Lett.* **53**(11):1057–1060 (1984).

259. L. B. W. Jolley, J. M. Waldram, and G. H. Wilson, *The Theory and Design of Illuminating Engineering Equipment,* John Wiley & Sons, New York, 1931.

260. D. G. Burkhard and D. L. Shealy, "Design of Reflectors Which Will Distribute Sunlight in a Specified Manner," *Solar Energy* **17**:221–227 (1975).

261. D. G. Burkhard and D. L. Shealy, "Specular Aspheric Surface to Obtain a Specified Irradiance from Discrete or Continuous Line Source Radiation: Design," *Appl. Opt.* **14**(6):1279–1284 (1975).

262. O. K. Kusch, *Computer-Aided Optical Design of Illuminating and Irradiating Devices,* ASLAN Publishing House, Moscow, 1993.

263. T. E. Horton and J. H. McDermit, "Design of Specular Aspheric Surface to Uniformly Radiate a Flat Surface Using a Nonuniform Collimated Radiation Source," *Journal of Heat Transfer,* 453–458 (1972).

264. J. S. Schruben, "Analysis of Rotationally Symmetric Reflectors for Illuminating Systems," *J. Opt. Soc. Am.* **64**(1):55–58 (1974).

265. J. Murdoch, *Illumination Engineering, from Edison's Lamp to the Laser,* MacMillan Publishing, 1985.

266. R. Winston, "Nonimaging Optics: Optical Design at the Thermodynamic Limit," *Proc. SPIE* **1528**:2–6 (1991).

267. D. Thackeray, "Reflectors for Light Sources," *J. Photog. Sci.* **22**:303–304 (1974).

268. P. W. Rhodes and D. L. Shealy, "Refractive Optical Systems for Irradiance Redistribution of Collimated Radiation: Their Design and Analysis," *Appl. Opt.* **19**:3545–3553 (1980).

269. W. A. Parkyn, "Design of Illumination Lenses via Extrinsic Differential Geometry," *Proc. SPIE* **3428**:154–162 (1998).

270. R. Winston and H. Ries, "Nonimaging Reflectors as Functionals of the Desired Irradiance," *J. Opt. Soc. Am. A* **10**(9):1902–1908 (1993).

271. W. B. Elmer, "The Optics of Reflectors for Illumination," *IEEE Trans. Industry Appl.* **IA-19**(5):776–788 (September/October 1983).

272. J. M. Gordon, P. Kashin, and A. Rabl, "Nonimaging Reflectors for Efficient Uniform Illumination," *Appl. Opt.* **31**(28):6027–6035 (1992).

273. J. M. Gordon and A. Rabl, "Nonimaging Compound Parabolic Concentrator-Type Reflectors with Variable Extreme Direction," *Appl. Opt.* **31**(34):7332–7338 (1992).

274. J. Gordon and H. Ries, "Tailored Edge-Ray Concentrators as Ideal Second Stages for Fresnel Reflectors," *Appl. Opt.* **32**(13):2243–2251 (1993).

275. H. R. Ries and R. Winston, "Tailored Edge-Ray Reflectors for Illumination," *J. Opt. Soc. Am. A* **11**(4):1260–1264 (1994).

276. A. Rabl and J. M. Gordon, "Reflector Design for Illumination with Extended Sources: The Basic Solutions," *Appl. Opt.* **33**(25):6012–6021 (1994).

277. P. T. Ong, J. M. Gordon, A. Rabl, and W. Cai, "Tailored Edge-Ray Designs for Uniform Illumination of Distant Targets," *Opt. Eng.* **34**(6):1726–1737 (1995).

278. P. T. Ong, J. M. Gordon, and A. Rabl, "Tailoring Lighting Reflectors to Prescribed Illuminance Distributions: Compact Partial-Involute Designs," *Appl. Opt.* **34**(34):7877–7887 (1995).

279. D. Jenkins and R. Winston, "Tailored Reflectors for Illumination," *Appl. Opt.* **35**(10):1669–1672 (1996).

280. D. Jenkins and R. Winston, "Integral Design Method for Nonimaging Concentrators," *J. Opt. Soc. Am. A* **13**(10):2106–2116 (1996).

281. J. M. Gordon, "Simple String Construction Method for Tailored Edge-Ray Concentrators in Maximum-Flux Solar Energy Collectors," *Solar Energy* **56**:279–284 (1996).

282. H. Schering and A. Merz, "Illuminating Device for Projectors," U.S. Patent 2,183,249, 1939.

283. Y. Kawamura, Y. Itagaki, K. Toyoda, and S. Namba, "A Simple Optical Device for Generating Square Flat-Top Intensity Irradiation from a Gaussian Laser Beam," *Opt. Comm.* **48**:44–46 (1983).

284. T. Sameshima and S. Usui, "Laser Beam Shaping System for Semiconductor Processing," *Opt. Comm.* **88**:59–62 (1992).

285. L. Austin, M. Scaggs, U. Sowada, and H.-J. Kahlert, "A UV Beam-Delivery System Designed for Excimers," *Photonics Spectra* 89–98 (May 1989).

286. A. Brunsting, "Redirecting Surface for Desired Intensity Profile," US Patent 4327972, 1982.

287. R. J. Bruno and K. C. Liu, "Laserbeam Shaping for Maximum Uniformity and Minimum Loss," *Laser Appl.* 91–94 (April 1987).

288. D. D. Dourte, C. Mesa, R. L. Pierce, and W. J. Spawr, "Optical Integration with Screw Supports," U.S. Patent 4,195,913, 1980.

289. S. L. Ream, "A Convex Beam Integrator," *Laser Focus:* 68–71 (November 79).

290. V. J. Doherty, "Design of Mirrors with Segmented Conical Surfaces Tangent to a Discontinuous Aspheric Base," *Proc. SPIE* **399**:263–271 (1983).

291. D. M. Dagenais, J. A. Woodroffe, and I. Itzkan, "Optical Beam Shaping of a High Power Laser for Uniform Target Illumination," *Appl. Opt.* **24**:671–675 (1985).

292. F. M. Dickey and B. D. O'Neil, "Multifaceted Laser Beam Integrators: General Formulation and Design Concepts," *Opt. Eng.* **27**:999–1007 (1988).

293. J. Geary, "Strip Mirror Integrator for Laser Beam Uniformity on a Target," *Opt. Eng.* **28**(8):859–864 (August 1989).

294. B. Lu, J. Zheng, B. Cai, and B. Zhang, "Two-Dimensional Focusing of Laser Beams to Provide Uniform Irradiation," *Opt. Comm.* **149**:19–26 (1998).

295. T. Henning, L. Unnebrink, and M. Scholl, "UV Laser Beam Shaping by Multifaceted Beam Integrators: Fundamentals and Principles and Advanced Design Concepts," *Proc. SPIE* **2703**:62–73 (1996).

296. L. Unnebrink, T. Henning, E. W. Kreutz, and R. Poprawe, "Optical System Design for Excimer Laser Materials Processing," *Proc. SPIE* **3779**:413–422 (1999).

297. S. R. David, C. T. Walker, and W. J. Cassarly, "Faceted Reflector Design for Uniform Illumination," *Proc. SPIE* **3482**:437–446 (June 12, 1998).

298. J. R. Coaton and A. M. Marsden, *Lamps and Lighting,* 4th ed., John Wiley & Sons, New York, 1997.

299. R. J. Donohue and B. W. Joseph, "Computer Design of Automotive Lamps with Faceted Reflectors," *IES* 36–42 (October 1972).

300. E. H. Wiley, "Projector Lamp Reflector," U.S. Patent 4,021,659, 1976.

301. W. P. Laudenschlarger, R. K. Jobe, and R. P. Jobe, "Light Assembly," U.S. Patent 4,153,929, 1979.

302. R. J. Heimer, "Multiple Apparent Source Optical Imaging System," U.S. Patent 4,241,389, 1980.

303. H. Hamada, K. Nakazawa, H. Take, N. Kimura, and F. Funada, "Lighting Apparatus," U.S. Patent 4,706,173, 1987.

304. M. Ronnelid, B. Perers, and B. Karlsson, "Optical Properties of Nonimaging Concentrators with Corrugated Reflectors," *Proc. SPIE* **2255**:595–602 (1994).

305. D. J. Lamb, J. F. Van Derlofske, and L. W. Hillman, "The Use of Aspheric Surfaces in Waveguide Illumination Systems for Automotive Displays," SAE Technical Paper 980874, 1998.

CHAPTER 3
VOLUME SCATTERING IN RANDOM MEDIA

Aristide Dogariu
School of Optics/The Center for Research and Education in Optics and Lasers (CREOL)
University of Central Florida

3.1 GLOSSARY

A	transversal area
C	correlation function
c	speed of light
D	diffusion constant
$\mathbf{E}$	field amplitude
$\mathbf{e}$	polarization direction
F	scattering form amplitude
f	forward scattering amplitude
G	pair correlation function
g	asymmetry parameter
I	field intensity
$\mathcal{I}$	specific intensity
$\mathbf{J}$	diffuse flux
K_B	Boltzmann constant
K_{K-M}	effective absorption coefficient
$\mathbf{k}$	wave vector
l_a	absorption length
l_s	scattering length
$l*$	transport mean free path
N	number of particles
n	refractive index
P	scattering form factor
$\mathbf{q}$	scattering wave vector

$\mathbf{r}$	vector position
S	static structure factor
S	scattering potential
$\mathbf{s}$	specific direction
S_{K-M}	effective scattering coefficient
T	absolute temperature
T	transmission coefficient
t	time
U	diffuse energy density rate
V	volume
v	volume fraction
ε	dielectric constant
η	shear viscosity
θ	scattering angle
λ	wavelength of light
μ_a	absorption coefficient
μ_s	scattering coefficient
ρ	number density
σ	scattering cross section
Ω	solid angle
ω	frequency

3.2 INTRODUCTION

Electromagnetic radiation impinging on matter induces oscillating charges that can be further regarded as secondary sources of radiation. The morphological details and the microscopic structure of the probed medium determine the frequency, intensity, and polarization properties of this re-emitted (scattered) radiation. This constitutes the basis of a long history of applications of light scattering as a characterization tool in biology, colloid chemistry, solid state physics, and so on.

A substantial body of applications deals with light scattering by particles. These smaller or larger ensembles of molecules have practical implications in many industries where they are being formed, transformed, or manipulated. Since the early works of Tyndall and Lord Rayleigh,[1] the study of light scattering by molecules and small particles has been consistently in the attention of many investigators. Classical reviews of the field are the books by Van de Hulst,[2] Kerker,[3] Bayvel and Jones,[4] and Bohren and Huffman;[5] we also note the recent survey of techniques and theoretical treatments by Jones.[6] Why and how the light is scattered by small particles has already been described in Vol. I, Chap. 6. Discussions on subjects related to light scattering can also be found in Vol. I, Chaps. 4, 5, and 43.

This chapter deals with how light scattering by individual particles constitutes the building block of more complicated physical situations. When the three-dimensional extent of the medium that scatters the light is much larger than the typical size of a local inhomogeneity (scattering center), the physical process of wave interaction with matter can be classified as volume scattering. In this regime, the measured radiation originates from many different locations dispersed throughout the volume. Depending upon the structural characteristics of the medium, various scattering centers can act as secondary, independent sources of radiation (*incoherent scattering*) or they can partially add their contributions in a collective manner (*coherent scattering*). Another situation of interest happens when, for highly disordered systems, light is scattered successively at many locations throughout the volume (*multiple scattering*). All these three aspects of volume scattering will be discussed in this chapter.

In practice, particles very rarely exist singly and, depending on the illuminated volume or the volume seen by the detection system, scattering by a large number of particles needs to be considered. The simplest situation is that of *incoherent scattering*. When the fields scattered by different centers are completely independent, the measured intensity results from an intensity-based summation of all individual contributions. The ensemble of particles is described by temporal and spatial statistics that do not show up in scattering experiments; one can say that the volume scattering does not resolve the spatial arrangement of scattering centers.

When the scattering centers are sufficiently close, the phases of wavelets originating from individual scattering centers are not independent. This is the case of collective or *coherent scattering*. One faces the problem of expanding the scattering theories to ensembles of particles that can have certain degrees of spatial or temporal correlations. A transition sets in from independent to coherent scattering regimes. The situation is common for gels or composites that scatter light due to local inhomogeneities of the refractive index with length scales of the order of wavelength and where spatial correlations between the scattering centers are encountered. This is the basis of one of the most successful application of volume scattering: the observation of structural characteristics of inhomogeneous systems.

In the case of highly disordered systems, the light propagation can be subject to scattering at many different locations within the probed volume and a *multiple scattering* regime sets in. For a long time, the intensity and phase fluctuations determined by multiple light scattering were regarded as optical "noise" that degrades the radiation by altering its coherence, broadening the beam, and decreasing its intensity. Experimentalists were trying to avoid this condition as much as possible and the development of comprehensive theories was not sufficiently motivated. In the 1980s and 1990s, however, remarkable advances in fundamental understanding and experimental methodologies proved that multiple scattering of waves is a source for unexplored physics leading to essentially new applications. New phenomena have been discovered and a series of experimental techniques has been implemented using particular coherent, polarization, temporal, and spectral properties of multiply scattered light. This revival of interest has been stimulated by the use of highly coherent sources in remote sensing and, especially, by considerable advances in solid-state physics. Many features of multiply scattered light are common to other classical waves like sound, heat, or microwaves, but several analogies with electron transport phenomena have been at the core of this renewed interest in the propagation of optical waves in random systems.[7]

There is also another situation, often encountered in optics, in which waves propagate through media with abrupt changes in their optical properties. Waves passing through inhomogeneous media with defined boundaries usually suffer surface scattering. In principle, scattering at rough surfaces can be considered as a limiting case of wave propagation and it is significant in various practical situations; this topic has been separately discussed in Vol. I, Chap. 7.

3.3 GENERAL THEORY OF SCATTERING

The schematic of a typical scattering experiment is depicted in Fig. 1, where a plane wave $\mathbf{E}^0$ with the wave vector $k = \omega/c$ is incident on a spatially random medium occupying a finite volume V. Light is scattered by local inhomogeneities of the dielectric constant $\varepsilon(\mathbf{r})$, and a basic theory of scattering aims at providing the link between the experimentally accessible intensity $I_s(\mathbf{R}) = |\mathbf{E}_s(\mathbf{R})|^2$ and the microscopic structure of the random medium.

The starting point of the theory is to describe the total electric field $\mathbf{E}(\mathbf{r})$ as a summation of the incoming and scattered fields and to consider that it satisfies the equation $(\nabla^2 + k^2)\mathbf{E}(\mathbf{r}) = -4\pi S(\mathbf{r})\mathbf{E}(\mathbf{r})$, where $S(\mathbf{r})$ represents a generic scattering potential. This equation can be con-

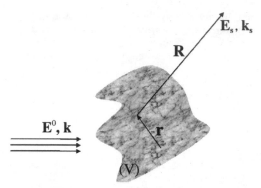

FIGURE 1 Schematic representation of the incident (**k**) and scattered (**k**$_s$) beams in a generic scattering experiment.

verted into an integral one and, for **R** sufficiently far from the scattering volume where the associated Green function simplifies, one eventually obtains the general result:[3, 8]

$$\mathbf{E}_s(\mathbf{R}) = \frac{e^{ikR}}{R} \frac{k^2}{4\pi} \int_{(V)} d\mathbf{r} \left[-\mathbf{k}_s \times \{\mathbf{k}_s \times [\varepsilon(\mathbf{r}) - 1] \cdot \mathbf{E}(\mathbf{r})\}\right] e^{-i\mathbf{k}_s \cdot \mathbf{r}} \tag{1}$$

This expression represents the scattered field as an outgoing spherical wave that depends on the direction and magnitude of the total field inside the scattering volume V.

Approximate solutions can be obtained for the case of weak fluctuations of the dielectric constant. One can expand the field $\mathbf{E}(\mathbf{r}) = \mathbf{E}_0(\mathbf{r}) + \mathbf{E}_1(\mathbf{r}) + \mathbf{E}_2(\mathbf{r}) + \ldots$ in terms of increasing orders of the scattering potential and use successive approximations of $\mathbf{E}(\mathbf{r})$ in Eq. (1) to obtain the so-called Born series. In the spirit of a first iteration, one replaces $\mathbf{E}(\mathbf{r})$ with $\mathbf{E}_0(\mathbf{r})$ and obtains the first Born approximation that describes the regime of single scattering.

Alternatively, one can write $\mathbf{E}(\mathbf{r}) = \exp[\Psi(\mathbf{r})]$ and develop the series solution for $\Psi(\mathbf{r})$ in terms of increasing orders of the scattering potential. This is the Rytov's series of exponential approximations, an alternative to the algebraic series representation of the Born method. The two approaches are almost equivalent; however, preference is sometimes given to Rytov's method because an exponential representation is believed to be more appropriate to describe waves in line-of-sight propagation problems.[9-11]

It is worth pointing out here that Eq. (1) can be regarded as an integral equation for the total field and, because the total field is a superposition of the incident field and contributions originating from scattering from all volume V, this generic equation includes all possible multiple scattering effects.[8, 12]

3.4 SINGLE SCATTERING

Incoherent Scattering

When a sparse distribution of scattering centers is contained in the volume V, all the scattering centers are practically exposed only to the incident field $\mathbf{E}(\mathbf{r}) = \mathbf{E}_0(\mathbf{r}) = E_0 \mathbf{e}_0 e^{i\mathbf{k} \cdot \mathbf{r}}$ where E_0 is the field magnitude, $\mathbf{e}_0$ is the polarization direction, and $\mathbf{k}$ is the wave vector. A considerable simplification is introduced when the magnitude of the scattered field is much smaller than that of the incident field, such that the total field inside the medium can be everywhere approximated with the incident field. This is the condition of the first Born approximation for tenuous media, which practically neglects multiple scattering effects inside the scattering volume V. From Eq. (1), it follows that:

$$\mathbf{E}_s(\mathbf{R}) = E_0 \frac{e^{ikR}}{R} \frac{k^2}{4\pi} \int_{(V)} d\mathbf{r} \, \{\mathbf{e}_s \cdot [\varepsilon(\mathbf{r}) - 1] \cdot \mathbf{e}_0\} \, e^{-i\mathbf{k}_s \cdot \mathbf{r}} \tag{2}$$

For instance, in the case of a system of N identical particles, Eq. (2) is evaluated to give

$$\mathbf{E}_s(\mathbf{R}) = E_0 \frac{e^{ikR}}{R} \frac{k^2}{4\pi} \sum_{j=1}^{N} e^{-i\mathbf{k}_s \cdot \mathbf{r}_j} \int_{(V_j)} d\mathbf{r} \, \{\mathbf{e}_s \cdot [\varepsilon(\mathbf{r}) - 1] \cdot \mathbf{e}_0\} \, e^{-i\mathbf{k}_s \cdot \mathbf{r}} \tag{3}$$

where V_j is the volume of the j-th particle located at $\mathbf{r}_j$. In terms of the scattering wave vector $\mathbf{q} = \mathbf{k}_s - \mathbf{k}$, the integral in Eq. (3) has the meaning of single scattering amplitude of the jth particle, $F(\mathbf{q}) = |f| \cdot B(\mathbf{q})$, and depends on the forward scattering amplitude f and a scattering amplitude normalized such that $B(0) = 1$. The ratio between the refractive index of the particles n_p and the suspending medium n_s determines the scattering strength f of an individual scatterer, while its shape and size are accounted for in $B(\mathbf{q})$.

For a collection of discrete scattering centers, the total scattered intensity of Eq. (3) factors out like this:[13]

$$I(\mathbf{q}) = \left(E_i \frac{e^{ikR}}{R} \frac{k^2}{4\pi} \right)^2 N \, F^2(\mathbf{q}) \sum_{i,j=1}^{N} e^{-i\mathbf{q} \cdot (\mathbf{r}_i - \mathbf{r}_j)} \propto N P(\mathbf{q}) \, S(\mathbf{q}) \tag{4}$$

where we separated the single scattering form factor $P(\mathbf{q})$ from the interference function or the static structure factor $S(\mathbf{q}) = \sum_{i,j=1}^{N} e^{-i\mathbf{q} \cdot (\mathbf{r}_i - \mathbf{r}_j)}$. The structure factor quantifies the phase-dependent contributions due to different locations of scattering centers. When an ensemble average is taken over the volume V and after separating out the diagonal terms, the static structure factor can be written as[13, 14]

$$S(\mathbf{q}) = 1 + [e^{-i\mathbf{q} \cdot (\mathbf{r}_i - \mathbf{r}_j)}] = 1 + \rho \int G(\mathbf{r}) e^{-i\mathbf{q} \cdot \mathbf{r}} d\mathbf{r} \tag{5}$$

in terms of the pair correlation function $G(\mathbf{r})$ (where $\mathbf{r}$ is the vectorial distance between two particles) describing the statistical properties of the spatial arrangement of scattering centers. It is through $S(\mathbf{q})$ that the link is made between the statistical mechanics description of the inhomogeneities and the measurable quantities in a scattering experiment.

As can be seen from Eq. (5), for the case of particles separated by distances much larger than the wavelength, $S(\mathbf{q})$ becomes unity; the situation corresponds to $G(\mathbf{r}) \equiv 1$, that is, constant probability of finding scattering centers anywhere in the scattering volume. This regime characterized by $S(\mathbf{q}) = 1$ is also called the *incoherent case of volume scattering* where $I(\mathbf{q})$ is a simple, intensity-based summation of individual contributions originating from different scattering centers.

Coherent Scattering

For higher volume fractions of particles, the pair correlation function depends on both the particle size and the concentration of particles. The estimation of the pair correlation functions—and, therefore, the evaluation of an explicit form for the structure factor—is a subject of highest interest and is usually approached through various approximations.[15]

Typical structure factors are shown in Fig. 2 for increasing volume fractions of spherical particles with radius r_0. These results are based on the Percus-Yevick approximation,[16, 17] which has the advantage of being available in a closed mathematical form, but the trend shown in Fig. 2 is rather general. Note that the strength of the interparticle interactions is practically measured by the magnitude of the first peak in $S(q)$.[13, 14]

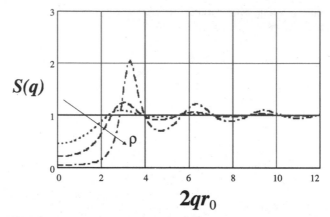

FIGURE 2 Typical structure factors corresponding to systems of identical spherical particles with increasing volume fractions as indicated. The horizontal line at $S(q) = 1$ corresponds to the independent scattering approximation.

In general, wave-scattering experiments can be used to infer the pair correlation function of the scattering centers in a random medium and also the characteristic of an individual scattering event. In an ideal experiment where one has access to all the values of $I(\mathbf{q})$ and the single-scattering phase function $P(\mathbf{q})$ is known, Eq. (3) can be inverted by standard methods.[17, 18] Capitalizing on conventional types of single scattering form factors, various inversion schemes have been implemented to extract the structure factor from the values of angular resolved intensity.[19] Convergence and stability in the presence of noise are the major requirements for a successful inversion procedure and a subsequent Fourier analysis to provide a description of the pair correlation function. Of course, for systems of nonspherical particles or when the system exhibits structural anisotropies, a fully vectorial inverse problem needs to be approached.

The factorization approximation of Eq. (4) has found numerous applications in optical scattering experiments for characterization of colloidal, polymeric, and complex micellar systems. Numerous static and dynamic techniques were designed to probe the systems at different length and time scales. For example, reaction kinetics or different phase transitions have been followed on the basis of angular and/or temporal dependence of the scattered intensity.[19] Another significant body of applications deals with light-scattering studies of aggregation phenomena.[20–22]

We should not conclude this section without mentioning here the similarities between volume light scattering and other scattering-based procedures such as x-ray and neutron scattering. Of course, the "scattering potentials" and the corresponding length scales are different in these cases, but the collective scattering effects can be treated in a similar manner. From a practical viewpoint, however, light scattering has the appealing features of being noninvasive and, most of the time, easier to implement.

The average power scattered by a single particle is usually evaluated using the physical concept of total scattering cross section $\sigma = k^{-4} \int_0^{2k} P(q)q\,dq$.[2, 3, 17] For a system with a number density $\rho = N/V$ of scattering centers, the regime of scattering is characterized by a scattering length $l_s = 1/\rho\sigma$. When the extent over which the wave encounters scattering centers is less than this characteristic scattering length l_s, we deal with the classical single scattering regime. Note that the system of particles can be in the single scattering regime and exhibit either independent or collective scattering. From the general theory of scattering it follows that the details of the scattering form factor depend on the size of the scattering particle compared to the wavelength.

The deviation of the scattering form factor from an isotropic character is characterized by the asymmetry parameter $g = \langle\cos(\theta)\rangle = 1 - 2\langle q^2\rangle/(2k)^2$, where $\langle q^2\rangle = k^{-2}\int_0^{2k} P(q)q^3 dq$.[2, 17]

As we have seen in the preceding section, when the particles are closely packed, the scattering centers are not independent—that is, they are sufficiently close that the fields scattered by different centers are partially in phase—and so collective scattering is considered through the static structure factor $S(q)$. The effect of wave coupling to an individual particle can therefore be isolated from the statistical mechanics description of the particle locations in the volume. In this regime, one can consider the dispersion of correlated particles as a collection of pseudoscattering centers—*equivalent particles*—that are characterized by a modified single scattering form factor $P(q)S(q)$. There is no interaction between these fictitious particles, and a corresponding single scattering phase function can also be defined to be:

$$\widetilde{P(q)} = \frac{k^2 |f|^2 \cdot |B(q)|^2 S(q)}{\displaystyle\int_0^{2k} |f|^2 \cdot |B(q)|^2 S(q)q\, dq} \tag{6}$$

The asymmetry parameter of these pseudoparticles can be written as

$$\tilde{g} = 1 - \frac{\displaystyle\int_0^{2k} |f|^2 \cdot |B(q)|^2 S(q)q^3\, dq}{2k^2 \displaystyle\int_0^{2k} |f|^2 \cdot |B(q)|^2 S(q)q\, dq} \tag{7}$$

The equivalent particle concept is illustrated in Fig. 3, where both the phase function and asymmetry parameters are presented for the specific case of silica particles suspended in water. This simplifying representation of coherent scattering effects is useful to further interpret complex multiple scattering phenomena.

Dynamic Scattering

So far we have limited the discussion to the case where the scattering potential varies across the volume V but, at a certain location, it remains constant in time. However, many physical

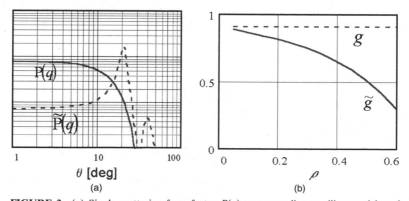

FIGURE 3   (*a*) Single scattering form factor $P(q)$ corresponding to silica particles of 0.476 μm placed in water and illuminated with $\lambda = 0.633$ μm and the form factor $\widetilde{P(q)}$ of an equivalent particle corresponding to a collection of such particles at the volume fraction $\rho = 0.5$. (*b*) Values of the asymmetry parameter for one silica particle and for a collection of particles with an increasing volume fraction as indicated.

systems are such that the volume distribution of scattering centers fluctuates in time and, therefore, gives rise to temporal fluctuations of the scattered radiation, that is, to dynamic scattering. A complete analysis of light scattering involves the autocorrelation function of the dielectric constant fluctuations $\langle \varepsilon^* (\mathbf{r},0),\varepsilon(\mathbf{r}, t)\rangle$, which manifests itself in the statistics of the temporal fluctuations of the measured light intensity. Evaluation of such correlation functions requires knowledge of the transport properties of the volume medium and is based on statistical mechanics and many-body theory. Despite the rather complex phenomenology, different photon correlation techniques have been successfully implemented in studies of reacting systems, molecular dynamics, or atmospheric scintillations.

The benchmarks of the dynamic light scattering were set in the 1970s.[23, 24] An interesting and useful particularity stems from the fact that, based on dynamic scattering, one can actually measure mechanical properties in the scattering volume without knowledge of the refractive index. Random motion of scattering centers induces small Doppler frequency shifts that, in turn, produce an overall broadening of the incident spectrum of light. The detection methods for such spectral changes depend primarily on the time scales of interest and range from high-resolution spectroscopy for very fast phenomena to various mixing or beating techniques for processes slower than about 1 μs.

Based on comparing the scattering signal with itself at increasing time intervals, photon correlation spectroscopy (PCS) has emerged as a successful technique for the study of volume distributions of small particles suspended in fluids.[25] Practically, one deals with the time-dependent fluctuations of the speckle pattern and the goal is to determine the temporal autocorrelation function $\langle E(0)E^*(t)\rangle$ that probes the microscopic dynamics. For instance, in the simple case of a monodispersed and noninteracting system of Brownian particles, $\langle E(0)E^*(t)\rangle$ is a single exponential that depends on the diffusion constant $D = k_B\mathbf{T}/6\pi\eta r_0$. Knowing the absolute temperature $\mathbf{T}$ and the shear viscosity η of the solvent, one can infer the particle radius r_0. Similar to the case of static scattering discussed previously, refinements can be added to the analysis to account for possible dynamic structuring, polydispersivity, and asphericity effects.[19, 23, 26]

3.5   MULTIPLE SCATTERING

When optical waves propagate through media with random distributions of the dielectric constant or when they encounter extended regions containing discrete scatterers or random continuum, one needs to solve a wave equation in the presence of large number of scattering centers; this is a difficult task and, as we will discuss, a series of simplifying approaches has been proposed. A survey of multiple scattering applications is presented by van de Hulst.[27]

Effective-Medium Representation

Some physical insight is given by a simple model that describes the wave attenuation due to scattering and absorption in terms of an effective dielectric constant. Without explicitly involving multiple scattering, one considers the wave propagation through a homogeneous effective medium that is defined in terms of averaged quantities. The effective permittivity ε_{eff} is calculated by simply considering the medium as a distribution of spheres with permittivity ε embedded in a continuum background of permittivity ε_0. If a wave with a wavelength much larger than the characteristic length scales (size of inhomogeneity and mean separation distance) propagates through a volume random medium, the attenuation due to scattering can be neglected; therefore, a frequency-independent dielectric constant will appropriately describe the medium. Based on an induced dipoles model, the Maxwell-Garnett mixing formula relates the effective permittivity to the volume fraction v of spheres $\varepsilon_{\text{eff}} = \varepsilon(1 + 2va)/(1 - va)$, where $a = (\varepsilon - \varepsilon_0)/(\varepsilon + 2\varepsilon_0)$.[28]

In recent developments, wave propagation through highly scattering media has been described by including multiple scattering interactions in a mean field approach. This is simply done by considering that the energy density is uniform when averaged over the correlation length of the microstructure.[29] Through this effective medium, a "coherent" beam propagates with a propagation constant that includes the attenuation due to both absorption and scattering. In general, the propagation of the coherent beam is characterized by a complex index of refraction associated with the effective medium and a nontrivial dispersion law can be determined by resonant scattering.[30] It is worth mentioning that, in the long-wavelength limit, the attenuation due to scattering is negligible and the classical mixture formula for the effective permittivity applies.

Analytical Theory of Multiple Scattering

A rigorous description of multiple light scattering phenomena can be made if statistical considerations are introduced for quantities such as variances and correlation functions for $\varepsilon(\mathbf{r})$ and general wave equations are subsequently produced. The advantage of an analytical theory is that the general formulation does not require a priori assumptions about the strength of individual scattering events or about the packing fraction of scattering centers. The drawback, however, is that, in order to deal with the complexity of the problem, one needs to use quite involved approximations and, sometimes, rather formal representations.

Multiple Scattering Equations.   As shown in Fig. 4, when the field $\mathbf{E}^0$ is incident on a random distribution of N scattering centers located at $\mathbf{r}_1, \mathbf{r}_2, \ldots, \mathbf{r}_N$ throughout the scattering volume V, the total field at one particular location inside V is the sum of the incident wave and the contributions from all the other particles

$$\mathbf{E} = \mathbf{E}^0 + \sum_{j=1}^{N} \mathbf{E}_j \tag{8}$$

the field scattered from the jth particle depends on the effective field incident on this particle and its characteristics (scattering potential) S_j

$$\mathbf{E}_j = \mathsf{S}_j \left(\mathbf{E}_j^0 + \sum_{i=1, i \neq j}^{N} \mathbf{E}_i^s \right) \tag{9}$$

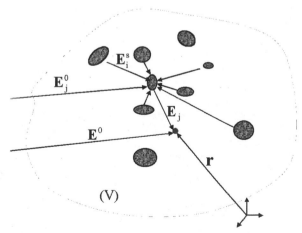

FIGURE 4   The field at $\mathbf{r}$ is a summation of the incident field $\mathbf{E}^0$ and contributions $\mathbf{E}_j$ from all other scattering centers; in turn, the effective field on each particle consists of an incident contribution $\mathbf{E}_j^0$ and contributions $\mathbf{E}_i^s$ from all other scattering centers.

It follows from Eqs. (8) and (9) that the total field can be formally written as

$$\mathbf{E} = \mathbf{E}^0 + \sum_{j=1}^{N} \mathsf{S}_j \left(\mathbf{E}_j^0 + \sum_{j=1}^{N} \sum_{i=1, i \neq j}^{N} \mathsf{S}_i \mathbf{E}_i^s \right) \tag{10}$$

which is a series of contributions from the incident field, single scattering, and increasing orders of multiple scattering.[18] In principle, knowing the scattering characteristics S_j of individual centers (this includes strength and location), one can develop the analytical approach by involving chains of successive scattering paths. From the summations in Eq. (10), by neglecting the scattering contributions that contain a scatterer more than once, Twersky has developed an expanded representation of multiple scattering that is practical only for cases of low-order scattering.[31]

Approximations of Multiple Scattering. Rigorous derivation of multiple scattering equations using Green's function associated with the multiple scattering process and a system transfer operator approach (the T-matrix formalism) can be found in Refs. 32–34. However, in many cases of practical interest, a very large number of scattering centers needs to be involved and it is impossible to obtain accurate descriptions for either the T-matrix or Green's function.

When large ensembles of scatterers are involved, a statistical description of the multiple scattering equations is appropriate. Probability density functions for finding scattering centers at different locations are determined by radial distribution function in a similar way as discussed in the context of correlated scattering. This expresses both the constraint on the particle locations and the fact that they are impenetrable. By using configuration averaging procedures, self-consistent integral equations can be obtained for the average and fluctuating parts of the field produced through multiple scattering. When such a procedure is applied to Eq. (10),

$$\langle \mathbf{E} \rangle = \mathbf{E}^0 + \int \mathsf{S}_j \mathbf{E}_j^0 G(\mathbf{r}_j) \, d\mathbf{r}_j + \int \int \mathsf{S}_i \mathsf{S}_j \mathbf{E}_i^0 G(\mathbf{r}_i) G(\mathbf{r}_j) \, d\mathbf{r}_i \, d\mathbf{r}_j + \ldots \tag{11}$$

a hierarchy of integral equations is generated and successive approximations are obtained by truncating the series at different stages. In fact, this expanded form has a more physically understandable representation in terms of the average field $\langle \mathbf{E}_j \rangle$ at the location of a generic particle j:

$$\langle \mathbf{E} \rangle = \mathbf{E}^0 + \int \mathsf{S}_j \langle \mathbf{E}_j \rangle \, G(\mathbf{r}_j) \, d\mathbf{r}_j \tag{12}$$

Foldy[35] was the first to introduce the concept of configurational averaging and used the joint probability distribution for the existence of a given configuration of scattering centers to average the resulting wave over all possible configurations. However, Foldy's approximation is appropriate for wave propagation in sparse media with a small fractional volume of scatterers.

A comprehensive discussion on multiple scattering equations and various approximations can be found in Tsang et al.,[34] including the quasi-crystalline approximation[36] and coherent potential[37] approximations.

Radiative Transfer

The drawback of an analytical theory of multiple scattering is that it is too complicated; for systems with volume disorder, often encountered in realistic situations, the scattering phenomena depend essentially on the ratio between the characteristic length scales of the system and the radiation wavelength. A statistical description in terms of such characteristic scattering lengths is usually sufficient. In general, the particular location, orientation, and size of a scattering center is irrelevant and the underlying wave character seems to be washed out.

Because energy is transported through multiple scattering processes, what matters is only the energy balance. Of course, this approach cannot account for subtle interference and correlation effects, but refinements can be developed on the basis of a microscopic interpretation of radiative transfer.[38]

A comprehensive mathematical description of the nonstationary radiative transport is given by Chandrasekhar.[39] The net effect of monochromatic radiation flow through a medium with a density ρ of scattering centers is expressed in terms of the specific intensity $I(\mathbf{r}, \mathbf{s}, t)$. This quantity is sometimes called *radiance* and is defined as the amount of energy that, at position $\mathbf{r}$, flows per second and per unit area in the direction $\mathbf{s}$. When radiation propagates over the distance ds, there is a loss of specific intensity due to both scattering and absorption $dI = -\rho(\sigma_{sc} + \sigma_{abs}) I\, ds$. In the meantime, there is a gain of specific intensity due to scattering from a generic direction $\mathbf{s}'$ into the direction $\mathbf{s}$ quantified by the scattering phase function $P(\mathbf{s}', \mathbf{s})$. Also, there could be an increase $\delta I(\mathbf{r}, \mathbf{s}, t)$ of specific intensity due to emission within the volume of interest; the net loss-gain balance illustrated in Fig. 5 represents the nonstationary radiative transfer equation:[18]

$$\left[\frac{1}{c} \frac{\partial}{\partial t} + \mathbf{s} \cdot \nabla + \rho(\sigma_{sc} + \sigma_{abs}) \right] I(\mathbf{r}, \mathbf{s}, t) = \rho \sigma_{sc} \int P(\mathbf{s}', \mathbf{s}) I(\mathbf{r}, \mathbf{s}, t)\, d\Omega + \delta I(\mathbf{r}, \mathbf{s}, t) \qquad (13)$$

No analytical solution exists for the transfer equation and, in order to solve specific problems, one needs to assume functional forms for the both the phase function and the specific intensity. Successive orders of approximation are obtained by spherical harmonic expansion of the specific intensity; for instance, the so-called P_1 approximation is obtained when the diffuse radiance is expressed as a linear combination of an isotropic radiance and a second term modulated by a cosine.[18]

Diffusion Approximation.   Perhaps one of the most widely used treatments for multiple light scattering is the diffusion approach. When (1) absorption is small compared to scattering, (2) scattering is almost isotropic, and (3) the radiance is not needed close to the source or boundaries, then the diffusion theory can be used as an approximation following from the general radiative transfer theory. To get insight into the physical meaning of this approxima-

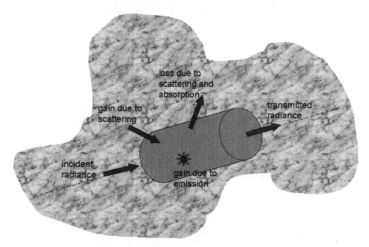

FIGURE 5   Loss-gain balance in a differential volume element; the incident radiance is attenuated by both absorption inside and scattering out of the volume element, but it can be increased by ambient scattering and emission within the volume.

tion it is convenient to define quantities that are directly measurable, such as the diffuse energy density rate (average radiance) $U(\mathbf{r}, t) = \int_{4\pi} I(\mathbf{r}, \mathbf{s}, t) \, d\Omega$ and the diffuse flux $\mathbf{J}(\mathbf{r}, t) = \int_{4\pi} I(\mathbf{r}, \mathbf{s}, t) \, \mathbf{s} \, d\Omega$. In the diffusion approximation, the diffuse radiance is approximated by the first two terms of a Taylor's expansion:[8, 40]

$$I(\mathbf{r}, \mathbf{s}, t) \simeq U(\mathbf{r}, t) + \frac{3}{4\pi} \mathbf{J}(\mathbf{r}, t) \cdot \mathbf{s} \tag{14}$$

and the following differential equation can be written for the average radiance

$$D\nabla^2 U(\mathbf{r}, t) - \mu_a U(\mathbf{r}, t) - \frac{\partial U(\mathbf{r}, t)}{\partial t} = \varphi(\mathbf{r}, t) \tag{15}$$

The isotropic source density is denoted by $\varphi(\mathbf{r}, t)$ and D is the diffusion coefficient, which is defined in units of length as

$$D = \frac{1}{3[\mu_a + \mu_s(1 - g)]} \tag{16}$$

in terms of the absorption scattering coefficients μ_a and μ_s. The diffusion equation is solved subject to boundary conditions and source specifics; most appealing is the fact that analytical solutions can be obtained for reflectance and transmittance calculations.

Because the phase function is characterized by a single anisotropy factor, the diffusion approximation provides mathematical convenience. Through renormalization, an asymmetry-corrected scattering cross section that depends only on the average cosine of scattering angle defines the diffusion coefficient in Eq. (16) and, therefore, an essentially anisotropic propagation problem is mapped into an almost isotropic (diffusive) model.

The photon migration approach based on the diffusion approximation has been very successful in describing the interaction between light and complex fluids[41] or biological tissues.[42, 43] It is instructive to note that three length scales characterize the light propagation in this regime: the absorption length $l_a = \mu_a^{-1}$, which is the distance traveled by a photon before it is absorbed; the scattering length $l_s = \mu_s^{-1}$, which is the average distance between successive scattering events; and the transport mean free path $l^* = l_s/(1 - g)$, which defines the distance traveled before the direction of propagation is randomized. In experiments that are interpreted in the frame of the diffusion approximation, l^* is the only observable quantity and, therefore, the spatial and temporal resolution are limited by l^* and l^*/c, respectively.

Low-Order Flux Models for Radiative Transfer. In an effort to describe the optical properties of highly scattering materials while reducing the computational difficulties, simplifying *flux models* have been designed for the radiative energy transport. A volume scattering medium consisting of a collection of scattering centers is described as homogeneous material characterized by effective scattering and absorption properties that are determined by its internal structure.

In this approach, the fundamental equation of radiative transfer is based on the balance between the net flux change, the flux input, and flux continuing out in an infinitesimal volume. Assuming two diffusing components, a one-dimensional model based on plane symmetry for unit cross section has been initially proposed by Schuster.[44] One of the most successful extensions of this model is the so-called Kubelka-Munk theory,[45] which relates the phenomenological, effective scattering (S_{K-M}) and absorption (K_{K-M}) coefficients to measurable optical properties such as diffuse reflectance or transmittance.

The two-flux model Kubelka-Munk theory is schematically illustrated in Fig. 6. Diffuse radiation is assumed to be incident on the slab; the diffuse radiant flux in the positive x direction is $\mathbf{J}_+$, while the one returning as a result of scattering is $\mathbf{J}_-$. The net flux balance at a distance x across an infinitesimal layer of thickness dx is

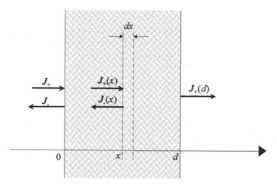

FIGURE 6   The two-flux approximation applied to radiative transfer through a diffusive layer.

$$d\mathbf{J}_+ = -(K_{K-M} + S_{K-M})\mathbf{J}_+ \, dx + s\mathbf{J}_+ \, dx$$

$$d\mathbf{J}_- = +(K_{K-M} + S_{K-M})\mathbf{J}_- \, dx - s\mathbf{J}_+ \, dx \tag{17}$$

where the coefficient K_{K-M} determines the flux attenuation due to absorption while S_{K-M} accounts for the net flux scattered between forward and backward directions. The solution of the simultaneous differential equations of the first order in one dimension is obtained by applying the boundary conditions for the layer shown in Fig. 6. The diffuse reflectance of the substrate at $x = d$ can also be accounted for and expressions for the total diffuse reflection and transmission are found for a specific application. In practical applications of the Kubelka-Munk theory, the effective scattering and absorption parameters are inferred by iteration from measurements of diffuse reflectance or transmission.

A considerable body of work was dedicated to relating the Kubelka-Munk parameters to microstructure and to incorporating both the single and multiple scattering effects. Refinements and higher-order flux models have also been developed. A more accurate model that accounts for the usual condition of collimated incident radiation was elaborated by Reichman.[46] A four-flux model has been developed that includes certain anisotropy of the scattered radiation.[47] A six-flux model was implemented to incorporate the effect of particle shape and interparticle correlation.[48] Despite the fact that it is based on empirical determination of coefficients and that its range of applicability is rather unclear, the simple-to-implement Kubelka-Munk theory makes a reasonably good description of experiments and has found applications in areas such as coatings, paper, paints, pigments, medical physics, and atmospheric physics.

Specific Effects in Multiple Light Scattering

The effects associated with light propagation through multiple scattering media depend on the scale of observation. An interesting analogy exists between optical and electron wave propagation in mesoscopic systems and, based on recent developments in solid-state physics, useful insights have been provided for a range of multiple scattering phenomena.[7,49] It is clear now that multiple light scattering in volume disorder does not merely scramble a coherent incident wave. In fact, the apparent randomness of the scattered field conceals intriguing and sometimes counterintuitive effects such as the enhancement of the backscattered intensity and a range of correlations and statistics of vector waves.

Weak Localization.   Until recently, the coherent light propagating through random media has been considered to be somehow *degraded,* losing its coherence properties. However, recent experiments have brought evidence of the enhanced backscattering of light due to interference between the waves taking time-reversed paths,[50] an effect that is associated with the more general phenomenon for the weak localization of waves in random media.[51, 52] First found in solid-state physics more than 30 years ago, this phenomenon is applicable to all classical waves. Surveys of the state of the art in the optical counterpart of weak localization can be found in Refs. 53 and 54.

When coherent light is scattered by a medium with volume randomness, interference effects occur between the scattered waves that travel through the medium along different paths. The result is a random pattern of interference called *laser speckle*. Because the correlations in this granular pattern extend over angular scales of typically 10^{-3} rad or less, when the individual scatterers are allowed to move over distances of the order of the wavelength or more, the distribution of intensities in the speckle pattern is rapidly averaged out and becomes essentially flat. So far, however, it is well understood and widely recognized that one kind of interference still survives in this average. This is the interference of the waves that emerge from the medium in directions close to exact backscattering and that have travelled along the same path but in opposite directions.

The pairs of time-reversed light paths have some particularities that can be easily understood in the general context of the waves scattered by random media. In Fig. 7, the main contributions to scattering from a dense distribution of scatterers are presented together with their angular dependence. In the narrow angle of interest, the single scattering $I^{(s)}$ and the incoherent ladder term of multiple scattering $I^{(ml)}$ are practically constant. The third, cyclical term $I^{(mc)}$, however, corresponds to the paired (or coherent) scattering channels and, being an interference term, has a definite angular structure. The existence of such a cyclical term (an idea originated from Watson[55]) is based on the fact that each scattering channel, involving a multitude of scattering centers, has its own coherent channel corresponding to the same sequence of scattering centers but time reversed. In the backward direction, the contribution of $I^{(mc)}$ equals that of $I^{(ml)}$ but its magnitude vanishes quickly away from this direction.

The angular profile of $I^{(mc)}$ can be qualitatively described by taking into account the interference between two light paths as shown in Fig. 7. The two time-reversed paths have the same initial and final wave vectors $\mathbf{k}_i$ and $\mathbf{k}_f$ and develop inside the medium through the same

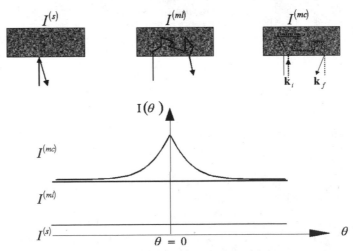

FIGURE 7 A schematic illustration of the origin of coherent backscattering. The classical contributions to backscattering are $I^{(s)}$ and $I^{(ml)}$; in addition, constructive interference occurs between reversed multiple scattering paths that have the same incident wave vector $\mathbf{k}_i$ and exit wave vector $\mathbf{k}_f$.

scattering centers. Under stationary conditions, the two outgoing waves corresponding to these paths are coherent and can interfere constructively for a special choice of $\mathbf{k}_i$ and $\mathbf{k}_f$. If the positions of the first and the last scatterer in the sequence are $\mathbf{r}_i$ and $\mathbf{r}_f$, respectively, the total phase shift between the two waves is $\mathbf{q}(\mathbf{r}_i - \mathbf{r}_f)$, where $\mathbf{q}$ is the momentum transfer $\mathbf{k}_i + \mathbf{k}_f$. Close to the backward direction ($\mathbf{k}_f = -\mathbf{k}_i$), the two waves add coherently and the interference may be described by a weighting factor $\cos[(\mathbf{k}_i + \mathbf{k}_f)(\mathbf{r}_i - \mathbf{r}_f)]$, which is controlled by the interparticle distance $|\mathbf{r}_i - \mathbf{r}_f|$. One can say that the coherence between the time-reversed sequences is lost for angles $\theta > \lambda/|\mathbf{r}_i - \mathbf{r}_f|$ and this actually sets the angular width of the cyclical term $I^{(mc)}$. The detailed angular profile of $I^{(mc)}$ is determined by the probability distribution function for $|\mathbf{r}_i - \mathbf{r}_f|$ and, based on a diffusive model for light propagation in a semi-infinite medium, an approximate formula was given in Akkermas et al.:[56]

$$I^{(mc)} = (1 - e^{-3.4ql*})/3.4ql* \tag{18}$$

It should be noted that the intensities that contribute to classical backscattering, that is, $I^{(s)}$ and $I^{(ml)}$, correspond to incoherent channels of scattering; they add up on an intensity basis; and, upon ensemble averaging, all angular dependences are washed out as seen in Fig. 7. As can be seen from Eq. (18), the angular shape of the coherent backscattering peak can be used to measure $l*$ for a specific multiple scattering medium.

Correlations in Speckle Patterns.   One of the significant discoveries in mesoscopic physics is the phenomenon of conductance fluctuations, which arise from correlations between the transmission probabilities for different output and input modes. In multiple light scattering, the same phenomenon shows up in the form of correlations between the intensities of light transmitted in different directions such as the case schematically depicted in Fig. 8. A nice feature of the optical scattering is that, in contrast with electronic conductance experiments, one has access to both angular dependence and angular integration of transmittance. A rich assortment of correlation functions of the different transmission quantities can be studied and their magnitudes, decay rates, and so on open up novel possibilities for multiple light-scattering-based characterization and tomographic techniques.[57, 58] These measurements exploit the spatial, spectral, and temporal resolution accessible in optical experiments.[59]

In a medium with volume disorder, optical waves can be thought to propagate through *channels* (propagating eigenmodes) defined angularly by one coherence area having the size of a speckle spot. The energy is coupled in and out of the medium through a number $(2\pi A/\lambda^2)$ of such channels where A is the transversal size of the medium. The transmission coefficient T_{ab} of one channel is defined as the ratio between the transmitted intensity in mode b and the incident intensity in mode a at a fixed optical frequency ω. An average transmission through

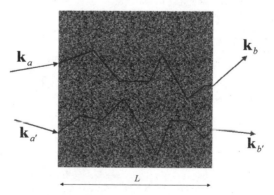

FIGURE 8   Scattering geometry containing two different wave trajectories.

the medium $\langle T_{ab} \rangle$ can be evaluated by ensemble averaging over different realizations of disorder and, for purely elastic scattering, it can be shown that $\langle T_{ab} \rangle \propto l^*/L$. The motion of scatterers, the frequency shift of the incident wave, and the variation of angle of incidence and/or detection introduce phase shifts (momentum differences) in different propagating channels. Surprising and novel features are found when one studies carefully how various T_{ab} channels are correlated with one another. A general function $C_{aba'b'} = \langle \delta T_{ab} T_{a'b'} \rangle$ can be designed to evaluate the correlation between changes in the transmission coefficients $\delta T_{ab} = T_{ab} - \langle T_{ab} \rangle$ and, to the lowest order in the disorder parameter $1/kl$, it can be shown to be a summation of three different terms.[60] The first term C^1, *short-range correlations,* represents the large local intensity fluctuations specific to speckle patterns and exhibits an angular *memory* effect: When the incident beam is tilted by a small angle, the transmitted speckle pattern will, on average, follow provided that the tilt angle is not too large.[61, 62] The second term C^2, *long-range correlations,* arises from paths that cross at a certain scattering site; it is smaller than C^1 by a factor $1/\Sigma T_{ab}$ and decays very slowly with the momentum difference.[63] Finally, a uniform positive correlation that is independent of momentum differences is included in C^3, *conductance correlations,* and it is determined by the less probable case of two crossings in propagation channels. This small correlation term [on the order of $1/\Sigma(T_{ab})^2$] causes just a shift in background, that is, the spatially averaged intensity in each speckle pattern is always a little darker or brighter than the total intensity averaged over many disorder realizations in the sample. These fluctuations do not decrease when averaging is done over larger and larger spatial regions and they are the optical analogue of the universal conductance fluctuations in electronic systems.[60, 64]

3.6 REFERENCES

1. Lord Rayleigh, *Phil. Mag.* **41**:107, 274, 447 (1871); Tyndall, *Phil. Mag.* **37**:156 (1869).

2. H. C. van der Hulst, *Light Scattering by Small Particles,* Wiley, New York, 1957. Reprinted by Dover, New York, 1981.

3. M. Kerker, *The Scattering of Light,* Academic Press, New York, 1969.

4. L. P. Bayvel and A. R. Jones, *Electromagnetic Scattering and Applications,* Elsevier, London, 1981.

5. C. F. Bohren and D. R. Huffman, *Absorption and Scattering of Light by Small Particles,* Wiley-Interscience, New York, 1983.

6. A. R. Jones, "Light Scattering for Particle Characterization," *Prog. Energy Combust. Sci.* **25**:1–53 (1999).

7. W. van Haeringen and D. Lenstra (eds.), *Analogies in Optics and Microelectronics,* Kluwer Academic Publishers, The Netherlands, 1990.

8. A. Ishimaru, *Electromagnetic Wave Propagation, Radiation, and Scattering,* Prentice Hall, Englewood Cliffs, NJ, 1991.

9. J. B. Keller, "Accuracy and Validity of the Born and Rytov Approximations," *J. Opt. Soc. Am.* **59**:1003 (1969).

10. B. Cairns and E. Wolf, "Comparison of the Born and Rytov Approximations for Scattering on Quasi-homogeneous Media," *Opt. Commun.* **74**:284 (1990).

11. T. M. Habashy, R. W. Groom, and B. R. Spies, "Beyond the Born and Rytov Approximations: A Nonlinear Approach to Electromagnetic Scattering," *J. Geophys. Res.* **98**:1759 (1993).

12. M. Born and E. Wolf, *Principles of Optics,* Cambridge University Press, Cambridge, 1999.

13. J. M. Ziman, *Models of Disorder,* Cambridge University Press, Cambridge, 1979.

14. N. E. Cusack, *The Physics of Structurally Disordered Matter,* Adam Hilger, Bristol, 1987.

15. Y. Waseda, *The Structure of Non-Crystalline Materials,* McGraw-Hill, New York, 1980.

16. J.-P. Hansen and I. McDonald, *Theory of Simple Liquids,* 2d ed., Academic Press, New York, 1986.

17. R. H. Bates, V. A. Smith, and R. D. Murch, *Phys. Rep.* **201**:185–277 (1991).

18. A. Ishimaru, *Wave Propagation and Scattering in Random Media,* Oxford University Press, Oxford, 1997.

19. W. Brown (ed.), *Light Scattering Principles and Development,* Clarendon Press, Oxford, 1996.

20. V. Degiorgio, M. Corti, and M. Giglio (eds.), *Light Scattering in Liquids and Macromolecular Solutions,* Plenum Press, New York, 1980.

21. J. Teixeira, "Experimental Methods for Studying Fractal Aggregates," p. 145, in *On Growth and Form,* H. E. Stanley and N. Ostrowski (eds.), Martinuus Nijhoff, Boston, 1986.

22. J. G. Rarity, R. N. Seabrook, and R. J. G. Carr, "Light-Scattering Studies of Aggregation," *Proc. R. Soc. Lond. A* **423**:89–102 (1989).

23. B. J. Berne and R. Pecora, *Dynamic Light Scattering,* John Wiley & Sons, New York, 1976.

24. H. Z. Cummins and E. R. Pike (eds.), *Photon Correlation and Light Beating Spectroscopy,* Plenum Press, New York, 1974.

25. W. Brown (ed.), *Dynamic Light Scattering,* Oxford University Press, New York, 1993.

26. B. Chu, *Laser Light Scattering, Basic Principles and Practice,* Academic Press, San Diego, 1991.

27. H. C. van de Hulst, *Multiple Light Scattering—Tables, Formulas, and Applications,* Academic Press, New York, 1980.

28. J. C. Maxwell-Garnett, *Philos. Trans. R. Soc. Lond. A* **203**:385 (1904).

29. E. N. Economou and C. M. Soukoulis, in *Scattering and Localization of Classical Waves in Random Media,* P. Sheng (ed.), World Scientific, Singapore, 1990.

30. A. Lagendijk and B. A. van Tiggelen, "Resonant Multiple Scattering of Light," *Phys. Reports* **270**:143–215 (1996).

31. V. Twerski, "On Propagation in Random Media of Discrete Scatterers," *Proc. Symp. Appl. Math.* **16**:84–116 (1964).

32. P. C. Waterman, *Phys. Rev. D* **3**:825 (1971).

33. V. K. Varadan and V. V. Varadan (eds.), *Acoustic, Electromagnetic and Elastic Wave Scattering— Focus on the T-Matrix Approach,* Pergamon, New York, 1980.

34. L. Tsang, J. A. Kong, and R. T. Shin, *Theory of Microwave Remote Sensing,* Wiley, New York, 1985.

35. L. L. Foldy, *Phys. Rev.* **67**:107 (1945).

36. M. Lax, *Phys. Rev.* **88**:621 (1952).

37. P. Soven, "Coherent-Potential Model of Substitutional Disordered Alloys," *Phys. Rev.* **156**:809–813 (1967).

38. M. C. W. van Rossum and T. M. Nieuwenhuizen, "Multiple Scattering of Classical Waves: Microscopy, Mesoscopy, and Diffusion," *Rev. Modern Phys.* **71**:313–371 (1999).

39. S. Chandrasekhar, *Radiative Transfer,* Dover Publishing, New York, 1960.

40. P. Sheng, *Introduction to Wave Scattering, Localization, and Mesoscopic Phenomena,* Academic Press, New York, 1995.

41. G. Maret, "Recent Experiments on Multiple Scattering and Localization of Light," in *Mesoscopic Quantum Physics,* E. Akkermanns, G. Montambaux, J.-L. Pichard, and J. Zinn-Justin (eds.), Elsevier Science, 1995.

42. For a collection of recent results and applications see: R. R. Alfano and J. M. Fujimoto (eds.), *OSA Trends in Optics and Photonics,* vol. 2, *Advances in Optical Imaging and Photon Migration,* Optical Society of America, Washington, DC, 1996; J. M. Fujimoto and M. S. Patterson (eds.), *OSA Trends in Optics and Photonics,* vol. 21, *Advances in Optical Imaging and Photon Migration,* Optical Society of America, Washington, DC, 1998; E. M. Sevick-Muraca and J. Izatt (eds.), *OSA Trends in Optics and Photonics,* vol. 22, *Biomedical Optical Spectroscopy and Diagnostics,* Optical Society of America, Washington, DC, 1998.

43. S. A. Arridge, "Optical Tomography in Medical Imaging," *Inverse Problems* **15**:41–93 (1999).

44. A. Schuster, "Radiation Through a Foggy Atmosphere," *Astrophys. J.* **21**(1) (1905).

45. P. Kubelka and F. Munk, "Ein Beitrag zur Optik der Farbenstriche," *Z. Tech. Phys.* **12**:593–601 (1931).

46. J. Reichman, "Determination of Scattering and Absorption Coefficients for Nonhomogeneous Media," *Appl. Opt.* **12**:1811 (1973).

47. B. Maheu, J. N. Letoulouzan, and G. Gouesesbet, "Four-Flux Models to Solve the Scattering Transfer Equations in Terms of Lorentz-Mie Parameters," *Appl. Opt.* **26**:3353–3361 (1984).

48. A. G. Emslie and J. R. Aronson, *Appl. Opt.* **12**:2563 (1973).

49. S. Datta, *Electronic Transport in Mesoscopic Systems,* Cambridge University Press, Cambridge, 1995.

50. K. Kuga and A. Ishimaru, "Retroreflectance from a Dense Distribution of Spherical Particles," *J. Opt. Soc. Am. A* **1**:831–835 (1984).

51. M. P. Albada and A. Lagendijk, "Observation of Weak Localization of Light in a Random Medium," *Phys. Rev. Lett.* **55**:2692–2695 (1985).

52. P. Wolf and G. Maret, "Weak Localization and Coherent Backscattering of Photons in Disordered Media," *Phys. Rev. Lett.* **55**:2696–2699 (1985).

53. M. P. Albada, M. B. van der Mark, and A. Lagendijk, "Experiments on Weak Localization and Their Interpretation," in *Scattering and Localization of Classical Waves in Random Media,* P. Sheng (ed.), World Scientific, Singapore, 1990.

54. Y. Barabenkov, Y. Kravtsov, V. D. Ozrin, and A. I. Saicev, "Enhanced Backscattering in Optics," in *Progress in Optics XXIX,* E. Wolf (ed.), North-Holland, Amsterdam, 1991.

55. K. Watson, "Multiple Scattering of Electromagnetic Waves in an Underdense Plasma," *J. Math. Phys.* **10**:688–702 (1969).

56. E. Akkermas, P. E. Wolf, R. Maynard, and G. Maret, "Theoretical Study of the Coherent Backscattering of Light in Disordered Media," *J. Phys. France* **49**:77–98 (1988).

57. R. Berkovits and S. Feng, "Correlations in Coherent Multiple Scattering," *Phys. Reports* **238**:135–172 (1994).

58. I. Freund, " '1001' Correlations in Random Wave Fields," *Waves Random Media* **8**:119–158 (1998).

59. A. Z. Genack, "Fluctuations, Correlation and Average Transport of Electromagnetic Radiation in Random Media," in *Scattering and Localization of Classical Waves in Random Media,* P. Sheng (ed.), World Scientific, Singapore, 1990.

60. S. Feng and P. A. Lee, "Mesoscopic Conductors and Correlations in Laser Speckle Patterns," *Science* **251**:633–639 (1991).

61. M. Rosenbluh, M. Hoshen, I. Freund, and M. Kaveh, "Time Evolution of Universal Optical Fluctuations," *Phys. Rev. Lett.* **58**:2754–2757 (1987).

62. J. H. Li and A. Z. Genack, "Correlation in Laser Speckle," *Phys. Rev. E* **49**:4530–4533 (1994).

63. M. P. van Albada, J. F. de Boer, and A. Lagendijk, "Observation of Long-Range Intensity Correlation in the Transport of Coherent Light Through a Random Medium," *Phys. Rev. Lett.* **64**:2787–2790 (1990).

64. S. Feng, C. Kane, P. A. Lee, and A. D. Stone, "Correlations and Fluctuations of Coherent Wave Transmission Through Disordered Media," *Phys. Rev. Lett.* **61**:834–837 (1988).

CHAPTER 4
SOLID-STATE CAMERAS

Gerald C. Holst
JCD Publishing
Winter Park, Florida

4.1 GLOSSARY

Quantity	Definition	Unit
A_D	Photosensitive area of one detector	m^2
c	Speed of light	3×10^8 m/s
C	Sense node capacitance	F
$E_{e\text{-faceplate}}(\lambda)$	Spectral radiant incidance	$W\,\mu m^{-1}\,m^{-2}$
$E_{q\text{-faceplate}}(\lambda)$	Spectral photon incidance	photons $s^{-1}\,\mu m^{-1}\,m^{-2}$
d	Detector size	mm
d_{CCH}	Detector pitch	mm
D	Optical diameter	m
DR_{array}	Array dynamic range	numeric
DR_{camera}	Camera dynamic range	numeric
fl	Optical focal length	m
F	f-number	numeric
G	Source follower gain	numeric
h	Planck's constant	6.626×10^{-34} J s^{-1}
M_e	Radiant exitance	W/m^2
$M_q(\lambda)$	Spectral photon exitance	photons $s^{-1}\,\mu m^{-1}\,m^{-2}$
m_{optics}	Optical magnification	numeric
M_v	Photometric exitance	lumen m^2
n_{dark}	Number of electrons created by dark current	numeric
n_{PE}	Number of photoelectrons	numeric
n_{well}	Charge well capacity	numeric
q	Electronic charge	1.6×10^{-19} C
R_{eq}	Equivalent resolution	mm

$\mathfrak{R}_e(\lambda)$	Detector responsivity	A/W
$\mathfrak{R}_{ave}$	Average detector responsivity	V/(J cm^{-2})
t_{int}	Integration time	s
U	Nonuniformity	numeric
V_{max}	Maximum output voltage	V
V_{signal}	Voltage created by photoelectrons	V
$<n_{floor}>$	Noise created by on-chip amplifier	rms electrons
$<n_{pattern}>$	Pattern noise	rms electrons
$<n_{PRNU}>$	Photoresponse nonuniformity noise	rms electrons
$<n_{shot}>$	Shot noise	rms electrons
$<n_{sys}>$	Total array noise	rms electrons
η	Image space horizontal spatial frequency	cycles/mm
η_C	Optical cutoff in image space	cycles/mm
η_N	Nyquist frequency in image space	cycles/mm
η_S	Sampling frequency in image space	cycles/mm
$\eta_{(\lambda)}$	Spectral quantum efficiency	numeric
λ	Wavelength	μm

4.2 INTRODUCTION

The heart of the solid-state camera is the solid-state array. It provides the conversion of light intensity into measurable voltage signals. With appropriate timing signals, the temporal voltage signal represents spatial light intensities. When the array output is amplified and formatted into a standard video format, a solid-state camera is created. Because charge-coupled devices (CCDs) were the first solid-state detector arrays, cameras are popularly called CCD cameras even though they may contain charge injection devices (CIDs) or complementary metal-oxide semiconductors (CMOS) as detectors.

Boyle and Smith[1] and Amelis et al.[2] invented CCDs in 1970. Since then, considerable literature[3-11] has been written on CCD physics, fabrication, and operation. *Charge-coupled devices* refers to a semiconductor architecture in which charge is transferred through storage areas. The CCD architecture has three basic functions: (1) charge collection, (2) charge transfer, and (3) the conversion of charge into a measurable voltage. The basic building block of the CCD is the metal-insulator semiconductor (MIS) capacitor. The most important MIS is the metal-oxide semiconductor (MOS). Because the oxide of silicon is an insulator, it is a natural choice.

Charge generation is often considered as the initial function of the CCD. With silicon photodetectors, each absorbed photon creates an electron-hole pair. Either the electrons or holes can be stored and transferred. For frame transfer devices, charge generation occurs under an MOS capacitor (also called a photogate). For some devices (notably interline transfer devices), photodiodes create the charge.

The CID does not use a CCD for charge transfer. Rather, two overlapping silicon MOS capacitors share the same row and column electrode. Column capacitors are typically used to integrate charge while the row capacitors sense the charge after integration. With the CID architecture, each pixel is addressable (i.e., it is a matrix-addressable device).

Active pixel sensors (APSs) are fabricated with CMOS technology. The advantage is that one or more active transistors can be integrated into the pixel. As such, they become fully addressable (can read selected pixels) and can perform on-chip image processing.

Devices may be described functionally according to their architecture (frame transfer, interline transfer, etc.) or by application. To minimize cost, array complexity, and electronic processing, the architecture is typically designed for a specific application. For example, astronomical cameras typically use full-frame arrays, whereas video systems generally use interline transfer devices. The separation between general imagery, machine vision, scientific devices, and military devices becomes fuzzy as technology advances.

4.3 IMAGING SYSTEM APPLICATIONS

Cameras for the professional broadcast television and consumer camcorder markets are designed to operate in real time with an output that is consistent with a standard broadcast format. The resolution, in terms of array size, is matched to the bandwidth that is recommended by the standard. An array that provides an output of 768 horizontal by 484 vertical pixels creates a satisfactory image for conventional (U.S.) television. Eight bits (256 gray levels) provides an acceptable image in the broadcast and camcorder industry.

In its simplest version, a machine vision system consists of a light source, camera, and computer software that rapidly analyzes digitized images with respect to location, size, flaws, and other preprogrammed data. Unlike other types of image analysis, a machine vision system also includes a mechanism that immediately reacts to an image that does not conform to parameters stored in the computer. For example, defective parts are taken off a production line conveyor belt.

For scientific applications, low noise, high responsivity, large dynamic range, and high resolution are dominant considerations. To exploit a large dynamic range, scientific cameras may digitize the signal into 12, 14, or 16 bits. Scientific arrays may have 5000×5000 detector elements.[12] Theoretically, the array can be any size, but manufacturing considerations may ultimately limit the array size.

Although low-light-level cameras have many applications, they tend to be used for scientific applications. There is no industrywide definition of a low-light-level imaging system. To some, it is simply a solid-state camera that can provide a usable image when the lighting conditions are less than 1 lux (lx). To others, it refers to an intensified camera and is sometimes called a low-light-level television (LLLTV) system. An image intensifier amplifies a low-light-level image that can be sensed by a solid-state camera. The image intensifier/CCD camera combination is called an intensified CCD (ICCD). The image intensifier provides tremendous light amplification but also introduces additional noise. The spectral response of the ICCD is governed by the image intensifier.

The military is interested in detecting, recognizing, and identifying targets at long distances. This requires high-resolution, low-noise sensors. Target detection is a perceptible act. A human determines if the target is present. The military uses the minimum resolvable contrast (MRC) as a figure of merit.[13]

4.4 CHARGE-COUPLED DEVICE ARRAY ARCHITECTURE

Array architecture is driven by the application. Full-frame and frame transfer devices tend to be used for scientific applications. Interline transfer devices are used in consumer camcorders and professional television systems. Linear arrays, progressive scan, and time delay and integration (TDI) are used for industrial applications. Despite an ever-increasing demand for color cameras, black-and-white cameras are widely used for many scientific and industrial applications.

The basic operation of linear, full-frame, frame transfer, and interline transfer devices is described in Chap. 22, "Visible Array Detectors," *Handbook of Optics,* Vol. I. This section describes some additional features.

Full-Frame Arrays

In full-frame arrays, the number of pixels is often based upon powers of 2 (e.g., 512×512, 1024×1024) to simplify memory mapping. Scientific arrays have square pixels and this simplifies image-processing algorithms.

Data rates are limited by the amplifier bandwidth and, if present, the conversion capability of the analog-to-digital converter. To increase the effective readout rate, the array can be divided into subarrays that are read out simultaneously. In Fig. 1, the array is divided into four subarrays. Because they are all read out simultaneously, the effective clock rate increases by a factor of 4. Software then reconstructs the original image. This is done in a video processor that is external to the CCD device where the serial data are decoded and reformatted.

Interline Transfer

The interline transfer array consists of photodiodes separated by vertical transfer registers that are covered by an opaque metal shield (Fig. 2). Although photogates could be used, photodiodes offer higher quantum efficiency. After integration, the charge that is generated by the photodiodes is transferred to the vertical CCD registers in about 1 microsecond (μs). The main advantage of interline transfer is that the transfer from the active sensors to the shielded storage is quick. There is no need to shutter the incoming light. The shields act like a venetian blind that obscures half the information that is available in the scene. The area fill factor may be as low as 20 percent. Because the detector area is only 20 percent of the pixel area, the output voltage is only 20 percent of a detector that would completely fill the pixel area. A microlens can optically increase the fill factor.

Because interline devices are most often found in general imagery products, most transfer register designs are based upon standard video timing. Figure 3 illustrates a four-phase transfer register that stores charge under two gates. With 2:1 interlace, both fields are collected

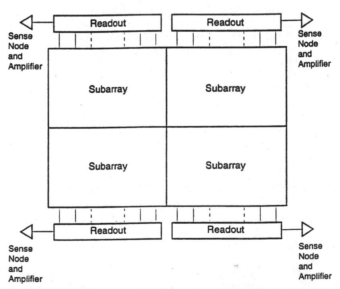

FIGURE 1   A large array divided into four subarrays. Each subarray is read out simultaneously to increase the effective data rate. Very large arrays may have up to 32 parallel outputs.

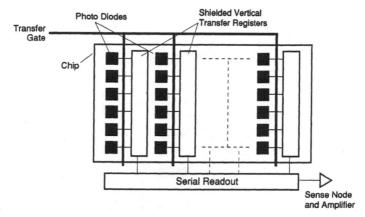

FIGURE 2 Interline transfer architecture. The charge is rapidly transferred to transfer registers via the transfer gate. Interline transfer devices can also have a split architecture similar to that shown in Fig. 1.

simultaneously but are read out alternately. This is called frame integration. With EIA 170 (formerly called RS 170), each field is read every 1/60 s. Because the fields alternate, the maximum integration time is 1/30 s for each field.

Pseudointerlacing (sometimes called field integration) is shown in Fig. 4. Changing the gate voltage shifts the image centroid by one-half pixel in the vertical direction. This creates 50 percent overlap between the two fields. The pixels have twice the vertical extent of standard interline transfer devices and therefore have twice the sensitivity. An array that appears to have 240 elements in the vertical direction is clocked so that it creates 480 lines. However,

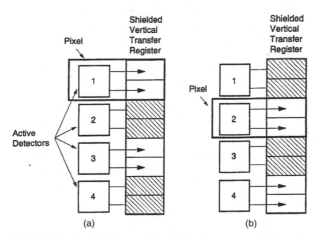

FIGURE 3 Detailed layout of the 2:1 interlaced array. (*a*) The odd field is clocked into the vertical transfer register, and (*b*) the even field is transferred. The vertical transfer register has four gates and charge is stored under two wells. The pixel is defined by the detector center-to-center spacing, and it includes the shielded vertical register area. The transfer gate is not shown.

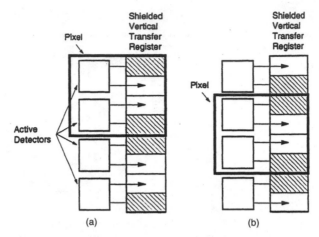

FIGURE 4 Pseudointerlace. By collecting charge from alternating active detector sites, the pixel centroid is shifted by one-half pixel. (*a*) Odd field and (*b*) even field.

this reduces the vertical modulation transfer function (MTF). With some devices, the pseudointerlace device can also operate in a standard interlace mode.

4.5 CHARGE INJECTION DEVICE

A CID consists of two overlapping MOS capacitors sharing the same row and column electrode. Figure 5 illustrates the pixel architecture. The nearly contiguous pixel layout provides a fill factor of 80 percent or greater. Charge injection device readout is accomplished by transferring the integrated charge from the column capacitors to the row capacitors. After this nondestructive signal readout, the charge moves back to the columns for more integration or is injected (discarded) back into the silicon substrate.

Although the capacitors are physically orthogonal, it is easier to understand their operation by placing them side by side. Figure 6 illustrates a functional diagram of an array, and Fig. 7 illustrates the pixel operation. In Fig. 7*a*, a large voltage is applied to the columns, and photogenerated carriers (usually holes) are stored under the column gate. If the column volt-

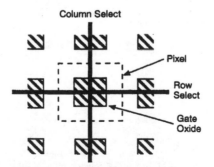

FIGURE 5 CID pixel architecture.

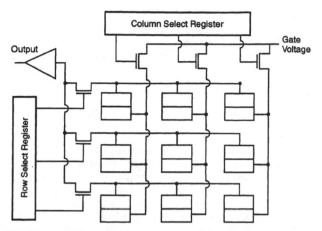

FIGURE 6   Functional layout of a 3 × 3 CID array. The row and column select registers are also called decoders. The select registers and readout electronics can be fabricated with CMOS technology.

age is brought to zero, the charge transfers to the row gate (Fig. 7*b*). The change in charge causes a change in the row gate potential that is then amplified and outputted. If V_1 is reapplied to the columns, the charge transfers back to the column gate. This is nondestructive readout and no charge is lost. By suspending charge injection, multiple-frame integration (time-lapse exposure) is created. In this mode, the observer can view the image on a display as the optimum exposure develops. Integration may proceed for up to several hours. Reset occurs by momentarily setting the row and column electrodes to ground. This *injects* the charge into the substrate (Fig. 7*c*).

The conversion of charge into voltage depends upon the pixel, readout line, and amplifier capacitance. The capacitance is high because all the pixels on a given row are tied in parallel. Therefore, when compared with a CCD, the charge conversion is small, yielding a small signal-to-noise ratio (SNR). Because the readout is nondestructive, it can be repeated numerous times. The multiple reads are averaged together to improve the SNR.

Because each pixel sees a different capacitance, CIDs tend to have higher pattern noise compared with CCDs. However, off-chip algorithms can reduce the amount of pattern noise. With no charge transfer, CIDs are not sensitive to charge transfer efficiency effects. Without

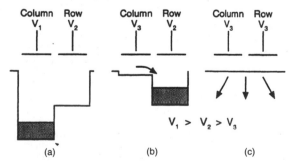

FIGURE 7   CID pixel operation. (*a*) Integration, (*b*) readout, and (*c*) injection.

multiple gates, CIDs have larger well capacities than comparably sized CCDs. Charge injection devices inherently have antibloom capability. Because charge is limited to a single pixel, it cannot overflow into neighboring pixels. Perhaps the greatest advantage of CIDs is random access to any pixel or pixel cluster. Subframes and binned pixels can be read out at high frame rates.

4.6   COMPLEMENTARY METAL-OXIDE SEMICONDUCTOR

With the APS approach, highly integrated image sensors are possible.[14] By placing processing on the chip, a CMOS camera can be physically smaller than a CCD camera that requires clocks, image reformatting, and signal processing in separate hardware. A sophisticated APS array could create[15] a "camera on a chip." It is possible to build a frame transfer device where pixels can be binned to enhance the SNR and provide variable resolution imaging.[16]

In 1993, Fossum[17] described state-of-the-art active pixel concepts. At that time, they were the double-gate floating surface transistor, charge modulation device, bulk charge modulation device, based-stored image sensor, and the static induction transistor. That review continued[14] in 1997.

The APS contains a photodiode, row select transistor, and a reset transistor. As shown in Fig. 8, by activating a row, the data from the pixels in that row are simultaneously copied into the columns. Each column will have a load transistor, a column select switch, and a sampling switch. The chip may also have an analog-to-digital converter and provide correlated double sampling. The pixels are then reset and a new integration is started. The APS does not rely upon charge transfer. Rather, the photodiode drives a capacitance line. The total capacitance limits the chip speed. Active pixel sensors can be fully addressable and subarrays can be read out at high frame rates just like CIDs.

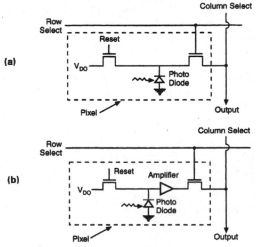

FIGURE 8   (a) Passive pixel device and (b) APS. Charge-coupled devices and CIDs are regarded as passive pixel sensors. Charge injection devices use photogates; CCDs use either photogates or photodiodes; CMOS devices typically use photodiodes.

Complementary metal-oxide semiconductor devices tend to have higher dark currents due to the highly doped silicon used. Therefore, it appears that CMOS sensors will not replace CCDs for low-noise scientific applications. Because the active devices often take up real estate, the area for the photosensor is reduced. This leads to reduced sensitivity that can be partially offset by a microlens. Because each pixel has its own amplifier, pattern noise is larger. However, more logic can be added to each pixel and it seems possible to use on-chip signal processing that suppresses pattern noise.[18] With charge limited to a single pixel, it cannot overflow into neighboring pixels and create blooming as seen with CCDs.

A further advantage is their low power consumption. Active pixel sensors can operate from a 5-volt (V) battery or less (compared with 10 to 15 V for a CCD). It appears at this time that CMOS will compete with CCDs in the general video marketplace where weight, size, and power consumption are factors.

4.7 ARRAY PERFORMANCE

The most common array performance measures are responsivity, read noise, and charge well capacity. From these the minimum signal, maximum signal, SNR, and dynamic range can be calculated. Full characterization includes quantifying the various noise sources, charge transfer efficiency, spectral quantum efficiency, linearity, and pixel nonuniformity.[19] These additional metrics are necessary for the most critical scientific applications.

Signal

Device specifications depend, in part, upon the application. Arrays for general video applications may have responsivity expressed in units of volts per lux. For scientific applications, the units may be in $V/(J\ cm^{-2})$ or, if a digital output is available, $DN/(J\ cm^{-2})$ where DN refers to a digital number. For example, in an 8-bit system, the digital numbers range from 0 to 255. These units are incomplete descriptors unless the device spectral response and source spectral characteristics are furnished.

The number of photoelectrons created by a detector is

$$n_{PE} = A_D \int_{\lambda_1}^{\lambda_2} E_{q\text{-faceplate}}(\lambda)\eta(\lambda)t_{INT}\,d\lambda \tag{1}$$

where A_D is the effective photosensitive area, t_{INT} is the integration time, $\eta(\lambda)$ is the spectral quantum efficiency, and $E_{q\text{-faceplate}}(\lambda)$ is the spectral photon incidance with unit of photons s^{-1} $\mu m^{-1}\ m^{-2}$ (further discussed in Sec. 4.8). Some arrays have a transparent window protecting the array. The faceplate is the front surface of that window. With this approach, the array quantum efficiency includes the transmittance of the window.

For CCDs, charge is converted to a voltage by a floating diode or floating diffusion. The diode, acting as a capacitor, is precharged at a reference level. The capacitance, or sense node, is partially discharged by the amount of negative charge transferred. The difference in voltage between the final status of the diode and its precharged value (reset level) is linearly proportional to the number of photoelectrons. The signal voltage after the source follower is

$$V_{signal} = V_{reset} - V_{out} = n_{PE}\frac{qG}{C} \tag{2}$$

The gain, G, of a source follower amplifier is approximately unity, and q is the electronic charge (1.6×10^{-19} C). The charge conversion is q/C. The output gain conversion is qG/C. It typically ranges from 0.1 $\mu V/e^-$ to 10 $\mu V/e^-$. The signal is then amplified and processed by electronics external to the CCD sensor.

Responsivity.   The spectral quantum efficiency is important to the scientific and military communities. When the array is placed into a general video or industrial camera, it is convenient to specify the output as a function of incident flux density or energy density averaged over the spectral response of the array.

The faceplate spectral photon incidance can be converted into the spectral radiant incidance by

$$E_{e\text{-faceplate}}(\lambda) = \frac{hc}{\lambda} \, E_{q\text{-faceplate}}(\lambda) \, \frac{w}{\mu m - m^2} \tag{3}$$

where h is Planck's constant (6.626×10^{-34} J s^{-1}) and c is the speed of light (3×10^8 m s^{-1}). Quantities associated with power (watts) have the subscript e, and those associated with photons have the subscript q. The quantum efficiency can be converted to amps per watt by

$$\Re_e(\lambda) = \frac{q\lambda}{hc} \, \eta(\lambda). \tag{4}$$

Then, the array output voltage (after the source follow amplifier) is

$$V_{\text{signal}} = \frac{G}{C} \, A_\text{D} \int_{\lambda_1}^{\lambda_2} E_{e\text{-faceplate}}(\lambda) \Re_e(\lambda) t_{\text{int}} \, d\lambda \tag{5}$$

It is desirable to express the responsivity in the form

$$V_{\text{signal}} = \Re_{\text{ave}} \left[\int_{\lambda_1}^{\lambda_2} E_{e\text{-faceplate}}(\lambda) t_{\text{INT}} \, d\lambda \right] \tag{6}$$

The value $\Re_{\text{ave}}$ is an average responsivity that has units of volts per joule per square centimeter and the quantity in the brackets has units of joules per square centimeter. Combining the two equations provides

$$\Re_{\text{ave}} = \frac{G}{C} \, A_\text{D} \, \frac{\displaystyle\int_{\lambda_1}^{\lambda_2} E_{e\text{-faceplate}}(\lambda) \Re_e(\lambda) \, d\lambda}{\displaystyle\int_{\lambda_1}^{\lambda_2} E_{e\text{-faceplate}}(\lambda) \, d\lambda} \quad \text{V/(J cm}^{-2}) \tag{7}$$

The value $\Re_{\text{ave}}$ is an average responsivity that depends upon the source characteristics and the spectral quantum efficiency. While the source can be standardized (e.g., CIE illuminant A or illuminant D$_{6500}$), the spectral quantum efficiency varies by device. Therefore, extreme care must be exercised when comparing devices solely by the average responsivity.

If a very small wavelength increment is selected, $E_{e\text{-faceplate}}(\lambda)$ and $\Re_e(\lambda)$ may be considered as constants and Eq. 7 can be approximated as

$$\Re_{\text{ave}} \approx \frac{G}{C} \, A_\text{D} \, \Re_e(\lambda_0) \tag{8}$$

Minimum Signal.   The noise equivalent exposure (NEE) is an excellent diagnostic tool for production testing to verify noise performance. NEE is a poor array-to-array comparison parameter and should be used cautiously when comparing arrays with different architectures. This is so because it depends on array spectral responsivity and noise. The NEE is the exposure that produces an SNR of one. If the measured root-mean-square (rms) noise on the analog output is V_{noise}, then the NEE is calculated from the radiometric calibration:

$$\mathrm{NEE} = \frac{V_{\mathrm{noise}}}{\mathfrak{R}_{\mathrm{ave}}} \ \mathrm{J/cm^2 \ rms} \tag{9}$$

When expressed in electrons, NEE is simply the noise value in rms electrons. The absolute minimum noise level is the noise floor, and this value is used most often to calculate the NEE. Although noise is an rms value, the notation "rms" is often omitted.

Maximum Signal.   The maximum signal is that input signal that saturates the charge well and is called the saturation equivalent exposure (SEE). It is

$$\mathrm{SEE} = \frac{V_{\mathrm{max}}}{\mathfrak{R}_{\mathrm{ave}}} \ \mathrm{J/cm^2} \tag{10}$$

The total number of electrons that can be stored is the well capacity, n_{well}. The well size varies with architecture, number of phases, and pixel size. It is approximately proportional to pixel area. Small pixels have small wells. If an antibloom drain is present, the maximum level is taken as the white clip level. The maximum signal is

$$V_{\mathrm{max}} = \frac{qG}{C} \ (n_{\mathrm{well}} - n_{\mathrm{dark}}) \tag{11}$$

For back-of-the-envelope calculations, the dark current is often considered negligible ($n_{\mathrm{dark}} \approx 0$).

Dynamic Range.   The array dynamic range is

$$DR_{\mathrm{array}} = \frac{n_{\mathrm{well}} - n_{\mathrm{dark}}}{\langle n_{\mathrm{sys}} \rangle} \tag{12}$$

where $\langle n_{\mathrm{sys}} \rangle$ is the overall rms noise. The array noise consists of dark current, shot, pattern, and readout noise (noise floor). In the absence of light and negligible dark current, the dynamic range is most often quoted as

$$DR_{\mathrm{array}} = \frac{n_{\mathrm{well}}}{\langle n_{\mathrm{floor}} \rangle} \tag{13}$$

Noise

Many books[4–8] and articles[20–23] have been written on noise sources. The level of detail used in noise modeling depends on the application. The noise sources include shot noise, reset noise, pattern noise, on-chip amplifier noise, and quantization noise. It is customary to specify all noise sources in units of equivalent rms electrons at the detector output.

Reset noise can be reduced to a negligible level with correlated double sampling (CDS). Correlated double sampling also reduces the source follower 1/f noise. The off-chip amplifier is usually a low-noise amplifier such that its noise is small compared with the on-chip amplifier noise. The use of an analog-to-digital converter that has more bits reduces quantization noise.

On-chip amplifier noise may be called readout noise, mux noise, noise-equivalent electrons, or the noise floor. The value varies by device and manufacturer. For most system analyses, it is sufficient to consider

$$\langle n_{\mathrm{sys}} \rangle = \sqrt{\langle n_{\mathrm{shot}}^2 \rangle + \langle n_{\mathrm{floor}}^2 \rangle + \langle n_{\mathrm{pattern}}^2 \rangle} \tag{14}$$

where $\langle n^2 \rangle$ is the noise variance and $\langle n \rangle$ is the standard deviation measured in rms electrons.

Shot Noise.   Both photoelectrons and dark current contribute to shot noise. These follow Poisson statistics so that the variance is equal to the mean:

$$\langle n_{\text{shot}}^2 \rangle = n_{\text{PE}} + n_{\text{dark}} \tag{15}$$

While the dark current average value can be subtracted from the output to provide only the signal due to photoelectrons, the dark current noise cannot. Cooling the array can reduce the dark current to a negligible value and thereby reduce dark current noise to a negligible level.

Pattern Noise.   Pattern noise refers to any spatial pattern that does not change significantly from frame to frame. Pattern noise is not noise in the usual sense. This variation appears as spatial noise to the observer. Fixed-pattern noise (FPN) is caused by pixel-to-pixel variations in dark current.[24, 25] As a signal-independent noise, it is additive to the other noise powers. Fixed-pattern noise is due to differences in detector size, doping density, and foreign matter getting trapped during fabrication.

Photoresponse nonuniformity (PRNU) is the variation in pixel-to-pixel responsivities and, as such, is a signal-dependent noise. This noise is due to differences in detector size, spectral response, and thickness in coatings. Photoresponse nonuniformity can be specified as a peak-to-peak value or an rms value referenced to an average value. This average value may either be full well or one-half full well value. That is, the array is uniformly illuminated and a histogram of responses is created. The PRNU can be the rms of the histogram divided by the average value or the peak-to-peak value divided by the average value. The definition varies by manufacturer so that the test conditions must be understood when comparing arrays. For this text, the rms value is used.

Because dark current becomes negligible when the array is sufficiently cooled, PRNU is the dominant pattern component for most arrays. As a multiplicative noise, PRNU is traditionally expressed as a fraction of the total number of charge carriers. If U is the fixed pattern ratio or nonuniformity, then

$$\langle n_{\text{pattern}} \rangle \approx \langle n_{\text{PRNU}} \rangle = U n_{\text{PE}} \tag{16}$$

Frame averaging will reduce all the noise sources except FPN and PRNU. Although FPN and PRNU are different, they are sometimes collectively called scene noise, pixel noise, pixel nonuniformity, or simply pattern noise.

Photon Transfer.   For many applications it is sufficient to consider photon shot noise, noise floor, and PRNU. The simplified noise model provides

$$\langle n_{\text{sys}} \rangle = \sqrt{n_{\text{PE}} + \langle n_{\text{floor}}^2 \rangle + (U n_{\text{PE}})^2} \tag{17}$$

Recall that the mean square photon fluctuation is equal to the mean photon rate. Either the rms noise or noise variance can be plotted as a function of signal level. The graphs are called the photon transfer curve and the mean-variance curve, respectively. Both graphs convey the same information.

For very low photon fluxes, the noise floor dominates. As the incident flux increases, the photon shot noise dominates. For very high flux levels, the noise may be dominated by PRNU. Figure 9 illustrates the rms noise as a function of photoelectrons when the dynamic range ($n_{\text{well}}/\langle n_{\text{floor}} \rangle$) is 60 dB. With large signals and small PRNU, the total noise is dominated by photon shot noise. When PRNU is large, U dominates the array noise at high signal levels. General video and industrial cameras tend to operate in high-signal environments, and cooling will have little effect on performance. A full SNR analysis is required before selecting a cooled camera.

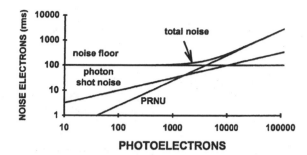

FIGURE 9   Photon transfer curve when $U = 2.5$ percent. The charge well capacity is 100,000 electrons and the noise floor is 100 e$^-$ rms to produce a dynamic range of 60 dB. The noise floor, photon shot noise, and PRNU have slopes of 0, 0.5, and 1, respectively. Dark noise is considered negligible.

4.8   CAMERA PERFORMANCE

Camera performance metrics are conceptually the same as array metrics. The camera is limited by the array noise and charge well capacity. The camera's FPN and PRNU may be better than the array pattern noise when correction algorithms are present. Frame averaging and binning can reduce the random noise floor and thereby appear to increase the camera's dynamic range. If the camera does not introduce any additional noise, modify the noise bandwidth, or minimize the maximum output, the camera SNR and DR_{camera} will be identical to the array values.

Camera Formula

The number of photoelectrons created by an object is

$$n_{PE} = \int_{\lambda_1}^{\lambda_2} \eta(\lambda)\, \frac{M_q(\lambda)}{4F^2(1 + m_{optics})^2}\, A_D t_{int}\tau_{optics}(\lambda)\, d\lambda \tag{18}$$

where $M_q(\lambda)$ is the object's spectral photon exitance in photons per second per micrometer per square meter and $\tau_{optics}(\lambda)$ is the lens system transmittance. The f-number has the usual definition ($F = fl/D$), and the optical magnification is $m_{optics} = R_2/R_1$. Here, R_1 and R_2 are related to the system's effective focal length, fl, by

$$\frac{1}{R_1} + \frac{1}{R_2} = \frac{1}{fl} \tag{19}$$

As the target moves to infinity ($R_1 \to \infty$), m_{optics} approaches zero.

Electronic still cameras are matched to conventional photographic cameras. In photography, shutter speeds (exposure times) vary approximately by a factor of two (e.g., 1/30, 1/60, 1/125, 1/250, etc.). Thus, changing the shutter speed by one setting changes n_{PE} approximately by a factor of 2. F-stops have been standardized to 1, 1.4, 2, 2.8, 4, 5.6, 8, ... The ratio of adjacent f-stops is $\sqrt{2}$. Changing the lens speed by one f-stop changes the f-number by a factor of $\sqrt{2}$. Here, also, the n_{PE} changes by a factor of 2.

The measurement of faceplate illumination (see Eq. 1) is usually performed with a calibrated lens. The value M_q is measured with a calibrated radiometer or photometer, and then $E_{q\text{-faceplate}}$ is calculated:

$$E_{q\text{-faceplate}} = \frac{M_q}{4F^2(1 + m_{\text{optics}})^2} \, \tau_{\text{optics}} \tag{20}$$

Minimum Signal

The maximum and minimum signals depend on the spectral output of the source and the spectral response of the detector. The source color temperature is not always listed but is a critical parameter for comparing systems. Although the CIE illuminant A is used most often, the user should not assume that this was the source used by the camera manufacturer.

Based on signal detection theory, the minimum illumination would imply that the SNR is one. However, the definition of minimum illumination is manufacturer dependent. Its value may be (a) when the video signal is, for example, 30 IRE units, (b) when the SNR is one, or (c) when an observer just perceives a test pattern. Because of its incredible temporal and spatial integration capability, the human visual system can perceive SNRs as low as 0.05. Therefore, comparing cameras based on "minimum" illumination should be approached with care.

The voltage signal (Eq. 5) exists at the output of the array. This voltage must be amplified by the camera electronics (gain = G_{camera}) to a value that is consistent with video standards. The minimum signal provided with gain "on" (G_{camera} greater than one) is usually calculated due to the difficulty of performing accurate, low-level radiometric and photometric measurements. These values may be provided at 30, 50, or 80 percent video levels. That is, the illumination that is given produces an output video that gives 30, 50, or 80 IRE units, respectively. Although a higher-gain amplifier could provide 100 percent video, the user can optimize the image by adjusting the gain and level of the display. That is, the display's internal amplifiers can be used for additional gain. In this context, the camera *system* consists of the camera and display.

If G_{camera} is expressed in decibels, it must be converted to a ratio. The scene illumination that creates a 30 percent video signal is

$$M_v (30\% \text{ video}) = \frac{0.3 M_v (\text{max video})}{G_{\text{camera}}} \tag{21}$$

where M_v (max video) is the scene illumination that produces the maximum output. The subscript v indicates photometric units, which are usually lux. For 50 and 80 percent video, the factor becomes 0.5 and 0.8, respectively. The input signal that produces an SNR of one is

$$M_v (\text{SNR} = 1) = M_v (\text{max video}) \, 10^{-\frac{DR_{\text{camera-dB}}}{20}} \tag{22}$$

where the camera's dynamic range is expressed in decibels (dB). Although photometric units are used most often, radiometric quantities can be used in Eqs. 21 and 22.

Camera Dynamic Range

Dynamic range depends on integration time, binning, and frame integration. In addition, the spectral content of the source and the array spectral responsivity affect the camera output voltage. Thus, the camera dynamic range can be quite variable depending on test conditions.

Although the theoretical maximum signal just fills the charge wells, the manufacturer may limit the maximum output voltage to a lower stored-charge value. Because the dynamic range is often expressed in decibels,

$$DR_{\text{camera-dB}} = 20 \log \left(\frac{V_{\text{max}}}{V_{\text{noise}}} \right) \tag{23}$$

Many camera manufacturers list the dynamic range as the signal-to-noise ratio. This value should not be confused with the actual SNR. With most cameras, the noise level is approximately equivalent to the array read noise. The electronics may increase the camera noise and image processing techniques may reduce it somewhat. Usually the dynamic range is calculated from the measured signal and noise. Because the read noise is amplifier white noise, it is appropriate to include the bandwidth as part of the measurements. For standard video-compatible cameras, the bandwidth is equal to the video format bandwidth.

4.9 MODULATION TRANSFER FUNCTION

The MTF of the camera is the product of all the subsystem MTFs. Usually, the electronics MTF does not significantly affect the overall MTF. Therefore, it is often adequate to consider only the optics and detector MTFs.

The MTF of a single rectangular detector in image space is

$$\text{MTF}_{\text{detector}}(u) = \left| \frac{\sin(\pi\, d\, \eta)}{(\pi\, d\, \eta)} \right| \tag{24}$$

where d is the horizontal extent of the photosensitive surface and η is the image-space horizontal spatial frequency variable (in cycles/mm). The detector size may be different in the horizontal and vertical directions, resulting in different MTFs in the two directions.

The MTF for a circular, clear-aperture, diffraction-limited lens is

$$\text{MTF}_{\text{optics}}(\eta) - \frac{2}{\pi}\left[\cos^{-1}\left(\frac{\eta}{\eta_C}\right) - \frac{\eta}{\eta_C}\sqrt{1 - \left(\frac{\eta}{\eta_C}\right)^2} \right] \tag{25}$$

The optical cutoff is $\eta_C = D/(\lambda_{\text{ave}}fl)$, where D is the aperture diameter and λ_{ave} is the average wavelength.

4.10 RESOLUTION

Detector arrays are specified by the number of pixels, detector size, and detector pitch. These are not meaningful until an optical system is placed in front of the array. The most popular detector resolution measure is the detector angular subtense. In object space, it is

$$\text{DAS} = \frac{d}{fl}\ \text{mrad} \tag{26}$$

and in image space, the detector size, d, becomes the measure of resolution.

Perhaps the most popular measure of optical resolution is Airy disk size. It is the bright center of the diffraction pattern produced by an ideal optical system. In the focal plane of the lens, the Airy disk diameter is

$$d_{\text{airy}} = 2.44\ \frac{\lambda}{D}\ fl = 2.44\lambda F \tag{27}$$

While often treated independently, the camera resolution depends upon both the optical and detector resolutions.

Shade created a metric for system performance. As reported by Lloyd,[26] Sendall modified Shade's equivalent resolution such that

$$R_{\text{eq}} = \frac{1}{2\displaystyle\int_0^\infty |\text{MTF}_{\text{sys}}(\eta)|^2\, d\eta} \tag{28}$$

As a summary metric, R_{eq} provides a better indication of system performance than a single metric, such as the detector size or blur diameter. R_{eq} cannot be directly measured. It is a mathematical construct simply used to express overall performance. As the MTF increases, R_{eq} decreases and the resolution "improves" (smaller is better). As an approximation, the system resolution may be estimated from the subsystem equivalent resolutions by

$$R_{eq} \approx \sqrt{R_{optics}^2 + R_{detector}^2} \qquad (29)$$

Placing the diffraction-limited MTF into Eq. 28 provides

$$R_{optics} = 1.845\lambda F \qquad (30)$$

Note that Shade's approach provides a value that is smaller than the Airy disk diameter. Recall that R_{eq} is only a mathematical construct used to analyze system performance. When R_{optics} dominates R_{eq}, we say the system is optics-limited.

Placing the detector MTF into Eq. 28 provides

$$R_{detector} = d \qquad (31)$$

Here, Shade's resolution matches the common method of describing detector performance: The smallest target that can be discerned is limited by the detector size. When $R_{detector}$ dominates R_{eq}, we say the system is detector-limited.

Using Eq. 29 to estimate the composite resolution in image space,

$$R_{eq} \approx d\sqrt{\left(\frac{1.845\lambda F}{d}\right)^2 + 1} \qquad (32)$$

As $\lambda F/d$ decreases, R_{eq} approaches d. For large values of $\lambda F/d$, the system becomes optics-limited and the equivalent resolution increases.

The more common ½-inch-format CCD arrays have detectors that are about 10 µm in size. Figure 10 illustrates R_{eq} as a function of f-number. Reducing the f-number below 5 does not improve resolution because the system is in the detector-limited region.

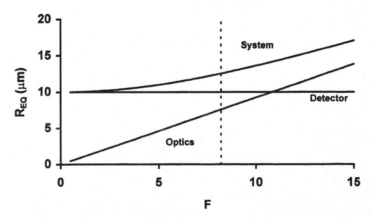

FIGURE 10 Resolution of a typical ½-inch-format CCD camera as a function of f-number. $\lambda = 0.5$ µm. The vertical line at $F = 8.20$ separates optics-limited from detector-limited operation. It occurs when the Airy disk size is equal to the detector size.

For most CCD camera applications, it is assumed that the camera is operating in the detector-limited region. This is only valid if $\lambda F/d$ is small. If $\lambda F/d$ is large, then the minimum discernable target size is definitely affected by the optics resolution.

We live in a world where "smaller is better." Detector sizes are shrinking. This allows the system designer to create physically smaller cameras. Replacing a ½-inch-format array with a ¼-inch-format array (typical detector size is 5 μm) implies a 2× improvement in resolution. However, this is only true if the system is operating in the detector-limited region. As d decreases, the f-number must also decrease to stay within the detector-limited region. Further, the f-number must also decrease to maintain the same signal intensity (Eq. 18). Reducing the f-number can place a burden on the optical designer.

4.11 SAMPLING

Sampling is an inherent feature of all electronic imaging systems. The scene is spatially sampled in both directions due to the discrete locations of the detector elements. The horizontal sample rate is

$$\eta_S = \frac{1}{d_{CCH}} \text{ cycles/mm} \tag{33}$$

Staring arrays can faithfully reproduce signals up to the Nyquist frequency:

$$\eta_N = \frac{\eta_S}{2} = \frac{1}{2d_{CCH}} \text{ cycles/mm} \tag{34}$$

The pitch in the horizontal and vertical directions, d_{CCH} and d_{CCV}, respectively, may be different and, therefore, the sampling rates will be different. Any input frequency above the Nyquist frequency will be aliased to a lower frequency (Fig. 11). After aliasing, the original signal can never be recovered. Diagonal lines appear to have jagged edges, or "jaggies," and periodic patterns create moiré patterns. Periodic structures are rare in nature and aliasing is seldom reported when viewing natural scenery, although aliasing is always present. Aliasing may become apparent when viewing periodic targets such as test patterns, picket fences, plowed fields, railroad tracks, and venetian blinds. It becomes bothersome when the scene

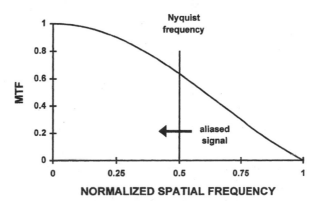

FIGURE 11 Signals above Nyquist frequency are aliased down to the base band. $\eta_S = 1$ and $\eta_N = 0.5$. The area bounded by the MTF above η_N may be considered as an "aliasing" metric.[27]

geometric properties must be maintained as with mapping. It affects the performance of most image-processing algorithms. While this is a concern to scientific and military applications, it typically is of little consequence to the average professional television broadcast and consumer markets.

We have become accustomed to the aliasing in commercial televisions. Periodic horizontal lines are distorted due to the raster. Cloth patterns, such as herringbones and stripes, produce moiré patterns. Cross-color effects occur in color imagery (red stripes may appear green or blue). Many videotape recordings are undersampled to keep the price modest, and yet the imagery is considered acceptable when observed at normal viewing distances.

Because the aliasing occurs at the detector, the signal must be band limited by the optical system to prevent it. Optical band limiting can be achieved by using small-diameter optics, blurring the image, or by inserting a birefringent crystal between the lens and array. The birefringent crystal changes the effective detector size and is found in almost all single-chip color cameras. Unfortunately, these approaches also degrade the MTF (reduce image sharpness) and are considered unacceptable for scientific applications.

If a system is Nyquist frequency–limited, then the Nyquist frequency is used as a measure of resolution. Because no frequency can exist above the Nyquist frequency, many researchers represent the MTF as zero above the Nyquist frequency (Fig. 12). This representation may be too restrictive for modeling purposes.

The fill factor can vary from 20 percent for interline transfer devices to nearly 100 percent for frame transfer devices. In the detector-limited region, the detector MTF determines the potential spatial frequencies that can be reproduced. The center-to-center spacing uniquely determines the Nyquist frequency (Fig. 13). A microlens will increase the effective detector size, but it does not affect the center-to-center spacing. The absolute value of the Nyquist frequency ($1/2d_{CCH}$) does not change. By increasing the detector size, the detector cutoff decreases. The *relative* locations of the sampling and Nyquist frequencies change. The figures in this text use relative (normalized) frequency scales.

Less-expensive color cameras contain only a single CCD chip. A color filter array (CFA) is placed over the chip to create red, green, and blue pixels. Figure 20 in Chap. 22 of *Handbook of Optics*, Vol. 1, provides a variety of CFA patterns. In many sensors, the number of detectors that are devoted to each color is different. The basic reason is that the human visual system (HVS) derives its detail information primarily from the green portion of the spectrum. That is, luminance differences are associated with green, whereas color perception is associ-

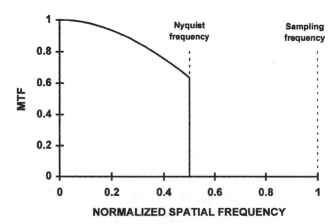

FIGURE 12   MTF representation of an undersampled system. $\eta_S = 1$ and $\eta_N = 0.5$. This restrictive representation erroneously suggests that there is no response above η_N.

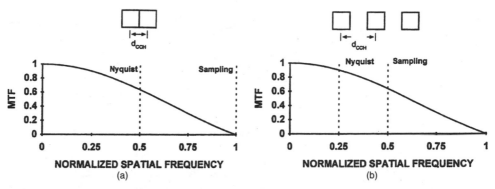

FIGURE 13 Two arrays with different horizontal center-to-center spacings. The detector size, d, is the same for both. (a) $d/d_{CCH} = 1$. This typifies frame transfer devices. (b) $d/d_{CCH} = 0.5$. This is representative of an ideal interline transfer CCD array. The detector MTF is plotted as a function of ηd.

ated with red and blue. The HVS requires only moderate amounts of red and blue to perceive color. Thus, many sensors have twice as many green as either red or blue detector elements. An array that has 768 horizontal elements may devote 384 to green, 192 to red, and 192 to blue. This results in an unequal sampling of the colors (Fig. 14). The output R, G, and B signals are created by interpolation of the sparse data (sparse pixels). The output signals *appear as if there are 768 red, green, and blue pixels.* This interpolation does not change the Nyquist frequency of each color. A birefringent crystal[28] inserted between the lens and the array effectively increases the detector sizes. A larger detector will have reduced MTF and this reduces aliasing. It also reduces edge sharpness.

Pseudointerlacing (Fig. 4) doubles the size of the detector. This reduces the detector cutoff and makes it equal to the Nyquist frequency (Fig. 15). From an aliasing point of view, aliasing has been significantly reduced. However, the MTF has also been reduced, and this results in reduced-edge sharpness.

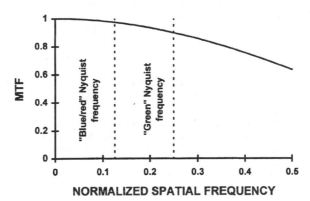

FIGURE 14 Horizontal detector MTF normalized to ηd. Unequally spaced red-, green-, and blue-sensitive detectors create different array Nyquist frequencies. This creates different amounts of aliasing and black-and-white scenes may break into color. The spacing between the red and blue detectors is twice the green detector spacing.

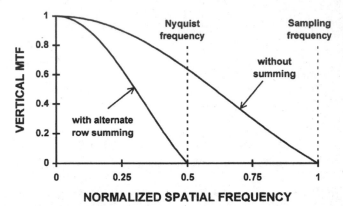

FIGURE 15 Alternate detector summing (pseudointerlacing) can reduce aliasing in the vertical direction. Vertical MTF with and without summing normalized to ηd.

4.12   STORAGE, ANALYSIS, AND DISPLAY

Chapter 22, "Visible Array Detectors," *Handbook of Optics,* Vol. I, describes the physics of CCD detectors. This chapter described some additional features of CCDs and introduced CID and CMOS detectors. These solid-state devices are the basic building blocks of the solid-state array. Certain array architectures lend themselves to specific applications. Once the camera is fabricated, the user selects an optical system. The minimum signal, maximum signal, and resolution discussed in this chapter included the lens *f*-number. It would appear that with all this knowledge, it would be easy to select a camera. A camera only becomes operational when its output is analyzed. Analysis is performed by an observer (general video) or a computer (machine vision).

For general video, the camera output must be formatted into a data stream consistent with the display device. The monochrome standard is often called EIA 170 (originally called RS 170) and the color format is simply known as NTSC (originally called EIA 170A or RS 170A). Worldwide, three color broadcast standards exist: NTSC, PAL, and SECAM. For higher vertical resolution, more lines are required. EIA 343A (originally RS 343A) is a high-resolution monochrome standard used for closed-circuit television cameras (CCTV). Although the standard encompasses equipment that operates from 675 to 1023 lines, the recommended values are 675, 729, 875, 945, and 1023 lines per frame.

These standards are commonplace. Monitors that display these formats are readily available. For computer analysis of this imagery, frame grabbers that accept multiple standards are easily obtained. The output of most general video cameras is an analog signal, and the frame grabber digitizes this signal for computer processing.

In principle, the clock rate of an analog-to-digital converter within the frame grabber can be set at any rate. However, to conserve on memory requirements and minimize clock rates, some frame grabbers tend to just satisfy the Nyquist frequency of a standard video signal. That is, if the highest frequency of the video bandwidth is f_{BW}, then the frame grabber sampling clock operates at $2f_{BW}$.

Some frame grabbers have an internal antialias filter. This filter ensures that the frame grabber does not produce any additional aliasing. The filter cutoff is linked to the frame grabber clock and is not related to the camera output. Once aliasing has occurred in the camera, the frame grabber antialias filter cannot remove it. If the filter does not exist, the frame grabber may create additional aliasing.[29] On the other hand, some frame grabbers have antialiasing filters that significantly limit the analog signal bandwidth and thereby reduce resolution.

The number of digital samples is simply related to the frame grabber clock rate and is not necessarily equal to the number of detector elements. Even if the number of digital samples matches the number of detector elements, phasing effects will corrupt the resolution. Imaging-processing algorithms operate on the digitized signal and the image-processing specialist must be aware of the overall resolution of the system. The frame grabber is an integral part of that system. These issues are mitigated when the camera output is in digital form and the frame grabber can accept digital signals.

As society moves toward digital television [high-definition television (HDTV) or advanced television system (ATS)], new demands are placed upon displays and frame grabbers. New standards also require new video-recording devices. This creates storage, frame grabber, and display problems. New standards do not impose any difficulties on camera design. The solid-state array can be made any size and appropriately designed electronics can support any video format.

Industrial and scientific applications require some forethought. The camera output is no longer a conventional format. Web inspection cameras may contain a linear array with as many as 4096 elements. Split architecture (Fig. 1) may have as many as 32 parallel outputs. Charge injection devices and CMOS cameras may offer variable-sized subframes at variable frame rates. Finding suitable frame grabbers is a challenge for these applications. Nonstandard formats (even after digitization) are not easy to display.

4.13 REFERENCES

1. W. S. Boyle and G. E. Smith, "Charge Coupled Semiconductor Devices," *Bell Systems Technical Journal* **49**:587–593 (1970).

2. G. F. Amelio, M. F. Tompsett, and G. E. Smith, "Experimental Verification of the Charge Coupled Concept," *Bell Systems Technical Journal* **49**:593–600 (1970).

3. M. J. Howes and D. V. Morgan (eds.), *Charge-Coupled Devices and Systems,* John Wiley & Sons, New York, 1979.

4. C. H. Sequin and M. F. Tompsett, *Charge Transfer Devices,* Academic Press, New York, 1975.

5. E. S. Yang, *Microelectronic Devices,* McGraw-Hill, New York, 1988.

6. E. L. Dereniak and D. G. Crowe, *Optical Radiation Detectors,* John Wiley & Sons, New York, 1984.

7. A. J. P. Theuwissen, *Solid-State Imaging with Charge-Coupled Devices,* Kluwer Academic Publishers, Dordrecht, The Netherlands, 1995.

8. G. C. Holst, *CCD Cameras, Arrays, and Displays,* second edition, JCD Publishing, Winter Park, FL, 1998.

9. J. Janesick, T. Elliott, R. Winzenread, J. Pinter, and R. Dyck, "Sandbox CCDs," *Charge-Coupled Devices and Solid State Optical Sensors V,* M. M. Blouke, ed., *Proc. SPIE* Vol. 2415, 1995, pp. 2–42.

10. J. Janesick and T. Elliott, "History and Advancement of Large Area Array Scientific CCD Imagers," *Astronomical Society of the Pacific Conference Series, Vol. 23, Astronomical CCD Observing and Reduction,* BookCrafters, 1992, pp. 1–39.

11. M. Kimata and N. Tubouchi, "Charge Transfer Devices," *Infrared Photon Detectors,* A. Rogalski, ed., SPIE Press, Bellingham, WA, 1995, pp. 99–144.

12. S. G. Chamberlain, S. R. Kamasz, F. Ma, W. D. Washkurak, M. Farrier, and P. T. Jenkins, "A 26.3 Million Pixel CCD Image Sensor," *IEEE Proceedings of the International Conference on Electron Devices,* pp. 151 155, Washington, D.C., December 10, 1995.

13. G. C. Holst, *CCD Cameras, Arrays, and Displays,* second edition, JCD Publishing, Winter Park, FL, 1998, pp. 338–370.

14. S. K. Mendis, S. E. Kemeny, and E. R. Fossum, "A 128×128 CMOS Active Pixel Sensor for Highly Integrated Imaging Systems," *IEEE IEDM Technical Digest* 583–586 (1993).

15. E. R. Fossum, "CMOS Image Sensors: Electronic Camera-on-a-chip," *IEEE Transactions on Electron Devices* **44**(10):1689–1698 (1997).

16. Z. Zhou, B. Pain, and E. R. Fossum, "Frame-Transfer CMOS Active Pixel Sensor with Pixel Binning," *IEEE Transactions on Electron Devices* **44**(10):1764–1768 (1997).

17. E. R. Fossum, "Active Pixel Sensors: Are CCD's Dinosaurs?, *Charge-Coupled Devices and Solid State Optical Sensors III,* M. M. Blouke, ed., *Proc. SPIE* Vol. 1900, 1993, pp. 2–14.

18. J. Nakamura, T. Nomoto, T. Nakamura, and E. R. Fossum, "On-Focal-Plane Signal Processing for Current-Mode Active Pixel Sensors," *IEEE Transactions on Electron Devices* **44**(10):1747–1757 (1997).

19. G. C. Holst, *CCD Cameras, Arrays, and Displays,* second edition, JCD Publishing, Winter Park, FL, 1998, pp. 102–144.

20. D. G. Crowe, P. R. Norton, T. Limperis, and J. Mudar, "Detectors," *Electro-Optical Components,* W. D. Rogatto, pp. 175–283. Volume 3 of *The Infrared & Electro-Optical Systems Handbook,* J. S. Accetta and D. L. Shumaker, eds., copublished by Environmental Research Institute of Michigan, Ann Arbor, MI, and SPIE Press, Bellingham, WA, 1993.

21. J. R. Janesick, T. Elliott, S. Collins, M. M. Blouke, and J. Freeman, "Scientific Charge-Coupled Devices," *Optical Engineering* **26**(8):692–714 (1987).

22. T. W. McCurnin, L. C. Schooley, and G. R. Sims, "Charge-Coupled Device Signal Processing Models and Comparisons," *Journal of Electronic Imaging* **2**(2):100–107 (1994).

23. M. D. Nelson, J. F. Johnson, and T. S. Lomheim, "General Noise Process in Hybrid Infrared Focal Plane Arrays," *Optical Engineering* **30**(11):1682–1700 (1991).

24. J. M. Mooney, "Effect of Spatial Noise on the Minimum Resolvable Temperature of a Staring Array," *Applied Optics* **30**(23):3324–3332 (1991).

25. J. M. Mooney, F. D. Shepherd, W. S. Ewing, J. E. Murguia, and J. Silverman, "Responsivity Nonuniformity Limited Performance of Infrared Staring Cameras," *Optical Engineering* **28**(11):1151–1161 (1989).

26. J. M. Lloyd, *Thermal Imaging,* Plenum Press, New York, 1975, p. 109.

27. G. C. Holst, *Sampling, Aliasing, and Data Fidelity,* JCD Publishing, Winter Park, FL, 1998, pp. 293–320.

28. J. E. Greivenkamp, "Color Dependent Optical Prefilter for the Suppression of Aliasing Artifacts," *Applied Optics* **29**(5):676–684 (1990).

29. G. C. Holst, *Sampling, Aliasing, and Data Fidelity,* JCD Publishing, Winter Park, FL, 1998, pp. 154–156.

CHAPTER 5
XEROGRAPHIC SYSTEMS

Howard Stark

Xerox Corporation
Corporate Research and Technology—Retired

5.1 INTRODUCTION AND OVERVIEW

The xerographic process was invented in the 1930s by Chester Carlson, who was looking for a simple process to copy office documents. The process consists of the creation of an electrostatic image on an *image receptor,* development of the image with dyed or pigmented charged particles referred to as *toner,* transfer of the toner from the image receptor to the paper, fusing the toner to the paper, cleaning the residual toner from the image receptor, and finally, erasing whatever is left of the original electrostatic image. The process is then repeated on the cleaned, electrostatically uniform image receptor. In the most common embodiment of the process, the electrostatic image is created optically from either a digital or a light lens imaging system on a charged *photoreceptor,* a material that conducts electric charge in the light and is an insulator in the dark. These steps are shown schematically in Fig. 1, in which the photoreceptor drum is shown to be rotating clockwise. In this review we summarize the more common ways in which these steps are carried out.

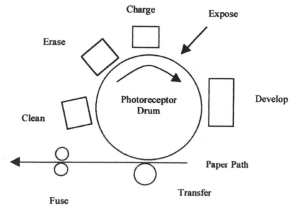

FIGURE 1 Schematic of xerographic process.

The process just outlined is the heart of copying and digital printing systems whose speeds can range from a few to 180 copies per minute. Repeating the process several times (once for each color and black if needed) can produce full-color images. Often the system contains means for either input or output collation and stapling or binding. The cost of these systems can range from hundreds of dollars to several hundred thousand dollars.

This review will not attempt to attempt to give complete references to the technical literature. There are several excellent books that do this.[1-4] In addition, there are older books that give an interesting historical perspective on the development of the technology.[5,6]

5.2 CREATION OF THE LATENT IMAGE

This section covers the creation of the electrostatic field image. First the more common optical systems are considered. Here, exposing a charged photoconductor to the optical image creates the latent electrostatic image. Then, ion writing systems, in which an insulator is charged imagewise with an ion writing head or bar to create the latent electrostatic image, are briefly discussed.

Optical Systems

We consider here ways in which the photoreceptor is charged, the required physical properties of the photoreceptor, and common exposure systems.

Charging.   Figure 2 schematically shows the charging and exposure of the photoreceptor. In this case the charging is shown to be positive. The two devices commonly used to charge the photoreceptor, the *corotron* and the *scorotron,* are shown schematically in Fig. 3. The corotron approximates a constant-current device, the scorotron a constant-voltage device.

The operational difference between the two devices is that the scorotron has a control screen. In both cases, a high potential, shown negative here, is applied to the corotron wires, creating a cloud of negative ions around the wires. In the case of the corotron, the negative ions drift under the influence of the electric field between the wires and the photoreceptor. Since the charging voltage of the photoreceptor is significantly less than that of the corona wires, the electric field and the resulting photoreceptor charging current remain roughly constant. The charge voltage of the photoreceptor is then simply determined from the current per unit length of the corotron, the photoreceptor velocity under the corotron, and the capacitance per unit area of the photoreceptor.

In the case of the scorotron, the photoreceptor voltage and the voltage on the control grid determine the charging field. Thus, when the photoreceptor reaches the grid voltage, the field and the current go to zero. Hence the constant-voltage-like behavior.

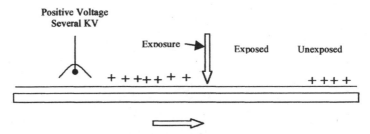

FIGURE 2   Charging and exposure of photoreceptor.

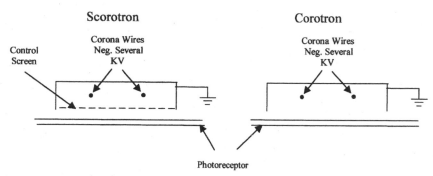

FIGURE 3   Scorotron and corotron.

Photoreceptor.   The discharge of the photoreceptor is accomplished by charge transport through the photoconductive medium. There are many materials that have been used as photoconductors. The first photoreceptors were films 50 or 60 μm thick of amorphous selenium on a metallic substrate. These were followed by amorphous films of various selenium alloys, which were panchromatic and in some cases more robust. Other materials that have been used include amorphous silicon, which is quite robust as well as being panchromatic. Organic photoreceptors are used in most of the recent designs. Here the photoreceptor consists of a photogeneration layer on the order of 1 μm and a transport layer on the order of 20 μm thick.

The photoreceptor discharge process is shown in Fig. 4. The photoconductor is negatively charged, creating an electric field between the deposited charge and the ground plane. Light is shown to be incident on the generator layer. A hole is released that drifts upward under the influence of the electric field. Ideally the hole reaches the surface and neutralizes the applied surface charge. The electron that remains in the generator layer neutralizes the positive charge in the ground plane. Important characteristics of this process include the *dark decay* of the photoreceptor—how well the photoreceptor holds its charge in the dark—the quantum efficiency of the generation process, the transit time of the hole across the transport layer, whether or not it gets trapped in the process, and whether or not there are any residual fields remaining across the generator layer.

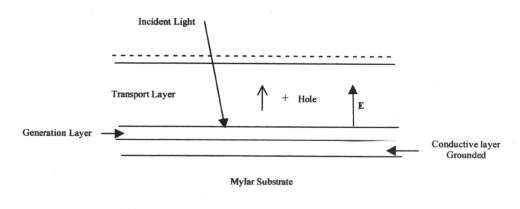

FIGURE 4   Photoreceptor discharge process.

In order for the photoreceptor to hold its charge in the dark (Ref. 2, pp. 104–112), the charge on the surface must not be injected into the transport layer and drift to the substrate. There must be no bulk generation of charge in the transport layer. Finally, there must be no injection and transport of charge from the conductive ground plane into the transport layer. Modern photoreceptors dark-decay at rates of less than a few volts per second.

The transport time of the photogenerated charge through the transport layer determines the rate at which the electrostatic latent image builds up. This limits the shortest time between exposure and development.

Charge trapped within the bulk of the photoreceptor can cause electrostatic ghosts of earlier images that may be developed. Proper erase procedures as well as careful photoreceptor processing are required to eliminate ghosts.

Exposure.   At present, both conventional light lens and digital exposure systems are in use. Conventional systems include both full-frame flash and slit scanning systems. Belt photoreceptors allow for a flat focal plane that permits a quite conventional exposure system; a full-frame flash exposure is used from a flat platen. More interesting is the system that is shown in Fig. 1. Here a slit is exposed at a fixed position on the rotating drum. To accommodate the movement of the drum, the image must move with it within the exposure slit. This is done with a moving platen for the document or a fixed platen with a moving exposure system and pivoting mirrors. Often a selfoc lens is used to conserve space.

The "original" in a digital imaging system is a document stored in computer memory. The idea includes both computer printers and digital copiers. The two most common means of optically writing the image on the photoreceptor are scanning lasers and image bars. In its simplest form an image bar exposes a line at a time across the photoreceptor. It consists of a full-width array of adjacent light-emitting diodes, one for each pixel. As the photoreceptor rotates under the image bar the diodes are turned on and off to write the image.

Laser scanning systems, also known as raster output scanning (ROS) systems, in their simplest embodiment use a laser diode that is focused and scanned across the photoreceptor by a rotating polygon. A so-called f-θ is used to achieve constant linear velocity of the spot across the photoreceptor. Often two or more diodes are focused several raster lines apart in order to write several lines at the same polygon speed. The laser diodes are modulated appropriately with the image information.

Prior to the development of laser diodes, HeNe lasers were used with acoustooptical modulators. In order to accommodate the slow response time of the acoustooptical crystal, the Scophony[7] system developed in the 1930s was used. The acoustooptic modulator consists of a piezoelectric transducer, which launches an acoustical wave in a transparent medium whose index of refraction is pressure sensitive. The acoustic wave, which is modulated with the image information, creates a phase-modulated diffraction pattern. The laser beam is expanded, passed through the crystal and by a stop that blocks the zeroth order of the diffraction pattern. The image of the acoustooptic modulator is then focused on the photoreceptor. Because of the phased imaging system, the resulting image is intensity modulated. However, the diffraction pattern is moving and thus the pixels are moving on the photoreceptor surface. To compensate for this motion, the image of the modulator is scanned in the opposite direction by the polygon at precisely the same speed at which the pixels are moving.

Ion Writing Systems

In an ion writing system the electrostatic image is created by depositing ions on a dielectric receiver in an imagewise fashion. It is typically used in high-speed applications. The requirements for the dielectric receiver are that it be mechanically robust and that it hold the charge through the development process. Transfer, fusing, and cleaning are essentially the same as in conventional xerography. Since (in principle at least) the photoreceptor can be replaced by a more durable dielectric receiver and since charging is eliminated, the process promises to be cheaper and more robust.

At least two techniques have been used commercially for writing the image: stylus writing and ion writing heads. In both cases the limitation appears to be resolution. A stylus writing head consists of an array of styli, one for each pixel. The dielectric receiver is moved under the array and a voltage greater than air breakdown is applied to each stylus as appropriate. The ion writing heads are an array of ion guns, which uses an electrode to control the ion flow to the receiver.

5.3 DEVELOPMENT

The role of the developer system is to apply toner to the appropriate areas of the photoreceptor. In the case of conventional exposure, these areas are the charged areas of the photoreceptor. This system is referred to as *charged area development* (CAD). For digital systems where lasers or image bars are employed, the designer has a choice; the regions to be developed can be left charged as in the conventional system. Alternatively, the photoreceptor can be discharged in the regions to be toned. This is referred to as *discharged area development* (DAD). Image quality and reliability drive the choice. For either CAD or DAD, charged toner is electrostatically attracted to the photoreceptor.

There are many different techniques for developing the electrostatic latent image. We consider first two-component magnetic brush development and, in that context, outline many of the more general considerations for all development systems. Other interesting systems will then be described.

Two-Component Magnetic Brush Development

The developer in two-component magnetic brush development consists of magnetized carrier beads and toner. Here the toner is typically 10 μm and the carrier 200 to 300 μm. The two components are mixed together and, by means of triboelectric charging, charge is exchanged between the toner and carrier. The much smaller toner particles remain attached to the carrier beads so that in a properly mixed developer there is little or no free toner. The role of the carrier is thus twofold: to charge the toner and, because of its magnetic properties, to enable the transport of the two-component developer. As will be seen, the conductivity of the carrier plays an important role in development. The carrier often consists of a ferrite core coated with a polymer chosen principally to control the charging characteristics of the developer.

The toner is a polymer containing pigment particles. For black systems the pigment is carbon black; for full color the subtractive primaries (cyan, magenta, and yellow) are used. In highlight color systems (black plus a highlight color) the pigment is the highlight color or perhaps a combination of pigments yielding the desired color. The choice of the polymer and, to some degree, the colorants is also constrained by the charging properties against the carrier and by the softening temperature, which is set by the fusing requirements. The covering power of the toner is determined by the concentration of the pigment. Typically a density of 1 is achieved with on the order of 1 mg/cm^2 of toner.

A typical magnetic brush development system is shown schematically in Fig. 5.

The developer roll transports the developer (beads and carrier) from the sump to the nip between the developer roll and the photoreceptor where development takes place. The magnetic fields hold the developer on the roll and the material is moved along by friction. A carefully spaced doctor blade is used to control and limit the amount of developer on the roll.

Key to the development process is the triboelectric charge on the toner particle. The charge exchange between the toner and carrier can be thought of in terms of the alignment of

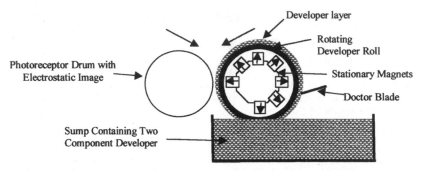

FIGURE 5 Magnetic brush development system.

the Fermi levels of the two materials in order to reach thermodynamic equilibrium. Thus a knowledge of the work functions (the energy required to lift an electron from the Fermi level to vacuum) gives a first-order estimate of the toner charge. Impurities and the pigments also play a large role in the triboelectric charging, as do the manufacturing processes. Often charge control agents are used to control the toner charging. The toner adheres to the carrier bead because of electrostatic attraction (they are of opposite polarity) and whatever other adhesion forces there are. Figure 6 schematically shows the details of the nip between the charged photoreceptor and the development roll.

The Development of Solid Areas. The principal driving force for development is the electric field $E = (V_{pr} - V_c)/d$ where it is assumed that the developer is an insulator. (The conductive case will be considered presently.) V_{pr} is the voltage on the photoreceptor and V_c is a small voltage used to suppress development in background regions by reversing the field. The toner, however, is attached to the carrier beads and must be detached before it can be deposited on the photoreceptor. The electric field plays a role in this, as do the mechanical forces that result from impaction with the photoreceptor and the mixing of the developer within the nip. The density of the developer in the nip, the toner concentration, the magnetic field, and the electric field thus all affect the rate of toner deposition. Developed toner may also be scavenged from the photoreceptor by, say, an oppositely charged carrier bead impacting on the developed area. Development proceeds until the two rates are equal or until the photoreceptor emerges from the nip.

The voltage V_c is used to provide a reverse electric field to prevent toner deposition in what should be toner-free regions. This is required to prevent toner from adhering to the photoreceptor for nonelectrostatic reasons. Developers may contain a small amount of wrong-

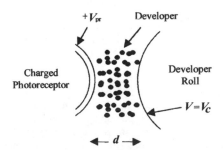

FIGURE 6 Details of nip between charged photoreceptor and developer roll.

sign toner for which this field is a development field. This requires careful formulation of the developer and as well as judicious sizing of the cleaning field.

As development proceeds and toner is deposited on the photoreceptor, current flows from the developer roll to the photoreceptor, neutralizing the charge on the photoreceptor. This process may be viewed as the discharging of a capacitor—the photoreceptor—through a resistor—the developer. Thus, as a first-order approximation, the time constant for development is simply determined from the capacitance per unit area of the photoreceptor and the resistivity of the developer. The nip design must be such that the photoreceptor is in the nip on the order of a time constant or more. Typically development takes place to 50 percent or more of complete neutralization of the photoreceptor and is roughly a linear function of the development field until saturation is reached.

The resistivity of the developer plays a large role in the rate of development. Two cases may be considered: the insulating magnetic brush (IMB) and the conductive magnetic brush (CMB). In the case of conductive development (CMB) the effective spacing to the development electrode or roller is smaller than d (see Fig. 6), thereby increasing the apparent electric field and the rate of development. Ideally development proceeds to neutralization for CMB. If the resistivity of the developer is large, the spacing is larger and the development is slower. In addition, space charge may develop in the nip, further slowing down development.

The Development of Lines. For solid-area development with a highly resistive developer, the electric field that controls development is given by $E = (V_{pr} - V_c)/d$ (Fig. 6). For lines this is no longer true. It was recognized early on that lines develop much faster than solids due to the fact that the electric field at the edge of a large solid area is quite a bit stronger than at the center. This edge-enhancing effect was quite prominent in early xerographic development systems. A general approach to understanding line development is to calculate the modulation transfer function (MTF) or sine-wave response. It is relatively straightforward to calculate the electric fields above a sinusoidal charge distribution (Ref. 2, pp. 25–37), as shown in Fig. 7. The question is what field to use and whether or not a linear analysis is appropriate in what would appear to be a very nonlinear system.

As development proceeds, the fields decrease due to the neutralization of the charge on the photoreceptor. Furthermore, the fields fall off approximately exponentially in distance from the surface of the photoreceptor. Finally, space charge can accumulate in the developer nip; thus the assumption of a well-defined dielectric constant is questionable. Shown in Fig. 8 (taken from Ref. 2) is the normal component of the initial electric field 27 μm above the photoreceptor surface. The photoreceptor is 60 μm thick. The dielectric constant is 6.6, and the photoreceptor is charged to an average field of 15 V/μm. The dielectric constant of the nip is assumed to be 21 and the thickness of the nip is assumed to be 1700 μm. Here it is seen that, at least initially, lines with a spatial frequency of, say, 5 lines per mm develop at a rate 4 times faster than a solid area. If development is designed to go close to completion, this ratio can be much reduced.

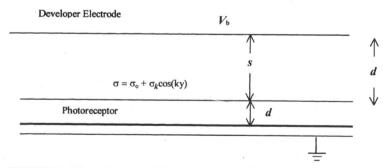

FIGURE 7 Calculating electric fields above a sinusoidal charge distribution.

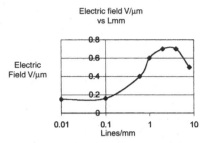

FIGURE 8 The normal electric field as a function of spatial frequency.

Measuring Toner Charge

The measurement of the charge on the toner is fundamental to characterization of a developer, that is, a toner and carrier bead mix. A Faraday cage is used with screens on either end (Fig. 9). The screen mesh is such that toner can pass through and the carrier cannot.

The developer is loaded into the cage. A jet of air is blown through the cage, removing the toner. Both the charge and the weight of the removed toner are measured. The quotient is referred to as the *tribo* and is measured in units of microcoulombs per gram. The results depend on how the developer is mixed. Useful, properly mixed developers have tribos ranging between 10 and, say, 30 μc/g.

The distribution of the charge can be obtained from what is called a *charge spectrograph.*[8] (See Fig. 10.) The charged toner is blown off the developer mixture and inserted into the laminar air stream flowing in the tube. An electric field is applied normal to the toner flow. Within the tube the toner is entrained in the air and drifts transversely in the direction of the electric field. It is collected on the filter. The displacement of the toner d can be calculated from the charge on the toner Q_t, the electric field E, and the viscous drag, which is proportional to the radius r_t of the particle and the viscosity of the air η. Thus

$$d = (Q_t/r_t)(E/6\pi\eta)$$

Using a computerized microscope to measure the number of toner particles as well as the radius and displacement, it is possible to obtain the charge and size distribution of the toner. This technique is particularly important as it yields the amount of wrong-sign toner in the developer.

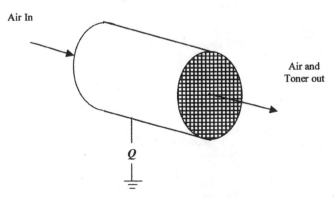

FIGURE 9 Faraday cage used to measure charge on toner.

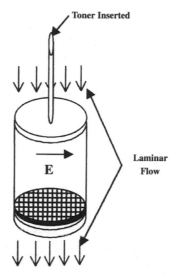

FIGURE 10 Charge spectrograph.

Other Development Systems

Among the other useful development systems are inductively charged single-component development, powder cloud development, and electrophoretic or liquid immersion development (LID).

Single-Component Development.   In inductively charged single-component development the toner is both conductive and magnetic. The system is shown schematically in Fig. 11. The toner is charged inductively in response to the electric field with charge injected from the developer roll. The charge is transferred to the toner closest to the developer roll. The toner then is attracted to the photoreceptor. The materials issues are to ensure charge injection from the developer roll while at the same time preventing charge transfer from the toner to the photoreceptor.

Powder Cloud Development.   A powder cloud development system is shown in Fig. 12. Here toner is injected above the grid, drifts through the grid, acquires a negative charge, and is transported to the photoreceptor by the electric field. As is seen, at the edge of a charged

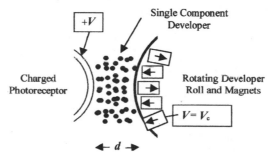

FIGURE 11 Single-component development.

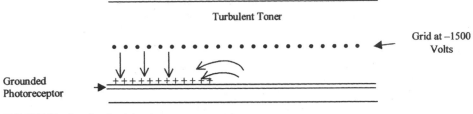

FIGURE 12   Powder cloud development system.

area, the fields are such as to prevent toner from entering into the region near the edge, thus diminishing edge development.

Liquid Immersion Development.   An electrophoretic developer consists of toner particles suspended in a nonconducting fluid. Charge exchange takes place between the fluid and the toner particles. The charged toner then follows the field lines to the photoreceptor. The development rates can be determined from the toner charge and the image-driven electric field. In commercial systems, the developer fluid is pumped between the photoreceptor and a development roll. Controlling the toner charge is a major materials problem. The details of preparation play an important role. Surface active agents also play an important role in the charging of the toner.

5.4   TRANSFER

After development, the toner is held on the photoreceptor by electrostatic forces. Electrostatic transfer is accomplished by applying an electric field either by means of a corotron or a bias transfer roll to attract the toner to the paper (Fig. 1). The paper is brought into contact with the image on the photoreceptor if nonelectrostatic forces are assumed negligible; it is possible to calculate[9] where the toner splits as a function of the applied field and the thickness of the photoreceptor, toner layer, and paper. The nonelectrostatic forces, say, Van der Waals forces, between toner particles and between the toner and the photoreceptor can play an important role. Transfer efficiencies can run over 90 percent.

The difficult engineering problem is to bring the possibly charged paper into and out of contact with the photoreceptor without causing toner disturbances due possibly to air breakdown or premature transfer resulting in a loss of resolution. Bias transfer rolls with a semiconductive coating having carefully controlled relaxation times are required.

5.5   FUSING

After transfer the toner must be permanently fixed to the paper. This can be accomplished with the application of heat and possibly pressure. The idea is to have the toner flow together as well as into the paper. Surface tension and pressure play important roles. Many different types of fusing systems exist, the simplest of which is to pass the paper under a radiant heater. Here the optical absorption of the toner must be matched to the output of the lamp. This usually requires black toner.

Roll fusing systems are quite common. Here the paper is passed between two rolls, with the heated roll on the toner side. The important parameters are the roll temperature and dwell time of the image in the nip of the rollers. Release agents are used to assist the release of the paper from the rollers.

The fusing system imposes material constraints on the toner. Low-melt toners are preferred for fusing, but they cause the developer to age faster.

5.6 CLEANING AND ERASING

There are many ways of removing the untransferred toner from the photoreceptor in preparation for the next imaging cycle. Vacuum-aided brushes are common. Here a fur brush is rotated against the photoreceptor; the toner is removed from the photoreceptor by the brush and from the brush by the vacuum. These systems tend to be noisy because of the vacuum assist. Electrostatic brushes have also been used. Here a biased conductive brush removes the toner from the photoreceptor and then "develops" it onto a conductive roller, which in turn is cleaned with a blade. A development system biased to remove toner from the photoreceptor has also been used. The simplest of the cleaning systems is a blade cleaner; it is compact, quiet, and inexpensive.

Along with the removal of untransferred toner, the photoreceptor must be returned to a uniform and preferably uncharged state. Ghosting from the previous image may result from trapped charge within the photoreceptor. Erasing is accomplished using strong uniform exposure.

5.7 CONTROL SYSTEMS

Proper operation of the xerographic system requires system feedback control to maintain excellent image quality. Among the things to be controlled are charging, exposure, and toner concentration and development. At a minimum, toner gets used and must be replaced. The simplest control system counts pages, assumes area coverage, and replenishes the toner appropriately. The photoreceptor potential after charging and after exposure can be measured with an electrostatic voltmeter. Changes to the charging and exposures can then be appropriately made. The toner optical density of a developed patch on the photoreceptor of known voltage contrast can be measured with a densitometer. In digital systems the actual area coverage can be determined by counting pixels. These two measurements allow the control of toner concentration.

5.8 COLOR

There are many full-color and highlight-color copiers and printers available commercially. All of the recent designs are digital in that the image is either created on a computer or read in from an original document and stored in a digital format. Thus in a full-color process the half-toned cyan, magenta, yellow, and black separations are written on the photoreceptor by the ROS or image bar. The process is then essentially repeated for each separation.

Full Color

The xerographic processor can be configured in several different ways for full color. In a cyclic process a system like the one shown in Fig. 13 is used. The paper is attached to the bias transfer roll and the cycle is repeated four times for full color. The significant new issues are the need for four developer housings, registration, and the fusing of the thicker toner layers. The cyan image must be developed with just cyan. It is also important not to disturb the cyan image on the photoreceptor as it passes through the other housings. The developer housings are therefore mechanically or electrically switched in or out depending on the image written.

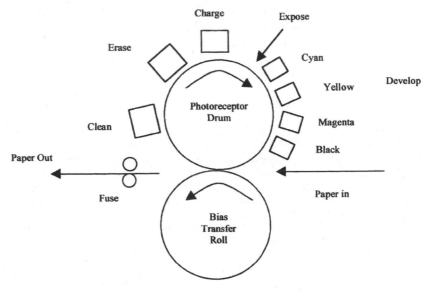

FIGURE 13   Cyclic full-color process.

In Fig. 13 the transfer roll and the photoreceptor are the same diameter. Registration is accomplished by writing the image on the same place on the photoreceptor each cycle. The temperature and dwell time in the fuser are controlled to achieve a good fuse.

Tandem configurations are also used. Here four separate processors are used as shown in Fig. 14. A belt transfer system is shown. The paper is attached to the belt and moved from station to station. The rollers shown behind the belt are used to apply the bias required for transfer. The order of development, shown here with yellow first, is chosen to minimize the effects of contamination. Other transfer systems are also possible. The image can be transferred directly to the belt and then, after the last station, to the paper.

Comparison of these two systems is interesting. The cyclic system has fewer parts and is therefore less expensive and likely more reliable. The tandem system has more parts but is a factor of 4 faster for full color than the cyclic system.

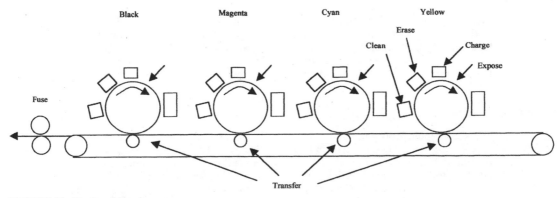

FIGURE 14   Tandem full-color process.

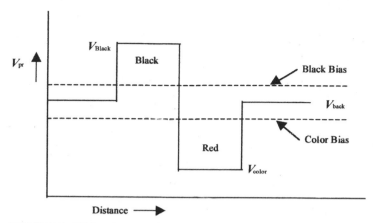

FIGURE 15   Tri-Level process.

Highlight Color

Highlight color is black plus one color, say red, which is used to accentuate or highlight important parts of the image. It can be achieved with the full-color systems just discussed. Also, the full-color system can be modified to contain just the black plus the highlight color. A single-pass highlight color system was developed at Xerox that retains much of the simplicity and throughput of a single-color system. Referred to as *Tri-Level*,[10,11] it encodes the black and highlight color on the photoreceptor as different voltages. The electrostatic arrangement is shown in Fig. 15. The photoreceptor is discharged to three levels. In this case full charge is black, the highlight is discharged to a low level, and white is maintained at some intermediate level.

Two developer housings are used, one for each color. In this case the black toner is negative and the color toner is positive. The housings are biased as shown. The image is passed sequentially through the two housings. The black region appears as a development field for black. Both the background and highlight color regions are cleaning fields for black and no development takes place. The same considerations apply to the highlight color. On the photoreceptor the resulting image contains opposite-polarity toner. The images are passed under a pretransfer corotron to reverse the sign of one of the toners. Thus, at the cost of an additional housing and a pretransfer charging device and no cost in throughput, a black-plus-one-color system can be designed.

The difficulties of the system are under consideration. Since the maximum charging voltage is limited, the voltage contrast is reduced by more than a factor of 2. The first image must not be disturbed when passing through the second developer housing. Both of these constraints require sophisticated developer housing design.

5.9 REFERENCES

1. L. B. Shein, *Electrophotography and Development Physics,* 2d ed., Springer-Verlag, 1992.

2. M. Scharfe, *Electrophotography Principles and Optimization,* Research Studies Press, 1983.

3. E. M. Williams, *The Physics and Technology of the Xerographic Process,* Wiley, 1984.

4. D. M. Pai and B. E. Springett, "Physics of Electrophotography," *Rev. Modern Phys.* **65**:163–211 (January 1983).

5. J. Dessauer and H. Clark (eds.), *Xerography and Related Processes,* Focal Press Limited, 1965.

6. R. M. Schaffert, *Electrophotography,* Focal Press Limited, 1975.

7. L. M. Myers, "The Scophony System: An Analysis of Its Possibilities," *TV and Shortwave World,* 201–294 (April 1936).

8. R. B. Lewis, E. W. Connors, and R. F. Koehler, U.S. Patent 4,375,673, 1983.

9. C. C. Yang, and G. C. Hartman, *IEEE Trans. ED* **23**:308 (1976).

10. W. E. Haas, D. G. Parker, and H. M. Stark, "The Start of a New Machine, IS&T's Seventh International Congress on Advances in Non-Impact Printing," Portland, OR, October 6–11, 1991.

11. W. E. Haas, D. G. Parker, and H. M. Stark, "Highlight Color Printing: The Start of a New Machine," *J. Imag. Sci. Technol.* **36**(4):366 (July/August 1992).

CHAPTER 6
PHOTOGRAPHIC MATERIALS

John D. Baloga
Eastman Kodak Company
Imaging Materials and Media, R&D

6.1 INTRODUCTION

Photographic materials are optical devices. Their fabrication and technical understanding encompass the field of optics in the broadest sense. Light propagation, absorption, scattering, and reflection must be controlled and used efficiently within thin multilayer coatings containing tiny light-detecting silver halide crystals and chemistry. Many subdisciplines within classical optics, solid-state theory, and photochemistry describe these processes.

In Chap. 20 of Ref. 1, Altman sweeps broadly through the basic concepts and properties of photographic materials. His brief, high-level descriptions are still generally valid for today's photographic products, and the reader will profit by reviewing that summary. This chapter focuses more sharply on four fundamental photographic technologies intimately related to the broad field of optics, then gives an overview of photographic materials available today.

Section 6.2 discusses the optical properties of multilayer color photographic films and papers. This section outlines the basic structure of the device and describes the principal function of each layer. Light absorption, reflection, scattering, and diffraction properties are discussed in the context of providing minimum optical distortion to the final image.

Silver halide light detection crystals are solid-state devices described by quantum solid-state photophysics and chemistry in addition to crystallography. These devices absorb light to create an electron-hole pair. Delocalization of the electron in the conduction band of the crystal, trapping, reduction of interstitial silver ions, nucleation and growth of clusters of silver atoms, and reversible regression of these phenomena must be controlled and optimized to produce the most sensitive light detectors that also possess long-term stability.

Section 6.3 describes the basic photophysics of silver halides according to our best understanding today. It outlines the solid-state properties most important to image detection and amplification. Surface chemical treatment and internal doping are discussed in the context of providing silver halide emulsions having the highest sensitivity to light.

A color photographic image is formed by high-extinction and high-saturation dyes structurally designed by chemists to optimize their color, permanence, and other useful characteristics. Their properties in both the ground state and photoexcited states are important to color and stability.

Section 6.4 briefly outlines the photochemistry of photographic image dyes. Excited-state properties are described that are important to the photostability of these dyes for image permanence.

Color science guides us to optimize all factors in photographic materials conspiring to render an accurate and pleasing image. These include spectral discrimination during the light detection phase and color correction to compensate for imperfect spectral detection and imperfect image dyes.

Section 6.5 sketches the photophysics and color science of photographic spectral sensitizers with an aim toward describing how modern photographic films sense the world in color nearly as the human eye sees it.

Today there exists a large diversity of photographic films. In addition to different manufacturers' brands, films differ by speed, type, color attributes, format, and a multitude of other factors. This can be confusing. And what are the differences between a consumer film bought in a drugstore and a more expensive professional film?

Section 6.6 gives an overview of the different types of films available today. Differences between high- and low-speed films are described with an understanding of the origins of these differences. Consumer films are compared to professional films and some of the special needs of each type of customer are described. Some general guidelines are offered for film choices among color reversal films, black-and-white films, and color negative films.

In addition to Chap. 20 in Ref. 1, several other texts contain useful information about photographic materials. Kapecki and Rodgers[2] offer a highly lucid and digestible sketch of basic color photographic film technology. Besides Chap. 20 in Ref. 1, this is a good place for the technically astute but nonpractitioner to gain a high level understanding of basic color photography. The technically detailed treatise by James[3] is perhaps the best single comprehensive source containing the science and technology of the photographic system outside photographic manufacturers' internal proprietary libraries. Hunt[4] provides a good comprehensive treatise on all aspects of color science and additionally gives a good technical review of color photographic materials. Chapter 6 in Ref. 1 contains a useful overview of the theory of light-scattering by particles, Chap. 9 contains a good overview of optical properties of solids, and Chap. 26 on colorimetry provides a good introduction to the basic concepts in that field.

6.2   THE OPTICS OF PHOTOGRAPHIC FILMS AND PAPERS

Introduction

Color photographic materials incorporate design factors that optimize their performance across all features deemed important to customers who use the product. These materials are complex in composition and structure. Color films typically contain over 12 distinct optically and chemically interacting layers. Some of the newest professional color reversal films contain up to 20 distinct layers.* Each layer contributes a unique and important function to the film's final performance. Careful design and arrangement of the film's various layers ultimately determine how well the film satisfies the user's needs.

One very important customer performance feature is film *acutance,* a measure of the film's ability to clearly record small detail and render sharp edges. Because images recorded on color film are frequently enlarged, image structure attributes such as acutance are very important. Magnifications such as those used to place the image on a printed page challenge the film's ability to clearly record fine detail.

The ability of a photographic material to record fine detail and sharp edges is controlled by two factors. The first includes light scatter, diffraction, and reflections during the exposure step. These are collectively called *optical* effects. This factor dominates the film's ability to record fine detail and sharp edges and is therefore critical to the film's design.

---

* Fujichrome Velvia 50 RVP professional film.

Chemical adjacency effects collectively known as *Eberhard effects* (see Chap. 21 in Ref. 3) are the second factor. In today's color and black-and-white film products these are present during film development and are caused by accumulating development by-products such as iodide ions or inhibitors released by development inhibitor–releasing (DIR) chemicals that restrain further silver development in exposed regions of the image, leading to a chemical unsharp mask effect (pp. 274 and 365 of Ref. 4). These chemical signals also give rise to the film's interlayer interimage effects (IIE; see p. 278 of Ref. 4) used for color correction. Film sharpness and IIE are strongly related.

This section describes multilayer factors that contribute to light scatter, reflection, diffraction, and absorption in photographic materials. Factors that influence the nonoptical part of film acutance, such as Eberhard effects, were briefly reviewed by Altman.[1] More information about these processes can be found in references cited therein.

Structure of Color Films

Color film structures contain image-recording layers, interlayers, and protective overcoat layers. These layers suspend emulsions and chemistry in a hardened gelatin binder coated over a polyacetate or polyester support. Figure 1 shows the principal layers in a color multilayer film structure.

The overcoat and ultraviolet (UV) protective layers contain lubricants; plastic beads 1 to 5 μm in size called *matte* to impart surface roughness that prevents sticking when the film is stored in roll form; UV-light-absorbing materials; and other ingredients that improve the film's handling characteristics and protect the underlying light-sensitive layers from damage during use and from exposure to invisible UV light. Ultraviolet light is harmful for two reasons: it will expose silver halide emulsions, thereby rendering an image from light invisible to humans, and it promotes photodecomposition of image dyes, leading to dye fade over time (p. 977 of Ref. 2). Visible light exposure must first pass through these overcoats, but they typically do little harm to acutance as they contain nothing that seriously scatters visible light.

Overcoat
UV Protective Layer
Fast Yellow Layer
Slow Yellow Layer
Yellow Filter Layer
Barrier Layer
Fast Magenta Layer
Mid Magenta Layer
Slow Magenta Layer
Optional Magenta Filter Layer
Barrier layer
Fast Cyan Layer
Mid Cyan Layer
Slow Cyan Layer
AHU Layer
Plastic Support
Optional Pelloid AHU Layer

FIGURE 1   Simplified diagram showing the key layers in a color photographic film (not drawn to scale).

The blue-light-sensitive yellow dye imaging layers appear next. Silver halides, with the exception of AgCl—used primarily in color papers—have an intrinsic blue sensitivity even when spectrally sensitized to green or red light. Putting the blue-light-sensitive layers on top avoids incorrect exposure leading to color contamination because a blue-light-absorbent yellow filter layer, located beneath the blue-sensitive imaging layers, absorbs all blue light that has passed completely through the blue-sensitive layers.

A collection of layers adjacent to one another that contain silver halide sensitized to a common spectral band is called a *color record.* Color records in films are always split into two or three separate layers. This arrangement puts the highest-sensitivity emulsion in the upper (fast) layer, where it gets maximum light exposure for photographic speed, and places low-sensitivity emulsions into the lower layer (or layers), where they provide exposure latitude and contrast control.

One or more interlayers separate the yellow record from the green-light-sensitive magenta record. These upper interlayers filter blue light to avoid blue light exposure color contaminating the red- and green-sensitive emulsion layers below. This behavior is called *punch through* in the trade. These interlayers also contain oxidized developer scavengers to eliminate image dye contamination between color records caused by oxidized color developer formed in one color record diffusing into a different color record to form dye.

The traditional yellow filter material is Carey Lea silver (CLS), a finely dispersed colloidal form of metallic silver, which removes most of the blue light at wavelengths less than 500 nm. The light absorption characteristics of this finely dispersed metallic silver are accounted for by Mie theory.[6] Unfortunately this material also filters out some green and red light, thus requiring additional sensitivity from the emulsions below. Bleaching and fixing steps in the film's normal processing remove the yellow CLS from the developed image.

Some films incorporate a yellow filter dye in place of CLS. Early examples contained yellow dyes attached to a polymer mordant in the interlayer. A *mordant* is a polymer that contains charged sites, usually cationic, that binds an ionized anionic dye by electrostatic forces. These dyes are removed during subsequent processing steps. Although these materials are free from red and green light absorption, it is a challenge to make them high in blue light extinction to avoid excessive chemical loads in the interlayer with consequent thickening.

Most recently, solid-particle yellow filter dyes[7] are seeing more use in modern films. These solid dyes are sized to minimize light scatter in the visible region of the spectrum and their absorption of blue light is very strong. They can be made with very sharp spectral cuts and are exceptionally well suited as photographic filter dyes. In some films solid-particle magenta filter dyes[8] are also used in the interlayers below the magenta imaging layers. Solid-particle filter dyes are solubilized and removed or chemically bleached colorless during the film's normal development process.

Because the human visual system responds most sensitively to magenta dye density, the green-light-sensitive image recording layers are positioned just below the upper interlayers, as close as practical to the source light exposure side of the film. This minimizes green light spread caused when green light passes through the turbid yellow record emulsions. It gives the maximum practical film acutance to the magenta dye record.

A lower set of interlayers separates the magenta record from the red-light-sensitive cyan dye record. These lower interlayers give the same type of protection against light and chemical contamination as do the upper interlayers. The magenta filter dye is absent in some films because red-light-sensitive emulsions are not as sensitive to green light exposure as to blue light exposure.

Located beneath the cyan record, just above the plastic film support, are antihalation undercoat (AHU) layers. The black absorber contained in this bottom layer absorbs all light that has passed completely through all imaging layers, thereby preventing its reflection off the gel-plastic and plastic-air interfaces back into the imaging layers. These harmful reflections cause a serious type of light spread called *halation,* which is most noticeable as a halo around bright objects in the photographic image.

The opacity required in the AHU is usually obtained by using predeveloped black filamentary silver, called *gray gel,* which is bleached and removed during the normal photographic processing steps. Alternatively, in many motion picture films a layer of finely divided carbon suspended in gelatin (*rem jet*) is coated on the reverse side of the film. When placed in this position the layer is called an *AHU pelloid.* It is physically removed by scrubbing just before the film is developed. The newest motion picture films incorporate a black solid-particle AHU filter dye that is solubilized and removed during normal development of the film and does not require a separate scrubbing step.

The overall thickness of a film plays an important role in minimizing harmful effects from light scatter. Because color films are structured with the yellow record closest to the light exposure source, it is especially important to minimize thickness of all film layers in the yellow record and below it, because light scattered in the yellow record progressively spreads as it passes to lower layers. The cyan record shows the strongest dependence on film thickness because red light passes through both yellow and magenta record emulsions en route to the cyan record. Because both emulsions often contribute to red light scatter, the cyan record suffers a stronger loss in acutance with film thickness.

Structure of Color Papers

The optical properties of photographic paper merit special consideration because these materials are coated in very simple structures on a highly reflective white Baryta-coated paper support. *Baryta* is an efficient diffuse reflector consisting of barium sulfate powder suspended in gelatin that produces isotropically distributed reflected light with little absorption.

Photographic papers generally contain about seven layers. On top are the overcoat and UV protective layers, which serve the same functions as previously described for film.

The imaging layers in color papers contain silver chloride emulsions for fast-acting development, which is important to the industry for rapid throughput and productivity. Generally only one layer is coated per color record, in contrast to films, which typically contain two or three layers per color record. The order of the color records in photographic papers differs from that in films due mainly to properties of the white Baryta reflective layer.

Because color paper is photographically slow, exposure times on the order of seconds are common. Most light from these exposures reflects turbidly off the Baryta layer. The imaging layers getting the sharpest and least turbid image are those closest to the reflective Baryta where light spread is least, not those closest to the top of the multilayer as is the case with film.

In addition, the Baryta layer as coated is not smooth. Its roughness translates into the adjacent layer, causing nonuniformities in that layer. Because the human visual system is most forgiving of physical imperfections in yellow dye, the yellow color record must be placed adjacent to the Baryta to take the brunt of these imperfections.

The magenta color record is placed in the middle of the color paper multilayer, as close to the Baryta layer as possible, since sharpness in the final image is most clearly rendered by magenta dye. This leaves the cyan record nearest to the top of the structure, just below the protective overcoats.

The magenta color record in the middle of the multilayer structure is spaced apart from the other two color records by interlayers containing oxidized developer scavengers to prevent cross-record color contamination, just as in films. However, no filter dyes are needed in color paper because silver chloride emulsions have no native sensitivity to light in the visible spectrum.

Light Scatter by Silver Halide Crystals

Because they consist of tiny particles, silver halide emulsions scatter light. Scattering by cubic and cubo-octahedral emulsions at visible wavelengths is most intense when the emulsion dimension ranges from 0.3 to 0.8 µm, roughly comparable to the wavelengths of visible light. This type of scattering is well characterized by Mie[9] theory and has been applied to silver halides.[10] We are often compelled to use emulsions having these dimensions in order to achieve photographic speed.

For photographic emulsions of normal grain size and concentration in gelatin layers of normal thickness, multiple scattering predominates. This problem falls within the realm of radiative transfer theory. Pitts[11] has given a rigorous development of radiative transfer theory to the problem of light scattering and absorption in photographic emulsions. A second approach that has received serious attention is the Monte Carlo technique.[12] DePalma and Gasper[13] were able to obtain good agreement between their modulation transfer functions (MTFs) determined by a Monte Carlo calculation and experimentally measured MTFs for a silver halide emulsion layer coated at various thicknesses.

Scattering by yellow record emulsions is especially harmful because red and green light must first pass through this record en route to the magenta and cyan records below. All other things being equal, the amount of light scattered by an emulsion layer increases in proportion to the total amount of silver halide coated in the layer.

Color films are constructed with low-scattering yellow record emulsions whenever possible. The amount of silver halide coated in the yellow record is held to a minimum consistent with the need to achieve the film's target sensitometric scale in yellow dye.

High-dye-yield yellow couplers[14] having high coupling efficiency are very useful in achieving silver halide mass reductions in the yellow record of color reversal films. These provide high yellow dye densities per unit silver halide mass, thereby reducing the amount of silver halide needed to achieve target sensitometry. Some upper-scale (high-density) increase in granularity often results from using these couplers in a film's fast yellow layer, but the benefits of reduced red and green light scatter overcome this penalty, especially because the human visual system is insensitive to yellow dye granularity.

Tabular emulsion grains offer a way to achieve typical photographic speeds using large-dimension emulsions, typically 1.0 to 3.0 µm in diameter, although smaller- and larger-diameter crystals are sometimes used. These do not scatter light as strongly at high angles from normal incidence with respect to the plane of the film as do cubic or octahedral emulsions having comparable photographic speeds. However, diffraction at the edges and reflections off the crystal faces can become a detrimental factor with these emulsions.

Silver halide tabular crystals orient themselves to lie flat in the plane of the gelatin layer. This happens because shear stress during coating of the liquid layer stretches the layer along a direction parallel to the support and also because the water-swollen thick layer compresses flat against the support after it dries by a ratio of roughly 20:1.

Reflections can become especially harmful with tabular emulsion morphologies since light reflecting from the upper and lower faces of the crystal interferes, leading to resonances in reflected light. The most strongly reflected wavelengths depend on the thickness of the tabular crystal. The thickness at which there is a maximum reflectance occurs at fractional multiples of the wavelength[15] given by:

$$t = \frac{\left(m + \frac{1}{2}\right)\lambda}{2n}$$

where t is the thickness of the tabular crystal, λ is the wavelength of light, n is the refractive index of the crystal, and m is an integer.

In an extreme case, tabular emulsions act like partial mirrors reflecting light from grain to grain over a substantial distance from its point of origin. This is called *light piping* by analogy with light traveling through an optical fiber (see Fig. 2). It is especially serious in the cyan record, where red light often enters at angles with respect to perpendicular incidence caused by scattering in upper layers.

Another cause of light piping is the presence of small-grain emulsion contaminants within a tabular grain population. A highly turbid contaminant emulsion, even if present in small amounts, can scatter light at sufficient angles to induce severe light piping through the tabular grain matrix. Modern emulsion science takes great pains to produce tabular emulsions free from highly scattering contaminants.

Gasper[16] treated the problem of scattering from thin tabular silver halide crystals in a uniform gelatin layer. For very thin and large-diameter tabular silver bromide crystals (thickness <0.06 µm, diameter >1.0 µm) light at normal incidence is scattered almost equally in the forward and backward hemispheres. As the grain thickness increases there appears increasing bias for forward scatter in a narrow cone. The efficiencies for scatter and absorption were found to be independent of grain diameter for crystal diameters larger than ~1.0 µm. For such crystals, the backscatter and absorption efficiencies are approximately equal to the specular reflectance and absorption of a semiinfinite slab of the same material and thickness.

But the forward scattering efficiency does not approximate the specular transmittance of the semiinfinite slab as a result of interference between the directly scattered forward field and the mutually coherent unscattered field, a process analogous to diffraction that does not occur for the semiinfinite slab with no edge. The specific turbidity depends on grain thickness, but not diameter for diameter >1.0 µm.

A light-absorbing dye is sometimes added to a film to reduce the mean free path of scattered and reflected light leading to an acutance improvement. This improvement happens at the expense of photographic speed, since light absorbed by the dye is not available to expose the silver halide emulsions. However, the trade-off is sometimes favorable if serious light piping is encountered.

Light-absorbing interlayers between color records are sometimes used in color films to help eliminate the harmful effects of reflected light. For example, some films have a magenta filter dye interlayer between the magenta and cyan records. In addition to its usefulness in reducing light punchthrough (green light passing through the magenta record to expose the red-sensitized layers), this filter layer helps eliminate harmful reflections of green light off cyan record tabular emulsions, which bounce back into the magenta record. These reflections, although potentially useful for improving photographic green speed, can be harmful to magenta record acutance.

6.3   THE PHOTOPHYSICS OF SILVER HALIDE LIGHT DETECTORS

Chapter 20 in Ref. 1 gave a very brief sketch of the general characteristics of silver halide crystals used for light detection and amplification in photographic materials. These crystals, com-

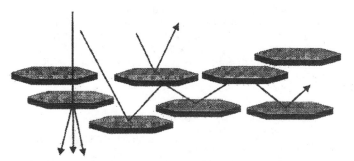

FIGURE 2   Light undergoes reflection, diffraction, and multiple reflections that may lead to light piping in film layers containing tabular silver halide crystals.

monly termed *emulsions,* are randomly dispersed in gelatin binder layers in photographic films and papers. For photographic applications the most commonly used halide types are AgCl, AgBr, and mixed halide solid solutions of $AgCl_xI_y$ and $AgBr_xI_y$. In addition, pure phases of AgCl and AgI are sometimes grown epitaxially on AgBr crystals to impart special sensitivity and development properties.

Upon exposure to light, silver halide crystals form a latent image (LI) composed of clusters of three to hundreds of photoreduced silver atoms either within the interior or most usefully on the surface of the crystal where access by aqueous developing agents is easiest.* Higher light exposures result in larger numbers of silver atoms per latent image cluster on average in addition to exposing a larger number of crystals on average.

The detection of light by a silver halide crystal, subsequent conversion to an electron hole pair, and ultimate reduction of silver ions to metallic silver atoms is in essence a solid-state photophysical process. The application of solid-state theory to the photographic process began with Mott and Gurney[17] and was extended by Seitz.[18] Additional reviews were published[19–21] as our understanding of these processes grew.

According to the nucleation and growth mechanism,[19] the photographic LI forms through a series of reversible electronic and ionic events (see Fig. 3). An electron is generated in the conduction band either by direct band gap photoexcitation of the crystal in its intrinsic spectral absorption region or by injection from a photoexcited spectral sensitizing dye on the crystal's surface. As this mobile electron travels through the crystal it becomes shallowly and transiently trapped at lattice imperfections carrying a positive charge. The ionized donor may be an interstitial silver ion or a surface defect. The kinked edge of a surface terrace, having a formal charge of $+\frac{1}{2}$, is one such site.

The electron may sample many such traps before becoming more permanently trapped at a preferred site. Trapping of the electron occurs in less than 0.1 ns at room temperature.[22] The shallowly trapped electron orbits about the trapping site in a large radius. At this stage the formal charge of the trap site becomes $-\frac{1}{2}$ and adjacent surface lattice ions relax, thereby increasing the depth of the trap.[22] In octahedral AgBr emulsions this shallow-deep transition occurs within 10 ns at room temperature.[22]

The negative charge formally associated with the trapped electron attracts a mobile interstitial Ag^+ ion, which the trapped electron reduces to an Ag^0 atom. The site's charge reverts to $+\frac{1}{2}$. A second photoelectron is then trapped at the site and reacts with a second interstitial Ag^+ ion to form Ag_2^0, and so on as additional photons strike the crystal. Frenkel defect concentrations—interior plus surface—in the range of $10^{14}/cm^3$ have been reported for AgBr crystals.[19] Latent image formation depends critically upon the existence of partially charged defects on the surface of the crystal.

Because the single-atom Ag^0 state is metastable and the Ag_2^0 state is not usefully detectable from gross fog by developing solutions (Chap. 4 of Ref. 3), a minimum of three photons must strike the crystal to form a detectable LI site, the second photon within the lifetime of the metastable trapped Ag^0 atom.† The transient existence of one or two silver atom sites has not been directly observed but is strongly inferred by indirect evidence.[21]

The highest photoefficiencies are achieved after silver halide emulsions are heated with aqueous sulfur and gold salts to produce Ag_2S and AgSAu species on the surface of the crystal.[23] Most of the evidence for sulfur sensitization suggests these sites improve electron trapping, perhaps by making trapping lifetimes longer at preexisting kink sites. Gold sensitization accomplishes at least two things: it reduces the size of the latent image center needed to cat-

---

* Some specialized developing agents can etch into the silver halide crystal to develop the internal latent image (LI), but the most commonly used color negative and color paper developers have little capability to do this. They are primarily surface developing agents. The color reversal black-and-white developer can develop slightly subsurface LI in color reversal emulsions.

† A metastable silver atom may dissociate back into an electron-hole pair, and the electron can migrate to another trapping site to form a new silver atom. This may happen many times in principle. The second electron must encounter the silver atom before electron-hole recombination or other irreversible electron loss takes place.

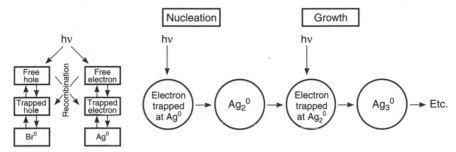

FIGURE 3 Diagram showing the basic features of the nucleation and growth mechanism for formation of a developable latent image site by light exposure of a silver halide crystal.

alyze development and it stabilizes silver atoms formed during the exposure process (Chap. 5 of Ref. 3).

Ideally only a single latent image site forms per exposed silver halide crystal. If more sites nucleate on a crystal, these can compete for subsequent photoelectrons, leading to losses in efficiency related to LI dispersity. One remarkable aspect of this process is that despite the large number of kink sites and chemically sensitized sites on a given crystal, in the most efficient silver halide emulsions only a single LI center forms per grain. The shallow-deep transition that occurs after the initial electron trapping selects one preferred defect out of the plethora of other possibilities for subsequent electron trapping.

The two predominant crystal faces that occur on photographically useful silver halide grains are [100] and [111]. Both surfaces have a negative charge[24] ranging from −0.1 to −0.3 V. A subsurface space charge layer rich in interstitial Ag+ ions compensates for the charged surface.[24] The [100] surfaces are flat with steps containing jogs that are either positively or negatively charged kink sites. The positive kinks on the surface form shallow electron traps.

In contrast, the [111] surface is not well characterized and is believed to be sensitive to surface treatment. Calculations of surface energies suggest that several arrangements may coexist, including microfaceted [100] planes and half layers of all silver or all halide ions in arrangements with hexagonal symmetry.[25,26]

All silver halides absorb light in the near-UV region of the spectrum. The absorption edge extends to longer wavelengths from AgCl to AgBr to AgI. AgCl and AgBr have indirect band gap energies of 3.29 and 2.70 eV, respectively.[20] Absorption of light produces electrons in a conduction band whose minimum lies at the zone center and holes in the valence band whose maximum is at an L point. Carriers produced at excess energy rapidly thermalize at room temperature since electron mobility is limited by phonon scattering.

Electron mobility in AgBr is about 60 cm^2/Vs at room temperature.[20] In AgCl the hole mobility is low, but in AgBr it is substantial—only about a factor of 30 lower than electron mobility.[21] Carriers generated by exposure will explore the crystal in a random walk as they scatter off phonons and transiently trap at charged defects. It is expected that electrons can fully sample a 1-μm crystal within less than 1 μs.

In most commercial photographic materials today, only about 25 percent to 50 percent of the electrons injected into the crystal's conduction band contribute to a developable LI site. The rest are wasted through electron-hole recombination, formation of internal LI that is inaccessible to the developing agent, or formation of multiple nucleation sites termed *dispersity*. The deliberate addition of dopants that act as shallow electron traps reduces the time electrons spend in the free carrier state and thereby limits their propensity to recombine with trapped holes.

Polyvalent transition metal ions are frequently doped into silver halide crystals to limit reciprocity effects, control contrast, and reduce electron hole recombination inefficiencies.[27]

They act as electron or hole traps and are introduced into the crystal from aqueous hexa-coordinated complexes during precipitation. They generally substitute for $(AgX_6)^{5-}$ lattice fragments. Complexes of Ru and Ir have been especially useful. Because the dopant's carrier-trapping properties depend on the metal ion's valence state and the steriochemistry of its ligand shell, there are many opportunities to design dopants with specific characteristics. Dopants incorporating Os form deep electron traps in AgCl emulsions with an excess site charge of +3. An effective electron residence lifetime of 550 s has been measured for these dopants at room temperature.[21] They are used to control contrast at high-exposure regions in photographic papers.

Quantum sensitivity is a measure of the average number of photons absorbed per grain to produce developability in 50 percent of an emulsion population's grains (Chap. 4 in Ref. 28). This microscopic parameter provides a measure of the photoefficiency of a given emulsion; the lower the quantum sensitivity value, the more efficient the emulsion. The quantum sensitivity measurement gives a cumulative measure of latent image formation, detection, and amplification stages of the imaging chain.

Quantum sensitivity has been measured for many emulsions. The most efficient emulsions specially treated by hydrogen hypersensitization yield a quantum sensitivity of about three photons per grain.[29] Although hydrogen hypersensitization is not useful for general commercial films because it produces emulsions very unstable toward gross fog, this sensitivity represents an ambitious goal for practical emulsions used in commercial photographic products. The most efficient practical photographic emulsions reported to date have a quantum sensitivity of about five to six photons per grain.[30] The better commercial photographic products in today's market contain emulsions having quantum sensitivities in the range of 10 to 20 photons per grain.

6.4 THE STABILITY OF PHOTOGRAPHIC IMAGE DYES TOWARD LIGHT FADE

Introduction

Azomethine dyes are formed in most photographic products by reaction between colorless couplers and oxidized p-phenylenediamine developing agents to form high-saturation yellow, magenta, and cyan dyes (Chap. 12 in Ref. 3). Typical examples are shown in Fig. 4. The diamine structural element common to all three dyes comes from the developing agent, where R_1 and R_2 represent alkyl groups. The R group extending from the coupler side of the chromophore represents a lipophilic ballast, which keeps the dye localized in an oil phase droplet. In some dyes it also has hue-shifting properties.

Heat, humidity, and light influence the stability of these dyes in photographic films and papers.[31] Heat and humidity promote thermal chemical reactions that lead to dye density loss. Photochemical processes cause dyes to fade if the image is displayed for extended periods of time. Ultraviolet radiation is especially harmful to a dye's stability, which is partly why UV absorbers are coated in the protective overcoat layers of a color film or paper.

Stability toward light is especially important for color papers, where the image may be displayed for viewing over many years. It is less important in color negative film, which is generally stored in the dark and where small changes in dye density can often be compensated for when a print is made. Light stability is somewhat important for color reversal (slide and transparency) film, since a slide is projected through a bright illuminant, but projection and other display times are short compared to values for color papers. A similar situation exists for movie projection films, where each frame gets a short burst of high-intensity light but the cumulative exposure is low.

Evaluation of the stability of dyes toward light is difficult because the time scale of the photochemical reactions, by design, is very slow or inefficient. The quantum yields of photochemical fading of photographic dyes, defined as the fraction of photoexcited dyes that fade,

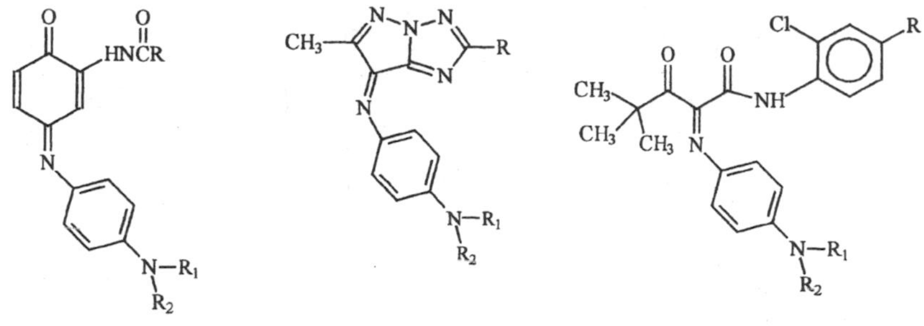

Cyan Dye   Magenta Dye   Yellow Dye

FIGURE 4 Typical examples of azomethine dyes used in photographic materials.

are on the order of 10^{-7} or smaller.[2,32] Accelerated testing using high-intensity illumination can sometimes give misleading results if the underlying photochemical reactions change with light intensity (p. 266 in Ref. 4).

Given that dyes fade slowly over time, it is best if all three photographic dyes fade at a common rate, thereby preserving the color balance of the image (p. 267 in Ref. 4). This rarely happens. The perceived light stability of color photographic materials is limited by the least stable dye. Historically, this has often been the magenta dye, whose gradual fade casts a highly objectionable green tint to a picture containing this dye. This is most noticeable in images containing memory colors, such as neutral grays and skin tones.

Photochemistry of Azomethine Dyes

Upon absorption of a photon, a dye becomes reversibly excited to the singlet state. For azomethine dyes, the lifetime of the excited state is estimated to be on the order of picoseconds.[33–36] Similarly, the lifetime of the triplet state of azomethine dyes, which is expected to form rapidly by intersystem crossing from the singlet excited state, has been estimated to be in the nanosecond range.[37] The short lifetime of these species seems to be at the basis of the observed low quantum yields of photochemical fading of azomethine dyes.

The nature of the initial elementary reactions involving excited or triplet states of dyes is not well understood. For some magenta dyes, the fading mechanism was reported to be photooxidative.[38] Cyan dye light fade appears to involve both reductive and oxidative steps[39] following photoexcitation of the dye.

Singlet oxygen has been traditionally postulated to be involved in the oxidative pathway.[40] This high-energy species can, in principle, be formed by energy transfer between a triplet dye and molecular oxygen, and can subsequently induce dye destruction, although reaction with singlet oxygen does not always lead to decomposition.[41]

$Dye + h\nu \star {}^{1}Dye^{*}$	Formation of an excited state by light absorption
${}^{1}Dye^{*} \rightarrow {}^{3}Dye$	Formation of triplet state by intersystem crossing
${}^{3}Dye + {}^{3}O_2 \star Dye + {}^{1}O_2$	Formation of singlet oxygen
${}^{1}O_2 + Dye \rightarrow$ Chemical reaction	Dye fade by singlet oxygen

Studies of the photophysical properties of photographic dyes, however, have shown that in general, azomethine dyes are good quenchers of singlet oxygen.[42] This implies that the formation of singlet oxygen by the triplet state of the dyes should be inefficient. An alternative feasible role for molecular oxygen in dye fade involves electron transfer mechanisms.

^{1}Dye* + O_2 ★ Dye$^+$ + O_2^- (Radical formation) $\Rightarrow$ subsequent reactions leading to dye destruction

^{3}Dye + O_2 ★ Dye$^+$ + O_2^- (Radical formation) $\Rightarrow$ subsequent reactions leading to dye destruction

The chemistry of dye fade may also be started by electron transfer between excited or triplet dye and a dye molecule in its ground state.

^{1}Dye* + Dye ★ Dye$^+$ + Dye$^-$ (Radical formation) $\Rightarrow$ subsequent reactions leading to dye destruction

^{3}Dye + Dye ★ Dye$^+$ + Dye$^-$ (Radical formation) $\Rightarrow$ subsequent reactions leading to dye destruction

Excited State Properties

The few papers that have been published about excited singlet-state properties of azomethine dyes have mainly focused on pyrazolotriazole magenta dyes.[33–36] These dyes have fluorescence quantum yields on the order of 10^{-4} at room temperature, but increase to ~1 in rigid organic glasses at 77K.[41] The fluorescence quantum yield is the fraction of photoexcited molecules that emit a quantum of light by fluorescence from an excited state to the ground state having the same multiplicity (usually singlet to singlet). The results at room temperature imply singlet-state lifetimes of only a few picoseconds.[41]

Room-temperature flash photolysis has identified a very short-lived fluorescent state plus a longer-lived nonfluorescent transient believed to be an excited singlet state whose structure is twisted compared to the ground state. This is consistent with the temperature-dependent results that allow rapid conformational change to the nonfluorescent singlet species at room temperature, in addition to intersystem crossing to the triplet excited state. But this conformational change is presumably restrained in a cold organic glass matrix, thereby allowing the excited singlet state time to fluoresce back to the ground state.

Investigation of the triplet states of azomethine dyes has proven difficult, and there are no definitive reports of direct observation of the triplet states. There are no reliable reports of phosphorescence from the triplet state of these dyes, although they have been studied by flash photolysis.[33–36,43–45] Using indirect energy transfer methods, triplet lifetimes have been estimated to be less than 10 ns.[37] Similar studies[44] gave triplet energies of 166 to 208 kJ mol^{-1} for yellow dyes, 90 to 120 kJ mol^{-1} for magenta dyes, and about 90 kJ mol^{-1} for cyan dyes.

Computational studies have been carried out recently on the ground and lowest triplet states of yellow azomethine dyes.[46] Consistent with crystallographic studies,[47,48]* the calculated lowest-energy ground state conformation S_0 is calculated to have the azomethine C$=$N bond coplanar with the anilide C$=$O carbonyl, but perpendicular to the C$=$O carbonyl of the ketone. The energy of the vertical transition triplet state having this conformation is quite high, calculated to be 230 kJ mol^{-1} above the ground state. Energy minimization on the lowest triplet energy surface starting with the geometry of S_0 results in a relaxation energy of 100 kJ mol^{-1} and leads to T_{1a}, characterized by a substantial twist around the C$=$N bond and by increased planarity of the keto carbonyl with the azomethine plane. These conformational changes offer an efficient mechanism for stabilizing the lowest triplet state in these dyes. A second triplet conformer T_{1b} with an energy 25 kJ mol^{-1} below T_{1a} was also calculated (see Fig. 5).

The low energy values calculated for T_{1a} and T_{1b} relative to S_{0a} (130 and 105 kJ mol^{-1}, respectively) have led to a new interpretation of the reported results[44] of energy transfer to S_{0a}. Using

---

* Other unpublished structures exhibiting the same conformation of S_{0a} have been solved at Kodak.

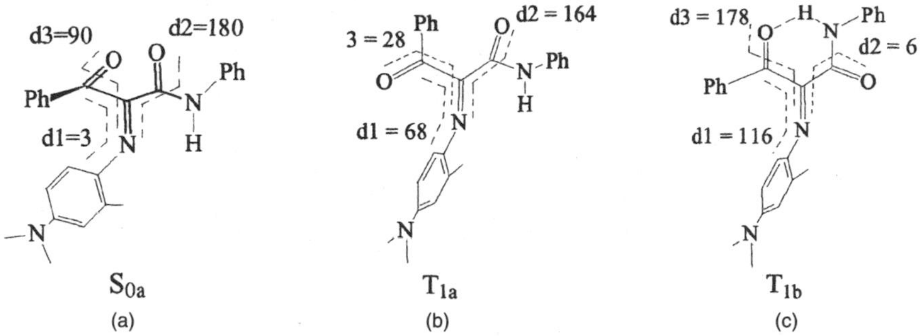

FIGURE 5 (*a*) Ground state conformation. (*b, c*) Two conformations of the lowest triplet state.

the nonvertical energy transfer model of Balzani,[49] it is shown that observed rates of energy transfer to S_{0a} can be consistent with formation of two distinct triplet states: one corresponding to a higher-energy excited triplet state (T_2) 167 kJ mol^{-1} above S_{0a}, and the second corresponding to one conformation of the lowest triplet state (T_1) with an energy of 96 kJ mol^{-1} above S_{0a}, in good agreement with the calculated T_{1b} configuration. The large structural differences between S_{0a} and the conformers of T_1 can explain the lack of phosphorescence in this system.

Light Stabilization Methods

Stabilization methods focus on elimination of key reactants (such as oxygen by various barrier methods), scavenging of reactive intermediates such as free radicals transiently formed during photodecomposition, elimination of ultraviolet light by UV absorbers, or quenching of photoexcited species in the reaction sequence.

Stabilizer molecules[50] have been added to photographic couplers to improve the stability of their dyes toward light fade. Although the exact mode by which these stabilizers operate has not been disclosed, it is probable that some quench the photoexcited state of the dyes, thereby preventing further reaction leading to fading; some scavenge the radicals formed during decomposition steps; and some may scavenge singlet oxygen.

Polymeric materials added to coupler dispersions also stabilize dyes. These materials are reported to improve thermal dye decomposition[51] by physically separating reactive components in a semirigid matrix or at least decreasing the mobility of reactive components within that matrix. They may provide benefits for light stability of some dyes by a similar mechanism.

Photographic dyes sometimes form associated complexes or microcrystalline aggregates (p. 322 of Ref. 4). These have been postulated to either improve light stability by electronic-to-thermal (phonon) quenching of excited states, or to degrade light stability by concentrating the dye to provide more available dye molecules for reaction with light-induced radicals. Both postulates may be correct depending upon the particular dyes.

Modern photographic dyes have vastly improved light stabilities over dyes of the past. For example, the magenta dye light stability of color paper dyes has improved by about two orders of magnitude between 1942 and the present time.[52] New classes of dyes[53,54] have proved especially resistant to light fade and have made memories captured by photographic materials truly archival.

6.5 *PHOTOGRAPHIC SPECTRAL SENSITIZERS*

Introduction

Spectral sensitizing dyes are essential to the practice of color photography. In 1873, Vogel[55] discovered that certain organic dyes, when adsorbed to silver halide crystals, extend their light sensitivity to wavelengths beyond their intrinsic ultraviolet-blue sensitivity. Since then, many thousands of dyes have been identified as silver halide spectral sensitizers having various degrees of utility, and dyes exist for selective sensitization in all regions of the visible and near-infrared spectrum. These dyes provide the red, green, and blue color discrimination needed for color photography.

The most widely used spectral sensitizing dyes in photography are known as *cyanine dyes.* One example is shown in Fig. 6. These structures are generally planar molecules as shown.

A good photographic spectral sensitizing dye must adsorb strongly to the surface of the silver halide crystal. The best photographic dyes usually self-assemble into aggregates on the crystal's surface. Aggregated dyes may be considered partially ordered two-dimensional dye crystals.[56] Blue-shifted aggregates are termed *H-aggregates,* while red-shifted ones, which generally show spectral narrowing and exitonic behavior, are termed *J-aggregates.* Dyes that form J-aggregates are generally most useful as silver halide spectral sensitizers.

A good sensitizing dye absorbs light of the desired spectral range with high efficiency and converts that light into a latent image site on the silver halide surface. The relative quantum efficiency[57] of sensitization is defined as the number of quanta absorbed at 400 nm in the AgX intrinsic absorption region to produce a specified developed density, divided by the number of like quanta absorbed only by dye within its absorption band. For the best photographic sensitizing dyes this number is only slightly less than unity.

Adsorption isotherms for J-aggregated cyanine dyes on AgX follow Langmuir behavior with a saturation coverage that indicates closely packed structures.[58] The site area per molecule derived from the Langmuir isotherms implies dye adsorption along its long axis with heterocyclic rings perpendicular to the surface. Recent scanning tunneling microscope work[59,60] has convincingly detailed the molecular arrangement within these aggregates, confirming general expectations.

The Photophysics of Spectral Sensitizers on Silver Halide Surfaces

Sensitizing dyes transfer an electron from the dye's excited singlet state to the conduction band of AgX. Evidence for this charge injection comes from extensive correlations of sensitization with the electrochemical redox properties of the dyes.[23,61,62] In the adsorbed state a good sensitizing dye has an ionization potential smaller than that of the silver halide being sensitized. This allows the energy of the dye excited state to lie at or above the silver halide conduction band even though the energy of the photon absorbed by the dye is less than the AgX band gap. Dye ionization potentials are generally well correlated with their electrochemical oxidation potentials.

The triplet energy level of a typical cyanine dye lies about 0.4 to 0.5 eV (about 35 to 50 kJ mol^{-1}) below the radiationally excited singlet level.[63] In most cases, this triplet state seems

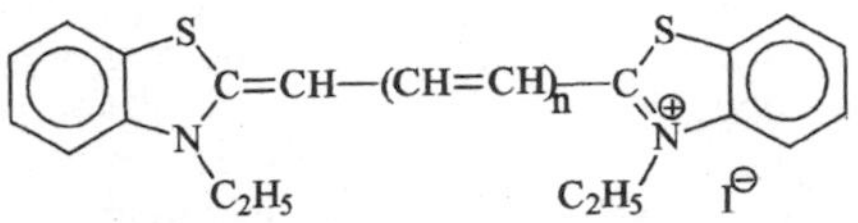

FIGURE 6 Typical cyanine spectral sensitizing dye used in photographic materials.

to have little effect on spectral sensitization at room temperature, although it may influence sensitization in special cases.[64]

Electron transfer takes place by tunneling from the excited-state dye to the silver halide conduction band. Simple calculations verify that penetration of the potential barrier will compete favorably with de-excitation of the dye by fluorescence emission (p. 253 of Ref. 3). The interaction time between an excited spectral sensitizer and silver bromide appears to occur between 10^{-13} and 10^{-10} (p. 237 of Ref. 3). Factors favoring irreversible flow of electrons from dye to silver halide are delocalization within the conduction band, the negative space charge layer on the surface of the crystal, trapping by remote sites on the silver halide surface, and irreversible trapping to form a latent image.

Free electrons from dye sensitization appear in the silver halide conduction band having the same mobility and lifetime as those formed by intrinsic absorption. These electrons participate in latent image formation by the usual mechanism.

After electron transfer, the oxidized dye radical cation becomes the hole left behind. Because there is evidence that a single dye molecule may function repeatedly,[65] the dye cation "hole" must be reduced again. This can occur by electron tunneling from an occupied site on the crystal's surface, whose energy level is favorable for that process. A bromide ion at a kink is one such site. This leaves a trapped hole on the crystal surface that may be thermally liberated to migrate through the crystal. This hole can lead to conduction band photoelectron loss by recombination. A supersensitizer molecule (discussed later) may also trap the hole.

The formation of the J-aggregate exciton band has significant effects on the excited-state dynamics of the dye. There is the possibility of coherent delocalization of the exciton over several molecules in the aggregate.[66] The exciton is mobile and can sample over 100 molecules within its lifetime.[67] This mobility means that the exciton is sensitive to quenching by traps within the aggregate structure. The overall yield of sensitization generally increases to maximum, then decreases somewhat as the aggregate size increases. Optimum sensitizations usually occur below surface monolayer coverage.

Sometimes the spectral sensitivity of a dye is increased by addition of a second substance. If the added sensitivity exceeds the sum of both sensitizers individually, the increase is superadditive and the second substance is called a *supersensitizer*. Maximum efficiency of a supersensitizer often occurs in the ratio of about 1:20 where the supersensitizer is the dilute component.

The supersensitizer may increase the spectral absorption of the sensitizer by inducing or intensifying formation of a J-aggregate. In some cases these changes are caused by a mutual increase in adsorption to the grain surface, as when the sensitizer and supersensitizer are ions of opposite charge.[68] However, the most important supersensitizers appear to increase the fundamental efficiency of spectral sensitization as measured by the relative quantum yield.

The low concentration of supersensitizer relative to sensitizer suggests trapping of an exciton moving through the quasicrystalline array of J-aggregated sensitizer molecules. One hypothesis is that the supersensitizer molecule traps the exiton, providing a site for facile electron transfer into the silver halide.[69]

Gilman and coworkers[70–73] proposed that the supersensitization of J-aggregating dyes takes place via hole trapping by the supersensitizer molecules. The exciton moving through the aggregate may be self-trapped at a site adjacent to a supersensitizer molecule. By hypothesis, the highest occupied (HO) electron energy level of the supersensitizer is higher than that of the partially empty HO level of the excited sensitizer molecule. An electron transfers from the supersensitizer to the sensitizer molecule. The sensitizer is reduced to an electron-rich, anion-free radical while the supersensitizer is oxidized to an electron-deficient cation radical of lower energy than the original hole, thereby localizing the hole on the supersensitizer (see Fig. 7).

Following electron transfer from the supersensitizer, the electron-rich free radical anion left on the sensitizer will possess an occupied electron level of higher energy than the excited level of the parent sensitizer. The reduction process thereby raises the level of the excited electron with respect to the conduction band of AgX, with a consequent increase in probability of electron tunneling into the silver halide conduction band.

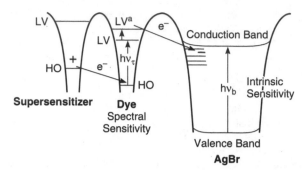

FIGURE 7 Simplified energy level diagram for a spectral sensitizer dye plus a supersensitizer dye adsorbed to a silver bromide surface.

The various proposed mechanisms of supersensitization are not mutually exclusive. Exciton trapping and hole trapping may operate together. Underlying all mechanisms are the roles of exciton migration throughout the aggregate, its interruption at or near the supersensitizer site, and a more facile transfer of the electron into the AgX conduction band at that localized state.

Spectral sensitizing dyes, especially at high surface coverage, desensitize silver halides in competition with their sensitization ability. Red spectral sensitizers generally desensitize more than green or blue spectral sensitizers do. Desensitization can be thought of as reversible sensitization. For example, a mobile conduction band electron can be trapped by a dye molecule to form a dye radical anion[74] or by reduction of a hole trapped on the dye.[64,75–77] Under normal conditions of film use (not in a vacuum), a dye radical anion may transfer the excess electron irreversibly to an O_2 molecule with formation of an O_2^- anion. Either postulate leads to irreversible loss of the photoelectron and consequent inefficiency in latent image formation.

Color Science of Photographic Spectral Sensitizers

Human vision responds to light in the range of 400 to 700 nm. The human eye is known to contain two types of light-sensitive cells termed *rods* and *cones,* so named because of their approximate shapes. Cones are the sensors for color vision. There is strong evidence for three types of cones sometimes termed long (L), middle (M), and short (S) wavelength receptors (Chap. 1 in Ref. 1). The normalized spectral responses of the human eye receptors (Chap. 26, Table 5 in Ref. 1) are shown in Fig. 8. The ability of humans to distinguish differences in spectral wavelength is believed to rely on difference signals created in the retina and optic nerve and interpreted by the brain from responses to light by the three cone detectors (Chap. 9 in Ref. 4).

The human eye's red sensitivity function peaks at around 570 nm and possesses little sensitivity at wavelengths longer than 650 nm. Also, there exists considerable overlap between red and green sensitivity functions, and to a lesser extent overlap between blue and green and blue and red sensitivity functions. This overlap in sensitivity functions, combined with signal processing in the retina, optic nerve, and brain, allows us to distinguish subtle color differences while simultaneously appreciating rich color saturation.

Photographic films cannot do this. The complex signal processing needed to interpret closely overlapping spectral sensitivity functions is not possible in a photographic material (Chap. 9 in Ref. 4). Some relief from this constraint might be anticipated by considering Maxwell's principle.[78]

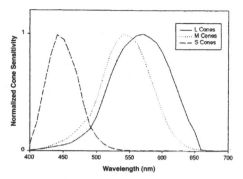

FIGURE 8 Normalized spectral sensitivity of the human eye.

Maxwell's principle states that any visual color can be reproduced by an appropriate combination of three independent color stimuli called *primaries*. The amount of each primary per wavelength needed to reproduce all spectral colors defines three color-matching functions (Chap. 6 in Ref. 1) for those primaries. If there exists a set of color-matching functions that minimize overlap between spectral regions, a photographic film with spectral sensitivity like that set of color-matching functions and imaging with the corresponding primary dyes would faithfully reproduce colors.

The cyan, magenta, and yellow dye primaries used in photographic materials lead to theoretical color-matching functions (Chap. 19-II in Ref. 3; Ref. 79)* having more separation than human visual sensitivity, but possessing negative lobes as shown in Fig. 9. This is impossible to achieve in practical photographic films (Chap. 9 in Ref. 4), although the negative feedback from IIE effects provides an imperfect approximation. Approximate color-matching functions have been proposed[80] that contain no negative lobes. But these possess a very short red sensitivity having a high degree of overlap with the green sensitivity, similar to that of the human eye.

It is therefore not possible to build a perfect photographic material possessing the simultaneous ability to distinguish subtle color differences, especially in the blue-green and orange spectral regions, and to render high-saturation colors. Compromises are necessary.

Most photographic materials intended to image colors at high saturation possess spectral sensitivities having small overlap, and in particular possess a red spectral sensitivity centered at considerably longer wavelengths than called for by theoretical color-matching functions. This allows clean spectral detection and rendering among saturated colors but sacrifices color accuracy.

For example, certain *anomalous reflection colors* such as the blue of the lobelia flower possess a sharp increase in spectral reflectance in the long red region not seen by the human eye but detected strongly by a photographic film's spectral sensitivity (see Fig. 10). This produces a serious color accuracy error (the flower photographs pink, not blue).[81] Among other such colors are certain green fabric dyes that image brown or gray. To correct for these hue errors, a special color reversal film was made whose general characteristic involves a short red spectral sensitivity.† This film does a considerably better job of eliminating the most serious hue errors, but unfortunately does not possess high color saturation. It is important in catalogue

---

* The theoretical color-matching functions were calculated using the analysis described in Ref. 79, Eq. (9a–c) for block dyes having trichromatic coefficients in Table II, and using the Judd-Vos modified XYZ color matching functions in Chap. 26, Table 2 of Ref. 1.
† Kodak Ektachrome 100 Film, EPN—a color reversal professional film.

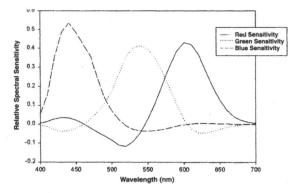

FIGURE 9 Spectral sensitivity for idealized photographic dyes. Note the large negative lobe around 525 nm.

and other industries where accurate color reproduction is of paramount importance, but most consumers prefer saturated colors.

More recently, color negative films possessing short red spectral sensitivities combined with powerful development inhibitor anchimeric releasing (DIAR) color correction chemistry and masking coupler technology have produced films* simultaneously having high color saturation plus accurate color reproduction, at least toward anomalous reflection colors. In addition, these films render improved color accuracy under mixed daylight plus (red-rich) incandescent lighting conditions. These films maintain the high color saturation capability important to consumers but also provide more accurate color hues—although they are not perfect.

A new type of color negative film has been described[82] that possesses a fourth color record whose function approximates the large negative lobe in the theoretical red color-matching function of photographic dyes. Several embodiments have been described and are employed in modern films. The most practical utilizes a fourth color record containing silver halide emulsions sensitive to short green light plus a magenta DIR color correction coupler.†

---

* Kodak Gold 100, Kodak Gold 200, and Kodak Max 400 color negative films.
† Fuji Reala 100 Film; Fuji Superia 200 and 400 film; Fuji Nexia 200 and 400 film; and Fuji 160 NPL, NPS, and NC professional films. These are all color negative films.

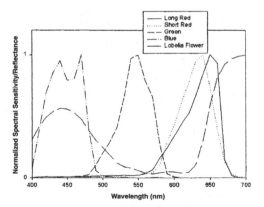

FIGURE 10 Spectral sensitivity for a real photographic film plus reflectance of the lobelia flower.

In simple terms, this special color record responds to short green light in the negative lobe region by releasing a development inhibitor that travels into the red-sensitive cyan color record, thereby reducing cyan dye in response to short green light, an effect opposite to that of normal red light exposure. This film's advantages reside in its ability to more accurately record colors in the blue-green spectral region, plus improved color accuracy under certain fluorescent lights that have a pronounced blue-green emission component.

Modern films are moving toward shorter-wavelength red spectral sensitivity. This allows improved color accuracy by shifting the red spectral sensitivity closer to that of the human eye. In addition, films have been designed to crudely approximate the principal negative lobe in the theoretical red color-matching function for photographic dye primaries, thereby producing a more accurate rendition of certain blue-green colors. Although not perfect, these techniques allow modern photographic films to record the world of color in all its richness and subtleties nearly as we ourselves see it.

6.6 GENERAL CHARACTERISTICS OF PHOTOGRAPHIC FILMS

Today there is available a large variety of films including different film speeds, types, and sizes. Many films are offered in both professional and consumer categories. This section describes some general characteristics of films as they relate to film speed and type. Some unique characteristics of professional films are contrasted to consumer films. Finally, color reversal, black-and-white, and color negative films are compared in more detail.

Low-Speed Films Compared to High-Speed Films

Higher granularity, higher sensitivity to ambient background radiation, and higher sensitivity to x rays are fundamental penalties that come with higher photographic speed. High-speed films often suffer additional penalties compared to their lower-speed counterparts, but these additional penalties are not intrinsic. They represent conscious choices made by the manufacturer nearly always traceable to the desire to minimize granularity.

As one example, 400-speed color slide film generally has lower color saturation than its 100-speed counterpart. The reduced color saturation comes as a result of lower interlayer interimage effects (IIE) caused by higher iodide content in the mixed silver bromo-iodide crystals, which, in turn, is needed to achieve higher speeds from a smaller sized crystal and is ultimately done to reduce the photographic granularity penalty.

Photographic Speed and Photographic Granularity.   To understand the intrinsic connection between photographic speed and photographic granularity, recall that light quanta striking a flat surface follow Poisson statistics (Chap. 1 of Ref. 28). Therefore, a uniform light exposure given to a photographic film delivers photon hits Poisson-distributed across the plane of the film. Silver halide crystals are on the order of 1 μm^2 in projected area. On average in a film exposure (the product of the intensity of the radiation and the duration) appropriate for an EI100-speed film, the most sensitive silver halide crystals receive:*

E_q(red light; 650 nm)	~48 photons per square micrometer
E_q(green light; 550 nm)	~40 photons per square micrometer
E_q(blue light; 450 nm)	~33 photons per square micrometer

---

* Analysis based on a typical EI100-speed color reversal film where the exposure falls near the photographically fastest, most sensitive part of the sensitometric curve. This analysis is described on p. 47 of Ref. 28.

Because the most efficient practical photographic emulsions to date have a quantum sensitivity of about five photons per grain,[30] these approximate numbers suggest there is enough light (on average) striking a film during an EI100-speed film exposure to form latent image on efficient 1-μm^2 silver halide grains. But an EI800-speed exposure at three stops down begins to push the limits of modern emulsion efficiency at five green photons per square micron.

Silver halide crystals within the film are light collectors. Just as large-area mirrors in telescopes collect more photons and are more sensitive detectors of starlight than smaller mirrors, all other things being equal, larger silver halide crystals likewise collect more photons to become more sensitive light detectors.

Strong correlates exist between silver halide surface area and photographic speed. Photographic speed often linearly follows the average surface area of the silver halide crystal population (p. 969 of Ref. 2): speed $\sim \bar{a}$.

According to the well-known Siedentopf formula (p. 60 in Ref. 28), the root-mean-square photographic granularity at density D is given by $G \sim D\hat{a}$. In this formula, the symbol $\hat{a}$ represents the average cross-sectional area of the image particle projected onto the film plane. To a good approximation, this is proportional (not necessarily linear) to the mass of developed silver regardless of whether the image particle is a dye cloud or silver deposit.

These considerations demonstrate the intrinsic and well-known correlation between photographic speed and photographic granularity. For a given crystal morphology, as the size of the silver halide crystal increases, both surface area and mass increase, leading to a speed plus granularity increase.

At a fundamental level, to lower the granularity of a higher-speed photographic film the sensitivity to light of the silver halide crystals must improve. Then smaller silver halide crystals can be used to attain speed. Factors that contribute to this problem were discussed in Sec. 6.2.

In general, larger silver halide crystals suffer more inefficiencies in the latent-image-forming process because latent-image dispersity, electron-hole recombination, and other irreversible photon waste processes become more problematic as the grain size increases (p. 993 in Ref. 2). Overcoming these inefficiencies at all speeds, but especially at high photographic speeds, is a major challenge of modern photographic science.

Photographic Speed and Sensitivity to High-Energy Radiation.   High-energy subatomic particles induce developable latent image formation in photographic films. The process by which high-energy charged particles produce latent image usually involves energy transfer to valence band electrons by inelastic scattering. Some of these scattered electrons are promoted into the silver halide conduction band and go on to produce latent image by the usual processes. Higher-speed films are more sensitive to this radiation because their larger silver halide crystals are hit (on average) more frequently and more collisions occur by passage through a larger crystal than through a smaller crystal.

Photographic films are slowly exposed by ambient background radiation (Chap. 23 in Ref. 3) caused by cosmic rays; ubiquitous stone and masonry building materials, which contain trace amounts of radioactive elements; and other low-level radiation sources. These radiation sources emit various types of charged and uncharged high-energy particles and photons including alpha particles, neutrons, γ rays, β particles, muons, and others. This results in a shorter ambient shelf life for very-high-speed films unless special storage precautions are taken.

Photographic emulsions are also sensitive to x rays. The formation of latent image silver by x rays (and γ rays) is due to the action of high-energy electrons emitted during the absorption or inelastic scattering of the electromagnetic radiation by matter (Chap. 23 in Ref. 3). These electrons may enter the conduction band of silver halide and proceed to latent image formation in the usual way. If the primary emitted electron is of sufficiently high energy, it may create secondary electrons of lower energy by inelastic scattering with other electrons. These secondary electrons enter the conduction band.

High-speed films have more sensitivity to x rays than do low-speed films. Some airline travelers ask airport security agents to hand-check their film rather than allowing it to pass through the baggage x-ray scanner. This precaution is prudent for higher-speed film because airport x-ray machines have become stronger emitters recently, especially at airports with special security concerns.

Despite the higher price, inherent penalties, and willful compromises that characterize high-speed films, they offer access to more picture space under low-light conditions. Among the advantages a high-speed film offers are the following:

1. Pictures under low light where flash is not permitted, such as theaters and museums.
2. Pictures under low light using slower lenses such as zoom and telephoto lenses.
3. Pictures under low light when the subject lies beyond the reach of flash.
4. Pictures under low light taken in rapid sequence without waiting for a flash to recharge.
5. Pictures under low light when flash disturbs the subject, as in some nature photography.
6. Pictures of people without flash. Flash is responsible for a major defect known as *redeye,* caused when light from a flash reflects off the retina of the eye.
7. Pictures using a stopped-down camera aperture for improved depth of field, so objects over a wider range of distance from the camera are imaged in focus.
8. Reduced dependence on flash for longer battery life.
9. Low-cost one-time-use cameras (OTUCs) when flash units are not needed.
10. Ability to freeze action using a rapid shutter speed.

For these and other reasons, a high-speed film may be the best choice for many film applications despite its shortcomings compared to lower-speed alternatives.

Professional Film Compared to Amateur Film

Most film manufacturers offer film in a professional category and an amateur (sometimes called *consumer*) category. These differ in many respects.

Consumer film is targeted for the needs of the amateur snap shooter. Drugstores, supermarkets, department stores, and other outlets frequented by nonprofessional photographers sell this film. Because most amateurs are price conscious, discount brands of film are available having some performance compromises that may not be important to some consumers.

To keep the price low, manufacturing tolerances for consumer film performance attributes are wider than those for professional film. In a well-manufactured film these departures from performance aims are small enough to go unnoticed by most average consumers but large enough to be objectionable to highly discerning professionals.

People want pleasing color in their pictures that renders the main subject attractive to the eye. For most amateurs this means saturated colors enhanced to some extent beyond their true appearance in the original scene. However, this preference is highly selective since oversaturated skin tones are objectionable. Other *memory colors*—those that we recognize and know what they should be, such as green grass and blue sky—must also be rendered carefully to produce the most pleasing color appearance (Chap. 5 in Ref. 4). The best consumer films accommodate these color preferences.

Exposures on a roll of consumer film are sometimes stored for long times. It is not unusual for amateurs to take pictures spanning two successive Christmas seasons on the same roll of film. Thus, latent image stability under a wide range of temperature and humidity conditions is very important to a good consumer film. On the other hand professional photographers develop their film promptly, making this requirement far less demanding in a professional film.

Most amateurs are not skilled photographers. It is advantageous for consumer film to be easy to use and provide good-looking pictures under a wide variety of conditions. Consumer film incorporates more tolerance toward over- and underexposures.

Film for amateurs is offered in three standard formats. The most common format is the 35-mm cassette. Consumer cameras that accept this film have become compact, lightweight, and fully automatic. The camera senses the film speed through the DX coding on the film cassette and adjusts its shutter, aperture, and flash accordingly. Autofocus, autorewind, and automatic film advance are standard on these cameras, making the picture-taking experience essentially point and shoot.

The new Advanced Photo System (APS) takes ease of use several steps further by incorporating simple drop-in film loading and a choice of three popular print aspect ratios from panoramic to standard size prints. A magnetic coating on the film records digital data whose use and advantages are just beginning to emerge.

The third format is 110 film. Although not as popular as 35-mm and APS, it is often found in inexpensive cameras for gift packages. Film for this format comes in a cartridge for easy drop-in loading into the camera.

Professional photographers demand more performance from their film than do typical amateurs. These people shoot pictures for a living. Their competitive advantage lies in their ability to combine technical skill and detailed knowledge of their film medium with an artistic eye for light, composition, and the color of their subject.

Film consistency is very important to professionals, and they pay a premium for very tight manufacturing tolerances on color balance, tone scale, exposure reciprocity, and other properties. They buy their film from professional dealers, catalogues, and in some cases directly from the manufacturer. Professionals may buy the same emulsion number in large quantity, then pretest the film and store it in a freezer to lock in its performance qualities. Their detailed knowledge about how the film responds to subtleties of light, color, filtration, and other characteristics contributes to their competitive advantage.

Professional film use ranges from controlled studio lighting conditions to harsh and variable field conditions. This leads to a variety of professional films designed for particular needs. Some photographers image scenes under incandescent lighting conditions in certain studio applications, movie sets, and news events where the lights are balanced for television cameras. These photographers need a tungsten-balanced film designed for proper color rendition when the light is rich in red component, but deficient in blue compared to standard daylight.

Catalogue photographers demand films that image colors as they appear to the eye in the original scene. Their images are destined for catalogues and advertisement media that must display the true colors of the fabric. This differs from the wants of most amateurs, who prefer enhanced color saturation.

Some professional photographers use high color saturation for artistic effect and visual impact. These photographers choose film with oversaturated color qualities. They may supplement this film by using controlled light, makeup on models, lens gels, or other filtration techniques to control the color in the image.

Professional portrait photographers need film that gives highly pleasing skin tones. Their films typically image at moderate to low color saturation to soften facial skin blotches and marks that most people possess but don't want to see in their portraits. In addition, portrait film features highly controlled upper and lower tone scale to capture the subtle texture and detail in dark suits and white wedding dresses and the sparkle in jewelry.

Professional photographers image scenes on a variety of film sizes ranging from standard 35-mm size to 120-size, and larger 4×5 to 8×10 sheet-film formats that require special cameras. Unlike cameras made for amateurs, most professional cameras are fully controllable and adjustable by the user. They include motor drives to advance the film rapidly to capture rapid sequence events. Professional photographers often take many pictures to ensure that the best shot has been captured.

Many specialty films are made for professional use. For example, some films are sensitized to record electromagnetic radiation in the infrared, which is invisible to the human eye. Note that these films are not thermal detectors. Infrared films are generally *false colored*, with the infrared-sensitized record producing red color in the final image, the red-light-sensitized record producing green color, and the green/blue-sensitized record producing blue color. Other professional films are designed primarily for lab use, such as duplicating film used to copy images from an original capture film.

Some film applications are important to a professional but are of no consequence to most amateurs. For example, professionals may knowingly underexpose their film and request *push processing,* whereby the film is held in the developing solution longer than normal. This brings up an underexposed image by rendering its midtone densities in the normal range, although some loss in shadow detail is inevitable. This is often done with color transparency film exposed under low-ambient-light conditions in the field. Professional films for this application are designed to provide invariant color balance not only under normal process times but also under a variety of longer (and shorter) processing times.

Multipop is another professional technique whereby film is exposed to successive flashes or *pops.* Sometimes the photographer moves the camera between pops for artistic effect. The best professional films are designed to record a latent image that remains nearly invariant toward this type of exposure.

Consumer film is made for amateur photographers and suits their needs well. Professional film comes in a wide variety of types to fit the varied needs of highly discerning professional photographers who demand the highest quality, specialized features, and tightest manufacturing tolerances—and are willing to pay a premium price for it.

Color Reversal Film

Color reversal films, sometimes called *color transparency* or *color slide* films, are sold in professional and amateur categories. This film is popular among amateurs in Europe and a few other parts of the world but has largely been replaced by color negative film among amateurs in North America. However, it is still very popular among advanced amateurs, who favor it for the artistic control this one-stage photographic system offers them, the ease of proofing and editing a large number of images, and the ability for large-audience presentations of their photographic work. Because the image is viewed by transmitted light, high amounts of image dye in these films can project intense colors on the screen in a dark auditorium for a very striking color impact.

Color reversal film is the medium of choice for most professional commercial photographers. It offers them the ability to quickly proof and select their most favored images from a large array of pictures viewed simultaneously on a light table. The colors are more intense compared to a reflection print, providing maximum color impact to the client. And professional digital scanners have been optimized for color reversal images, making it convenient to digitally manipulate color reversal images and transcribe them to the printed page. Today's professional imaging chain infrastructure has been built around color reversal film, although digital camera image capture is making inroads, especially for low-quality and low-cost applications.

Large-format color reversal film, from 120 size to large sheet formats, is often shot by professionals seeking the highest-quality images destined for the printed page. A larger image needs little or no magnification in final use and avoids the deterioration in sharpness and grain that comes from enlarging smaller images. Additionally, art directors and other clients find it easier to proof larger images for content and artistic appeal.

EI400-speed color reversal films generally possess low color saturation, low acutance, high grain, and high contrast compared to lower-speed color reversal films. The high contrast can lead to some loss in highlight and shadow detail in the final image. Except for very low-light situations, EI100 is the most popular speed for a color reversal film. It offers the ultrahigh image quality professionals and advanced amateurs demand.

Modern professional color reversal films maintain good color balance under push processing conditions. They also perform very well under a variety of exposure conditions including multipop exposures and very long exposure times.

When professional photographers believe nonstandard push or pull process times are needed, they may request a "clip test" (sometimes also called a "snip test") whereby some images are cut from the roll of film and processed. Adjustments to the remainder of the roll are based on these preprocessed images. Some photographers anticipate a clip test by deliberately shooting sacrificial images at the beginning of a roll.

Professional color reversal films can be segregated into newer modern films and older *legacy films*. Modern films contain the most advanced technology and are popular choices among all photographers. Legacy films are popular among many professional photographers who have come to know these films' performance characteristics in detail after many years of experience using them.

Legacy films continue to enjoy significant sales because the huge body of personal technical and artistic knowledge about a film accumulated by a professional photographer over the years contributes to his or her competitive advantage. This inertia is a formidable barrier to change to a new, improved, but different film whose detailed characteristics must be learned.

In some cases, the improved features found in modern films are not important to a particular professional need. For example, a large 4 × 5 sheet film probably needs little or no enlargement, so the poorer grain and sharpness of a legacy film may be quite satisfactory for the intended application. The detailed knowledge the professional has about how the legacy film behaves in regard to light and color may outweigh the image structure benefits in a modern film. Also, all films possess a unique tone scale and color palette. A professional may favor some subtle and unique feature in the legacy film's attributes.

Kodachrome is a special class of color reversal legacy films. Unlike all other types of color reversal films, this film contains no dye-forming incorporated couplers. The dyes are imbibed into the film during processing so the film's layers are coated very thin. This results in outstanding image sharpness. Moreover, many photographers prefer Kodachrome's color palette for some applications.

Kodachrome dyes are noted for their image permanence. Images on Kodachrome film have retained their colors over many decades of storage, making this film attractive to amateurs and professionals alike who want to preserve their pictures. However, modern incorporated coupler color reversal films also have considerably improved dark storage image stability compared to their predecessors.

Table 1 lists some color reversal films available today. This is not a comprehensive list. Most film manufacturers improve films over time, and portfolios change frequently.

TABLE 1   Color Reversal Capture Films

Modern Films	Legacy Films	Kodachrome Films	Tungsten Films
Kodak Ektachrome 100VS (very saturated color)	Fuji Velvia 50 RVP (very saturated color)	Kodachrome 25 PKM	Kodak Ektachrome 64T EPY
Kodak Ektachrome E100S (standard color, good skin tones)	Kodak Ektachrome 64 EPR (standard color)	Kodachrome 64 PKR	Fujichrome 64T RTP
Fuji Astia 100 RAP (standard color, good skin tones)	Kodak Ektachrome 100 Plus EPP (standard color)	Kodachrome 200 PKL	Agfachrome 64T
Fuji Provia 100 RDPII (standard color)	Kodak Ektachrome 100 EPN (accurate color)		Kodak Ektachrome 160T EPT
Agfachrome 100 (standard color)	Kodak Ektachrome 200 EPD (lower color saturation)		Kodak Ektachrome 320T EPJ
Kodak Ektachrome E200 (can push 3 stops)	Agfachrome 200		
	Kodak Ektachrome 400 EPL		
	Fuji Provia 400		
	Agfachrome 400		

Unless a photographer has some particular reason for choosing a legacy color reversal film, the modern films are generally a better overall choice. More detailed information about these films' uses and characteristics can be found in the film manufacturers' Web pages on the Internet.

Black-and-White Film

Black-and-white (B&W) films and papers are sold through professional product supply chains. Color films and papers have largely displaced B&W for amateur use, with the exception of advanced amateurs who occasionally shoot it for artistic expression.

The image on a B&W film or print is composed of black metallic silver. This image has archival stability under proper storage conditions and is quite stable even under uncontrolled storage. Many remarkable and historically important B&W images exist that date from 50 to over 100 years ago. This is remarkable considering the primitive state of the technology and haphazard storage conditions.

Pioneer nature photographer Ansel Adams reprinted many of his stored negatives long after he stopped capturing images in the field. This artistic genius recreated new expressions in prints from scenes captured many years earlier in his career because his negatives were well preserved.

Ansel Adams worked with B&W film and paper. Many photographic artists work in B&W when color would distract from the artistic expression they wish to convey. For example, many portraits are imaged in B&W.

B&W films pleasingly image an extended range of tones from bright light to deep shadow. They are panchromatically sensitized to render colors into grayscale tones. A B&W film's tone scale is among its most important characteristics.

Photographers manipulate the contrast of a B&W image by push- or pull-processing the negative (varying the length of time in the developer solution) or by printing the negative onto a select contrast paper. Paper contrast grades are indexed from 1 through 5, with the higher index giving higher contrast to the print. Multigrade paper is also available whereby the contrast in the paper is changed by light filtration at the enlarger.

It is generally best to capture the most desired contrast on the negative because highlight or shadow information lost on the negative cannot be recovered at any later printing stage. Paper grades are designed for pleasing artistic effect, not for highlight or shadow recovery.

Legacy films are very important products in B&W photography. Although typically a bit grainier and less sharp than modern B&W films, their tone scale characteristics, process robustness, and overall image rendition have stood the test of time and are especially favored by many photographers. Besides, the B&W photographic market is small so improvements to this line of films occur far less frequently than do improvements to color films.

Unlike the case with color films, which are designed for a single process, a photographer may choose among several developers for B&W films. Kodak Professional T-Max developer is a good choice for the relatively modern Kodak T-Max B&W film line. Kodak developer D-76 is also very popular and will render an image with slightly less grain and slightly less speed. It is a popular choice for legacy films. Kodak Microdol-X developer is formulated to give very low grain at the expense of noticeable speed loss. Kodak developer HC110 is popular for home darkrooms because of its low cost and ease of use, but it renders a grainier image compared to other developers. Other manufacturers, notably Ilford Ltd., carry similar types of B&W developers.

A new type of B&W film has emerged that is developed in a standard color negative process. This incorporated-coupler 400-speed film forms a black-and-white image from chemical dyes instead of metallic silver. Among its advantages are very fine grain at normal exposures in the commonly available color negative process. Its shortcomings are the inability to control contrast on the negative by push or pull processing and its objectionable high grain when underexposed compared to a comparably underexposed 400-speed legacy film. And many photographers prefer the tone scale characteristics found in standard B&W films.

Table 2 lists some B&W films available today. This is not a comprehensive list. Modern films generally feature improved grain and sharpness compared to legacy films. However, the

TABLE 2   Black and White Films

Modern Films	Legacy Films	Specialty Films
Kodak T-Max 100 Professional (fine grain, high sharpness)	Ilford Pan F Plus 50	Kodak Technical Pan 25 (fine grain and very high sharpness)
Ilford Delta 100 Pro (fine grain, high sharpness)	Kodak Plus-X 125	Kodak IR (infrared sensitive)
Kodak T-Max 400 Professional (finer grain compared to legacy films)	Ilford FP4 Plus 125	Ilford SFX 200 (extended long red sensitivity but not IR)
Ilford Delta 400 Pro (finer grain compared to legacy films)	Kodak Tri-X 400	Ilford Chromogenic 400 (incorporated couplers, color negative process)
Fuji Neopan 400 (finer grain compared to legacy films)	Ilford HP5 Plus 400	Kodak Professional T400 CN (incorporated couplers, color negative process)
Kodak T-Max P3200 Professional (push process to high speed, high grain)		
Ilford Delta P3200 Pro (push process to high speed, high grain)		

legacy films are favorite choices among many photographers because of their forgiving process insensitivity, the experience factor of knowing their detailed behavior under many conditions, and the characteristic look of their images. More detailed information about these films' uses and characteristics can be found in the film manufacturers' Web pages on the Internet.

Color Negative Film

Today's color negative film market is highly segmented. It can be most broadly divided into consumer films and professional films. The consumer line is further segmented into 35-mm film, 110 format film (a minor segment whose image size is 17×13 mm), single-use cameras, and the new Advanced Photo System (APS). Each segment serves the needs of amateur photographers in different ways. The basic films are common across segments, with the main difference being camera type.

Consumer color negative film is a highly competitive market with many film and camera manufacturers offering products. In general, films for this marketplace offer bright color saturation that is most pleasing to amateurs. The higher-speed films show progressively more grain than do the lower-speed films. The highest-speed films are less often used when enlargement becomes significant, such as in APS and 110 format, but are prevalent in single-use cameras.

Consumer 35-mm Films.   Films in 35-mm format form the core of all consumer color negative products. These same basic films are also found in single-use cameras, 110 format cameras, and APS cameras. Film speeds include 100, 200, 400, and 800. All manufacturers offer films at 100, 200, and 400 speed, but only a few offer films at speeds of 800 and above.

The 400-speed film is most popular for indoor shots under low light and flash conditions. Lower-speed films are most often used outdoors when light is plentiful.

The best consumer films feature technology for optimized color rendition including the high color saturation pleasing to most consumers, plus accurate spectral sensitization for realistic color rendition under mixed lighting conditions of daylight plus fluorescent and incandescent light, which is often present inside homes. Technology used for accurate spectral sensitization was briefly described for Fujicolor Superia and Kodak Gold films in Sec. 6.5.

Kodak consumer films segment into Kodak Gold and Kodak Royal Gold films. Kodak Gold consumer films emphasize colorfulness, while Kodak Royal Gold films emphasize fine grain and high sharpness.

Films made in 110 format generally fall into the 100- to 200-speed range because the enlargement factors needed to make standard-size prints place a premium on fine grain and high sharpness.

Single-Use Cameras.   The single-use camera marketplace is extremely competitive because of the high popularity of this film and camera system among amateurs. These low-cost units can be bought at all consumer outlets and are especially popular at amusement parks, theme parks, zoos, and similar family attractions. After the film is exposed, the photographer returns the entire film plus camera unit for processing. Prints and negatives are returned to the customer while the camera body is returned to the manufacturer, repaired as needed, reloaded with film, and repackaged for sale again. These camera units are recycled numerous times from all over the world.

Single-use cameras come in a variety of styles including a waterproof underwater system, low-cost models with no flash, regular flash models, an APS style offering different picture sizes, and a panoramic style. Single-use cameras typically contain 800-speed film, except for APS and panoramic models, which usually contain 400-speed film due to enlargement demands for finer grain.

Advanced Photo System (APS).   APS is the newest entry into consumer picture taking. The size of the image on the negative is 29×17 mm, about 60 percent of the image size of standard 35-mm film (image size 36×24 mm). This puts a premium on fine grain and high sharpness for any film used in this system. Not all film manufacturers offer it. Speeds offered are 100, 200, and 400, with the higher speeds being the most popular.

Some manufacturers offer a B&W film in APS format. The film used is the incorporated-coupler 400-speed film that forms a black-and-white image from chemical dyes and is developed in a standard color negative process.

The APS camera system features simple drop-in cassette loading. Unlike the 35-mm cassette, the APS cassette contains no film leader to thread into the camera. When loaded, the camera advances the film out of the cassette and automatically spools it into the camera, making this operation mistake-proof. Three popular print sizes are available; panoramic, classic, and high-definition television (HDTV) formats. These print sizes differ in their width dimension.

A thin, nearly transparent magnetic layer is coated on the APS film's plastic support. Digital information recorded on this layer, including exposure format for proper print size plus exposure date and time, can be printed on the back of each print. Higher-end cameras offer prerecorded titles—for example, "Happy Birthday"—that can be selected from a menu and placed on the back of each print. Additional capabilities are beginning to emerge that take more advantage of this magnetic layer and the digital information it may contain.

The APS system offers easy mid-roll change on higher-end cameras. With this feature, the last exposed frame is magnetically indexed so the camera "remembers" its position on the film roll. A user may rewind an unfinished cassette to replace it with a different cassette (different film speed, for example). The original cassette may be reloaded into the camera at a later time. The camera will automatically advance the film to the next available unexposed frame.

Markings on the cassette indicate whether the roll is unexposed, partially exposed, fully exposed, or developed. There is no uncertainty about whether a cassette has been exposed or not. Unlike the 35-mm system, where the negatives are returned as cut film, the negatives in the APS system are rewound into the cassette and returned to the customer for compact storage. An index print is given to the customer with each processed roll so that each numbered frame in the cassette can be seen at a glance when reprints are needed.

Compact film scanners are available to digitize pictures directly from an APS cassette or 35-mm cut negatives. These digitized images can be uploaded into a computer for the full range of digital manipulations offered by modern photo software. Pictures can be sent over the Internet, or prints can be made on inkjet printers. For the best photo-quality prints, special photographic ink cartridge assemblies are available with most inkjet printers to print onto special high-gloss photographic-quality paper.

Table 3 summarizes in alphabetical order many brands of 35-mm consumer film available today worldwide. Some of these same basic films appear in single-use cameras, APS format, and 110 format, although not all manufacturers listed offer these formats. This is not a comprehensive list. Because this market is highly competitive, new films emerge quickly to replace existing films. It is not unusual for a manufacturer's entire line of consumer films to change within three years. More detailed information about these films' uses and characteristics can be found in the film manufacturers' Web pages on the Internet.

Professional Color Negative Film.   Professional color negative film divides roughly into portrait and wedding film and commercial and photojournalism film. Portrait and wedding films feature excellent skin tones plus good neutral whites, blacks, and grays. These films also incorporate accurate spectral sensitization technology for color accuracy and excellent results under mixed lighting conditions. The contrast in these films is about 10 percent lower than in most consumer color negative films for softer, more pleasing skin tones, and its tone scale captures highlight and shadow detail very well.

The most popular professional format is 120 size, although 35-mm and sheet sizes are also available. Kodak offers natural color (NC) and vivid color (VC) versions of Professional Portra film, with the vivid color having a 10 percent contrast increase for colorfulness and a sharper look. The natural color version is most pleasing for pictures having large areas of skin tones, as in head and shoulder portraits.

Commercial films emphasize low grain and high sharpness. These films offer color and contrast similar to those of consumer films. Film speeds of 400 and above are especially popular for photojournalist applications, and the most popular format is 35 mm. These films are often push processed.

Table 4 summarizes in alphabetical order many brands of professional color negative film available today worldwide. This is not a comprehensive list. Because this market is highly competitive, new films emerge quickly to replace existing films. More detailed information about these films' uses and characteristics can be found in the film manufacturers' Web pages on the Internet.

Silver halide photographic products have enjoyed steady progress during the past century. This chapter has described most of the technology that led to modern films. Most noteworthy among these advances are:

1. Efficient light management in multilayer photographic materials has reduced harmful optical effects that caused sharpness loss and has optimized beneficial optical effects that lead to efficient light absorption.

2. Proprietary design of silver halide crystal morphology, internally structured halide types, transition metal dopants to manage the electron-hole pair, and improved chemical surface treatments has optimized the efficiency of latent image formation.

TABLE 3 Consumer 35-mm Color Negative Films

Manufacturer and Brand Name					
Agfacolor HDC Plus	100	200	400		
Fujicolor Superia	100	200	400	800	
Ilford Colors	100	200	400		
Imation HP	100	200	400		
Kodak Gold	100	200	Max400	Max800	
Kodak Royal Gold	100	200	400		1000
Konica Color Centuria	100	200	400	800	
Polaroid One Film	100	200	400		

TABLE 4   Professional Color Negative Films

Manufacturer and brand name (commercial film)					
Agfacolor Optima II Prestige	100	200	400		
Fujicolor Press			400	800	
Kodak Professional Ektapress	PJ100		PJ400	PJ800	
Konica Impresa	100	200			3200 SRG

Manufacturer and brand name (portrait film)			
Agfacolor Portrait	160 XPS		
Fujicolor	160 NPS	400 NPH	800 NHGII
Kodak Professional Portra	160 NC	400 NC	Pro 1000
	160 VC	400 VC	
Konica Color Professional	160		

3. Transition metal dopants and development-modifying chemistry have improved process robustness, push processing, and exposure time latitude.

4. New spectral sensitizers combined with powerful chemical color correction methods have led to accurate and pleasing color reproduction.

5. New image dyes have provided rich color saturation and vastly improved image permanence.

6. New film and camera systems for consumers have made the picture-taking experience easier and more reliable than ever before.

Photographic manufacturers continue to invest research and product development resources in their silver halide photographic products. Although digital electronic imaging options continue to emerge, consumers and professionals can expect a constant stream of improved silver halide photographic products for many years to come.

6.7   REFERENCES

1. M. Bass, E. Van Stryland, D. Williams, and W. Wolfe (eds.), *Handbook of Optics,* vol. 1, 2d ed., McGraw-Hill, New York, 1995.

2. J. Kapecki and J. Rodgers, *Kirk-Othmer Encyclopedia of Chemical Technology,* vol. 6, 4th ed., John Wiley & Sons, New York, 1993.

3. T. H. James (ed.), *The Theory of the Photographic Process,* 4th ed., Macmillan, New York, 1977.

4. R. W. G. Hunt, *The Reproduction of Colour,* Fountain Press, Tolworth, England, 1987.

5. C. R. Barr, J. R. Thirtle, and P. W. Vittum, *Photogr. Sci. Eng.* **13**:214 (1969).

6. E. Klein and H. J. Metz, *Photogr. Sci. Eng.* **5**:5 (1961).

7. R. E. Factor and D. R. Diehl, U. S. Patent 4,940,654, 1990.

8. R. E. Factor and D. R. Diehl, U. S. Patent 4,855,221, 1989.

9. G. Mie, *Ann. Physik.* **25**:337 (1908); M. Kerker, *The Scattering of Light and Other Electromagnetic Radiation,* Academic Press, New York, 1969.

10. D. H. Napper and R. H. Ottewill, *J. Photogr. Sci.* **11**:84 (1963); *J. Colloid Sci.* **18**:262 (1963); *Trans. Faraday Soc.* **60**:1466 (1964).

11. E. Pitts, *Proc. Phys. Soc. Lond.* **67B**:105 (1954).

12. J. M. Hammersley and D. C. Handscomb, *Monte Carlo Methods,* John Wiley & Sons, New York, 1964.

13. J. J. DePalma and J. Gasper, *Photogr. Sci. Eng.* **16**:181 (1972).

14. M. Motoki, S. Ichijima, N. Saito, T. Kamio, and K. Mihayashi, U. S. Patent 5,213,958, 1993.

15. Miles V. Klein, *Optics,* John Wiley & Sons, New York, 1970, pp. 582–585.

16. J. Gasper, Eastman Kodak Company, unpublished data.

17. N. F. Mott and R. W. Gurney, *Electronic Processes in Ionic Crystals,* Oxford University Press, London, 1940.

18. F. Seitz, *Rev. Mod. Phys.* **23**:328 (1951).

19. J. F. Hamilton, *Adv. Phys.* **37**:359 (1988).

20. A. P. Marchetti and R. S. Eachus, *Advances In Photochemistry,* John Wiley & Sons, New York, 1992, **17**:145.

21. R. S. Eachus, A. P. Marchetti, and A. A. Muenter, *Ann. Rev. Phys. Chem.* **50**:117 (1999).

22. R. J. Deri, J. P. Spoonhower, and J. F. Hamilton, *J. Appl. Phys.* **57**:1968 (1985).

23. T. Tani, *Photographic Sensitivity,* Oxford University Press, Oxford, UK, 1995, p. 235.

24. L. M. Slifkin and S. K. Wonnell, *Solid State Ionics* **75**:101 (1995).

25. J. F. Hamilton and L. E. Brady, *Surf. Sci.* **23**:389 (1970).

26. R. C. Baetzold, Y. T. Tan, and P. W. Tasker, *Surf. Sci.* **195**:579 (1988).

27. R. S. Eachus and M. T. Olm, *Annu. Rep. Prog. Chem., Sect. C Phys. Chem.* **83**:3 (1986).

28. J. C. Dainty and R. Shaw, *Image Science,* Academic Press, London, 1974.

29. R. K. Hailstone, N. B. Liebert, M. Levy, R. T. McCleary, S. R. Girolmo, D. L. Jeanmaire, and C. R. Boda, *J. Imaging Sci.* **3**(3) (1988).

30. J. D. Baloga, "Factors in Modern Color Reversal Films," *IS&T 1998 PICS Conference,* p. 299 (1998).

31. R. J. Tuite, *J. Appl. Photogr. Eng.* **5**:200 (1979).

32. W. F. Smith, W. G. Herkstroeter, and K. I. Eddy, *Photogr. Sci. Eng.* **20**:140 (1976).

33. P. Douglas, *J. Photogr. Sci.* **36**:83 (1988).

34. P. Douglas, S. M. Townsend, P. J. Booth, B. Crystall, J. R. Durrant, and D. R. Klug, *J. Chem. Soc. Faraday Trans.* **87**:3479 (1991).

35. F. Wilkinson, D. R. Worrall, and R. S. Chittock, *Chem. Phys. Lett.* **174**:416 (1990).

36. F. Wilkinson, D. R. Worrall, D. McGarvy, A. Goodwin, and A. Langley, *J. Chem. Soc. Faraday Trans.* **89**:2385 (1993).

37. P. Douglas, S. M. Townsend, and R. Ratcliffe, *J. Imaging Sci.* **35**:211 (1991).

38. P. Egerton, J. Goddard, G. Hawkins, and T. Wear, *Royal Photographic Society Color Imaging Symposium,* Cambridge, UK, September 1986, p. 128.

39. K. Onodera, T. Nishijima, and M. Sasaki, *Proceedings of the International Symposium on the Stability and Conservation of Photographic Images,* Bangkok, Thailand, 1986.

40. Y. Kaneko, H. Kita, and H. Sato, *Proceedings of IS&T's 46th Annual Conference,* 1993, p. 299.

41. R. J. Berry, P. Douglas, M. S. Garley, D. Clarke, and C. J. Winscom, "Photophysics and Photochemistry of Azomethine Dyes," *IS&T 1998 PICS Conference,* p. 282 (1998).

42. W. F. Smith, W. G. Herkstroeter, and K. I. Eddy, *J. Am. Chem. Soc.* **97**:2164 (1975).

43. W. G. Herkstroeter, *J. Am. Chem. Soc.* **95**:8686 (1973).

44. W. G. Herkstroeter, *J. Am. Chem. Soc.* **97**:3090 (1975).

45. P. Douglas and D. Clarke, *J. Chem. Soc. Perkin Trans.* **2**:1363 (1991).

46. F. Abu-Hasanayn, *Book of Abstracts, 218th ACS National Meeting,* New Orleans, LA, August 22–26, 1999.

47. K. Hidetoshi et al. *International Congress of Photographic Science meeting,* Belgium, 1998.

48. N. Saito and S. Ichijima, *International Symposium on Silver Halide Imaging,* 1997.

49. V. Balzani, F. Bolletta, and F. Scandola, *J. Am. Chem. Soc.* **102**:2152 (1980).

50. R. Jain and W. R. Schleigh, U. S. Patent 5,561,037, 1996.

51. O. Takahashi, H. Yoneyama, K. Aoki, and K. Furuya, "The Effect of Polymeric Addenda on Dark Fading Stability of Cyan Indoaniline Dye," *IS&T 1998 PICS Conference,* p. 329 (1998).

52. R. L. Heidke, L. H. Feldman, and C. C. Bard, *J. Imag. Tech.* **11**(3):93 (1985).

53. S. Cowan and S. Krishnamurthy, U. S. Patent 5,378,587 (1995).

54. T. Kawagishi, M. Motoki, and T. Nakamine, U. S. Patent 5,605,788 (1997).

55. H. W. Vogel, *Berichte* **6**:1302 (1873).

56. J. E. Maskasky, *Langmuir* **7**:407 (1991).

57. J. Spence and B. H. Carroll, *J. Phys. Colloid Chem.* **52**:1090 (1948).

58. A. H. Hertz, R. Danner, and G. Janusonis, *Adv. Colloid Interface Sci.* **8**:237 (1977).

59. M. Kawasaki and H. Ishii, *J. Imaging Sci. Technol.* **39**:210 (1995).

60. G. Janssens, J. Gerritsen, H. van Kempen, P. Callant, G. Deroover, and D. Vandenbroucke, *The Structure of H-, J-, and Herringbone Aggregates of Cyanine Dyes on AgBr(111) Surfaces.* Presented at ICPS 98 International Conference on Imaging Science, Antwerp, Belgium (1998).

61. P. B. Gilman, *Photogr. Sci. Eng.* **18**:475 (1974).

62. J. Lenhard, *J. Imaging Sci.* **30**:27 (1986).

63. W. West, "Scientific Photography," in *Proceedings of the International Conference at Liege, 1959,* H. Sauvenier (ed), Pergamon Press, New York, 1962, p. 557.

64. T. Tani, *Photogr. Sci. Eng.* **14**:237 (1970).

65. J. Eggert, W. Meidinger, and H. Arens, *Helv. Chim. Acta.* **31**:1163 (1948).

66. J. M. Lanzafame, A. A Muenter, and D. V. Brumbaugh, *Chem. Phys.* **210**:79 (1996).

67. A. A. Muenter and W. Cooper, *Photogr. Sci. Eng.* **20**:121 (1976).

68. W. West, B. H. Carroll, and D. H. Whitcomb, *J. Phys. Chem.* **56**:1054 (1952).

69. R. Brunner, A. E. Oberth, G. Pick, and G. Scheibe, *Z. Elektrochem.* **62**:146 (1958).

70. P. B. Gilman, *Photogr. Sci. Eng.* **11**:222 (1967).

71. P. B. Gilman, *Photogr. Sci. Eng.* **12**:230 (1968).

72. J. E. Jones and P. B. Gilman, *Photogr. Sci. Eng.* **17**:367 (1973).

73. P. B. Gilman and T. D. Koszelak, *J. Photogr. Sci.* **21**:53 (1973).

74. H. Sakai and S. Baba, *Bull. Soc. Sci. Photogr. Jpn.* **17**:12 (1967).

75. B. H. Carroll, *Photogr. Sci. Eng.* **5**:65 (1961).

76. T. Tani, *Photogr. Sci. Eng.* **15**:384 (1971).

77. T. A. Babcock, P. M. Ferguson, W. C. Lewis, and T. H. James, *Photogr. Sci. Eng.* **19**:49 (1975).

78. E. J. Wall, *History of Three Color Photography,* American Photographic Publishing Company, Boston, MA, 1925.

79. A. C. Hardy and F. L. Wurzburg Jr., "The Theory of Three Color Reproduction," *J. Opt. Soc. Am.* **27**:227 (1937).

80. M. L. Pearson and J. A. C. Yule, *J. Color Appearance* **2**:30 (1973).

81. S. G. Link, "Short Red Spectral Sensitizations for Color Negative Films," *IS&T 1998 PICS Conference,* p. 308 (1998).

82. Y. Nozawa and N. Sasaki, U.S. Patent 4,663,271 (1987).

CHAPTER 7

RADIOMETRY AND PHOTOMETRY: UNITS AND CONVERSIONS

James M. Palmer
Research Professor
Optical Sciences Center
University of Arizona
Tucson, Arizona

7.1 GLOSSARY *

A	area, m^2
A_{proj}	projected area, m^2
E, E_e	radiant incidence (irradiance), W m^{-2}
E_p, E_q	photon incidence, s^{-1} m^{-2}
E_v	illuminance, lm m^{-2} [lux (lx)]
I, I_e	radiant intensity, W sr^{-1}
I_p, I_q	photon intensity, s^{-1} sr^{-1}
I_v	luminous intensity, candela (cd)
K_m	absolute luminous efficiency at λ_p for photopic vision, 683 lm/W
K'_m	absolute luminous efficiency at λ_p for scotopic vision, 1700 lm/W
$K(\lambda)$	absolute spectral luminous efficiency, photopic vision, lm/W
$K'(\lambda)$	absolute spectral luminous efficiency, scotopic vision, lm/W
L, L_e	radiance, W m^{-2} sr^{-1}
L_p, L_q	photon radiance (photonance), s^{-1} m^{-2} sr^{-1}
L_v	luminance, cd sr^{-1}
M, M_e	radiant exitance, W m^{-2}
M_p, M_q	photon exitance, s^{-1} m^{-2}
Q, Q_e	radiant energy, joule (J)

---

* *Note.* The subscripts are used as follows: *e* (energy) for radiometric, *v* (visual) for photometric, and *q* (or *p*) for photonic. The subscript *e* is usually omitted; the other subscripts may also be omitted if the context is unambiguous.

Q_p, Q_q	photon energy, J or eV
Q_v	luminous energy, lm s^{-1}
$\Re(\lambda)$	spectral responsivity, A/W or V/W
$V(\lambda)$	relative spectral luminous efficiency, photopic vision
$V'(\lambda)$	relative spectral luminous efficiency, scotopic vision
$V_q(\lambda)$	relative spectral luminous efficiency for photons
λ	wavelength, nm or µm
λ_p	wavelength at peak of function, nm or µm
Φ, Φ_e	radiant power (radiant flux), Watt (W)
Φ_p, Φ_q	photon flux, s^{-1}
Φ_v	luminous power (luminous flux), lumen (lm)
ω	solid angle, steradian (sr)
Ω	projected solid angle, sr

7.2 *INTRODUCTION AND BACKGROUND*

After more than a century of turmoil, the symbols, units, and nomenclature (SUN) for radiometry and photometry seem somewhat stable at present. There are still small, isolated pockets of debate and even resistance; by and large, though, the community has settled on the International System of Units (SI) and the recommendations of the International Organization for Standardization [ISO (ISO, 1993)] and the International Union of Pure and Applied Physics [IUPAP (IUPAP, 1987)].

The seed of the SI was planted in 1799 with the deposition of two prototype standards, the meter and the kilogram, into the Archives de la République in Paris. Noted physicists Gauss, Weber, Maxwell, and Thompson made significant contributions to measurement science over the next 75 years. Their efforts culminated in the Convention of the Meter, a diplomatic treaty originally signed by representatives of 17 nations in 1875 (currently there are 48 member nations). The Convention grants authority to the General Conference on Weights and Measures (CGPM), the International Committee for Weights and Measures (CIPM), and the International Bureau of Weights and Measures (BIPM). The CIPM, along with a number of subcommittees, suggests modifications to the CGPM. In our arena, the subcommittee is the Consultative Committee on Photometry and Radiometry (CCPR). The BIPM, the international metrology institute, is the physical facility that is responsible for realization, maintenance, and dissemination of standards.

The SI was adopted by the CGPM in 1960 and is the official system in the 48 member states. It currently consists of seven base units and a much larger number of derived units. The base units are a choice of seven well-defined units that, by convention, are regarded as independent. The seven base units are as follows:

1. Meter
2. Kilogram
3. Second
4. Ampere
5. Kelvin
6. Mole
7. Candela

The derived units are those that are formed by various combinations of the base units.

International organizations involved in the promulgation of SUN include the International Commission on Illumination (CIE), the IUPAP, and the ISO. In the United States, the American National Standards Institute (ANSI) is the primary documentary (protocol) standards organization. Many other scientific and technical organizations publish recommendations concerning the use of SUN for their scholarly journals. Several examples are the Illuminating Engineering Society (IESNA), the International Astronomical Union (IAU), the Institute for Electrical and Electronic Engineering (IEEE), and the American Institute of Physics (AIP).

The terminology employed in radiometry and photometry consists principally of two parts: (1) an adjective distinguishing between a radiometric, photonic, or photometric entity, and (2) a noun describing the underlying geometric or spatial concept. In some instances, the two parts are replaced by a single term (e.g., radiance).

There are some background concepts and terminology that are needed before proceeding further.

Projected area is defined as the rectilinear projection of a surface of any shape onto a plane normal to the unit vector. The differential form is $dA_{\text{proj}} = \cos(\beta)dA$, where β is the angle between the local surface normal and the line of sight. Integrate over the (observable) surface area to get

$$A_{\text{proj}} = \int_A \cos \beta dA \tag{1}$$

Some common examples are shown in Table 1.

Plane angle and solid angle are both derived units in the SI system. The following definitions are from National Institute of Standards and Technology (NIST) SP811 (NIST, 1995).

The radian is the plane angle between two radii of a circle that cuts off on the circumference an arc equal in length to the radius.

The abbreviation for the radian is *rad*. Since there are 2π rad in a circle, the conversion between degrees and radians is 1 rad = $(180/\pi)$ degrees. (See Fig. 1.)

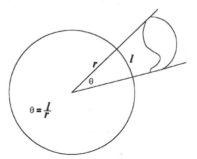

$$\theta = \frac{l}{r}$$

FIGURE 1 Plane angle.

TABLE 1 Projected Areas of Common Shapes

Shape	Area	Projected area
Flat rectangle	$A = L \times W$	$A_{\text{proj}} = L \times W \cos \beta$
Circular disc	$A = \pi r^2 = \pi d^2/4$	$A_{\text{proj}} = \pi r^2 \cos \beta = (\pi d^2 \cos \beta)/4$
Sphere	$A = 4\pi r^2 = \pi d^2$	$A_{\text{proj}} = A/4 = \pi r^2$

A solid angle is the same concept that is extended to three dimensions.

> One steradian (sr) is the solid angle that, having its vertex in the center of a sphere, cuts off an area on the surface of the sphere equal to that of a square with sides of length equal to the radius of the sphere.

The solid angle is the ratio of the spherical area (A_{proj}) to the square of the radius, r. The spherical area is a projection of the object of interest onto a unit sphere, and the solid angle is the surface area of that projection, as shown in Fig. 2. Divide the surface area of a sphere by the square of its radius to find that there are 4π sr of solid angle in a sphere. One hemisphere has 2π sr. The accepted symbols for solid angle are either lowercase Greek omega (ω) or uppercase Greek omega (Ω). I recommend using ω exclusively for solid angle, and reserving Ω for the advanced concept of projected solid angle ($\omega \cos \theta$). The equation for solid angle is $d\omega = dA_{proj}/r^2$. For a right circular cone, $\omega = 2\pi(1 - \cos \theta)$, where θ is the half-angle of the cone.

Both plane angles and solid angles are dimensionless quantities, and they can lead to confusion when attempting dimensional analysis. For example, the simple inverse square law, $E = I/d^2$ appears dimensionally inconsistent. The left side has units W m^{-2} while the right side has W sr^{-1} m^{-2}. It has been suggested that this equation be written $E = I\Omega_0/d^2$, where Ω_0 is the unit solid angle, 1 sr. Inclusion of the term Ω_0 will render the equation dimensionally correct, but will far too often be considered a free variable rather than a constant equal to 1, leading to erroneous results.

Isotropic and lambertian both carry the meaning "the same in all directions" and, regrettably, are used interchangeably.

Isotropic implies a spherical source that radiates the same in all directions [i.e., the intensity (W/sr or cd) is independent of direction]. The term *isotropic point source* is often heard. No such entity can exist, for the energy density would necessarily be infinite. A small, uniform sphere comes very close. Small in this context means that the size of the sphere is much less than the distance from the sphere to the plane of observation, such that the inverse square law is applicable. An example is a globular tungsten lamp with a milky white diffuse envelope some 10 cm in diameter, as viewed from a distance greater than 1 m. From our vantage point, a distant star is considered an isotropic point source.

Lambertian implies a flat radiating surface. It can be an active emitter or a passive, reflective surface. The intensity decreases with the cosine of the observation angle with respect to the surface normal (Lambert's law). The radiance (W m^{-2} sr^{-1}) or luminance (cd m^{-2}) is independent of direction. A good approximation is a surface painted with a quality matte, or flat, white paint. The intensity is the product of the radiance, L, or luminance, L_v, and

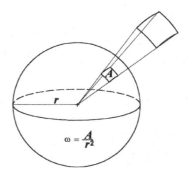

FIGURE 2 Solid angle.

the projected area A_{proj}. If the surface is uniformly illuminated, it appears equally bright from whatever direction it is viewed. Note that the flat radiating surface can be used as an elemental area of a curved surface.

The ratio of the radiant exitance (power per unit area, W m^{-2}) to the radiance (power per unit projected area per unit solid angle, W m^{-2} sr^{-1}) of a lambertian surface is a factor of π and not 2π. This result is not intuitive, as, by definition, there are 2π sr in a hemisphere. The factor of π comes from the influence of the cos (θ) term while integrating over a hemisphere.

A sphere with a lambertian surface illuminated by a distant point source will display a radiance that is maximum at the point where the local normal coincides with the incoming beam. The radiance will fall off with a cosine dependence to zero at the terminator. If the intensity (integrated radiance over area) is unity when viewing from the direction of the source, then the intensity when viewing from the side is $1/\pi$. Think about this and ponder whether our Moon has a lambertian surface.

7.3   SYMBOLS, UNITS, AND NOMENCLATURE IN RADIOMETRY

Radiometry is the measurement of optical radiation, defined as electromagnetic radiation within the frequency range from 3×10^{11} to 3×10^{16} hertz (Hz). This range corresponds to wavelengths between 0.01 and 1000 micrometers (μm) and includes the regions commonly called *ultraviolet, visible,* and *infrared.*

Radiometric units can be divided into two conceptual areas: (1) those having to do with power or energy, and (2) those that are geometric in nature. In the first category are:

Energy is an SI-derived unit, measured in joules (J). The recommended symbol for energy is Q. An acceptable alternate is W.

Power (also called *radiant flux*) is another SI-derived unit. It is the derivative of energy with respect to time, dQ/dt, and the unit is the watt (W). The recommended symbol for power is uppercase Greek phi (Φ). An accepted alternate is P.

Energy is the integral of power over time, and it is commonly used with integrating detectors and pulsed sources. Power is used for continuous sources and nonintegrating detectors. The radiometric quantity power can now be combined with the geometric spatial quantities area and solid angle.

Irradiance (also referred to as *flux density* or *radiant incidance*) is an SI-derived unit and is measured in W m^{-2}. Irradiance is power per unit area *incident* from all directions within a hemisphere onto a surface that coincides with the base of that hemisphere. A related quantity is *radiant exitance,* which is power per unit area *leaving* a surface into a hemisphere whose base is that surface. The symbol for irradiance is E, and the symbol for radiant exitance is M. Irradiance (or radiant exitance) is the derivative of power with respect to area, $d\Phi/dA$. The integral of irradiance or radiant exitance over area is power. There is no compelling reason to have two quantities carrying the same units, but it is convenient.

Radiant intensity is an SI-derived unit and is measured in W sr^{-1}. Intensity is power per unit of solid angle. The symbol is I. *Intensity* is the derivative of power with respect to solid angle, $d\Phi/d\omega$. The integral of radiant intensity over solid angle is power.

A great deal of confusion surrounds the use and misuse of the term *intensity.* Some use it for W sr^{-1}; some use it for W m^{-2}; others use it for W m^{-2} sr^{-1}. It is quite clearly defined in the SI system, in the definition of the base unit of luminous intensity—the candela. Attempts are often made to justify these different uses of intensity by adding adjectives like *optical* or *field* (used for W m^{-2}) or *specific* (used for W m^{-2} sr^{-1}). In the SI system, the underlying geometric concept for intensity is quantity per unit of solid angle. For more discussion, see Palmer (1993).

Radiance is an SI-derived unit and is measured in $W\ m^{-2}\ sr^{-1}$. Radiance is a directional quantity, power per unit of projected area per unit of solid angle. The symbol is *L*. Radiance is the derivative of power with respect to solid angle and projected area, $d\Phi/d\omega\ dA\ \cos(\theta)$, where θ is the angle between the surface normal and the specified direction. The integral of radiance over area and solid angle is power.

Photon quantities are also common. They are related to the radiometric quantities by the relationship $Q_p = hc/\lambda$, where Q_p is the energy of a photon at wavelength λ, *h* is Planck's constant, and *c* is the velocity of light. At a wavelength of 1 μm, there are approximately 5×10^{18} photons per second in a watt. Conversely, one photon has an energy of about 2×10^{-19} J (W/s) at 1 μm.

7.4   SYMBOLS, UNITS, AND NOMENCLATURE IN PHOTOMETRY

Photometry is the measurement of light, electromagnetic radiation detectable by the human eye. It is thus restricted to the wavelength range from about 360 to 830 nanometers (nm; 1000 nm = 1 μm). Photometry is identical to radiometry *except* that everything is weighted by the spectral response of the nominal human eye. *Visual photometry* uses the eye as a comparison detector, while *physical photometry* uses either optical radiation detectors constructed to mimic the spectral response of the nominal eye, or spectroradiometry coupled with appropriate calculations to do the eye response weighting.

Photometric units are basically the same as the radiometric units, except that they are weighted for the spectral response of the human eye and have strange names. A few additional units have been introduced to deal with the amount of light that is reflected from diffuse (matte) surfaces. The symbols used are identical to the geometrically equivalent radiometric symbols, except that a subscript *v* is added to denote *visual*. Table 2 compares radiometric and photometric units.

The SI unit for light is the *candela* (unit of luminous intensity). It is one of the seven base units of the SI system. The candela is defined as follows (BIPM, 1998):

> The candela is the luminous intensity, in a given direction, of a source that emits monochromatic radiation of frequency 540×10^{12} hertz and that has a radiant intensity in that direction of 1/683 watt per steradian.

The candela is abbreviated as *cd*, and its symbol is I_v. This definition was adopted by the 16th CGPM in 1979. The candela was formerly defined as the luminous intensity, in the perpendicular direction, of a surface of $1/600,000\ m^2$ of a blackbody at the temperature of freezing platinum under a pressure of 101,325 newtons per square meter ($N\ m^{-2}$). This earlier definition was initially adopted in 1948 and later modified by the 13th CGPM in 1968. It was abrogated in 1979 and replaced by the current definition.

The 1979 definition was adopted for several reasons. First, the freezing point of platinum ($\approx$2042 K) was tied to another base unit, the kelvin, and was therefore not independent. The

TABLE 2   Comparison of Radiometric and Photometric Units

Quantity	Radiometric	Photometric
Power	Φ: watt (W)	Φ_v: lumen (lm)
Power per area	*E, M:* $W\ m^{-2}$	E_v: $lm\ m^{-2}$ = lux (lx)
Power per solid angle	*I:* $W\ sr^{-1}$	I_v: $lm\ sr^{-1}$ = candela (cd)
Power per area per solid angle	*L:* $W\ m^{-2}\ sr^{-1}$	L_v: $lm\ m^{-2}\ sr^{-1}$ = $cd\ m^{-2}$ = nit

uncertainty of the thermodynamic temperature of this fixed point created an unacceptable uncertainty in the value of the candela. If the best estimate of this freezing point were to change, it would then impact the candela. Second, the realization of the platinum blackbody was extraordinarily difficult; only a few were ever built. Third, if the platinum blackbody temperature were slightly off, possibly because of temperature gradients in the ceramic crucible or contamination of the platinum, the freezing point might change or the temperature of the cavity might differ. The sensitivity of the candela to a slight change in temperature is significant. At a wavelength of 555 nm, a change in temperature of only 1 K results in a luminance change approaching 1 percent. Fourth, the relative spectral radiance of blackbody radiation changes drastically (some three orders of magnitude) over the visible range. Finally, recent advances in radiometry offered new possibilities for the realization of the candela.

The value of 683 lm/W was selected based upon the best measurements with existing platinum freezing point blackbodies at several national standards laboratories. It has varied over time from 620 to nearly 700 lm/W, depending largely upon the assigned value of the freezing point of platinum. The value of $1/600,000$ m^2 was chosen to maintain consistency with prior standards. Note that neither the old nor the new definition of the candela say anything about the spectral responsivity of the human eye. There are additional definitions that include the characteristics of the eye, but the base unit (candela) and those SI units derived from it are "eyeless."

Note also that in the definition of the candela, there is no specification for the spatial distribution of intensity. Luminous intensity, while often associated with an isotropic point (i.e., small) source, is a valid specification for characterizing any highly directional light source, such as a spotlight or an LED.

One other issue: since the candela is no longer independent but is now defined in terms of other SI-derived quantities, there is really no need to retain it as an SI base quantity. It remains so for reasons of history and continuity and perhaps some politics.

The *lumen* is an SI-derived unit for luminous flux (power). The abbreviation is *lm,* and the symbol is Φ_v. The lumen is derived from the candela and is the luminous flux that is emitted into unit solid angle (1 sr) by an isotropic point source having a luminous intensity of 1 cd. The lumen is the product of luminous intensity and solid angle (cd · sr). It is analogous to the unit of radiant flux (watt), differing only in the eye response weighting. If a light source is isotropic, the relationship between lumens and candelas is 1 cd = 4π lm. In other words, an isotropic source that has a luminous intensity of 1 cd emits 4π lm into space, which is 4π sr. Also, 1 cd = 1 lm sr^{-1}, which is analogous to the equivalent radiometric definition.

If a source is not isotropic, the relationship between candelas and lumens is empirical. A fundamental method used to determine the total flux (lumens) is to measure the luminous intensity (candelas) in many directions using a goniophotometer, and then numerically integrate over the entire sphere. Later on, this "calibrated" lamp can be used as a reference in an integrating sphere for routine measurements of luminous flux.

The SI-derived unit of luminous flux density, or illuminance, has a special name: *lux.* It is lumens per square meter (lm m^{-2}), and the symbol is E_v. Most light meters measure this quantity, as it is of great importance in illuminating engineering. The IESNA's *Lighting Handbook* (IESNA, 1993) has some 16 pages of recommended illuminances for various activities and locales, ranging from morgues to museums. Typical values range from 100,000 lx for direct sunlight to between 20 and 50 lx for hospital corridors at night.

Luminance should probably be included on the list of SI-derived units, but it is not. Luminance is analogous to radiance, giving the spatial and directional dependences. It also has a special name, *nit,* and is candelas per square meter (cd m^{-2}) or lumens per square meter per steradian (lm m^{-2} sr^{-1}). The symbol is L_v. Luminance is most often used to characterize the "brightness" of flat-emitting or -reflecting surfaces. A common use is the luminance of a laptop computer screen. They typically have between 100 and 250 nits, and the sunlight-readable ones have more than 1000 nits. Typical CRT monitors have luminances between 50 and 125 nits.

Other Photometric Units

There are other photometric units, largely historical. The literature is filled with now obsolete terminology, and it is important to be able to properly interpret these terms. Here are several terms for illuminance that have been used in the past.

1 meter-candle = 1 lx

1 phot (ph) = 1 lm cm^{-2} = 10^4 lx

1 footcandle (fc) = 1 lm ft^{-2} = 10.76 lx

1 milliphot = 10 lx

Table 3 is useful to convert from one unit to another. Start with the unit in the leftmost column and multiply it by the factor in the table to arrive at the unit in the top row.

There are two classes of units that are used for luminance. The first is conventional, directly related to the SI unit, the cd m^{-2} (nit).

1 stilb = 1 cd cm^{-2} = 10^4 cd m^{-2} = 10^4 nit

1 cd ft^{-2} = 10.76 cd m^{-2} = 10.76 nit

The second class was designed to "simplify" characterization of light that is reflected from diffuse surfaces by incorporating within the definition the concept of a perfect diffuse reflector (lambertian, reflectance $\rho = 1$). If 1 unit of illuminance falls upon this ideal reflector, then 1 unit of luminance is reflected. The perfect diffuse reflector emits $1/\pi$ units of luminance per unit of illuminance. If the reflectance is ρ, then the luminance is ρ/π times the illuminance. Consequently, these units all incorporate a factor of $(1/\pi)$.

1 lambert (L) = $(1/\pi)$ cd cm^{-2} = $(10^4/\pi)$ cd m^{-2} = $(10^4/\pi)$ nit

1 apostilb = $(1/\pi)$ cd m^{-2} = $(1/\pi)$ nit

1 foot-lambert (ft-lambert) = $(1/\pi)$ cd ft^{-2} = 3.426 cd m^{-2} = 3.426 nit

1 millilambert = $(10/\pi)$ cd m^{-2} = $(10/\pi)$ nit

1 skot = 1 milliblondel = $(10^{-3}/\pi)$ cd m^{-2} = $(10^{-3}/\pi)$ nit

Table 4 is useful to convert from one unit to another. Start with the unit in the leftmost column and multiply it by the factor in the table to arrive at the unit in the top row.

Human Eye.   The SI base unit and units derived therefrom have a strictly physical basis; they have been defined monochromatically at a single wavelength of 555 nm. But the eye does not see all wavelengths equally. For other wavelengths or for band or continuous-source spectral distributions, the spectral properties of the human eye must be considered. The eye has two general classes of photosensors: *cones* and *rods*.

Cones.   The cones are responsible for light-adapted vision; they respond to color and have high resolution in the central foveal region. The light-adapted relative spectral response of the eye is called the *spectral luminous efficiency function for photopic vision, V(λ),* and is

TABLE 3   Illuminance Unit Conversions

	fc	lx	phot	milliphot
1 fc (lm/ft^2) =	1	10.764	0.0010764	1.0764
1 lx (lm/m^2) =	0.0929	1	0.0001	0.1
1 phot (lm/cm^2) =	929	10,000	1	0.001
1 milliphot =	0.929	10	0.1	1

TABLE 4 Luminance Unit Conversions*

	nit	stilb	cd/ft²	apostilb	lambert	ft-lambert
1 nit (cd/m²) =	1	10^{-4}	0.0929	π	$\pi/10000$	0.0929π
1 stilb (cd/cm²) =	10,000	1	929	$10^4\pi$	π	929π
1 cd/ft² =	10.764	1.0764×10^{-3}	1	10.764π	$\pi/929$	π
1 apostilb =	$1/\pi$	$10^4/\pi$	$0.0929/\pi$	1	10^{-4}	0.0929
1 lambert =	$10^4/\pi$	$1/\pi$	$929/\pi$	10^4	1	929
1 ft-lambert =	$10.76/\pi$	$1/(929\pi)$	$1/\pi$	10.764	1.076×10^4	1

*Note: Photometric quantities are the result of an integration over wavelength. It therefore makes no sense to speak of spectral luminance or the like.

published in tabular form (CIE, 1983). This empirical curve, shown in Fig. 3, was first adopted by the CIE in 1924. It has a peak that is normalized to unity at 555 nm, and it decreases to levels below 10^{-5} at about 370 and 785 nm. The 50 percent points are near 510 nm and 610 nm, indicating that the curve is slightly skewed. A logarithmic representation is shown in Fig. 4.

More recent measurements have shown that the 1924 curve may not best represent typical human vision. It appears to underestimate the response at wavelengths shorter than 460 nm. Judd (1951), Vos (1978), and Stockman and Sharpe (1999) have made incremental advances in our knowledge of the photopic response.

Rods.   The rods are responsible for dark-adapted vision, with no color information and poor resolution when compared with the foveal cones. The dark-adapted relative spectral response of the eye is called the *spectral luminous efficiency function for scotopic vision, V(λ)*, also published in tabular form (CIE, 1983). Figures 3 and 4 also show this empirical curve, which was adopted by the CIE in 1951. It is defined between 380 and 780 nm. The $V(\lambda)$ curve has a peak of unity at 507 nm, and it decreases to levels below 10^{-3} at about 380 and 645 nm. The 50 percent points are near 455 and 550 nm.

Photopic (light-adapted cone) vision is active for luminances that are greater than 3 cd m^{-2}. Scotopic (dark-adapted rod) vision is active for luminances that are lower than 0.01 cd m^{-2}. In between, both rods and cones contribute in varying amounts, and in this range the vision is called *mesopic*. There are currently efforts under way to characterize the composite spectral response in the mesopic range for vision research at intermediate luminance levels. Definitive

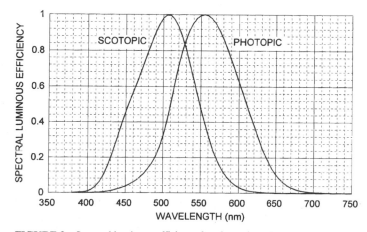

FIGURE 3 Spectral luminous efficiency for photopic and scotopic vision.

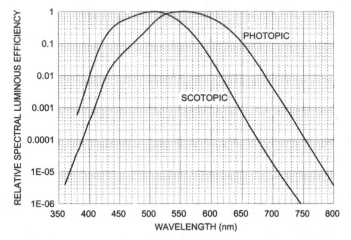

FIGURE 4   Spectral luminous efficiency for photopic and scotopic vision (log scale).

values at 1-nm intervals for both photopic and scotopic spectral luminous efficiency functions may be found in CIE (1983). Values at 5-nm intervals are given by Zalewski (1995).

The relative spectral luminous efficiency functions can be converted for use with photon flux (s^{-1}) by multiplying by the spectrally dependent conversion from watts to photons per second. The results are shown in Fig. 5. The curves are similar to the spectral luminous efficiency curves, with the peaks shifted to slightly shorter wavelengths, and the skewness of the curves is different. This function can be called $V_q(\lambda)$ for photopic response or $V'_q(\lambda)$ for scotopic response. The conversion to an absolute curve is made by multiplying by the response at the peak wavelength. For photopic vision ($\lambda = 550$ nm), $K_{mp} = 2.45 \times 10^{-16}$ lm / photon s^{-1}. There are, therefore, 4.082×10^{15} photon s^{-1} lm^{-1} at 550 nm, and more at all other wavelengths. For scotopic vision ($\lambda_p = 504$ nm), $K'_{mp} = 6.68 \times 10^{-16}$ lm / photon s^{-1}. There are 1.497×10^{15} photon s^{-1} lm^{-1} at 504 nm, and more at all other wavelengths.

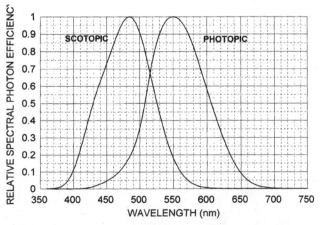

FIGURE 5   Spectral photon efficiency for scotopic and photopic vision.

Approximations.   The $V(\lambda)$ curve appears similar to a Gaussian (normal) function. A non-linear regression technique was used to fit the Gaussian shown in Eq. 2 to the $V(\lambda)$ data

$$V(\lambda) \cong 1.019e^{-285.4(\lambda - 0.559)^2} \tag{2}$$

The scotopic curve can also be fit with a Gaussian, although the fit is not quite as good as the photopic curve. My best fit is

$$V'(\lambda) \cong 0.992e^{-321.0(\lambda - 0.503)^2} \tag{3}$$

The results of the curve fitting are shown in Figs. 6 and 7. These approximations are satisfactory for application with continuous spectral distributions, such as sunlight, daylight, and incandescent sources. Calculations have demonstrated errors of less than 1 percent with blackbody sources from 1500 K to more than 20,000 K. The equations must be used with caution for narrow-band or line sources, particularly in those spectral regions where the response is low and the fit is poor.

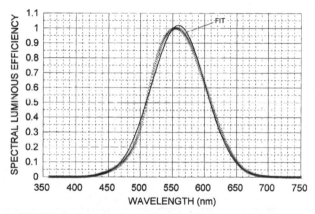

FIGURE 6   Gaussian fit to photopic relative spectral efficiency curve.

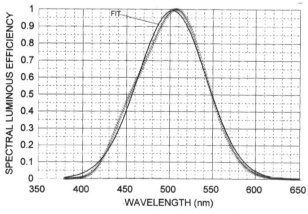

FIGURE 7   Gaussian fit to scotopic relative spectral efficiency curve.

Usage. The SI definition of the candela was chosen in strictly physical terms at a single wavelength. The intent of photometry, however, is to correlate a photometric observation to the visual perception of a human observer. The CIE introduced the two standard spectral luminous efficiency functions $V(\lambda)$ (photopic) and $V'(\lambda)$ (scotopic) as spectral weighting functions, and they have been approved by the CIPM for use with light sources at other wavelengths. Another useful function is the CIE $V_M(\lambda)$ Judd-Vos modified $V(\lambda)$ function (Vos, 1978), which has increased response at wavelengths that are shorter than 460 nm. It is identical to the $V(\lambda)$ function for wavelengths that are longer than 460 nm. This function, while not approved by CIPM, represents more realistically the spectral responsivity of the eye. More recently, studies on cone responses have led to the proposal of a new, improved luminous spectral efficiency curve, with the suggested designation $V_2^*(\lambda)$ (Stockman and Sharpe, 1999).

7.5 CONVERSION OF RADIOMETRIC QUANTITIES TO PHOTOMETRIC QUANTITIES

The definition of the candela states that there are 683 lm W^{-1} at a frequency of 540 terahertz (THz), which is very nearly 555 nm (vacuum or air), the wavelength that corresponds to the maximum spectral responsivity of the photopic (light-adapted) human eye. The value 683 lm W^{-1} is K_m, the absolute luminous efficiency at λ_p for photopic vision. The conversion from watts to lumens at any other wavelength involves the product of the power (watts), K_m, and the $V(\lambda)$ value at the wavelength of interest. For example, a 5-mW laser pointer has 0.005 W $\times$ 0.032 $\times$ 683 lm $W^{-1} = 0.11$ lm. $V(\lambda)$ is 0.032 at 670 nm. At 635 nm, $V(\lambda)$ is 0.217, and a 5-mW laser pointer has 0.005 W $\times$ 0.217 $\times$ 683 lm $W^{-1} = 0.74$ lm. The shorter-wavelength laser pointer will create a spot that has nearly seven times the luminous power as the longer-wavelength laser.

Similar calculations can be done in terms of photon flux at a single wavelength. As was shown previously, there are 2.45×10^{-16} lm in 1 photon s^{-1} at 550 nm, the wavelength that corresponds to the maximum spectral responsivity of the light-adapted human eye to photon flux. The conversion from lumens to photons per second at any other wavelength involves the product of the photon flux (s^{-1}) and the $V_p(\lambda)$ value at the wavelength of interest. For example, again compare laser pointers at 670 and 635 nm. As shown before, a 5-mW laser at 670 nm [$V_p(\lambda) = 0.0264$] has a luminous power of 0.11 lm. The conversion is $0.11 \times 4.082 \times 10^{15}/0.0264 = 1.68 \times 10^{16}$ photon s^{-1}. At 635 nm [$V_p(\lambda) = 0.189$], the 5-mW laser has 0.74 lm. The conversion is $0.74 \times 4.082 \times 10^{15}/0.189 = 1.6 \times 10^{16}$ photon s^{-1}. The 635-nm laser delivers just 5 percent more photons per second.

In order to convert a source with nonmonochromatic spectral distribution to a luminous quantity, the situation is decidedly more complex. The spectral nature of the source must be known, as it is used in an equation of the form

$$X_v = K_m \int_0^\infty X_\lambda V(\lambda) d\lambda \tag{4}$$

where X_v is a luminous term, X_λ is the corresponding spectral radiant term, and $V(\lambda)$ is the photopic spectral luminous efficiency function. For X, luminous flux (lm) may be paired with spectral power (W nm^{-1}), luminous intensity (cd) with spectral radiant intensity (W sr^{-1} nm^{-1}), illuminance (lx) with spectral irradiance (W m^{-2} nm^{-1}), or luminance (cd m^{-2}) with spectral radiance (W m^{-2} sr^{-1} nm^{-1}). This equation represents a weighting, wavelength by wavelength, of the radiant spectral term by the visual response at that wavelength. The constant K_m is the maximum spectral luminous efficiency for photopic vision, 683 lm W^{-1}. The wavelength limits can be set to restrict the integration to only those wavelengths where the product of the spectral term X_λ and $V(\lambda)$ is nonzero. Practically, the limits of integration need only extend from 360 to 830 nm, limits specified by the CIE $V(\lambda)$ function. Since this $V(\lambda)$ function is defined by a table of empirical values (CIE, 1983), it is best to do the integration numerically. Use of the Gaussian equation (Eq. 2) is only an approximation.

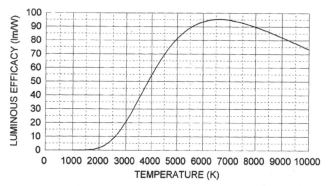

FIGURE 8   Luminous efficacy of blackbody radiation versus temperature (K).

For source spectral distributions that are blackbody-like (thermal source, spectral emissivity constant between 360 and 830 nm) and of known source temperature, it is straightforward to convert from power to luminous flux and vice versa. Equation 4 is used to determine a scale factor for the source term X_λ. Figure 8 shows the relationship between total power and luminous flux for blackbody (and graybody) radiation as a function of blackbody temperature. The most efficient temperature for the production of luminous flux is near 6630 K.

There is nothing in the SI definitions of the base or derived units concerning the eye response, so there is some flexibility in the choice of the weighting function. The choice can be made to use a different spectral luminous efficiency curve, perhaps one of the newer ones. The equivalent curve for scotopic (dark-adapted) vision can also be used for work at lower light levels. The $V'(\lambda)$ curve has its own constant, K'_m, the maximum spectral luminous efficiency for scotopic vision. K'_m is defined as 1700 lm/W at the peak wavelength for scotopic vision (507 nm). This value was deliberately chosen such that the absolute value of the scotopic curve at 555 nm coincides with the photopic curve, 683 lm/W at 555 nm. Some researchers are referring to "scotopic lumens," a term that should be discouraged because of the potential for misunderstanding. In the future, expect to see spectral weighting to represent the mesopic region as well.

The CGPM has approved the use of the CIE $V(\lambda)$ and $V'(\lambda)$ curves for determination of the value of photometric quantities of luminous sources.

7.6 CONVERSION OF PHOTOMETRIC QUANTITIES TO RADIOMETRIC QUANTITIES

The conversion from watts to lumens in the previous section required only that the spectral function, X_λ, of the radiation be known over the spectral range from 360 to 830 nm, where $V(\lambda)$ is nonzero. Attempts to go in the other direction, from lumens to watts, are far more difficult. Since the desired quantity was inside of an integral, weighted by a sensor spectral responsivity function, the spectral function, X_λ, of the radiation must be known over the entire spectral range where the source emits, not just the visible.

For a monochromatic source in the visible spectrum (between the wavelengths of 380 and 860 nm), if the photometric quantity (e.g., lux) is known, apply the conversion $K_m \times V(\lambda)$ and determine the radiometric quantity (e.g., W m^{-2}). In practice, the results that one obtains are governed by the quality of the $V(\lambda)$ correction of the photometer and the knowledge of the wavelength of the source. Both of these factors are of extreme importance at wavelengths where the $V(\lambda)$ curve is steep (i.e., other than very close to the peak of the $V(\lambda)$ curve).

Narrowband sources, such as LEDs, cause major problems. Typical LEDs have spectral bandwidths ranging from 10 to 40 nm full width at half-maximum (FWHM). It is intuitive that in those spectral regions where the $V(\lambda)$ curve is steep, the luminous output will be greater than that predicted using the $V(\lambda)$ curve at the peak LED wavelength. This expected result increases with wider-bandwidth LEDs. Similarly, it is also intuitive that the luminous output is less than that predicted using the $V(\lambda)$ curve when the peak LED wavelength is in the vicinity of the peak of the $V(\lambda)$ curve. Therefore, there must be two wavelengths where the conversion ratio (lm/W) is largely independent of LED bandwidth. An analysis of this conversion ratio was done using a Gaussian equation to represent the spectral power distribution of an LED and applying Eq. 4. Indeed, two null wavelengths were identified (513 and 604 nm) where the conversion between radiometric and photometric quantities is constant (independent of LED bandwidth) to within 0.2 percent up to an LED bandwidth of 40 nm. These wavelengths correspond (approximately) to the wavelengths where the two maxima of the first derivative of the $V(\lambda)$ curve are located.

At wavelengths between these two null wavelengths (513 and 604 nm), the conversion ratio (lm/W) decreases slightly with increasing bandwidth. The worst case occurs when the peak wavelength of the LED corresponds with the peak of $V(\lambda)$. It is about 5 percent lower for bandwidths up to 30 nm, increasing to near 10 percent for 40-nm bandwidth. At wavelengths outside of the null wavelengths, the conversion (lm/W) increases with increasing bandwidth, and the increase is greater when the wavelength approaches the limits of the $V(\lambda)$ curve. Figure 9 shows that factor by which a conversion ratio (lm/W) should be multiplied as a function of LED bandwidth, with the peak LED wavelength as the parameter. Note that the peak wavelength of the LED is specified as the radiometric peak and not as the dominant wavelength (a color specification). The dominant wavelength shifts with respect to the radiometric peak, the difference increasing with bandwidth.

Most often, LEDs are specified in luminous intensity [cd or millicandela (mcd)]. The corresponding radiometric unit is watts per steradian (W/sr). In order to determine the radiometric power (watts), the details of the spatial distribution of the radiant intensity must be known prior to integration.

For broadband sources, a photometric quantity cannot in general be converted to a radiometric quantity unless the radiometric source function, Φ_λ, is known over all wavelengths. However, if the source spectral distribution is blackbody-like (thermal source, spectral emissivity constant between 360 and 830 nm), and the source temperature is also known, then an illuminance can be converted to a spectral irradiance curve over that wavelength range. Again, use Eq. 4 to determine a scale factor for the source term, X_λ. Figs. 10 and 11 show the calculated spectral irradiance versus wavelength for blackbody radiation with source temperature as the parameter.

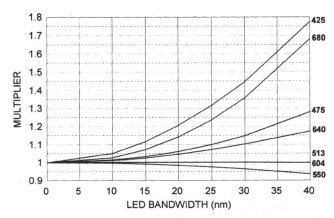

FIGURE 9   Multiplier for converting LED luminous intensity to radiant intensity.

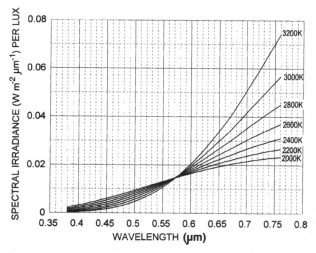

FIGURE 10 Spectral irradiance versus wavelength of blackbody radiation versus temperature.

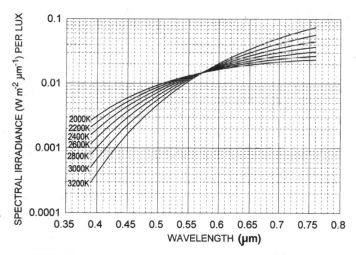

FIGURE 11 Spectral irradiance versus wavelength of blackbody radiation versus temperature.

7.7 RADIOMETRIC/PHOTOMETRIC NORMALIZATION

In radiometry and photometry, there are two mathematical activities that fall into the category called *normalization*. The first is bandwidth normalization. The second doesn't have an official title but involves a conversion between photometric and radiometric quantities, used to convert detector responsivities from amperes per lumen to amperes per watt.

The measurement equation relates the output signal from a sensor to the spectral responsivity of the sensor and the spectral power that is incident upon the sensor.

$$SIG = \int_0^\infty \Phi_\lambda \Re(\lambda)\mathrm{d}\lambda \tag{5}$$

An extended discourse on bandwidth normalization (Nicodemus, 1973) showed that the spectral responsivity of a sensor (or detector) can be manipulated to yield more source information than is immediately apparent from the measurement equation. The sensor weights the input spectral power according to its spectral responsivity, $\Re(\lambda)$, such that only a limited amount of information about the source can be deduced. If either the source or the sensor has a sufficiently small bandwidth such that the spectral function of the other term does not change significantly over the passband, the equation simplifies to

$$SIG = \Phi_\lambda \cdot \Re(\lambda) \cdot \Delta\lambda \qquad (6)$$

where $\Delta\lambda$ is the passband. Spectroradiometry and multifilter radiometry, using narrow-bandpass filters, take advantage of this simplified equation. For those cases where the passband is larger, the techniques of bandwidth normalization can be used. The idea is to substitute for $\Re(\lambda)$ an equivalent response that has a uniform spectral responsivity, $\Re_n$, between wavelength limits λ_1 and λ_2 and zero response elsewhere. Then, the signal is given by

$$SIG = \Re_n \int_{\lambda_1}^{\lambda_2} \Phi_\lambda d\lambda \qquad (7)$$

and now the integrated power between wavelengths λ_1 and λ_2 is determined. There are many ways of assigning values for λ_1, λ_2, and $\Re_n$ for a sensor. Some of the more popular methods were described by Nicodemus (1973) and Palmer (1980). An effective choice is known as the *moments method* (Palmer and Tomasko, 1980), an analysis of the zeroth, first, and second moments of the sensor spectral responsivity curve. The derivation of this normalization scheme involves the assumption that the source function is exactly represented by a second-degree polynomial. If this condition is met, the moments method of determining sensor parameters yields exact results for the source integral. In addition, the results are completely independent of the source function. The errors encountered are related to deviation of the source function from said second-degree polynomial.

Moments normalization has been applied to the photopic spectral luminous efficiency function, $V(\lambda)$, and the results are given in Table 5 and shown in Fig. 12. These values indicate the skewed nature of the photopic and scotopic curves as the deviation from the centroid and the peak wavelengths. The results can be applied to most continuous sources, like blackbody and tungsten radiation, which are both continuous across the visible spectrum. To demonstrate the effectiveness of moments normalization, the blackbody curve was multiplied by the $V(\lambda)$ curve for temperatures ranging from 1000 K to 20,000 K to determine a photometric function [e.g., lumens per square meter (or lux)]. Then, the blackbody curve was integrated over the wavelength interval between λ_1 and λ_2 to determine the equivalent (integrated between λ_1 and λ_2) radiometric function (e.g., in-band watts per square meter). The ratio of lux to watt per square meter is 502.4 ± 1.0 (3σ) over the temperature range from 1600 K to more than 20,000 K. This means that the in-band (487.6 to 632.8 nm) irradiance for a continuous blackbody-like source can be determined using a photometer that is properly calibrated

TABLE 5   Bandwidth Normalization on Spectral Luminous Efficiency

	Photopic	Scotopic
Peak wavelength (λ_p)	555 nm	507 nm
Centroid wavelength (λ_C)	560.19 nm	502.40 nm
Short wavelength (λ_1)	487.57 nm	436.88 nm
Long wavelength (λ_2)	632.81 nm	567.93 nm
Moments bandwidth	145.24 nm	131.05 nm
Normalized $\Re_n$	0.7357	0.7407
Absolute $\Re_n$	502.4 lm/W	1260 lm/W

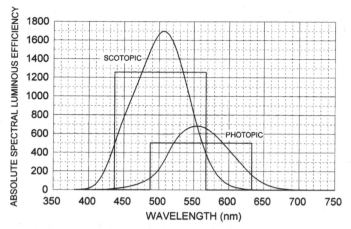

FIGURE 12 Absolute spectral luminous efficiency functions and equivalent normalized bandwidths.

in lux and that is well corrected for $V(\lambda)$. Simply divide the reading in lux by 502.4 to get the in-band irradiance in watts per square meter between 487.6 and 632.8 nm.

If a photometer is available with a $V'(\lambda)$ correction, calibrated for lux (scotopic) with K_m' of 1700 lm/W, a similar procedure is effective. The integration takes place over the wavelength range from 436.9 to 567.9 nm. The ratio of lux (scotopic) to watts per square meter is 1260 ± 2 (3σ) over the temperature range from 1800 K to more than 20,000 K. This means that the in-band (436.9 to 567.9 nm) irradiance for a continuous blackbody-like source can be determined using a $V'(\lambda)$-corrected photometer that is properly calibrated in lux (scotopic). Simply divide the reading in lux (scotopic) by 1260; the result is the in-band irradiance in watts per square meter between 436.9 and 567.9 nm.

A common problem is the interpretation of specifications for photodetectors, which are given in photometric units. An example is a photomultiplier with an S-20 photocathode, which has a typical responsivity of 200 μA/lm. Given this specification and a curve of the relative spectral responsivity, the problem is to determine the output when exposed to a known power from an arbitrary source.

Photosensitive devices, in particular vacuum photodiodes and photomultiplier tubes, are characterized using CIE Illuminant A, a tungsten source at 2854 K color temperature. The illuminance is measured using a photopically corrected photometer, and this illuminance is applied to the device under scrutiny. This technique is satisfactory only if the source that you are using is spectrally comparable to Illuminant A. If you wish to use a source with a different spectral distribution, a photometric normalization must be done. Eberhart (1968) generated a series of conversion factors for various sources and standardized detector spectral responsivities (S-1, S-11, S-20, etc.).

The luminous flux from any source is given by

$$\Phi_v = \int_{380}^{830} \Phi_\lambda V(\lambda) d\lambda \qquad (8)$$

and the output of a detector when exposed to said luminous flux is

$$S = \int_0^\infty \Phi_\lambda \Re(\lambda) d\lambda \qquad (9)$$

where Φ_v is the illuminance, Φ_λ is spectral irradiance, $V(\lambda)$ is the spectral luminous efficiency of the photopic eye, $\Re(\lambda)$ is the absolute spectral responsivity of the photodetector, and S is the photodetector signal. The total responsivity of the detector when exposed to this source is

$$\Re = \frac{\int_0^\infty \Phi_\lambda \Re(\lambda) d\lambda}{\int_{380}^{830} \Phi_\lambda V(\lambda) d\lambda} \quad \text{A/lm} \tag{10}$$

This total responsivity is specific to the source that is used to make the measurement, and it cannot be applied to other sources with differing spectral distributions. To determine the total responsivity (amperes per watt) for a second source, four integrals are required.

$$\Re = \frac{\int_0^\infty \Phi_\lambda \Re(\lambda) d\lambda}{\int_{380}^{830} \Phi_\lambda V(\lambda) d\lambda} \cdot \frac{\int_{380}^{830} \Phi'_\lambda V(\lambda) d\lambda}{\int_0^\infty \Phi'_\lambda d\lambda} \quad \text{A/W} \tag{11}$$

The left ratio of integrals is Eq. 10, the responsivity for source, Φ_λ. The right ratio of integrals is the photometric-to-radiometric conversion for the second source, Φ'_λ. If the two sources have identical spectral power curves ($\Phi_\lambda = \Phi'_\lambda$), two of the integrals cancel, indicating the desirability of using a calibration source that is substantially identical to the application source.

7.8   OTHER WEIGHTING FUNCTIONS AND CONVERSIONS

The general principles that are outlined here can be applied to action spectra other than those already defined for the human eye (photopic and scotopic). Some action spectra take the form of defined spectral regions, such as UVA (315–400 nm), UVB (280–315 nm), UVC (100–280 nm), IR-A (770–1400 nm), IR-B (1400–3000 nm), and IR-C (3000–10^6 nm). Others are more specific. $A(\lambda)$ is for aphakic hazard, $B(\lambda)$ is for photochemical blue-light hazard, $R(\lambda)$ is for retinal thermal hazard, and $S(\lambda)$ is an actinic ultraviolet action spectrum (ANSI, 1996). PPF (a.k.a. PhAR) is a general action spectrum for plant growth. Many others have been defined, including those for erythema (sunburn), skin cancer, psoriasis treatment, mammalian and insect vision, and other generalized chemical and biological photoeffects. Various conversion factors from one action spectrum to another are scattered throughout the popular and archival literature. Ideally, they have been derived via integration over the appropriate spectral regions.

7.9   BIBLIOGRAPHY

ANSI, *Photobiological Safety of Lamps,* American National Standards Institute RP27.3-96 (1996).

BIPM (Bureau International des Poids et Mesures), *The International System of Units (SI),* 7th edition, 1998.

CIE, *The Basis of Physical Photometry,* CIE 18.2, Vienna, 1983.

E. H. Eberhart, "Source-Detector Spectral Matching Factors," *Appl. Opt.* **7**:2037 (1968).

IESNA, *Lighting Handbook: Reference and Application,* M. S. Rea, ed., Illuminating Engineering Society of North America, New York, 1993.

ISO, "Quantities and Units—Part 6. Light and Related Electromagnetic Radiations," *ISO Standards Handbook, Quantities and Units,* 389.15 (1993).

IUPAP, *Symbols, Units, Nomenclature and Fundamental Constants in Physics,* prepared by E. Richard Cohan and Pierre Giacomo, Document IUPAP-25 (SUNAMCO 87-1), International Union of Pure and Applied Physics, 1987.

D. B. Judd, "Report of U.S. Secretariat Committee on Colorimetry and Artificial Daylight," *Proceedings of the Twelfth Session of the CIE (Stockholm)*, CIE, Paris, 1951. [The tabular data is given in Wyszecki and Stiles (1982).]

F. E. Nicodemus, "Normalization in Radiometry," *Appl. Opt.* **12**:2960 (1973).

NIST, *Guide for the Use of the International System of Units (SI)*, prepared by Barry N. Taylor, NIST Special Publication SP811 (1995). (Available in PDF format from *http://physics.nist.gov/cuu/*.)

J. M. Palmer, "Radiometric Bandwidth Normalization Using r.m.s. Methods," *Proc. SPIE* **256**:99 (1980).

J. M. Palmer and M. G. Tomasko, "Broadband Radiometry with Spectrally Selective Detectors," *Opt. Lett.* **5**:208 (1980).

J. M. Palmer, "Getting Intense on Intensity," *Metrologia* **30**:371 (1993).

A. Stockman and L. T. Sharpe, "Cone Spectral Sensitivities and Color Matching," *Color Vision: From Genes to Perception*, K. Gegenfurtner and L. T. Sharpe, eds., Cambridge, 1999. (This information is summarized at the Color Vision Lab at UCSD Web site located at *cvision.uscd.edu/*.)

J. J. Vos, "Colorimetric and Photometric Properties of a 2-Degree Fundamental Observer," *Color Research and Application* **3**:125 (1978).

G. Wyszecki and W. S. Stiles, *Color Science*, 2nd ed., Wiley, New York, 1982.

E. F. Zalewski, "Radiometry and Photometry," Chap. 24, *Handbook of Optics*, Vol. II, McGraw-Hill, New York, 1995.

7.10   FURTHER READING

Books, Documentary Standards, Significant Journal Articles

American National Standard Nomenclature and Definitions for Illuminating Engineering, ANSI Standard ANSI/IESNA RP-16 96.

W. R. Blevin and B. Steiner, "Redefinition of the Candela and the Lumen," *Metrologia* **11**:97 (1975).

C. DeCusatis, *Handbook of Applied Photometry*, AIP Press, 1997. (Authoritative, with pertinent chapters written by technical experts at BIPM, CIE, and NIST. Skip chapter 4!)

J. W. T. Walsh, *Photometry*, Constable, London, 1958. (The classic!)

Publications Available on the World Wide Web

All you ever wanted to know about the SI is contained at BIPM and at NIST. Available publications (highly recommended) include the following:

"The International System of Units (SI)," 7th ed. (1998), direct from BIPM. This is the English translation of the official document, which is in French. Available in PDF format at *www.bipm.fr/*.

NIST Special Publication SP330, "The International System of Units (SI)." The U.S. edition (meter rather than metre) of the above BIPM publication. Available in PDF format from *http://physics.nist.gov/cuu/*.

NIST Special Publication SP811, "Guide for the Use of the International System of Units (SI)," Available in PDF format from *http://physics.nist.gov/cuu/*.

Papers published in recent issues of the NIST Journal of Research are also available on the Web in PDF format from *nvl.nist.gov/pub/nistpubs/jres/jres.htm*. Of particular relevance is "The NIST Detector-Based Luminous Intensity Scale," Vol. 101, p. 109 (1996).

Useful Web Sites

AIP (American Institute of Physics): *www.aip.org*
ANSI (American National Standards Institute): *www.ansi.org/*
BIPM (International Bureau of Weights and Measures): *www.bipm.fr/*

CIE (International Commission on Illumination): *www.cie.co.at/cie/*
Color Vision Lab at UCSD: *cvision.uscd.edu/*
CORM (Council for Optical Radiation Measurements): *www.corm.org*
IESNA (Illuminating Engineering Society of North America): *www.iesna.org/*
ISO (International Standards Organization): *www.iso.ch/*
IUPAP (International Union of Pure and Applied Physics): *www.physics.umanitoba.ca/iupap/*
NIST (National Institute of Standards and Technology): *physics.nist.gov/*
OSA (Optical Society of America): *www.osa.org*
SPIE (International Society for Optical Engineering): *www.spie.org*

P · A · R · T · 2

VISION OPTICS

Jay M. Enoch **Part Editor**

School of Optometry, University of California at Berkeley
Berkeley, California, and
Department of Ophthalmology,
University of California at San Francisco
San Francisco, California

CHAPTER 8

UPDATE TO PART 7 ("VISION") OF VOLUME I OF THE *HANDBOOK OF OPTICS*

Theodore E. Cohn
School of Optometry and Department of Bioengineering
University of California
Berkeley, California

8.1 INTRODUCTION

Part 7 of Volume I (Bass et al., eds., 1995) contained six chapters dealing, respectively, with the optics of the eye, visual performance, colorimetry, displays for vision research, optical generation of the vision stimulus, and with psychophysical methods.

In the five years since publication of that volume, much new information has been added to the literatures underlying these topic areas. In this time period, some things have changed but others did not. For example, Charman (p.24.8) observed "Surgical methods of correction for both spherical and astigmatic refractive errors are currently at an experimental stage . . . and their long-term effectiveness remains to be proven." Today, readers will already know, from ubiquitous advertisement on radio and elsewhere, that some such surgical methods have progressed beyond the experimental stage and are available in clinical settings. Other, alternative strategies have entered the experimental stage [for updates, see the chapters by Liang, by Long, and by Schor in the present volume (Liang et al., 2000; Long, 2000; Schor, 2000)]. Ironically, Charman's observation concerning long-term effectiveness needs no updating.

Authors of chapters in Volume I have been consulted as to the question of whether there is information of which readers should be aware to make best use of the earlier volume. This section summarizes their observations plus those of the editor.

Optics of the eye. Possibly the most significant recent advance regarding optics of the eye has been the application of the Hartmann-Shack wavefront sensor to the measurement of the ocular aberration. This method (Dreher et al., 1989; Bille et al., 1990; Liang et al., 1994; Liang et al., 2000) allows the ocular wavefront aberration to be very rapidly mapped and has been shown to produce results which are in good agreement with psychophysical methods (Salmon et al., 1998). The most exciting aspect of this work is the use of this aberration information to control compensatory adaptive optical systems, thereby potentially enhancing the optical performance of the eye at larger pupil diameters so that it approaches diffraction-limited performance. Some success has been achieved with systems based on liquid crystal spatial light modulators (Vargas-Martin et al., 1998) but to date the best results have been given by adaptive systems based on deformable mirrors (Liang et al., 1997). Use of this approach in a retinal camera has already allowed individual receptors in the retina of the living human eye to be photographed (Miller et al., 1996). Additional information will be found in Chapter 10 by Miller in the present volume.

Visual performance. Erratum: Caption to Fig. 11 (p. 25.23) should read "open circles," not "open squares."

Colorimetry. The CIE (CIE, 1995) has updated its recommendations for uniform color spaces. These recommendations, which extend the CIELAB color space, are described in a technical report of the CIE.

Displays for vision research. Vision research benefits increasingly from the commodity nature of vision displays. Displays of very high resolution and especially speed (up to 150 Hz) with commensurately plunging prices have enabled vision researchers to accomplish things previously unachievable. At the same time, however, the media industry has become able to develop visually interesting scenes for entertainment with the result that vision researchers are likely to gain increasing attention from industry as to safety and related issues. The infamous Pokémon incident of December 16, 1997 (Nomura and Takahashi, 1999), is a case in point. At roughly 6:50 P.M. that day, thousands of viewers of this popular children's show were exposed to an alternating blue and red pattern flickering at about 12 Hz that lasted only about 4 s. Hospitals in Japan recorded over 650 admissions for the treatment of seizures in the period shortly thereafter. (Later surveys established that nearly 1 percent of viewers experienced minor symptoms such as headache and nausea during that particular incident.) The consequence of widespread display of this computer-generated pattern was not surprising to neurologists who routinely use similar provocative visual stimuli to test a patient's susceptibility to epileptic seizures. The lesson for vision science is that visual effects are not the only consequence of visual stimuli, and enhanced ability to produce stimuli of arbitrary spatial, temporal, and chromatic complexity, as enabled by modern display technology, requires a sort of vigilance that has not heretofore seemed important.

Optical generation of the vision stimulus. No update.

Psychophysical methods. For a tutorial on the advantages of using visual noise and efficiency calculations to supplement standard threshold studies, see Pelli and Farell (1999).

8.2   BIBLIOGRAPHY

M. Bass et al. (eds.), *Handbook of Optics,* McGraw-Hill, New York, vol. I, 1995.

J. F. Bille, A. Dreher, and G. Zinser. "Scanning Laser Tomography of the Living Human Eye," in *Noninvasive Diagnostic Techniques in Ophthalmology,* B. Masters, ed., Springer, New York, Chap. 28, pp. 528–547, 1990.

Commission Internationale de L'Eclairage, "Industrial Colour-Difference Evaluation," CIE Technical Report 116-1995, CIE Central Bureau, Vienna, Austria, 1995. http://www.cie.co.at/cie/home.html.

A. W. Dreher, J. F. Bille, and R. N. Weinreb, "Active Optical Depth Resolution Improvement of the Laser Tomographic Scanner," *Applied Optics* **28**:804–808 (1989).

J. Liang, B. Grimm, S. Goelz, and J. F. Bille, "Objective Measurement of the Wave Aberrations of the Human Eye with the Use of a Hartmann-Shack Wavefront Sensor," *J. Opt. Soc. Am. A* **11**:1949–1957 (1994).

J. Liang, D. R. Williams, and D. T. Miller, "Supernormal Vision and High Resolution Imaging Through Adaptive Optics," *J. Opt. Soc. Am. A* **14**:2884–2892 (1997).

D. T. Miller, "Adaptive Optics in Retinal Microscopy and Vision," in *Handbook of Optics,* vol. III, Chap. 10, McGraw-Hill, New York, 2000, pp. 10.1–10.16.

W. F. Long, R. Garzia, and J. L. Weaver, "Assessment of Refraction and Refractive Errors," in *Handbook of Optics,* second ed., vol. III, chap. 11, pp. 11.1–11.24, 2000.

D. T. Miller, D. R. Williams, G. M. Morris, and J. Liang, "Images of Cone Receptors in the Living Human Eye," *Vision Res.* **36**:1067–1079 (1996).

M. Nomura and T. Takahashi, "The Pokémon Incident," Information Display, No. 8, 26–7, Society for Information Display, 1999.

D. G. Pelli and B. Farell, "Why Use Noise?" *J. Opt. Soc. Am. A* **16**:647–653 (1999).

T. O. Salmon, L. N. Thibos, and A. Bradley, "Comparison of the Eye's Wave-Front Aberration Measured Psychophysically and with the Hartman-Shack Wave-Front Sensor," *J. Opt. Soc. Am. A* **15**:2457–2465 (1998).

C. Schor, "Binocular Vision Factors That Influence Optical Design," in *Handbook of Optics,* second ed., vol. III, chap. 12, pp. 12.1–12.40, 2000.

F. Vargas-Martin, P. M. Prieto, and P. Artal, "Correction of the Aberrations in the Human Eye with a Liquid-Crystal Spatial Light Modulator: Limits to Performance," *J. Opt. Soc. Am. A* **15**:2552–2562 (1998).

CHAPTER 9
BIOLOGICAL WAVEGUIDES

Vasudevan Lakshminarayanan
School of Optometry and Department of Physics and Astronomy
University of Missouri—St. Louis
St. Louis, Missouri

Jay M. Enoch
School of Optometry
University of California—Berkeley
Berkeley, California

9.1 GLOSSARY

Apodization.   Modification of the pupil function. In the present context, the Stiles-Crawford directional effect can be considered as an apodization, i.e., variable transmittance across the pupil.

Born approximation.   If an electromagnetic wave is expressed as the sum of an incident wave and the diffracted secondary wave, the scattering of the secondary wave is neglected. This neglect represents what is known as Born's first order approximation.

Bruch's membrane.   Inner layer of the choroid.

Cilia.   Hair.

Cochlea.   The spiral half of the labyrinth of the inner ear.

Coleoptile.   The first true leaf of a monocotyledon.

Copepod.   A subclass of crustacean arthropods.

Cotyledons.   The seed leaf—the leaf or leaves that first appear when the seed germinates.

Ellipsoid.   The outer portion of the inner segment of a photoreceptor (cone or rod). It is located between the myoid and the outer segment. It contains mitochondria for metabolic activity.

Fluence.   A measure of the time-integrated energy flux usually given in units of Joules/cm^2. In biology, it is used as a measure of light exposure.

Helmholtz's reciprocity theorem.   Also known as the reversion theorem. It is a basic statement of the reversibility of optical path.

Henle's fibers.   Slender, relatively uniform fibers surrounded by more lucent Müller cell cytoplasm in the retina. They arise from cone photoreceptor bodies and expand into cone pedicles.

Hypocotyl.   The part of the axis of the plant embryo that lies below the cotyledons.

Interstitial matrix.   Space separating photoreceptors. It is of a lower refractive index than the photoreceptors and, hence, can be considered as "cladding."

Mesocotyl.   The node between the sheath and the cotyledons of seedling grasses.

Mode.   The mode of a dielectric waveguide is an electromagnetic field that propagates along the waveguide axis with a well defined phase velocity and that has the same shape in any arbitrary plane transverse along that axis. Modes are obtained by solving the source-free Maxwell's equations with suitable boundary conditions.

Myoid.   It is the inner portion of the inner segment of a photoreceptor (cone or rod), located between the ellipsoid and the external limiting membrane of the retina.

Organ of Corti.   The structure in the cochlea of the mammals which is the principal receiver of sound; contains hair cells.

Pedicle.   Foot of the cone photoreceptor; it is attached to a narrow support structure.

Phototropism.   Movement stimulated by light.

Spherule.   End bulb of a rod photoreceptor.

V-parameter.   Waveguide parameter (Equation 3). This parameter completely describes the optical properties of a waveguide and depends upon the diameter of the guide, the wavelength of light used, and the indices of refraction of the inside (core) and outside (cladding) of the guide.

Equation 1.   This is a statement of Snell's law.

n_1 = refractive index of medium 1; light is incident from this medium.

n_2 = refractive index of medium 2; light is refracted into this medium.

θ_1 = angle of incidence of light measured with respect to the surface normal in medium 1.

θ_2 = angle of refraction of light in medium 2 measured with respect to the surface normal.

Equation 2.   This is an expression for the numerical aperture.

n_1 = refractive index of the inside of the dielectric cylinder (the "core").

n_2 = refractive index of the surround (the "cladding").

n_3 = refractive index of medium light is incident from before entering the cylinder.

θ_L = limiting angle of incidence.

Equation 3.   Expression for the waveguide V-parameter.

d = diameter of the waveguide cylinder.

λ = wavelength of incident light.

n_1 = refractive index of core.

n_2 = refractive index of cladding.

9.2   INTRODUCTION

In this chapter, we will be studying two different types of biological waveguide model systems in order to relate the models' structures to the waveguide properties initiated within the light-guide construct. In these models, we will be discussing how these biological waveguides follow the Stiles-Crawford effect of the first kind, wavelength sensitivity of transmission, and Helmholtz's reciprocity theorem of optics. Furthermore, we will be investigating the different sources of biological waveguides such as vertebrate photoreceptors, cochlear hair cells (similar to cilia), and fiber-optic plant tissues, as well as considering their applications and comparing their light-guiding effects to the theoretical models. The emphasis will be on waveguiding in vertebrate retinal photoreceptors, though similar considerations apply to other systems.

9.3 WAVEGUIDING IN RETINAL PHOTORECEPTORS AND THE STILES-CRAWFORD EFFECT

Photoreceptor optics is defined as the science that investigates the effects of the optical properties of the retinal photoreceptors—namely, their size, shape, refractive index, orientation, and arrangement—on the absorption of light by the photopigment, as well as the techniques and instrumentation necessary to carry out such studies. It also explores the physiological consequences of the propagation of light within photoreceptor cells. The *Stiles-Crawford effect of the first kind*[1] [*SCE I*], discovered in 1933, represents a major breakthrough in our understanding of retinal physiology and the modern origin of the science of photoreceptor optics. The SCE I refers to the fact that visual sensitivity is greatest for light entering near the center of the eye pupil, and the response falls off roughly symmetrically from this peak. The SCE I underscores the fact that the retina is a complex optical processing system whose properties play a fundamental role in visual processing. The individual photoreceptors (rods and cones) behave as light collectors which capture the incident light and channel the electromagnetic energy to sites of visual absorption, the photolabile pigments, where the transduction of light energy to physicochemical energy takes place. The photoreceptors act as classic fiber-optic elements and the retina can be thought of as an enormous fiber bundle (a typical human retina has about 130×10^6 receptors). Some aspects of the behavior of this fiber optic bundle can be studied using the SCE I function, a psychophysical measurement of the directional sensitivity of the retina. The SCE I is an important index reflecting waveguide properties of the retina and is a psychophysical measurement of the directional sensitivity of the retina. The SCE I can be used to elucidate various properties, and the photodynamic nature, of the photoreceptors. The SCE I can also be used to indicate the stage and degree of various retinal abnormalities. The reader is referred to the literature for detailed discussion of various aspects of photoreceptor optics.[2,3,4,5] The books by Dowling[6] and Rodieck[7], for example, give good basic information on the physiology of the retina as well as discussions of the physical and chemical events involved in the early stages of visual processing.

9.4 WAVEGUIDES AND PHOTORECEPTORS

It appears as though there are two major evolutionary lines associated with the development of visual receptors: the ciliary type (vertebrates) and the rhabdomeric type (invertebrates). Besides being different in structure, the primary excitation processes also differ between these two types of detectors. Ciliary type (modified hair) detectors are found in all vertebrate eyes. They have certain common features with other sensory systems, e.g., the hair cells of the auditory and vestibular systems. It is possible to say that all vertebrate visual detectors have waveguides associated with the receptor element. The incident light is literally guided or channeled by the receptor into the outer segment where the photosensitive pigments are located. In other words, the fiber transmits radiant energy from one point (the entrance pupil or effective aperture of the receptor) to other points (the photolabile pigment transduction sites). Given the myriad characteristics found in photoreceptors in both vertebrate and invertebrate eyes (including numerous superposition and apposition eyes), what is common in all species is the presence of the fiber-optic element. It is apparent that fiber-optic properties evolved twice in the vertebrate and invertebrate receptors. This fact emphasizes the importance of these fiber-optic properties.

Like any optical device, the photoreceptor as a waveguide can accept and contain light incident within a solid angle about its axis. Therefore, all waveguides are directional in their acceptance of light and selection of only a certain part of the incident electromagnetic wavefront for transmittance to the transducing pigment.

The detailed analysis of optics and image formation in various forms of vertebrate and invertebrate eyes is beyond the scope of this chapter (see references 8 through 10 for excellent reviews). However, all such eyes must share a common characteristic. That is, each detecting element must effectively collect light falling within its bound and each must "view" a slightly different aspect of the external world, the source of visual signal. Because of limitations of ocular imagery, there must be a degree of overlap between excitation falling on neighboring receptors; there is also the possibility of optical cross talk. It has been postulated that there is an isomorphic correspondence between points on the retina and a direction in (or points in) real (the outside world) space. A number of primary and secondary sources of light stimulus for the organism are located in the outside world. The organism has to locate those stimuli falling within its field of view and to relate those sensed objects to its own egocentric (localization) frame of reference. Let us assume that there is a fixed relationship between sense of direction and retinal locus. The organism not only needs to differentiate between different directions in an orderly manner, but it must also localize objects in space. This implies that a pertinent incident visual signal is identifiable and used for later visual processing. Further, directionality-sensitive rods and cones function most efficiently only if they are axially aligned with the aperture of the eye, i.e., the source of the visual signal. Thus, it becomes necessary to relate the aperture of the receptor as a fiber-optic element relative to the iris aperture. The retinal receptor-waveguide-pupillary aperture system should be thought of as an integrated unit that is designed for optimization of light capture of quanta from the visual signal in external space and for rejection of stray light noise in the eye/image. A most interesting lens-aperture-photoreceptor system in the invertebrate world is that of a copepod, *Copilia*. The *Copilia* has a pair of image-forming eyes containing two lenses, a corneal lens and another lens cylinder (with an associated photoreceptor), separated by a relatively enormous distance. Otherwise, the photoreceptor system is similar to the insect eye. *Copilia* scans the environment by moving the lens cylinder and the attached photoreceptor. The rate of scan varies from about five scans per second to one scan per two seconds. This is like a mechanical television camera. There are other examples of scanning eyes in nature.[8]

The retina in the vertebrate eye is contained within the white, translucent, diffusing, integrating-sphere-like eye. It is important to realize that the retinal receptor "field of view" is not at all limited by the pupillary aperture. The non-imaged light may approach the receptor from virtually any direction. In Fig. 1 the larger of the two converging cones of electromagnetic energy is meant to portray (in a limited sense) the extent of the solid angle over which energy may impinge on the receptor. While stray light from any small part of this vast solid angle may be small, the total, when integrated over the sphere surrounding the receptor, may be quite a significant sum. To prevent such noise from destroying image quality, the organism has evolved specific mechanisms., including screening dark pigments in the choroid and pigment epithelium, morphological designs, photomechanical changes, aligned photolabile pigment and fiber-optic properties.

Since the pupillary aperture is the source of the pertinent visual signal, the primary purpose of the retinal fiber-optic element, the receptor, is to be a directionally selective mechanism, which accepts incident light from only within a rather narrow solid angle. Given that the receptor has a rather narrow, limited aperture, exhibits directionality, and favors absorption of light passing along its axis, this optical system will be most effective if it is aligned with the source of the signal, the pupillary aperture. This argument applies equally for cones and rods.

In order for a material to serve as a fiber-optic element, the material must be reasonably transparent and have a refractive index that is higher than the immediately surrounding medium. To reduce cross-talk in an array of such fiber-optic elements, they must be separated by a distance approximately equal to or greater than a wavelength of light by a medium having a lower index of refraction than the fiber. The inner and outer segments of retinal photo-

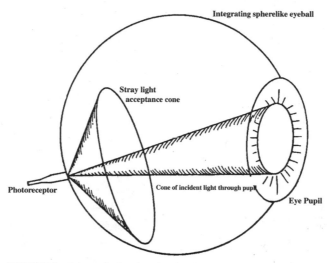

FIGURE 1 Schematic drawing showing light incidence at a single photoreceptor. Light incident on the retina has a maximum angle of incidence of about ±10° for a receptor at the posterior pole of the eye. Stray light may enter the receptor over a much larger acceptance angle (shown by the larger cone).

receptors and the separating interstitial matrix satisfy such condition. The index of refraction of biological materials is generally dependent on the concentration of proteins, lipids, and lipoproteins. The receptor has a high concentration of such solids. These materials are in low concentrations in the surrounding interstitial matrix space. As examples, Sidman[9] has reported refractive index values of 1.4 for rod outer segments, 1.419 for foveal cone outer segments, and a value between 1.334 and 1.347 for the interstitial matrix in the primate retina. Enoch and Tobey[10] found a refractive index difference of 0.06 between the rod outer segment and the interstitial matrix in frogs. Photoreceptors are generally separated by about 0.51 μm. A minimum of about 0.5 μm (1 wavelength in the green part of the spectrum) is needed. Separation may be disturbed in the presence of pathological processes. This can alter waveguiding effects, transmission as a function of wavelength, directionality, cross-talk, stray light, resolution capability, etc.

9.5 PHOTORECEPTOR ORIENTATION AND ALIGNMENT

As noted above, Stiles and Crawford[1] found that radiant energy entering the periphery of the dilated pupil was a less effective stimulus than a physically equal beam entering the pupil center. Under photopic foveal viewing conditions, it was found that a beam entering the periphery of a pupil had to have a radiance of from 5 to 10 times that of a beam entering the pupil center to have the same subjective brightness. Stiles and Crawford plotted a parameter η, defined as the ratio of the luminance of the standard beam (at pupil center) to that of the displaced beam at photometric match point as a function of the entry portion of the displaced beam. The resultant curve has essentially a symmetric fall-off in sensitivity (Fig. 2). This cannot be explained as being due to preretinal factors and this directional sensitivity of the retina is known as the Stiles-Crawford Effect of the first kind (SCE I). Later a chromatic effect (alteration of perceived hue and saturation of displaced beam) was also discovered that is called the *Stiles-Crawford Effect of the second kind*. SCE I is unaffected by polarized light. Normative values of

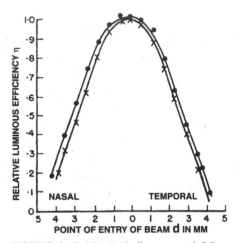

FIGURE 2 Psychophysically measured Stiles-Crawford function for left and right eyes of an observer. Nasal and temporal refer to direction with respect to pupil center. The observer's task is to match the brightness of a beam whose point of entry is varied in the pupil to a fixed beam coming in through the center of the pupil.

the parameters of this function describing the effect (as well as various mathematical descriptions) can be found in Applegate and Lakshminarayanan.[13] Histological and X-ray diffraction studies show that in many species, including human, photoreceptors tend to be oriented not toward the center of the eye, but toward an anterior point. This pointing tendency is present at birth. The directional characteristics of an individual receptor are difficult to assess, and cannot, as yet, be directly related to psychophysical SCE I data. If one assumes that there is a retinal mechanism controlling alignment and that the peak of the SCE I represents the central alignment tendencies of the receptors contained within the sampling area under consideration, it is possible to investigate where the peak of the receptor distribution is pointed at different loci tested across the retina and to draw conclusions relative to the properties of an alignment system or mechanism. Results of the SCE I functions studied over a wide area of the retina show that the overall pattern of receptor orientation is toward the approximate center of the exit pupil of the eye. This alignment is maintained even up to 35 degrees in the periphery. There is evidence that receptor orientation is extremely stable over time.[14]

Photoreceptor orientation disturbance and realignment of receptors following retinal disorders have also been extensively studied. A key result of these studies is that with remission of the pathological process, the system is capable of recovering orientation throughout life. Another crucial conclusion from these studies is that alignment is *locally controlled* in the retina. The stability of the system, as well as corrections in receptor orientation after pathology, implies the presence of at least one mechanism (possibly phototropic) for alignment. Even though the center-of-the-exit-pupil-of-the-eye receptor alignment characteristic is predominant, exceptions to this have been found. These individuals exhibit an approximate center-of-the-retinal-sphere-pointing receptor alignment tendency (Fig. 3). Though certain working hypotheses have been made, what maintains the receptor is an open unanswered question and is the subject of current research that is beyond the scope of this chapter. To summarize, the resultant directionality must reflect the properties of the media, the waveguide properties of receptors, the alignment properties of photolabile pigments, and so forth. Mechanical tractional effects also affect receptor alignment.[15] Other factors could include stress due to microscars, traction or shear forces due to accommodation, inertial forces during saccades,

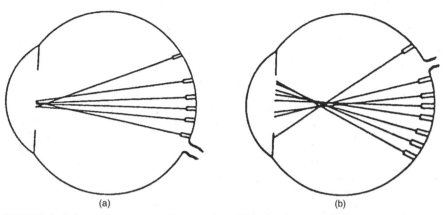

FIGURE 3   Inferred photoreceptor alignment from SCE measurements. (*a*) The more usual center-of-the-exit-pupil pointing tendency. (*b*) Center-of-the-retinal-sphere pointing tendency (rare).

etc. The SCE I function obtained experimentally depends upon all these factors as well as the sum of photoreceptor alignment properties sampled in the retinal test area. The response is dominated by the receptor units most capable of responding in the test area and is affected by the level of light adaptation.

What are the functional visual consequences of SCE I? Recent studies have dealt with this question.[16] SCE is important in photometry and retinal image quality. When light from the edge of a large pupil plays a significant role in degrading retinal image quality, attenuating this light by the effective SCE I apodization will, of course, improve image quality. In general, the magnitude of this improvement will depend on the magnitude of the SCE and the level of blur caused by marginal rays. The presence of aberrations influences the impact of the SCE apodization on defocused image quality. Since defocus of the opposite sign to that of the aberration can lead to well-focused marginal rays, attenuating these well-focused rays with apodization will not improve image quality. SCE will likely improve image quality significantly for positive defocus and not for negative defocus in eyes with significant positive spherical aberration. In addition to generally attenuating the impact of defocus on image contrast, this apodization will effectively remove some of the phase reversals (due to zero crossings of the modulation transfer function) created by positive defocus. The displacement of phase reversals to higher spatial frequencies will have a direct impact on defocused visual resolution. Spatial tasks in which veridical phase perception is critical will be significantly more resistant to positive defocus because of the SCE. It has been argued that measurements of the *transverse chromatic aberration* (*TCA*) can be affected due to SCE. Recent unpublished results (S. Burns, Boston) show that the effect of the SCE on the amount of TCA varies strongly across individuals, and between eyes in the same individual. In conclusion, aberrations and SCE modify the amount and direction of TCA, and SCE does not necessarily reduce the impact of TCA. Longitudinal chromatic aberration, on the other hand, is only slightly affected by SCE I.

9.6   *INTRODUCTION TO THE MODELS AND THEORETICAL IMPLICATIONS*

First, we need to explore the theoretical constructs or models of the biological waveguide systems based on the photoreceptors to get a full appreciation of how these models transmit light through their structure. There are two models of biological waveguide systems that will be discussed here. The first model, which is the retinal layer of rods and cones, is defined as beginning approximately at the *external limiting membrane* (*ELM;* Fig. 4), and it is from that boundary that the photoreceptors are believed to take their orientation relative to the pupillary aperture. In any local area within that layer, the photoreceptors are roughly parallel cylindrical structures of varying degrees of taper, which have diameters of the order of magnitude of the wavelength of light, and which are separated laterally by an organized lower refractive index substance, called the *interphotoreceptor matrix* (*IPM*). The photoreceptors thus form a highly organized array of optical waveguides whose longitudinal axes are, beginning at the ELM, aligned with the central direction of the pupillary illumination. The IPM serves as a cladding for those waveguides. We will assume that the ELM approximates the inner bound of the receptor fiber bundle for light that is incident upon the retina in the physiological direction.[17] Portions of the rod and cone cells that lie anterior to the ELM are not necessarily aligned with the pupillary aperture, and might be struck on the side, rather than on the axis, by the incident illumination. Some waveguiding is observed for a short distance into this region. However, these portions of the rod and cone cells anterior to the ELM would only be weakly excited by the incoming light.

Figure 5 shows the three-segment model that is generally used, denoting the idealized photoreceptor. All three segments are assumed to have a circular cross section, with uniform myoid and outer segment and a smoothly tapered ellipsoid. For receptors with equal myoid and outer segment radii, the ellipsoid is untapered. In this model, the sections are taken to be homogeneous, isotropic, and of higher refractive index than the surrounding near homogeneous and isotropic medium, which is the IPM. Both the myoid and the ellipsoid are assumed to be lossless, or nonabsorbing, since the cytochrome pigments in the mitochondria of the ellipsoids are not considered in this model. Absorption in the outer segments by the photolabile pigment molecules aligned within the closely packed disk membranes is described by the *Naperian absorption coefficient* (α). The aperture imposed at the ELM facilitates calculation of the equivalent illumination incident upon the receptor.

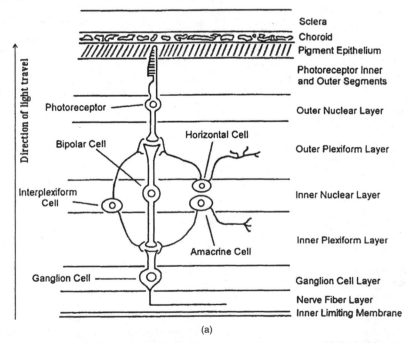

(a)

FIGURE 4   Schematic diagram of the retina and the photoreceptor. (*a*) Retinal layers. Note: the direction of light travel from the pupil is shown by the arrow.

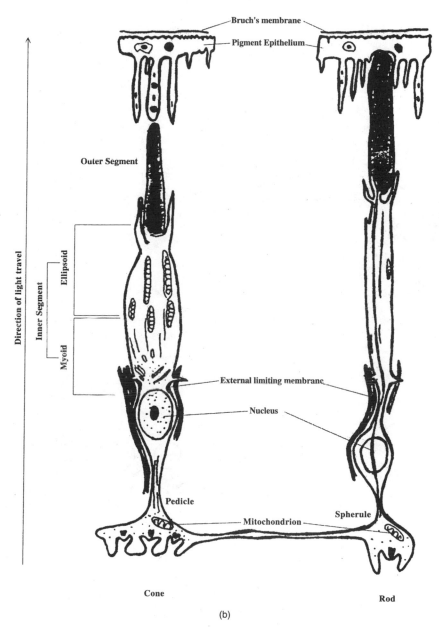

FIGURE 4 (*Continued*) (*b*) Structure of the cone and rod photoreceptor cells.

A number of assumptions and approximations have been made for these two models described above:

1. The receptor cross section is approximately circular and its axis is a straight line. This appears to be a reasonable assumption for the freshly excised tissue that is free of ocular diseases or pathology.

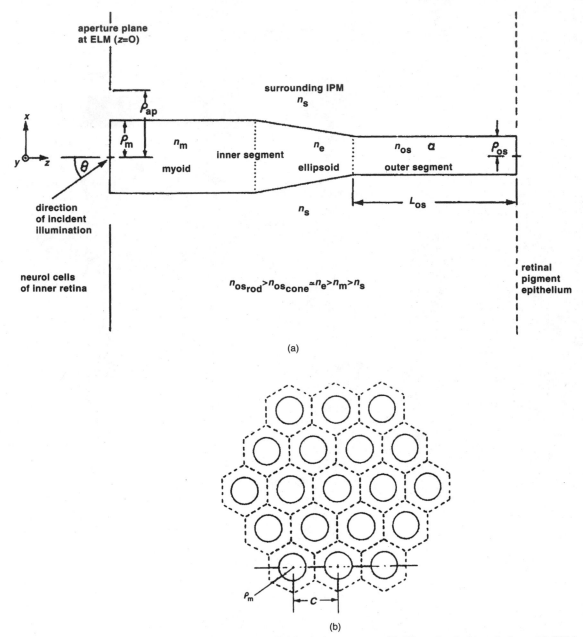

(a)

(b)

FIGURE 5 (*a*) Generalized three-segment model of an idealized photoreceptor. (*b*) View of retinal mosaic from the ELM (External Limiting Membrane), which is the aperture plane of the biological waveguide model for photoreceptors. The solid circles represent myoid cross sections.

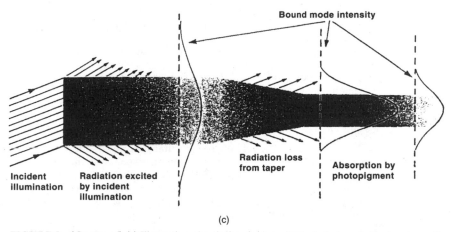

FIGURE 5 (*Continued*) (*c*) Illustration of optical excitation of retinal photoreceptor showing radiation loss, funneling by ellipsoid taper, and absorption in the outer segment. Regions of darker shading represent greater light intensity.

2. Ellipsoid taper is assumed to be smooth and gradual. Tapers that deviate from these conditions may introduce strong mode coupling and radiation loss.

3. Individual segments are assumed to be homogeneous. This is a first-order approximation, as each of the photoreceptor sections has different inclusions, which produce local inhomogeneities.

4. Individual segments are assumed to be isotropic as a first-order approximation. However, in reality, the ellipsoid is packed with highly membranous mitochondria whose dimensions approach the wavelength of visible light, and which are shaped differently in different species. The mitochondrion-rich ellipsoid introduces light scattering and local inhomogeneity. The outer segment is composed of transversely stacked disks, having a repetition period an order of magnitude smaller than the wavelength of light; in addition, these outer segments contain photolabile pigment.

5. Linear media are assumed. Nonlinearities may arise from a number of sources. These could include the following:

   • Absorption of light is accompanied by bleaching of the photopigments and the production of absorbing photoproducts that have different absorption spectra from those of the pigments. Bleaching is energy- and wavelength-dependent and the bleached pigment is removed from the pool of available pigment, thus affecting the absorption spectrum. Under physiological conditions in the intact organism where pigment is constantly renewed, the relatively low illumination levels that result in a small percent bleach can be expected to be consistent with the assumed linearity. However, with very high illumination levels where renewal is virtually nonexistent, researchers must utilize special procedures to avoid obtaining distorted spectra (e.g., albino retina).

   • Nonlinearity may arise in the ellipsoid owing to illumination-dependent activity of the mitochondria, since it has been shown that mitochondria of ellipsoid exhibit metabolically linked mechanical effects. If such effects occurred in the retina, then illumination-driven activity could alter inner segment scattering.

   • Some species exhibit illumination-dependent photomechanical responses that result in gross movement of retinal photoreceptors. That movement is accompanied by an elongation or contraction of the myoid in the receptor inner segment and has been analyzed in terms of its action as an optical switch.[18] In affected species, rod myoids elongate and cone myoids shorten in the light, while rod myoids contract and the cone myoids elongate in the dark. In the light, then, the cones lie closer to the outer limiting membrane and away from the shielding effect of the pigment epithelial cell processes, and contain absorbed pigment, while the pigment granules migrate within the pigment epithelial processes.

6. The refractive indices of the segment are assumed to be independent of wavelength, which completely contradicts the definition of the *refractive index*, which is $n = c/v$ where v (velocity) $= f$ (frequency) $\cdot x$ (wavelength), and v is proportional to wavelength because f is constant for all media. Therefore, a change in the velocity dependent on wavelength also changes the refractive index in real life.

7. The medium to the left of the external limiting membrane is assumed to be homogeneous. However, in reality, although the ocular media and inner (neural) retina are largely transparent, scattering does occur within the cornea, aqueous, lens, vitreous, and inner retina. Preparations of excised retina will be devoid of the ocular media, but the inner retina will be present. In turn, the preparations would account for light scattering upon light incidence.

8. It is assumed that the medium surrounding the photoreceptors is homogeneous and non-absorbing. Again, this an approximation. Microvilli originating in the Müller cells extend into the space between the inner segments, and microfibrils of the *retina pigment epithelium (RPE)* surround portions of the outer segments. Also, the assumption of homogeneity of the surround neglects the reflection and backscatter produced by the RPE, choroid, and sclera. Note that preparations of excised retinae that are used for observing waveguiding are usually (largely) devoid of interdigitating RPE.

9. In some species the outer segment is tapered. Gradual taper can be accommodated in the model. Sharp tapers, however, must be dealt with separately.

10. The exact location of the effective input plane to the retinal fiber optics bundle is not known; however, in early studies it was assumed that the ELM was the effective inner bound of the retinal fiber bundle. To date, though, universal agreement does not exist regarding the location of the effective input plane; some investigators assume it to be located at the outer segment level.

11. In species with double cones (e.g., goldfish), the two components may not be of equal length or contain identical photolabile pigments.

Next, we will evaluate the electromagnetic validity of the preceding models based on the assumptions discussed above. An additional assumption is that meaningful results relating to the in situ properties of a retinal photoreceptor can be obtained by means of calculations performed upon a single photoreceptor model. Involved here are considerations regarding optical excitation, optical coupling between neighboring receptors, and the role of scattered light within the retina. These assumptions may be summarized as follows:

1. The illumination incident upon the single photoreceptor through the aperture in Fig. 5 is taken to be representative of that illumination available to an individual receptor located in the retinal mosaic. Figure 5b illustrates the cross section of a uniform receptor array as seen from the ELM. For a uniformly illuminated, infinitely extended array, the total illumination available to each receptor would be that falling on a single hexagonal area.

2. The open, or unbounded, transverse geometry of the dielectric waveguide supports two classes of modes, referred to as bound, guided, or trapped modes, and unbound, unguided, or radiation modes. Different behavior is exhibited by the two mode species. As their names suggest, the bound modes carry power along the waveguide, whereas the unbound or radiation modes carry power away from the guide into the radiation field (Fig. 5c). From a ray viewpoint, the bound modes are associated with rays that undergo total internal reflection and are trapped within the cylinder; unbound modes are associated with rays that refract or tunnel out of the cylinder. These modes are obtained as source-free solutions of Maxwell's equations and are represented in terms of Bessel and Hankel functions

(for a cylindrical geometry). These mode shapes show a high degree of symmetry. A full discussion of waveguide theory can be found in Snyder and Love.[19]

3. Coherent illumination of the photoreceptors is assumed. Here, the equivalent aperture of the photoreceptor is defined to be the circle whose area is equal to that of the hexagon seen in Fig. 5b specified by the appropriate intercellular spacing and inner segment diameter.

The earliest analyses of light propagation within the retinal photoreceptors employed geometric optics. As early as 1843, Brucke discussed the photoreceptor's trapping of light via total internal reflection (as quoted by von Helmholtz),[20] but stopped just short of attributing a directionality to vision. In fact, Hannover had observed waveguide modes in photoreceptors,[21] but interpreted them as being cellular fine structure (Fig. 6). Interest resumed after the discovery of the Stiles-Crawford effect of the first kind (SCE I). O'Brien[22] suggested a light-funneling effect produced by the tapered ellipsoids, and Winston and Enoch[23] employed geometric optics to investigate the light-collecting properties of the retinal photoreceptor.

Recall the behavior of rays incident upon the interface separating two dielectric media (see Fig. 7a). Rays incident from medium 1 at an angle θ_1 to the interface normal, will be refracted at an angle θ_2 in medium 2; the angles are related by Snell's law where

$$n_1 \sin \theta_1 = n_2 \sin \theta_2, \tag{1}$$

where n_1 and n_2 are the refractive indices of media 1 and 2, respectively. As shown in Fig. 6a, the refracted rays are bent toward the interface, since the incidence is from a medium of greater refractive index toward a medium of lesser refractive index. In this situation, the limiting angle of incidence angle θ_L equals the critical angle of the total internal reflection, for which the refracted ray travels along the interface (angle $\theta_2 = 90$ degrees); from Snell's law, the critical angle = arcsin (n_2/n_1). Rays incident at angles smaller than critical angle will undergo partial reflection and refraction; rays incident at angles equal to or larger than critical angle will undergo total internal reflection.

Let us now consider the pertinent geometry involved with a dielectric cylinder. A circular dielectric cylinder of refractive index n_1 is embedded in a lower-refractive-index surround n_2. Light is incident upon the fiber end face at an angle θ to the z or cylinder axis from a third

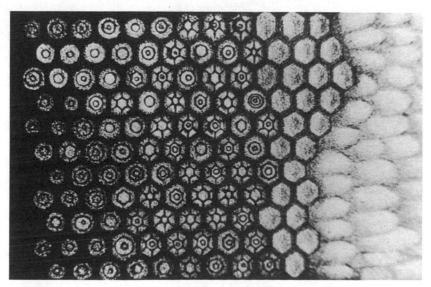

FIGURE 6   Drawing of frog outer segment viewed end-on. (*From Ref. 21.*)

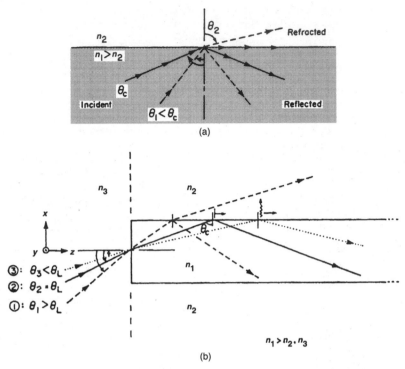

FIGURE 7   (*a*) Reflection and refraction of light rays incident upon an interface separating two different media. (*b*) Ray optics analysis of light incident on an idealized cylindrical photoreceptor (see text for details).

medium of refractive index $n_3 < n_1$ (Fig. 7*b*). We consider the meridional rays, i.e., rays which intersect the fiber axis. There are three possibilities for meridional rays, shown by rays 1 through 3. Once inside the cylinder, ray 1 is incident upon the side wall at an angle smaller than the critical angle, is only partially reflected, and produces, at each reflection, a refracted ray that carries power away from the fiber. Ray 2 is critically incident and is thus totally reflected; this ray forms the boundary separating meridional rays that are refracted out of the fiber from meridional rays that trapped within the fiber. Ray 3 is incident upon the side wall at an angle greater than the critical angle, and is trapped within the fiber by total internal reflection.

Ray optics thus predicts a limiting incidence angle θ_L, related to the critical angle via Snell's law, given by

$$n_3 \sin \theta_L = [(n_1)^2 - (n_2)^2]^{1/2} \tag{2}$$

for which incident meridional rays may be trapped within the fiber. Only those rays incident upon the fiber end face at angles $\theta <$ limiting angle θ_L are accepted; those incident at greater angles are quickly attenuated owing to refraction. The numerical aperture (NA) of the fiber is defined to be $n_3 \sin \theta_L$.

Application of meridional ray optics thus yields an acceptance property wherein only rays incident at $\theta < \theta_L$ are trapped within the fiber. One thing to note is that the greater the angle of incidence is, the faster the attenuation of light will be through the fiber. This result is consistent with results obtained with electromagnetic waveguide modal analysis, which show that the farther away from a cutoff a mode is, the more tightly bound to the cylinder it is, and the closer its central ray is to the cylinder axis.

Geometric optics predicts that an optical fiber can trap and guide light; however, it does not provide an accurate quantitative description of phenomena on fibers having diameters of the order of the wavelength of visible light, where diffraction effects are important. Also, taking a simplified electromagnetic analysis of the photoreceptor waveguide model, illumination is assumed to be monochromatic, and the first Born approximation[24]—which ignores the contribution of waves that are scattered by the apertures—is used to match the incident illumination in the equivalent truncating aperture to the modes of the lossless guide that represents the myoid. Modes excited on the myoid are then guided to the tapered ellipsoid. Some of the initially excited radiation field will have leaked away before the ellipsoid is reached. Each segment is dealt with separately; thus, the ellipsoid will guide or funnel the power received from the myoid to the outer segment. Since each segment is dealt with separately, and because the modes of those segments are identical to those of an infinitely extended cylinder of the same radius and refractive index, the dielectric structure is considered to have an infinite uniform dielectric cylinder of radius p and refractive index n_1 embedded in a lower-refractive-index surround n_2.

9.7 QUANTITATIVE OBSERVATIONS OF SINGLE RECEPTORS

In order to treat the optics of a single receptor theoretically, it is modeled as a cylindrical waveguide of circular symmetry, with two fixed indices of refraction inside and outside the guide. Note that light is propagated along the guide in one or more of a set of modes, each associated with a characteristic three-dimensional distribution of energy density in the guide. In the ideal guide their distributions may be very nonuniform but will show a high degree of symmetry. Modes are grouped into orders, related to the number of distinct modal surfaces on which the electromagnetic fields go to zero. Thus, in Fig. 8, the lowest (first) order mode (designated HE_{11}) goes to zero only at an infinite distance from the guide axis. Note that for all modes the fields extend to infinity in principle. At the terminus of the guide the fiber effectively becomes an antenna, radiating light energy into space. If one observes the end of an optical waveguide with a microscope, one sees a more or less complex pattern of light which roughly approximates a cross section through the modes being transmitted.

The optical properties of the ideal guide are completely determined by a parameter called the *waveguide parameter V,* defined by

$$V = (\pi d/\lambda) \, [(n_1)^2 - (n_0)^2]^{1/2} \tag{3}$$

where d is the diameter of the guide, λ the wavelength in vacuum, and n_1 and n_0 are the indices of refraction of the inside (core) and outside (cladding) of the fiber, respectively. For a given guide, V is readily changed by changing wavelength as shown above. With one exception, each of the various modes is associated with a cutoff value, V_c, below which light cannot propagate indefinitely along the fiber in that mode. Again, the lowest order, HE_{11}, mode does not have a cutoff value. V may be regarded as setting the radial scale of the energy distribution; for $V \gg V_c$, the energy is concentrated near the axis of the guide with relatively little transmitted outside the core. As V approaches V_c, the energy spreads out with proportionally more being transmitted outside the guide wall. A quantity η (not to be confused with the relative luminous efficiency defined for SCE I) has been defined that represents the fraction of transmitted energy which propagates inside the guide wall. For a given mode or combination of modes and given V, η can be evaluated from waveguide theory. Figure 8a shows schematic representations of the mode patterns commonly observed in vertebrate receptors, cochlear hair cells, cilia (various), etc., together with V_c for the corresponding modes. Figure 8b shows values of η as a function of V for some of the lower order modes.[25]

It is evident that an understanding of receptor waveguide behavior requires a determination of V for a range of typical receptors. An obvious straightforward approach is suggested, where if one could determine V at any one wavelength, it could be found for other wave-

MODAL PATTERNS

DESIGNATIONS	NEGATIVE FORM	CUTOFF
HE_{11}		0.0
TE_{01}, TM_{01}		2.405
HE_{21}		2.4+
$(TE_{01}$ OR TM_{01}) + HE_{21}		2.4+
HE_{12}		3.812
EH_{11}		3.812
HE_{31}		3.8+
HE_{12} + (HE_{31} OR EH_{11})		3.8+
HE_{31} + EH_{11}		3.8+
EH_{21} OR HE_{41}		5.2
TE_{02}, TM_{02}, HE_{22}		5.52
$(TE_{02}$ OR TM_{02}) + HE_{22}		5.52

(a)

FIGURE 8 (a) Some commonly observed modal patterns found in human photorecep-tors; their designations and V values.

lengths by simple calculation. How is this done? If one observes a flat retinal preparation illu-minated by a monochromator so that the mode patterns are sharp, then by rapidly changing the wavelength throughout the visible spectrum, some of the receptors will show an abrupt change of the pattern from one mode to another. This behavior suggests that for these recep-tors V is moving through the value of V_c for the more complex mode resulting in a change in the dominant mode being excited. To be valid, this approach requires that the fraction of energy within the guide (η) should fall abruptly to zero at cutoff. In experimental studies, it has been shown that η does go to zero at cutoff for the second-order modes characterized by bilobed or thick annular patterns as shown in Fig. 8. These modes include HE_{12}, TE_{01}, TM_{01},

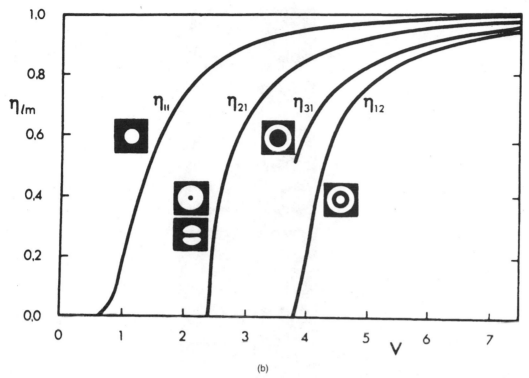

FIGURE 8   (*Continued*) (*b*) The fraction of transmitted energy as a function of the V parameter for some commonly observed modal patterns.

$TM_{01}(TE_{01} + HE_{21})$, and $(TM_{01} + HE_{21})$. Figure 9 displays some waveguide modes obtained in humans and monkeys, as well as variation with wavelength. In turn, it has been discovered that these second-order modal patterns are by far the most frequently observed in small-diameter mammalian receptors. Also, a detailed understanding of receptor-waveguide behavior requires accurate values for guide diameters. Another fact to note concerning parameter V is that the indices of refraction of the entire dielectric cylinder are fairly constant with no pronounced change in waveguide behavior as determined from past experimental studies. For reinforcement, considerable power is carried outside the fiber for low (above-cutoff) values of V. In the limit as $V \rightarrow$ infinity, the mode's total power is carried within the fiber.

The most complete study of the electrodynamics of visible-light interaction with the outer segment of the vertebrate rod based on detailed, first-principles computational electromagnetics modeling using a direct time integration of Maxwell's equation in a two-dimensional space grid for both transverse-magnetic and transverse-electric vector field modes was presented by Piket-May, Taflove, and Troy.[26] Detailed maps of the standing wave within the rod were generated (Fig. 10). The standing-wave data were Fourier analyzed to obtain spatial frequency spectra. Except for isolated peaks, the spatial frequency spectra were found to be essentially independent of the illumination wavelength. This finding seems to support the hypothesis that the electrodynamic properties of the rod contribute little if at all to the wavelength specificity of optical absorption.[27] Frequency-independent structures have found major applications in broad band transmission and reception of radio-frequency and microwave signals. As Piket-May et al.[26] point out, it is conceivable that some engineering application of frequency-independent, retinal-rod-like structures may eventually result for optical signal processing.

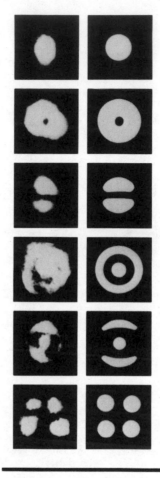

(a)

FIGURE 9 (*a*) Human waveguide modal patterns and their theoretical, idealized equivalents. The bar represents 10 microns.

9.8 WAVEGUIDE MODAL PATTERNS FOUND IN MONKEY/HUMAN RETINAL RECEPTORS

When a given modal pattern is excited, the wavelength composition of the energy associated with the propagation of that pattern in the retinal receptor (acting as a waveguide) may differ from the total distribution of energy at the point of incidence; for it is important to realize that the energy which is not transmitted is not necessarily absorbed by the receptor. It may be reflected and refracted out of the receptor. A distinct wavelength separation mechanism is seen to exist in those retinal receptors observed. In some instances, one type of modal pattern was seen having a dominant hue. Where multiple colors were seen in a given receptor unit, those colors in some instances were due to changes in modal pattern with wavelength. The appearance of multiple colors, observed when viewing the outer segments, occurred in some receptors where mode coupling or interactive effects were present. There are differences in

MONKEY WAVE-GUIDE MODAL PATTERNS

HE₁₁ ⊢—TE₀₁ ᴏʀ TM₀₁ —⊣ HE₁₂

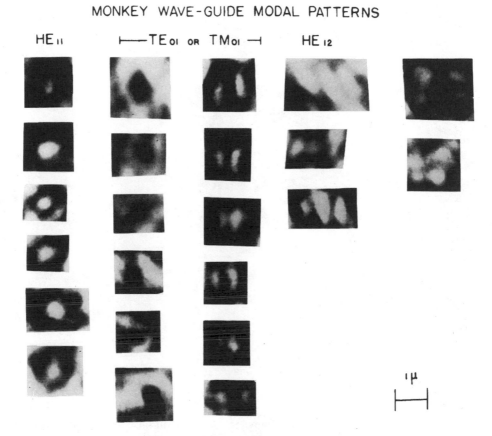

(b)

FIGURE 9   (*Continued*) (*b*) Monkey waveguide modal patterns. The bar represents 1 micron.

the wavelength distribution of the energy transmitted to the terminal end of the receptor. In addition, there are[28] differences in the spatial or regional distribution of that energy within the outer segments as a function of wavelength due to the presence of different modal patterns and interactions governed by parameter V.

If the same well-oriented central foveal receptors of humans are illuminated at three angles of obliquity of incident light, other changes are observed. It is observed subjectively that increased obliquity results in more red and blue being transmitted, or less yellow and green being seen. Why does this occur? This phenomenon is dependent on the angle of incidence of light upon the receptor and the spectral sensitivity of the outer segment. First, the angle of incidence of light upon the receptor determines the amount of light transmitted through the receptor, as shown. Therefore, different values of V result in different amounts of light being collected for a fixed angle of incidence, as well as in different shapes, hence, in different widths, of the respective selectivity curves. Also, as the angle of incidence approaches the limiting angle, there is a decrease in the amount of bound-mode power of light transmitted through the receptor. In turn, absorption is greater for the larger-diameter receptors which, having higher V values, are more efficient at collecting light. So for higher amounts of light transmitted through the receptor with less contrast of different hues, the receptor has a

MODAL PATTERN CHANGES INDUCED BY VARYING WAVELENGTH

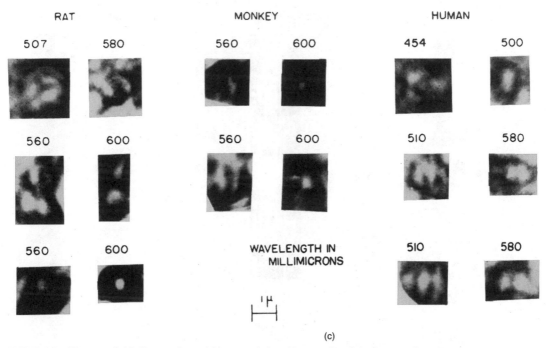

(c)

FIGURE 9 (*Continued*) (*c*) Changes in modal patterns induced by wavelength in three species.

high *V* value (large myoid diameter) and the light incident upon it is closer to normal to the myoid's surface versus its limiting angle. To explain the phenomenon of increased obliquity of incident rays upon the outer segment and/or myoid, one can theorize that this could be due to a higher incidence of striking possibly more pigment present in those segments. In turn, these pigments in the outer segment of the receptor absorb higher amounts of light of wavelength in the range of 460 to 560 nm for transduction and allow higher amounts of shorter wavelength light (blue light—below 460 nm) and longer wavelengths of light (red light—above 580 nm) to pass through the receptor acting as a waveguide. Another fact to note is that the absorption peak occurs at decreasing wavelengths for decreasing receptor radius, and the absorption curves broaden for decreasing radius of outer segment.

Also, since wavelength transmissivity, as well as total transmissivity, changes with obliquity of incidence, it may be further assumed that receptors that are not oriented properly in the retina will not respond to stimuli in the same manner as receptors that are oriented properly. In other words, disturbance in the orientation of receptors should result in anomalous color vision.[29]

It has been suggested that Henle fibers found in the retina may act as optical fibers directing light toward the center of the fovea.[30] Henle fibers arise from cone cell bodies, and turn horizontally to run radially outward from the center of the fovea, parallel to the retinal surface. At the foveal periphery, the Henle fibers turn again, perpendicular to the retinal surface, and expand into cone pedicules. Pedicules of foveal cones form an annulus around the fovea. Cone pedicules are located between 100 to 1000 microns from the center of the fovea and the Henle fibers connecting them to cone photoreceptor nuclear regions can be up to 300 microns long. If this hypothesis is correct, then Henle fiber-optic transmission could increase foveal irradiance, channeling short wavelength light from perifoveal pedicules to the central fovea and increasing the

FIGURE 10 Computed figure showing the magnitude of the optical standing wave within a retinal rod at three different wavelengths (475, 505, and 714 nm). White areas are standing wave peaks, dark areas standing wave nulls. (*From Ref. 26.*)

fovea's risk of photochemical damage from intense light exposure. This property has also been suggested as a reason for foveomacular retinitis burn without direct solar observation and could also be considered as a previously unsuspected risk in perifoveal laser photocoagulation therapy.

Modern theories of the Stiles-Crawford Effect I (SCE I) have used waveguide analysis to describe the effect. The most noteworthy is that of Snyder and Pask.[31] Snyder and Pask hypothesized an ideal "average" foveal cone with uniform dimensions and refractive indices. The cone is thought to be made up of two cylinders representing inner and outer segments. The response of a single cone is assumed to be a monotonically increasing function of absorbed light. The psychophysical response is assumed to be a linear combination of single-cone responses. Values of various parameters in the associated equations were adjusted to fit the wavelength variation of the SCE I reported by Stiles.[32] The model predicts the magnitude and trends of the experimental data (Fig. 11). The predictions, however, are highly dependent on the choice of refractive indices. Even small variations (~0.5%) produces enormous changes in the results. The model is based not upon the absorption of light, but upon transmitted modal power. Other waveguide models include those of, for example, Alpern,[33] Starr,[34] and Wijngaard et al.[35] These papers deal with SCE II, the color effect. Additionally, Goyal et al.[36] have proposed an inhomogeneous waveguide model for cones. A caveat when considering waveguide models is that attention must be given to the physical assumptions made. These, along with uncertainties in the values of various physical parameters (e.g., refractive index), can seriously affect the results and hence the conclusions drawn from these models.

A formalism for modal analysis in absorbing photoreceptor waveguides has been presented by Lakshminarayanan and Calvo.[37,38,39] Here, an exact expression for the fraction of total energy confined in a waveguide was derived and applied to the case of a cylindrically symmetric waveguide supporting both the zeroth and first order sets of excited modes. The fraction of energy flow confined within the cylinder can be decomposed into two factors, one containing a Z-dependence (Z is direction of cylinder axis) as a damping term and the second as a Z-dependent oscillatory term. It should be noted that because of non-negligible absorption, Bessel and Hankel functions with complex arguments have to be dealt with. Using this formalism, if it is found that, using physiological data, the attenuation of the fraction of confined power increased with distance, the energy is completely absorbed for a penetration distance $Z > 60$ microns.

More interesting results are obtained when two neighboring waveguides separated by a distance of the order of the wavelength (a situation found in the foveal photoreceptor mosaic) are

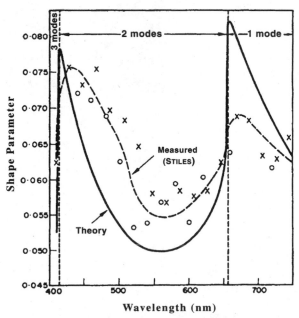

FIGURE 11 The idealized cone model of Snyder and Pask (reference 31). Here the magnitude of the shape parameter (the curvature of the SCE function) is plotted as a function of the wavelength. The solid curve is the theoretical prediction and the dashed curve is plotted through Stiles's experimental points. Note the qualitative agreement between theory and experiment.

considered. An incoherent superposition of the intensity functions in the two waveguides is assumed. The results are shown in Fig. 12. The radial distribution of transmitted energy for two values of core radius (0.45 microns and 0.75 microns) is shown in Fig. 13. It is seen that 70% of the energy is still confined within the waveguide boundary, with some percentage flowing into the cladding, producing an "efficient" waveguide with radius of the order of 0.75 microns and 0.95 microns in the two cases. This effect together with the incoherent superposition implies a mechanism wherein as light propagates along the waveguide, there is a two-way transverse distribution of energy from both sides of the waveguide boundary, providing a continuous energy flow while avoiding complete attenuation of the transmitted signal. Analysis also showed that for specific values of the modal parameters, some energy can escape in a direction transverse to the waveguide axis at specific points of the boundary. This mechanism, a result of the Z-dependent oscillatory term, also helps to avoid complete attenuation of the transmitted signal.

The spatial impulse response of a single-variable cross-section photoreceptor has been analyzed by Lakshminarayanan and Calvo.[40] The modal field propagating under strict confinement conditions can be shown to be written in terms of a superposition integral which describes the behavior of a spatially invariant system. Using standard techniques of linear systems analysis, it is possible to obtain the modulation transfer functions of the inner and outer segments. This analysis neglects scattering and reflection at the inner-outer segment interface. The transfer function depends on both the modal parameter defining the waveguide regime and spatial frequency (Fig. 14). It is seen that both inner and outer segments behave as low-pass filters, although for the outer segment a wide frequency range is obtained, especially for the longer wavelengths. Using a different analysis, Staccy and Pask[41,42] have shown similar results and conclude that photoreceptors themselves contribute to visual acuity through a wavelength-dependent response at each spatial frequency and that the response is dependent on the coherence of the source.

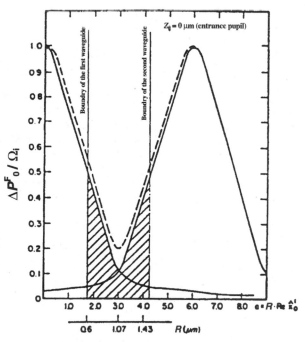

FIGURE 12   Effect of two neighboring absorbing waveguide photoreceptors on the fraction of confined power per unit area. The shaded area denotes the region in which the energy of both waveguides is "cooperative."

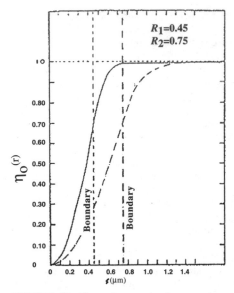

FIGURE 13 Flow of transmitted energy in a monomode absorbing waveguide for two value of the core radius (unbroken lines—0.45 microns; dashed line—0.75 microns) (see text for details).

9.9 LIGHT GUIDE EFFECT IN COCHLEAR HAIR CELLS AND HUMAN HAIR

The vibration of the organ of Corti in response to sound is becoming evident at the cellular level using the optical waveguide property of cochlear hair cells to study the motion of single hair cells. It has been reported that conspicuous, well-defined bright spots can be seen on the transilluminated organ of Corti. The spots are arranged in the mosaic fashion of hair cells known from surface views of the organ of Corti.[43] Thus, the hair cells appear to act as light guides. Also, a definite property of biological waveguide requires the extracellular matrix or fluid to be lower in refractive index as compared to the fiber (in this case, cochlear hair cell), and this property has been elucidated by phase-contrast micrographs of cochlear hair cells.[44]

Cochlear hair cells are transparent, nearly cylindrical bodies with a refractive index higher than the surrounding medium. Inner hair cells are optically denser than the surrounding supporting cells, and outer hair cells are optically denser than the surrounding cochlear fluid. Therefore, in optical studies, hair cells may be considered as optical fibers and the organ of Corti as a fiber-optics array. Even though the significance of the function of the organ of Corti in audition is understood, the role of waveguiding in hair cells is not clearly established.

To check for light-guide action, a broad light is usually used so that many hair cells show up simultaneously as light guides. For broad beams the conditions of light entrance into individual cells are not specified. Light can enter hair cells along the cylindrical wall. Part of such light is scattered at optical discontinuities inside the cells. To a large extent scattered light is trapped in cells by total internal reflection. Thus, to maximize contrast for transillumination with broad beams, no direct light is allowed to enter the microscope. Areas with hair cells appear dark. However, the heads of hair cells appear brightly illuminated owing to the light-collecting effect of hair cells on scattered light. Also, the diameter of cochlear hair cells is from 5 to 9 μm, that is, 10 to 20 times the wavelength of light. In fiber optics, fibers with diameters larger than 10 to 20 times the wavelength are considered multimodal because they sup-

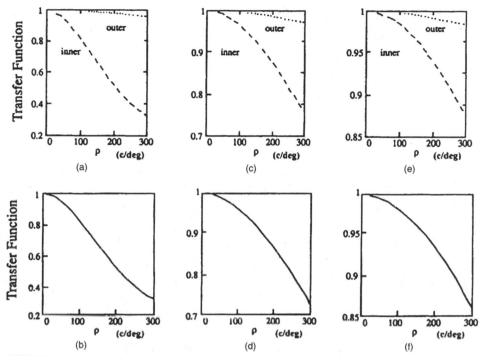

FIGURE 14   Transfer function of the inner and outer segments (top) and the total transfer function of the photoreceptor for three wavelengths: 680, 550 and 460 nm. The values of the modal parameters are fixed. Horizontal axis is in units of cycles/degree.

port a large number of characteristic modes of light-energy propagation. Fibers with diameters smaller than 10 to 20 times the wavelength support fewer modes, and discrete modes of light-energy propagation can be launched more readily. Because of this division, cochlear hair cells fall in between typically single-mode and multimode optical fibers.[45] Also, an important parameter for the evaluation of fibers with small diameters is the parameter V. Recall that the smaller parameter V, the smaller the number of discrete modes a given fiber is capable of supporting. In addition, parameter V is greater for cochlear hair cells than for retinal receptor cells mainly because they have larger diameters than retinal rods and cones (see above for parameter V).

Gross features of surface preparations of the organ of Corti can be recognized with the binocular dissecting microscope. When viewed, the rows of hair cells can be seen as strings of bright pearls. Bright spots in the region of hair cells are due to light transmission through hair cells. Bright spots disappear when hair cells are removed.

Helmholtz's reciprocity theorem of optics[46] (reversibility of light path) also applies to hair cells where exit radiation pattern is equivalent to entrance radiation pattern. There is no basic difference in light-guide action when the directions of illumination and observation are reversed compared to the normal directions. Obviously, Helmholtz's theorem applies to visual photoreceptors too, since they are just modified cilia.

It is also seen that the detection of light guidance is shown to strongly correlate with the viewing angle, where incident light hits the numerical aperture of the hair cell, which is 30 to 40 degrees. When the incident light is normal to the numerical aperture of the hair cell, higher transmittance of light and contrast is observed; however, at more oblique angles to the numerical aperture of the hair cell, less contrast is observed because the angle of incident light is approaching the limiting angle, which is a limit to total internal reflection and more refraction or scattering of rays occurs through the fiber.

A useful tool for studying the quality and preservation of preparations of organ of Corti utilizes the concept that hair cells show little change in appearance under the light microscope for up to 20 to 40 minutes after the death of an animal. Consequently, poor fixation and preservation of the organ of Corti can be recognized immediately at low levels of magnification.

To study the modal light intensity distributions on single hair cells, researchers use a monocular microscope. Hair cells exercise relatively little mode selection, which is to be expected because of their relatively large diameters (large parameter V). As the monocular microscope is focused up and down individual hair cells, it is seen that the intensity distributions in different cross sections of hair cells may differ. Focusing above the exit of hair cells shows that the near-field radiation pattern may differ from the intensity distribution on the cell exit. Furthermore, other studies have shown that the modal distribution of light intensity can be changed drastically for individual cells by changing the angle of light incidence relative to the axis of hair cells. Also, several low-order patterns on the exit end of single hair cells are easily identifiable and are designated according to the customary notation.

As a note, since cochlear hair cells show relatively little mode selection, it is not always possible to launch a desired mode pattern on a given cell with ordinary light. In order to restrict the spectrum of possible modes, it is advantageous to work with monochromatic, coherent light, i.e., laser light. However, the identification of patterns relative to the cells was found to be difficult. This is due to the fact that laser stray light is highly structured (speckled), which makes it difficult to focus the microscope and to recognize cell boundaries. Light guidance in cochlear hair cells may be considered to be a general research tool for cochlear investigations in situations where the organ of Corti can be transilluminated and where relatively low levels of magnification are desirable.

Furthermore, biological waveguide effects have also been show to occur in human hair,[47] which are similar to cochlear hair cell waveguide effects. There is a reduction in light intensity with increased hair length, but by far the most important factor was hair color. Little or no light was transmitted by brown hair, while gray hair acts a natural fiber optic which can transmit light to its matrix, the follicular epithelium, and to the dermis. Interestingly, whether light transmission down hairs affects skin and hair needs investigation. As pointed out by Enoch and Lakshminarayanan,[48] many major sensory systems incorporate cilia in their structure. In the eye, the invagination of the neural groove of the surface ectoderm of the embryo leads to the development and elaboration of the nervous system. The infolding and incorporation of surface tissue within the embryo accounts for the presence of ciliated cells within nervous tissue neurons, which after maturation form photoreceptors containing cilia, lining the outer wall of the collapsed brain ventricle that will form the retina (see Oyster[49] for a description of the embryology and growth of the retina). A study of the evolution of the relationships between cilia and specific sensory systems would be of great interest.

9.10   FIBER-OPTIC PLANT TISSUES

Dark-grown (etiolated) plant tissues are analogous to multiple fiber-optic bundles capable of coherent transfer of light over at least 20 mm; hence, they can transmit a simple pattern faithfully along their length. Each tissue examined can accept incident light at or above certain angles relative to normal as is expected of any optical waveguide. The peak of the curve that describes this angular dependence, the acceptance angle, and the overall shape of the curve appear to be characteristic of a given plant species and not of its individual organs.[50] Knowledge of this optical phenomenon has permitted localization of the site of photoreception for certain photomorphogenic responses in etiolated oats. Before the discussion of the fiber-optic properties of etiolated plant tissues, it is important to have a general idea of what plant structures are presently being discussed to get a full understanding of the structure to function relationship (Fig. 15). To define, the *epicotyl* is a region of shoot above the point of attach-

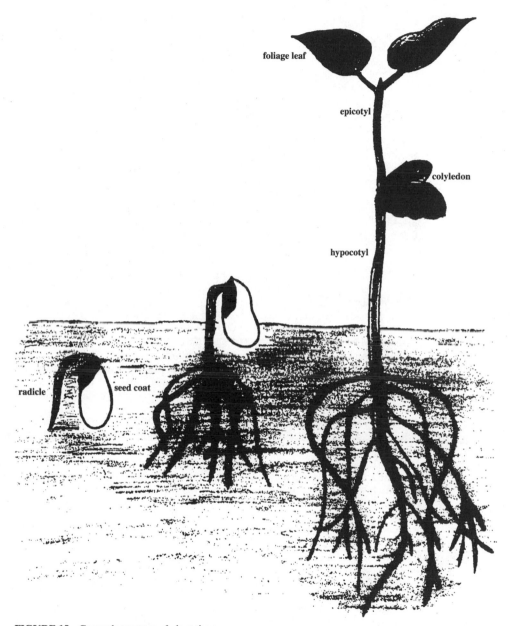

FIGURE 15   General structure of plant tissues.

ment of the cotyledons, often bearing the first foliage leaves. The *hypocotyl* is the part of the embryo proper below the point of attachment of the cotyledons; it will form the first part of the stem of the young plant. The *coleoptile* is a cylindrical sheath that encloses the first leaves of seedlings of grasses and their relatives.

Total internal reflection of light should be independent of fluence rate since this phenomenon is simply a special case of light scattering. The amount of light emerging from the tissue segment was measured first at the highest fluence rate available and then again when the incident light beam had been attenuated with one or more calibrated neutral density filters over a range of 4.5 absorbance units. The amount of light axially transmitted at a given fluence rate (expressed as a percentage of the amount transmitted at the highest fluence rate for individual tissue segments) decreased log-linearly as filters of increasing absorbance were used for oat, mung bean, and corn tissues. Therefore, light guiding in etiolated tissues is clearly fluence-rate-independent over at least 4.5 orders of magnitude.[51]

Light guiding through segments of etiolated root, stem, and coleoptiles, plus primary leaves from several species, all display the spectral dependence expected of any light scattering agent, with low transmission in the blue relative to that in the far-red regions of the spectrum. As tissue length increases, the difference between transmissions in the blue and far-red regions of the spectrum increases.

It has been found that the spectra obtained when the incident beam of white light is applied to the side of the intact plant are indistinguishable from those obtained when the cut end of a tissue is irradiated. Spectra taken on tissue segments excised from different regions of the plant may show traces of certain pigments present only in those regions. Whereas there are no differences between spectra of the apical and basal portions of oat mesocotyls, the hook regions of mung bean hypocotyls contain small amounts of carotenoids, which is indicated as a small dip in the transmittance curve that is absent in spectra from the lower hypocotyl or hypocotyl-root transition zone. Oat coleoptilar nodes, which join the mesocotyl to the coleoptile, decrease light transmission from one organ to another when light is passed from the coleoptile to the mesocotyl or from the mesocotyl to the coleoptile, respectively, but show a spectral dependence which is independent of the direction of light transmission based on Helmholtz's reciprocity theorem of light. Older etiolated leaves do not transmit red light axially relative to the cylindrical coleoptile which sheaths them. Consequently, the presence or absence of the primary leaves does not alter the spectrum of the etiolated coleoptile.

Plant tissues mainly consist of vertical columns of cells. Light internally reflected through them passes predominantly along these columns of cells rather than across columns when the incident beam is at or above the tissue acceptance angle.[50] As optical fibers, the tissues are about 10% as effective as glass rods of comparable diameter, and about 1% as effective as commercial fiber optics.[51] Within a given cell, light could be traveling predominantly in the cell wall, the cytoplasm, the vacuole, or any combination of these three compartments. Observations of the cut ends of tissue segments that are light-guiding indicate that the cell wall does not transmit light axially but that the cell interior does. Also, plant cells contain pigments that are naturally restricted to the vacuole or to the cytoplasmic organelles. Mung bean hypocotyls contain carotenoids (absorption maxima near 450 nm), which are located in the etioplasts, organelles found only in the cytoplasm, whereas etiolated beet and rhubarb stems are composed of cells with clear cytoplasm and vacuoles containing betacyanin and/or anthocyania (absorption maxima near 525 to 530 nm). Spectra of the light guided through these tissues show that light guiding occurs both through the cytoplasm and the vacuole. The relative size of each compartment, the pigment concentration, and the refractive indices involved will largely determine the amount of light that a given compartment will transmit. As the seedlings are exposed to light and become green, troughs due to absorption by chlorophylls (near 670 nm) appear. Completely greened tissue has very little light-guiding capacity except in the far-red region of the spectrum.

Also, the red to far-red (R/FR) ratio determines the fraction of phytochrome in the active state in the plant cell.[52] This ratio plays an important role in the regulation of growth.[52] Similarly, the blue to red (B/R) ratio may be important for responses mediated by phytochrome alone when transmission is low in the red relative to the blue, or for those mediated by both phytochrome and a pigment absorbing blue light. The R/FR ratio of the light emitted from several curved, etiolated plant tissues decreases as the tissue length increases. However, the patterns of change in the B/R ratio differ with each tissue examined. All B/R ratios presumably decrease along the first 20 mm of tissues but in mung hypocotyl (stem) this ratio then increases, whereas in oat mesocotyl it increases and then decreases further. Changes in the R/FR and B/R ratios also occur in a given length of tissue as it becomes green. The R/FR ratio does not change in tissues that remain white (i.e., oat mesocotyl) but decreases dramatically in those that synthesize pigments (i.e., oat coleoptile or mung bean, beet, and rhubarb stems). The B/R ratios, unlike the R/FR ratios, tend to increase as the tissues becomes green, reflecting a smaller decrease in transmission at 450 nm than 660 nm as pigments accumulate.

All of these results demonstrate certain properties of light-guiding effects in plants: the fluence-rate independence of light guiding in plants; an increase in tissue length showing a more pronounced difference between transmission in the blue and far-red regions of the spectra; and the effect of the greening status on the pertinent R/FR and B/R ratios for several species of plants grown under a variety of conditions. Certainly, both the B/R and the R/FR ratios change dramatically in light that is internally reflected through plants as etiolated tissue length is increased and as greening occurs. Finally, the sensitivity of etiolated plant tissues to small fluences of light and the inherent capacity of plants to detect and respond to small variations in the R/FR ratio indicate that light guiding occurring in etiolated plants beneath the soil and the concomitant spectral changes that occur along the length of the plant may be of importance in determining and/or fine-tuning the plant's photomorphogenic response.[53,54]

9.11 SUMMARY

In short, we have pointed out some similar relationships between the theoretical waveguide models and the actual biological waveguides such as vertebrate photoreceptors, cochlear hair cells of the inner ear, human hair (cilia), and fiber-optic plant tissues. The similar relationships between the theoretical model of wave guides and actual biological waveguides were demonstrated as (a) almost entirely following the Stiles-Crawford effect of the first kind (SCE-I) for vertebrate photoreceptors, (b) exhibiting wavelength sensitivity of transmission, and (c) conforming to Helmholtz's reciprocity theorem of optics. Of special interest is the fact that visual photoreceptors exhibit self-orientation. Similar phototropic self-orientation is exhibited by growing and fully-grown plants (e.g., sunflowers). Lastly, for theoretical or possible applications, the light-guiding properties of biological waveguides can be used not only to determine the structure of such guides, but also to provide the basis for determining the functioning of other structures related to these guides.

9.12 REFERENCES

1. W. S. Stiles and B. H. Crawford, "The Luminous Efficiency of Rays Entering the Eye Pupil at Different Points," *Proc. R. Soc. Lond. B.* **112**:428–450 (1933).

2. A. W. Snyder and R. Menzel, *Photoreceptor Optics,* Springer-Verlag, New York, 1975.

3. J. M. Enoch and F. L. Tobey Jr., *Vertebrate Photoreceptor Optics,* vol. 23, Springer Series in Optical Science. Springer-Verlag, Berlin, 1981.

4. J. M. Enoch and V. Lakshminarayanan, "Retinal Fiber Optics," in W. N. Charman (ed.), *Vision and Visual Dysfunction: Visual Optics and Instrumentation,* vol. 1, MacMillan Press, London, pp. 280–309, 1991.

5. V. Lakshminarayanan, "Waveguiding in Retinal Photoreceptors: An Overview," in J. P. Raina and P. R. Vaya (eds.), *Photonics—96: Proceedings of the International Conference on Photonics and Fiber Optics,* vol. 1, McGraw-Hill, New Delhi, pp. 581–592, 1996.

6. J. Dowling, *The Retina: An Approachable Part of the Brain,* Harvard University Press, Cambridge, MA, 1987.

7. R. W. Rodieck, *The First Steps in Seeing,* Sinauer, Sunderland, MA, 1998.

8. W. N. Charman, "Optics of the Eye," in M. Bass et al. (eds.), *Handbook of Optics,* 2nd ed., vol. I, chap. 24, McGraw-Hill, New York, 1995.

9. M. F. Land, "The Optics of Animal Eyes," *Contemp. Physiol.* **29**:435–455 (1988).

10. M. F. Land, "Optics and Vision in Invertebrates," in H. Autrum (ed.), *Handbook of Sensory Physiology,* vol. VII/6B, Springer-Verlag, Berlin, pp. 471–592, 1981.

11. R. Sidman, "The Structure and Concentration of Solids in Photoreceptor Cells Studied by Refractometry and Interference Microscopy," *J. Biophys. Biochem. Cytol.* **3**:15–30 (1957).

12. J. M. Enoch and F. L. Tobey Jr., "Use of the Waveguide Parameter V to Determine the Differences in the Index of Refraction Between the Rat Rod Outer Segment and Interstitial Matrix," *J. Opt. Soc. Am.,* **68**(8):1130–1134 (1978).

13. R. A. Applegate and V. Lakshminarayanan, "Parametric Representation of Stiles-Crawford Function: Normal Variation of Peak Location and Directionality," *J. Opt. Soc. Am. A.,* **10**(7):1611–1623 (1993).

14. J. M. Enoch, J. S. Werner, G. Haegerstrom-Portnoy, V. Lakshminarayanan, and M. Rynders, "Forever Young: Visual Functions Not Affected or Minimally Affected by Aging: A Review," *J. Gerontol. Biol. Sci.* **54**(8):B336–351 (1999).

15. V. Lakshminarayanan, J. E. Bailey, and J. M. Enoch, "Photoreceptor Orientation and Alignment in Nasal Fundus Ectasia," *Optom. Vis. Sci.* **74**(12):1011–1018 (1997).

16. X. Zhang, Ye. Ming, A. Bradley, and L. Thibos, "Apodization of the Stiles Crawford Effect Moderates the Visual Impact of Retinal Image Defocus," *J. Opt. Soc. Am.* **A.16**(4):812–820 (1999).

17. J. M. Enoch, "Optical Properties of Retinal Receptors," *J. Opt. Soc. Am.* **53**:71–85 (1963).

18. W. H. Miller and A. W. Snyder, "Optical Function of Myoids," *Vision Res.* **12**(11):1841–1848 (1972).

19. A. W. Snyder and J. D. Love, *Optical Waveguide Theory,* Chapman and Hall, London, 1983.

20. H. von Helmholtz, *Treatise in Physiological Optics,* Dover, New York, 1962, p. 229.

21. A. Hannover, *Vid. Sel. Naturv. Og Math. Sk.* **X** (1843).

22. B. O'Brien, "A Theory of the Stiles Crawford Effect," *J. Opt. Soc. Am.* **36**(9):506–509 (1946).

23. R. Winston and J. M. Enoch, "Retinal Cone Receptors as an Ideal Light Collector," *J. Opt. Soc. Am.* **61**(8):1120–1122 (1971).

24. M. Born and E. Wolf, "Elements of the Theory of Diffraction," in *Principles of Optics,* 6th ed., chap. 8, Pergamon Press, Oxford, 1984, p. 453.

25. K. Kirschfeld and A. W. Snyder, "Measurement of Photoreceptors Characteristic Waveguide Parameter," *Vision Res.* **16**(7):775–778 (1976).

26. M. J. May Picket, A. Taflove, and J. B. Troy, "Electrodynamics of Visible Light Interaction with the Vertebrate Retinal Rod," *Opt. Letters* **18**:568–570 (1993).

27. A. Fein and E. Z. Szuts, *Photoreceptors: Their Role in Vision,* Cambridge University Press, Cambridge, UK, 1982.

28. J. M. Enoch, "Receptor Amblyopia," *Am. J. Ophthalmol.* **48**:262–273 (1959).

29. F. W. Campbell and A. H. Gregory, "The Spatial Resolving Power of the Human Retina with Oblique Incidence," *J. Opt. Soc. Am.* **50**:831 (1960).

30. M. A. Mainster, "Henle Fibers May Direct Light Toward the Center of the Fovea," *Lasers and Light in Ophthalmol.* **2**:79–86 (1988).

31. A. W. Snyder and C. Pask, "The Stiles Crawford Effect—Explanation and Consequences," *Vision Res.* **13**(6):1115–1137 (1973).

32. W. S. Stiles, "The Directional Sensitivity of the Retina and the Spectral Sensitivities of the Rods and Cones," *Proc. R. Soc. Lond.* B **127**:64–105 (1939).

33. M. Alpern, "A Note on Theory of the Stiles Crawford Effects." In J. D. Mollon and L. T. Sharpe (eds.), *Color Vision: Physiology and Psychophysics,* Academic Press, London, 1983, pp. 117–129.

34. S. J. Starr, *Effects of Luminance and Wavelength on the Stiles Crawford Effect in Dichromats,* Ph.D dissertation, Univ. of Chicago, Chicago, 1977.

35. W. Wijngaard, M. A. Bouman, and F. Budding, "The Stiles Crawford Color Change," *Vision Res.* **14**(10):951–957 (1974).

36. I. C. Goyal, A. Kumar, and A. K. Ghatak, "Stiles Crawford Effect: An Inhomogeneous Model for Human Cone Receptor," *Optik* **49**:39–49 (1977).

37. V. Lakshminarayanan and M. L. Calvo, "Initial Field and Energy Flux in Absorbing Optical Waveguides II. Implications," *J. Opt. Soc. Am.* **A4**(11):2133–2140 (1987).

38. M. L. Calvo and V. Lakshminarayanan, "Initial Field and Energy Flux in Absorbing Optical Waveguides I. Theoretical formalism," *J. Opt. Soc. Am.* **A4**(6):1037–1042 (1987).

39. M. L. Calvo and V. Lakshminarayanan, "An Analysis of the Modal Field in Absorbing Optical Waveguides and Some Useful Approximations," *J. Phys. D: Appl. Phys.* **22**:603–610 (1989).

40. V. Lakshminarayanan and M. L. Calvo, "Incoherent Spatial Impulse Response in Variable-Cross Section Photoreceptors and Frequency Domain Analysis," *J. Opt. Soc. Am.* **A12**(10):2339–2347 (1995).

41. A. Stacey and C. Pask, "Spatial Frequency Response of a Photoreceptor and Its Wavelength Dependence I. Coherent Sources," *J. Opt. Soc. Am.* **A11**(4):1193–1198 (1994).

42. A. Stacey and C. Pask, "Spatial Frequency Response of a Photoreceptor and Its Wavelength Dependence II. Partial Coherent Sources," *J. Opt. Soc. Am.* **A14**:2893–2900 (1997).

43. H. Engstrom, H. W. Ades, and A. Anderson, *Structural Pattern of the Organ of Corti,* Almiquist and Wiskell, Stockholm, 1966.

44. R. Thalmann, L. Thalmann, and T. H. Comegys, "Quantitative Cytochemistry of the Organ of Corti. Dissection, Weight Determination, and Analysis of Single Outer Hair Cells," *Laryngoscope* **82**(11):2059–2078 (1972).

45. N. S. Kapany and J. J. Burke, *Optical Waveguides,* New York, Academic Press, 1972.

46. M. Born and E. Wolf, "Elements of the Theory of Diffraction," in *Principles of Optics,* 6th ed., chap. 8, Pergamon Press, Oxford, 1984, p. 381.

47. J. Wells, "Hair Light Guide" (letter), *Nature* **338**(6210):23 (1989).

48. J. M. Enoch and V. Lakshminarayanan, "Biological Light Guides" (letter), *Nature* **340**(6230):194 (1989).

49. C. Oyster, *The Human Eye: Stucture and Function,* Sinauer, Sunderland, MA, 1999.

50. D. F. Mandoli and W. R. Briggs, "Optical Properties of Etiolated Plant Tissues," *Proc. Natl. Acad. Sci. USA* **79**:2902–2906 (1982).

51. D. F. Mandoli and W. R. Briggs, "The Photoreceptive Sites and the Function of Tissue Light-Piping in Photomorphogenesis of Etiolated Oat Seedlings," *Plant Cell Environ.* **5**(2):137–145 (1982).

52. W. L. Butler, H. C. Lane, and H. W. Siegelman, "Nonphotochemical Transformations of Phytochrome, In Vivo," *Plant Physiol.* **38**(5):514–519 (1963).

53. D. C. Morgan, T. O'Brien, and H. Smith, "Rapid Photomodulation of Stem Extension in Light Grown Sinapin Alba: Studies on Kinetics, Site of Perception and Photoreceptor," *Planta* **150**(2):95–101 (1980).

54. W. S. Hillman, "Phytochrome Conversion by Brief Illumination and the Subsequent Elongation of Etiolated Pisum Stem Segments," *Physiol. Plant* **18**(2):346–358 (1965).

CHAPTER 10
ADAPTIVE OPTICS IN RETINAL MICROSCOPY AND VISION*†

Donald T. Miller

Graduate Program in Visual Science
School of Optometry
Indiana University
Bloomington, Indiana 47405-3680

10.1 GLOSSARY

Adaptive optics (AO). Optoelectromechanical method for closed-loop compensation of wavefront distortions before the wavefront is focused into an image.

Aliasing. Misrepresentation of image content due to undersampling.

Angular frequency. Number of periods of a sinusoidal pattern—located at the retina (or object plane)—that subtends 1° at the nodal point of the eye. Units are in cycles per degree.

Contrast sensitivity. Reciprocal of the minimum perceptible contrast.

Deformable mirror. Mirror whose surface profile can be accurately modified at high speed in response to applied electrical signals.

Detection acuity. Highest detectable spatial frequency.

Ganglion cells. Retinal cells that connect the neural tissue of the eye to the region of the brain termed the *lateral geniculate nucleus.*

Hartmann-Shack wavefront sensor (HSWS). Instrument consisting of a point light source, lenslet array, and CCD camera that is used to noninvasively measure the wave aberrations of the eye.

Higher-order aberrations. Aberrations higher than second order.

Low-order aberrations. Aberrations of zero through second order.

Microaneurysm. Small localized bulge or enlargement of a blood vessel. Usually a result of infection, injury, or degeneration.

Nerve fiber. Axon of a ganglion cell.

* *Acknowledgments.* The author thanks David Williams, Larry Thibos, Arthur Bradley, and Pablo Artal for their helpful comments about the manuscript. This work was supported in part by the National Science Foundation Center for Adaptive Optics.

† The majority of the information in this article appeared in the journal *Physics Today* (January 2000), pp. 31–36.

Ophthalmoscope. Optical instrument for viewing and recording images of the back of the eye through the eye's intact optics.

Photoreceptor. Cell that initiates the visual process in the retina by transducing the captured photons into an encoded neural signal.

Psychophysics. A branch of science that deals with the relationship between physical stimuli and sensory response—for example, using an eye chart to measure a patient's visual acuity.

Retinal microscopy. Observation of microstructures in the retina (e.g., photoreceptors) made through the optics of the eye.

Segmented mirror. An array of small closely spaced mirrors having independent piston and/or tilt movement that are controlled by electrostatic attraction (or repulsion).

Single-sheeted mirror. A single continuous reflective surface whose profile is controlled by an array of actuators that push and pull on the mirror's back surface.

Spectacles. Ophthalmic appliance that corrects prism, defocus, and astigmatism.

Strehl ratio. Ratio of the maximum light intensity in the aberrated point spread function to that in the aberration-free point spread function with the same pupil diameter. Values range from 0 to 1.

Visual acuity. Capacity for distinguishing the details of an object. In units of the Snellen fraction, visual acuity with spectacle correction is typically between 20/15 and 20/20.

Wave aberration. Distortions in the wavefront at the pupil that are incurred by defects in the optics of the system (e.g., optics of the eye).

Zernike polynomials. Two-dimensional polynomials defined to be orthonormal on a circle of unit radius and commonly used to mathematically represent the wave aberration of the eye.

10.2 INTRODUCTION

Vision is a most acute human sense. Yet, one needs only to examine the first step in the visual process—the formation of an image on the retina—to realize that it is often far from perfect. One reason is that the human eye has significant optical defects (i.e., aberrations) that distort the passing optical wavefront. This distortion blurs the retinal image, thereby degrading our visual experience. Diffraction, which is caused by the finite size of the eye's pupil, is the other reason for blurriness. Together, aberrations and diffraction limit not only what the eye sees looking out, but also determine the smallest internal structures that can be observed when looking into the eye with a microscope (see Fig. 1). Spectacles and contact lenses can correct the eye's major aberrations, but if all the aberrations could be quantified and corrected while, at the same time, minimizing diffraction, high-resolution retinal microscopy could become routinely feasible—and we might eventually achieve supernormal vision.

The low-order aberrations of defocus, astigmatism (cylinder), and prism (tilt, in which the eyes do not look in the same direction) are the only optical defects of the eye that are routinely corrected by the use of spectacles or contact lenses. Spectacles have been used to correct defocus since at least as early as the thirteenth century. They began to be used to correct astigmatism shortly after 1801, when Thomas Young discovered that condition in the human eye. In the past two centuries, however, little progress has occurred in correcting additional, higher-order aberrations. There are three reasons for this lack of success.

First, the most significant and troublesome aberrations in the eye are defocus and astigmatism. Correcting these two usually improves vision to an acceptable level. Second, until recently, defocus and astigmatism were the only aberrations of the eye that could be easily measured. Third, even when higher-order aberrations could be measured by the cumbersome techniques that were available in the past, there was no simple or economical way to correct them. In 1961, for instance, M. S. Smirnov of the Soviet Union's Academy of Sciences in Moscow measured many of the higher-order aberrations in the human eye for the first time.[1,2] Unfortunately, his psychophysical method required between 1 and 2 h of measurements per eye, followed by an

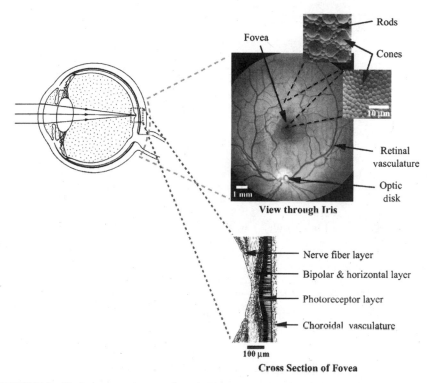

FIGURE 1 The human eye is approximately 25 mm in diameter. The cornea and the ocular lens refract light to form an inverted image on the retina. The retina is a delicate 200- to 400-μm-thick structure, made of highly organized layers of neurons, that is sandwiched between the retinal and choroidal vasculatures. A mosaic of light-sensitive single cells called rod and cone photoreceptors composes one of the retinal layers. The 5 million cones and 100 million rods in each eye sample the optical image at the retina, transducing the captured photons into an encoded neural image. Rods provide low-acuity monochrome vision at night. Cones provide high-acuity color vision during the day. The other retinal layers, as shown in the cross section of the foveal part of the retina, contain a dense arrangement of neurons including bipolar, horizontal, and amacrine cells that preprocess the neural image. The nerve fiber layer transports the final retinal signal through the optics disk opening and to the region of the brain termed the *lateral geniculate nucleus.*

additional 10 to 12 h of calculations. (*Psychophysics* is a branch of science that deals with the relationship between physical stimuli and sensory response—for example, using an eye chart to measure a patient's visual acuity.) On the strength of his results, Smirnov suggested that it would be possible to manufacture custom lenses to compensate for the higher-order aberrations of individual eyes. Now, nearly 4 decades later, no such lens has yet been made.

Recent technological advances, notably in adaptive optics (AO),[3] have provided solutions to the problems that Smirnov encountered. The concept of AO was proposed in 1953 by astronomer Horace Babcock[4]—then director of the Mount Wilson and Palomar Observatories—as a means of compensating for the wavefront distortion induced by atmospheric turbulence.[5]

Transforming AO into a noninvasive vision tool fundamentally consists of measuring the aberrations of the eye and then compensating for them. But the technique has to be able to deal with the problem of variation of ocular and retinal tissue for a given person, as well as from person to person. Also, it must take human safety into account. Despite the difficulties, adaptive optics has met those challenges. Furthermore, as described in this chapter, it has the potential for tackling a wide range of exciting clinical and scientific applications.

10.3 THE MATHEMATICS OF THE EYE'S ABERRATIONS

In an eye without any aberrations whatsoever, the cone of rays emanating from a point source on the retina is refracted by the eye's optics to form parallel rays and a planar wavefront. In a normal eye, though—even one belonging to someone with 20/20 or 20/15 vision—optical defects in the refractive surfaces skew the rays and distort the originally planar wavefront. The distortions blur the retinal image and diminish the eye's ability to see fine detail.

Mathematically speaking, the wave aberration of the eye, as in most optical systems, is usually described by a series of polynomials—a convenient approach in which an individual polynomial represents a particular type of aberration of known characteristics. For example, with Zernike polynomials, which are the preferred choice for the eye, the first-order polynomials represent tilt, the second-order correspond to defocus and astigmatism, the third-order are coma and coma-like aberrations, the fourth-order are spherical and spherical-like aberrations, and the higher orders are known collectively as *irregular aberrations*. Although the Zernike series represents the aberrations of the eye effectively, its popularity is undoubtedly due in part to its long use in astronomy for representing atmospheric turbulence.[6] For the eye, other series could be as good, or even better.

Although we have just begun to accurately measure higher-order aberrations, their presence in the human eye has been recognized for some time.[2] They vary considerably among subjects, and many do not correspond to the classical aberrations of man-made optical systems (e.g., Seidel aberrations). Indeed, Hermann von Helmholtz, commenting on the eye 150 years ago, put it as follows: "Now, it is not too much to say that if an optician wanted to sell me an instrument which had all these defects, I should think myself quite justified in blaming his carelessness in the strongest terms, and giving him back his instrument."[7]

10.4 THE EFFECT OF DIFFRACTION AND ABERRATIONS

Retinal blur is caused not only by ocular aberrations, but also by diffraction, which, for pupil diameters of less than about 3 mm, is actually the largest source of retinal image blur. A pupil about 3 mm in diameter—which is a typical pupil size under bright viewing conditions—generally provides the best optical performance in the range of spatial frequencies that are important for normal vision (i.e., 0–30 cycles per degree). Increasing the pupil size does increase the high-spatial-frequency cutoff, allowing finer details to be discerned, but at the high cost of increasing the deleterious effects of aberrations. Clearly, the very best optical performance would come from correcting all the eye's aberrations across the dilated pupil.

Quantitatively, the effects of diffraction and aberrations on corrected and uncorrected pupils of various diameters are best illustrated by the modulation transfer function (MTF). The MTF is a popular image-quality metric that reflects an optical system's amplitude response to sinusoidal patterns. In vision science, the sinusoids are described in terms of angular frequency and are measured in cycles per degree. Using an angular, as opposed to a linear, measure avoids the need to measure the length of the eye, and the same frequency holds for both object and image space. For example, 1 cycle per degree corresponds to one period of a sinusoid located at the retina (or object plane) that subtends exactly 1° at the eye's nodal point. Most of the visual information we use occupies the range of 0 to 30 cycles per degree.

Figure 2 shows the MTF for normal pupils that are 3 and 7.3 mm in diameter, as well as for a large corrected 8-mm pupil. The large corrected pupil has a higher optical cutoff frequency and an increase in contrast across all spatial frequencies. Peering inward through normal-sized pupils, only ganglion cell bodies and the largest photoreceptors and capillaries are of sufficient size to be resolved with reasonable contrast. However, with the corrected 8-mm pupil, all structures except the smallest nerve fibers can be clearly discerned. Having a large corrected pupil is therefore critical for making use of the spatial frequencies that define single cells in the retina.

Many of the higher-order aberrations in the eye have been measured in research laboratories by means of a variety of psychophysical and objective methods.[2] The optometric and ophthalmic professions, however, have largely ignored such methods because they have not proven robust and efficient enough for practical use in a clinical setting.

To address these limitations, in 1994, Junzhong Liang, Bernhard Grimm, Stefan Goelz, and Josef Bille at the University of Heidelberg[8] developed an outstanding technique based on a Hartmann-Shack wavefront sensor (HSWS), an example of which is shown and described in Fig. 3. The technique is analogous to the artificial guide-star approach that is now being used to measure the wave aberration of atmospheric turbulence.[9]

The HSWS was a watershed. It provided—for the first time—a rapid, automated, noninvasive, and objective measure of the wave aberration of the eye. The method required no feedback from the patient; its measurements could be collected in a fraction of a second; and its near-infrared (near-IR) illumination was less intrusive to the subject than earlier techniques had been. In 1997, Liang, with David Williams at the University of Rochester, developed the technique further and measured the first 10 radial orders (corresponding to the first 65 Zernike modes) to compile what is probably the most complete description to date of the eye's wave aberration.[10]

The left column of Fig. 4 displays surface plots of the point spread function (PSF), measured on two subjects by means of the Rochester HSWS. The PSF, defined as the light distribution in the image of a point source, can be obtained from the measured wave aberration (see Fig. 5) by means of a Fourier transform. In Fig. 4, the average Strehl ratio—defined in the figure caption—for the two (normal) subjects is 0.05. That number is substantially smaller than the diffraction limit of 0.8, reflecting the severity of the aberrations in the normal eye for large pupils. The corresponding wave aberrations, which are shown in Fig. 5, have an average rms of 0.73 μ.

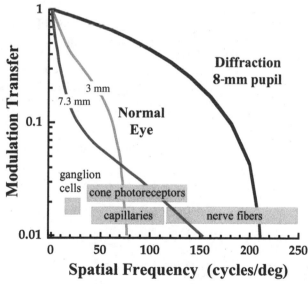

FIGURE 2   Modulation transfer function of the eye under normal viewing, pupils that are 3 and 7.3 mm in diameter and with the best refraction achievable using trial lenses average across 14 eyes.[10,19] Also shown is the diffraction-limited MTF for the 8-mm pupil. The area between the normal and diffraction curves represents the range of contrast and spatial frequencies inaccessible with a normal eye, but reachable with AO. Also indicated is the range of fundamental frequencies defining various-sized ganglion cell bodies, photoreceptors (within 2.5° of the foveal center), foveal capillaries, and nerve fibers.

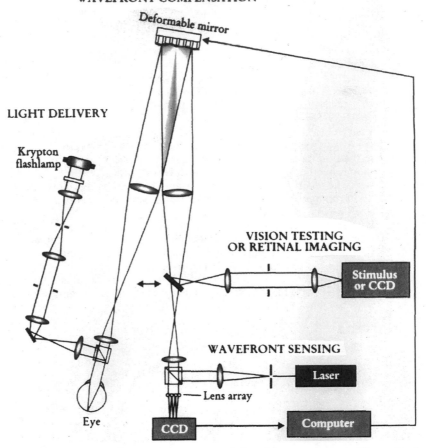

FIGURE 3 Adaptive optics retina camera consisting of four components: (1) wavefront sensing, (2) wavefront compensation, (3) light delivery system, and (4) vision testing and retinal imaging. The HSWS works by focusing a laser beam onto the subject's retina. The reflection from the retinal spot is distorted as it passes back through the refracting media of the eye. A two-dimensional lenslet array, placed conjugate with the eye's pupil, samples the exiting wavefront forming an array of images of the retinal spot. A CCD sensor records the displacement of the spots, from which first local wavefront slopes and then global wavefront shape are determined. Wavefront compensation is realized with a 37-actuator single-sheeted deformable mirror (made by Xinetics, Inc). The mirror lies in a plane conjugate with the subject's pupil and the lenslet array of the HSWS. The light source was a Krypton flashlamp that illuminated a 1° patch of retina with 4-ms flashes. The flashlamp output is filtered to 10 nm, with a center wavelength typically between 550 and 630 nm to eliminate the chromatic aberrations of the eye. A scientific-grade CCD camera was positioned conjugate to the retina to record the reflected aerial image of the retina.

10.5 CORRECTING THE EYE'S ABERRATIONS

There have been various attempts to correct aberrations in the eye beyond defocus and astigmatism. In 1994, Dirk-Uwe Bartsch and Gerhard Zinser at UC San Diego and William Freeman at Heidelberg Engineering attempted to increase the axial resolution of a confocal laser scanning ophthalmoscope by nulling the cornea with a specially designed contact lens.[11] Contact with

Best refraction   **Adaptive correction**

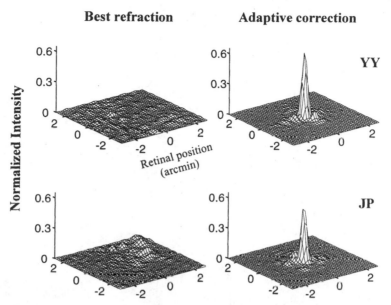

FIGURE 4   The PSF of the eye for two normal subjects (YY and JP) without and with adaptive compensation. The subjects' pupils were 6 mm in diameter. The best refraction was obtained by means of trial lenses. The normalized-intensity ordinate corresponds to the Strehl ratio. The Strehl ratio, which ranges from 0 to 1, is defined as the ratio of the maximum light intensity in the aberrated PSF to that in the aberration-free PSF with the same pupil diameter. (*Source: Measurements courtesy of Geun-Young Yoon and David Williams, University of Rochester.*)

the cornea is required in this approach and aberrations at other ocular surfaces are not corrected. This was noted by Thomas Young when he nulled the refraction of his cornea by submerging his head in water and observed that the astigmatism of his eye remained essentially unchanged.[12] In addition, nulling the cornea disturbs the natural balancing of aberrations that may occur between the cornea and the eye's crystalline lens.[13-15] It may be for these reasons that correcting the cornea alone has produced only modest improvement. Psychophysicists avoid the optics of the eye altogether by producing interference fringes on the retina, using two coherent beams focused in the pupil plane of the eye.[16] This is directly analogous to Young's two-slit experiment. Although a useful tool in the laboratory for probing the visual system without optical blurring, it is not applicable for improving vision or retinal imaging.

Adaptive correction of the eye was first attempted in 1989 by Andreas Dreher, Josef Bille, and Robert Weinreb at UC San Diego.[17] They used a 13-actuator segmented mirror. Their attempt succeeded, but only to the extent that they could correct the astigmatism in one subject's eye by applying his conventional prescription for spectacles.

At the University of Rochester in 1997, Liang, Williams, Michael Morris, and I constructed the adaptive optics (AO) camera shown in Fig. 3.[18,19] The AO channel consisted of an HSWS and a 37-actuator deformable mirror. The system included an additional channel for collecting images of the living retina and measuring the visual performance of subjects. A short description of the system is given in the figure caption and for the interested reader a more detailed one can be found in the preceding two references. With a pupil that is 6 mm in diameter, the system substantially corrected aberrations up through the fourth Zernike order, where most of the significant aberrations in the eye reside. Thus, we demonstrated for the first time that AO allows coma, spherical aberration, and—to a lesser extent—irregular aberra-

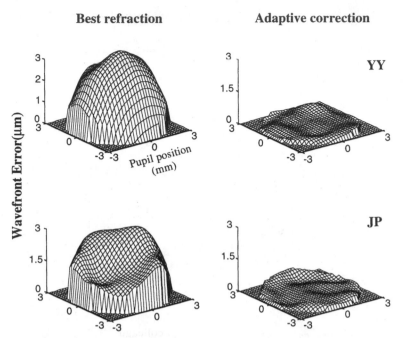

FIGURE 5 The wave aberration of the eye for two normal subjects (YY and JP) without and with adaptive compensation. The subjects' pupils were 6 mm in diameter. The best refraction was obtained by means of trial lenses. The wave aberrations were used to compute the corresponding PSFs displayed in Fig. 4. (*Source: Measurements courtesy of Geun-Young Yoon and David Williams, University of Rochester.*)

tions to be corrected in the eye. The right column of Fig. 4 shows the substantial improvement in retinal image quality that adaptive compensation can facilitate. For the two subjects, the Strehl ratio increased from an average value of 0.05 (uncompensated) to 0.58 (compensated), a factor of 11 increase. Figure 5 shows the corresponding wave aberrations that were used to compute the four PSFs in Fig. 4. Compensation in the two subjects decreased the average wave aberration rms from 0.73 to 0.074 μ, a 10-times improvement. Fig. 5 visually illustrates the effectiveness of AO to flatten the wave aberrations of the eye.

Deformable mirrors, such as the one used in Rochester, are the technology of choice for wavefront correction in AO. Their expense and size, however, have prohibited many vision scientists from taking advantage of this technology and have also precluded the development of commercial ophthalmic instruments that are equipped with AO. Significant reduction in cost and size can be obtained with other corrector technologies, among them microelectromechanical membranes, segmented micromirrors, and liquid crystal spatial light modulators (LC-SLMs), which are based on the same technology as LCD wristwatches. An LC-SLM has been evaluated in conjunction with the eye by Larry Thibos, Xiaofeng Qi, and Arthur Bradley at Indiana University for correction of prism, defocus, and astigmatism.[20,21] Using a similar LC-SLM, Fernando Vargas-Martín, Pedro Prieto, and Pablo Artal at Universidad de Murcia (Spain) attempted to correct the full-wave aberrations of the eye and collect retinal images.[22] Xin Hong, Thibos, and myself at Indiana University have investigated the requirements of segmented piston correctors (e.g., LC-SLM and micromirror devices) that will enable diffraction-limited imaging in normal eyes in both mono- and polychromatic light.[23] Membrane mirrors are just beginning to be explored with recent results on artificial eyes by Lijun Zhu, Pan-Chen Sun, Bartsch, Freeman, and Yeshaiahu Fainman at UC San Diego.[24]

10.6   RETINAL MICROSCOPY WITH ADAPTIVE OPTICS

Imaging the retina through the eye's intact optics dates back to Helmholtz (see Fig. 6), who devised the first ophthalmoscope. Although retina cameras have advanced substantially since the nineteenth century, only recently have they been able to observe retinal features with a size as small as a single cell. Optical blur had been a major factor in precluding retina camera observations, but other characteristics of the eye also impede high-resolution retinal imaging—namely, the factor of 10^{-4} attenuation of light inside the eye, eye motion [with temporal fluctuations of up to 100 Hz and spatial displacements of 2.3 arcmin root mean square (rms), even in the fixating eye], a low damage threshold for light, severe ocular chromatic aberration [2 diopters (D) across the visible spectrum], microfluctuations in ocular power (0.1 D rms), and a light-sensitive iris that can change diameter from 1.5 to 8 mm.

During the last 10 years, research groups in Canada,[25] United Kingdom,[26] Spain,[27,28] and the United States[18,19,29] have developed high-resolution retina cameras that address these imaging problems. The designs incorporate such methods as a dilated pupil, quasi-monochromatic illumination, short exposure times of a few milliseconds, and precise correction of defocus and astigmatism. Those retinal images, containing in some cases spatial frequencies well above 60 cycles per degree (normal vision makes use of spatial frequencies of 0–30 cycles per degree), have been good enough to give researchers a tantalizing first-ever glimpse of single cells in the retina of the living human eye. However, the microscopic structure revealed in the images typically has been of low contrast, while also being noisy and limited to subjects with good optics. To achieve better imaging will require the correction of aberrations beyond defocus and astigmatism.

Using the AO camera in Fig. 3, my colleagues and I obtained the first images of the living retina with correction to the higher-order aberrations. Figure 7 shows three images, collected with the camera, of the same patch of retina both with and without adaptive correction. A regular array of small bright spots fills the images. Each bright spot corresponds to light exiting an individual cone photoreceptor after reflection from retinal tissue lying behind the receptor (compare it with the photoreceptor micrographs shown in Fig. 1). The center image was collected with adaptive correction and has noticeably higher contrast and resolution, with individual photoreceptors more clearly defined. To reduce the effect of noise present in these images, multiple images of the same patch of retina were registered and averaged. The rightmost image is a registered stack of 61 corrected images. The improvement in image clarity that is made possible by AO is striking.

10.7   ADAPTIVE OPTICS AND VISION

The correction of higher-order aberrations also offers the possibility of better vision. Because diffraction, which cannot be mitigated, is more deleterious for small pupils than aberrations are, the largest benefits will accrue when the pupil is large—under indoor lighting conditions, for example. Under bright outdoor conditions, the natural pupil of most normal eyes is sufficiently small that diffraction dominates and monochromatic aberrations beyond defocus and astigmatism are relatively unimportant.

To illustrate what the retinal image would be like with supernormal optics, imagine yourself viewing the Statue of Liberty from a boat that is 3 km away in New York Harbor. The statue subtends almost 0.9°—about the same as a U.S. quarter at a distance of 5 ft. Under optimal viewing conditions and with defocus and astigmatism removed, your retinal image of the statue with a normal 3-mm pupil would look like the image in Fig. 8a. If you viewed the statue through AO programmed to fully correct all ocular aberrations across your 3-mm pupil, the retinal image of the statue would look like the image in Fig. 8b. Notice the finer detail and higher contrast. The comparison illustrates that retinal image quality can be improved even for pupil sizes as small as 3 mm.

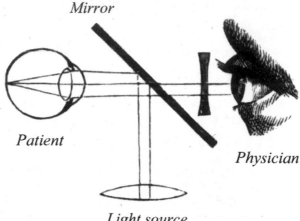

Mirror

Patient

Physician

Light source

FIGURE 6 Schematic of Helmholtz's original ophthalmoscope. Imaging the inside of the human eye began with Helmholtz's invention of the ophthalmoscope in 1850. For the first time the inner structure of the living eye could be viewed, giving the physician a tremendous aid in diagnosing the eye's health. Though quite simple, it is still the basis for today's modern ophthalmoscopes. It consisted of a mirror and a light source. The mirror enabled the light source to illuminate the patient's retina at the same time the physician viewed the retinal reflection through a small hole in the mirror. The hole physically separated the illumination and viewing paths, preventing corneal reflections from obstructing the physician's view. The physician's observations were initially recorded in hand paintings and then replaced by photographic film, with the first successful photograph of the human retina taken in 1886 using a 2.5-min exposure. Eye motion was reduced by fixing the camera to the patient's head. The next 100 years produced a steady advance in better light sources (in particular, the electronic flashlamp), film quality, development techniques, procedures to collect retinal photographs, and now the use of large-format CCD cameras and automated computer-driven systems. It is interesting to note, however, that as successful as these advanced retina cameras have been, none correct for the aberrations of the eye beyond defocus and astigmatism. Therefore, their optical resolution is still fundamentally limited by the same ocular aberrations that limited Helmholtz 150 years ago. (*Source: Reproduced from Helmholtz, 1924, with labels modified.*[31])

The best retinal image quality, however, is obtained with the largest physiological pupil diameter (8 mm) and with full correction of all ocular aberrations. This case is depicted in Fig. 8c, which shows that the theoretical maximum optical bandwidth that can be achieved with the human eye is substantially higher than what we are endowed with.

Improving the quality of the retinal image is an important first step toward achieving supernormal vision—but it is only the first step. In an eye with perfect optics, visual performance becomes constrained by neural factors, specifically, the spacing between retinal photoreceptors, which represents a neural limitation to visual resolution that is only slightly higher than the normal optical limit. In addition, there is one expected penalty with supernormal visual optics. If the optical quality of the retinal image improves dramatically, the photoreceptor mosaic will appear relatively coarse by comparison, as shown in Fig. 9. As a result of that mismatch, very fine spatial details in the retinal image will be smaller than the distance between

Uncorrected　　Corrected　　Stack of Corrected

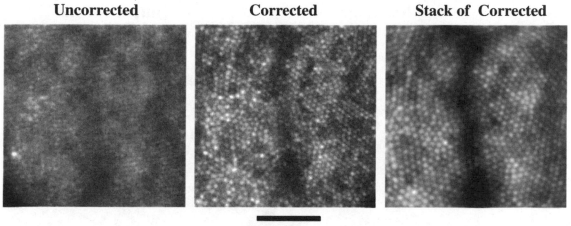

10-min. arc
(48.6 μm)

FIGURE 7 Retinal images, before and after adaptive compensation. All three images are of the same 0.5° patch of retina, obtained with 550 nm of light through a 6-mm pupil. The dark vertical band across each image is an out-of-focus shadow of a blood vessel. The leftmost image shows a single snapshot taken when only defocus and astigmatism had been corrected. The middle image shows additional aberrations having been corrected with AO. It also shows noticeably higher contrast and resolution, with individual photoreceptors more clearly defined. To reduce the effect of noise present in the images, the rightmost image demonstrates the benefit of registering and then averaging 61 images of a single retinal patch. (*Source: Images courtesy of Austin Roorda and David Williams, University of Rochester.*)

neighboring cones, so those details will not be properly registered in the neural image. This misrepresentation of the image due to neural undersampling by a relatively coarse array of photoreceptors is called *aliasing*. For everyday vision, however, the penalty of aliasing is likely to be outweighed by the reward of heightened contrast sensitivity and detection acuity. Therefore, we anticipate that correcting the eye's optical aberrations will yield a net increase in the quality of the patient's visual experience and, therefore, is worth pursuing. Indeed, laboratory observations indicate that stimuli seen through AO have the strikingly crisp appearance that is expected of an eye with supernormal optical quality.[18,19] This is qualitatively consistent with the six- and threefold increase in contrast sensitivity recently measured in Williams's laboratory (University of Rochester) for quasi-monochromatic and white-light stimuli, respectively.

It is unclear how well supernormal vision achieved under controlled conditions in the laboratory will transfer to everyday use. Nor is it clear just what the benefits of acquiring such vision will be. We can be fairly sure, however, that supernormal vision will not be realized with traditional spectacles, because the rotation of the eye in relation to the fixed spectacle lens severely limits the types of aberrations that can be corrected.

To counteract the effects of eye rotation, engineers could set out to devise a sophisticated visor (like the one worn by *Star Trek*'s Geordi LaForge) that dynamically adapts the correction to the wearer's focus and direction of gaze. Contact lenses, though, are a more realistic and inexpensive option. They move with the eye, and with current technology, they could be tailored to correct the higher-order aberrations in individual eyes. Another option is refractive surgery. The leading surgical technique today, laser-assisted in situ keratomileusis, corrects the refractive errors of both defocus and astigmatism by ablating corneal tissue with an excimer laser. That technology could be extended to compensate for the higher-order aberrations.

Adaptive optics has the potential to realize these corrective procedures by means of providing substantially faster and more sensitive measurements than previously possible. For example, measuring the eye's wave aberration before and after refractive surgery might help improve the surgery and optimize retinal image quality—perhaps making the surgical option

**Normal
Vision** ⟶ **Supernormal
Vision?**

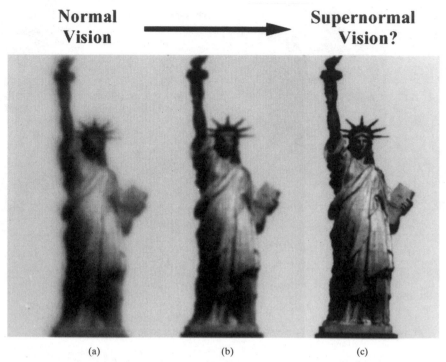

(a)   (b)   (c)

FIGURE 8   Predicted retinal image of the Statue of Liberty when viewed at a distance of 3 km with
(*a*) normal optics and a 3-mm-diameter pupil, (*b*) corrected optics and a 3-mm pupil, and (*c*) corrected optics and an 8-mm pupil. A narrow-band chromatic filter is placed in front of the viewer's
eye to prevent ocular chromatic aberrations. Optical microfluctuations in the eye are ignored. The
wave aberration used in the simulation is that of the author's eye (he has normal vision with spectacle correction). (*Source: Wave aberration measurement courtesy of Xin Hong and Larry Thibos, Indiana
University.*)

superior to contact lenses and spectacles. The same approach may be applicable to improving
other corrective procedures, such as the use of intracorneal polymer rings that are surgically
implanted in the cornea to alter corneal shape. Wave aberration measurements might also
assist in the design and fitting of contact lenses. That application would be of particular benefit to patients with irregularly shaped corneas, such as in kerataconus (where cone-shaped
bulges in the cornea form after abnormal corneal thinning).

10.8   MEDICAL AND SCIENTIFIC APPLICATIONS

The medical and scientific benefits of AO—other than improving vision—are just beginning
to be explored. Among the numerous potential applications are detection of local pathological changes and investigation of the functionality of the various retinal layers. A few specific
examples follow; more can be found in the literature.[19]

Glaucoma is a disease in which blindness ensues from the gradual loss of optic nerve fibers.
With conventional techniques, it can be detected only after significant damage has occurred.
By exploiting the greater sensitivity of AO, eye care specialists could make more accurate
measurements of the thickness of the nerve fiber layer around the head of the optic nerve,
and, conceivably, individual fibers or ganglion cell bodies could be monitored. Such detailed
information would permit glaucoma to be detected earlier. Another example is age-related
macular degeneration, which causes a progressive loss of central vision in 10 percent of those

FIGURE 9 Enlargement of the Statue of Liberty image of Fig. 8c, overlaid with a hexagonally packed mosaic of circles that represent the foveal cone mosaic. This neural mosaic is relatively coarse compared with the detail of the retinal image, a phenomenon that may introduce artifacts into the neural image and ultimately cause a kind of misperception called *aliasing*.

over age 60. It has been suggested that photoreceptor death is an early indicator of this disease, yet this loss is not detectable in the retina with current techniques. Adaptive optics could reveal the early stages of this disease by directly monitoring receptor loss over time.

Diabetes, another disease that affects the retina, produces microaneurysms in the retinal vasculature that gradually grow larger and leak blood. Detection at a small size is critical for successful retinal surgery. With AO, much smaller microaneurysms could be detected than can be now, possibly without the use of the invasive fluorescent dyes required by the current detection methods.

It is now possible to study the living retina at a microscopic scale. Unfortunately, studies of the retina at the single-cell scale require removing the retina from the eye. That procedure has two great drawbacks: (1) It drastically limits the types of experiment that can be conducted, and (2) it irreversibly changes the tissue in ways that confound experiments.

It is now possible to collect whole temporal sequences in individual eyes to track retinal processes that vary over time at the microscopic spatial scale. Those data could be used, for example, to study both the development of the normal eye and the progression of retinal disease. Studying the retina at the single-cell level would improve our understanding not only of the light-collecting properties of the retina, but also of the spatial arrangement and relative numbers of the three types of cone photoreceptor that mediate color vision. Recently, Austin Roorda and Williams at Rochester used the AO camera shown in Fig. 3 as a retinal densitometer to obtain the first images showing the arrangement of the blue, green, and red cones in the living human eye.[30] Fig. 10 displays their false-grayscale images from two subjects. In spite of the striking differences in the proportion and pattern of cone types in these two subjects, each had normal color vision.

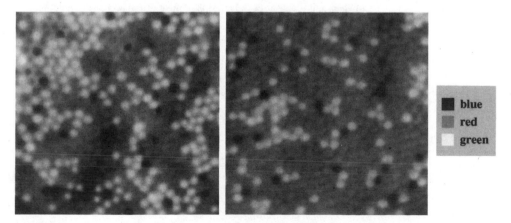

FIGURE 10 False-color images showing the arrangement of the blue, green, and red cone photoreceptors in the eyes of two different people. Both subjects have normal color vision in spite of the striking differences in the pattern and proportion of the three cone types. (*Source: Courtesy of Roorda and Williams, University of Rochester.*)

The realization of all these future applications will depend heavily on the continuing development of less expensive and smaller devices for AO. Such advances have traditionally excited the astronomy and military communities. Today, that excitement is also shared by those of us who conduct research in optometry, ophthalmology, and vision science.

10.9 REFERENCES

1. M. S. Smirnov, "Measurement of the Wave Aberration of the Human Eye," *Biophysics* **7**:766–795 (1962).

2. For reviews, see W. N. Charman, "Optics of the Eye," *Handbook of Optics,* M. Bass, E. W. Van Stryland, D. R. Williams, W. L. Wolfe (eds.), McGraw-Hill, New York, 1995. See also *Journal of the Optical Society of America* **A15**:2444–2596 (1998), published as a special issue, "Measurement and Correction of the Optical Aberrations of the Human Eye."

3. R. K. Tyson, *Principles of Adaptive Optics,* Academic Press, San Diego, 1998.

4. H. W. Babcock, "The Possibility of Compensating Astronomical Seeing," *Publications of the Astronomical Society of the Pacific* **65**:229–236 (1953).

5. L. A. Thompson, "Adaptive Optics in Astronomy," *Physics Today* 24–31 (December 1994).

6. M. C. Roggemann and B. Welch, *Imaging through Turbulence,* M. J. Weber (ed.), CRC Press, Boca Raton, 1996.

7. H. von Helmholtz, *Popular Scientific Lectures,* M. Kline (ed.), Dover Publications, New York, 1962.

8. J. Liang et al., "Objective Measurement of the Wave Aberrations of the Human Eye Using a Shack-Hartmann Wavefront Sensor," *Journal of the Optical Society of America* **A11**:1949–1957 (1994).

9. R. Q. Fugate et al., "Measurement of Atmospheric Wavefront Distortion Using Scattered Light from a Laser Guide-Star," *Nature* **353**:144–146 (1991).

10. J. Liang and D. R. Williams, "Aberrations and Retinal Image Quality of the Normal Human Eye," *Journal of the Optical Society of America* **A14**:2873–2883 (1997).

11. D. Bartsch et al., "Resolution Improvement of Confocal Scanning Laser Tomography of the Human Fundus," in *Vision Science and Its Applications,* 1994 Technical Digest Series (Optical Society of America, Washington, DC), **2**:134–137 (1994).

12. T. Young, "On the Mechanism of the Eye," *Philosophical Transactions of the Royal Society of London* **91**:23–88 (1801).

13. S. G. El Hage and F. Berny, "Contribution of the Crystalline Lens to the Spherical Aberration of the Eye," *Journal of the Optical Society of America* **63**:205–211 (1973).

14. P. Artal and A. Guirao, "Contributions of the Cornea and the Lens to the Aberrations of the Human Eye," *Optics Letters* **23**:1713–1715 (1998).

15. T. O. Salmon and L. N. Thibos, "Relative Contribution of the Cornea and Internal Optics to the Aberrations of the Eye," *Optometry and Vision Science* **75**:235 (1998).

16. F. W. Campbell and D. G. Green, "Optical and Retinal Factors Affecting Visual Resolution," *J. Physiol. (London)* **181**:576–593 (1965).

17. A. W. Dreher et al., "Active Optical Depth Resolution Improvement of the Laser Tomographic Scanner," *Applied Optics* **24**:804–808 (1989).

18. J. Liang et al., "Supernormal Vision and High Resolution Retinal Imaging Through Adaptive Optics," *Journal of the Optical Society of America* **A14**:2884–2892 (1997).

19. D. R. Williams et al., "Wavefront Sensing and Compensation for the Human Eye," *Adaptive Optics Engineering Handbook,* R. K. Tyson (ed.), Marcel Dekker, New York, 2000.

20. L. N. Thibos and A. Bradley, "Use of Liquid-Crystal Adaptive Optics to Alter the Refractive State of the Eye," *Optometry and Vision Science* **74**:581–587 (1997).

21. L. N. Thibos et al., "Vision Through a Liquid-Crystal Spatial Light Modulator," in G. D. Love (ed.), *Proceedings for Adaptive Optics for Industry and Medicine,* World Scientific, Singapore, 1999.

22. F. Vargas-Martín et al., "Correction of the Aberrations in the Human Eye with a Liquid-Crystal Spatial Light Modulator: Limits to Performance," *Journal of the Optical Society of America A* **15**:2552–2562 (1998).

23. D. T. Miller et al., "Requirements for Segmented Spatial Light Modulators for Diffraction-Limited Imaging Through Aberrated Eyes," in G. D. Love (ed.), *Proceedings for Adaptive Optics for Industry and Medicine,* World Scientific, Singapore, 1999.

24. L. Zhu et al., "Adaptive Control of a Micromachined Continuous Membrane Deformable Mirror for Aberration Compensation," *Applied Optics* **38**:168–176 (1999).

25. A. Roorda et al., "Confocal Scanning Laser Ophthalmoscope for Real-Time Photoreceptor Imaging in the Human Eye," in *Vision Science and Its Applications,* Technical Digest Series (Optical Society of America, Washington, DC), **1**:90–93 (1997).

26. A. R. Wade and F. W. Fitzke, "*In vivo* Imaging of the Human Cone-Photoreceptor Mosaic Using a Confocal Laser Scanning Ophthalmoscope," *Lasers and Light* **8**:129–136 (1998).

27. P. Artal and R. Navarro, "High-Resolution Imaging of the Living Human Fovea: Measurement of the Intercenter Cone Distance by Speckle Interferometry," *Optics Letters* **14**:1098–1100 (1989).

28. S. Marcos et al., "Coherent Imaging of the Cone Mosaic in the Living Human Eye," *Journal of the Optical Society of America A* **13**:897–905 (1996).

29. D. T. Miller et al., "Images of the Cone Mosaic in the Living Human Eye," *Vision Research* **36**:1067–1079 (1996).

30. A. Roorda and D. R. Williams, "The Arrangement of the Three Cone Classes in the Living Human Eye," *Nature* **397**:520–522 (1999).

31. H. von Helmholtz, *Helmholtz's Treatise on Physiological Optics,* J. P. C. Southall (ed.), Optical Society of America, New York, 1924.

CHAPTER 11
ASSESSMENT OF REFRACTION AND REFRACTIVE ERRORS*

William F. Long
Ralph Garzia
University of Missouri—St. Louis
St. Louis, Missouri

Jeffrey L. Weaver
American Optometric Association
St. Louis, Missouri

11.1 GLOSSARY

A	constant specific to each intraocular lens
f	spectacle lens focal length
F	spectacle lens power
F_V	back vertex power of a spectacle lens
F_x	effective back vertex power of a spectacle lens displaced a distance x
K	keratometry reading
L	axial length of the eye
P	power of intraocular implant lens
x	displacement distance of a spectacle lens

11.2 INTRODUCTION

At the receiving end of most optical devices is a human eye and chances are good that that eye suffers from refractive error. That refractive error and its correction can have important implications for optical designers: e.g., vertex separation due to spectacles can limit field through vignetting; instrument accommodation requires an allowance that will benefit younger users but blur presbyopes; Polaroid® sunglasses blot out a liquid crystal display; pro-

* *Acknowledgments:* Thanks to Bruce S. Frank, M.D., Raymond I. Myers, O.D., and Kathleen Haywood, Ph.D., for their suggestions in preparing the manuscript for this chapter.

gressive addition bifocal wearers get stiff necks looking at a video display. Fifty years ago patients with refractive error had only glass spectacles that might have one of three or four bifocal designs and a choice of a handful of tints. Now they have a bewildering range of options, each with advantages and disadvantages that optical designers and vision care professionals need to bear in mind. At any given time, one or both eyes of the observer may not be optimally or fully corrected, and this issue needs consideration by the designer of optical instruments. Here, we detail some of the approaches and problems encountered. For example, see Appendix in Chap. 18.

11.3   REFRACTIVE ERRORS

The refractive elements of the eye are the *cornea* which provides about ⅔ of the eye's optical power and the *crystalline lens* which provides the remainder. The shape of the crystalline lens may be changed in response to the action of the intraocular muscles in order to vary the eye's optical power. This is called *accommodation*. These elements bring light to a focus on the *retina* (Fig. 1).[1]

Most eyes suffer from greater or lesser degrees of *ametropia,* the failure to accurately focus light from a remote object on the retina.[2] The degree of ametropia or *refractive error* is quantified clinically by the optical power F of the spectacle lens that corrects the focus of the eye. The dioptric power of a lens is the reciprocal of its focal length f, so $F = 1/f$. The mks unit of optical power is the *diopter*, abbreviated D, which dimensionally is a reciprocal meter. Optical power is especially convenient in ophthalmic work since most ophthalmic lenses may be treated as thin lenses and the power of superimposed thin lenses is simply the algebraic sum of the powers of the individual lenses.

Spectacle lenses are usually manufactured in quarter diopter steps. By ophthalmic convention, spectacle prescriptions are written with a plus or minus sign and two places beyond the decimal, e.g., +3.75 D.

Types of Refractive Errors

An unaccommodated eye which focuses collimated light perfectly is *emmetropic* and, of course, requires no spectacle correction to see distant objects (Fig. 1).

A *myopic* or "near sighted" eye focuses light in front of the retina (Fig. 1). For the myope, remote objects are blurred, but closer objects are in focus. Myopia is corrected by lenses of negative power that diverge light.

A *hyperopic* or "far sighted" eye focuses light behind the retina (Fig. 1). By exercising the accommodation of the eye, the hyperope may be able to focus remote objects clearly on the retina, but the nearer the object, the greater the hyperopia and the less the accommodative

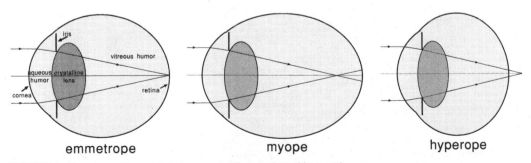

FIGURE 1   Focusing collimated light in emmetropic, myopic, and hyperopic eyes.

ability of the eye and the more difficult this is to do. Hyperopia is corrected by lenses of positive power that converge light.

As the name implies, the *astigmatic* eye does not bring light to a point focus at all. Instead, the eye forms two line foci perpendicular to one another. Astigmats experience some blur at all object distances—the greater the astigmatism, the greater the blur. The effect of astigmatism on vision is far more serious when the two line foci are oriented obliquely instead of being more or less vertical and horizontal.

Astigmatism can be corrected with a *sphero-cylinder lens* (Fig. 2), a lens having a toric surface. Optically, a sphero-cylinder correcting lens acts like a cylinder lens that brings the two line foci of the eye together, superimposed on a spherical lens that moves the superimposed line foci to the retina. In clinical practice lens prescriptions are written in terms of *sphere, cylinder,* and *axis.* For example, the prescription −2.00/−1.00 × 015 corresponds to a −2.00 D spherical lens (sometimes designated −2.00 DS) superimposed on a −1.00 D cylindrical lens with its axis 15° (sometimes designated −1.00 DC × 015). Cylinder axes are designated with respect to the usual trigonometric coordinate system superimposed on the eye as seen by the examiner, i.e., with 0° to the examiner's right and the patient's left. Sometimes for clinical reasons astigmatism may be corrected by a *spherical equivalent* lens, a spherical lens that causes the lines foci to bracket the retina. For example, the spherical equivalent lens for the previous prescription would be −2.50 D.

Refractive errors usually assume one of these three simple types because the curvatures of the various refractive surfaces of the eye vary slowly over the area of the entrance pupil.[3,4] In various pathological conditions, however, *irregular refraction* may be encountered in which no simple spherical or toric lens, nor indeed any lens of any type, can produce sharp vision. Conditions in which this occurs include distortion of the cornea in keratoconus, a dystrophic deformity of the cornea, or subsequent to rigid contact lens wear; disruption of the optics of the crystalline lens by cataract; displacement of the crystalline lens; or tilting of the retina.[5,6]

A final refractive anomaly is *presbyopia,* an age-related reduction of the ability of the eye to accommodate due to sclerosing of the crystalline lens. While the crystalline lens becomes progressively more rigid throughout life, the effect of presbyopia only becomes clinically important in a patient's early 40s.[7] By the late 50s, the crystalline lens has become completely rigid and the patient experiences *absolute presbyopia* with no accommodation at all. Presbyopia almost always proceeds at the same rate in both eyes and at very similar rates among the population as a whole. Presbyopia is corrected by adding positive power to the distance prescription, either in the form of reading glasses or bifocals. Myopes may be able to compensate for presbyopia by simply removing their distance glasses.

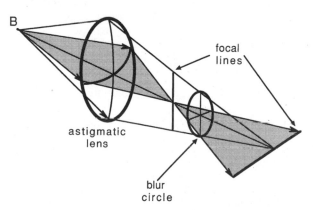

FIGURE 2   The astigmatic pencil formed by a sphero-cylinder lens.

11.4   *ASSESSING REFRACTIVE ERROR*

There are two basic problems in determining refractive error or *ocular refraction:*

1. It is not possible to measure the prescription parameters of sphere, cylinder, and axis separately and sequentially.
2. Clinical tests are typically accurate to within ±0.25 D, roughly the depth of field of the eye,[8] but the final results of a refraction must be accurate to within a somewhat smaller tolerance.

In practice, a refractionist employs a variety of testing procedures iteratively, repeating tests and refining results until a satisfactory end point is reached. Normally testing starts with *objective* tests in which the patient is passive, proceeding to the more sensitive *subjective* tests which require patient responses. Some of the most common procedures are discussed below.

Objective Tests

Objective tests require no judgment on the part of the patient, but with classic instrumentation require considerable skill on the part of the refractionist. In recent years, however, many of these tests have been adapted to automated equipment which simplifies the examiner's task. Three common tests are keratometry, direct ophthalmoscopy, and retinoscopy.

Keratometry measures the curvature of the front surface of the cornea by determining the magnification of the image of a bright object, the *keratometer mire,* reflected by the cornea. The total power of the cornea may be estimated from the curvature measurement. The astigmatism of the cornea may be used to predict the probable astigmatism of the whole eye. In an eye from which the crystalline lens has been removed, for example, corneal astigmatism *is* the total ocular astigmatism. Various rules of thumb have evolved for estimating total astigmatism from keratometric measurements in normal eyes, of which the best known is that of Javal and the simplest to use is that of Grosvenor.[9] All these rules reflect the fact that large degrees of astigmatism are almost invariably due to the cornea.

The *direct ophthalmoscope* is used primarily to view the retina, but can also be used to estimate refractive error. The instrument sends light through the patient's pupil to the retina which the examiner may view through the instrument. A variable lens in the ophthalmoscope enables the examiner to visualize the internal eye. The power of that lens which gives a clear view of the retina added algebraically to the examiner's refractive error equals the patient's refractive error. This is a rather crude estimate of refraction allowing at best a measurement of the equivalent sphere of the patient's refractive error to within a diopter or so. Most doctors make little use of this method, but it can be of value with children and other noncommunicative patients.[10]

The most common objective measurement of refractive error is *retinoscopy.* The optics of the retinoscope are somewhat like those of the direct ophthalmoscope, but the purpose of the retinoscope is not to see the retina but the red reflex in the pupil due to the retina, the same red eye reflex which so often spoils flash photos.

In practice, the retinoscope is held at some known distance—the *working distance*—from the eye. The refractionist observes the red reflex while the retinoscope beam is moved across the patient's eye. Because of vignetting of the patient's pupil and the examiner's pupil, the reflex will appear to move across the patient's eye. The direction and speed of that motion depend on the patient's refractive error[11].

In the most common configuration, collimated light is sent to the eye by a retinoscope held ⅔ meter from the eye. If the reflex moves in the same direction as the retinoscope beam—*with motion*—the retina is conjugate to a point closer than the working distance. If the reflex moves in the opposite direction to the retinoscope beam—*against motion*—the retina is conjugate to a point farther than the working distance. The closer the conjugacy is to the working distance, the faster the motion. The examiner interposes lenses until an end point is reached when the reflex moves infinitely fast or when motion reverses. The patient's refractive error is the value of the interposed lens minus 1.50 DS—the working distance correction.

A retinoscopist may judge the axis of astigmatism from the shape and motion of the reflex. The astigmatic refraction is then determined by examining each of the two principal meridians separately. This is facilitated by using a retinoscope with a streak-shaped beam.

Retinoscopy can provide a very good estimate of refractive error in a very short time, but requires a lot of skill. And even with great skill, there are a few patients whose retinoscopic findings vary considerably from their true refractive state.

Auto-refractors determine a patient's refraction without the examiner's skill and judgment. Instruments have been developed using a variety of optical principles, including those of the retinoscope and the direct ophthalmoscope.[12]

Subjective Techniques

Subjective techniques provide the final refinement of refractive findings. Countless subjective techniques have been developed over the years,[13,14,15] but in practice most refractionists rely on only two or three of them.

The most important tools for subjective refraction are an acuity chart and a set of lenses. An acuity chart consists of rows of progressively smaller letters or other symbols. The chart may simply be hung on the wall or, more commonly, projected onto a screen. The symbols on the chart are rated according to size. For example, a patient with a two minute minimum angle of resolution could just read a letter rated 12 meters or 40 feet when it is 6 meters or 20 feet away. In Snellen notation, the patient would have 6/12, or 20/40, visual acuity—the testing distance being at the top of the fraction, the letter rating at the bottom. A very rough estimate of the magnitude of refractive error may be obtained by dividing the denominator of the Snellen fraction in English units by one hundred.[16] For example, a 20/50 patient would have about a ±0.50 D refractive error.

Classically, refraction was carried out with a *trial lens set,* a box of labeled spherical and cylindrical lenses. These lenses could be held before the patient or placed in a special spectacle frame called a *trial frame.* Nowadays, most refractionists employ a *phoropter,* a device mounted before the patient which contains a range of spherical and cylindrical lenses and other optical devices designed to make rapid change of lenses possible.

Refractions can be carried out with or without cycloplegia with a pharmaceutical agent like Cyclopentolate which eliminates most accommodation. While cycloplegia may be important in certain patients, refractions without cycloplegia are generally preferable since they permit a more realistic assessment of lens acceptance as well as an evaluation of the patient's accommodative and binocular status.[17]

Refraction starts with an estimate determined from an objective technique, acuity rule of thumb, or the previous spectacle prescription. The usual first step in refraction is determination of the patient's equivalent sphere by adding plus spherical power to the patient's approximate prescription until the acuity chart is blurred (termed *"fogging* the patient"), then reducing plus spherical power (*"unfogging* the patient") until the best possible acuity is just achieved. The end point may also be determined or confirmed using the red-green test in which the patient simultaneously views black letters on red and green backgrounds. Due to the eye's chromatic aberration, letters on the red and green side will appear equally sharp and clear with the right equivalent sphere in place.

The next step is to determine the patient's astigmatism. One test for this is the *astigmatic dial,* a chart with radial lines arranged at 30° intervals (Fig. 3). With the approximate cylinder removed, the patient is fogged and asked to identify the line that seems clearest. That line corresponds to the line focus nearest the retina, from which the axis of astigmatism may be calculated. The magnitude of astigmatism may be determined by adding cylinder lenses at the axis of astigmatism until all the lines appear equally sharp and clear.

Estimates of astigmatism are refined to a final prescription with the *Jackson Cross cylinder,* a sphero-cylinder lens with equal and opposite powers along its principal meridians, typically ±0.25 D. The cross cylinder lens is designed to mechanically flip around one of its principal meridians to check astigmatism power and around an axis halfway between the prin-

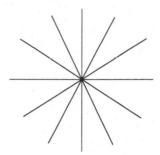

FIGURE 3 The astigmatic dial.

cipal meridians to check astigmatism orientation. The patient looks through the equivalent sphere noting changes in clarity of small letters on the acuity chart while the refractionist flips the cross cylinder lens, adjusting the power and orientation of the astigmatic portion of the prescription according to the patient's opinion of which flip cylinder orientation is more clear.

The final step in refraction is to increase the plus power or decrease the minus power of the refracting lens until the patient can just achieve best acuity.

11.5 *CORRECTING REFRACTIVE ERROR*

Philosophies of Correcting Refractive Error

In determining spectacle prescription the doctor must take into account the patient's refraction and binocular status. Elaborate schemes have been developed for making these clinical decisions,[18,19,20] but they come down to a few basic concepts.

First, astigmatic corrections can cause adaptation problems due to spatial distortion, hence the minimum cylindrical correction consistent with clear vision and comfort is given. In some cases, that might just be the equivalent sphere.

The distance prescription should allow patients to see remote objects clearly. For an absolute presbyope this means prescribing the exact results of the refraction. Young myopes often like to accommodate a bit through their spectacles, so they should receive at a minimum the full refraction, possibly with a quarter or half diopter over correction. Young hyperopes are often blurred by their full prescription until they learn to more completely relax accommodation, so it is wise to give them less than a full prescription at first.

Likewise, the near prescription should provide clear, single vision for reading and other near tasks. For the presbyope this means bifocal lenses, reading glasses, or some other optical arrangement to supplement the eye's accommodative power. Typically, the power of the near correction is chosen so that the patient uses about half the available accommodation for reading. Patients with an especially strong link between accommodation and convergence may also need bifocal lenses or reading glasses to prevent eyestrain and even double vision at near distances. Myopes with minimal astigmatism may be comfortable doing near work with no spectacle correction at all.

Spectacle Correction

Spectacle Lens Optics.   Throughout this discussion, we've assumed that spectacle lenses can be treated as thin lenses and that their exact placement before the eyes is not critical. Fortunately, that is true for the spectacle lenses of that vast majority of patients whose refractive errors are of modest size, but for high refractive errors, we must deal with spectacle optics more precisely.

First of all, the power specified for a spectacle lens is actually its *back vertex power,* the reciprocal of distance from the *ocular surface* of the lens (the surface next to the eye) to the secondary focal point of the lens. The back vertex power of the ocular refraction is read directly from the phoropter. The *lensometer* or *focimeter,* an instrument commonly used in determining spectacle power, actually measures back vertex power of a lens.[21]

The back vertex power of the correction depends on the *vertex distance* of the spectacle lens. The vertex distance is the distance between the ocular surface of the lens and the cornea, typically from 12 to 15 mm. If refraction at a given vertex distance gives a spectacle lens of back vertex power F_V, but the patient will actually wear the prescription at a vertex distance x millimeters from the refraction's spectacle plane (moving toward the eye corresponding to $x > 0$), the back vertex power of the spectacle prescription should be

$$F_x = \frac{F_V}{1 - 0.01xF_V} \tag{1}$$

From this equation it can be shown that changes of vertex of a millimeter or two aren't significant until the magnitude of the prescription power exceeds eight diopters.[22]

The only significant degree of freedom available to the spectacle lens designer is the *base curve* of the lens, the reference curve of the lens form. The base curve is adjusted so that the eye encounters as nearly as possible the same spectacle prescription in all fields of gaze.[23] In terms of the Seidel aberrations, this corresponds to minimizing unwanted marginal astigmatism and regulating curvature of field. A variety of designs, called *corrected curve lens series,* have evolved, each based on slightly different assumptions about vertex distance and the relative importance of the aberrations. In modern lens designs, the front surface of the lens is spherical, and the back surface toric, the so-called *minus cylinder* form. To achieve adequate performance in high powered lenses, especially positive powered lenses, it is necessary to use aspheric curves.

Spectacle Lens Materials.   Ophthalmic lenses may be made of several kinds of glass or plastic.[24] The final choice of material depends on lens weight, thickness, impact resistance, scratch resistance, and chromaticity.

For most of this century, the preferred material has been glass, typically ophthalmic Crown glass of index 1.523, with higher index glasses used to reduce thickness in high powers. The introduction of impact resistance standards in North America in the early 1970's, which required thicker lenses, and the trend to larger eye sizes in spectacle frames soon led to uncomfortably heavy glass lenses. Plastic lens materials which weigh about half as much as glass lenses have since become standard. The most common material is CR39 which has an index of 1.498. Another common material is polycarbonate, a very lightweight material of index 1.586 which is highly impact-resistant. Its toughness makes polycarbonate useful in safety lenses and in lenses for high refractive errors, but the material is very soft and suffers from considerable chromatic aberration.[25]

Presbyopic Corrections.   Presbyopia may be managed by using different spectacles for near and distance, but bifocal lenses are a more practical solution. The top part of the bifocal lens is for distance viewing, and an area of the lower part of the bifocal lens, the *segment,* is for near tasks. The *add* of a bifocal is the difference between segment and distance power. Adds typically range from +1.00 D to +2.50 D, the power increasing with age.

Bifocal segments come in a variety of shapes, the most common being the *straight top, Executive,* and *round* styles illustrated in Fig. 4. Plastic bifocals are made in one piece. In glass bifocals, Executive lenses are also one piece, but straight top and round lenses are made of a higher index glass, fused and countersunk into the distance lens. The choice of lens depends on the field of view the patient needs for near tasks and on cosmetic factors. Whatever the shape, bifocals are fit with the top of the segment at the margin of the patient's lower lid.

As the patient progresses toward absolute presbyopia, a multifocal may be necessary to see adequately at all distances. In trifocals a band with power intermediate between the dis-

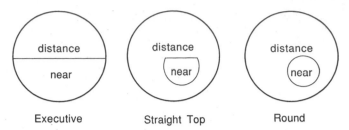

FIGURE 4 The most common bifocal styles are the Executive in which the segment covers the entire bottom of the lens, the straight top bifocal, and the round bifocal.

tance and near correction sits at the top of the bifocal segment. Trifocals are set somewhat higher than bifocals (usually at the lower margin of the patient's pupil) to allow easy access of the intermediate and near portions of the segment.

An increasingly popular alternative to bifocals and trifocals is the *progressive addition lens* or *PAL*. In PALs an aspheric front surface allows power to vary continuously from the distance portion of the lens at the top to the near portion at the bottom. Distant and near portions are connected by a narrow optical channel (Fig. 5). Outside the channel and the wider segment, the optics of the bottom portion of the PAL provides only distorted and blurred vision. Dozens of PAL designs have been brought to market in recent years, each representing a different solution to the problem of designing the aspheric surface.[26]

PALs have been marketed principally as lineless bifocals that don't betray the age of the wearer. But PALs' transition corridors provide useful vision at intermediate distances, eliminating the need for trifocals. The chief disadvantages of PALs are expense, adaptation, and a reading area smaller than that of conventional bifocals. Nonetheless, most patients quickly learn to use these lenses and prefer them to conventional bifocals.[27]

A couple of specialty multifocals are worth mentioning. One is the *double segment bifocal* that has an intermediate segment above the eye in addition to the regular segment below. This configuration is useful for presbyopic mechanics, electricians, painters, plumbers, and others who need to see near objects overhead. The power of the upper segment can be chosen according to occupational needs.

Another is the computer lens. Regular bifocals are prescribed on the assumption that near tasks will be at a typical reading distance 40 cm from the eye. But computer users must see at arm's length, typically about 70 cm from the eye. The intermediate corridor of a conventional

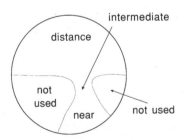

FIGURE 5 A progressive addition lens (PAL) has a large area of distance prescription power that changes to the addition power through a narrow corridor. The areas on the lower temporal and nasal side of the lens become less and less usable the farther they are from the intermediate corridor and near portion.

TABLE 1 Tint Numbers and Transmission Ranges

Tint number	Transmission range
0	80%–100%
1	43%–80%
2	18%–43%
3	8%–18%
4	3%–8%

PAL is too narrow for prolonged use, and a single vision lens prescribed for this working distance may leave the user with too narrow a range of clear vision. Some companies have produced special progressive addition lenses for computer use which have a wide intermediate area with a small near zone at the bottom and distance zone at the top. Such lenses are also helpful with a number of occupations not connected with computers, for example musicians.[28]

Tints in Spectacle Lenses.   Tints are used in spectacle lenses for comfort, safety, and cosmesis. Color tints in glass lenses are produced by adding a compound to the glass as it is made or by coating the completed glass lens. Plastic lenses are tinted by dipping them into a dye. In ophthalmic practice, tints are classified numerically based on their light transmission as shown in Table 1.[29]

The most common application of spectacle tints is, of course, sunglasses. Optimal transmittance for comfort is about 10 percent,[30] a #3 tint, which also helps preserve the eye's capacity for dark adaptation.[31] The most common sunglass tints are gray, brown, yellow, and green. Typical transmittance curves are shown in Fig. 6. Gray lenses block light without affecting color perception. Other tints can, potentially, alter color perception, but sunglasses following the ANSI Z80.3 Sunglass Standard should cause no difficulties, at least in the important task of recognizing traffic signals.[32]

Photochromic lenses have become a popular alternative to conventional tinted lenses. In sunlight a clear photochromic lens darkens to about a #2 tint in five minutes or so. Darkened lenses will lighten to near transparency in about 10 minutes.[33] Somewhat greater darkening occurs at lower temperatures. The darkest tint is inadequate for a true sunglass lens and photochromics are, accordingly, marketed as "comfort" lenses. The darkening is a response to ultraviolet (UV) light, so photochromic lenses work poorly in cars where the windows transmit little UV radiation. Photochromic lenses lighten in response to infrared (IR) radiation.

Photochromic lenses come in glass, ophthalmic plastic, and even polycarbonate. They can be had in single vision and bifocal formats, including progressive addition lenses.[34]

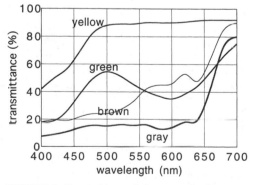

FIGURE 6 Spectral transmission curves for four common plastic lens tints.

Polaroid® sunglasses are made with a layer of Polaroid material sandwiched within spectacle material. The axis of polarization is vertical to block glare from horizontally polarized light scattered from the road or water. Polaroid lenses can be tinted for extra density and can be made in most prescriptions, including bifocals and progressive addition lenses.

Antireflective (AR) coatings use thin film interference to diminish or eliminate reflections off spectacle surfaces. AR coatings improve lens cosmesis and can eliminate or at least minimize unwanted ghost images formed by lens reflection that some patients find distracting.

IR and UV spectacle coatings block light at long and short wavelengths that have been associated with a variety of abiotic effects including keratitis and cataract.[35,36] Because the crystalline lens is the site of most of the eye's UV absorption,[37] it is especially important to provide UV protection in visual corrections worn subsequent to cataract surgery.

Standards for spectacle prescriptions in the United States are spelled out in the ANSI Z80.1 standards.[38] In North America all lenses are expected to have sufficient impact resistance to pass the drop ball test in which a half ounce steel ball of ⅝ in diameter is dropped from a height of 50 inches onto the lens. To satisfy this requirement, glass lenses must be hardened by thermal air tempering or chemical tempering. Plastic lenses require no special hardening treatment.

11.6   CONTACT LENSES

Both kinds of contact lenses in common use—hydrophilic or "soft" contact lenses and rigid or "hard" contact lenses—provide refractive correction while resting on the cornea. Contact lenses are prescribed chiefly for cosmetic reasons, but they provide real optical benefits for two kinds of patients—those with high refractive errors and those with irregular corneas. Contact lenses allow patients with high refractive errors to avoid using the optically inferior peripheral portions of high powered spectacle lenses, plus providing high myopes a slightly magnified retinal image. Corneal irregularities can be masked by the tear layer between the contact lens and the cornea, which may help in keratoconus and other conditions.

Soft Contact Lenses

"Soft" or hydrophilic contact lenses are the most common nowadays. These lenses are made of hydrogels, polymers able to absorb a significant amount of water.[39] Hydrogels are brittle and fragile when dry, but soft and flexible when hydrated. When saturated with ophthalmic saline solution that has the pH of ordinary tears, a properly fitted soft contact lens drapes smoothly over the cornea, providing, in effect, a new cornea with the right power to eliminate refractive error. Once in place, the lens is not easy to dislodge, making soft contacts ideal for sports. A typical lens is about 14 mm in diameter, wide enough to cover the cornea and overlap the sclera (the white of the eye) somewhat. Around the central *optic zone* there is a *peripheral zone* about a millimeter wide with a flatter curvature (Fig. 7). Hydrogel materials transmit atmospheric oxygen to the cornea well, enabling the cornea to sustain its metabolism and transparency. To provide the tear exchange which carries additional oxygen to the eye, the lens should move slightly when the patient blinks, typically about a half millimeter.[40] The patient is unaware of the movement.

The power of the soft contact lens is just that of the spectacle prescription adjusted for vertex distance using equation (1). Most soft lenses prescribed are spherical, small amounts of astigmatism being corrected by using the equivalent sphere. Soft astigmatic lenses are available using a variety of ingenious methods to stabilize the lens to prevent lens rotation so that the astigmatic correction retains its proper orientation.[41] The most commonly used method is that of *dynamic stabilization* which orients the lens through the action of the lids on the superior and inferior thickness gradient of the lens during blinking.

While there is some initial foreign body sensation with soft lenses, they quickly become quite comfortable. After an adaptation period of a couple of weeks, most patients can wear

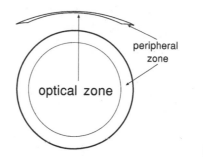

FIGURE 7 The hydrogel lens seen from the front and in cross section.

the lenses from 10 to 12 hours per day with no trouble. Well-fitted lenses cause little or no change in the shape of the cornea, so it is easy to move back and forth between contact lenses and glasses. Occasional contact lens wear is also practical for sports or other special occasions.

Soft contact lenses must be cleaned and disinfected between uses to eliminate infectious agents and mucous deposits. Daily-wear soft contact lenses are to be removed after each day's wear. They usually last about a year. Disposable and frequent replacement lenses intended for shorter periods of use ranging from a day to a couple of weeks offer advantages to many patients, particularly those prone to rapid deposit build-up or without ready access to cleaning and sterilizing solutions.

Extended-wear lenses are intended to be worn for longer periods, typically a few days to a week. Extended-wear hydrophilic lenses have a high water content and/or reduced thickness to increase oxygen permeability,[42] which also makes the lenses more fragile. New materials with increased oxygen transmissibility seem likely to extend the wearing period even more. Extended-wear lenses are usually disposable.

Soft contact lenses are usually clear or with a faint blue "finding tint" which makes the lenses easier to see and handle without affecting the patient's appearance in the lenses. Cosmetic soft contact lenses can change the apparent color of the iris. Overall transparent tints can make light colored irides assume a variety of colors. Patients with dark irides can use lenses with an array of tinted dots surrounding the pupil to change apparent eye color. The tints from the dots along with the iris seen between the dots produce a very convincing illusion. Surprisingly, this arrangement produces no real effect on patients' visual fields.[43]

Novelty soft contact lenses have been marketed with a variety of ghoulish designs intended to elicit parental ire. More serious variants of this idea are cosmetic lenses designed to disguise unsightly damage to eyes subsequent to injury or disease.[44]

Rigid Contact Lenses

The first rigid corneal contact lenses widely used clinically were made of polymethylmethacrylate (PMMA) which had good optical and mechanical properties but which transmitted no oxygen from the air to the cornea. Contact lens fitters had to adjust the lens diameter and the curves on the ocular surface of the contact lens, the lens's *base curve*, to allow the lens to move in order to provide sufficient tear exchange to maintain corneal physiology.

Modern rigid lenses are made of a variety of gas permeable materials.[45] The greater biocompatibility of these materials makes it possible to fit larger lenses that move rather less than the earlier PMMA lenses. A gas permeable lens is typically from 9 to 10 mm in diameter. The curvature of the ocular surface is close to that of the central cornea, as measured by keratometry, with a flatter peripheral zone about 1 mm wide. The lenses are quite thin, on the order of 0.15 mm, which makes them more comfortable as well as improving oxygen transmissibility. The lens is picked up by the upper lid on the blink, and typically slides about a millimeter down the cornea while the eye is open. The patient is unaware of the movement so long as the optic zone covers the pupil.

The meniscus of tears between the cornea and ocular surface of a rigid contact lens forms the so-called *tear lens*. The power of the tear lens plus that of the contact lens itself must equal the vertex adjusted power of the spectacle prescription. The tendency of the very thin gas permeable lenses to flex and change shape on the eye makes it difficult to predict the power of the tear lens accurately, and most fitters rely on trial lenses to determine the contact lens power they will prescribe.

Because the optical index of tears approximates that of the cornea, the power of the anterior surface of the tear pool behind a rigid lens effectively replaces the power of the cornea. This means that corneal astigmatism effectively disappears behind a rigid contact lens with a spherical posterior surface. Since refractive astigmatism is so often due mainly to corneal astigmatism, astigmatic refractive errors can frequently be managed satisfactorily with simple spherical rigid lenses. When the astigmatism is too great or there are mechanical problems in fitting a spherical lens to the astigmatic cornea, rigid lenses with a front and/or back toric surface can be used. The orientation of toric lenses is most commonly achieved with *prism ballast* in which the bottom part of the lens is made thicker and heavier by incorporating a prismatic power in the lens.[46]

Rigid contact lenses are, initially at least, much less comfortable than soft contact lenses. A significant period of adaptation to the lenses is required, starting from a few hours a day. Most patients can eventually wear the lenses from 10 to 12 hours per day. There have been attempts to use certain rigid gas permeable lenses as extended-wear lenses. Once established, wearing schedules should be retained to preserve the eyes' adaptation to the lenses.

Rigid lenses are usually clear or with a light finding tint of blue, brown, or green. A dot is usually placed on the right lens to help the patient identify it.

All rigid lenses, especially those made of PMMA, tend to change the shape of the cornea.[47] Fluctuations in corneal shape can make it difficult to determine an appropriate spectacle prescription for a patient to use after a day of contact lens wear, a phenomenon known as *spectacle blur*. *Orthokeratology* attempts to use corneal molding by rigid contact lenses to reduce or eliminate refractive error.[48] A variety of procedures have been advocated, all of them using a series of contact lenses with base curves flatter than that of the cornea to reduce corneal curvature and, with it, the patient's myopia. With months of effort on the part of doctor and patient, modest reductions in myopia—a diopter or two—are possible. The gains are transient, however, and patients must wear "retainer" lenses several hours daily to prevent regression.

Rigid lenses must be cleaned and sterilized with special solutions each time they are removed. A rigid gas permeable lens can be used for several years.

The chief contraindications for both soft or hard lenses are dry eye and allergy.

Contact Lenses in Presbyopia

The presbyope represents a special challenge to the contact lens fitter. All approaches compromise the convenience of contacts and/or the quality of vision.

An obvious solution to the presbyopia problem is simply to wear reading glasses over the contact lenses. But many contact lens patients don't want to give up their spectacle-free existence.

Two kinds of bifocal contact lenses have been marketed. In some rigid bifocal contact lenses, so-called *alternating vision* lenses, a segment is polished onto the lens. The lens is fit so that when the patient looks down the lower eyelid moves the lens up placing the segment over the pupil. Such lenses present substantial fitting problems.[49]

Other bifocal contact lenses are designed so the segment and distance portion of the lens are simultaneously within the pupil. These are called *simultaneous vision* lenses. The lens wearer sees superimposed clear and blurred images of either near or remote objects so that there is degradation of clarity and contrast.[50] Simultaneous vision contact lens designs have been made for both rigid and soft contact lenses with distant and near corrections arranged in a variety of geometric configurations. One ingenious design is the diffractive contact lens which makes use of the diffractive principle of the Airy disk.[51]

The most common contact lens approach to presbyopia is *monovision* in which one eye is corrected for distance and one for near.[52,53] Most patients adapt surprisingly well to having one eye blurred, especially if the difference between the eyes is small. There are, of course, consequences to visual function.

Scleral Contact Lenses

Scleral or *haptic* contact lenses rest on the sclera of the eye, vaulting the cornea. Outside of cosmetic uses as prostheses[54] or in special effects lenses for the movies,[55] scleral lenses are almost never encountered nowadays.

11.7 CATARACT, APHAKIC, AND PSEUDOPHAKIC CORRECTIONS

Cataract and Cataract Surgery

Cataract is partial or complete loss of transparency of the crystalline lens of the eye. Cataract may be associated with a number of factors, including trauma, certain kinds of medications, UV radiation, and diabetes and other diseases, but the great majority of cases are due to imperfectly understood age-related changes.[56] The disruption of the optics of the crystalline lens in cataract produces irregular optics with increased light scatter that lowers visual acuity and image contrast, increases glare sensitivity, and alters color perception.[57]

Treatment for cataract is surgical removal of the crystalline lens. An eye from which the lens has been removed is termed *aphakic*. In *intracapsular cataract extraction (ICCE)* the lens and the lens capsule are both removed. Intracapsular surgery has now been almost completely replaced by *extracapsular cataract extraction (ECCE)* in which the lens is removed but the capsule left behind.[58] *Phacoemulsification* carries this one step further by breaking the lens up with an ultrasonic probe before removal, thus permitting the surgery to proceed through an incision at the corneal margin as small as 3 millimeters.[59] Postoperative opacification of the posterior capsule occurs within months or years in about a third of patients. When this occurs, a hole may be opened in the capsule with a YAG laser, an in-office procedure.[60]

Post-Surgical Correction in Cataract

The crystalline lens accounts for about a third of the eye's optical power,[61] so the aphakic eye requires a substantial optical correction, typically on the order of +10 D.[62] This can be provided with spectacles, contact lenses, or intraocular lens implants.

Aphakic spectacle corrections are heavy and thick with serious optical problems, among them a retinal image magnification of around 35 percent, aberrations and distortion, sensitivity to precise vertex placement, unsightly magnification of the patient's eyes, and a ring shaped blind spot at the periphery of the lens due to prismatic effects.[63] Aspheric lenses and specially designed frames make aphakic spectacles optically and cosmetically a bit more tolerable but their near complete elimination in the era of the intraocular lens implant is unlamented.

Optically, contact lenses provide a much better aphakic visual correction than spectacles.[64,65] They provide vastly improved cosmesis, give an image size very close to that of the phakic eye, and eliminate the distortions and prismatic effects of aphakic spectacle lenses.

There are serious practical problems in aphakic contact lens fits, however, and contact lenses are rarely used in aphakia anymore.[66]

Intraocular lens implants (IOLs) are lenses placed in the eye during cataract surgery to replace the power of the crystalline lens. While initial experiments placed the IOL in the anterior chamber using angle-fixated lenses or iris support, the posterior chamber has been found to be the safest, most effective site for primary implantation.[67] IOL implantation in the capsule after extracapsular extraction has become the standard procedure. A patient with an intraocular implant is said to be *pseudophakic*.

In addition to its optical and mechanical properties, the optical material for an IOL should be biocompatible, sterilizable, and able to withstand the laser beam in YAG capsulotomy. Polymethylmethacrylate (PMMA), the same material used in early corneal contact lenses, is the most common material used in intraocular lenses.[68,69] Ultraviolet light absorbers may be added to PMMA to replace the UV absorption of the extracted crystalline lens. "Foldable" lenses made of hydrogel, silicone, or foldable acrylic materials permit insertion of the implant through the narrow incision in phacoemulsification.[70] A great variety of supports have been devised for IOLs. The most popular now are the *modified C loop* (Fig. 8) with PMMA loops which seat it within the capsule or ciliary sulcus.

The power of the IOL to be implanted is calculated from the corneal curvature as measured with the keratometer, and from the axial length of the eye determined by ultrasonography. A number of formulas have been developed to do this.[71] A typical one that is often used, called the SRK formula after the initials of its developers, is

$$P = A - 2.5L - 0.9K,$$

where P is implant power necessary to achieve emmetropia, L axial length, K the keratometer reading, and A a constant specific to each IOL which may be further adjusted for surgeon factors. The SRK formula is a regression formula based on 2068 eyes. A newer version of the formula, the SRK II formula, adds or subtracts a factor C dependent on axial length.

The refraction actually targeted by the surgeon depends on the patient's needs. Typically, a surgeon will try to achieve slight myopia, using the depth of field of the eye to get a range of vision, although myopia sufficient to work without glasses at near or even monovision are also possibilities.

Spectacle prescriptions are determined clinically for the pseudophakic patient just as they would be for any other absolute presbyope. Since, of course, these patients have no accommodation, all will need some sort of bifocal or other provision for close work.

Bifocal IOLs have been produced in designs using the same optical principles as simultaneous vision bifocal contact lenses. As with the bifocal contact lenses, there is increased depth of field at the expense of decreased contrast sensitivity and lowered acuity.[72,73]

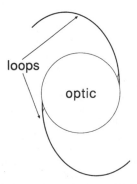

FIGURE 8 Intraocular lens with C loops.

11.8   ANISEIKONIA AND ANISOPHORIA

The visual correction which focuses an image on the retina also changes shape and magnification of the retinal image. A relative difference in size and shape of retinal images between the two eyes is called *aniseikonia*. Aniseikonia has been associated with visual discomfort, especially when there is a significant difference in the refraction of the two eyes or when there has been a sudden change in the eyes' optics.[74] Aniseikonia is most frequently associated with *anisometropia,* a condition in which there is a difference in refractive error between the two eyes. Clinically significant differences are one diopter or more in the sphere or cylinder.[75]

Diagnosis of aniseikonic problems is done mainly by process of elimination: if a patient sees sharply with each eye and still complains of headache, eye strain, or glare, aniseikonia may be the underlying cause. The degree of aniseikonia can be measured with a number of clinical test charts or devices, of which the Space Eikonometer is the gold standard.[76] Few doctors have such devices—the Space Eikonometer hasn't been manufactured since the 1970's— so they are reduced to estimating the degree of aniseikonia from optical rules of thumb.[77,78] Unfortunately, these optical estimates don't take into account "neural" magnification which has been shown to have significant effect. For example, Knapp's Law, derived from purely optical considerations, predicts that the aniseikonia due to anisometropia in which the eyes have different lengths but similar optical components is best corrected with spectacles, while experimental work suggests contact lens correction works better.[79]

Image magnification can be controlled to some extent by the prescribing doctor through specification of spectacle parameters or by opting for contact lenses or, in cataract cases, an intraocular lens implant. While aniseikonia has been a vexed topic in the ophthalmic literature, in practice few doctors concern themselves with it except in extreme cases, mainly unilateral cataract or aphakia.[80] Most patients adapt quickly to changes in retinal images,[81,82] a fact confirmed in clinical practice by the success of progressive addition bifocals and contact lenses, including monovision. Aniseikonia management has, indeed, become a "lost art."[83,84]

Optical anisophoria is a tendency for the prismatic differences in the two spectacle lenses to dissociate the eyes of an anisometropic patient outside of center gaze. Vertical prismatic dissociation is by far the most troublesome. Most patients can avoid anisophoria by simply looking through the optical center of the lens. Presbyopes, however, have to look a centimeter or so below the optical center to use the bifocal segment. A number of strategies exist to reduce or eliminate prism at the reading level, including *bicentric grinding* which grinds additional prism into the lower part of the lens.[85] Fortunately, most patients adapt to the prismatic differences with surprising ease.[86] Of course, contact lenses are the definitive remedy for anisophoric problems.

11.9   REFRACTIVE SURGERY

The correction of refractive error by surgical means has attracted considerable attention in the last couple of decades. An alphabet soup of techniques has evolved and new ones are being developed regularly. Almost all involve changing the curvature of the cornea, the most powerful optical element of the eye as well as the most accessible.

As with any surgery, refractive surgery carries with it certain risks. In particular, there is a degree of uncertainty in the resulting optical correction. Studies typically quote results of a technique in terms of patients' postsurgical visual acuities. In evaluating these results, however, it is important to remember that the relation between visual acuity and refractive error is not straightforward.[87] Though surgery may reduce the need for glasses, many post-surgical patients will still require them at least some of the time, especially for visually demanding tasks. Moreover, visual acuity measurements which involve bright, high-contrast targets provide no insight into other aspects of optical performance such as diminished image contrast and disability glare.

The discussion below deals with the methods most common at this time.

Incisional Procedures

Any incision of the cornea creates a relaxing or flattening effect at right angles to the direction of the incision.[88] *Radial keratotomy* (*RK*) involves making a series of radial incisions in the cornea, sparing a central optical zone. This process results in peripheral steepening and central flattening of the cornea. The amount of flattening is dependent upon the number of incisions, the depth of the incisions, and the size of the optical zone.[89]

In the National Eye Institute–funded Prospective Evaluation of Radial Keratotomy (PERK) study, 38 percent of eyes had a post-operative refractive error within 0.50 D of the intended correction, and 60 percent were within 1.00 D. Uncorrected visual acuity was 20/20 or better in 53 percent of the eyes and 20/40 or better in 85 percent. With spectacle correction, 97 percent of the patients were able to attain or nearly attain their presurgical visual acuity.[90]

Since the incisions are from 90 to 95 percent of the corneal thickness, there are real surgical risks of perforating the cornea. Even without perforation, there is evidence that the ocular integrity is permanently weakened with this procedure.[91] Postoperatively, there tends to be a shift towards hyperopia, often in excess of one diopter.[92] Diurnal variations in vision are common, characterized by shifts in visual acuity and refraction.[93] Depending upon the pupil size and the size of the optical zone, there can be significant complications of glare—in the form of halos, starburst patterns around light sources, and discomfort.[94]

The incidence of severe, vision-threatening complications in RK is low, but the popularity of this surgery has dropped dramatically since the availability of alternative laser procedures.

Astigmatic keratotomy (*AK*) is a variation of RK used to correct mild-to-moderate astigmatism. This procedure creates a flattening of the steepest meridian by making transverse (tangential or "T") incisions perpendicular to the steep meridian. There are other variations of incisional procedures that are rarely used.

Laser Procedures

Photorefractive keratectomy (*PRK*) uses a 193 nm argon fluoride excimer laser to precisely photoablate a thin layer of the central corneal stroma, the thick layer of connective tissue constituting most of the cornea. Each pulse of this laser can remove as little as 0.25 µ of corneal tissue.[95] For low to moderate myopia, 6.0 mm or greater diameter single zone or multizone ablation zones have become the standard because they result in superior refractive results and fewer complaints related to night vision.[96]

In selected studies of the treatment of myopia using FDA-approved lasers, the percentage of eyes with uncorrected post-operative visual acuity of 20/20 or better ranged from 2 to 100 percent, and those 20/40 or better ranged from 27 to 100 percent. Clinically significant loss of best corrected visual acuity occurred in from 0 to 18 percent of the eyes in these studies. Much of the variability between studies can be attributed to the amount of correction attempted, since eyes with low to moderate myopia typically can expect a very good outcome, while high amounts of myopia have markedly poorer results.

Corneal haze is common during early months of the healing phase. Low levels of haze have minimal effect on vision performance, but persistent or late onset corneal haze, often leading to regression and decreased visual acuity, has been observed in moderate to high myopes.[97]

Short-term refractive shifts are rare with PRK, but myopic regression is common, particularly within the first year and with small ablation zones. Ablation decentration of greater than 1 mm will result in poorer visual outcomes, including glare, halos, decreased visual acuity, and monocular diplopia.[98] Refinements in lasers, algorithms, and surgical skills should continue to improve the procedure.

Laser in-situ keratomileusis (*LASIK*) is a technique that uses an automated surgical knife or microkeratome to slice a hinged corneal flap, after which stromal tissue is removed as in PRK. The flap is then repositioned without sutures, and the eye is allowed to heal. Since there is no removal of the surface corneal epithelium, visual acuity is restored rapidly, and pain is minimal. This procedure has become the most widely performed refractive surgical procedure.

Uncorrected visual acuity after this procedure is 20/20 or better in from 38 to 63 percent and 20/40 or better in from 82.5 to 100 percent of selected studies. Less than 9.5 percent of eyes suffered clinically significant loss of best corrected acuity.[99,100,101] Medical complications of LASIK occur in approximately 5 percent of cases.[102]

Intrastromal Rings

Intrastromal corneal rings (*ICRs*) are circular arcs implanted in the stroma of the peripheral cornea, concentric to the optic axis, which act mechanically to flatten the curvature of the entire cornea. Current procedures use two 150° arcs, with thicknesses of a few tenths millimeter—the greater the thickness, the greater the reduction in curvature and the greater the refractive correction. FDA studies have yielded acuities of 20/15 or better in 13 percent of patients; 20/20 or better in 34 percent; 20/25 or better in 55 percent; and 20/40 or better in 81 percent.[103] The procedure is intended for patients with modest degrees of myopia (less than three diopters) and small astigmatism (less than a diopter). Unlike laser procedures, ICR doesn't work directly on the optical axis of the eye and is at least somewhat reversible through removal of the rings.

Intraocular Lenses

Cases of high myopia may be treated by extracting the crystalline lens, just as for a cataract patient, and inserting an implant of suitable power. This is called a *clear lens exchange* (*CLE*). Phakic lens implants leave the patient's crystalline lens intact. *Anterior chamber phakic lens implants* (*ACPLIs*) are placed in the anterior chamber and *posterior chamber phakic lens implants* (*PCPLIs*) vault the crystalline lens in the posterior chamber.

11.10 *MYOPIA RESEARCH AND VISUAL TRAINING*

Myopia results from the lack of precise correlation between the optical components of the eye and its axial length. Of the ocular components, axial length correlates most closely with refractive error, on the order of –0.76.[104] The continued eye elongation through puberty is compensated by concurrent changes in the crystalline lens. This *emmetropization* process occurs in the form of a decrease in crystalline lens refractive power, crystalline lens surface flattening, and crystalline lens thinning.[105]

Recent research has expounded on the influences of heredity and environmental factors on the onset and development of myopia. A family history of myopia is associated with an increased prevalence of its occurrence. Monozygotic twins have a higher correlation between refractive error and ocular and optical components than dizygotic twins. A higher prevalence of myopia is found among children of myopic parents than nonmyopic parents. Even before the appearance of myopia, children with two myopic parents already have longer axial lengths than children with only one myopic parent or no myopic parents.[106]

It was noted during investigations of anomalous visual experience on visual neurophysiology that a significant amount of myopia developed in monkey eyes deprived of form vision by lid suturing.[107] Subsequent experiments have confirmed and expanded these results. The distribution of refractive errors created in form-deprived eyes is broad with a large number of high myopic refractive errors represented.[108] The effect of deprivation on refractive error development also occurs in a variety of animal species. This phenomenon occurs in human infants with severe congenital drooping of the upper eyelid.[109]

Form deprivation myopia is axial in nature, and is attributable primarily to an increase in vitreous chamber depth. In fact, the increase in vitreous chamber depth correlates well with the magnitude of induced myopia.[110] The earlier the onset of form deprivation and the longer the duration of deprivation, the higher the degree of myopia produced. Form deprivation does not produce myopia in adults, but only during a sensitive period of ocular growth early in life. The fact that eyelid suture does not cause deprivation myopia in dark reared animals supports the notion that visual experience is necessary for this process.[111]

Diffusing spectacle lenses that reduce image contrast also produce form deprivation myopia. The degree of deprivation myopia is correlated with the amount of image degradation. The contrast threshold for triggering from deprivation myopia is relatively modest— approximately 0.5 log units at high spatial frequencies, enough to reduce visual acuity to 6/15. Optically induced blur results in modifications of axial growth to neutralize the direction of retinal defocus.[112] Myopia also has been induced in monkeys reared in environments restricted to near viewing.[113] Atropine, a parasympathetic inhibiting cycloplegic agent, prevents myopia development, indicating that ocular accommodation may be a contributing factor.[114]

The evidence from environmentally induced refractive error in animals suggests that the visual environment has an active influence on ocular growth and therefore refractive error by disrupting the emmetropization process. Image degradation associated with relatively small amounts of chronic retinal defocus is capable of producing alterations in refractive error.

In humans, ocular accommodation may contribute to myopia development through its influence on retinal image quality. It has been found that accommodative responses to dioptric blur are lower in myopes compared to emmetropes across a range of accommodative stimuli.[115] Lower stimulus-response gradients reflect reduced blur-induced accommodation and reduced sensitivity to retinal defocus in myopia. In fact, reduced accommodative response to dioptric blur has been found to occur in children prior to the onset of myopia compared to children that remain emmetropic.[116]

After a short period of sustained near fixation accommodative adaptation occurs.[117] This is the product of a transient increase in the magnitude of tonic accommodation—the response of the accommodative system in the absence of visual stimuli such as blur—and results in an increase in the overall accuracy of the aggregate accommodative response and a reduction in the demand for blur-driven accommodation.[118] Following near fixation, tonic accommodation should regress to pre-fixation levels. It has been found that myopes have a shallower regression gradient than emmetropes, indicating a relatively prolonged rate of decay.[119] This slower recovery could produce transient periods of myopic retinal defocus on the order of 0.2 D to 1.00 D.[120] Individuals with poor adaptive responses are likely to demonstrate an increased amount of retinal defocus during nearwork tasks and during recovery from sustained near fixation.

This process could be mediated by sympathetic (adrenergic) nervous system innervation of accommodation. The innervation to the accommodative system is mediated by the autonomic nervous system. The parasympathetic branch of the autonomic system produces positive or increased levels of accommodation. The sympathetic branch, which is inhibitory in nature and mediated by beta-adrenergic receptors, produces negative or decreased accommodation. Timolol maleate, a pharmaceutical agent with adrenergic antagonistic properties, produces a significant myopic shift in tonic accommodation (0.87 D) and a shallower regression gradient after a period of near work, leading to increased magnitude and duration of post-nearwork retinal defocus.[121]

These research results suggest that reduced sensitivity to retinal blur and poor and/or prolonged adaptive responses may create a form of deprivation myopia by deregulating the emmetropization process. These environmental influences on myopia development may be modifiable. Bifocal lens therapy in myopes with clinically diagnosed deficiencies in blur-driven accommodation has been shown to reduce the rate of myopia progression by approximately 0.25 D/yr compared to single vision lenses.[122,123,124] Conventional optometric accommodative vision therapy has been used to reduce abnormal nearwork-induced transient myopia in adults.[125] In the future, pharmacological treatment through manipulation of the sympathetic nervous system may be available.

11.11 REFERENCES

1. W. N. Charman, "Optics of the Eye," chap. 24, in Michael Bass (ed.), *Handbook of Optics*, vol. I, McGraw-Hill, New York, 1995, p. 24.5–24.6.

2. W. N. Charman, "Optics of the Eye," chap. 24, in Michael Bass (ed.), *Handbook of Optics*, vol. I, McGraw-Hill, New York, 1995, p. 24.8.

3. W. Long, "Why Is Ocular Astigmatism Regular?", *Am. J. Optom. Physiol. Opt.* **59**(6):520–522 (1982).

4. W. Charmin and G. Walsh, "Variations in the Local Refractive Correction of the Eye Across Its Entrance Pupil," *Opt. Vis. Sci.* **66**(1):34–40 (1989).

5. C. Sheard, "A Case of Refraction Demanding Cylinders Crossed at Oblique Axes, Together with the Theory and Practice Involved," *Ophth. Rec.* **25**:558–567 (1916).

6. T. D. Williams, "Malformations of the Optic Nerve Head," *Am. J. Optom. Physiol. Optics*, **55**(10):706–718 (1978).

7. W. N. Charman, "Optics of the Eye," chap. 24, in Michael Bass (ed.), *Handbook of Optics*, Vol. I, McGraw-Hill, New York, 1995, pp. 24.33–24.34.

8. W. N. Charman, "Optics of the Eye," chap. 24, in Michael Bass (ed.), *Handbook of Optics*, vol. I, McGraw-Hill, New York, 1995, p. 24.27.

9. T. Grosvenor, S. Quintero, and D. Perrigin, "Predicting Refractive Astigmatism, A Suggested Simplification of Javal's Rule," *Am. J. Optom. Physiol. Opt.* **65**(4):292–297 (1988).

10. J. E. Richman and R. P. Garzia, "The Use of an Ophthalmoscope for Objective Refraction on Non-Cooperative Patients," *Am. J. Optom. Physiol. Optics* **60**(4):329–334 (1983).

11. Theodore Grosvenor, *Primary Care Optometry*, third ed., Butterworth-Neinemann, Boston, 1996, pp. 263–265.

12. Charles E. Campbell, William J. Benjamin, and Howard C. Howland, in William J. Benjamin (ed.), *Borish's Clinical Refraction*, W. B. Saunders, Philadelphia, 1998, pp. 590–628.

13. Irvin M. Borish and William J. Benjamin, "Monocular and Binocular Subjective Refraction," chap. 19, in William J. Benjamin (ed.), *Borish's Clinical Refraction*, W. B. Saunders, Philadelphia, 1998, pp. 629–723.

14. Theodore Grosvenor, "Subjective Refraction," chap. 9, in *Primary Care Optometry*, third ed., Butterworth-Neinemann, Boston, 1996, pp. 285–307.

15. Michael Polasky, "Monocular Subjective Refraction," chap. 18, in J. Boyd Eskridge, John F. Amos, and Jimmy D. Bartlett (eds.), *Clinical Procedures in Optometry*, Lippincott, Philadelphia, 1991, pp. 174–193.

16. M. J. Hirsch, "Relation of Visual Acuity to Myopia," *Arch. Ophthalmol.* **34**(5):418–421 (1945).

17. Theodore Grosvenor, *Primary Care Optometry*, third ed., Butterworth-Neinemann, Boston, 1996, pp. 301–302.

18. David A. Goss, *Ocular Accommodation, Convergence, and Fixation Disparity: A Manual of Clinical Analysis*, Professional Press Books, New York, 1986.

19. Homer Hendrickson, *The Behavioral Optometry Approach to Lens Prescribing: 1984 Revised Edition*, Optometric Extension Program, Santa Ana, CA, 1984.

20. S. K. Lesser, *Introduction to Modern Analytical Optometry: 1974 Revised Edition*, Optometric Extension Program, Duncan, OK, 1974.

21. D. F. C. Loran and C. Fowler, "Instrumentation for the Verification of Spectacle and Contact Lens Parameters," chap. 19, in W. Neil Charman (ed.), *Vision and Visual Dysfunction, Vol. 1, Visual Optics and Instrumentation*, MacMillan, London, 1991, pp. 418–419.

22. W. F. Long, "Paraxial Optics of Visual Correction," chap. 4, in W. Neil Charman (ed.), *Vision and Visual Dysfunction, Vol. 1, Visual Optics and Instrumentation*, MacMillan, London, 1991, p 55.

23. M. Jalie, "Spectacle Lens Design," chap. 18, in *The Principles of Ophthalmic Lenses*, third ed., Association of Dispensing Opticians, London, 1980, pp. 406–431.

24. C. Fowler, "Correction by Spectacle Lenses," chap. 5, in W. Neil Charman (ed.), *Vision and Visual Dysfunction, Vol. 1, Visual Optics and Instrumentation,* MacMillan, London, 1991, pp. 64–66.

25. Troy E. Fannin and Theodore Grosvenor, *Clinical Optics,* Butterworths, Boston, 1987, p. 16.

26. *Seventh Annual Ophthalmic Lens Desk Reference,* Optometric Management, 6S–47S (1998).

27. H. J. Boroyan, M. H. Cho, B. C. Fuller, et al., "Lined Multifocal Wearers Prefer Progressive Addition Lenses," *J. Am. Optom. Assn.* **66**(5):296–300 (1995).

28. W. F. Long, "Recorder with Music Glasses Obbligato", *The American Recorder* **33**:25–27 (1992).

29. *Ophthalmic Optics Files, Coatings,* Essilor, p. 6.

30. S. M. Luria, "Preferred Density of Sunglasses," *Amer. J. Optom. Physiol. Opt.* **61**(6):379–402 (1984).

31. R. H. Peckham and R. D. Harley, "The Effect of Sunglasses in Protecting Retinal Sensitivity," *American Journal of Ophthalmology* **34**(11):1499–1507 (1951).

32. J. K. Davis, "The Sunglass Standard and Its Rationale," *Optometry and Vision Science* **67**(6):414–430 (1990).

33. D. G. Pitts and B. R. Chou, "Prescription of Absorptive Lenses," chap. 24, in William J. Benjamin (ed.), *Borish's Clinical Refraction,* W. B. Saunders, Philadelphia, 1998, pp. 932–934.

34. Alicia Burrill, "Seeing Photochromics in a Whole New Light," *Review of Optometry,* Feb. 15:35–36 (1999).

35. Troy E. Fannin and Theodore Grosvenor, *Clinical Optics,* Butterworths, Boston, 1987, pp. 185–186.

36. D. G. Pitts and B. R. Chou, "Prescription of Absorptive Lenses," chap. 24, in William J. Benjamin (ed.), *Borish's Clinical Refraction,* W. B. Saunders, Philadelphia, 1998, pp. 949–953.

37. W. N. Charman, "Optics of the Eye," chap. 24, in Michael Bass (ed.), *Handbook of Optics,* vol. I, McGraw-Hill, New York, 1995, p. 24.10.

38. ANSI Z80.1-1995, *American National Standard for Ophthalmics—Prescription Ophthalmic Lenses—Recommendations,* Optical Laboratories Association, Merrifield, 1995.

39. M. F. Refojo, "Chemical Composition and Properties," chap. 2, in Montague Ruben and Michel Guillon (eds.), *Contact Lens Practice,* Chapman & Hall, London, 1994, pp. 36–40.

40. Harold A. Stein, Melvin I. Freeman, Raymond M. Stein, and Lynn D. Maund, *Contact Lenses, Fundamentals and Clinical Use,* SLACK, Thorofare, NJ, 1997, p. 111.

41. Richard G. Lindsay and David J. Westerhout, "Toric Contact Lens Fitting," chap. 14, in Anthony J. Phillips and Lynne Speedwell (eds.), *Contact Lenses,* fourth ed., Butterworth-Heineman, Oxford, 1997, pp. 483–485.

42. M. Guillon and M. Ruben, "Extended or Continuous Wear Lenses," chap. 43, in Montague Ruben and Michel Guillon (eds.), *Contact Lens Practice,* Chapman & Hall, London, 1994, pp. 1007–1008.

43. L. R. Trick and D. J. Egan, "Opaque Tinted Contact Lenses and the Visual Field," *ICLC* **17**(July/August):192–196 (1990).

44. M. J. A. Port, "Cosmetic and Prosthetic Contact Lenses," chap. 20.6, in Anthony J. Phillips and Lynne Speedwell (eds.), *Contact Lenses,* fourth ed., Butterworth-Heineman, Oxford, 1997, pp. 752–754.

45. M. F. Refojo, "Chemical Composition and Properties," chap. 2, in Montague Ruben and Michel Guillon (eds.), *Contact Lens Practice,* Chapman & Hall, London, 1994, pp. 29–36.

46. Richard G. Lindsay and David J. Westerhout, "Toric Contact Lens Fitting," chap. 14, in Anthony J. Phillips and Lynne Speedwell (eds.), *Contact Lenses,* fourth edition, Butterworth-Heineman, Oxford, 1997, p. 477.

47. G. E. Lowther, "Induced Refractive Changes," chap. 44, in Montague Ruben and Michel Guillon (eds.), *Contact Lens Practice,* Chapman & Hall, London, 1994, pp. 1035–1044.

48. J. G. Carney, "Orthokeratology," chap. 37, in Montague Ruben and Michel Guillon (eds.), *Contact Lens Practice,* Chapman & Hall, London, 1994, pp. 877–888.

49. J. T. de Carle, "Bifocal and Multifocal Contact Lenses," chap. 16, in Anthony J. Phillips and Lynne Speedwell (eds.), *Contact Lenses,* fourth ed., Butterworth-Heineman, Oxford, 1997, pp. 549–559.

50. W. J. Benjamin and I. M. Borish, "Presbyopia and the Influence of Aging on Prescription of Contact Lenses," chap. 33, in Montague Ruben and Michel Guillon (eds.), *Contact Lens Practice,* Chapman & Hall, London, 1994, pp. 783–787.

51. A. L. Cohen, "Bifocal Contact Lens Optics," *Cont. Lens Spectrum* **4**(June):43–52 (1989).

52. P. Erickson and C. Schor, "Visual Function with Presbyopic Contact Lens Correction," *Optom. Vis. Sci.* **67**(1):22–28 (1990).

53. C. Schor, L. Landsman, and P. Erickson, "Ocular Dominance and the Interocular Suppression of Blur in Monovision," *Am. J. Optom. Physiol. Opt.* **64**(10):723–730 (1987).

54. Devendra Kumar and Krishna Gopal, *Cosmetic Contact Lenses and Artificial Eyes,* Delhi University Press, Delhi, 1981, pp. 72–88.

55. M. K. Greenspoon, "Optometry's Contribution to the Motion Picture Industry," *Journal of the American Optometric Association* **58**(12):993–99 (1987).

56. Karla J. Johns et al., *Lens and Cataract,* American Academy of Ophthalmology, San Francisco, 1997, pp. 40–63.

57. S. Masket, "Preoperative Evaluation of the Patient with Visually Significant Cataract," chap. 2, in Roger F. Steinert (ed.), *Cataract Surgery: Technique, Complications, and Management,* Saunders, Philadelphia, 1995, pp. 11–18.

58. D. V. Leaming, "Practice Styles and Preferences of ASCRS Members—1992 Survey," *J. Cataract Refract. Surg.* **19**:600–606 (1993).

59. Jared Emery, "Extracapsular Cataract Surgery, Indications and Techniques," chap. 8, in Roger F. Steinert (ed.), *Cataract Surgery: Technique, Complications, and Management,* Saunders, Philadelphia, 1995, p. 108.

60. Claudia U. Richter and Roger F. Steinert, "Neodymium:Yttrium-Aluminum-Garnet Laser Posterior Capsulotomy," chap. 32, in Roger F. Steinert (ed.), *Cataract Surgery: Technique, Complications, and Management,* Saunders, Philadelphia, 1995, pp. 378–396.

61. H. H. Emsley, *Visual Optics, Fifth Ed., Vol. 1: Optics of Vision,* Butterworths, London, 1953, pp. 343–348.

62. A. E. A Ridgway, "Intraocular Lens Implants," chap. 7, in W. Neil Charman (ed.), *Vision and Visual Dysfunction, Vol. 1, Visual Optics and Instrumentation,* MacMillan, London, 1991, pp. 126.

63. Troy E. Fannin and Theodore Grosvenor, *Clinical Optics,* Butterworths, Boston, 1987, pp. 360–377.

64. Troy F. Fannin and Theodore Grosvenor, *Clinical Optics,* Butterworths, Boston, 1987, pp. 448–449.

65. Montague Ruben and Michel Guillon, *Contact Lens Practice,* Chapman & Hall, London, 1994, pp. 503–505.

66. A. E. A. Ridgway, "Intraocular Lens Implants," chap. 7, in W. Neil Charman (ed.), *Vision and Visual Dysfunction, Vol. 1, Visual Optics and Instrumentation,* MacMillan, London, 1991, p. 127.

67. Ehud I. Assia, Victoria E. Castaneda, Ulrich F. C. Legler, Judy P. Hoggatt, Sandra J. Brown, and David J. Apple, "Studies on Cataract Surgery and Intraocular Lenses at the Center for Intraocular Lens Research," *Ophthalmology Clinics of North America* **4**(2):251–266 (1991).

68. Richard L. Lindstrom, "The Polymethylmethacrylate (PMMA) Intraocular Lenses," chap. 22, in Roger F. Steinert (ed.), *Cataract Surgery: Technique, Complications, and Management,* Saunders, Philadelphia, 1995, pp. 271–278.

69. A. E. A. Ridgway, "Intraocular Lens Implants," chap. 7, in W. Neil Charman (ed.), *Vision and Visual Dysfunction, Vol. 1, Visual Optics and Instrumentation,* MacMillan, London, 1991, pp. 127–129.

70. Richard L. Lindstrom, "Foldable Intraocular Lenses," chap. 23, in Roger F. Steinert (ed.), *Cataract Surgery: Technique, Complications, and Management,* Saunders, Philadelphia, 1995, pp. 279–294.

71. John A. Retzlaff, Donald R. Sanders, and Manus Kraff, *Lens Implant Power Calculation,* Slack, NJ, 1990, pp. 19–22.

72. W. N. Charman, I. M. Murray, M. Nacer, and E. P. O'Donoghue, "Theoretical and Practical Performance of a Concentric Bifocal Intraocular Implant Lens," *Vision Research* **38**:2841–2853 (1998).

73. Jack T. Holladay, "Principles and Optical Performance of Multifocal Intraocular Lenses," *Ophthalmology Clinics of North America* **4**(2):295–311 (1991).

74. H. M. Burian, "History of Dartmouth Eye Institute," *Arch. Ophthal.* **40**:163–175 (1948).

75. M. Scheiman and B. Wick, *Clinical Management of Binocular Vision,* Lippincott, Philadelphia, 1994, p. 86.

76. J. Boyd Eskridge, "Eikonometry," chap. 77, in J. Boyd Eskridge, John F. Amos, and Jimmy D. Bartlett (eds.), *Clinical Procedures in Optometry,* Lippincott, Philadelphia, 1991, pp. 732–737.

77. H. H. Emsley, *Visual Optics, Fifth Ed., Vol. 1: Optics of Vision,* Butterworths, London, 1953, pp. 361–369.

78. Theodore Grosvenor, *Primary Care Optometry,* third ed., Butterworth-Heinemann, Boston, 1996, pp. 399–401.

79. A. Bradley, J. Rabin and R. D. Freeman, "Nonoptical Determinants of Aniseikonia," *Invest. Ophthalmol. Vis. Sci.* **24**(4):507–512 (1983).

80. J. M. Enoch, "Aniseikonia and Intraocular Lenses: A Continuing Saga," *Optom. Vision Sci.* **71**(1):67–68 (1994).

81. Knox Laird, "Anisometropia," chap. 10, in Theodore Grosvenor and Merton C. Flom (eds.), *Refractive Anomalies: Research and Clinical Applications,* Butterworth Heinemann, Boston, 1991, p. 189.

82. D. B. Carter, "Symposium on Conventional Wisdom in Optometry," *Am. J. Optom. Arch. Am. Acad. Optom.* **44**:731–745 (1967).

83. Leonard R. Achiron, Ned Witkin, Susan Primo, and Geoffrey Broocker, "Contemporary Management of Aniseikonia," *Survey of Ophthalmology* **41**(4):321–330 (1997).

84. J. A. M. Jennings, "Binocular Vision Through Correcting Lenses: Aniseikonia," chap. 9, in W. Neil Charman (ed.), *Vision and Visual Dysfunction, Vol. 1, Visual Optics and Instrumentation,* MacMillan, London, 1991, p. 179.

85. Troy E. Fannin and Theodore Grosvenor, *Clinical Optics,* Butterworths, Boston, 1987, pp. 323–328.

86. D. Henson and B. G. Dharamshi, "Oculomotor Adaptation to Induced Heterophoria and Anisometropia," *Invest. Ophthalmol. Vis. Sci.* **22**(2):234–240 (1982).

87. H. L. Blum, H. B. Peters, and J. W. Bettman, *Vision Screening for Elementary Schools, The Orinda Study,* University of California Press, Berkeley, 1959, pp. 99–100.

88. D. S. Holsclaw, "Surgical Correction of Astigmatism," *Ophthalmol. Clin. N. Am.* **10**(4):555–576 (1997).

89. D. K. Talley, *Refractive Surgery Update,* American Optometric Association, 1998.

90. G. O. Waring, M. J. Lynn, P. J. McDonnell, et al., "Results of the Prospective Evaluation of Radial Keratotomy (PERK) Study 10 Years After Surgery," *Arch. Ophthalmol.* **112**(10):1298–1308 (1994).

91. M. Campos et al., "Ocular Integrity After Refractive Surgery: Photorefractive Keratectomy, Phototherapeutic Keratectomy, and Radial Keratotomy," *Ophthalmic Surg.* **23**(9):598–602 (1992).

92. H. Sawelson and R. G. Marks, "Ten Year Refractive and Visual Results of Radial Keratotomy," *Ophthalmology* **102**(12):1892–1901 (1995).

93. P. J. McDonnell, A. Nizam, M. J. Lynn, et al., "Morning-to-Evening Change in Refraction, Corneal Curvature, and Visual Acuity 11 Years After Radial Keratotomy in the Prospective Evaluation of Radial Keratotomy Study," *Ophthalmol.* **103**(2):233–239 (1996).

94. G. O. Waring, M. J. Lynn, H. Gelender, et al., "Results of the Prospective Evaluation of Radial Keratotomy (PERK) Study One Year After Surgery," *Ophthalmology* **92**(2):177–198 (1985).

95. C. R. Munnerlyn, S. J. Koons, and J. Marshall, "Photorefractive Keratectomy: A Technique for Laser Refractive Surgery," *J. Cataract Refract. Surg.* **14**(1):46–52 (1988).

96. American Academy of Ophthalmology, "Excimer Laser Photorefractive Keratectomy (PRK) for Myopia and Astigmatism," *Ophthalmol.* **102**(2):422–437 (1999).

97. I. Lipshitz, A. Lowenstein, D. Varssano, et al., "Late Onset Corneal Haze After Photorefractive Keratectomy for Moderate and High Myopia," *Ophthalmol.* **104**(3):369–374 (1997).

98. C. R. Ellerton and R. R. Krueger, "Postoperative Complications of Excimer Laser Photorefractive Keratectomy for Myopia," *Ophthalmol. Clin. N. Am.* **11**(2):165–181 (1998).

99. D. J. Salchow, M. E. Zirm, C. Stieldorf, and A. Parisi, "Laser In Situ Keratomileusis for Myopia and Myopic Astigmatism," *J. Cataract Refract. Surg.* **24**(Feb):175–182 (1998).

100. A. El-Maghraby, T. Salah, G. O. Waring, III, et al., "Radomized Bilateral Comparison of Excimer Laser In Situ Keratomileusis and Photorefractive Keratectomy for 2.50 to 8.00 Diopters of Myopia," *Ophthalmol.* **106**(3):447–457 (1999).

101. G. O. Waring, III, J. D. Carr, R. D. Stulting, et al., "Prospective Randomized Comparison of Simultaneous and Sequential Bilateral Laser In Situ Keratomileusis for the Correction of Myopia," *Ophthalmol.* **106**(4):732–738 (1999).

102. R. D. Stulting, J. D. Carr, K. P. Thompson, et al., "Complications of Laser In Situ Keratomileusis for the Correction of Myopia," *Ophthalmol.* **106**(1):13–20 (1999).

103. D. J. Schanzlin, "KeraVision®Intacs™: History, Theory, and Surgical Technique," in *KeraVision Intacs: A New Refractive Technology,* Supplement to Refractive EyeCare™ for Ophthalmologists, May, pp. 6–10 (1999).

104. G. W. H. M. Van Alphen, "On Emmetropia and Ametropia," *Ophthalmoligica* **142**(suppl):1–92 (1961).

105. K. Zadnik, "Myopia Development in Childhood," *Optom. Vision Sci.* **74**(8):603–608 (1997).

106. K. Zadnik, W. A. Satariano, D. O. Mutti, R. I. Sholtz, and A. J. Adams, "The Effect of Parental History of Myopia on Children's Eye Size," *J.A.M.A.* **271**:1323–1327 (1994).

107. T. N. Wiesel and E. Raviola, "Myopia and Eye Enlargement After Neonatal Lid Fusion," *Nature* **266**(5597):66–68 (1977).

108. E. L. Smith III, R. S. Harwerth, M. L. J. Crawford, and G. K. von Noorden, "Observations on the Effects of Form Deprivation on the Refractive Status of the Monkey," *Invest. Ophthalmol. Vis. Sci.* **28**(8):1236–1245 (1987).

109. C. S. Hoyt, R. D. Stone, C. Fromer, and F. A. Billdon, "Monocular Axial Myopia Associated with Neonatal Eyelid Closure in Human Infants," *Am. J. Ophthalmol.* **91**(2):197–200 (1981).

110. E. L. Smith III and L.-F. Hung, "Optical Diffusion Disrupts Emmetropization and Produces Axial Myopia in Young Monkeys," *Invest. Ophthalmol. Vis. Sci.* **36**(4)(Suppl.):S758 (1995).

111. E. Raviola and T. N. Wiesel, "Effect of Dark-Rearing on Experimental Myopia in Monkeys," *Invest. Ophthalmol. Vis. Sci.* **17**(6):485–488 (1978).

112. E. L. Smith III, L.-F. Hung, and R. S. Harwerth, "Effects of Optically Induced Blur on the Refractive Status of Young Monkeys," *Vis. Res.* **34**:293–301 (1994).

113. F. A. Young, "The Effect of Restricted Visual Space on the Refractive Error of the Young Monkey Eye," *Invest. Ophthalmol.* **27**(Dec):571–577 (1963).

114. F. A. Young, "The Effect of Atropine on the Development of Myopia in Monkeys," *Am. J. Optom. Arch. Am. Acad. Optom.* **42**(Aug):439–449 (1965).

115. J. Gwiazda, F. Thorn, J. Bauer, and R. Held, "Myopic Children Show Insufficient Accommodative Response to Blur," *Invest. Ophthalmol. Vis. Sci.* **34**(3):690–694 (1993).

116. J. Gwiazda, J. Bauer, F. Thorn, and R. Held, "Dynamic Relationship Between Myopia and Blur-Driven Accommodation in School-Aged Children," *Vis. Res.* **35**:1299–1304 (1995).

117. M. Rosenfield and B. Gilmartin, "Temporal Aspects of Accommodative Adaptation," *Optom. Vis. Sci.* **66**:229–234 (1989).

118. M. Rosenfield and B. Gilmartin, "Accommodative Adaptation During Sustained Near-Vision Reduces Accommodative Error," *Invest. Ophthalmol. Vis. Sci.* **39**(4):S639 (1998).

119. B. Gilmartin and M. A. Bullimore, "Adaptation of Tonic Accommodation to Sustained Visual Tasks in Emmetropia and Late-onset Myopia," *Optom. Vis. Sci.* **68**(1):22–26 (1991).

120. E. Ong and K. J. Ciuffreda, "Nearwork-Induced Transient Myopia," *Doc. Ophthalmol.* **91**(1):57–85 (1995).

121. B. Gilmartin and N. R. Winfield, "The Effect of Topical β-Adrenoceptor Antagonists on Accommodation in Emmetropia and Myopia," *Vis. Res.* **35**(9):1305–1312 (1995).

122. D. A. Goss and T. Grosvenor, "Rates of Childhood Myopia Progression with Bifocals As a Function of Nearpoint Phoria: Consistency of Three Studies," *Optom. Vis. Sci.* **67**(8):637–640 (1990).

123. D. A. Goss and E. F. Uyesugi, "Effectiveness of Bifocal Control of Childhood Myopia Progression As a Function of Near Point Phoria and Binocular Cross Cylinder," *J. Optom. Vis. Dev.* **26**(1):12–17 (1995).

124. D. A. Goss, "Effect of Bifocal Lenses on the Rate of Childhood Myopia Progression," *Am. J. Optom. Physiol. Opt.* **63**(2):135–141 (1986).

125. K. J. Ciuffreda and X. Ordonez, "Vision Therapy to Reduce Nearwork-Induced Transient Myopia," *Optom. Vision Sci.* **75**(5):311–315 (1998).

CHAPTER 12

BINOCULAR VISION FACTORS THAT INFLUENCE OPTICAL DESIGN

Clifton Schor
University of California
School of Optometry
Berkeley, California

12.1 GLOSSARY

Accommodation.   Change in focal length or optical power of the eye produced by change in power of the crystalline lens as a result of contraction of the ciliary muscle. This capability decreases with age.

Ametrope.   An eye with a refractive error.

Aniseikonia.   Unequal perceived image sizes from the two eyes.

Baseline.   Line intersecting the entrance pupils of the two eyes.

Binocular disparity.   Differences in the perspective views of the two eyes.

Binocular fusion.   Act or process of integrating percepts of retinal images formed in the two eyes into a single combined percept.

Binocular parallax.   Angle subtended by an object point at the nodal points of the two eyes.

Binocular rivalry.   Temporal alternation of perception of portions of each eye's visual field when the eyes are stimulated simultaneously with targets composed of dissimilar colors or different contour orientations.

Center of rotation.   A pivot point within the eye about which the eye rotates to change direction of gaze.

Concomitant.   Equal amplitude synchronous motion or rotation of the two eyes in the same direction.

Conjugate.   Simultaneous motion or rotation of the two eyes in the same direction.

Corresponding retinal points.   Regions of the two eyes that when stimulated result in identical perceived visual directions.

Convergence.   Inward rotation of the two eyes.

Cyclopean eye.   A descriptive term used to symbolize the combined viewpoints of the two eyes into a single location midway between them.

Cyclovergence.   Unequal torsion of the two eyes.

Disconjugate. Simultaneous motion or rotation of the two eyes in opposite directions.

Divergence. Outward rotation of the two eyes.

Egocenter. Directional reference point for judging direction relative to the head from a point midway between the two eyes.

Emmetrope. An eye with no refractive error.

Entrance pupil. The image of the aperture stop formed by the portion of an optical system on the object side of the stop. The eye pupil is the aperture stop of the eye.

Extraretinal cues. Nonvisual information used in space perception.

Eye movement. A rotation of the eye about its center of rotation.

Fixation. The alignment of the fovea with an object of interest.

Focus of expansion. The origin of velocity vectors in the optic flow field.

Fronto-parallel plane. Plane that is parallel to the face and orthogonal to the primary position of gaze.

Haplopia. Perception of a single target by the two eyes.

Heterophoria. Synonymous with phoria.

Horopter. Locus of points in space whose images are formed on corresponding points of the two retinas.

Hyperopia. An optical error of the eye in which objects at infinity are imaged behind the retinal surface while the accommodative response is zero.

Midsagittal plane. Plane that is perpendicular to and bisects the baseline. The plane vertically bisecting the midline of the body.

Motion parallax. Apparent relative displacement or motion of one object or texture with respect to another that is usually produced by two successive views by a moving observer of stationary objects at different distances.

Myopia. An optical error of the eye in which objects at infinity are imaged in front of the retinal surface while the accommodative response is zero.

Nonconcomitant. Unequal amplitudes of synchronous motion or rotation of the two eyes in the same direction.

Optic Flow. Pattern of retinal image movement.

Percept. That which is perceived.

Perception. The act of awareness through the senses, such as vision.

Perspective. Variations of perceived size, separation, and orientation of objects in 3-D space from a particular viewing distance and vantage point.

Phoria. An error of binocular alignment revealed when one eye is occluded.

Primary position of gaze. Position of the eye when the visual axis is directed straight ahead and is perpendicular to the frontoparallel plane.

Retinal disparity. Difference in the angles formed by two targets with the entrance pupils of the eyes.

Shear. Differential displacement during motion parallax of texture elements along an axis perpendicular to the meridian of motion.

Skew movement. A vertical vergence or vertical movement of the two eyes in opposite directions.

Stereopsis. The perception of depth stimulated by binocular disparity.

Strabismus. An eye turn or misalignment of the two eyes during attempted binocular fixation.

Tilt. Amount of angular rotation in depth about an axis in the frontoparallel plane—synonymous with orientation.

Torsion. Rotation of the eye around the visual axis.

Vantage point.   Position in space of the entrance pupil through which 3-D space is transformed by an optical system with a 2-D image or view plane.

Visual field.   The angular region of space or field of view limited by the entrance pupil of the eye, the zone of functional retina, and occlusion structures such as the nose and orbit of the eye.

Visual plane.   Any plane containing the fixation point and entrance pupils of the two eyes.

Yoked movements.   Simultaneous movement or rotation of the two eyes in the same direction.

12.2 COMBINING THE IMAGES IN THE TWO EYES INTO ONE PERCEPTION OF THE VISUAL FIELD

The Visual Field

Optical designs that are intended to enhance vision usually need to consider limitations imposed by having two eyes and the visual functions that are made possible by binocular vision. When developing an optical design it is important to tailor it to specific binocular functions that you wish to enhance or at least not to limit binocular vision. Binocular vision enhances visual perception in several ways. First and foremost, it expands the *visual field* (Walls, 1967). The horizontal extent of the visual field depends largely on the placement and orientation of the eyes in the head. Animals with laterally placed eyes have panoramic vision that gives them a full 360 degrees of viewing angle. The forward placement of our eyes reduces the visual field to 190 degrees. Eye movements let us expand our field of view. The forward placement of the eyes adds a large region of binocular overlap so that we can achieve another binocular function, stereoscopic depth perception. The region of binocular overlap is 114 degrees and the remaining monocular portion is 37 degrees for each eye. Each eye sees slightly more of the temporal than the nasal visual field and more of the ipsilateral than the contralateral side of a binocularly viewed object.

Visual perception is not uniform throughout this region. The fovea or central region of the retina is specialized to allow us to resolve fine detail in the central 5 degrees of the visual field. Eye movements expand the regions of available space. They allow expansion of the zone of high resolution, and accommodation expands the range of distances within which we can have high acuity and clear vision from viewing distances as near as 5 cm in the very young to optical infinity (Hofstetter, 1983). The peripheral retina is specialized to aid us in locomotor tasks such as walking so that we can navigate safely without colliding with obstacles. It is important that optical designs allow users to retain the necessary visual field to perform tasks aided by the optical device.

Space Perception

Space perception includes our perception of direction; distance; orientation and shape of objects; object trajectories with respect to our location; body orientation, location, and motion in space; and heading. These percepts can be derived from a 2-D image projection on the two retinas of a 3-D object space. Reconstruction of a 3-D percept from a 2-D image requires the use of information within the retinal image that is geometrically constrained by the 3-D nature of space. Three primary sources of visual or vision-related information are used for this purpose, namely, monocular information, binocular information, and extraretinal or motor information. Monocular visual or retinal cues include familiarity with the size and shape of objects, linear perspective and shape distortions, texture density, shading, partial image occlusion or overlap, size expansion and optic flow patterns, and motion parallax (Dember & Warm, 1979). Binocular information is also available, including stereoscopic depth sensed from combinations of horizontal and vertical disparity, and motion in depth (Cumming, 1994). Extraretinal cues include accommodation, convergence, and gaze direction of the eyes. Many of these cues are redundant and provide ways to check consistency and to sense errors that can be corrected by adaptively reweighting their emphasis or contribution to the final 3-D percept.

Monocular Cues.   Linear perspective refers to the distortions of the retinal image that result from the projection or imaging of a 3-D object located at a finite viewing distance onto a 2-D surface such as the retina. Texture refers to a repetitive pattern of uniform size and shape such as a gravel bed. Familiarity with the size and shape of a target allows us to sense its distance and orientation. As targets approach us and change their orientation, even though the retinal image expands and changes shape, the target appears to maintain a constant perceived size and rigid shape. Retinal image shape changes are interpreted as changes in object orientation produced by rotations about three axes. Texture density can indicate which axes the object is rotated about. Rotation about the vertical axis produces tilt and causes a compression of texture along the horizontal axis. Rotation about the horizontal axis produces slant and causes a compression of texture along the vertical axis. Rotation about the z axis produces roll or torsion and causes a constant change in orientation of all texture elements. Combinations of these rotations cause distortions or foreshortening of images; however, if the distortions are decomposed into three rotations, the orientation of rigid objects can be computed. The gradient of compressed texture varies inversely with target distance. If viewing distance is known, the texture density gradient is a strong cue to the amount of slant about a given axis.

Perspective cues arising from combinations of observer's viewpoint, object distance, and orientation can contribute to perception of distance and orientation. Perspective cues utilize edge information that is extrapolated until it intersects another extrapolated edge of the same target (Fig. 1) (Epstein, 1977). When the edges of an object are parallel, they meet at a point in the retinal image plane called the vanishing point. Normally, we do not perceive the vanishing point directly, and it needs to be derived by extrapolation. It is assumed that the line of sight directed at the vanishing point is parallel to the edges of the surface (such as a roadway) that extrapolates to the same vanishing point. As with texture density gradients, perspective distortion of the retinal image increases with object proximity. If the viewing distance is known, the amount of perspective distortion is an indicator of the amount of slant or tilt about a given axis.

Depth ordering can be computed from image overlap. This is a very powerful cue and can override any of the other depth cues, including binocular disparity. A binocular version of form overlap is called DaVinci stereopsis where the overlap is compared between the two eyes' views (Nakayama & Shimojo, 1990). Because of their lateral separation, each eye sees a slightly different view of 3-D objects. Each eye perceives more of the ipsilateral side of the object than the contralateral side. Thus for a given eye, less of the background is occluded by a near object on the ipsilateral than the contralateral side. This binocular discrepancy is sufficient to provide a strong sense of depth in the absence of binocular parallax.

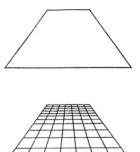

FIGURE 1   When the edges of an object are parallel, they meet at a point in the retinal image plane called the vanishing point.

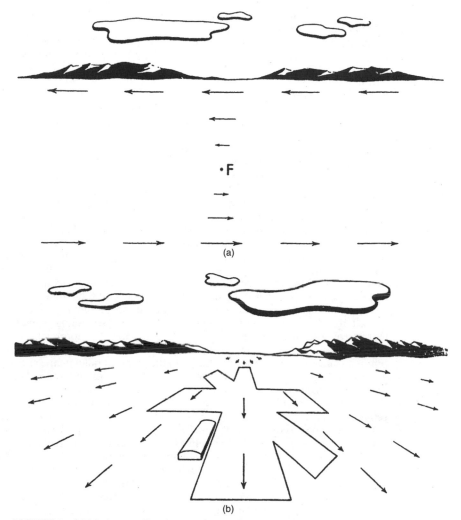

FIGURE 2 (*a*) Motion parallax. Assume that an observer moving toward the left fixates a point at F. Objects nearer than *F* will appear to move in a direction opposite to that of the movement of the observer; objects farther away than *F* will appear to move in the same direction as the observer. The length of the arrows signifies that the apparent velocity of the optic flow is directly related to the distance of objects from the fixation point. (*b*) Motion perspective. The optical flow in the visual field as an observer moves forward. The view is that as seen from an airplane in level flight. The direction of apparent movement in the terrain below is signified by the direction of the motion vectors (arrows); speed of apparent motion is indicated by the length of the motion vectors. The expansion point in the distance from which motion vectors originate is the heading direction.

Kinetic Cues. Kinetic monocular motion cues include size expansion and optic flow patterns and motion parallax. The distance of targets we approach becomes apparent from the increased velocity of radial flow of the retina (loom) (M. G. Harris, 1994). The motion of texture elements across the whole image contributes to our perception of the shape and location of objects in space. The relative motion of texture elements in the visual field is referred to as optic flow and this flow field can be used to segregate the image into objects at different depth planes, as well as to perceive the shape or form of an object. Motion parallax is the shear of optic flow in opposite directions resulting from the translation of our viewpoint (Rogers & Graham, 1982). Movies that pan the horizon yield a strong sense of 3-D depth from the motion parallax they produce between near and far parts of 3-D objects. Two sequential views resulting from a lateral shift in the viewpoint (motion parallax) are analogous to two separate and simultaneous views from two eyes with separate viewpoints (binocular disparity). Unlike horizontal disparity, the shear between sequential views by itself is ambiguous, but once we know which way our head is translating, we can correctly identify the direction or sign of depth (Ono, et al., 1986). The advantage of motion parallax over binocular disparity is that we can translate in any meridian, horizontal or vertical, and get depth information, while stereo only works for horizontal disparities (Rogers & Graham, 1982).

Motion perception helps us to determine where we are in space over time. Probably the most obvious application is to navigation and heading judgments (Fig. 2). Where are we headed, and can we navigate to a certain point in space? Generally, we perceive a focus of expansion in the direction we are heading as long as our eyes and head remain fixed (Crowell & Banks, 1996; L. R. Harris, 1994; M. G. Harris, 1994; Regan & Beverley, 1979; Warren & Hannon, 1988). These types of judgments occur whenever we walk, ride a bicycle, or operate a car. Navigation also involves tasks to intercept or avoid moving objects. For example, you might want to steer your car around another car or avoid a pedestrian walking across the crosswalk. Or you may want to intercept something, as when catching or striking a ball. Another navigation task is anticipation of time of arrival or contact. This activity is essential when pulling up to a stop sign or traffic light. If you stop too late, you enter the intersection and risk a collision. These judgments are also made while walking down or running up a flight of stairs. Precise knowledge of time to contact is essential to keep from stumbling. Time to contact can be estimated from the distance to contact divided by the speed of approach. It can also be estimated from the angular subtence of an object divided by its rate of expansion (Lee, 1976). The task is optimal when looking at a surface that is in the frontoparallel plane, and generally we are very good at it. It looks like we use the rate of expansion to predict time of contact since we make unreliable judgments of our own velocity (McKee & Watamaniuk, 1994).

Another class of motion perception is self- or ego-motion (Brandt, et al., 1973). It is related to the vestibular sense of motion and balance and tells us when we are moving. Self-motion responds to vestibular stimulation, as well as visual motion produced by translational and rotational optic flow of large fields. Finally, we can use binocular motion information to judge the trajectory of a moving object and determine if it will strike us in the head or not. This motion in depth results from a comparison of the motion in one eye to the other. The visual system computes the direction of motion from the amplitude ration and relative direction of horizontal retinal image motion (Beverley & Regan, 1973) (Fig. 3). Horizontal motion in the two eyes that is equal in amplitude and direction corresponds to an object moving in the frontoparallel plane. If the horizontal retinal image motion is equal and opposite, the object is moving toward us in the midsagittal plane and will strike us between the eyes. If the motion has any disconjugacy at all, it will strike us. A miss is indicated by yoked, albeit nonconcomitant, motion.

Binocular Cues and Extraretinal Information from Eye Movements.   Under binocular viewing conditions, we perceive a single view of the world as though seen by a single cyclopean eye, even though the two eyes receive slightly different retinal images. The binocular percept is the average of monocularly sensed shapes and directions (allelotropia). The benefit of this binocular combination is to allow us to sense objects as single with small amounts of binocular disparity so that we can interpret depth from the stereoscopic sense. Interpretation of the effects of prism and magnification upon perceived direction through an instrument needs to consider how the visual directions of the two eyes are combined. If we had only one eye, direction could be judged from the nodal point of the eye, a site where viewing angle in space equals visual angle in the eye, assuming the nodal point is close to the radial center of the retina. However, two eyes present a problem for a system that operates as though it has

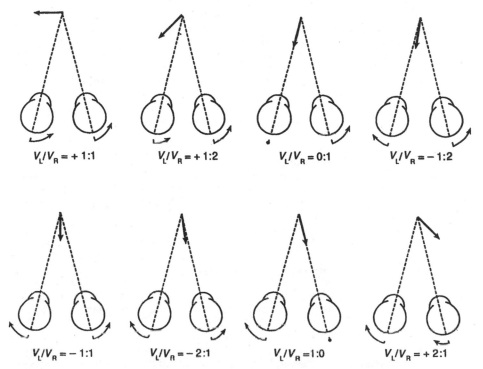

$V_L/V_R = +1{:}1$ 　　 $V_L/V_R = +1{:}2$ 　　 $V_L/V_R = 0{:}1$ 　　 $V_L/V_R = -1{:}2$

$V_L/V_R = -1{:}1$ 　　 $V_L/V_R = -2{:}1$ 　　 $V_L/V_R = 1{:}0$ 　　 $V_L/V_R = +2{:}1$

FIGURE 3 Motion in depth. Relative velocities of left and right retinal images for different target trajectories. When the target moves along a line passing between the eyes, its retinal images move in opposite directions in the two eyes; when the target moves along a line passing wide of the head, the retinal images move in the same direction, but with different velocities. The ratio (V_l/V_r) of left- and right-eye image velocities provides an unequivocal indication of the direction of motion in depth relative to the head.

only a single cyclopean eye. The two eyes have viewpoints separated by approximately 6.5 cm. When the two eyes converge accurately on a near target placed along the midsagittal plane, the target appears straight ahead of the nose, even when one eye is occluded. In order for perceived egocentric direction to be the same when either eye views the near target monocularly, there needs to be a common reference point for judging direction. This reference point is called the cyclopean locus or egocenter, and is located midway on the interocular axis. The egocenter is the percept of a reference point for judging visual direction with either eye alone or under binocular viewing conditions.

Perceived Direction

Direction and distance can be described in polar coordinates as the angle and magnitude of a vector originating at the egocenter. For targets imaged on corresponding retinal points, this vector is determined by the location of the retinal image and by the direction of gaze that is determined by the average position of the two eyes (conjugate eye position). The angle the two retinal images form with the visual axes is added to the conjugate rotational vector component of binocular eye position (the average of right and left eye position). This combination yields the perceived egocentric direction. Convergence of the eyes, which results from disconjugate eye movements, has no influence on perceived egocentric direction. Thus, when the two eyes fixate near objects to the left or right of the midline in asymmetric convergence, only the conjugate component of the two eyes' positions contributes to perceived direction. These facets of egocentric direction were summarized by Hering (1879) as five laws of visual direction, and they have been restated by Howard (1982). The laws are mainly concerned with targets imaged on corresponding retinal regions (i.e., targets on the horopter).

We perceive space with two eyes as though they were merged into a single cyclopean eye. This merger is made possible by a sensory linkage between the two eyes that is facilitated by the anatomical superposition or combination of homologous regions of the two retinas in the visual cortex. The binocular overlap of the visual fields is very specific. Unique pairs of retinal regions in the two eyes (corresponding points) must receive images of the same object so that these objects can be perceived as single and at the same depth as the point of fixation. This requires that retinal images must be precisely aligned by the oculomotor system with corresponding retinal regions in the two eyes. As described in Sec. 12.8, binocular alignment is achieved by yoked movements of the eyes in the same direction (version) and by movements of the eyes in opposite direction (vergence). Slight misalignment of similar images from corresponding points is interpreted as depth. Three-dimensional space can be derived geometrically by comparing the small differences between the two retinal images that result from the slightly different vantage points of the two eyes caused by their 6.5-cm separation. These disparities are described as horizontal, vertical, and torsional, as well as distortion or shear differences between the two images. The disparities result from surface shape, depth, and orientation with respect to the observer, as well as direction and orientation (torsion) of the observer's eyes (van Ee & Erkelens, 1996). These disparities are used to judge the layout of 3-D space and to sense the solidness or curvature of surfaces. Disparities are also used to break through camouflage in images such as those seen in tree foliage.

Binocular Visual Direction-Corresponding Retinal Points and the Horopter.   Hering (1879) defined binocular correspondence by retinal locations in the two eyes that, when stimulated, resulted in a percept in identical visual directions. For a fixed angle of convergence, projections of corresponding points along the equator and midline of the eye converge upon real points in space. In other cases, such as oblique eccentric locations of the retina, corresponding points have visual directions that do not intersect in real space. The horopter is the locus in space of real objects or points whose images can be formed on corresponding retinal points. It serves as a reference throughout the visual field for the same depth or disparity as at the fixation point. To appreciate the shape of the horopter, consider a theoretical case in which corresponding points are defined as homologous locations on the two retinas. Thus, corresponding points are equidistant from their respective foveas. Consider binocular matches between the horizontal meridians or equators of the two retinas. Under this circumstance, the visual directions of corresponding points intersect in space at real points that define the longitudinal horopter. This theoretical horopter is a circle whose points will be imaged at equal eccentricities from the two foveas on corresponding points except for the small arc of the circle that lies between the two eyes (Ogle, 1964). While the theoretical horopter is always a circle, its radius of curvature increases with viewing distance. This means that its curvature decreases as viewing distance increases. In the limit, the theoretical horopter is a straight line at infinity that is parallel to the interocular axis. Thus, a surface representing zero disparity has many different shapes depending on the viewing distance. The consequence of this spatial variation in horopter curvature is that the spatial pattern of horizontal disparities is insufficient information to specify depth magnitude or even depth ordering or surface shape. It is essential to know viewing distance to interpret surface shape and orientation from depth-related retinal image disparity. Viewing distance could be obtained from extraretinal information such as the convergence state of the eyes, or from retinal information in the form of vertical disparity as described below.

The empirical or measured horopter differs from the theoretical horopter in two ways (Ogle, 1964). It can be tilted about a vertical axis and its curvature can be flatter or steeper than the Vieth-Müller circle. This is a circle that passes through the fixation point in the midsagittal plane and the entrance pupils of the two eyes. Points along this circle are imaged at equal retinal eccentricities from the foveas of the two eyes. The tilt of the empirical horopter

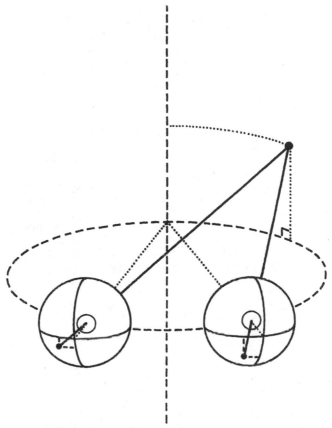

FIGURE 4 The horopter. The theoretical horopter or Vieth-Müller circle passes through the fixation point and two entrance pupils of the eyes. The theoretical vertical horopter is a vertical line passing through the fixation point in the midsagittal plane. Fixation on a tertiary point in space produces vertical disparities because of greater proximity to the ipsilateral eye. (*Reprinted by permission from Tyler and Scott, 1979.*)

can be the result of a horizontal magnification of the retinal image in one eye. Image points of a frontoparallel surface subtend larger angles in the magnified images. A similar effect occurs when no magnifier is worn and the fixation plane is tilted toward one eye. The tilt causes a larger image to be formed in one eye than the other. Thus, magnification of a horizontal row of vertical rods in the frontoparallel plane causes the plane of the rods to appear tilted to face the eye with the magnifier.

How does the visual system distinguish between a surface fixated in eccentric gaze that is in the frontoparallel plane (parallel to the face), but tilted with respect to the Vieth-Müller circle, and a plane fixated in forward gaze that is tilted toward one eye? Both of these planes project identical patterns of horizontal retinal image disparity, but they have very different physical tilts in space. In order to distinguish between them, the visual system needs to know the horizontal gaze eccentricity. One way is to register the extraretinal eye position signal, and the other is to compute gaze eccentricity from the vertical disparity gradient associated with eccentrically located targets. It appears that the latter method is possible, since tilt of a frontoparallel plane can be induced by a vertical magnifier before one eye (the induced effect). The effect is opposite to the tilt produced by placing a horizontal magnifier before the same eye (the geometric effect). If both a horizontal and vertical magnifier are placed before one eye, in the form of an overall magnifier, no tilt of the plane occurs. Interestingly, in eccentric gaze, the proximity to one eye causes an overall magnification of the image in the nearer eye, which could be sufficient to allow observers to make accurate frontoparallel settings in eccentric gaze (Fig. 4). Comparison of vertical and horizontal magnification seems to be sufficient to disambiguate tilt from eccentric viewing.

Flattening of the horopter from a circle to an ellipse results from nonuniform magnification of one retinal image. If the empirical horopter is flatter than the theoretical horopter, corresponding retinal points are more distant from the fovea on the nasal than the temporal hemiretina. Curvature changes in the horopter can be produced with nonuniform magnifiers such as prisms (Fig. 5). A prism magnifies more at its apex than its base. If a base-out prism is placed before the two eyes, the right half of the image is magnified more in the left eye than in the right eye and the left half of the image is magnified more in the right eye than in the left eye. This causes flat surfaces to appear more concave and the horopter to be less curved or more convex.

The Vertical Horopter.   The theoretical vertical point-horopter for a finite viewing distance is limited by the locus of points in space where visual directions from corresponding points will intersect real objects (Tyler & Scott, 1979). These points are described by a vertical line in the midsagittal plane that passes through the Vieth-Müller Circle. Eccentric object points

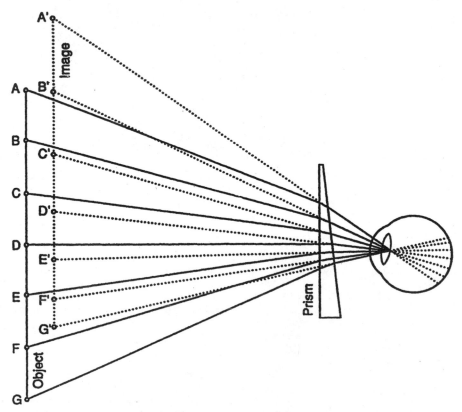

FIGURE 5   Figure showing nonuniform magnification of a prism. Disparateness of retinal images, producing stereopsis.

in tertiary gaze (points with both azimuth and elevation) lie closer to one eye than the other eye. Because they are imaged at different vertical eccentricities from the two foveas, tertiary object points cannot be imaged on theoretically corresponding retinal points. However, all object points at an infinite viewing distance can be imaged on corresponding retinal regions, and at this infinite viewing distance, the vertical horopter becomes a plane.

The empirical vertical horopter is declined (top slanted away from the observer) in comparison to the theoretical horopter. Helmholtz (1909) reasoned that this was because of a horizontal shear of the two retinal images which causes a real vertical plane to appear inclined toward the observer. Optical infinity is the only viewing distance that the vertical horopter becomes a plane. It is always a vertical line at finite viewing distances. Targets that lie away from the midsagittal plane always subtend a vertical disparity due to their unequal proximity and retinal image magnification in the two eyes. The pattern of vertical disparity varies systematically with viewing distance and eccentricity from the midsagittal plane. For a given target height, vertical disparity increases with horizontal eccentricity from the midsagittal plane and decreases with viewing distance. Thus, a horizontally extended target of constant height will produce a vertical disparity gradient that increases with eccentricity. The gradient will be greater at near than at far viewing distances. It is possible to estimate viewing distance from the vertical disparity gradient or from vertical disparity at a given point if target eccentricity is known. This could provide a retinal source of information about viewing distance to allow the visual system to scale disparity to depth and to compute depth ordering. Several investigators have shown that modifying vertical disparity can influence the magnitude of depth

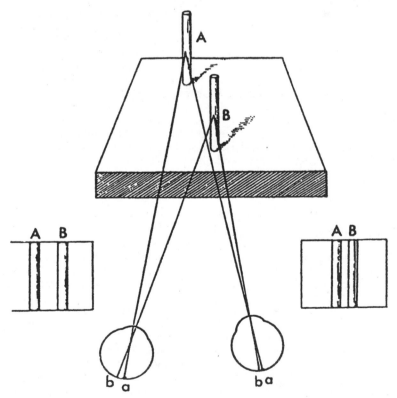

FIGURE 6   The differences in the perspective views of the two eyes produce binocular vision that can be used to perceive depth.

(scaling) and surface slant, such that vertical disparity is a useful source of information about viewing distance that can be used to scale disparity and determine depth ordering.

Stereopsis

Three independent variables involved in the calculation of stereodepth are retinal image disparity, viewing distance, and the separation in space of the two viewpoints (i.e., the baseline or interpupillary distance) (Fig. 6). In stereopsis, the relationship between the linear depth interval between two objects and the retinal image disparity that they subtend is approximated by the following expression:

$$\Delta d = \eta \times \frac{d^2}{2a}$$

where η is retinal image disparity in radians, d is viewing distance, $2a$ is the interpupillary distance, and Δd is the linear depth interval. $2a$, d, and Δd are all expressed in the same units (e.g., meters). The formula implies that in order to perceive depth in units of absolute distance (e.g., meters), the visual system utilizes information about the interpupillary distance and the viewing distance. The viewing distance could be sensed from the angle of convergence (Foley, 1980) or from other retinal cues such as oblique or vertical disparities. These disparities occur naturally with targets in tertiary directions from the point of fixation (Garding, et al., 1995; Gillam and Lawergren, 1983; Liu, et al., 1994; Mayhew and Longuet-Higgins, 1982; Rogers & Bradshaw, 1993; Westheimer & Pettet, 1992).

The equation illustrates that for a fixed retinal image disparity, the corresponding linear depth interval increases with the square of viewing distance and that viewing distance is used to scale the horizontal disparity into a linear depth interval. When objects are viewed through base-out prisms that stimulate additional convergence, perceived depth should be reduced by underestimates of viewing distance. Furthermore, the pattern of zero retinal image disparities described by the curvature of the longitudinal horopter varies with viewing distance. It can be concave at near distances and convex at far distances in the same observer (Ogle, 1964). Thus, without distance information, the pattern of retinal image disparities across the visual field is insufficient to sense either depth ordering (surface curvature) or depth magnitude (Garding, et al., 1995). Similarly, the same pattern of horizontal disparity can correspond to different slants about a vertical axis presented at various horizontal gaze eccentricities (Ogle, 1964). Convergence distance and direction of gaze are important sources of information used to interpret slant from disparity fields associated with slanting surfaces (Backus, et al., 1999). Clearly, stereodepth perception is much more than a disparity map of the visual field.

Binocular Fusion and Suppression

Even though there is a fairly precise point-to-point correspondence between the two eyes for determining depth in the fixation plane, images in the two eyes can be combined into a single percept when they are anywhere within a small range or retinal area around corresponding retinal points. Thus, a point in one eye can be combined perceptually with a point imaged within a small area around its corresponding retinal location in the other eye. These ranges are referred to as Panum's fusional area (PFA) and they serve as a buffer zone to eliminate diplopia for small disparities near the horopter. PFA allows for the persistence of single binocular vision in the presence of constant changes in retinal image disparity caused by various oculomotor disturbances. For example, considerable errors of binocular alignment (>15 arc min) may occur during eye tracking of dynamic depth produced either by object motion or by head and body movements (Steinman & Collewijn, 1980). Stereopsis could exist without singleness but the double images near the fixation plane would be a distraction. The depth of focus of the human eye serves a similar function. Objects that are nearly conjugate to the retina appear as clear as objects focused precisely on the retina. The buffer for the optics of the eye is much larger than the buffer for binocular fusion. The depth of focus of the eye is approximately 0.75 diopters. Panum's area, expressed in equivalent units, is only 0.08 meter angle or approximately one tenth the magnitude of the depth of focus. Thus, we are more tolerant of focus errors than we are of convergence errors.

Fusion is only possible when images have similar size, shape, and contrast polarity. This similarity ensures that we only combine images that belong to the same object in space. Natural scenes contain many objects that are at a wide range of distances from the plane of fixation. Some information is coherent, such as images formed within Panum's fusional area. Some information is fragmented, such as partially occluded regions of space resulting in visibility to only one eye. Finally, some information is uncorrelated because it is either ambiguous or in conflict with other information, such as the superposition of separate diplopic images arising from objects seen by both eyes behind or in front of the plane of fixation. One objective of the visual system is to preserve as much information from all three sources as possible to make inferences about objective space without introducing ambiguity or confusion of space perception. In some circumstances conflicts between the two eyes are so great that conflicting percepts are seen alternately every 4 seconds (binocular rivalry suppression) or in some cases one image is permanently suppressed, as when you look through a telescope with one eye while the other remains open. The view through the telescope tends to dominate over the view of the background seen by the other eye. For example, the two ocular images may have unequal clarity or blur, as in asymmetric convergence, or large unfusable disparities originating from targets behind or in front of the fixation plane may appear overlapped with other large diplopic images. Fortunately, when dissimilar images are formed within a fusable range of disparity, that perception of the conflicting images is suppressed.

Four classes of stimuli evoke what appear to be different mechanisms of interocular suppression. The first is unequal contrast or blur of the two retinal images which causes interocular blur suppression. The second is physiologically diplopic images of targets in front of or behind the singleness horopter which result in suspension of one of the redundant images (Cline, et al., 1989). The third is targets of different shape presented in identical visual directions. Different size and shape result in an alternating appearance of the two images referred to as either binocular retinal rivalry or percept rivalry suppression. The fourth is partial occlusion that obstructs the view of one eye such that the portions of the background are only seen by the unoccluded eye and the overlapping region of the occluder that is imaged in the other eye is permanently suppressed.

12.3 DISTORTION OF SPACE BY MONOCULAR MAGNIFICATION

The magnification and translation of images caused by optical aids will distort certain cues used for space perception while leaving other cues unaffected. This will result in cue misrepresentation of space as well as cue conflicts that will produce errors of several features of space perception. These include percepts of direction, distance, trajectory or path of external moving targets, heading and self-motion estimates, and the shape and orientation (curvature, slant, and tilt) of external objects. Fortunately, if the optical distortions are long-standing, the visual system can adapt and make corrections to their contributions to space perception.

Magnification and Perspective Distortion

Most optical systems magnify or minify images, and this can produce errors in perceived direction. When objects are viewed through a point other than the optical center of a lens, they appear displaced from their true direction. This is referred to as a prismatic displacement. Magnification will influence most of the monocular cues mentioned above except overlap. Magnification will produce errors in perceived distance and produce conflicts with cues of texture density and linear perspective. For example, the retinal image of a circle lying on a horizontal ground plane has an elliptical shape when viewed at a remote distance, and a more circular shape when viewed from a shorter distance. When a remote circle is viewed through a telescope, the uniformly magnified image appears to be closer and squashed in the vertical direction to form an ellipse (shape distortion). Uniform magnification also distorts texture-spacing gradients such that the ground plane containing the texture is perceived as inclined or tilted upward toward vertical. Normally, the change in texture gradient is greatest for near objects (Fig. 7). When a distant object is magnified, it appears nearer but its low texture density gradient is consistent with a plane inclined toward the vertical. Magnification also affects slant derived from perspective cues in the same way. For this reason, a tilted rooftop appears to have a steeper pitch when viewed through binoculars or a telescope, the pitcher-to-batter distance appears reduced in a telephoto view from the outfield, and the spectators may appear larger than the players in front of them in telephoto lens shots.

Magnification and Stereopsis

Magnification increases both retinal image disparity and image size. The increased disparity stimulates greater stereodepth while increased size stimulates reduced perceived viewing distance (Westheimer, 1956). Greater disparity will increase perceived depth while reduced per-

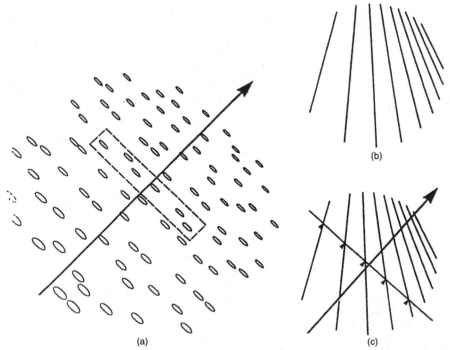

(a)   (b)

(c)

FIGURE 7   Texture gradients. The tilt of a surface is the direction in which it is slanted away from the gaze normal of the observer. (*a*) If the surface bears a uniform texture, the projection of the axis of tilt in the image indicates the direction in which the local density of the textures varies most, or, equivalently, it is perpendicular to the direction that the texture elements are most uniformly distributed (dotted rectangle). Either technique can be used to recover the tilt axis, as illustrated. Interestingly, however, the tilt axis in situations like (*b*) can probably be recovered most accurately using the second method, i.e., searching for the line that is intersected by the perspective lines at equal intervals. This method is illustrated in (*c*). (*Reprinted by permission from K. Stevens, "Surface Perception from Local Analysis of Texture and Contour," Ph.D. thesis, Department of Electrical Engineering and Computer Science, Massachusetts Institute of Technology, Cambridge, MA, 1979.*)

ceived viewing distance will scale depth to a smaller value. Because stereoscopic depth from disparity varies with the square of viewing distance, the perceptual reduction of viewing distance will dominate the influence of magnification on stereoscopic depth. Thus, perceived depth intervals sensed stereoscopically from retinal image disparity appear smaller with magnified images because depth derived from binocular retinal image disparity is scaled by the square of perceived viewing distance. Objects appear as flat surfaces with relative depth, but without thickness. This perceptual distortion is referred to as cardboarding. Most binocular optical systems compensate for this reduction in stereopsis by increasing the baseline or separation of the two ocular objectives. For example, the prism design of binoculars folds the optical path into a compact package and also expands the distance between the objective lenses creating an expanded interpupillary distance (telestereoscope). This increase of effective separation between the two eyes' views exaggerates disparity, and if it exceeds the depth reduction caused by perceived reductions in viewing distance, objects can appear in hyperstereoscopic depth. The depth can be predicted from the formula in the preceding section that computes disparity from viewing distance, interpupillary distance, and physical linear depth interval.

Magnification and Motion Parallax—Heading

Magnification will produce errors in percepts derived from motion cues. Motion will be exaggerated by magnification and perceived distance will be shortened. As with binocular disparity, depth associated with motion parallax will be underestimated by the influence of magnification on perceived distance. Heading judgements in which gaze is directed away from the heading direction will be in error by the amount the angle of eccentric gaze is magnified. Thus, if your car is headed north and your gaze is shifted 5 degrees to the left, heading errors will be the percentage the 5-degree gaze shift is magnified by the optical system and you will sense a path that is northeast. Heading judgements that are constantly changing, such as along a curved path, might be affected in a similar way if the gaze lags behind the changing path of the vehicle. In this case, changes of perceived heading along the curved path would be exaggerated by the magnifier. Unequal magnification of the two retinal images in anisometropia will cause errors in sensing the direction of motion in depth by changing the amplitude ratio of the two moving images.

Bifocal Jump

Most bifocal corrections have separate optical centers for the distance lens and the near addition. One consequence is that objects appear to change their direction or jump as the visual axis crosses the top line of the bifocal segment. This bifocal jump occurs because prism is introduced as the eye rotates downward behind the lens away from the center of the distance correction and enters the bifocal segment (Fig. 8). When the visual axis enters the optical zone of the bifocal, a new prismatic displacement is produced by the displacement of the visual axis from the optical center of the bifocal addition. All bifocal additions are positive so that the prismatic effect is always the same: objects in the upper part of the bifocal appear displaced upward compared to views just above the bifocal. This results in bifocal jump. A consequence of bifocal jump is that a region of space seen near the upper edge of the bifocal is not visible. It has been displaced out of the field of view by the vertical prismatic effect of the bifocal. The invisible or missing part of the visual field equals the prismatic jump that can be calculated from Prentice's rule [distance (cm) between the upper edge and center of the bifocal times the power of the add]. Thus for a 2.5 D bifocal add with a top-to-center distance of 0.8 cm, approximately one degree of the visual field is obscured by bifocal jump. This loss may seem minor; however, it is very conspicuous when paying attention to the ground plane while walking. There is a clear horizontal line of discontinuity that is exaggerated by the unequal speed of optic flow above and below the bifocal boundary. Optic flow is exaggerated by the magnification of the positive bifocal segment. It can produce errors of estimated walking speed and some confusion if you attend to the ground plane in the lower field while navigating toward objects in the upper field. This can be a difficult problem when performing such tasks as walking down steps. It is best dealt with by maintaining the gaze well above the bifocal segment while walking, which means that the user is not looking at the ground plane in the immediate vicinity of the feet. Some bifocal designs eliminate jump by making the pole of the bifocal and distance portion of the lens concentric. The bifocal line starts at the optical centers of both lenses. This correction always produces some vertical displacement of the visual field since the eyes rarely look through the optical center of the lens; however, there is no jump.

Discrepant Views of Objects and Their Images

Translation errors in viewing a scene can lead to errors in space perception. A common problem related to this topic is the interpretation of distance and orientation of objects seen on video displays or in photographs. The problem arises from discrepancies between the camera's vantage point with respect to a 3-D scene and the station point of the observer who is viewing the 2-D projected image after it has been transformed by the optics of the camera lens. Errors of observer distance and tilt of the picture plane contribute to perceptual errors of perspective distortion, magnification, and texture density gradients expected from the station point. The station point represents the position of the observer with respect to the view plane or projected images. To see the picture in the correct perspective, the station point should be the same as the location of the camera lens with respect to the film plane. Two violations of these requirements are errors in viewing distance and translation errors (i.e., horizontal or vertical displacements). Tilt of the view screen can be decomposed into errors of distance and translation errors. In the case of the photograph, the correct distance for the station point is the camera lens focal length multiplied by the enlargement scaling factor. Thus,

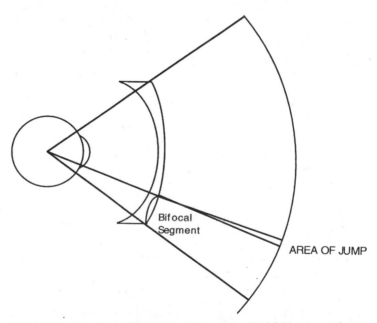

FIGURE 8 Jump figure. The missing region of the visual field occurs at the top edge of the bifocal segment as a result of prism jump.

when a 35-mm negative taken through a 55-mm lens is enlarged 7 times to a 10-in print, it should be viewed at a distance of 38.5 cm or 15 in (7×5.5 cm).

12.4 DISTORTION OF SPACE PERCEPTION FROM INTEROCULAR ANISO-MAGNIFICATION (UNEQUAL BINOCULAR MAGNIFICATION)

Optical devices that magnify images unequally or translate images unequally in the two eyes produce errors of stereoscopic depth and perceived direction, as well as eye alignment errors that the oculomotor system is not accustomed to correcting. If these sensory and motor errors are long-standing, the visual system adapts perceptual and motor responses to restore normal visual performance. Patients who receive optical corrections that have unequal binocular magnification will adapt space percepts in 10 days to 2 weeks.

Lenses and Prisms

Magnification produced by lenses is uniform compared to the nonuniform magnification produced by prisms, where the magnification increases toward the apex of the prism. Uniform magnification differences between the horizontal meridians of the two eyes produce errors in perceived surface tilt about a vertical axis (geometric effect). A frontoparallel surface appears to face the eye with the more magnified image. This effect is offset by vertical magnification of one ocular image that causes a frontoparallel surface to appear to face the eye with the less magnified image. Thus, perceptual errors of surface tilt occur mainly when magnification is greater in the horizontal or vertical meridian. These meridional magnifiers are produced by cylindrical corrections for astigmatism. Nonuniform magnification caused by horizontal prism, such as base-out or base-in (Fig. 5), cause surfaces to appear more concave or convex, respectively. Finally, cylindrical corrections for astigmatism cause scissor or rotational distortion of line orientations (Fig. 9). When the axes of astigmatism are not parallel in the two eyes,

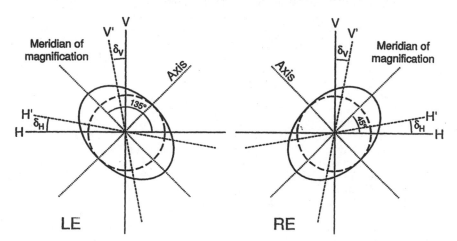

FIGURE 9 Magnification ellipse. Meridional magnification of images in the two eyes in oblique meridians cause "scissors" or rotary deviations of images of vertical (and horizontal) lines and affect stereoscopic spatial localization in a characteristic manner. [*Reprinted from K. N. Ogle and P. Boder, "Distortion of Stereoscopic Spatial Localization," J. Opt. Soc. Am. 38:723–733 (1948).*]

cyclodisparities are introduced. These disparities produce slant errors and cause surfaces to appear inclined or declined. These spatial depth distortions are adapted to readily so that space appears veridical; however, some people are unable to cope with the size differences of the two eyes and suffer from an anomaly referred to as aniseikonia. While aniseikonic errors are tolerated by many individuals, they can cause spatial distortions which, in given situations, may create meaningful problems.

Aniseikonia

Unequal size of the two retinal images in anisometropia can precipitate a pathological mismatch in perceived size of the two ocular images (aniseikonia). Aniseikonia is defined as a relative difference in perceived size and/or shape of the two ocular images. Most difficulties with aniseikonia occur when anisometropia is corrected optically with spectacles. Patients experience a broad range of symptoms, including spatial disorientation, depth distortion, diplopia, vertigo, and asthenopia. These symptoms result from several disturbances, including spatial distortion and interference with binocular sensory and motor fusion.

Two optical distortions produced by spectacle corrections of anisometropia are called axial and lateral aniseikonia. Axial aniseikonia describes size distortions produced by differences in the magnification at the optical centers of ophthalmic spectacle corrections for anisometropia. The magnification difference produced by the spectacle correction depends in part on the origin of the refractive error. Retinal images in the uncorrected axial and refractive ametropias are not the same. In axial myopia, the uncorrected retinal image of the elongated eye is larger than that of an emmetropic eye of similar refractive power. An uncorrected refractive myope has a blurred image of similar size to an emmetropic eye with the same axial length. Knapp's law states that image size in an axial ametropic eye can be changed to equal the image size in an emmetropic eye having the same refractive power by placing the ophthalmic correction at the anterior focal plane of the eye. The anterior focal plane of the eye is usually considered to correspond approximately to the spectacle plane. However, this is only a very general approximation. The anterior focal point will vary with power of the eye and, in anisometropia, it will be unequal for the two eyes. Assuming the anterior focal point is the spectacle plane, then the abnormally large image of an uncorrected axial-myopic eye can be reduced to near normal with a spectacle correction, whereas a contact lens correction would leave the retinal image of the myopic eye enlarged. If the ametropia is refractive in nature (i.e., not axial), then the uncorrected retinal images are equal in size and a contact lens correction will produce less change in image size than a spectacle correction. If the refractive myope is corrected with spectacles, the retinal image becomes smaller than that in an emmetropic eye of similar refractive power. This change in image size of refractive anisometropes corrected with spectacle lenses is one of several factors leading to ocular discomfort in aniseikonia.

Paradoxically, spectacle correction of axial myopes often produces too drastic a reduction in ocular image size. The ocular image of the axially myopic eye, corrected with spectacle lenses, appears smaller than that of the less ametropic eye in anisometropia (Arner, 1969; Awaya & von Noorden, 1971; Bourdy, 1968; Bradley, et al., 1983; Rose & Levenson, 1972; Winn, et al., 1987). As mentioned earlier, ocular image size depends not only on retinal image size, but also on the neural factor (Ogle, 1964). For example, in cases of axial myopia, the elongation of the eye produces stretching of the retina and pigment epithelium. Accordingly, retinal images of normal size occupy a smaller proportion of the stretched retina than they would in a retina with normal retinal element density. Indeed, anisometropes with axial myopia report minification of the ocular image in their more myopic eye after the retinal images have been matched in size (Bradley, et al., 1983; Burnside & Langley, 1964; Engle, 1963). Thus, in the uncorrected state, myopic anisometropic individuals have optical magnification and neural minification. As a result, a spectacle correction would overcompensate for differential size of ocular images.

Surprisingly, not all patients with a significant amount of aniseikonia (from 4 to 6 percent) complain of symptoms. This could be due in part to suppression of one ocular image. However, normally we are able to fuse large size differences of up to 15 percent of extended targets (Julesz, 1971). Sinusoidal gratings in the range from 1 to 5 cpd can be fused with grating differing in spatial frequency by from 20 to 40 percent (Blakemore, 1970; Fiorentini & Maffei, 1971; Tyler & Sutter, 1979) and narrow band pass filtered bars can be fused with from 100 to 400 percent difference in interocular size (Schor & Heckmen 1989; Schor, et al., 1984). These enormous size differences demonstrate the remarkable ability of the visual system to fuse large size differences in axial aniseikonia. However, in practice, it is difficult to sustain fusion and stereopsis with greater than from 8 to 10 percent binocular image size difference.

With extended targets, magnification produces an additional lateral displacement due to cumulative effects such that nonaxial images are enlarged and imaged on noncorresponding points. Given our tolerances for axial disparity gradients in size, it appears that most of the difficulty encountered in aniseikonia is lateral. Von Rohr (1911) and Erggelet (1916) considered this prismatic effect as the most important cause of symptomatic discomfort in optically induced aniseikonia.

Differential magnification of the two eyes' images can be described as a single constant; however, the magnitude of binocular-position disparity increases proportionally with the eccentricity with which an object is viewed from the center of a spectacle lens. Prentice's rule approximates this prismatic displacement from the product of distance from the optical center of the lens in cm and the dioptric power of the lens. In cases of anisometropia corrected with spectacle lenses, this prismatic effect produces binocular disparity along the meridian of eccentric gaze. When these disparities are horizontal, they result in distortions of stereoscopic depth localization, and when they are vertical, they can produce diplopia and eye strain. Fortunately, there is an increasing tolerance for disparity in the periphery provided by an increase in the binocular sensory-fusion range with retinal eccentricity (Ogle, 1964). In cases of high anisometropia, such as are encountered in monocular aphakia, spectacle corrections can cause diplopia near the central optical zone of the correction lens. In bilateral aphakia, wearing glasses or contact lenses does not interfere with the normal motor or sensory fusion range to horizontal disparity; however, stereo-thresholds are elevated (Nolan, et al., 1975). In unilateral aphakia, contact lenses and intraocular lens implants support a normal motor fusion range; however, stereopsis is far superior with the interocular lens implants (Nolan, et al., 1975). Tolerance of size difference errors varies greatly with individuals and their work tasks. Some may tolerate from 6 to 8 percent errors, while others may complain with only from .25 to .5 percent errors.

Interocular Blur Suppression with Anisometropia

There is a wide variety of natural conditions that present the eyes with unequal image contrast. These conditions include naturally occurring anisometropia, unequal amplitudes of accommodation, and asymmetric convergence on targets that are closer to one eye than the other. This blur can be eliminated in part by a limited degree of differential accommodation of the two eyes (Marran & Schor, 1998) and by interocular suppression of the blur. The latter mechanism is particularly helpful for a type of contact lens patient who can no longer accommodate (presbyopes) and who prefer to wear a near contact lens correction over one eye and a far correction over the other (monovision) rather than wear bifocal spectacles. For most people (66 percent), all of these conditions result in clear, nonblurred, binocular percepts with a retention of stereopsis (Schor, et al., 1987), albeit with the stereothreshold elevated by approximately a factor of two. Interocular blur suppression is reduced for high contrast targets composed of high spatial frequencies (Schor, et al., 1987). There is an interaction between interocular blur suppression and binocular rivalry suppression. Measures of binocular rivalry reveal a form of eye dominance defined as the eye that is suppressed least when viewing dichoptic forms of different shape. When the dominant eye for rivalry and aiming or sighting is the same, interocular suppression is more effective than when dominance for sighting and rivalry are crossed (i.e., in different eyes) (Collins & Goode, 1994).

When the contrast of the two eyes' stimuli is reduced unequally, stereoacuity is reduced more than if the contrast of both targets is reduced equally. Lowering the contrast in one eye reduces stereoacuity twice as much as when contrast is lowered in both eyes. This contrast paradox only occurs for stereopsis and is not observed for other disparity-based phenomena, including binocular sensory fusion. The contrast paradox occurs mainly with low spatial frequency stimuli (<2.0 cycles/degree). Contrast reduction caused by lens defocus is greater at high spatial frequencies, and perhaps this is why stereopsis is able to persist in the presence of interocular blur caused by anisometropia.

12.5 *DISTORTIONS OF SPACE FROM CONVERGENCE RESPONSES TO PRISM*

Horizontal vergence responses to prism influence estimates of viewing distance (Ebenholtz 1981; Foley, 1980; Owens & Leibowitz, 1980; Wallach & Frey, 1972). The resulting errors in perceived distance will influence depth percepts that are scaled by perceived viewing distance, such as stereopsis and motion parallax. Time of arrival or contact with external objects can be estimated from the distance-to-contact divided by the speed of approach. If distance is underestimated while wearing convergent or base-out (temporal) prism, then time-to-contact, computed from velocity and distance, will also be underestimated. In addition, prism stimuli to convergence will produce conflicts in distance estimates from vergence and retinal image size. When vergence responds to prism, retinal image size remains constant instead of increasing as it would in response to real changes in viewing distance. The conflict is resolved by a decrease in perceived size. This illusion is referred to as convergence micropsia. An average magnitude of shrinkage is 25 percent for an accommodation and convergence increase equivalent to 5 diopters or 5 meter angles (McCready, 1965). Very large prism stimuli to vergence present major conflicts between the fixed stimulus to accommodation and the new vergence stimulus. The oculomotor system compromises by partially responding to the vergence stimulus. If physical size of a target in a virtual display has been adjusted to correspond to an expected full vergence response (increased retinal image size with convergence stimuli), but a smaller vergence response actually occurs, the perceptual compensation for size is based on the smaller vergence response. As a result, a paradoxical increase of perceived size accompanies convergence (convergence macropsia) (Ellis, et al., 1995; Morita & Hiruma, 1996; Yeh, 1993). The incompletely fused images are perceived at the distance of convergence alignment of the eyes rather than at the intended distance specified by physical size and disparity. Thus, aniseikonic and anisophoric differences in the two eyes, which are made worse by anisometropia, can cause difficulty when binocular tasks associated with the use of optical instruments are performed.

12.6 *EYE MOVEMENTS*

There are two general categories of eye movements. Once class stabilizes the motion of the retinal image generated by our own body and head movements and the other class stabilizes the retinal image of an object moving in space. In the former case, the whole retinal image is moving in unison, whereas in the latter case only the retinal image of the moving object moves relative to the stationary background, and when that image is tracked it remains stabilized on the fovea during target inspection. In all, there are five types of eye movements (VOR, OKN, saccades, pursuits, and vergence) and they are classified as either stabilizing or tracking movements.

Two classes of reflex eye movements perform the image stabilization task. The first class of stabilizing eye movement compensates for brief head and body rotation and is called the vestibulo-ocular reflex (VOR). During head movements in any direction, the semicircular canals of the vestibular labyrinth signal how fast the head is rotating and the oculomotor system responds reflexively to this signal by rotating the eyes in an equal and opposite rotation. During head rotation, this reflex keeps the line of sight fixed in space and visual images stationary on the retina. Body rotation in darkness causes a repetitious sequence of eye movements which move the line of sight slowly in the direction opposite to head rotation (slow phase) followed by a rapid (saccadic) movement in the direction the head moves (fast phase). This reflex is almost always active and without it we would be unable to see much of anything due to constant smear of the retinal image. The vestibulo-ocular reflex is extremely accurate for active head motion (Collewijn, et al., 1983) because it has an adaptive plasticity that makes it easily calibrated in response to altered visual feedback (Gauthier & Robinson, 1975; Gonshor & Melvill-Jones, 1976). The VOR may be inhibited or suppressed by the visual fixation mechanism (Burde, 1981) as a result of the system's attempt to recalibrate to the head-stationary image.

The optokinetic reflex is an ocular following response that stabilizes whole field motion when we move about in a visual scene. Whereas the vestibulo-ocular reflex operates in total darkness, the optokinetic reflex requires a visible retinal image. Rotation or translation of the visual field causes the eyes to execute a slow following and fast reset phase sequence of eye movements that have the same pattern as the VOR. The optokinetic reflex supplements the vestibulo-ocular reflex in several ways. The vestibulo-ocular reflex responds to acceleration and deceleration of the head, but not to constant velocity. In contrast, the optokinetic reflex responds to constant retinal image velocity caused by constant body rotation or translation. The vestibulo-ocular reflex controls initial image stabilization and the optokinetic reflex maintains the stabilization.

Tracking eye movements place the retinal image of the object of regard on the fovea and keep it there, even if the object moves in space. The class of eye movements that places the image on the fovea is called a saccade. This is a very fast eye movement that shifts the image in a steplike motion. Saccades can be made voluntarily, in response to visual, tactile, and auditory stimuli, and even in darkness in response to willed directions of gaze. Their chief characteristic is that they are fast, reaching velocities of nearly 1000 deg/sec. Once the image of the target is on the fovea, slow following eye movements called pursuits maintain it there. They can keep images stabilized that are formed of objects moving as fast as 30 deg/sec in humans. Generally, pursuits cannot be made voluntarily in the absence of a visual moving stimulus. However, they can respond to a moving sound or tactile sensation in darkness.

12.7   COORDINATION AND ALIGNMENT OF THE TWO EYES

If you observe the motion of the two eyes, you will notice that they either move in the same or in opposite directions. When they move in the same direction, they are said to be yoked, or conjugate, like a team of horses. These are also called version movements. At other times the eyes move in opposite directions, especially when we converge from far to near. These are called disjunctive or vergence movements. Both conjugate and disjunctive movements can be either fast or slow, depending on whether the eyes are shifting fixation or maintaining foveation (stable fixation) of a target. A century ago, Ewald Hering described the movements of the two eyes as equal and symmetrical (Hering's law). Our eyes are yoked when we are born, but the yoking can be modified in response to anisometropic spectacles that produce unequal image sizes and require us to make unequal eye movements to see bifoveally. This image inequality leads to the binocular anomalies of aniseikonia (unequal perceived image size) and anisophoria (nonconcomitant phoria, or a phoria that varies with direction of gaze).

Targets are brought into alignment with corresponding points by movements of the eyes in opposite directions. These vergence movements have three degrees of freedom (horizontal, vertical, and torsional). Horizontal vergence can rotate the eyes temporalward (divergence) or nasalward (convergence). Since near targets only require convergence of the eyes, there is limited need for divergence other than to overcome postural errors of the eyes. Consequently, the range of divergence is less than half the range of convergence. The eyes are able to diverge approximately 4 degrees and converge approximately from 8 to 12 degrees in response to disparity while focusing accurately on a target at a fixed viewing distance. Vergence movements serve the same function as conjugate movements of the eyes, which is to preserve the alignment of the two retinal images on corresponding retinal points. Vergence is required when changing viewing distance, whereas versions accomplish the task when changing the direction of gaze.

The stimulus to vergence depends on both the viewing distance and the distance between the two eyes (interpupillary distance). It is equal to the angle subtended by a target in space at the entrance pupils of the two eyes (binocular parallax). As will be described in Sec. 12.9 ("Prism-Induced Errors of Eye Alignment"), the stimulus to vergence can be changed by stimulating accommodation with lenses, or changing disparity (binocular parallax) with prisms. The exact stimulus value of lenses and prisms to vergence depends upon the placement of the prism relative to the eye and this relationship will be described in Sec. 12.8 ("Effects of Lenses and Prism on Vergence and Phoria"). Vergence is measured in degrees of ocular rotation and prism diopters. The meter angle is also used in order to get a comparable metric for accommodation and convergence. Meter angles equal the reciprocal of the viewing distance in meters, and the value is approximately equal to the accommodative stimulus in diopters.

$$MA = \frac{1}{\text{target distance in meters}}$$

Meter angles (MA) can be converted into prism diopters simply by multiplying MA by the interpupillary distance (IPD) in cm. Prism diopters equal the 100 tan vergence angle in degrees. Vergence eye movements respond to four classes of stimuli described nearly a century ago by Maddox.

Intrinsic Stimuli to Vergence

There are four classes of stimuli to align the eyes with horizontal vergence at various viewing distances. The first is an intrinsic innervation that develops postnatally. During the first 6 weeks of life, our eyes diverge during sleep. This divergence posture is known as the anatomical position of rest and it represents the vergence angle in the absence of any innervation. As we develop during the first 1.5 months of life, the resting vergence angle is reduced to nearly zero by the development of tonic vergence which reduces the resting position from the anatomical position of rest to the physiological position of rest. The anatomical position of rest refers to eye alignment in the absence of all innervation, such as in deep anesthesia. The physiological position of rest refers to the angle that vergence assumes when we are awake and alert, in the absence of any stimulus to accommodation, binocular fusion, or perceived distance. Usually, this physiological position of rest differs from the vergence demand (i.e., the amount of vergence needed to fuse a distance target). The discrepancy between the vergence demand at a far viewing distance and the physiological position of rest is called the phoria. The phoria is an error of binocular alignment, but the error is only manifest during monocular viewing or any condition that disrupts binocular vision. If binocular vision is allowed (not disrupted), other components of vergence reduce the error nearly to zero. Adaptation of the phoria continually calibrates eye alignment and keeps the phoria constant throughout life.

There are two components of phoria adaptation that compensate for eye misalignment. One is a conjugate process called prism adaptation that refers to our ability to adjust our phoria in response to a uniform prism. This prism is constant throughout the visual field and produces a conjugate change in phoria in all directions of gaze. The conjugate adaptation is very rapid and can be completed in several minutes in a normal patient. The eyes are able to adapt horizontal vergence several degrees over a 1-h time span and larger amounts for unlimited adaptation periods. The slow-anomalous fusional movements reported by Bagolini (1976) for strabismus patients are similar in time course to adaptive responses to prism, indicating that the potential range of adaptation must exceed 10 degrees.

The second component of phoria adaptation is a nonconjugate process that adjusts the phoria separately for each direction and distance of gaze. For example, anisometropic spectacle corrections disrupt the normal alignment pattern needed for fusion. A person wearing a plus (converging) lens over the right eye needs to overconverge in left gaze and overdiverge in right gaze. Our oculomotor system is able to make these noncomitant adjustments such that the phoria varies with the direction and distance of gaze so as to compensate for the problems encountered with anisometropic spectacle corrections. This disconjugate adaptation occurs in a matter of hours and persists when the spectacle lenses are removed. It provides a way to align the eyes without retinal cues of binocular disparity. The adapted variations of phoria are guided by a synergistic link with the eye positions in which they were learned. For example, when an anisometropic person first wears a plus lens spectacle over the right eye, the right eye is hypophoric in upgaze and hyperphoric in downgaze. After wearing this spectacle correction for several hours, the vergence errors are adapted to so that there is no longer a phoria or vergence error while wearing the spectacles. However, when the spectacles are removed, the right eye is hyperphoric in upgaze and hypophoric in downgaze. This demonstrates a synergistic link between vertical eye position and vertical phoria where eye position guides the vertical vergence alignment of the two eyes, even when only one eye is open and the other is occluded.

Binocular Parallax

Horizontal vergence also responds to misalignment of similar images from corresponding retinal points (retinal image disparity). It reduces the disparity to a value within Panum's fusional area and causes images to appear fused or single. This fusional or disparity vergence is analogous to optical reflex accommodation. It serves as a rapid fine-adjustment mechanism that holds or maintains fixation of a given target and corrects for minor changes of disparity caused by small movements of the head or the target being inspected. It has a range of approximately 3 degrees of convergence and divergence. Rapid adaptations of the phoria account for increased ranges of fusion in response to disparity.

Vertical and cyclovergence eye movements also respond to binocular disparity. Vertical vergence can rotate one eye up or down relative to the other eye (skew movements) over a range of 2 degrees. These movements respond to vertical disparity that is always associated with near targets in tertiary directions of gaze (targets at oblique eccentricities). Vertical disparity is a consequence of spatial geometry. Tertiary targets are closer to one eye than the other so that they subtend unequal image sizes in the two eyes (Fig. 4). Unequal vertical eccentricities resulting from unequal image magnification require a vertical vergence to align tertiary targets onto corresponding retinal regions. At viewing distances greater than 10 in, vertical disparities are very small and only require a limited range of movement.

Torsion or cyclovergence can twist one eye about its visual axis relative to the orientation of the other eye. Normally, torsional vergence keeps the horizontal meridians of the retinas aligned when the eyes are converged on a near target in the upper or lower visual field. Torsional vergence stimulated by cyclodisparities is limited to a range of approximately 10 degrees. Like vertical vergence, torsional vergence responds to unequal orientations or torsional disparities that result from the perspective distortions of near targets viewed by the two eyes. Ocular torsion is mainly stimulated by lateral head roll and amounts to approximately

10 percent of the amount of head roll, up to a maximum of 6 degrees (60-degree head roll). The head roll can produce vertical disparities of the near fixation targets if the eyes are not accurately converged. With the head tilted toward the right shoulder, a divergence fixation error can cause the right eye to be depressed from the target and a convergence fixation error will cause the right eye to be elevated with respect to the fixation target. However, no vertical fixation errors result from head roll if the eyes are accurately converged on the fixation target.

Defocus and Efforts to Clear Vision

Stimulation of either convergence or accommodation causes synergistic changes in convergence, accommodation, and pupil size. This synergy is called the near triad. During monocular occlusion, when the eyes focus from optical infinity to 1 meter (1 diopter) they converge approximately from 2 to 3 degrees without disparity cues. This accommodative/vergence response brings the eyes into approximate binocular alignment, and any residual alignment errors produce binocular disparity that is corrected by fusional vergence. The activation of convergence when accommodation is stimulated while one eye is covered is called accommodative vergence. The gain or strength of this interaction is called the accommodative convergence/accommodation (AC/A) ratio, which describes the amount of convergence activated per diopter of accommodation. Similarly, if convergence is stimulated while wearing pinhole pupils, accommodation also responds without blur cues. This reflex is called convergence-accommodation and the gain or strength of this interaction is the convergence-accommodation/convergence (CA/C) ratio. Normally, these ratios are tuned appropriately for the geometry of our eye separation so that efforts of either accommodation or convergence bring the eyes into focus and alignment at nearly the same viewing distance. Occasionally, because of uncorrected refractive errors or large abnormal phorias, or because of optical distortions caused by instruments, there is an imbalance of accommodation and vergence causing them to respond to different distances. The mismatch may result in fatigue and associated symptoms if the visual task is extended. The resulting mismatch needs to be corrected by fusional vergence and optical reflex accommodation. If the mismatch is beyond the range of the compensation, then vision remains blurred or double.

Perceived Distance

The eyes align and focus in response to perceived distance. The term proximal refers to perceived distance. Perceived distance is classified as a spatiotopic cue because it results in a conscious percept of distance and direction. Our perception of space is made up of an infinite number of different targets in different directions and distances. Our eyes can only select and fuse one of these target configurations at a time. One of the main tasks of the oculomotor system is to select one fixation target from many alternatives. This selection process usually results in a shift of binocular fixation from one distance and direction to another. Proximal vergence serves as a course adjustment mechanism and accounts for nearly 100 percent of the eye movement when gaze shifts are too large for us to sense due to blur and disparity. We perform these large gaze shifts without needing to see the new intended target clearly or even with both eyes (Schor, et al., 1992). We only need to have an estimate of target location using monocular cues for distance and direction. The majority of binocular shifts in fixation are stimulated by proximal vergence. Once the shift has been made, another component of vergence (fusional vergence) takes over to maintain the response at the new target location. Errors in perceived distance can result in errors of eye alignment and focus. Mislocalization of large print or symbols in a head-mounted display, or of the distance of a microscope slide, can cause the eyes to accommodate excessively and produce a blurred image of a target that is imaged at optical infinity.

The Zone of Clear and Single Binocular Vision

These components of vergence add linearly to form a range of accommodative and disparity stimuli that the observer is able to respond to with single and clear binocular vision (Hofstetter, 1983) (Fig. 10). The range is determined by the convergence and divergence fusional vergence range from the resting phoria position, plus the linkage between accommodation and vergence (described in an earlier subsection, "Defocus and Efforts to Clear Vision") that increases with accommodative effort until the amplitude of accommodation is reached. Geometrically matched stimuli for convergence and accommodation that correspond to common viewing distances normally lie within this range of clear single binocular vision unless there is an alignment error such as an eye turn or strabismus. Mismatches between stimuli for accommodation and convergence produced by lenses, prism, or optical instruments can be compensated for as long as the combined stimuli for accommodation and convergence lie within this stimulus operating zone for binocular vision. The ability to respond to unmatched stimuli for convergence and accommodation does not result from decoupled independent responses of the two systems. Accommodation and convergence always remain cross-coupled, but fusional vergence and optical reflex accommodation are able to make responses to overcome differences in phorias produced by accommodative vergence and prism.

12.8 EFFECTS OF LENSES AND PRISM ON VERGENCE AND PHORIA

Errors of horizontal and vertical vergence can result from differences in the two ocular images caused by differential magnification and translation. Optical power differences will cause magnification differences. Translation of the image is caused by prism or by decentration of the line of sight from the optical center of the lens system.

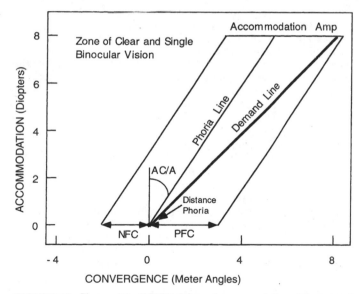

FIGURE 10   Figure of the ZCSBV. Schematic representation of the zone of clear single binocular vision in terms of five fundamental variables (amplitude of accommodation, distance phoria, AC/A ratio, positive fusional convergence, and negative fusional convergence).

Magnification-Induced Errors of Eye Alignment

Magnification of the ocular image is influenced by several factors, including index of refraction, lens thickness, front surface power, overall power of the lens, and the ocular distance from the eyepiece. For a spectacle lens the magnification is described by the product of a shape factor and a power factor. Magnification described by the shape factor (M_s) equals

$$M_s = \frac{1}{1 - F_1 t/n}$$

and by the power factor (M_p) equals

$$M_p = \frac{1}{1 - P_v h}$$

where F_1 is the front surface power, t is the lens thickness in cm, n is the index of refraction, h is the distance of the spectacle plane from the cornea in cm, and P_v is the thin lens power. Percent angular magnification caused by the power factor is approximately 1.5 percent per diopter power. The percent magnification ($M\%$) can be approximated by the following formula (Ogle, 1964):

$$M\% = \frac{F_1 t}{n + P_v h}$$

Differences between the lenses before the two eyes will produce image size differences, depending on the source of refractive error. Two basic sources are an abnormal axial length (axial ametropia) or abnormal corneal and lenticular power (refractive ametropia). Magnification differences between the two eyes caused by interocular differences in axial length can be eliminated largely by correcting axial errors with spectacle lenses located in the primary focal plane (the spectacle plane) and placing the lenses not too near to the eyes. Magnification differences between the two eyes caused by unequal powers of the two corneas can be prevented by correcting refractive errors with contact lenses instead of spectacle lenses (Knapp's law). If Knapp's law is violated by correcting a refractive myope with spectacle lenses placed in the primary focal plane of the eye (16.67 mm), the myopic eye will have a smaller image than an emmetropic eye with a normal corneal power. Similarly, if an axial myope is corrected with a contact lens, the myopic eye will have a larger image than an emmetropic eye with a normal axial length. However, the perceived image size in axial refractive errors also depends on whether the retina is stretched in proportion to the axial length. If it is, then the contact lens correction for an axial ametrope would produce a perceived image size equal to that of an emmetropic eye with a normal axial length. This analysis only applies to perceived image size and it is an attempt to correct perceived image size differences between the two eyes (aniseikonia). However, it does not address the unequal ocular movements stimulated by anisometropic spectacle corrections (anisophoria) resulting [as described in Sec. 12.10, "Head and Eye Responses to Direction (Gaze Control)"] from the separation of the entrance pupil and center of rotation.

Anisophoria is defined by Friedenwald (1936) as a heterophoria which varies in magnitude with direction of gaze. It is produced by spectacle corrections for anisometropia independently of whether refractive errors are axial or refractive in nature, and only concerns the refractive power of the spectacle lens correction. The lateral prismatic effect occurs even when size distortions of an axial refractive error have been properly corrected with a spectacle lens. A prismatic deviation will always exist whenever the visual axis of the eye and the optical axis of a lens do not coincide. As a result, the initial movement of an ametropic eye corrected with spectacle lenses will be too small in hypermetropia and too large in myopia. These optical factors produce differential prism in ansiometropia corrected with spectacles. Initially, anisometropes corrected with spectacles will make vergence errors during versional movements to eccentric targets because the eyes follow Hering's law of equal innervation to the extra ocular muscles. Interocular differences in power of spectacle lenses can stimulate vergence eye movements in all meridians of anisometropia (vertical, horizontal, and oblique). This optically induced anisophoria can be compensated for easily in the horizontal meridian by disparity vergence eye movements. The vertical errors will be the most troublesome because of the limited range and slow speed of vertical vergence (Houtman, et al., 1977; Perlmutter & Kertesz, 1978). Difficulty in compensating for optically induced anisophoria is believed to contribute to the symptoms of aniseikonia corrected with spectacle lenses (Friedenwald, 1936; Romole, 1984). Anisophoria may explain why both refractive and axial anisometropes prefer contact lens corrections that do not produce anisophoria. Contact lenses move with the eye, and the line of sight always passes through the optical center such that fixation errors of targets at infinity always equal the correct motor error to guide the oculomotor system during gaze shifts. Optically induced anisophoria can be very large. For example, Ogle (1964) computes a 4.2-prism-diopter phoria produced by a +4-diopter anisometropia spectacle correction during a 20-degree eye movement. This constitutes a 12 percent error. Normally, the oculomotor system is capable of recalibration of Hering's law to compensate for optically induced anisophoria. For example, anisometropic patients wearing bifocal spectacle corrections do not exhibit the predicted change in vertical phoria while viewing in downgaze (Cusick and Hawn, 1941; Ellerbrock and Fry, 1942). Adjustments of heterophoria in all directions of gaze occur in approximately 2.5 h for differential afocal magnification of up to 10 percent (Henson and Dharamski, 1982; Schor, et al., 1990). Similar oculomotor adjustments occur normally in asymmetric convergence whereby the abducted eye makes larger saccades than the adducted eye in order to compensate for geometrically induced aniseikonia (Morrison, 1977). Unequal prism before the two eyes produced by a 10 percent magnification difference, as observed in monocular aphakia corrected with spectacles or spectacle and contact lens combinations (Enoch, 1968), is extremely difficult to adapt to. Presumably a 10 percent magnification difference sets an upper limit to the degree of anisometropia that will not disrupt binocular versional eye movements and binocular alignment. It is possible that individuals who have symptoms of aniseikonia when corrected with spectacle lenses, are simply unable to make the necessary adjustments in conjugacy to overcome their optically induced anisophoria. Indeed, Charnwood (1950) observed that aniseikonia patients preferred prism incorporated into their spectacle corrections to reduce anisophoria, and Field (1943) reported that nearly all of 25 patients with clinically significant aniseikonia had oculomotor imbalance. Clearly, anisophoria is an important biproduct of anisometropia corrected with spectacles.

Torsional vergence errors can also result from optical corrections that have cylindrical corrections for astigmatism (Fig. 9). The power of a cylindrical lens varies with the square of the cosine of the angular separation from the major power meridian. This power variation produces a meridian-dependent magnification of the ocular image described as a magnification ellipse. The ellipse produces rotational deviations of lines at different orientations toward the meridian of maximum magnification. The rotary deviation between a line and its image meridian (δ) depends on the angle between the orientation of the line (γ) and the major axis of the magnification ellipse (θ). The rotary deviation is approximated by the following equation:

$$100 \tan \delta = \tfrac{1}{2} f \sin 2(\gamma\theta)$$

where f equals the percent of magnification produced by the cylinder (Ogle, 1950). This equation demonstrates that most of the shear or scissor distortion of lines will occur in meridians between the major and minor axes of the magnification ellipse. These rotary deviations produce lines imaged on noncorresponding retinal meridians or torsional disparities because the axis of astigmatism is normally not parallel in the two eyes. Rather, astigmatism tends to be mirror symmetric about oblique meridians such that an axis of 45 degrees in one eye is often paired with an axis of 135 degrees in the other eye. As a result, the shear or rotary deviations of vertical meridians tend to be in opposite directions. The resulting cyclodisparities produce perceptual declination errors of stereo surface slant (tilt of a surface about a vertical or horizontal axis). Cyclodisparities about horizontal axes stimulate cyclovergence movements of the two eyes (torsion of the two eyes in opposite directions). If the disparities are long-lasting, the cyclovergence response adapts and persists even in the absence of binocular stimulation (Maxwell & Schor, 1999).

12.9 *PRISM-INDUCED ERRORS OF EYE ALIGNMENT*

Errors of binocular eye alignment also result from translation of one ocular image with respect to the other. The translation can be caused by prism or by displacement of the optical center of the lens from the line of sight. When an object is viewed through a prism, it appears in a new direction that is deviated toward the prism apex and the eye must move in that direction to obtain foveal fixation. The amount the eye must rotate to compensate for the prismatic displacement of targets at finite viewing distances decreases as the distance between the eye and prism increases. This is because the angle that the displaced image subtends at the center of eye rotation decreases with eye-prism separation. This reduction is referred to as prism effectivity (Keating, 1988). It does not occur for targets at infinity since the target distance is effectively the same to both the prism and the center of rotation. The effective prism magnitude at the eye (Z_e) is described by the following equation:

$$Z_e = \frac{Z}{1 - C_{rot} \times V}$$

where Z is prism value, C_{rot} is center of rotation to prism separation in meters, and V is the reciprocal of the object-to-lens separation in meters which corresponds to the dioptric vergence leaving the prism (Keating, 1988).

Binocular disparities induced by a prism or lens determine the direction of the phoria or vergence error. For example, base-out prism produces a crossed disparity and divergence error (exophoria) which is corrected under binocular conditions by fusional convergence. Minus lenses stimulate accommodation and accommodative convergence, which under binocular conditions produce a convergence error (esophoria) that is corrected by fusional divergence. Similarly, base-in prism produces an uncrossed disparity and convergence error (esophoria) which is corrected under binocular conditions by fusional divergence, and plus lenses stimulate a relaxation of both accommodation and accommodative convergence, which produces a divergence error (exophoria) that is corrected by fusional convergence.

The prismatic translation caused by optical displacement is described by Prentice's rule, where the prism displacement (Z) equals the power of the lens in diopters (P) times the displacement of the lens center from the line of sight (d) in centimeters ($Z = dP$). The angular deviation of any paraxial ray that passes through a lens at a distance from the optical center is described by this rule. Prentice's rule can also be applied to cylindrical lenses where the magnification occurs in a power meridian. The axis of a cylindrical lens is perpendicular to its power meridian. The prism induced can be calculated from the vector component of the displacement that is perpendicular to the axis of the cylinder. This distance (Hp) equals the sine of the angle (θ) between the displacement direction and the axis of the cylinder times the magnitude of the displacement (h):

$$Hp = h \sin \theta$$

Thus, according to Prentice's rule, prism induced by displacing a cylindrical lens equals

$$Z = P h \sin \theta$$

The magnitude of prism induced by lens displacement also depends on the separation between the lens and the center of rotation of the eye. Effective prism displacement (Ze) decreases as the distance of the lens from the eye increases according to the following equation (Keating, 1988):

$$Z_e = \frac{Zc + Z}{1 - C_{rot} \times V}$$

Where $Zc = 100 \tan c$, $c =$ the angle formed between the object point and the displaced point from the optical center (major reference point) and the straight ahead line, $V =$ the reciprocal of the image distance to the lens in meters, which corresponds to the dioptric vergence leaving the prism, Z = the prism calculated from Prentice's rule, and C_{rot} = the separation between the lens and the center of rotation of the eye. The reduction of prismatic displacement with eye-to-lens separation is referred to as prism effectivity with lenses.

When the two eyes view targets through lenses that are displaced from the optical axes, a disparity stimulus is introduced. Most typically, when the optical centers of a binocular optical system are separated by an amount greater or less than the interpupillary distance, disparate horizontal prism is introduced that requires the eyes to make a horizontal vergence adjustment to achieve binocular alignment. For example, when looking through a pair of plus lenses that are farther apart than the interpupillary distance, each eye is viewing through a point that is nasalward from the optical axis. This introduces base-out prism that stimulates convergence of the two eyes. This prism vergence stimulus may produce conflicts with the accommodative stimulus that is not affected by lens displacement so that the eyes need to focus for one distance and converge for another. Accommodation and convergence are linked synergistically so that their responses are closely matched to similar target distances. There is a limited degree of disparity vergence compensation available to overcome any discrepancy between the two systems, and these efforts can result in headaches and other ocular discomfort.

Anisometropic ophthalmic lens prescriptions with bifocal corrections in the lower segment of the lens can also introduce vertical prism while viewing near objects through the reading addition. The amount of vertical prism is predicted by Prentice's rule applied to the amount of anisometropia. Thus, for a 2-diopter anisometropic individual, a patient looking downward 1 cm through their bifocal has a vertical phoria of 2 prism diopters. The vertical phoria can be corrected with a slab-off prism (a base-down prism that is cemented onto the bifocal over the lens with the most plus power). However, this procedure is rarely necessary, since the patient usually is able to adapt the vertical phoria to vary nonconjugately with vertical eye position as described earlier.

12.10 HEAD AND EYE RESPONSES TO DIRECTION (GAZE CONTROL)

When objects attract our attention in the peripheral visual field, we are able to fixate or shift our central gaze toward them using head and eye movements. The eyes move comfortably over a range of 15 degrees starting from primary position. Larger gaze shifts require combinations of head and eye movements. Normally, the eyes begin the large movements, then the head moves toward the target while the eye movements are reduced by the VOR, and when the head stops the eyes refine their alignment. The reduction of eye motion during head rotation results from a gaze stabilization reflex that allows us to keep our eyes fixed on a stationary object as we move our body and/or head. This reflex operates with head motion sensors called the vestibular organs.

The manner in which we shift our gaze influences optical designs. Optical systems usually are mounted on the head. They need to provide a large enough viewing angle to accommodate the normal range of eye movements, and to remain stable on the head as it moves. They also need to provide peak resolution in all directions so that the oculomotor system is able to shift gaze while the head remains stationary. Gaze shifts can be inaccurate because of interactions between spectacle refractive corrections and movements of the eyes.

The accuracy of eye movements is affected by stationary optical systems such as spectacles that are mounted in front of the eyes, as opposed to optical systems such as contact lenses that move with the eye. The problem occurs when viewing distant targets because the entrance pupil of the eye and its center of rotation are at different distances from the spectacle plane. The entrance pupil is at a distance of about 15 mm from the spectacle lens and the ocular center of rotation lies about 27 mm behind the spectacle plane. Thus the center of rotation and entrance pupil are separated by about 12 mm, and a target in space subtends different angles at these two points. The angle at the entrance pupil determines the visual error and the angle at the center of rotation determines the motor error. Whether a spectacle lens is worn or not, this difference alters retinal image feedback during eye rotations for any target imaged at a finite viewing distance from the eye. Images at infinity don't have this problem because they subtend equal angles at both the entrance pupil and center of rotation. Thus, eye movements cause motion parallax between targets viewed at near and far viewing distances, and some animals like the chameleon use this parallax to judge distance (Land, 1995).

Problems occur when the spectacle lens forms an image of an object at infinity at some finite distance from the eye. This image of the distant target subtends different angles at the entrance pupil and center of rotation because of their 12-mm separation. Thus, perceived visual angle and motor error are unequal. Depending on whether the image is formed in front of (myopic error) or behind (hyperopic error) the eye, this discrepancy requires that a fixational eye movement must be smaller or larger, respectively, than the perceived eccentricity or retinal angular error of the target sensed prior to the eye movement. The amount of eye rotation can be computed from the equation describing prism effectivity with lenses (Z_e) (Keating, 1988):

$$Z_e = \frac{Z_c}{1 - C_{rot} \times V}$$

where Z_c equals 100 tan c (c equals angular eccentricity of the target measured at the lens plane), V equals the dioptric vergence of the image distance from the lens plane, and C_{rot} equals the lens to center of rotation separation in meters. When different power spectacle lenses are worn before the two eyes to correct unequal refractive errors (anisometropia), a target that would normally require equal movements of the two eyes now requires unequal binocular movements. The oculomotor system learns to make these unequal movements, and it retains this ability even when one eye is occluded after wearing anisometropic spectacle corrections for only 1 h.

During head movements, magnification of the retinal image by spectacles increases retinal image motion during head rotation and causes the stationary world to appear to move opposite to the direction the head rotates. This can produce vertigo, as well as stimulate a change in the vestibular stabilization reflex. When a magnifier is worn for only 1 h, the reflex can be exaggerated to compensate for the excessive retinal image motion. Following this adaptation, when the optical aid is removed, the exaggerated VOR causes the world to appear to move in the same direction that the head rotates. If you wear glasses, you can demonstrate this adaptation to yourself by removing your glasses and shaking your head laterally. If your wear negative lenses to correct for myopia, the minification they produce has caused a reduction in image stabilization and without your glasses, the world will appear to move opposite to the direction of head shake.

12.11   *FOCUS AND RESPONSES TO DISTANCE*

Our visual sense of space requires that we be able to perceive objects accurately at different distances. Single and clear views of targets at finite viewing distances require that the visual system align the similar images of the targets on corresponding retinal regions and that the optics of the eye adjust the focus to accommodate the near target (accommodation). The range of accommodation varies with age, starting at birth at about 18.5 D and decreasing 1 D every 3 years until approximately age 55 when it is no longer available (absolute presbyopia). The amplitude of accommodation available to an individual of a given age is remarkably stable across the population if extremes of refractivity are not considered.

Spectacle corrections and other optical systems can influence both aspects of the near response. Accommodation declines with age until eventually the near point is beyond the near working distance (functional presbyopia). In this situation, near targets can be imaged farther away from the eye with plus lenses so that they lie within the available range of remaining accommodation. The accommodative system is only able to maintain ⅔ of its full amplitude for long periods of time and when this range is exceeded by the optical demands of working distance, bifocals are typically prescribed. The magnitude of the bifocal correction depends upon the accommodative amplitude and the near working distance. For a typical 16-in or 40-cm near working distance that requires 2.5 D of accommodation, a bifocal correction is needed when the amplitude of accommodation is less than 3.75 D. Typically, this amplitude is reached at about 45 years of age.

If spectacles are worn, the amplitude of accommodation is influenced by the sign (plus or minus) of the spectacle correction. Accommodation demand equals the difference in the dioptric values of the far point of the eye and the image distance of a near target formed by optics placed before the eye. These dioptric values can be described relative to the front surface of the eye or the spectacle plane. The reciprocal of a specific viewing distance to the spectacle plane describes the spectacle accommodative demand, and this value is independent of the lens correction. However, the ocular accommodative demand (amount the eye needs to change power) required to focus targets at this distance does depend on the power and sign of the spectacle correction. Ocular accommodation is computed by the difference between the dioptric vergence subtended by the near target at the cornea after it is imaged by the spectacle lens, and the vergence subtended at the cornea by the far point of the eye. For myopes corrected with minus spectacle lenses, ocular accommodation through spectacles is greater than the spectacle accommodative demand, and the reverse is true for hyperopes. For example, the ocular accommodative response to a target placed 20 cm in front of the spectacle plane 14 mm from the eye is 4.67 D for an emmetrope, 4 D for a myope wearing a −6-D corrective spectacle lens, and 5.56 D for a hyperope wearing a +6-D corrective spectacle lens. Thus, near targets present unequal accommodative stimuli to the two eyes of anisometropes corrected with spectacles. Although the accommodative response is consensual (equal in the two eyes), the difference in the two stimuli will go unnoticed as long as it lies within the depth of focus of the eye (approximately 0.25 D). However, in the example given above, a 6-D anisometrope would have about a 1-D difference in the stimulus to accommodation presented by a target at a 20-cm viewing distance. There is a limited amount of differential binocular accommodation to compensate for this unequal stimulus (Marran and Schor, 1998). Finally, this lens effectivity causes spectacle lens wearers who are hyperopic to become presbyopic before myopes.

12.12 *VIDEO HEADSETS, HEADS-UP DISPLAYS, AND VIRTUAL REALITY: IMPACT ON BINOCULAR VISION*

A variety of head-mounted visual display systems (HMD) have been developed to assist performance of a variety of tasks. Telepresence is an environment in which images from remote cameras are viewed on video screens to perform tasks at distal sites. In other applications, information in the form of symbology is added to a natural image to augment reality, such as in a heads-up aviation application, and in still another application virtual environments are simulated and displayed for entertainment or instructional value. The head-mounted video systems usually display information in one of three configurations. These are monocular or monoscopic that present information to one eye, biocular where both eyes view an identical image, and binocular where slight differences in the two eyes' images stimulate stereoscopic depth perception. The biocular view avoids image conflicts between the two eyes that arise in the monoscopic systems, and stereopsis in the binocular systems helps to sort out or distinguish elevation and distance information in the perspective scene. Stereopsis also provides the user with fine depth discrimination and shape sensitivity for objects located within an arm's length. Each of these systems attempts to present coherent cues to space perception; however, due to physical limitations this is not always possible, and cue conflicts result.

Distance Conflicts

The current designs of HMD devices introduce three stimulus conflicts. They present conflicting information about target distance to accommodation and convergence, conflicting visual-vestibular information, and conflicts concerning the spatial location of information to the two eyes. For example, the symbology presented to aviators in heads-up displays is set for infinity, but large symbols appear to be near and this can initiate an inappropriate accommodative response. However, given sufficient time, subjects will accommodate correctly (Schor and Task, 1996). Currently, all HMD units present a fixed optical viewing distance that is not always matched to the convergence stimulus in biocular units, or to the many states of convergence stimulated in binocular units. This situation is exacerbated by the presence of uncorrected refractive errors that can cause images to be blurred in the case of myopia, force extra accommodation in the case of hyperopia, or present unequal blur and accommodative stimuli in the case of anisometropia. Normally, observers with uncorrected refractive errors are able to adjust target distance to clear images, but the HMD has a fixed target distance that is not adjustable. Units that provide self-adjusting focus run the risk of users' introducing larger errors or mismatching the clarity of the two ocular images.

As described above, accommodation and convergence are tightly coupled to respond to approximately the same viewing distance. There is some flexibility to compensate for small differences between stimuli to accommodation and convergence, but long periods of usage—particularly those requiring fusional divergence to overcome an esophoria caused by excessive accommodation—can produce eye strain. Tolerance for accommodative errors can be increased by extending the depth of focus with small exit pupils or with screens that have reduced resolution. It is also helpful to set the viewing distance at an intermediate value of approximately 2 m (Southard, 1992) to allow uncorrected mild myopes to see clearly while allowing a 3-degree range of stereodepth disparities. In biocular units where disparity is not varied, the optical distance can be set closer to 1 m to further compensate for uncorrected myopia and to place images near arm's length. Prolonged viewing at close distances can cause instrument myopia, a form of excessive accommodation, and postimmersion short-term aftereffects such as accommodative spasm (Rosenfield & Ciuffreda, 1994). Accommodative fatigue also can result after 30 min of stereo-display use (Inoue & Ohzu, 1990). If refractive corrections are worn, the optimal image distance will depend on the tasks performed. Simulated computer screens should be at arm's length, whereas views for driving simulators should be remote to provide more realism in the display.

Errors of perceived distance can occur when vergence distance is not matched to image size or perceptual task. These errors can result in perceptual image expansion (macropsia) or constriction (micropsia), particularly in binocular displays that stimulate a range of convergence values. These effects can be minimized by limiting the range of binocular disparities to a few degrees for near objects and to use a biocular display when presenting views of distant objects simultaneously with foreground detail. Perspective cues, along with motion parallax, overlap or superposition, and texture density gradients, provide powerful depth cues for remote or complex scenes containing a wide range of depth information, and binocular disparity adds little to these displays. Stereopsis is most useful for facilitating manual dexterity tasks with fine depth discrimination between objects and their 3-D shape located within an arm's length of the user. Static displays that don't utilize motion parallax will benefit more from the addition of stereoscopic cues.

Some HMDs and other virtual environments combine real and virtual information with a see-through visor or with remote rear-projection systems that surround the viewer panoramically as in the Computer-Augmented Virtual Environment (CAVE) (Kenyon, et al., 1995). These mixed environments can introduce multiple and sometimes conflicting accommodative stimuli. When real variable focus and virtual fixed-focus images are at different image distances, both cannot be cleared at the same time, though they may appear to be in similar locations in space. Conflicting near virtual targets can interfere with recognition of real distant targets (Norman & Ehrlich, 1986). Several optical solutions have been suggested to minimize these focus conflicts (Marran & Schor, 1998). The depth of focus can be extended with pinhole pupils. A pinhole exit pupil projected to the eye pupil in a Maxwellian view system will preserve field of view. The virtual image is viewed through the Maxwellian pupil imaged via a beam splitter while the real environment is viewed through the natural pupil. Eye tracking might be necessary in Maxwellian systems to shift the exit pupil to follow eye movements (de Wit & Beek, 1997). Monovision is another possible solution in which a near lens is worn over one eye and a far lens over the other. A target that is in focus for one eye will be blurred for the other. The visual system is able to suppress the blurred image and retain the focused images of both eyes, and also to retain a large functional range of stereopsis (Erickson & Schor, 1990). Monovision has been used successfully in presbyopes with contact lens corrections, and it has been shown to work on pre-presbyopes in most individuals (Schor, et al., 1987). The limitation of monovision is that it provides clear vision only for the two focal depths, plus and minus the depth of focus. The range of the two clear vision distances might be extended by aniso-accommodation: the ability of the two eyes to focus independently of each other while remaining binocular (Marran & Schor, 1998). Another solution is a chromatic bifocal that takes advantage of the longitudinal chromatic aberration of the eye. Wavelengths of light at opposite ends of the visual spectrum have a 2-D difference in dioptric power for a given object distance. Given that nearer objects tend to be in the lower visual field, different monochromatic wavelengths combined with field segregation could be used in the virtual image control panel. Users viewing near objects could view long-wavelength images in the lower half of the virtual control panel. Thus, both real and red virtual images would stimulate an increase in accommodation. For distant real objects, the user could attend to the upper short-wavelength images in the virtual display panel, and real and blue virtual targets would stimulate a relaxation of accommodation. A simpler solution is to place a bifocal lens segment in the lower portion of the see-through window, bringing near real images to the same focal distance as the images seen on the virtual display.

Visual-Vestibular Conflicts

Virtual systems that are worn on the head or remain fixed in space both present conflicting visual-vestibular information. Normally, when the head rotates or translates, the eyes rotate in the opposite direction to keep the images of objects in space stabilized on the retina. The relationship between retinal image motion and head and eye motion is disrupted by the virtual display. In the case of an external virtual projection system like CAVE (Kenyon, et al., 1995), simulated movements of the visual scene that correspond to body movements are not accompanied by the appropriate vestibular signals. For example, a driving simulator displays changes in the road view while the virtual car is being steered. When the driver makes a turn, the visual field appears to rotate but there is no corresponding vestibular signal. As a result, drivers of virtual systems tend to oversteer the turns because they underestimate the curved path of the automobile. Perceptual disturbances during similar phenomena in flight simulators are referred to as simulator sickness. In contrast, head-mounted video displays can present a head-fixed visual stimulus while the head rotates such that vestibular signals are not accompanied by the appropriate visual motion stimuli. These conflicts can also result in symptoms of nausea, vertigo, and general malaise even when very small head movements occur (Howarth, 1996; Howarth & Costello, 1997; Hughes, et al., 1973). Visual-vestibular conflicts also occur with telepresence systems that present images that are either magnified or minified as a result of differences between the camera distance-to-objects and the viewer distance to the viewscreen. As a result, head movements that cause camera movements will stimulate excessive retinal image motion with magnification or too little retinal image motion with image minification. The mismatch will cause adaptive adjustments of the VOR and possible simulator sickness.

The oculomotor system responds to this visual-vestibular conflict during steady fixation by suppressing the VOR and adaptively reducing its gain. Thus, a user viewing a 2-h motion picture on an HMD while making a normal range of head movements during conversation with other persons will adapt their VOR to be less responsive to head motion. This adaptation can produce postimmersion symptoms where the adapted VOR remains suppressed such that head movements without the HMD will not stimulate a sufficient reflex movement to stabilize the retinal image. Users will perceive the world to move with their head, which can cause nausea, vertigo, and malaise. Attempts to move images with the head using a head tracking system that updates image position during head rotation have been unsuccessful (Piantanida, et al., 1992) because of small delays in image processing that desynchronize expected image motion and head movements. It is possible that an image could be blanked during large head movements and stabilized with a head tracker during small head movements. The blanking would be similar to the normal suppression of vision during saccadic eye movements. Other solutions are to use see-through and see-around virtual displays so that real objects in the periphery are always visible during normal head movements so that some visual and vestibular stimuli remain coherent.

Spatial Location Conflicts

No existing HMD has the wide field of view of the human visual system in a natural environment. The horizontal field of view in currently available HMDs ranges from 22.5 to 155 degrees (Davis, 1997). The larger fields of view in binocular systems are expanded by partially overlapping the two monocular images. Note that the entire monocular and binocular portions of the field of view lie within the from 114- to 120-degree binocular overlap of the normal visual system. Partial overlap in HMDs can be produced by displacing the right field to the right and the left field to the left. Each eye has a monocular flanking component in its temporal visual field. It is also possible to reverse the displacement so that the monocular components are in the nasal-visual field of each eye (Fig. 11). These two organizations are consistent with two different occlusion aperture configurations. The former has the same organization as a binocular field seen on a near screen in front of a background. Thus, the total image is seen as a discontinuous surface in depth with the central binocular portion at a proximal distance compared to the monocular portion that is seen farther away (Nakayama & Shimojo, 1990). The latter case has the same organization as a single continuous surface viewed through an aperture (porthole effect) (Fig. 11). The former system is more likely to cause binocular rivalry in the monocular regions of the field than is the latter approach. This suppression can cause a moonlike crescent shape (luning) in the flanking monocular region that corresponds to the border of the image seen by the contralateral eye, which blocks visibility of the monocular field crescent seen by the ipsilateral eye. Luning can be reduced by blurring the edges of each eye's image borders or by using the crossed-view configuration. Interestingly, the latter crossed-view approach is preferable in head-mounted displays (Melzer & Moffitt, 1991).

Visual suppression and binocular rivalry suppression are also stimulated in monocular image displays. Rivalry occurs between the image presented to one eye and the natural scene presented to the other. Even if one eye is occluded, the dark image of the patched eye will rival with the visible image of the open eye. This rivalry can be minimized by presenting the patched eye with a uniform field with the same mean luminance and color as the open eye image. This can be accomplished easily by placing a diffusing filter over the nonseeing eye. In cases where the two eyes view independent images, it might be possible to control which eye is suppressed by moving the image in the seeing eye and keeping it stationary in the nonseeing eye. This technique is based on the observation that motion is a dominant feature for binocular rivalry.

Optical Errors

The success of any binocular viewing instrument depends in large part on the precision of alignment and quality of distortion-free optics. Alignment errors can produce vertical, horizontal, and torsional disparities. Of these, the vertical errors are the most troublesome. There is limited range for vertical vergence and slow response times (Houtman, et al., 1977; Perlmutter & Kertesz, 1987). The oculomotor system can adapt its vertical alignment posture (vertical phoria) if the vertical disparity is constantly present; however, if the HMD is worn

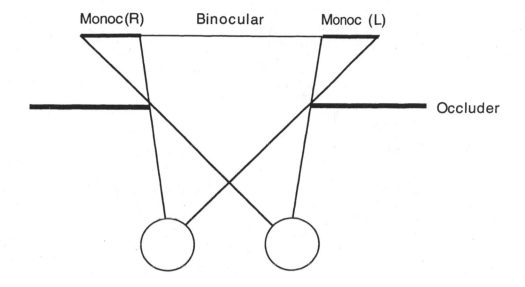

Porthole   occlusion   configuration
Contralateral   monocular   crescents

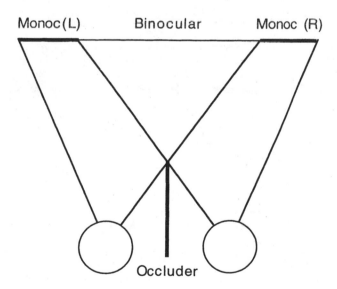

Septum   occlusion   configuration
Ipsilateral   monocular   crescents

FIGURE 11 Occlusion figure with porthole and septum. Partial overlap of the visual field simulates obstructed views by a porthole aperture or septum in the midsagittal plane.

intermittently, then no adaptation will occur. Horizontal errors, especially in the convergence direction, can be overcome easily by vergence eye movements. Torsional errors cause overall torsional disparities that can cause perception of slant about the horizontal axis if the virtual scene is superimposed by see-through optics on the natural visual environment. However, if an opaque visor is used so that only the virtual display is visible, slant distortion will not be perceived with cyclotorsion disparity (Howard & Kaneko, 1994). This is because the visual system only interprets slant when there is horizontal shear disparity between vertical lines but no vertical shear disparity between horizontal lines. When both forms of shear are present, this is interpreted as resulting from eye rotation rather than real object slant. Cyclodisparity in large field displays will stimulate cyclovergence responses that can have postimmersion aftereffects in the form of cyclophoria (Maxwell & Schor, 1999). Errors in the baseline separation of the entrance pupils of the optical system can produce divergent disparities if the separation is set wider than the interpupillary distance. The oculomotor system has limited divergence ability and this stimulus can cause headaches and visual discomfort. Finally, low-quality optics that have distortions, such as barrel, pincushion, and keystone, will introduce horizontal, vertical, and torsional disparities in both biocular and binocular displays if the distortions are not identical for the two eyes or if the visual fields are displaced slightly in the two eyes. Displacements could result from alignment errors or from partial field overlap used to expand the field of view. The disparities produced by these distortions can cause idiosyncratic warpage of the virtual image and vertical disparities that stress the oculomotor system.

12.13 REFERENCES

R. S. Arner, "Eikonometer Measurements in Anisometropes with Spectacles and Contact Lenses," *J. Am. Optom. Assn.* **40**:712–715 (1969).

S. Awaya and G. K. von Noorden, "Aniseikonia Measurement by Phase Difference Haploscope in Myopia Anisometropia and Unilateral Aphakia (with Special Reference to Knapp's Law and Comparison Between Correction with Spectacle Lenses and Contact Lenses)," *J. Jpn. Contact Lens Soc.* **13**:131 (1971).

B. T. Backus, M. S. Banks, R. van Ee, and J. A. Crowell, "Horizontal and Vertical Disparity, Eye Position, and Stereoscopic Slant Perception," *Vis. Research* **39**:1143–1170 (1999).

B. Bagolini, "Part II: Sensorial-Motor Anomalies in Strabismus (Anomalous Movements)," *Doc. Ophthalmol.* **41**:23–41 (1976).

K. I. Beverley and D. Regan, "Evidence for the Existence of Neural Mechanisms Selectively Sensitive to the Direction of Movement in Space," *J. Physiol. Lond.* **235**:17–29 (1973).

C. B. Blakemore, "A New Kind of Stereoscopic Vision," *Vision Res.* **10**:1181–1199 (1970).

C. Bourdy, "Aniseikonia and Dioptric Elements of the Eye," *J. Am. Opt. Assn.* **39**:1085–1093 (1968).

A. Bradley, J. Rabin, and R. D. Freeman, "Nonoptical Determinants of Aniseikonia," *Invest. Ophthalmol. Vis. Sci.* **24**:507–512 (1983).

T. Brandt, J. Dichgans, and W. Koenig, "Differential Central and Peripheral Contributions to Self-Motion Perception," *Exp. Brain Res.* **16**:476–491 (1973).

R. M. Burde, "The Extraocular Muscles," in R. A. Moses (ed.), *Adler's Physiology of the Eye: Clinical Applications*, St. Louis, CV Mosby, pp. 84–183, 1981.

R. M. Burnside and C. Langley, "Anisometropia and the Fundus Camera," *Am. J. Ophthalmol.* **58**:588–594 (1964).

J. R. B. Charnwood, "An Essay on Binocular Vision," London: Hatton, pp. 73–88, 1950.

D. Cline, H. W. Hofstetter, and J. R. Griffin, *Dictionary of Visual Science*, fourth ed., Chilton, Radnor, PA, 1989.

H. Collewijn, A. J. Martins, and R. M. Steinman, "Compensatory Eye Movements During Active and Passive Head Movements: Fast Adaptation to Changes in Visual Magnifications," *J. Physiology* **340**:259–286 (1983).

M. J. Collins and A. Goode, "Interocular Blur Suppression and Monovision," *Acta Ophthalmologica* **72**:376–380 (1994).

J. A. Crowell and M. S. Banks, "Ideal Observer for Heading Judgments," *Vision Res.* **36**:471–490 (1996).

B. Cumming, "Motion—In Depth," Chapter 12, in A. T. Smith and R. J. Snowden (eds.), *Visual Detection of Motion,* Academic Press, San Diego, CA, 1994.

P. L. Cusick and H. W. Hawn, "Prism Compensation in Cases of Anisometropia," *Arch. Ophth.* **25**:651–654 (1941).

E. T. Davis, "Visual Requirements in HMDs: What Can We See and What Do We Need to See?," Chapter 8, in J. E. Melzer and K. Moffett (eds.), *Head Mounted Displays,* Optical and Electro-Optical Engineering Series, McGraw-Hill, New York, 1997.

W. Dember and J. Warm, *Psychology of Perception,* Holt, Rinehart, and Winston, New York, 1979.

G. C. de Wit and R. A. E. W. Beek, "Effects of a Small Exit Pupil in a Virtual Reality Display System," *Opt. Eng.* **36**:2158–2162 (1997).

S. M. Ebenholtz, "Hysteresis Effects in the Vergence Control System: Perceptual Implications," in D. F. Fisher, R. A. Monty, and J. W. Senders (eds.), *Eye Movements: Visual Perception and Cognition,* Erlbaum, Hillsdale, NJ, pp. 83–94, 1981.

V. Ellerbrock and G. Fry, "Effects Induced by Anisometropic Corrections," *Am. J. Optom.* **19**:444–459 (1942).

S. R. Ellis, U. J. Bucher, and B. M. Menges, "The Relationship of Binocular Convergence and Errors in Judged Distance to Virtual Objects," in *Proceedings of the International Federation of Automatic Control,* Boston, 1995.

E. Engle, "Incidence and Magnitude of Structurally Imposed Retinal Image Size Differences," *Percept. Motor Skills* **16**:377–384 (1963).

J. M. Enoch, "A Spectacle–Contact Lens Combination Used as a Reverse Galilean Telescope in Unilateral Aphakia," *Am. J. Optom.* **45**:231–240 (1968).

W. Epstein (ed.), *Stability and Constancy in Visual Perception: Mechanisms and Processes,* John Wiley & Sons, New York, 1977.

H. Erggelet, "Über brillenwirkungen," *Ztschr. f. Ophth. Optik* **3**:170–183 (1916).

P. Erickson and C. M. Schor, "Visual Function with Presbyopic Contact Lens Correction," *Optometry and Visual Science* **67**:22–28 (1990).

H. B. Field, "A Comparison of Ocular Imagery," *Arch. Oph.* **29**:981–988 (1943).

A. Fiorentini and L. Maffei, "Binocular Depth Perception Without Geometrical Cues," *Vision Res.* **11**:1299–1305 (1971).

J. Foley, "Binocular Distance Perception," *Psychol. Rev.* **87**:411–434 (1980).

J. S. Friedenwald, "Diagnosis and Treatment of Anisophoria," *Arch. Ophth.* **15**:283–304 (1936).

J. Garding, J. Porrill, J. E. W. Mayhew, and J. P. Frisby, "Stereopsis, Vertical Disparity and Relief Transformations," *Vision Res.* **35**(5):703–722 (1995).

G. M. Gauthier and D. A. Robinson, "Adaptation of the Human Vestibular Ocular Reflex to Magnifying Lenses," *Brain Res.* **92**:331–335 (1975).

B. Gillam and B. Lawerren, "The Induced Effect, Vertical Disparity, and Stereoscopic Theory," *Perception and Psychophysics* **34**:121–130 (1983).

A. Gonshor and G. Melvill-Jones, "Extreme Vistibulo-Ocular Adaptation Induced by Prolonged Optical Reversal of Vision," *J. Physiol. (Lond.)* **256**:381–414 (1976).

L. R. Harris, "Visual Motion Caused by Movements of the Eye, Head and Body," Chapter 14, in A. T. Smith and R. J. Snowden (eds.), *Visual Detection of Motion,* Academic Press, San Diego, pp. 397–435, 1994.

M. G. Harris, "Optic and Retinal Flow," Chapter 11, in A. T. Smith and R. J. Snowden (eds.), *Visual Detection of Motion,* Academic Press, San Diego, pp. 307–332, 1994.

D. B. Henson and B. Dharamski, "Oculomotor Adaptation to Induced Heterophoria and Anisometropia," *Invest. Ophthalmol. Vis. Sci.* **22**:234–240 (1982).

E. Hering, *Spatial Sense and Movements of the Eye* (in German), C. A. Radde (trans.), American Academy of Optometry, Baltimore, MD, 1942.

H. W. Hofstetter, "Graphical Analysis," Chapter 13, in C. M. Schor and K. Ciuffreda (eds.), *Vergence Eye Movements: Clinical and Basic Aspects,* Butterworth, Boston, pp. 439–464, 1983.

W. A. Houtman, T. H. Roze, and W. Scheper, "Vertical Motor Fusion," *Doc. Ophthalmol.* **44**:179–185 (1977).

I. P. Howard, *Human Visual Orientation,* John Wiley & Sons, New York, 1982.

I. P. Howard and H. Kaneko, "Relative Shear Disparities and the Perception of Surface Inclination," *Vision Res.* **34**:2505–2517 (1994).

P. A. Howarth, "Empirical Studies of Accommodation, Convergence and HMD Use," in *Proceedings of the Hoso-Bunka Foundation Symposium: The Human Factors in 3D Imaging,* Hoso-Bunka Foundation, Tokyo, Japan, 1996.

P. A. Howarth and P. J. Costello, "The Occurrence of Virtual Simulation Sickness Symptoms When an HMD Was Used as a Personal Viewing System," *Displays* **18**:107–116 (1997).

R. L. Hughes, L. R. Chason, and J. C. H. Schwank, *Psychological Considerations in the Design of Helmet-Mounted Displays and Sights: Overview and Annotated Bibliography,* U.S. Government Printing Office, AMRL-TR-73-16, National Technical Information Service and U.S. Dept. of Commerce, Washington, DC, 1973.

T. Inoue and H. Ohzu, "Accommodation and Convergence When Looking at Binocular 3D Images," in K. Noro and O. Brown, Jr. (eds.), *Human Factors in Organizational Design and Management—III,* Elsevier Science Publishers, Amsterdam, pp. 249–252, 1990.

B. Jules, *Foundations of Cyclopean Perception,* University of Chicago Press, Chicago, 1971.

M. P. Keating, *Geometric, Physical and Visual Optics,* Butterworths, Boston, 1988.

R. V. Kenyon, T. A. DeFanti, and J. T. Sandin, "Visual Requirements for Virtual-Environment Generation," *Society for Information Display Digest* **3**:357–360 (1995).

M. F. Land, "Fast-Focus Telephoto Eye," *Nature* **373**:658 (1975).

D. N. Lee, "A Theory of Visual Control of Braking Based on Information About Time to Collision," *Perception* **5**:437–459 (1976).

L. Liu, S. B. Stevenson, and C. M. Schor, "A Polar Coordinate System for Describing Binocular Disparity," *Vision Res.* **34**(9):1205–1222 (1994).

L. Marran and C. M. Schor, "Lens Induced Aniso-Accommodation," *Vision Res.* **38**(**22**):3601–3619 (1998).

J. S. Maxwell and C. M. Schor, "Adaptation of Ocular Torsion in Relation to Head Position," *Vision Res., 1999* (in press).

J. E. W. Mayhew and H. C. Longuet-Higgins, "A Computational Model of Binocular Depth Perception," *Nature* **297**:376–378 (1982).

D. W. McCready, "Size, Distance Perception and Accommodation Convergence Micropsia. A Critique," *Vision Res.* **5**:189–206 (1965).

S. P. McKee and S. Watamaniuk, "The Psychophysics of Motion Perception," Chapter 4, in A. T. Smith and R. J. Snowden (eds.), *Visual Detection of Motion,* Academic Press, San Diego, CA, pp. 85–114, 1994.

J. E. Melzer and K. W. Moffitt, "Ecological Approach to Partial Binocular Overlap," in *Large Screen Projection, Avionic and Helmet-Mounted Displays, Proceedings of the SPIE,* H. M. Assenheim, R. A. Flasck, T. M. Lippert, and J. Bentz (eds.), vol. 1456, Optical Engineering Press, Bellingham, WA, pp. 175–191, 1991.

T. Morita and N. Hiruma, "Effect of Vergence Eye Movements for Size Perception of Binocular Stereoscopic Images," in *Proceedings of the Third International Display Workshops,* SID, Kobe, Japan, 1996.

L. C. Morrison, "Stereoscopic Localization with the Eyes Asymmetrically Converged," *Am. J. Optom.* **54**:556–566 (1977).

K. Nakayama and S. Shimojo, "DaVinci Stereopsis: Depth and Subjective Occluding Contours from Unpaired Image Points," *Vision Research* **30**:1811–1825 (1990).

J. Nolan, A. Hawkswell, and S. Becket, "Fusion and Stereopsis in Aphakia," in S. Moore, J. Mein, and L. Stockbridge (eds.) *Orthoptics: Past, Present and Future,* Shatten Int. Cont. Med. Book Co., New York, pp. 523–529, 1975.

J. Norman and S. Ehrlich, "Visual Accommodation and Virtual Image Displays: Target Detection and Recognition," *Hum. Factors* **28**:135–151 (1986).

K. N. Ogle, *Researches in Binocular Vision,* Hafner, New York, 1964.

M. E. Ono, J. Rivest, and H. Ono, "Depth Perception as a Function of Motion Parallax and Absolute-Distance Information," *J. Exp. Psychol.: Human Perception and Performance,* **12**:331–337 (1986).

D. A. Owens and H. W. Leibowitz, "Accommodation, Convergence and Distance Perception in Low Illumination," *Am. J. Optom. Physiol. Opt.* **57**:540–550 (1980).

A. L. Perlmutter and A. Kertesz, "Measurement of Human Vertical Fusional Response," *Vision Res.* **18**:219–223 (1978).

T. Piantanida, D. K. Bowman, J. Larimer, J. Gille, and C. Reed, "Studies of the Field of View/Resolution Trade-Off in Virtual-Reality Systems," *Hum. Vis. Visual Proc. Digital Display III, Proc. SPIE* **1666**:448–456 (1992).

D. Regan and K. J. Beverley, "Visually Guided Locomotion: Psychophysical Evidence for a Neural Mechanism Sensitive to Flow Patterns," *Science* **205**:311–313 (1979).

B. J. Rogers and M. F. Bradshaw, "Vertical Disparities, Differential Perspective and Binocular Stereopsis," *Nature* **361**:253–255 (1993).

B. Rogers and B. J. Graham, "Similarities Between Motion Parallax and Stereopsis in Human Depth Perception," *Vision Res.* **22**:261–270 (1982).

A. Romole, "Dynamic Versus Static Aniseikonia," *Aust. J. Optom.* **67**:108–113 (1984).

L. Rose and A. Levenson, "Anisometropia and Aniseikonia," *Am. J. Optom.* **49**:480–484 (1972).

M. Rosenfield and K. J. Ciuffreda, "Cognitive Demand and Transient Nearwork-Induced Myopia," *Optoml. Vision Sci.* **71**:381–385 (1994).

C. M. Schor and T. Heckmen, "Interocular Differences in Contrast and Spatial Frequency: Effects on Stereopsis and Fusion," *Vision Res.* **29**:837–847 (1989).

C. M. Schor and L. Task, "Effects of Overlay Symbology in Night Vision Goggles on Accommodation and Attention Shift Reaction Time," *Aviation, Space and Environmental Medicine* **67**:1039–1047 (1996).

C. M. Schor, J. Alexander, L. Cormack, and S. Stevenson, "A Negative Feedback Control Model of Proximal Convergence and Accommodation," *Ophth. and Physiol. Optics* **12**:307–318 (1992).

C. M. Schor, J. Gleason, and D. Horner, "Selective Nonconjugate Binocular Adaptation of Vertical Saccades and Pursuits," *Vision Res.* **30**(11):1827–1845 (1990).

C. M. Schor, L. Landsman, and P. Erickson, "Ocular Dominance and the Interocular Suppression of Blur in Monovision," *Am. J. Optom. and Physiol. Optics* **64**:723–736 (1987).

C. M. Schor, I. C. Wood, and J. Ogawa, "Spatial Tuning of Static and Dynamic Local Stereopsis," *Vision Res.* **24**:573–578 (1984).

D. A. Southard, "Transformations for Stereoscopic Visual Stimuli," *Computers & Graphics* **16**:401–410 (1992).

R. M. Steinman and H. Collewijn, "Binocular Retinal Image Motion During Active Head Rotation," *Vision Res.* **20**:415–429 (1980).

C. W. Tyler and A. B. Scott, "Binocular Vision," in *Duane's Foundations of Ophthalmology,* W. H. Tasman and E. A. Jaegar (eds.), vol. 2, chap. 24, J. B. Lippincott, Philadelphia, pp. 1–24, 1979.

C. W. Tyler and E. Sutter, "Depth from Spatial Frequency Difference: An Old Kind of Stereopsis," *Vision Res.* **19**:858–865 (1979).

R. van Ee and C. J. Erkelens, "Stability of Binocular Depth Perception with Moving Head and Eyes," *Vision Res.* **36**:3827–3842 (1996).

H. von Helmholtz, *Handbuch der physiologischen Optik,* third German edition (Helmholtz's Treatise on Physiological Optics), J. P. C. Southall (trans.), George Banta Publishing, Menasha, WI, 1962.

M. Von Rohr, *Die Brille als optisches Instrument.* Wilhelm Englemann, Leipzig, p. 172, 1911.

H. Wallach and K. J. Frey, "Adaptation in Distance Perception Based on Oculomotor Cues," *Perception and Psychophysics* **11**:77–83 (1972).

G. L. Walls, *The Vertebrate Eye and Its Adaptive Radiation,* Hafner, New York, 1967.

W. H. Warren and D. J. Hannon, "Direction of Self-Motion Is Perceived from Optical Flow," *Nature* **336**:162–163 (1988).

G. Westheimer, "Effect of Binocular Magnification Devices on Stereoscopic Depth Resolution," *J. Opt. Soc. Am.* **46**:278–280 (1956).

G. Westheimer and M. W. Pettet, "Detection and Processing of Vertical Disparity by the Human Observer," *Proc. R. Soc. Lond. B Biol. Sci.* **250**(1329):243–247 (1992).

B. Winn, C. A. Ackerley, E. K. Brown, J. Murray, J. Prars, and M. F. St. John, "Reduced Aniseikonia in Axial Anisometropia with Contact Lens Correction," *Ophthalmic Physiol. Opt.* **8**:341–344 (1987).

Y. Y. Yeh and L. D. Silverstein, "Spatial Judgements with Monoscopic and Stereoscopic Presentations of Perspective Displays," *Human Factors* **34**:583–600 (1992).

CHAPTER 13
OPTICS AND VISION OF THE AGING EYE*

John S. Werner
Brooke E. Schefrin
University of California, Davis
Department of Ophthalmology and
Section of Neurobiology, Physiology, and Behavior
Sacramento, California

13.1 GLOSSARY

Aphakic.   A person without a lens, nearly always as a result of cataract surgery.

Age-related Macular Degeneration (AMD).   A disease that results in loss of vision in the central retina or macula due to damage to the receptors and retinal pigment epithelium of the retina, often secondary to changes in the choroid (blood vessel layer) and Bruch's membrane. There are two major forms, *wet* or exudative form and *dry* or atrophic form. The dry form is more common, but the wet form is responsible for the majority of cases involving severe vision loss.

Cataract.   An opacity or clouding of the lens resulting in a reduction in the amount of light transmitted to the retina and an increase in light scatter. This is the leading cause of blindness worldwide, but can be treated successfully by removal of the lens.

Diabetic Retinopathy.   A disease of the eye caused by diabetes that causes a proliferation of blood vessels. These blood vessels may swell, become tortuous, and/or bulge and leak.

Glaucoma.   A group of diseases often, but not always, associated with high intraocular pressure, resulting in death of ganglion cells.

Intraocular Lens (IOL).   An artificial lens usually implanted in the eye to replace the natural lens following its removal. IOLs may be placed in front of the pupillary plane (anterior chamber lenses) or behind the pupil (posterior chamber lenses). They are manufactured from a variety of materials, the most common of which are polymethylmethacrylate, silicone, and hydrogels. Today, most are placed in the residual lens capsule, which is located behind the pupil.

---
* *Acknowledgments:* The authors are grateful for the helpful comments of Arthur Bradley, John L. Barbur, James D. Brandt, James T. Handa, Mark J. Mannis, Lawrence S. Morse, David S. Pfoff, and Vittorio Porciatti. This research was supported by NIA grant AG04058 and a Jules and Doris Stein Research to Prevent Blindness Professorship.

LASIK (Laser In Situ Keratomileusis). A procedure for surgically reshaping the cornea to reduce or eliminate the need for external refractive lenses.

Phakic. A person with a natural eye lens, from the Greek *phakos* or lens.

Presbyopia. Greek for "old sight." The condition of an individual who has lost the ability to accommodate sufficiently for near work. This results in the need for reading glasses around the age of 40 or 50 years.

PRK (Photorefractive Keratectomy). A procedure for surgically modifying the cornea to reduce or eliminate the need for external refractive lenses.

Pseudophakic. An individual with an artificial or intraocular lens implanted to replace the natural lens following its removal.

13.2 INTRODUCTION

Senescent changes in the optics of the human eye produce changes in the geometric and spectral distribution of light comprising the retinal image. The impact of these changes for vision is often profound for the individual. They may have enormous economic consequences for health care and society. It is essential to study such changes scientifically in order to understand the neural basis of visual senescence. Before describing the optical and visual changes that characterize normal aging, it is useful to consider the scope of the problem in light of demographic changes in the world's aging population.

13.3 THE GRAYING OF THE PLANET

The proportion of the world's population that is elderly is higher now than at any other point in recorded history. In the last two decades, the number of individuals above age 65 years has increased by more than twice the rate for individuals below 65 years.[1] At present, every month another 1 million of the world's population enters its seventh decade of life. As a result, the number of individuals age 60 and above is likely to change from one in ten today to one in five by 2050, and one in three by the year 2150. As illustrated by Figure 1, the global growth rate for individuals over age 80 is occurring at an even faster pace. Indeed, the number of centenarians in the United States has doubled every ten years since 1960. At the same time, there has been a corresponding decline in the proportion of the population below 15 years of age. It is not clear how long these demographic shifts will continue into the future or when they will reach equilibrium.

Many factors have contributed to this remarkable demographic shift in the age distribution, including a change in the fertility rate and improved hygiene, nutrition, disease control, and education. These trends are generally more apparent in less developed countries, and the demographic shift in the elderly population is now more rapid in these parts of the world. For example, in 1995 there was a global increase of 12 million persons at or above age 60 and nearly 80 percent of it was due to a change in less developed countries. The countries with the fewest resources at present will thus have less time to cope with the social and economic consequences of an elderly population.

Some Implications of Living Longer

Some theoretical calculations place the maximum human life span at about 120 years, but there are reasons to think that this may not apply to the eye. The demographic shifts in longevity with respect to the eye are apparently not keeping pace with those for the rest of the

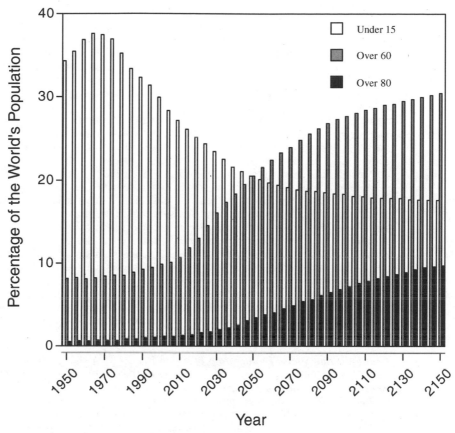

FIGURE 1   Percentage of the global population projected from 1950 to 2150 for three age groups based on statistics provided by the United Nations.[2]

body. The organization Research to Prevent Blindness estimates that the number of blind individuals will double between 1999 and 2030. If true, the social, health, and economic consequences will be profound. One eminent authority put it this way: "Mathematically speaking, it is permissible to say that those who prolong life promote blindness" (Ref. 3, p. 306).

Health Implications and the Problem of Defining Normal Aging.   The current projection for the future is that as the baby boomers (the cohort of 76 million Americans born between 1946 and 1964) start to turn age 65 in the year 2011, they will be better educated and will have better health and more financial resources than their parents did when they retired. All indications are that retiring baby boomers will read more and want to spend more time in outdoor leisure activities than their parents did at comparable ages. For the age group over 65 years of age, there are also some indications that the rates of disability and disease are decreasing compared to previous generations. However, this is not so for vision. Older adults experience a higher rate of visual disability and ocular disease than younger individuals, and the rate accelerates after about 40 years of age. For example, glaucoma is about eight times more likely for individuals above age 65 than in the general population. And in the Framingham Study[4] 50 percent of participants age 75 and over had at least one of four visual diseases: age-related macular degeneration, cataract, diabetic retinopathy, or glaucoma.

There is little agreement about where to draw the line between senescent degradation of vision and some forms of ocular pathology. For many researchers, normal aging is defined by exclusion—the changes in vision that occur when overt ocular disorder and disease are not detected. An alternative distinction between aging and disease is based on the amount of tissue changed or when visual loss exceeds a criterion. Both of these definitions are used to define cataract, a clouding of the lens resulting from cumulative changes in products of oxidation that result in loss of transparency and an increase in intraocular light scatter. There are several different types of cataract, but all degrade the optical quality of images formed on the retina and consequently undermine the quality of the neural image transmitted by the retina. The biochemical events leading up to these lenticular changes occur continuously across the life span, so the presence of this disorder is largely a matter of where the criterion is set on the continuum of normal aging.

Most individuals will still be able to approach their golden years without frank visual pathology. However, no individual should expect to have the quality of vision experienced in youth. Indeed, by the time of retirement, most individuals already will have experienced the need for reading glasses to correct near vision due to the reduction in or loss of accommodation. Some problems of optical origin can be corrected or compensated by optical means, but additional losses in vision may be expected due to neural changes for which there is no known remedy. Some current research has focused on factors that might contribute to normal aging of the eye that could be controlled by life style choices such as diet[5] or exposure to ultraviolet[6] and short-wave visible[7] radiation. Whatever the cause, any impairment of our dominant sensory system will inevitably have behavioral consequences. In some cases, this may be an inconvenience such as slower reading or the need for better lighting, while in other cases it may lead to more serious problems such as accidents, falls, or greater dependence on others for daily activities.

Economic and Social Implications of Aging Eyes.   There is a gradual loss in accommodation throughout adulthood that becomes complete by the end of the sixth decade of life. Combining this information with the fact that each day 10,000 people in the United States reach the age of 50, there is little wonder that ophthalmic corrections for reading and near work will account for approximately half the revenues in the multibillion dollar ophthalmic lens industry.

Visual impairment secondary to diseases of the eye also creates a sizeable economic impact on our society. For example, visual impairment for individuals above age 40 already costs the U.S. federal government in excess of $4 billion per annum.[8] In addition, according to Health Products Research, Inc., approximately 2.5 million cataract surgeries are performed each year in the United States at an annualized cost of over $2.5 billion.[9] The costs of presbyopic spectacle correction is even higher, reaching $8 billion in 1998. These numbers are almost certain to increase due to the increasing proportion of the population that is elderly.

Additional economic consequences of our aging population will be created because the elderly generally require more special services, housing needs, and medical care. For example, one third of medical expenditures in the United States are spent on the oldest 12 percent of the population. There are also likely to be new burdens on public transportation systems in view of the fact that about half the world's elderly population today live in cities, and that proportion is expected to increase to about three-fourths, at least in more developed countries. On average, every residential dwelling will likely have at least one elderly person.[10] New demands will be made for housing these elderly individuals in order to promote independent living. Being able to care for oneself is not only important for self-esteem, it has enormous economic impact. Architects and lighting designers will have to take into account age-related changes to the visual system in planning for new environments inhabited by the elderly. Current standards for lighting are almost entirely based on college-age students, although there have been new calls for explicit attention to aging visual functions.[11,12] The goal of such standards would be to find conditions that can compensate for sensory losses to be described in the pages that follow, while not introducing other problems such as disability glare. Such lighting considerations might promote greater safety in the home to facilitate independent mobility and support communication through computers and reading, as well as other visual tasks involving hobbies and work. And with the world becoming ever more visually oriented through computer use, VDJs, and so forth, blindness will often be more limiting.

13.4 THE SENESCENT EYE AND THE OPTICAL IMAGE

The globe of the eye grows rapidly in the first year of life and then more slowly until about age five years when it reaches an asymptotic sagittal length of about 23.5 mm. Thereafter, there is suggestive evidence that it may shrink slightly with age, by about 1 mm between age 20 and 80 years.[3] Provided no other changes occur, this alone would result in about a 2-diopter (D) change in the direction of hypermetropia (farsightedness).

While the globe of the eye changes relatively little over the adult years, the lens continues to grow. These and other structural changes alter the refractive state of the eye throughout the life span. The properly refracted eye is said to be emmetropic, but emmetropia is not the norm during all stages of development and senescence. Generally speaking, in early childhood, the eye is often hypermetropic and in time will shift in an emmetropic or myopic (nearsighted) direction, whereas in later life there is an overall tendency for the refractive error of the eye to shift in a hypermetropic direction.

To understand the aging of the visual system, it is necessary to consider how the stimulus available for vision changes over the life span. Figure 2 shows the optical density* of the human ocular media, typical of younger persons. Because light itself is thought to facilitate some aspects of ocular and retinal aging, the figure also presents the relative quantal distribution of sunlight. The spectral distribution of sunlight reaching different parts of the earth's surface will differ somewhat from that shown in Figure 2 due to such factors as latitude, altitude, cloud cover, and aerosol content. The solar quantum spectrum incident on the cornea will vary with surface reflectance (e.g., snow and sand are better reflectors of ultraviolet, or UV, radiation than are grass and leaves) and a variety of other factors (e.g., sunglasses and brimmed hats may decrease UV light reaching the cornea). While it is straightforward to document surface UV radiation, it is very difficult to calculate ocular exposure. For example, while the sunlight stays constant when moving from a snowless to a snow-covered field, ocular UV irradiance increases substantially due to the high reflectivity of the snow, since reflected sunlight is not blocked by the brow ridge and superior lid like light from directly overhead.[13] Most of the ocular media contribute to a reduction in UV radiation (wavelengths below 400 nm) reaching the retina. The most effective UV absorbers of the ocular media are found in the crystalline lens, whose optical density increases from infancy to old age.

Senescent changes influencing image formation are most prevalent in three principal structures of the eye (cornea, pupil, and lens), and each of these changes may result in degradations of optical quality. The next sections describe some of these changes in more detail.

The Cornea

The human cornea consists of three discrete layers separated by two basement membranes. The outermost epithelial layer composed of basal, wing, and surface cells acts as a barrier to the external environment and is subject to a rapid rate of renewal, with the result that its cells are replaced every 5 to 7 days.[17] The middle layer, or stroma, is a matrix of collagen fibers as well as fibroblast-like cells called keratocytes that accounts for approximately 90 percent of the corneal thickness. The stroma is sandwiched between two membranes. Anterior and posterior to the stroma are Bowman's and Decemet's membranes, respectively. The transparency of the stroma is due to the uniform density of collagen fibers and any cause of disruption of

---

* Optical density = $\log (1/T)$, where T is the light transmission.

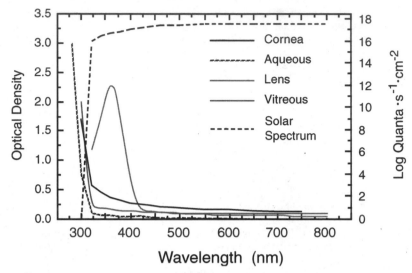

FIGURE 2 Optical density of human ocular media is plotted as function of wavelength (left ordinates; after Boettner and Wolter[14]). Log quanta in sunlight reaching the surface of the earth from an angle of 30° from the horizon is plotted as a function of wavelength (right ordinates; after Brandhorst, et al.[15]). (*After Werner.*[16])

this order such as increased hydration may cause light scatter and loss of transparency. Because of the tendency of the stroma to imbibe water, there must be a way to regulate corneal hydration. The extent of corneal hydration is largely controlled by energy-dependent ion-coupled fluid transport mechanisms located at the inner most layer of the cornea, the corneal endothelium. The endothelium is made up of a single layer of cells. At birth, the corneal endothelium has about 4500 cells/mm^2, but the number is reduced to about 2000 to 2500 cells/mm^2 by age 70 or 80. Although the surviving cells become larger to maintain approximately constant total tissue volume, a smaller total number of fluid transport mechanisms remain in the corneal endothelium. As a result, proper corneal hydration is sometimes a problem in the elderly.

The cornea is the most powerful refractive element in the human eye, accounting for about 40 to 45 D of optical power. Although it has a lower refractive index than the lens, the ratio of refractive indices, which is directly proportional to dioptric power, is highest at the air–cornea interface. The cornea is not spherical; its greater curvature in the center provides about 4 D more power than at the periphery. The refractive index of the cornea[18] and curvature of its anterior surface[19] appear to undergo little change with age.

As can be seen from Figure 2, the cornea absorbs much of the UV radiation below about 320 nm. Absorption of this radiation in sufficient intensity can damage the cornea (e.g., actinic keratitis), but long-term effects are rare because the corneal epithelium is capable of rapid molecular renewal. This is probably why the absorption spectrum of the cornea undergoes only small changes with advancing age.[20] Cumulative absorption of UV light may, however, lead to corneal abnormalities such as pterygium, a condition that involves the advancing growth of abnormal tissue from the limbus (the border between the cornea and sclera) onto the clear cornea.

The anterior surface of the cornea and globe of the eye is bathed by tears, but tear production by the main accessory and lacrimal glands decreases with age.[3] This may produce a condition called *dry eye* that is further exacerbated by incomplete lid closure, loss of tonicity in the lids, and a reduction in the number of blinks in the elderly. Corneal desiccation may affect the optical quality of the eye by changing its outermost portion from a smooth to an irregular optical surface and, in extreme cases, lead to scarring. Either of these sequelae may diminish visual acuity in individuals afflicted with this condition.

The cornea and lens are separated by the anterior chamber, which is filled with a fluid, the aqueous humor. This fluid maintains the metabolism of the cornea and the lens, as these structures have no blood supply. Separating the lens from the retina is a colorless gelatinous substance called the vitreous humor. Both the aqueous and vitreous humors absorb UV radiation below about 300 nm. Their optical density changes negligibly with age[14] and any damage sustained by the aqueous humor resulting from absorption of light is likely to be unimportant for senescence because this fluid is rapidly replaced.[17]

The Pupil

The dilator and sphincter muscles of the iris control the size of the pupillary aperture that for young adults varies in diameter from approximately 2 to 8 mm with the level of the prevailing illumination. As in any optical system, the size of the aperture has important consequences for image formation. Smaller pupils in the human eye diminish the effects of spherical aberration and increase depth of focus by reducing the size of the blur circle on the retina. If the pupillary diameter becomes too small, however, it will decrease optical quality through diffraction effects. The optimal pupil diameter from the point of view of image quality is about 2.4 mm.[21] Larger pupils under scotopic conditions represent a compromise to improve sensitivity because more light is incident on the retina at the cost of diminished image quality.

Maximal pupil diameter occurs in adolescence and then progressively decreases as a function of increasing age. As shown in Figure 3, the age-related change in diameter differs for dark-adapted and light-adapted conditions. Changes in pupil diameter with light level provide a mechanism of adaptation through adjustment of visual sensitivity with illumination. However, the change from maximum to minimum size influences light level by only about a factor ten, so the pupil plays only a small role in adaptation mechanisms that allow us to function over a range of ambient illumination covering about 12 log units.

Age-related changes in pupil size are noticeable in the elderly, but actually begin in the early adult years. For example, under darkened conditions there is essentially a monotonic decrease in pupil diameter beyond 10 years of age.[23,24] Spatial vision depends on light level,

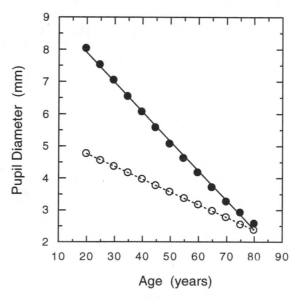

FIGURE 3   Mean diameter of the human eye pupil is plotted as a function of age. Filled and unfilled symbols show dark- and light-adapted values, respectively. (*Data from Kornzweig.*[22])

and although a smaller pupillary diameter allows less incident light to fall onto the retina, it is unlikely that a smaller pupil accounts for many age-related changes in spatial vision. For example, it has been demonstrated that age-related differences in contrast sensitivity under photopic and mesopic ambient light levels cannot be explained solely by differences in pupillary diameter.[25] In fact, there are some benefits associated with an age-related reduction in pupil diameter. The smaller pupil of older persons diminishes age-related changes in the optical transfer function of the eye at low luminance levels.[26] In addition, the smaller pupil in the elderly reduces the diameter of the retinal blur circle, thereby increasing the depth of focus by about 0.5 D between the ages of 30 and 70 years.[27] However, this improvement in depth of focus is far too small to compensate for the approximately 7 D age-related loss in amplitude of accommodation over this same age span.

The Lens

The lens consists of only one cell type, but its thickness increases throughout life due to the accumulation of cells to form a series of concentric layers. The oldest cells are thus found in the center or nucleus, while the youngest cells comprise the outer or cortical layers. As the cells of the lens age, they lose their organelles and become essentially metabolically inert.

Molecules that absorb UV and short-wave visible light are present in the newborn lens and their numbers increase continuously over the life span. Thus, the curve for the lens shown in Figure 2 can be, to a first approximation, scaled multiplicatively with age. A number of studies have measured the optical density of the human lens as a function of age. Some have described these changes in density by a linear function of age,[28,29] while other evidence demonstrates a bilinear function due to an acceleration in the aging function after about 60 years.[30] The precise form of the aging function is difficult to determine because of substantial individual differences at every age (on the order of 1.0 log unit for light at 400 nm). However, it should be noted that the average infant lens transmits at least 25 times more light at 400 nm to the retina than the average 70-year-old eye!

An acceleration in the rate of change in lens density beyond approximately 60 years of age could be due to the inclusion of subjects approaching cataract.[31] However, it is not clear where normal aging ends and cataract begins since the change in density is continuous across life. In his comprehensive monograph on this subject, Young[32] summarized a substantial literature, which led him to conclude that: "No discontinuity between senescent and cataractous changes can be detected at the molecular level in the human lens. . . . The most sophisticated techniques of modern biophysical and biochemical analysis have so far failed to uncover any feature of the cataractous lens that suggests cataract is anything more than an advanced stage of normal aging" (p. 56).

It should be added that there are several forms of cataract depending on the part of the lens affected (nucleus, cortex, posterior capsule) or etiology (congenital, age-related, chemically induced, trauma-induced, disease-induced). Losses of lenticular transparency can sometimes occur through occupational hazards associated with infrared light exposure such as in "glassblowers' cataract" due to the transfer of heat absorbed by the iris.[33] Most often, changes in the transparency of the lens are associated with cumulative exposure to UV radiation. Weale[34] has proposed that the specific form of cataract depends on genetic factors, while the most common etiology is cumulative exposure to solar radiation. Epidemiological evidence provides support for this view. For example, there is a clear correlation between incidence of cataract and other diseases (e.g., pterygium and skin cancer) known to be related to sunlight exposure. In addition, the onset of cataract and its incidence in the population increases from the poles to the equator. Young's[32] summary of this literature indicates prevalence rates of about 14 percent for countries above 36° latitude, 22 percent for mid-latitude (28° to 36°) countries, and 28 percent for low-latitude (10° to 26°) countries. While epidemiological studies are complex due to the many factors that vary among groups, UV radiation appears to be the only factor that changes with latitude *and* has been shown to play a significant role in the aging of the lens.

Cataract is treated by surgical removal of the lens and in more than 90 percent of surgeries performed in the United States an intraocular lens (IOL) is implanted in its place. A properly selected IOL obviates the need to wear thick spectacle corrections or contact lenses for the hypermetropia that would otherwise occur. Most of these lenses now contain UV-absorbing chromophores, but this is a relatively recent design improvement. A study composed of pseudophakic patients who had undergone bilateral cataract extraction and implantation of two IOLs differing in the amount of UV absorption suggested that increased UV exposure reduced visual sensitivity of short-wavelength-sensitive cones. It was estimated that 5 years of exposure to light through an IOL without UV-absorbing chromophores produced the equivalent of more than 30 years of normal aging.[6] Thus, the lens pigmentation responsible for UV absorption may be an adaptation to protect the retina from photochemical effects that are damaging.

Accommodation and Presbyopia. Accommodation refers to the ability of the eye to adjust its focus for objects located at varying distances in the environment. Contraction of the ciliary muscle located within the ciliary body of the eye allows the anterior surface of the lens capsule to change shape, thereby altering the dioptric power of the lens. The amplitude of accommodation refers to the *dioptric* difference between the farthest and nearest distances in which an object remains in focus. Age-related losses in elasticity of the lens capsule[35] produce a loss in accommodative amplitude, as shown in Figure 4. The open symbols are from the Basel longitudinal study of individuals followed for 20 years, with different subjects starting the study at different ages.[36] These data agree well with the classic data of Donders[37] shown by closed symbols.

As the amplitude of accommodation decreases, a near correction for tasks such as reading may be required. This typically occurs within the fifth decade of life and is known as presbyopia (meaning "old sight"). Because the accommodative amplitude changes continuously, the definitions of presbyopia and its onset are somewhat arbitrary. The two sets of data in Figure 4, although separated in time by about a century, are from similar geographic locations. Age of onset of presbyopia depends on many factors, including geographical latitude and average ambient temperature.[3]

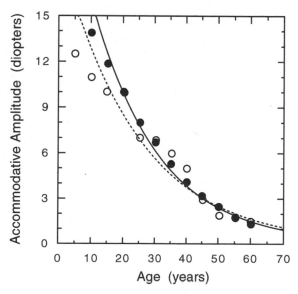

FIGURE 4 Amplitude of accommodation (diopters) is plotted as a function of age. Open symbols (dashed curve) are from Brückner et al.[36]; closed symbols (solid curve) are from Donders.[37] Curves are based on best-fitting exponential functions.

The Macular Pigment

The central retina contains a yellow pigment called the macular pigment (MP). It is deposited in the receptor fiber and inner plexiform layers of the retina[38] where it selectively absorbs short-wave light en route to the photoreceptors. Figure 5 shows the spectral absorbance of the foveal macular pigment for an average observer. The peak density is at 460 nm, near the peak sensitivity of the short-wave-sensitive cones (440 nm).

Several different methods (fundus photography, entoptic visualization, and psychophysics) indicate that the peak density of the MP is at the fovea and follows an exponential decay to negligible levels at from approximately 5 to 10° retinal eccentricity.[40,41] There are large individual differences in the absorption of the macular pigment (from about 0 to 1.0 log unit at 460 nm), but there is little age dependency after early childhood.[42,43]

While the function of the MP is still uncertain, it has been suggested that the presence of this pigment improves the optical image quality at the retina and may aid in maintaining the health of the retina. Because MP selectively absorbs short-wave light, it is thought to minimize the effects of chromatic aberration,[44] thereby improving form vision in the central retina. With regard to the health of the eye, MP is made up of carotenoids that tend to neutralize photosensitized molecules[45] and inhibit free radical reactions that are toxic to retinal cells. Thus, MP may protect the retina from actinic damage that may lead to age-related changes in visual sensitivity and the development of age-related macular degeneration.[7,46]

Retinal Image Quality

Many factors act to alter the retinal image throughout the life span. In this section, the combined effects of these factors are described, along with one aspect of the retinal image (chromatic aberration) that does not seem to be age dependent.

Intraocular Scatter.   The spatial distribution of light on the retina from a point source is given by the point spread function (PSF). The fraction of light transmitted from a point source

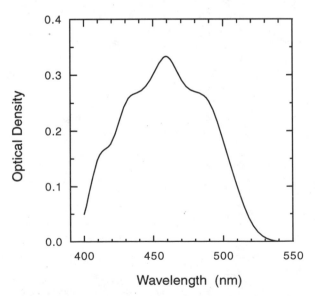

FIGURE 5   Optical density of the human macular pigment is plotted as a function of wavelength. (*Data from Vos.*[39])

in the forward direction is the total transmittance, while the fraction transmitted within a cone of a particular angle α is the direct transmittance. The value of α is arbitrary, but typically is about 1°. The difference between the total and direct transmittances produces a "tail" in the PSF and is called *stray light*. However, because the photoreceptors are directionally selective (see Chapter 9 by Lakshminarayanan and Enoch), the optical PSF is not necessarily equivalent to the functional PSF. Westheimer and Liang[47] have used psychophysical methods to measure the PSF of younger and older observers for a broadband source. The total scattered light from 7′ to 90° was about five times larger in a 69-year-old observer compared to a young adult. Put another way, 90 percent of the light from a point source fell within a region of 7′ for a young subject, but only 50 percent of the light fell within 7′ for the older subject. (This analysis of the PSF does not take into account differences in the density of the ocular media that would further increase the difference in retinal images between young and old.)

A number of studies have shown that light scatter by the human ocular media increases across the life span.[48–50] The relation between light scatter and age has been described as nonlinear due to an acceleration in the function in middle age.[49] A more recent study[50] measured the integral of the scatter function of the eye from 2.2° to 90°. The results show little or no increase in scatter from 16 to 40 years of age, but a rapid increase afterwards (i.e., a factor of 5 increase at 65 years compared to young adults).

Stray light in the human eye is not strongly related to wavelength,[51] implying that it is not due to scatter by small particles (on the order of the wavelength of light). A small wavelength dependence, greater at long than middle wavelengths (opposite to Rayleigh scatter, λ^{-4}), has been attributed to fundal reflections, especially in lightly pigmented eyes.[52] This can be important in evaluating stray light in the elderly because iris pigmentation decreases with age. The lens seems to be the main source of age-related change in intraocular stray light.[48] The cornea produces light scatter as well, but the amount does not change significantly with age.[14]

Another source of scattered light in the elderly may be due to the age-related accumulation of two fluorophores in the lens.[53] These are molecules that emit light at one wavelength (e.g., 520 nm) when irradiated by another (e.g., 410 nm). As these emissions are scattered randomly, they would be expected to degrade visual resolution. Elliott et al.[54] demonstrated a small effect of fluorescence on the resolution of low contrast targets.

Intraocular stray light reduces the contrast of the retinal image. Low luminance and low contrast visual acuity exhibit marked deficits with aging, and this is further exacerbated by glare (veiling, auto headlights, etc.). Normal levels of scattered light do not affect color detection thresholds significantly, but high levels can affect chromatic saturation and estimates of color constancy.[55] The primary effect of intraocular stray light is to impair visual performance when lighting varies dramatically within a visual scene, such as when driving at night, because scatter from a high intensity source can dramatically reduce the contrast of low intensity targets. The effects of light scatter are sometimes referred to as *disability glare*. *Discomfort glare*, on the other hand, describes a subject's visual discomfort when high intensity glare sources are present. The level of discomfort glare shows large intersubject variability, but it is generally correlated with the amount of disability glare. The older eye is not necessarily more sensitive to glare, but it receives more of it due to greater lenticular scatter.[56]

Modulation Transfer Function.   A quantitative description of the quality of an optical system is provided by its modulation transfer function (MTF), the function describing the relation between contrast attenuation and spatial frequency. The eye's MTF can be derived from *in vivo* measurements of the optical point spread function on the retina. Recently, Guirao et al.[26] derived the MTFs from the point spread functions of 60 observers ranging in age from 20 to 70 years, while controlling for such important factors as pupil size and the refractive state of the eye. In Figure 6, the solid and dashed curves represent MTFs for observers with mean ages of 24 and 63 years, respectively. Optical performance decreases with increasing spatial frequency for both age groups, and there is an age-related loss in performance at all spatial frequencies. Results similar to these have also been reported by Artal et al.[57] The results in Figure 6 may be attributed to optical aberrations but not ocular scatter, because the latter is factored out in the

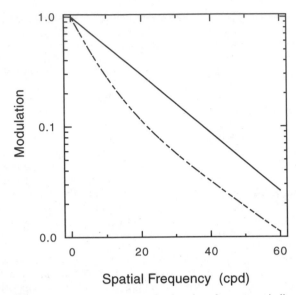

FIGURE 6 Modulation transfer functions for average individuals at 23 (solid curve) and 63 (dashed curve) years of age for a 3-mm pupil, 543 nm stimulus, calculated using equation and parameters from Guirao et al.[26]

measurement procedure. Inclusion of intraocular scatter would tend to produce a more or less uniform halo in the retinal image that further reduces the contrast of stimuli in the elderly eye.

Further analysis of the MTF reveals a linear degradation in the quality of the retinal image throughout the adult years. However, the smaller pupil of the elderly tends to diminish the differences between young and old MTFs under normal viewing conditions, especially at low luminance levels. Indeed, Guirao et al.[26] indicate that the average MTF is approximately constant with age for natural pupils at low luminance levels.

Chromatic Aberration. The refractive index of the ocular media is wavelength-dependent so that the spectral components of a broadband stimulus will not be imaged in the same plane within the eye. This is an example of chromatic aberration. Chromatic aberration can be appreciated from the colored fringes that can be observed in many telescopes and microscopes. The sum of the refractive indices of the optical elements of the eye declines as wavelength increases so that the image plane for short-wave light lies closer to the cornea than long-wave light. The degree of chromatic aberration can be quantified in terms of the lens power required to image a particular wavelength on the retina when the eye is focused for a reference wavelength. With a 578 nm reference, the aberration varies from −1.70 D at 405 nm to 0.40 D at 691 nm.[58] Howarth et al.[59] measured the longitudinal chromatic aberration for four younger (from 27 to 33 years) and six older (from 48 to 72 years) participants using three different methods. All methods failed to show a change in axial chromatic aberration with age.

13.5 SENESCENT CHANGES IN VISION

Age-related changes in visual performance are due to both optical and neural factors. Changes in performance that cannot be explained solely by optical factors are assumed to be due to senescent changes in neural pathways.

Senescent Changes in Sensitivity Under Scotopic and Photopic Conditions

Visible radiation is normally limited to the band between 400 and 700 nm. The limit at short wavelengths is due to the absorption characteristics of the ocular media; the limit at long wavelengths is due to the absorption spectrum of the photopigments contained in the rod and cone photoreceptors.

Various studies using psychophysical procedures have demonstrated age-related losses in the sensitivity of scotopic and photopic mechanisms. Dark adaptation functions (absolute threshold versus time in the dark) have two scalloped-shaped branches, the first branch being mediated by cone photoreceptors and the second by rods. A number of early studies have shown that the asymptotes of each of these branches become elevated as a function of increasing age. In addition, it has been noted that the rate of photopic and scotopic dark adaptation decreases with increasing age.[60-62] It was not clear from some of the early studies whether these effects were entirely due to age-related changes in retinal illuminance caused by, among other factors, smaller pupils (see Section 13.4, subsection entitled "The Pupil"). Some,[62-64] but not all,[65] studies that have taken age-related changes in pupil size and ocular media density into account have demonstrated age-related losses in scotopic sensitivity. Based on those studies that have shown losses in sensitivity that cannot be explained solely by preretinal changes, it can be inferred that losses in sensitivity reflect neural changes that take place at or beyond the level of the photoreceptors.

A variety of studies suggests that changes at the receptor level may partially account for these age-related changes in light and dark adaptation, as well as overall sensitivity under scotopic and photopic conditions. *In vivo* measurements of the photopigment kinetics indicate that a change in the rate of dark adaptation is due, at least in part, to a slowing down of photopigment regeneration.[61,66] Other changes at the receptor level involve the relative ability of "aging" rods to efficiently capture light quanta. Although there are substantial losses in numbers of rod photoreceptors with age,[67] the amount of rhodopsin photopigment in the retina decreases only slightly, if at all, with increasing age.[68] This implies that individual surviving rods must contain more photopigment in the older retina. However, some outer segments of these aged photoreceptors undergo morphological changes such that they appear to be more convoluted in shape.[69] These relatively deformed outer segments may contribute to reductions in scotopic sensitivity because the rhodopsin molecules now possess less than optimal orientations for absorbing light.

Photopic vision is mediated by three different classes of cone photoreceptors, with each class of receptor having peak sensitivity at either short (S), medium (M), or long (L) wavelengths. Several studies have shown that the sensitivity of the S-cones decreases with age.[70-72] In addition, the results of Werner and Steele[72] (Figure 7) indicate that M- and L-cones also sustain significant sensitivity losses with age, and that the rate of loss is similar for all three cone types, on the order of approximately 0.13 log units, or 26 percent, per decade of life.

One complication in specifying the rates of sensitivity loss for individual cone types as a function of age is that the magnitudes of these losses depend on the level of light adaptation. For example, it has been shown for an S-cone mechanism that age-related differences are greatest near the absolute threshold and gradually diminish at higher light levels.[73] Recent measurements of the absolute threshold of cone mechanisms in the fovea, 4° and 8° eccentricity along the horizontal meridian in the temporal retina, show linear declines in sensitivity that could not be accounted for by age-related changes in preretinal structures of the eye.[41] Thus, losses in sensitivity that occur over the life span reflect, in part, neural changes taking place in both scotopic and photopic mechanisms.

Color Vision

The ability to discriminate hue declines with age beginning in the early adult years as assessed by color matching[74,75] and arrangement tasks.[76-78] In general, these losses in discrimination may be due to age-related changes in retinal illuminance caused by smaller pupils and

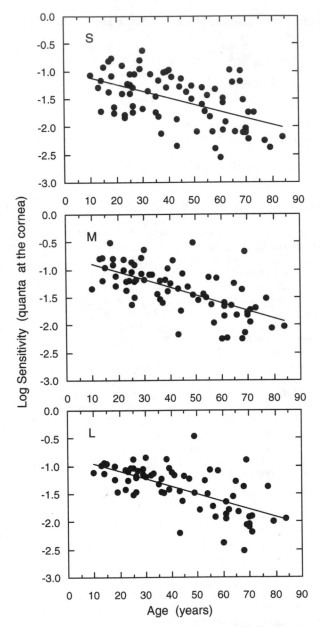

FIGURE 7 Log relative sensitivity (on a quantal basis) of S-, M-
and L-cone mechanisms is plotted as a function of age. (*Data from
Werner and Steele.*[72])

increased short-wave light absorption of the lens. Results from the Farnsworth-Munsell 100-
hue test indicate that losses in discrimination mediated by S-cones occur earlier in life than
losses in discrimination that depend on M- and L-cones. However, this effect is due, at least in
part, to the construction of the FM-100 test.[79] When chromatic discriminations are measured
under conditions that equate the stimuli individually for retinal illuminance, losses occur not

only for S-cone-mediated discriminations,[80] but throughout color space for both spectral[81] and nonspectral stimuli.[82] These latter results indicate that age-related losses in chromatic discrimination are, in part, due to neural processes.

Despite substantial senescent losses in sensitivity of cone mechanisms and poorer chromatic discrimination, our perceptions of brightness,[83] saturation,[84] and hue are less affected by age. For example, Schefrin and Werner[85] presented younger and older subjects with a set of broadband chromatic surfaces (like paint chips) and asked them to scale the hues using the fundamental hue names (red, green, yellow, blue). The subject's task was to describe the percentage of each hue contained in the test stimulus. Despite different retinal stimuli, due to lenticular senescence and smaller pupils, young and old did not differ significantly in their responses. Similarly, the wavelengths that appear pure blue or yellow do not change with age.[86] Even the position in color space of a stimulus that appears purely white, when created by an additive mixture of short- and long-wave monochromatic light, appears stable from 11 to 78 years.[87,88] In terms of the spectral composition of the light mixture reaching the retina, the progressive yellowing of the lens will transmit less short-wave relative to long-wave light. Thus, one would expect that the locus of a purely white stimulus within color space would shift with age. Results showing that the locus of this white point is independent of age implies that one or more neural processes compensate, at least in part, for age-related variations in lenticular senescence and any small differences in the rate of aging for the different cone classes.[89]

Spatial Vision

Visual acuity, the minimum angle of resolution, is the most time-honored measure of human visual performance, and in many clinical settings it is the only measure of visual ability. It can be measured quickly and reliably with a variety of test patterns such as Snellen letters or Landolt Cs. Traditionally, visual acuity is measured with letters having high contrast and moderate luminances. On average, such data suggest some stability in acuity until about age 50 and then a linear decrease with increasing age.[90] The results have been quite variable between studies, however, and the failure to screen subjects for disease and to optimize their refractions for the test distance makes it difficult to know the true course of age-related change in visual acuity. Figure 8 shows results from several recent studies plotted in terms of log minimum angle of resolution (left axis) or equivalent Snellen and metric acuities (right axis). Data from Owsley et al.[91] resemble earlier studies, although participants were screened for visual disease and refracted for the test distance. However, with more stringent screening and optimized refraction, the change in visual acuity follows a somewhat different course according to more recent studies.[92,93] As shown in Figure 8, these latter studies show better overall visual acuity with a progressive decline after about age 25 years. Which data sets represent "normal" aging is unclear, but the better acuity values indicate that most older individuals can often reliably resolve finer details, or higher spatial frequencies, than suggested by earlier research.

A more general characterization of spatial vision is provided by the contrast sensitivity function (CSF), defined by the reciprocal of the minimum contrast required to detect sinusoidal gratings that vary in spatial frequency (the number of cycles per degree of visual angle, cpd). Figure 9 shows CSFs for nonflickering stimuli interpolated from a large data set by linear regression for hypothetical 20- (closed symbols) and 75-year-old (open symbols) observers. These functions could be computed in this way because age-related changes in contrast sensitivity appear to be linear over this age range. A comparison of the photopic CSFs (circles) shows little, if any, change in sensitivity with age at low spatial frequencies, but notable declines in sensitivity are found at middle and high spatial frequencies. These results are consistent with other studies that measured contrast sensitivity in subjects refracted for the test distance and free from ocular disease.[94-97] Currently, it is unclear whether age-related differences in contrast sensitivity, especially at high spatial frequencies, become greater if the temporal modulation of luminance-varying sinusoidal gratings is increased.[91,96,97]

A combination of evidence from a number of studies suggests that both preneural and neural factors are required to explain the decline in contrast sensitivity with age. Preretinal

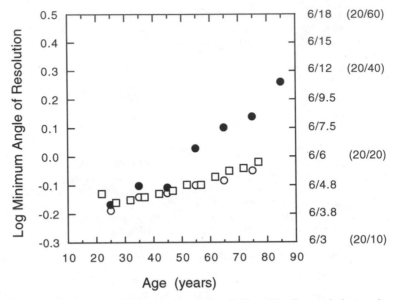

FIGURE 8 Log minimum angle of resolution (left) and Snellen equivalent acuity (right) plotted as a function of age. Filled circles are from Owsley et al.[91]; open squares and open circles are from Frisén and Frisén,[92] and Elliott et al.[93] respectively.

factors, such as smaller pupils and increases in ocular media density and intraocular scatter, can partially account for these results, especially since contrast sensitivity at high spatial frequencies decreases in proportion to the square root of decreases in luminance.[99] However, losses in contrast sensitivity are still demonstrated by studies that have controlled for pupil size[95] and have attempted to mimic age-related reductions in retinal illuminance due to smaller pupils and increases in lens density.[97] In addition, losses of contrast sensitivity at high spatial frequencies are found when measured with laser interferometry, a technique that for the most part bypasses the optics of the eye and reduces intraocular light scatter.[100] It has been suggested that age-related changes in visual pathways account for less than half of the total decline in photopic contrast sensitivity at medium and high spatial frequencies, with the remainder of the loss in sensitivity due to preretinal factors.[101] However, this point needs to be more thoroughly investigated.

Measurements of the CSF under scotopic conditions show a pattern of loss quite different from those made under photopic conditions, as illustrated by the squares in Figure 9. These data were obtained following 30 min dark adaptation, using an artificial pupil and stimulus wavelength to minimize age-related variation in retinal illuminance.[98] The largest sensitivity loss to horizontally oriented sinusoidal gratings is at low spatial frequencies, a result that implies age-related change in neural mechanisms.

In Figure 9, extrapolation of the CSFs to zero at high spatial frequencies provides a measure of visual resolution. At both photopic and scotopic light levels, age-related changes in the high frequency cut-off are noted even though participants were included only if they had retinae without pathology when examined by ophthalmoscopy.

Temporal Vision

The critical fusion frequency (CFF), the lowest frequency of flicker that is indiscriminable from a steady light, decreases with age.[102] Part of this loss in temporal resolution is due to reti-

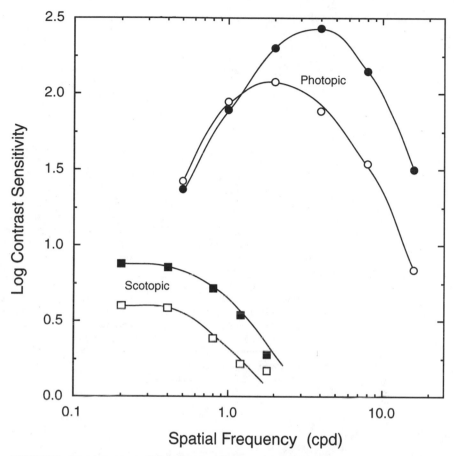

FIGURE 9 Log contrast sensitivity is plotted as a function of spatial frequency for static sinusoidal gratings. Closed and open circles represent average observers age 20 and 75 years, respectively, determined from regression equations for stimuli centered on the fovea. (*Data from Owsley et al.*[91]) Closed and open squares represent average observers age 20 and 75 years, respectively, determined from regression equations for stimuli centered at 8° nasal. (*Data from Schefrin et al.*[98])

nal illuminance differences between younger and older observers. Lower luminance levels are associated with a loss in temporal resolution. However, it seems likely that neural factors also contribute to a loss in sensitivity to flicker.[103]

Zhang and Sturr[104] measured temporal summation functions in younger and older observers; contrast thresholds for a 0.5 cpd grating were measured at six stimulus durations and four illuminance levels. Contrast thresholds were higher in older observers, but the shapes of the contrast thresholds versus stimulus duration functions were similar. Thus, under these conditions, the critical duration over which temporal information is summed is similar for younger and older observers.

While the CFF measures the high temporal resolution limit, a more complete description of temporal processing is provided by the temporal contrast sensitivity function, the amount of contrast needed for detection as a function of temporal frequency. Tyler[105] reported small age-related declines in temporal sensitivity for long-wavelength sinusoidal modulation, with the magnitude of the loss increasing with temporal frequency. The decrease in sensitivity began around age 20 years. Similar results were reported by Kim and Mayer,[106] but sensitivity losses were not observed until about the mid-forties, after which there was a loss in temporal sensitivity of about 0.78 decilog per decade. It was argued in both studies that these losses in sensitivity are greater than those expected from age-related changes in optical factors alone. Based on simultaneous recordings of pattern electroretinograms and visually evoked potentials in young and elderly observers, Porciatti et al.[107] have suggested that at least some of the loss in temporal sensitivity with age is due to changes at postretinal sites.

The Visual Field

The extent of the visual field narrows over adulthood by several degrees each decade until about age 50 to 60, and then somewhat more rapidly with increasing age, according to Wolf.[108] Other research has suggested that the volume of the visual field decreases starting at about 30 to 40 years of age and continues for the remainder of the life span.[109]

Senescent constriction of, and reduced light sensitivity within, the visual field are due in part to age-related reductions in pupil diameter and increases in light absorbance by the ocular media. Johnson et al.[110] performed static perimetry using experimental conditions that notably lessen the effects of lenticular senescence and reduced pupil size. They demonstrated age-related losses in sensitivity that varied from approximately 0.5 to 1 dB per decade within the central 30° visual field. Their results suggest that at least part of the visual field loss must be ascribed to neural factors.

Another approach to perimetry is to define the functional or "useful field of view," or the area of the visual field over which information can be processed within a single fixation. The size of the useful field of view decreases when observers are engaged in additional tasks.[111] Ball et al.[112] compared sensitivity of groups of younger and older observers within the central 30°. Test stimuli were presented in the presence and absence of distracting stimuli to measure the useful field of view. The older group was more impaired by the distracting stimuli, particularly at greater retinal eccentricities. This restriction of the functional visual field is apparently not due to optical or retinal factors.

Depth and Stereovision

Our ability to perceive the distances of objects depends on several different kinds of information. One of the most important depth cues is stereopsis. Stereoscopic depth information is provided by retinal disparity that occurs when a single object is imaged at different retinal locations between the two eyes. There is a paucity of investigations that have examined age-related changes in stereovision. This is unfortunate because the use of retinal disparity to extract three-dimensional depth could be used to probe cortical changes that may occur with aging.[113,114] In addition, based on the increased number of falls and other accidents that occur among the elderly, measures of stereoscopic depth perception across adulthood might be valuable.

Jani[115] used a standardized task of stereovision (Flashlight Diastereo Test) to determine the percentage of 1207 individuals who failed the test at various ages. This percentage decreased from his 0- to 9-year-old group and remained stable from 10 to 19 through about 40 to 49 years. With further increases in age, there was a monotonic increase in the percentage of individuals failing the test. A more sensitive indicator of stereovision requires measurement of thresholds. Bell et al.[116] using a Verhoeff stereopter (based on the farthest distance at which the relative distance between three rods can be detected) showed that stereo resolution decreases slowly from 30 to 40 years of age and then more rapidly after about age 40. It is not clear whether these changes in stereovision are secondary to changes in the retinal illuminance of the stimuli and/or the spatial CSF. The precise course and mechanisms for senescent changes in stereovision remain moot.

Some Visual Functions That Change Minimally with Age

Senescent changes in visual function are widespread, as might be expected given the optical and early neural changes that take place. It is perhaps surprising, therefore, that a number of visual functions change relatively little with age. The stability of color perception across the life span has already been described, but Enoch et al.[88] have summarized a number of other visual functions that also appear resistant to aging. For example, vernier acuity tasks have been used with subjects between 10 and 94 years of age. They were asked to align two or three lines or several points. This can be done with great precision (hence the name *hyperacuity*) and it does not change with age. The directional sensitivity of the retina (measured by the Stiles-Crawford effect of the first kind) also does not appear much affected by senescence. This task depends on the integrity of both physical and physiological mechanisms (see Chap. 9).

These and perhaps other (undiscovered) constancies in visual function across adulthood may provide insight into the types of neural changes that accompany age. In addition, such results might also serve as useful indices to distinguish between normal aging and changes in the same functions due to disease.

13.6 AGE-RELATED OCULAR DISEASES AFFECTING VISUAL FUNCTION

There are several ocular and retinal disorders that are so strongly correlated with increasing age that *age* has become a part of their name, e.g., age-related cataracts, age-related macular degeneration, etc. Age, of course, represents only the passage of time, not a process, but the name does accurately imply that these problems are more likely to occur in the elderly.

Cataract

Cataract is an opacity of the lens that interferes with vision by reducing the amount of light transmitted while at the same time increasing its scatter within the eye. There are a number of different forms of cataract, and each form exhibits its own characteristic effects on vision and has different stages of development. For example, cataracts can form within the lens nucleus or cortex, and they may originate in either the axial or equatorial regions of the lens. Cataracts, depending on their type, can cause a host of visual problems, including loss of visual acuity, changes in refractive error, aniseikonia, perceived spatial distortions, disability glare, and the formation of multiple images. Oftentimes, more than one form is present in a given eye, and the lens of each eye may develop cataract at different rates.

Cataract is found in about one in seven individuals above the age of 40, and the proportion increases with age. The rate is much higher, and typically occurs at an earlier age, in individuals with diabetes. It is known that UV and infrared radiation can cause cataracts. Most individuals exhibit changes in their crystalline lenses with advancing age and senescent cataract appears to be an extension of this process.[32] In developing countries, cataracts occur much earlier in life (e.g., from 14 to 18 years earlier in India than in the United States), perhaps due to nutritional factors or greater UV exposure as many developing countries lie at lower latitudes.

Cataract is the leading worldwide cause of blindness. Fortunately, it is treated with a very high success rate by surgically removing the cataractous lens. In most cases in developed countries, an IOL is implanted. Cataract surgery in the West is one of the most successful therapies known, with visual improvement in nearly all cases. Approximately 2.5 million cataract surgeries were performed in the United States in 1998.

Postcataract surgery, the patient describes a somewhat "bluer world." That is, the yellowing, aging lens is replaced with the clearer IOL or other visual correction.

Age-Related Macular Degeneration

Age-related macular degeneration (AMD) is generally a progressive, bilateral disease that compromises the integrity and functioning of the central retina. More than 13 million persons over the age of 40 in the United States have signs of macular degeneration. An individual in the early stages of AMD may experience declines in visual acuity, losses in sensitivity for central photopic and scotopic mechanisms,[117,118] an inability to make fine distinctions between colors,[119] and metamorphopsia (a distortion in the appearance of objects in the visual scene). As the disease progresses, individuals afflicted with this condition undergo further declines in visual performance, including the onset of central scotomas (defined areas of absolute or relative vision loss within the visual field) making it difficult to recognize readily faces of individuals encountered. However, because the disease does not spread past the macula or central retina, individuals can still navigate about in the environment and even regain some function through training to use unaffected adjacent portions of the macular region. Unfortunately, about 1.2 million persons in the United States have reached sight-threatening late stages of AMD. This latter group is composed primarily of Caucasians aged 60 years and above, as the disease affects African-Americans with lower frequency. The numbers of individuals having AMD are increasing rapidly with the aging of the population, affecting more than 30 percent of individuals over the age of 75.

AMD is divided clinically into the wet and dry forms of the disease.[120] In the more devastating wet form of this disease, choroidal neovascular membranes grow through cracks in Bruch's membrane, the layer separating the choroid from the retinal pigment epithelium (RPE). Among other things, the RPE is responsible for regulating and providing nourishment to, and removing waste products from, the retina. Death of retinal cells occurs when the retina is separated for lengthy time periods from the underlying RPE, as can happen when the leaky blood vessels comprising the neovascular membranes begin to exude their contents into the potential space between the RPE and the receptor layer of the retina. In addition, the presence of these subretinal neovascular membranes leads to scar formation beneath the macula, which is the primary cause of vision loss in the wet form of AMD. The wet form comprises only 10 percent of all cases of AMD, but accounts for 90 percent of legal blindness from AMD.

In contrast to the wet form of AMD that may produce a sudden loss of vision, dry (or atrophic) AMD results in a slow progressive vision loss that usually takes place over many years. The dry form of AMD is characterized by the formation of drusen, deposits of accumulated waste products from the retina and atrophy of the RPE cells. Presently, it is unclear what causes retinal atrophy in the dry form of AMD, but it is thought to reflect a metabolic disturbance of the receptor/RPE complex.

Currently, clinical advances have been made to retard the progress of the wet form of AMD. Fluorescein and indocyanine green angiography have been effective tools for both diagnosing wet AMD in its early stages and defining the subretinal sites of neovascular membranes and exudative leakage from these blood vessels. By taking advantage of this valuable knowledge, laser photocoagulation is now used to destroy these vascular membranes, thereby retarding the progress of the disease. Beyond these forms of treatment, visual rehabilitation has concentrated on the use of low-vision aids and training AMD patients to use parafoveal and peripheral portions of their retina that are less affected by the disease.

Diabetic Retinopathy

Diabetic retinopathy is also a common disorder associated with aging. Juvenile-onset (Type 1) diabetes is generally diagnosed during the first two decades of life, whereas adult-onset (Type 2) diabetes is diagnosed later in life. Type 2 diabetes is associated with increasing age, obesity, lack of exercise, and a family history of the disease; Mexican-Americans and African-Americans are affected relatively more often than Caucasian-Americans. A patient with diabetes (either Type 1 or Type 2) develops diabetic retinal changes over time; at least 40 percent of diabetic patients will eventually develop detectable retinopathy. The incidence of diabetic retinopathy is related to both blood sugar control and the duration of the disease, with most estimates suggesting clinical evidence of retinopathy occurs about 10 years after the onset of hyperglycemia. Approximately 8000 people are blinded each year by this disease.[8]

There are several stages to diabetic retinopathy. Early in the disease there is patchy loss of the endothelial cells lining the retinal capillaries, which leads to increased permeability of these blood vessels and the formation of microaneurysms. The appearance of small punctate hemorrhages, microaneurysms, and proteinaceous deposits within the retina are hallmarks of the early or *background* stage of diabetic retinopathy. As the disease progresses, blood vessels may become occluded resulting in retinal hypoxia. In response to disrupted blood flow, the hypoxic retina releases growth factors that induce the proliferation of new blood vessels. These blood vessels formed in the *proliferative* stage of diabetic retinopathy are both relatively more fragile and more permeable than normal vessels, thus continuing the cascade of increased presence of blood and protein exudate within the retina. In the latter stages of the disease there is death of ganglion cells from intraretinal hypoxia, and as the hemorrhages start to resolve, scar tissue forms leading to retinal traction and subsequent retinal detachment in areas adjacent to relatively high concentrations of neovascularization. Tractional retinal detachments may cause visual distortion and blindness.

Initial treatment of diabetic retinopathy involves controlling blood sugar levels via dietary changes, oral agents, and/or insulin to facilitate the uptake of glucose into body tissue. Poorly controlled blood sugar levels can lead to fluctuations in vision due to changes in the refractive power of the crystalline lens, and the early formation of cataracts. The Diabetes Control and Complications Trial[121] demonstrated that tight control of blood glucose reduced the incidence of diabetic retinopathy. Additionally, fluctuations in blood glucose levels have been shown to influence normal functioning of S-cone pathways[122,123] perhaps accounting for decreased chromatic discrimination along a tritan axis demonstrated by diabetic individuals.[124]

In the later stages of diabetic retinopathy, laser photocoagulation is used in an attempt to maintain the integrity of the central retina. Laser photocoagulation is utilized to control the complications of neovascularization and reduced vision secondary to macular edema. These treatments have been shown to decrease severe vision loss by approximately 50 percent. Because laser photocoagulation is a destructive form of treatment, newer strategies are being investigated that target specific growth factors that cause neovascularization.

Glaucomas

The glaucomas are a group of chronic diseases that result in the progressive degeneration of retinal ganglion cells whose axons comprise the optic nerve. In most but not all cases of glaucoma, elevated intraocular pressure (IOP) is believed to be one of the primary risk factors for ganglion cell death. Ganglion cell death occurs through the process of apoptosis, brought on, it is believed, by the loss of trophic factors normally transported to the ganglion cell body by retrograde axoplasmic flow. In glaucoma, cell death is thought to occur following a disruption in axoplasmic flow of the retinal ganglion cell axons comprising the optic nerve; the site of this disruption is believed to be at the optic nerve head, where the insult may be mechanical due to elevated IOP and/or a disruption in the vascular supply to the nerve head.[125]

Aqueous humor serves as the optically clear "blood supply" to the metabolically active but avascular structures of the lens and cornea. IOP is determined by a balance between production of aqueous humor (by the ciliary body, a secretory tissue that lies directly behind the iris) and drainage through the trabecular meshwork, a circumferential drainage system located just anterior to the iris at an angle formed by the cornea and the iris root. Abnormalities in the trabecular meshwork are thought to underlie the elevated IOP in many forms of open-angle glaucoma.

Of the approximately 67 million persons affected by glaucoma worldwide, about equal numbers suffer from open-angle and angle-closure types, with the former being more prevalent in the United States and Europe and the latter being more prevalent in Asia and the Pacific Rim. Glaucoma accounts for about 10 percent of blindness in the United States. The frequencies of glaucomatous disorders increase with increasing age, particularly after the age of 40. By age 65 they affect about 10 percent of African-Americans and about 3 percent of Caucasian-Americans.

Angle-closure glaucoma typically occurs when a forward protruding iris covers the trabecular meshwork, thus impeding the flow of aqueous humor from the eye. In this form of glaucoma, there can be sudden dramatic increases in IOP resulting in an abnormally hydrated cornea that leads to diminished visual acuity and accompanying ocular pain. In contrast, open-angle glaucoma is a slow, progressive condition in which the IOP is too high for a particular eye, resulting in ganglion cell death. Visual loss due to glaucoma occurs generally in the peripheral visual field and affects central vision only late in the disease process. Because the loss is gradual and usually occurs in areas overlapping 'normal' vision in the fellow eye, patients are generally unaware of their visual loss until a great deal of damage has occurred. In addition to IOP, other risk factors associated with glaucoma include age, ethnicity, and family history of glaucoma. Because increased IOP is not invariably associated with glaucoma, a complete eye examination is needed to make an accurate diagnosis. Oftentimes, these examinations will reveal changes to the retina such as an abnormal appearance of the optic nerve head and, on visual field testing, enlarged blind spots and arcuate scotomas.

Because damage to optic nerve fibers is irreversible and the progress of glaucoma can, in most instances, be successfully controlled, early detection is of great importance. Declines in visual performance at earlier stages in the disease process can be detected through novel forms of visual field testing. For example, psychophysical measurements of the sensitivity of an S-cone pathway can detect glaucomatous changes in the visual field[126] that can occur 5 to 7 years before conventional visual field testing.[127] Studies by Quigley[128] suggest that ganglion cells having large diameter axons that provide inputs to the magnocellular layers of the lateral geniculate body are affected more by this disease than ganglion cells having smaller diameter axons that provide input to the parvocellular layers of the lateral geniculate body (cf. Ref. 129). The efficacy of visual field testing using stimulus conditions that are thought to tap into the magnocellular pathways is under active investigation.[130]

Current treatment regimens for glaucoma involve the use of pharmacological agents and surgical procedures aimed at controlling IOP. Drug therapy includes the use of topical and systemic agents, either singly or in combination, to inhibit the formation of aqueous and promote increased drainage of aqueous from the eye. Surgical procedures involving lasers or incisions typically aid in aqueous outflow and are used when drug therapy, by itself, fails to adequately retard the progress of the disease. In addition, ongoing research is currently investigating possible therapies involving neuroprotective agents designed to reduce ganglion cell death after mechanical or chemical damage and to promote new growth.[131]

13.7 THE AGING WORLD FROM THE OPTICAL POINT OF VIEW

The most common problems of aging requiring optical correction are presbyopia and aphakia (removal of the lens in cataract surgery). (Also, see Chap. 11.)

Presbyopic Corrections

Spectacles. Presbyopia refers to a loss of accommodation resulting in the near point, the closest distance on which the eye can focus, receding from the eye. Presbyopes who are not emmetropic typically require different optical corrections for optimal vision at near (from approximately 0.3 to 0.5 m) and far (beyond approximately 6 m). Since the invention of the bifocal lens by Benjamin Franklin, several ingenious approaches to this problem have been developed. With traditional bifocal lenses, the individual looks through different portions of a lens, each portion having a different dioptric power, in order to clearly see objects at near or far. Extensions of this concept include trifocal lenses having essentially three dioptric powers confined to three distinct zones, and aspheric lens designs in which the dioptric power continually changes in a prescribed manner through the inferior portion of the lens. Peripheral zones of the latter designs often exhibit aberrations, but are nevertheless chosen frequently for cosmetic reasons.

Contact Lenses.   The principles of bifocal and multifocal corrections have been incorporated into contact lenses and intraocular lenses (see next section). Like their spectacle counterparts, bifocal and aspheric contact lenses have been designed to provide clearer vision at varying distances. Some designs of bifocal contact lenses consist of concentric annuli that alternate in power between near and far optical corrections. Thus, some of the light forms a focused image, but at the same time a fraction of the light is always defocused. In these annular designs, the proportion of light in the focused image varies with lens design and pupil size.

Another technique used to optically correct the presbyopic individual is monovision. Instead of offering two different corrections in the same eye, a monovision correction consists of different unifocal optical corrections in each eye, one for near and the other for far. Monovision and multifocal contact lenses of annular design share the same feature that the entire visual scene cannot be simultaneously in focus. Contact lens wearers who can tolerate or suppress the portion of the image that is blurred while attending to the focused portion of the image are typically successful in either of these two fitting modalities. It should be noted that the retinal blur produced by monovision and simultaneous bifocal contact lens corrections leads to losses in contrast sensitivity, especially at high spatial frequencies, resulting in decreases in visual acuity, and hyperacuity.[132] Stereopsis can also be affected adversely by blurred imagery along with aniseikonic and anisometropic viewing conditions that accompany monovision corrections.[132,133]

Intraocular Lenses.   Following extraction of a cataractous lens, a refractive error of about 17 D remains to be corrected. A person without a natural lens is called an aphake or aphakic. In most cases, an IOL is inserted and the wearer is said to be pseudophakic. IOLs have traditionally been constructed from polymethylmethacrylate, but a variety of other materials have been developed (e.g., silicone, hydrogels) so that the lens can be folded and inserted through smaller incisions than are required to insert a hard plastic lens. Many IOLs in the past 15 years have incorporated UV-absorbing chromophores, but older IOLs were virtually transparent to UV radiation from 320 to 400 nm.

Bifocal and multifocal corrections have been incorporated into IOLs. While the optical principles are mostly the same as presbyopic contact lens corrections, some different problems may emerge because of the placement of the optic and maintenance of its position. For example, IOL tilt is associated with astigmatism, while decentration of a bifocal IOL having separate refractive zones could result in one zone being vignetted by the eye pupil.

Diffractive optics have been developed to avoid problems of decentration in bifocal IOLs by having near and far refractive corrections employed over most of the lens surface. As illustrated by Figure 10, this is made possible by a series of annular zones (usually about 30) separated by small discontinuities specially designed to produce constructive interference patterns on the retina for light from optical infinity and a prescribed near point. Still, since about half the light goes for near correction and half for far correction,[134] the focused image of a near or distant object is superimposed on a defocused image of the same object. Reduced image contrast in any bifocal or multifocal optic can be expected due to the superposition of this second out-of-focus image. The consequences of these problems may be exacerbated in older patients due to their lower contrast sensitivity and smaller pupil sizes. Problems with glare and halos around light sources are greater in patients with bilateral multifocal compared to bilateral monofocal IOLs, but the former patients sometimes achieve less dependence on additional spectacle corrections.[135]

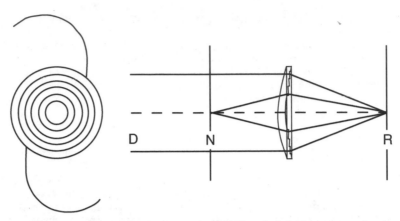

FIGURE 10 Schematic of an intraocular lens (IOL) on the left. The prongs, or haptic, attached to the central optic are used to maintain the position of the lens when inserted in the eye. The concentric rings represent ridges of a bifocal diffractive optic. Ray traces (right) illustrate the principle that results in distant (D) and near (N) images focused in the plane of the retina (R).

Finally, a new class of IOL, a phakic IOL (sometimes called an implantable contact lens) is being developed. This is a correction for extreme myopia or hyperopia provided to individuals without removal of their natural lens.[136] Extreme myopia may sometimes be corrected by removal of the natural lens, but with the loss of accommodation in young pre-presbyopic individuals. Phakic IOLs have been designed for implantation in either the anterior or posterior chamber in order to preserve accommodation. It is not clear how the long-term efficacy of this procedure will compare with other invasive procedures such as reshaping the cornea by refractive surgery.

Refractive Surgeries

Refractive surgeries are performed on individuals wishing to reduce or eliminate their need for external optical corrections. These types of surgeries modify the shape of the cornea, resulting in a change in its refractive power. While refractive surgery has recently become popular, the Japanese ophthalmologist Sato tried at mid-century to alter the curvature of the cornea using a series of radial cuts on the inside of the cornea to attempt to allow the periphery to bulge and to flatten the center. This approach was unsuccessful because it produced irreparable damage to the endothelial cell layer of the cornea. As discussed earlier, corneal hydration is regulated primarily by endothelial cells.

Radial keratotomy (RK) was developed by Fyodorov in the former Soviet Union and brought to the United States in the late 1970s. RK is the inverse of Sato's earlier technique in that incisions are made in the epithelial and stromal layers of the cornea. The number, orientation, and depth of these incisions depend on the patient's refractive error. Radial cuts are made in the cornea that spare a central *optical zone* through which the patient sees. The treatment for astigmatism includes making either linear or arcuate incisions in the mid-periphery of the cornea. Numerous optical problems with this procedure have been reported due to technical difficulties with the number and centration of the incisions, and corneal scarring. These problems include glare, fluctuations in refractive error throughout the day and with changes in ambient light level, and often a long-term shift toward hypermetropia.[137] Because of these problems and other issues, RK is no longer the standard surgical procedure used to correct refractive errors.

Currently, surgical techniques that reshape the cornea through the use of a high-energy laser (approximately 200 nm) are preferred by most refractive surgeons. Photorefractive keratectomy (PRK) and laser in situ keratomileusis (LASIK) procedures focus energy onto the corneal stroma for the purpose of vaporizing tissue at that level. In PRK, either the central corneal epithelium is mechanically scraped away prior to applying laser energy to the eye or energy is allowed to pass through the intact epithelium on its way to the stromal bed. Either of these methods leads to a disruption of the corneal surface, resulting in varying degrees of postoperative pain and fluctuating vision while a new layer of epithelium is laid down on the corneal surface. For these reasons, LASIK is currently the preferred surgical procedure, especially when correcting for relatively high amounts of refractive error. In this procedure, a microkeratome is passed over almost the entire extent of the cornea, creating an approximately 180-μm-thick flap of tissue. Following the application of laser energy to the stroma, the flap is carefully laid back down over the cornea. Initially the epithelial flap is held in place by an adhesion between stromal tissue coating the inner layer of the flap and the stromal bed beneath the flap. Also, the flap is pulled toward the cornea by the action of transport mechanisms within the corneal endothelium that pump fluid back into the aqueous humor, thereby creating an intracorneal vacuum. Later, the adherence of the flap to the cornea is increased when the edges of the flap are sealed off via epithelial cell migration and the slight formation of scar tissue.

Recent studies demonstrate advantages of LASIK over PRK. In a randomized study of myopic patients (from −2.5 to −8.0 D), surgery was performed in one eye with LASIK and the other with PRK.[138] Comparable refraction was obtained with the two procedures at two years, but uncorrected visual acuity was better with the LASIK procedure. The authors suggest that this was due to smoother and clearer central cornea with the LASIK procedure.

13.8 CONCLUSIONS

The proportion of the population above age 65 years is increasing and there is a parallel rise in the proportion of the population that is blind or in need of refractive corrections. Senescent deterioration in the eye's optics can lead to a wide range of changes in visual function due to reductions in retinal illuminance and degradations in image formation. Age-related changes in vision also result from neural losses. A few visual functions are spared, but substantial senescent changes in the spatial, spectral, and temporal analyses of the retinal image have been well documented. The demographic shift in the proportion of the population that is elderly has not yet reached an equilibrium, so the social and economic consequences can be expected to be even more profound as we enter the new millennium.

13.9 REFERENCES

1. U. S. Senate Special Committee on Aging, *Aging America,* U.S. Department of Health and Human Services, Washington, DC, 1985–86.

2. United Nations, Population Division, *World Population Projections to 2150,* New York, 1998.

3. R. A. Weale, *A Biography of the Eye,* Lewis, London, 1982.

4. H. M. Leibowitz, D. E. Krueger, L. R. Maunder, R. C. Milton, M. M. Kini, H. A. Kahn, R. J. Nickerson, J. Pool, T. L. Colton, J. P. Ganley, J. I. Loewenstein, and T. R. Dawber, "The Framingham Eye Study Monograph," *Surv. Ophthalmol.* (Supplement) **24**:335–610 (1980).

5. B. R. Hammond, Jr., E. J. Johnson, R. M. Russell, N. I. Krinsky, K-J. Yeum, R. B. Edwards, and D. M. Snodderly, "Dietary Modification of Human Macular Pigment Density," *Invest. Ophthalmol. Visual Sci.* **38**:1795–1801 (1997).

6. J. S. Werner, V. G. Steele, and D. S. Pfoff, "Loss of Human Photoreceptor Sensitivity Associated with Chronic Exposure to Ultraviolet Radiation," *Ophthalmology* **96**:1552–1558 (1989).

7. G. Haegerstrom-Portnoy, "Short-Wavelength-Sensitive-Cone Sensitivity Loss with Aging: A Protective Role for Macular Pigment?," *J. Opt. Soc. Am.* **A5**:2140–2144 (1988).

8. National Society to Prevent Blindness, *Prevent Blindness America,* Schaumburg, IL, 1994.

9. Health Products Research, Inc., *Vision Information Services (VIS)*^{SM}—*The IOL Report,* Health Products Research, Inc., Whitehouse, NJ, 1999.

10. S. Brink, "Greying of Our Communities Worldwide," *Ageing Internat.* **23**:13–31 (1997).

11. W. Adrian (ed), *Lighting for Aging and Visual Health,* Lighting Research Institute, New York, 1995.

12. H. Bouma and K. Sagawa, "Ageing Effects on Vision," in *24th Session of the CIE—Warsaw,* Commission Internationale de l'Eclairage, Publication No. 133, Warsaw, Poland, 1999, pp. 368–370.

13. D. Sliney and M. Wolbarsht, *Safety with Lasers and Other Optical Sources,* Plenum Press, New York, 1980.

14. E. A. Boettner and J. R. Wolter, "Transmission of the Ocular Media," *Invest. Ophthalmol.* **1**:776–783 (1962).

15. H. Brandhorst, J. Hickey, H. Curtis, and E. Ralph, "Interim Solar Cell Testing Procedures for Terrestrial Applications," *NASA Report TM X-7177,* Lewis Research Center, Cleveland, OH, 1975.

16. J. S. Werner, "The Damaging Effects of Light on the Eye and Implications for Understanding Changes in Vision Across the Life Span," in P. Bagnoli and W. Hodos (eds.), *The Changing Visual System: Maturation and Aging in the Central Nervous System,* Plenum, New York, 1991, pp. 295–309.

17. R. W. Young, "The Bowman Lecture, 1982: Biological Renewal. Applications to the Eye," *Trans. Ophthal. Soc. U.K.* **102**:42–75 (1982).

18. M. Millodot and I. A. Newton, "A Possible Change of Refractive Index with Age and Its Relevance to Chromatic Aberration," *Graefe's Arch. Clin. Exp. Ophthalmol.* **201**:159–167 (1976).

19. M. W. Morgan, "Changes in Visual Function in the Aging Eye," in A. A. Rosenbloom, Jr., and M. W. Morgan (eds.), *Vision and Aging: General and Clinical Perspectives,* Fairchild Publications, New York, 1986, pp. 121–134.

20. S. Lerman, "Biophysical Aspects of Corneal and Lenticular Transparency," *Current Eye Res.* **3**:3–14 (1984).

21. F. W. Campbell and R. W. Gubisch, "Optical Quality of the Human Eye," *J. Physiol.* (*London*) **186**:558–578 (1966).

22. A. L. Kornzweig, "Physiological Effects of Age on the Visual Process," *Sight-Saving Rev.* **24**:130–138 (1954).

23. V. Kadlecová, M. Peleška and A. Vaško, "Dependence on Age of the Diameter of the Pupil in the Dark," *Nature* (*London*) **18**:1520–1521 (1958).

24. I. E. Loewenfeld, "Pupillary Changes Related to Age," in H. S. Thompson, R. Daroff, L. Frisén, J. S. Glasen, and M. D. Sanders (eds.), *Topics in Neuroophthalmology,* Williams and Wilkins, Baltimore, MD, 1979.

25. M. E. Sloane, C. Owsley, and S. L. Alvarez, "Aging, Senile Miosis and Spatial Contrast Sensitivity at Low Luminance," *Vision Res.* **28**:1235–1246 (1988).

26. A. Guirao, C. González, M. Redondo, E. Gevaghty, S. Norrby, and P. Artal, "Average Optical Performance of the Human Eye As a Function of Age in a Normal Population" *Invest. Ophthalmol. Visual Sci.* **40**:203–213 (1999).

27. R. A. Weale, *The Senescence of Human Vision,* Oxford University Press, Oxford, 1992.

28. J. S. Werner, "Development of Scotopic Sensitivity and the Absorption Spectrum of the Human Ocular Media," *J. Opt. Soc. Am.* **72**:247–258 (1982).

29. R. A. Weale, "Age and the Transmittance of the Human Crystalline Lens," *J. Physiol.* (*London*) **395**:577–587 (1988).

30. J. Pokorny, V. C. Smith, and M. Lutze, "Aging of the Human Lens," *Appl. Opt.* **26**:1437–1440 (1987).

31. P. A. Sample, F. D. Esterson, R. N. Weinreb, and R. M. Boynton, "The Aging Lens: In Vivo Assessment of Light Absorption in 84 Eyes," *Invest. Ophthalmol. Visual Sci.* **29**:1306–1311 (1988).

32. R. W. Young, *Age-Related Cataract,* Oxford University Press, New York, 1991.

33. D. G. Pitts, L. L. Cameron, J. G. Jose, S. Lerman, E. Moss, S. D. Varma, S. Zigler, S. Zigman, and J. Zuclich, "Optical Radiation and Cataracts," in M. Waxler and V. M. Hitchins (eds.), *Optical Radiation and Visual Health,* CRC Press, Boca Raton, Florida, 1986, pp. 5–41.

34. R. A. Weale, "Senile Cataract. The Case Against Light," *Ophthalmology* **90**:420–423 (1983).

35. R. F. Fisher, "The Elastic Constants of the Human Lens," *J. Physiol. (London)* **212**:147–180 (1971).

36. R. Brückner, E. Batschelet, and F. Hugenschmidt, "The Basel Longitudinal Study on Aging (1955–1978)," *Doc. Ophthalmol.* **64**:235–310 (1987).

37. F. C. Donders, *On the Anomalies of Accommodation and Refraction of the Eye,* London, 1864.

38. D. M. Snodderly, J. D. Auran, and F. C. Delori, "The Macular Pigment. II. Spatial Distribution in Primate Retinas," *Invest. Ophthalmol. Visual Sci.* **25**:674–685 (1984).

39. J. J. Vos, *Tabulated Characteristics of a Proposed 2° Fundamental Observer,* Institute for Perception, Soesterberg, The Netherlands, 1978.

40. D. Moreland and P. Bhatt, "Retinal Distribution of Macular Pigment," in G. Verriest (ed.), *Colour Vision Deficiencies VII,* Dr. W. Junk, The Hague, 1984, pp. 127–132.

41. J. S. Werner, B. E. Schefrin, and M. L. Bieber, "Senescence of Foveal and Parafoveal Cone Sensitivities and Their Relations to Macular Pigment Density," *J. Opt. Soc. Am. A.* (in press)

42. J. S. Werner, S. K. Donnelly, and R. Kliegl, "Aging and Human Macular Pigment Density; Appended with Translations from the Work of Max Schultze and Ewald Hering," *Vision Res.* **27**:257–268 (1987).

43. R. A. Bone, J. T. Landrum, L. Fernandez, and S. L. Tarsis, "Analysis of the Macular Pigment by HPLC: Retinal Distribution and Age Study," *Invest. Ophthalmol. Visual Sci.* **29**:843–849 (1988).

44. V. M. Reading and R. A. Weale, "Macular Pigment and Chromatic Aberration," *J. Opt. Soc. Am.* **64**:231–234 (1974).

45. K. Kirschfield, "Carotenoid Pigments: Their Possible Role in Protecting Against Photooxidation in Eyes and Photoreceptor Cells," *Proc. R. Soc. London* **B216**:71–85 (1982).

46. B. R. Hammond, Jr., B. R. Wooten, and D. M. Snodderly, "Preservation of Visual Sensitivity of Older Subjects: Association with Macular Pigment Density," *Invest. Ophthalmol. Visual Sci.* **39**:397–406 (1998).

47. G. Westheimer and J. Liang, "Influence of Ocular Light Scatter on the Eye's Optical Performance," *J. Opt. Soc. Am.* **A12**:1417–1424 (1995).

48. E. Wolf, "Glare and Age," *Arch. Ophthalmol.* **64**:502–514 (1960).

49. J. K. Ijspeert, P. W. T. de Waard, T. J. T. P. van den Berg, and P. T. V. M. de Jong, "The Intraocular Straylight Function in 129 Healthy Volunteers; Dependence on Angle, Age and Pigmentation," *Vision Res.* **36**:699–707 (1990).

50. M. L. Hennelly, J. L. Barbur, D. F. Edgar, and E. G. Woodward, "The Effect of Age on the Light Scattering Characteristics of the Eye," *Ophthal. Physiol. Opt.* **18**:197–203 (1998).

51. B. R. Wooten and G. A. Geri, "Psychophysical Determination of Intraocular Light Scatter As a Function of Wavelength," *Vision Res.* **27**:1291–1298 (1987).

52. T. J. T. P. van den Berg, J. K. Ijspeert, and P. W. T. de Waard, "Dependence of Intraocular Straylight on Pigmentation and Light Transmission Through the Ocular Wall," *Vision Res.* **31**:1361–1367 (1991).

53. S. Lerman and R. F. Borkman, "Spectroscopic Evaluation and Classification of the Normal, Aging, and Cataractous Lens," *Ophthal. Res.* **8**:335–353 (1976).

54. D. B. Elliott, K. C. H. Yang, K. Dumbleton, and A. P. Cullen, "Ultraviolet-Induced Lenticular Fluorescence: Intraocular Straylight Affecting Visual Function," *Vision Res.* **33**:1827–1833 (1993).

55. J. L. Barbur, A. J. Harlow, and C. Williams, "Light Scattered in the Eye and Its Effect on the Measurement of the Colour Constancy Index," in C. R. Cavonius (ed.), *Colour Vision Deficiencies XIII,* Kluwer Academic Publishers, Dordrecht, The Netherlands, 1997, pp. 439–438.

56. J. J. Vos, "Age Dependence of Glare Effects and Their Significance in Terms of Visual Ergonomics," in W. Adrian (ed.), *Lighting for Aging Vision and Health.* Lighting Research Institute, New York, 1995, pp. 11–25.

57. P. Artal, M. Ferro, I. Miranda, and R. Navarro, "Effects of Aging in Retinal Image Quality," *J. Opt. Soc. Am.* **A10**:1656–1662 (1993).

58. G. Wald and D. R. Griffin, "The Change in Refractive Power of the Human Eye in Dim and Bright Light," *J. Opt. Soc. Am.* **37**:321–336 (1947).

59. P. A. Howarth, X. X. Zhang, A. Bradley, D. L. Still, and L. N. Thibos, "Does the Chromatic Aberration of the Eye Vary with Age?," *J. Opt. Soc. Am.* **A5**:2087–2092 (1988).

60. R. G. Domey, R. A. McFarland, and E. Chadwick, "Dark Adaptation As a Function of Age and Time: II. A Derivation," *J. Gerontol.* **15**:267–279 (1960).

61. C. D. Coile and H. D. Baker, "Foveal Dark Adaptation, Photopigment Regeneration, and Aging," *Vis. Neurosci.* **8**:27–29 (1992).

62. G. R. Jackson, C. Owsley, and G. McGwin, "Aging and Dark Adaptation," *Vision Res.* **39**:3975–3982 (1999).

63. J. F. Sturr, L. Zhang, H. A. Taub, D. J. Hannon, and M. M. Jackowski, "Psychophysical Evidence for Losses in Rod Sensitivity in the Aging Visual System," *Vision Res.* **37**:475–481 (1997).

64. B. E. Schefrin, M. L. Bieber, R. McLean, and J. S. Werner, "The Area of Complete Scotopic Spatial Summation Enlarges with Age," *J. Opt. Soc. Am.* **A15**:340–348 (1998).

65. E. Pulos, "Changes in Rod Sensitivity Through Adulthood," *Invest. Ophthalmol. Visual Sci.* **30**:1738–1742 (1989).

66. J. E. E. Keunen, D. V. Norren, and G. J. V. Meel, "Density of Foveal Cone Pigments at Older Age," *Invest. Ophthalmol. Visual Sci.* **28**:985–991 (1987).

67. C. A. Curcio, C. L. Milican, K. A. Allen, and R. E. Kalina, "Aging of the Human Photoreceptor Mosaic: Evidence for Selective Vulnerability of Rods in Central Retina," *Invest. Ophthalmol. Visual Sci.* **34**:3278–3296 (1993).

68. J. J. Plantner, H. L. Barbour, and E. L. Kean, "The Rhodopsin Content of the Human Eye," *Current Eye Res.* **7**:1125–1129 (1988).

69. J. Marshall, J. Grindle, P. L. Ansell, and B. Borwein, "Convolution in Human Rods: An Ageing Process," *Br. J. Ophthalmol.* **63**:181–187 (1979).

70. A. Eisner, S. A. Fleming, M. L. Klein, and W. M. Mauldin, "Sensitivities in Older Eyes with Good Acuity: Cross-Sectional Norms," *Invest. Ophthalmol. Visual Sci.* **28**:1824–1831 (1987).

71. C. A. Johnson, A. J. Adams, J. D. Twelker, and J. M. Quigg, "Age-Related Changes of the Central Visual Field for Short-Wavelength-Sensitive Pathways," *J. Opt. Soc. Am.* **A5**:2131–2139 (1988).

72. J. S. Werner and V. G. Steele, "Sensitivity of Human Foveal Color Mechanisms Throughout the Life Span," *J. Opt. Soc. Am.* **A5**:2122–2130 (1988).

73. B. E. Schefrin, J. S. Werner, M. Plach, N. Utlaut, and E. Switkes, "Sites of Age-Related Sensitivity Loss in a Short-Wave Cone Pathway," *J. Opt. Soc. Am.* **A9**:355–363 (1992).

74. R. Lakowski, "Is the Deterioration of Colour Discrimination with Age Due to Lens or Retinal Changes?," *Die Farbe* **11**:69–86 (1962).

75. Y. Ohta and H. Kato, "Colour Perception Changes with Age," *Mod. Probl. Ophthalmol.* **17**:345–352 (1976).

76. G. Verriest, "Further Studies on Acquired Deficiency of Color Discrimination," *J. Opt. Soc. Am.* **53**:185–195 (1963).

77. V. C. Smith, J. Pokorny, and A. S. Pass, "Color-Axis Determination on the Farnsworth-Munsell 100-Hue Test," *Am. J. Ophthalmol.* **100**:176–182 (1985).

78. K. Knoblauch, F. Saunders, M. Kusuda, R. Hynes, M. Podor, K. E. Higgins, and F. M. de Monasterio, "Age and Illuminance Effects in the Farnsworth-Munsell 100-Hue Test," *Appl. Opt* **26**:1441–1448 (1987).

79. J. Birch, *Diagnosis of Defective Colour Vision,* Oxford University Press, New York, 1993.

80. B. E. Schefrin, K. Shinomori, and J. S. Werner, "Contributions of Neural Pathways to Age-Related Losses in Chromatic Discrimination," *J. Opt. Soc. Am.* **A12**:1233–1241 (1995).

81. K. Shinomori and J. S. Werner, "Individual Variation in Wavelength Discrimination: Task and Model Analysis," in *Proceedings of the 8th Congress of the International Colour Association,* Color Science Association of Japan, Tokyo, 1997, vol. I, pp. 195–198.

82. J. M. Kraft and J. S. Werner, "Aging and the Saturation of Colors: 1. Colorimetric Purity Discrimination," *J. Opt. Soc. Am.* **A16**:223–230 (1999).

83. J. M. Kraft and J. S. Werner, "Spectral Efficiency Across the Life Span: Flicker Photometry and Brightness Matching," *J. Opt. Soc. Am.* **A11**:1213–1221 (1994).

84. J. M. Kraft and J. S. Werner, "Aging and the Saturation of Colors: 2. Scaling of Color Appearance," *J. Opt. Soc. Am.* **A16**:231–235 (1999).

85. B. E. Schefrin and J. S. Werner, "Age-Related Changes in the Color Appearance of Broadband Surfaces," *Color Res. Appl.* **18**:380–389 (1993).

86. B. E. Schefrin and J. S. Werner, "Loci of Spectral Unique Hues Throughout the Life Span," *J. Opt. Soc. Am.* **A7**:305–311 (1990).

87. J. S. Werner and B. E. Schefrin, "Loci of Achromatic Points Throughout the Life Span," *J. Opt. Soc. Am.* **A10**:1509–1516 (1993).

88. J. M. Enoch, J. S. Werner, G. Haegerstrom-Portnoy, V. Lakshminarayanan, and M. Rynders, "Forever Young: Visual Functions Not Affected or Minimally Affected by Aging," *J. Gerontol: Bio. Sci.* **A54**:B336–B351 (1999).

89. J. S. Werner, "Visual Problems of the Retina During Ageing: Compensation Mechanisms and Colour Constancy Across the Life Span," *Prog. Retin. Eye Res.* **15**:621–645 (1996).

90. D. G. Pitts, "The Effects of Aging on Selected Visual Functions: Dark Adaptation, Visual Acuity, Stereopsis, and Brightness Contrast," in R. Sekuler, D. Kline, and K. Dismukes (eds.), *Aging and Human Visual Functions* A. R. Liss, New York, 1982, pp. 131–159.

91. C. Owsley, R. Sekuler, and D. Siemsen, "Contrast Sensitivity Throughout Adulthood," *Vision Res.* **23**:689–699 (1983).

92. L. Frisén and M. Frisén, "How Good is Normal Acuity? A Study of Letter Acuity Thresholds As a Function of Age," *Graefe's Arch. Clin. Exp. Ophthalmol* **215**:149–157 (1981).

93. D. B. Elliott, K. C. H. Yang, and D. Whitaker, "Visual Acuity Changes Throughout Adulthood in Normal, Healthy Eyes: Seeing Beyond 6/6," *Optom. Vis. Sci.* **72**:186–191 (1995).

94. G. Derefeldt, G. Lennerstrand, and B. Lundh, "Age Variations in Normal Human Contrast Sensitivity," *Acta Ophthalmol.* **57**:679–690 (1979).

95. K. E. Higgins, M. J. Jafe, R. C. Caruso, and F. M. deMonasterio, "Spatial Contrast Sensitivity: Effects of Age, Test Retest, and Psychophysical Method," *J. Opt. Soc. Am.* **A5**:2173–2180 (1988).

96. U. Tulunay-Keesey, J. N. Ver Hoeve, and C. Terkla-McGrane, "Threshold and Suprathreshold Spatiotemporal Response Throughout Adulthood," *J. Opt. Soc. Am.* **A5**:2191–2200 (1988).

97. D. Elliott, D. Whitaker, and D. MacVeigh, "Neural Contribution to Spatiotemporal Contrast Sensitivity Decline in Healthy Ageing Eyes," *Vision Res.* **30**:541–547 (1990).

98. B. E. Schefrin, S. J. Tregear, L. O. Harvey, Jr., and J. S. Werner, "Senescent Changes in Scotopic Contrast Sensitivity," *Vision Res.* **39**:3728–3736 (1999).

99. D. H. Kelly, "Adaptation Effects on Spatiotemporal Sine Wave Thresholds," *Vision Res.* **12**:89–101 (1972).

100. J. D. Morrison and C. McGrath, "Assessment of the Optical Contributions to the Age-Related Deterioration in Vision," *Q. J. Exp. Psychol.* **70**:249–269 (1985).

101. K. B. Burton, C. Owsley, and M. E. Sloane, "Aging and Neural Spatial Contrast Sensitivity: Photopic Vision," *Vision Res.* **33**:939–946 (1993).

102. N. W. Coppinger, "The Relationship Between Critical Flicker Frequency and Chronological Age for Varying Levels of Stimulus Brightness," *J. Gerontol.* **10**:48–52 (1955).

103. D. Kline and H. Scheiber, "Visual Persistence and Temporal Resolution," in R. Sekuler, D. Kline, and K. Dismuke (eds.), *Aging and Human Visual Functions,* A. R. Liss, New York, 1982, pp. 231–244.

104. L. Zhang and J. F. Sturr, "Aging, Background Luminance, and Threshold Duration Functions for Detection of Low Spatial Frequency Sinusoidal Gratings," *Optom. Vis. Sci.* **72**:198–204 (1995).

105. C. W. Tyler, "Two Processes Control Variations in Flicker Sensitivity over the Life Span," *J. Opt. Soc. Am.* **A6**:481–490 (1989).

106. C. B. Y. Kim and M. J. Mayer, "Flicker Sensitivity in Healthy Aging Eyes. II. Cross-Sectional Aging Trends from 18 Through 77 Years of Age," *J. Opt. Soc. Am.* **A11**:1958–1969 (1994).

107. V. Porciatti, D. C. Burr, C. Morrone, and A. Fiorentini, "The Effects of Ageing on the Pattern Electroretinogram and Visual Evoked Potential in Humans," *Vision Res.* **32**:1199–1209 (1992).

108. E. Wolf, "Studies on the Shrinkage of the Visual Field with Age," *Highway Res. Rec.* **167**:1–7 (1967).

109. A. Iwase, Y. Kitazawa, and Y. Ohno, "On Age-Related Norms of the Visual Field," *Jpn. J. Ophthal.* **32**:429–437 (1988).

110. C. A. Johnson, A. J. Adams, and R. A. Lewis, "Evidence for a Neural Basis of Age-Related Visual Field Loss in Normal Observers," *Invest. Ophthalmol. Visual Sci.* **30**:2056–2064 (1989).

111. M. Ikeda and T. Takeuchi, "Influence of Foveal Load on the Functional Visual Field," *Percept. Psychophys.* **18**:255–260 (1975).

112. K. K. Ball, B. L. Beard, D. L. Roenker, R. L. Miller, and D. S. Griggs, "Age and Visual Search: Expanding the Useful Field of View," *J. Opt. Soc. Am.* **A5**:2210–2219 (1988).

113. K. O. Devaney and H. A. Johnson, "Neuron Loss in the Aging Visual Cortex of Man," *J. Gerontol.* **35**:836–841 (1980).

114. P. D. Spear, "Neural Bases of Visual Deficits During Aging," *Vision Res.* **33**:2589–2609 (1993).

115. S. N. Jani, "The Age Factor in Stereopsis Screening," *Am. J. Optom.* **43**:653–657 (1966).

116. M. D. Bell, E. Wolf, and C. D. Bernholz, "Depth Perception As a Function of Age," *Aging Human Develop.* **3**:77–81 (1972).

117. G. Haegerstrom-Portnoy and B. Brown, "Two-Color Increment Thresholds in Early Age Related Maculopathy," *Clin. Vis. Sci.* **4**:165–172 (1989).

118. J. S. Sunness, R. W. Massof, M. A. Johnson, N. M. Bressler, S. B. Bressler, and S. L. Fine, "Diminished Foveal Sensitivity May Predict the Development of Advanced Age-Related Macular Degeneration," *Ophthalmology* **96**:375–381 (1989).

119. R. A. Applegate, A. J. Adams, J. C. Cavender, and F. Zisman, "Early Color Vision Changes in Age-Related Maculopathy," *Appl. Opt.* **26**:1458–1462 (1987).

120. A. C. Bird, N. M. Bressler, S. B. Bressler, I. H. Chisholm, G. Coscas, M. D. Davis, P. T. V. M. de Jong, C. C. W. Klaver, B. E. K. Klein, R. Klein, P. Mitchell, J. P. Sarks, S. H. Sarks, G. Soubrane, H. R. Taylor, and J. R. Vingerling, "An International Classification and Grading System for Age-Related Maculopathy and Age-Related Macular Degeneration," *Surv. Ophthalmol.* **39**:367–374 (1995).

121. Diabetes Control and Complications Trial Research Group, "The Effect of Intensive Treatment of Diabetes on the Development and Progression of Long-Term Complications in Insulin-Dependent Diabetes Mellitus," *New Eng. J. Med.* **329**:977–986 (1993).

122. V. J. Volbrecht, M. E. Schneck, A. J. Adams, J. A. Linfoot, and E. Ai, "Diabetic Short-Wavelength Sensitivity: Variations with Induced Changes in Blood Glucose Level," *Invest. Ophthalmol. Visual Sci.* **35**:1243–1246 (1994).

123. M. E. Schneck, B. Fortune, E. Switkes, M. Crognale, and A. J. Adams, "Acute Effects of Blood Glucose on Chromatic Visually Evoked Potentials in Persons with Diabetes and in Normal Persons," *Invest. Ophthalmol. Visual Sci.* **38**:800–810 (1997).

124. R. Lakowski, P. A. Aspinall, and P. R. Kinnear, "Association Between Colour Vision Losses and Diabetes Mellitus," *Ophthalmol. Res.* **4**:145–159 (1972/73).

125. J. Flammer and S. Orgül, "Optic Nerve Blood-Flow Abnormalities in Glaucoma," *Prog. Retin. Eye Res.* **17**:267–289 (1998).

126. G. Heron, A. Adams, and R. Husted, "Central Visual Fields for Short Wavelength Sensitive Pathways in Glaucoma and Ocular Hypertension," *Invest. Ophthalmol. Visual Sci.* **29**:64–72 (1988).

127. C. A. Johnson, A. J. Adams, E. J. Casson, and J. D. Brandt, "Blue-on-Yellow Perimetry Can Predict the Development of Glaucomatous Visual Field Loss," *Arch. Ophthalmol.* **111**:645–650 (1993).

128. H. A. Quigley, "Neuronal Death in Glaucoma," *Prog. Retin. Eye Res.* **18**:39–57 (1998).

129. V. Porciatti, E. D. Bartolo, M. Nardi, and A. Fiorentini, "Responses to Chromatic and Luminance Contrast in Glaucoma: A Psychophysical and Electrophysiological Study," *Vision Res.* **37**:1975–1987 (1997).

130. C. A. Johnson and S. J. Samuels, "Screening for Glaucomatous Visual Field Loss with Frequency-Doubling Perimetry," *Invest. Ophthalmol. Visual Sci.* **38**:413–425 (1997).

131. S. J. Chew and R. Ritch, "Neuroprotection: The Next Breakthrough in Glaucoma? Proceedings of the Third Annual Optic Nerve Rescue and Restoration Think Tank," *J. Glaucoma* **6**:263–266 (1997).

132. P. Erickson and C. Schor, "Visual Function with Presbyopic Contact Lens Correction," *Optom. Vis. Sci.* **67**:22–28 (1990).

133. R. L. Woods, "The Aging Eye and Contact Lenses—A Review of Visual Performance," *J. Br. Contact Lens Assoc.* **15**:31–43 (1992).

134. S. P. B. Percival and S. S. Setty, "Comparative Analysis of Three Prospective Trials of Multifocal Implants," *Eye* **5**:712–716 (1991).

135. J. C. Javitt, F. Wang, D. J. Trentacost, M. Rowe, and N. Tarantino, "Outcomes of Cataract Extraction with Multifocal Intraocular Lens Implantation," *Ophthalmology* **104**:589–599 (1997).

136. E. Rosen and C. Gore, "Staar Collamer Posterior Chamber Phakic Intraocular Lens to Correct Myopia and Hyperopia," *J. Cataract Refract. Surg.* **24**:596–606 (1998).

137. G. O. Waring, III, M. J. Lynn, P. J. McDonnell, and the PERK Study Group, "Results of the Prospective Evaluation of Radial Keratotomy (PERK) Study 10 Years After Surgery," *Arch. Ophthalmology* **112**:1298–1308 (1994).

138. A. El-Maghraby, T. Salah, G. O. Waring, S. Klyce, and O. Ibrahim, "Randomized Bilateral Comparison of Excimer Laser In Situ Keratomileusis and Photorefractive Keratectomy for 2.50 and 8.00 Diopters of Myopia," *Ophthalmology* **106**:447–457 (1999).

CHAPTER 14

RADIOMETRY AND PHOTOMETRY REVIEW FOR VISION OPTICS*

Yoshi Ohno
Optical Technology Division
National Institute of Standards and Technology
Gaithersburg, Maryland

14.1 INTRODUCTION

Radiometry is the measurement of optical radiation, which is electromagnetic radiation in the frequency range between 3×10^{11} Hz and 3×10^{16} Hz. This range corresponds to wavelengths between 10 nm and 1000 μm, and includes the regions commonly called the ultraviolet, the visible, and the infrared. Typical radiometric units include watt (radiant flux), watt per steradian (radiant intensity), watt per square meter (irradiance), and watt per square meter per steradian (radiance).

Photometry is the measurement of light, which is defined as electromagnetic radiation detectable by the human eye. It is thus restricted to the visible region of the spectrum (wavelength range from 360 nm to 830 nm), and all the quantities are weighted by the spectral response of the eye. Photometry uses either optical radiation detectors constructed to mimic the spectral response of the eye, or spectroradiometry coupled with appropriate calculations for weighting by the spectral response of the eye. Typical photometric units include lumen (luminous flux), candela (luminous intensity), lux (illuminance), and candela per square meter (luminance).

The difference between radiometry and photometry is that radiometry includes the entire optical radiation spectrum (and often involves spectrally resolved measurements), while photometry deals with the visible spectrum weighted by the response of the eye. This chapter provides some guidance in photometry and radiometry (Refs. 1 through 6 are available for further details). The terminology used in this chapter follows international standards and recommendations.[7–9]

14.2 BASIS OF PHYSICAL PHOTOMETRY

The primary aim of photometry is to measure visible optical radiation, light, in such a way that the results correlate with the visual sensation of a normal human observer exposed to that

---

* Chapter 24 in Vol. I of this Handbook gives further treatment of issues in radiometry, and Chap. 25 in Vol. II of this Handbook describes measurement of optical properties of materials.

radiation. Until about 1940, visual comparison measurement techniques were predominant in photometry.

In modern photometric practice, measurements are made with photodetectors. This is referred to as *physical photometry.* In order to achieve the aim of photometry, one must take into account the characteristics of human vision. The relative spectral responsivity of the human eye was first defined by the CIE (Commission Internationale de l'Éclairage) in 1924,[10] and redefined as part of colorimetric standard observers in 1931.[11] Called the *spectral luminous efficiency for photopic vision,* or $V(\lambda)$, it is defined in the domain from 360 nm to 830 nm, and is normalized at its peak, 555 nm (Fig. 1). This model has gained wide acceptance. The values were republished by CIE in 1983,[12] and published by CIPM (Comité International des Poids et Mesures) in 1982[13] to supplement the 1979 definition of the candela. (The tabulated values of the function at 1 nm increments are available in Refs. 12 through 15.) In most cases, the region from 380 nm to 780 nm suffices for calculation with negligible errors because the value of the $V(\lambda)$ function falls below 10^{-4} outside this region. Thus, a photodetector having a spectral responsivity matched to the $V(\lambda)$ function replaced the role of human eyes in photometry.

Radiometry concerns physical measurement of optical radiation in terms of optical power, and in many cases, as a function of its wavelength. As specified in the definition of the candela by CGPM (Conférence Générale des Poids et Mesures) in 1979[16] and by CIPM in 1982,[13] a photometric quantity X_v is defined in relation to the corresponding radiometric quantity $X_{e,\lambda}$ by the equation:

$$X_v = K_m \int_{360\,nm}^{830\,nm} X_{e,\lambda}\, V(\lambda)\mathrm{d}\lambda \tag{1}$$

The constant, K_m, relating the photometric quantities and radiometric quantities, is called the *maximum spectral luminous efficacy (of radiation) for photopic vision.* The value of K_m is given by the 1979 definition of candela that defines the spectral luminous efficacy of light at the frequency 540×10^{12} Hz (at the wavelength 555.016 nm in standard air) to be 683 lm/W. The value of K_m is calculated as $683 \times V(555.000\,nm)/V(555.016\,nm) = 683.002$ lm/W.[12] K_m is normally rounded to 683 lm/W with negligible errors.

It should be noted that $V(\lambda)$ is defined for the *CIE standard photometric observer for photopic vision,* which assumes additivity of sensation and a 2° field of view at relatively high luminance levels (higher than approximately 1 cd/m²). The human vision in this level is called

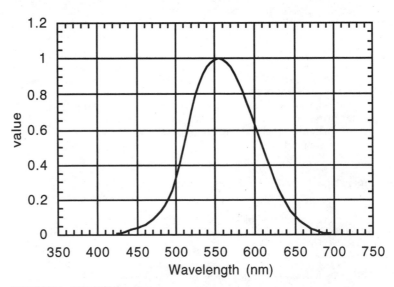

FIGURE 1 CIE $V(\lambda)$ function.

photopic vision. The spectral responsivity of human vision deviates significantly at very low levels of luminance (less than approximately 10^{-3} cd/m^2). This type of vision is called scotopic vision. Its spectral responsivity, peaking at 507 nm, is designated by $V'(\lambda)$, which was defined by CIE in 1951,[17] recognized by CIPM in 1976,[18] and republished by CIPM in 1982.[13] Human vision in the region between photopic vision and scotopic vision is called mesopic vision. While there have been active researches in this area,[19] there is no internationally accepted spectral luminous efficiency function for the mesopic region yet. In current practice, almost all photometric quantities are given in terms of photopic vision, even at low light levels. Quantities in scotopic vision are seldom used except for special calculations for research purposes. (Further details of the contents in this section are given in Ref. 12.)

14.3 *PHOTOMETRIC BASE UNIT—THE CANDELA*

The history of photometric standards dates back to the early nineteenth century, when the intensity of light sources was measured in comparison with a standard candle using visual bar photometers. At that time, the flame of a candle was used as a unit of luminous intensity that was called the *candle*. The old name for luminous intensity *candle power* came from this origin. Standard candles were gradually superseded by flame standards of oil lamps, and in the early twentieth century investigations on platinum point blackbodies began at some national laboratories. An agreement was first established in 1909 among several national laboratories to use such a blackbody to define the unit of luminous intensity, and the unit was recognized as the *international candle*. This standard was adopted by the CIE in 1921. In 1948, it was adopted by the CGPM[16] with a new Latin name *candela* with the following definition:

> The candela is the luminous intensity, in the perpendicular direction, of a surface of 1/600000 square meter of a blackbody (full radiator) at the temperature of freezing platinum under a pressure of 101325 newton per square meter.

Although the 1948 definition served to establish the uniformity of photometric measurements in the world, difficulties in fabricating the blackbodies and in improving accuracy were addressed. Beginning in the mid-1950s, suggestions were made to define the candela in relation to the unit of optical power, watt, so that complicated source standards would not be necessary. Finally, in 1979, a new definition of the candela was adopted by the CGPM[16] as follows:

> The candela is the luminous intensity, in a given direction, of a source that emits monochromatic radiation of frequency 540×10^{12} hertz and that has a radiant intensity in that direction of (1/683) watt per steradian.

The value of K_m (683 lm/W) was determined in such a way that the consistency from the prior unit was maintained, and was determined based on the measurements by several national laboratories. (Technical details on this redefinition of the candela are reported in Refs. 20 and 21.) This 1979 redefinition of the candela has enabled photometric units to be derived from radiometric units using a variety of techniques. (The early history of photometric standards is described in greater detail in Ref. 22.)

14.4 *QUANTITIES AND UNITS IN PHOTOMETRY AND RADIOMETRY*

In 1960, the SI (Système International) was established, and the candela became one of the seven SI base units.[23] (For further details on the SI, Refs. 23 through 26 may be consulted.) Several quantities and units, defined in different geometries, are used in photometry and radiometry. Table 1 lists the photometric quantities and units, along with corresponding quantities and units for radiometry.

While the candela is the SI base unit, the luminous flux (lumen) is perhaps the most fundamental photometric quantity, as the other photometric quantities are defined in terms of lumen with an appropriate geometric factor. The definitions of these photometric quantities are given in the following sections. (The descriptions given here are somewhat simplified from the rigorous definitions for ease of understanding. Refer to Refs. 7 through 9 for official rigorous definitions.)

Radiant Flux and Luminous Flux

Radiant flux (also called *optical power* or *radiant power*) is the energy Q (in joules) radiated by a source per unit of time, expressed as

$$\Phi = \frac{dQ}{dt} \tag{2}$$

The unit of radiant flux is the *watt* (W = J/s).

Luminous flux (Φ_v) is the time rate of flow of light as weighted by $V(\lambda)$. The unit of luminous flux is the *lumen* (lm). It is defined as

$$\Phi_v = K_m \int_\lambda \Phi_{e,\lambda}\, V(\lambda) d\lambda \tag{3}$$

where $\Phi_{e,\lambda}$ is the spectral concentration of radiant flux as a function of wavelength λ. The term luminous flux is often used in the meaning of total luminous flux in photometry (see the following subsection entitled "Total Radiant Flux and Total Luminous Flux").

Radiant Intensity and Luminous Intensity

Radiant intensity (I_e) or *luminous intensity* (I_v) is the radiant flux (luminous flux) from a point source emitted per unit solid angle in a given direction, as defined by

$$I = \frac{d\Phi}{d\Omega} \tag{4}$$

TABLE 1   Quantities and Units Used in Photometry and Radiometry

Photometric quantity	Unit	Relationship with lumen	Radiometric quantity	Unit
Luminous flux	lm (lumen)		Radiant flux	W (watt)
Luminous intensity	cd (candela)	lm sr^{-1}	Radiant intensity	W sr^{-1}
Illuminance	lx (lux)	lm m^{-2}	Irradiance	W m^{-2}
Luminance	cd m^{-2}	lm sr^{-1} m^{-2}	Radiance	W sr^{-1} m^{-2}
Luminous exitance	lm m^{-2}		Radiant exitance	W m^{-2}
Luminous exposure	lx·s		Radiant exposure	W m^{-2}·s
Luminous energy	lm·s		Radiant energy	J (joule)
Total luminous flux	lm (lumen)		Total radiant flux	W (watt)
Color temperature	K (kelvin)		Radiance temperature	K (kelvin)

where $d\Phi$ is the radiant flux (luminous flux) leaving the source and propagating in an element of solid angle $d\Omega$ containing the given direction. The unit of radiant intensity is W/sr, and that of luminous intensity is the *candela* (cd = lm/sr). (See Fig. 2.)

Solid Angle.   The solid angle (Ω) of a cone is defined as the ratio of the area (A) cut out on a spherical surface (with its center at the apex of that cone) to the square of the radius (r) of the sphere, as given by

$$\Omega = \frac{A}{r^2} \tag{5}$$

The unit of solid angle is *steradian* (sr), which is a dimensionless unit. (See Fig. 3.)

Irradiance and Illuminance

Irradiance (E_e) or *illuminance* (E_v) is the density of incident radiant flux or luminous flux at a point on a surface, and is defined as radiant flux (luminous flux) per unit area, as given by

$$E = \frac{d\Phi}{dA} \tag{6}$$

where $d\Phi$ is the radiant flux (luminous flux) incident on an element dA of the surface containing the point. The unit of irradiance is W/m^2, and that of illuminance is *lux* (lx = lm/m^2). (See Fig. 4.)

Radiance and Luminance

Radiance (L_e) or *luminance* (L_v) is the radiant flux (luminous flux) per unit solid angle emitted from a surface element in a given direction, per unit projected area of the surface element perpendicular to the direction. The unit of radiance is W sr^{-1} m^{-2}, and that of luminance is cd/m^2. These quantities are defined by

$$L = \frac{d^2\Phi}{d\Omega \cdot A \cdot \cos\theta} \tag{7}$$

where $d\Phi$ is the radiant flux (luminous flux) emitted (reflected or transmitted) from the surface element and propagating in the solid angle $d\Omega$ containing the given direction. dA is the

Point source

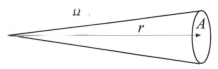

FIGURE 2   Radiant intensity and luminous intensity.

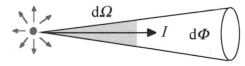

FIGURE 3   Solid angle.

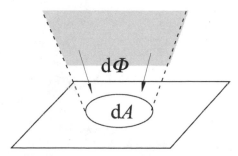

FIGURE 4   Irradiance and illuminance.

area of the surface element, and θ is the angle between the normal to the surface element and the direction of the beam. The term dA cos θ gives the projected area of the surface element perpendicular to the direction of measurement. (See Fig. 5.)

Radiant Exitance and Luminous Exitance

Radiant exitance (M_e) or *luminous exitance* (M_v) is defined to be the density of radiant flux (luminous flux) leaving a surface at a point. The unit of radiant exitance is W/m^2 and that of luminous exitance is lm/m^2 (but it is not lux). These quantities are defined by

$$E = \frac{d\Phi}{dA} \tag{8}$$

where dΦ is the radiant flux (luminous flux) leaving the surface element. Luminous exitance is rarely used in the general practice of photometry. (See Fig. 6.)

Radiant Exposure and Luminous Exposure

Radiant exposure (H_e) or *luminous exposure* (H_v) is the time integral of irradiance $E_e(t)$ or illuminance $E_v(t)$ over a given duration Δt, as defined by

$$H = \int_{\Delta t} E(t) dt \tag{9}$$

The unit of radiant exposure is J m^{-2}, and that of luminous exposure is *lux·second* (lx·s).

Radiant Energy and Luminous Energy

Radiant energy (Q_e) or *luminous energy* (Q_v) is the time integral of the radiant flux or luminous flux (Φ) over a given duration Δt, as defined by

$$Q = \int_{\Delta t} \Phi(t) dt \tag{10}$$

The unit of radiant energy is *joule* (J), and that of luminous energy is *lumen·second* (lm·s).

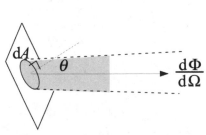

FIGURE 5 Radiance and luminance.

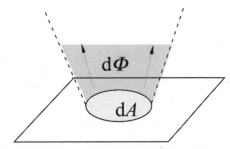

FIGURE 6 Radiant exitance and luminous exitance.

Total Radiant Flux and Total Luminous Flux

Total radiant flux or *total luminous flux* (Φ_v) is the geometrically total radiant (luminous) flux of a light source. It is defined as

$$\Phi = \int_{\Omega} I \, d\Omega \tag{11}$$

or

$$\Phi = \int_{A} E \, dA \tag{12}$$

where I is the radiant (luminous) intensity distribution of the light source and E is the irradiance (illuminance) distribution over a given closed surface surrounding the light source. If the radiant (luminous) intensity distribution or the irradiance (illuminance) distribution is given in polar coordinates (θ, ϕ), the total radiant (luminous) flux of the source Φ is given by

$$\Phi = \int_{\phi=0}^{2\pi} \int_{\theta=0}^{\pi} I(\theta, \phi) \sin\theta \, d\theta \, d\phi \tag{13}$$

or

$$\Phi = r^2 \int_{\phi=0}^{2\pi} \int_{\theta=0}^{\pi} E(\theta, \phi) \sin\theta \, d\theta \, d\phi \tag{14}$$

For example, the total luminous flux of an isotropic point source having luminous intensity of 1 cd would be 4π lumens.

Radiance Temperature and Color Temperature

Radiance temperature (unit: kelvin) is the temperature of the Planckian radiator for which the radiance at the specified wavelength has the same spectral concentration as for the thermal radiator considered.

Color temperature (unit: kelvin) is the temperature of a Planckian radiator with radiation of the same chromaticity as that of the light source in question. This term is commonly used to specify the colors of incandescent lamps whose chromaticity coordinates are practically on the blackbody locus. The next two terms are also important in photometry.

Distribution temperature (unit: kelvin) is the temperature of a blackbody with a spectral power distribution closest to that of the light source in question, and is used for quasi-Planckian sources such as incandescent lamps (refer to Ref. 27 for details).

Correlated color temperature (unit: kelvin) is the temperature of the Planckian radiator whose perceived color most closely resembles that of the light source in question. Correlated color temperature is used for sources with a spectral power distribution significantly different from that of Planckian radiation (e.g., discharge lamps; refer to Ref. 28 for details).

Relationship Between SI Units and English Units

The SI units as described previously should be used in all radiometric and photometric measurements according to international standards and recommendations on SI units. However, some English units are still rather widely used in some countries, including the United States. The use of these non-SI units is discouraged. The definitions of these English units are given in Table 2 for conversion purposes only.

TABLE 2   English Units and Definition

Unit	Quantity	Definition
Footcandle (fc)	Illuminance	Lumen per square foot (lm ft^{-2})
Footlambert (fL)	Luminance	1/π candela per square foot (π^{-1} cd ft^{-2})

The definition of footlambert is such that the luminance of a perfect diffuser is 1 fL when illuminated at 1 fc. In the SI unit, the luminance of a perfect diffuser would be 1/π (cd/m^2) when illuminated at 1 lx. For convenience of changing from English units to SI units, the conversion factors are listed in Table 3. For example, 1000 lx is the same illuminance as 92.9 fc, and 1000 cd/m^2 is the same luminance as 291.9 fL. (Conversion factors to and from some more units are given in Ref. 5.)

Troland

This unit is not an SI unit, not used in metrology, and is totally different from all other photometric units mentioned previously. It is introduced here because this unit is commonly used by vision scientists. Troland is defined as the retinal illuminance when a surface of luminance one candela per square meter is viewed through a pupil at the eye (natural or artificial) of area one square millimeter. Thus, the troland value, T, for the luminance, L (cd/m^2), of an external field and the pupil size, p (mm^2), is given by

$$T = L \cdot p \tag{15}$$

or, for pupil size p(m^2),

$$T = L \cdot 10^6 \times p \tag{16}$$

The troland value is not the real illuminance in lux on the retina, but is a quantity proportional to it. Since the natural pupil size changes with luminance level, luminance changes do not have a proportional visual effect. Thus, troland value rather than luminance is often useful in visual experiments. There is no simple or defined conversion between troland value (for natural pupil) and luminance (cd/m^2), without knowing the actual pupil size. (Further details of this unit can be found in Ref. 29.)

14.5 PRINCIPLES IN PHOTOMETRY AND RADIOMETRY

Several important theories in practical photometry and radiometry are introduced in this section.

TABLE 3   Conversion Between English Units and SI Units

To obtain the value in	Multiply the value in	By
lx from fc	fc	10.764
fc from lx	lx	0.09290
cd/m^2 from fL	fL	3.4263
fL from cd/m^2	cd/m^2	0.29186
m (meter) from feet	feet	0.30480
mm (millimeter) from inch	inch	25.400

Inverse Square Law

Illuminance E (lx) at a distance d (m) from a point source having luminous intensity I (cd) is given by

$$E = \frac{I}{d^2}$$ (17)

For example, if the luminous intensity of a lamp in a given direction is 1000 cd, the illuminance at 2 m from the lamp in this direction is 250 lx. Note that the inverse square law is valid only when the light source is regarded as a point source. Sufficient distances relative to the size of the source are needed to assume this relationship.

Lambert's Cosine Law

The luminous intensity of a Lambertian surface element is given by

$$I(\theta) = I_n \cos \theta$$ (18)

(See Fig. 7.)

Lambertian Surface.   A surface whose luminance is the same in all directions of the hemisphere above the surface.

Perfect (Reflecting/Transmitting) Diffuser.   A Lambertian diffuser with a reflectance (transmittance) equal to 1.

Relationship Between Illuminance and Luminance

The luminance L (cd/m²) of a Lambertian surface of reflectance ρ, illuminated by E (lx) is given by

$$L = \frac{\rho \cdot E}{\pi}$$ (19)

(See Fig. 8.)

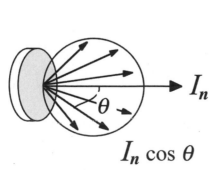

FIGURE 7   Lambert's cosine law.

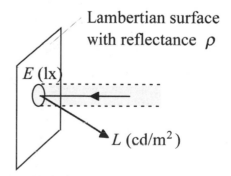

FIGURE 8   Relationship between illuminance and luminance.

Reflectance (ρ).   The ratio of the reflected flux to the incident flux in a given condition. The value of ρ can be between 0 and 1.

In the real world, there is no existing perfect diffuser nor perfectly Lambertian surfaces, and Eq. 19 does not apply. For real object surfaces, the following terms apply.

Luminance Factor (β).   Ratio of the luminance of a surface element in a given direction to that of a perfect reflecting or transmitting diffuser, under specified conditions of illumination. The value of β can be larger than 1. For a Lambertian surface, reflectance is equal to the luminance factor. Eq. 19 for real object is restated using β as

$$L = \frac{\beta \cdot E}{\pi} \tag{20}$$

Luminance Coefficient (q).   Quotient of the luminance of a surface element in a given direction by the illuminance on the surface element, under specified conditions of illumination,

$$q = \frac{L}{E} \tag{21}$$

Using *q*, the relationship between luminance and illuminance is thus given by

$$L = q \cdot E \tag{22}$$

Luminance factor corresponds to radiance factor, and luminance coefficient corresponds to radiance coefficient in radiometry. BRDF (bidirectional reflectance distribution function) is also used for the same concept as radiance coefficient.

Integrating Sphere

An integrating sphere is a device to make a spatial integration of luminous flux (or radiant flux) generated (or introduced) in the sphere and to detect it with a single photodetector. In the case of measurement of light sources, the spatial integration is made over the entire solid angle (4π).

In Fig. 1, assuming that the integrating sphere wall surfaces are perfectly Lambertian, the illuminance E on any part of the sphere wall created by luminance L of an element Δa is given by

$$E = \frac{L\Delta a}{4R^2} \tag{23}$$

where R is the radius of the sphere. This equation holds no matter where the two surface elements are. In other words, the same amount of flux incident anywhere on the sphere wall will create an equal illuminance on the detector port. In the case of actual integrating spheres, the surface is not perfectly Lambertian, but due to interreflections of light in the sphere, the distribution of reflected light will be uniform enough to assume the relationship of Eq. 23. (See Fig. 9.)

The direct light from an actual light source is normally not uniform; thus, it must be shielded from the detector. When a light source with luminous flux Φ is operated in a sphere having reflectance ρ, the flux created by interreflections is given by

$$\Phi(\rho + \rho^2 + \rho^3 + \cdots) = \Phi \cdot \frac{\rho}{1 - \rho} \tag{24}$$

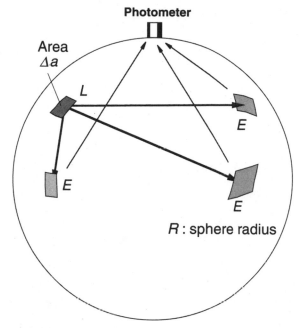

FIGURE 9 Flux transfer in a sphere.

Then, the illuminance E_d created by all the interreflections is given by

$$E_d = \frac{\Phi \cdot \rho}{1 - \rho} \cdot \frac{1}{4\pi \cdot R^2} \tag{25}$$

The sphere efficiency (E_d/Φ) is strongly dependent on reflectance ρ due to the term $1 - \rho$ in the denominator. For example, the detector signal at $\rho = 0.98$ is 10 times larger than at $\rho = 0.8$.

Planck's Law

The spectral radiance of a blackbody at a temperature T (K) is given by

$$I_e(\lambda, T) = c_1 n^{-2} \pi^{-1} \lambda^{-5} \left[\exp\left(\frac{c_2}{n\lambda T} \right) - 1 \right]^{-1} \tag{26}$$

where $c_1 = 2\pi h c^2 = 3.7417749 \times 10^{-16}$ W $\cdot$ m^2, $c_2 = hc/k = 1.438769 \times 10^{-2}$ m $\cdot$ K (1986 CODATA from Ref. 9), h is Planck's constant, c is the speed of light in vacuum, k is the Boltzmann constant, n (=1.00028) is the refractive index of standard air (12),[30] and λ is the wavelength.

Wien's Displacement Law

Taking the partial derivative of the Planck's equation with respect to temperature T, and setting the result equal to zero, the solution yields the relationship between the peak wavelength λ_m for Planck's radiation and temperature T (K), as given by

$$\lambda_m T = 2897.8 \ \mu m \cdot K \tag{27}$$

This shows that the peak wavelength of blackbody radiation shifts to shorter wavelengths as the temperature of the blackbody increases.

Stefan-Boltzmann's Law

The (spectrally total) radiant exitance M_e from a blackbody in a vacuum is expressed in relation to the temperature T (K) of the blackbody, in the form

$$M_e(T) = \int_0^\infty M_e(\lambda, T)\mathrm{d}\lambda = \sigma T^4 \tag{28}$$

where $M_e(\lambda, T)$ is the spectral radiant exitance of the blackbody, and σ is the Stefan-Boltzmann constant, equal to 5.67051×10^{-8} W $\cdot$ m^{-2} $\cdot$ K^{-4} (1986 CODATA from Ref. 9). Using this value, the unit for M_e is W $\cdot$ m^{-2}.

14.6   PRACTICE IN PHOTOMETRY AND RADIOMETRY

Photometry and radiometry are practiced in many different areas and applications, dealing with various light sources and detectors, and cannot be covered in this chapter. Various references are available on practical measurements in photometry and radiometry.

Further references in practical radiometry include books on absolute radiometry,[31] optical detectors,[32] spectroradiometry,[33] photoluminescence,[34] radiometric calibration,[35] etc. There are a number of publications from CIE that are regarded as international recommendations or standards. CIE publications in radiometry include reports on absolute radiometry,[36] reflectance,[37] spectroradiometry,[38] detector spectral response,[39] photobiology and photochemistry,[40] etc. There are also a number of publications from the National Institute of Standards and Technology (NIST) in radiometry, on spectral radiance,[41] spectral irradiance,[42] spectral reflectance,[43] spectral responsivity,[44] and so on (Ref. 45 provides greater depths of knowledge in radiometry).

For practical photometry, ref. 4 provides the latest information on standards and practical measurements of photometry in many aspects. A recent publication from NIST[46] is also available. CIE publications are also available on many subjects in photometry, including characterization of illuminance meters and luminance meters,[47] luminous flux measurement,[48] measurements of LEDs,[49] characteristics of displays,[50] and many others. A series of measurement guide documents are published from the Illuminating Engineering Society of North America (IESNA) for operation and measurement of particular types of lamps[51–53] and luminaires. The American Society for Testing and Materials (ASTM) provides many useful standards and recommendations on optical properties of materials and color measurements.[54] Colorimetry is a branch of radiometry and is becoming increasingly important among color imaging industry and multimedia applications. The basis of colorimetry is provided by CIE publications[28,55,56] and many other authoritative references are available[29,57].

14.7   REFERENCES

1. F. Grum and R. J. Becherer, *Optical Radiation Measurements, Vol. 1 Radiometry,* Academic Press, San Diego, CA, 1979.

2. R. McCluney, *Introduction to Radiometry and Photometry,* Artech House, Norwood, MA, 1994.

3. W. L. Wolfe, *Introduction to Radiometry,* SPIE—The International Society for Optical Engineering, P.O. Box 10, Bellingham, WA 98227-0010, 1998.

4. Casimer DeCusatis (ed.), *OSA/AIP Handbook of Applied Photometry,* AIP Press, Woodbury, NY, 1997.

5. *IES Lighting Handbook, 8th Edition, Reference and Application,* Illuminating Engineering Society of North America, New York, 1993.

6. J. M. Palmer, Radiometry and Photometry FAQ, http://www.optics.Arizona.EDU/Palmer/rpfaq/rpfaq.htm

7. *International Lighting Vocabulary,* CIE Publication No. 17.4, 1987.

8. *International Vocabulary of Basic and General Terms in Metrology,* BIPM, IEC, IFCC, ISO, IUPAC, IUPAP, and OIML, 1994.

9. *Quantities and Units,* ISO Standards Handbook, third edition, 1993.

10. CIE Compte Rendu, p. 67 (1924).

11. CIE Compte Rendu, Table II, pp. 25–26 (1931).

12. *The Basis of Physical Photometry,* CIE Publication No. 18.2 (1983).

13. CIPM, *Comité Consultatif de Photométrie et Radiométrie 10e Session—1982,* BIPM, Pavillon de Breteuil, F-92310 Sèvres, France (1982).

14. *Principles Governing Photometry,* Bureau International Des Poids et Mesures (BIPM) Monograph, BIPM, F-92310 Sèvres, France (1983).

15. CIE Disk D001, Photometric and Colorimetric Tables (1988).

16. CGPM, *Comptes Rendus des Séances de la 16e Conférence Générale des Poids et Mesures,* Paris 1979, BIPM, F-92310 Sèvres, France (1979).

17. CIE Compte Rendu, Vol. 3, Table II, pp. 37–39 (1951).

18. CIPM Procès-Verbaux 44, 4 (1976).

19. *Mesopic Photometry: History, Special Problems and Practical Solutions,* CIE Publication No. 81 (1989).

20. W. R. Blevin and B. Steiner, *Metrologia* **11**:97 (1975).

21. W. R. Blevin, "The Candela and the Watt," *CIE Proc.* P-79-02 (1979).

22. J. W. T. Walsh, *Photometry,* Constable, London, 1953.

23. *Le Système International d'Unité (SI), The International System of Units (SI),* 6th Edition, Bur. Intl. Poids et Mesures, Sèvres, France, 1991.

24. B. N. Taylor, *Guide for the Use of the International System of Units (SI),* Natl. Inst. Stand. Technol. Spec. Publ. 811, 1995.

25. B. N. Taylor (ed.), *Interpretation of the SI for the United States and Metric Conversion Policy for Federal Agencies,* Natl. Inst. Stand. Technol. Spec. Publ. 814, 1991.

26. *SI Units and Recommendations for the Use of Their Multiples and of Certain Other Units,* ISO 1000: 1992, International Organization for Standardization, Geneva, Switzerland, 1992.

27. *CIE Collection in Photometry and Radiometry,* Publication No. 114/4, 1994.

28. *Colorimetry,* second edition, CIE Publication No. 15.2, 1986.

29. G. Wyszecki and W. S. Stiles, *Color Science,* John Wiley and Sons, Inc., New York, 1982.

30. W. R. Blevin, "Corrections in Optical Pyrometry and Photometry for the Refractive Index of Air," *Metrologia* **8**:146 (1972).

31. F. Hengstberger (ed.), *Absolute Radiometry,* Academic Press, San Diego, CA, 1989.

32. W. Budde, *Optical Radiation Measurements, Vol. 4—Physical Detectors of Optical Radiation,* Academic Press, Orlando, FL, 1983.

33. H. Kostkowski, *Reliable Spectroradiometry,* Spectroradiometry Consulting, P.O. Box 2747, La Plata, MD 20646-2747, USA.

34. K. D. Mielenz (ed.), *Optical Radiation Measurements, Vol. 3, Measurement of Photoluminescence,* Academic Press, New York, 1982.

35. C. L. Wyatt, *Radiometric Calibration: Theory and Models,* Academic Press, New York, 1978.

36. *Electrically Calibrated Thermal Detectors of Optical Radiation (Absolute Radiometers),* CIE Publication 65, 1985.

37. *Absolute Methods for Reflection Measurements,* CIE Publication 44, 1979.

38. *The Spectroradiometric Measurement of Light Sources,* CIE Publication 63, 1984.

39. *Determination of the Spectral Responsivity of Optical Radiation Detectors,* CIE Publication 64, 1984.

40. *CIE Collection in Photobiology and Photochemistry,* CIE Publication 106, 1993.

41. J. H. Walker, R. D. Saunders, and A. T. Hattenburg, *Spectral Radiance Calibrations,* NBS Special Publication 250-1, 1987.

42. J. H. Walker, R. D. Saunders, J. K. Jackson, and D. A. McSparron, *Spectral Irradiance Calibrations,* NBS Special Publication 250-20, 1987.

43. P. Y. Barnes, E. A. Early, and A. C. Parr, *Spectral Reflectance,* NIST Special Publication 250-48, 1998.

44. T. C. Larason, S. S. Bruce, and A. C. Parr, *Spectroradiometric Detector Measurements,* NIST Special Publication 250-41, 1998.

45. F. Nicodemos (ed.), *Self-Study Manual on Optical Radiation Measurements,* NBS Technical Note 910 Series, Parts 1–12, 1978–1985.

46. Y. Ohno, *Photometric Calibrations,* NIST Special Publication 250–37, 1997.

47. *Methods of Characterizing Illuminance Meters and Luminance Meters,* CIE Publication 69, 1987.

48. *Measurement of Luminous Flux,* CIE Publication 84, 1989.

49. *Measurement of LEDs,* CIE Publication 127, 1997.

50. *The Relationship Between Digital and Colorimetric Data for Computer-Controlled CRT Displays,* CIE Publication 122, 1996.

51. *IES Approved Method for the Electric and Photometric Measurement of Fluorescent Lamps,* IESNA LM-9.

52. *Electrical and Photometric Measurements of General Service Incandescent Filament Lamps,* IESNA LM-45.

53. *Electrical and Photometric Measurements of Compact and Fluorescent Lamps,* IESNA LM-66.

54. *ASTM Standards on Color and Appearance Measurement,* fifth edition, 1996.

55. *Colorimetric Illuminants,* ISO/CIE 10526, 1991.

56. *Colorimetric Observers,* ISO/CIE 10527, 1991.

57. F. Grum and C. J. Bartleson (eds.), *Optical Radiation Measurements, Vol. 2, Color Measurement,* Academic Press, New York, 1980.

CHAPTER 15
OCULAR RADIATION HAZARDS*

David H. Sliney, Ph.D.
*U.S. Army Center for Health Promotion
and Preventive Medicine,
Laser/Optical Radiation Program,
Aberdeen Proving Ground, Maryland*

15.1 GLOSSARY

Bunsen-Roscoe law of photochemistry. The reciprocal relation between irradiance (dose-rate) and exposure duration to produce a constant radiant exposure (dose) to produce an effect.

Coroneo effect. See *Ophthalmoheliosis*.

Erythema. Skin reddening (damage) caused by UV radiation; sunburn.

Macula lutea. Yellowed-pigment central region (macular area) of the retina of the human eye.

Maxwellian view. A method for directing a convergent beam into the eye's pupil to illuminate a large retinal area.

Ophthalmoheliosis. The concentration of UV radiation upon the nasal region of the cornea (the limbus) and nasal equatorial region of the lens, thus contributing to the development of pterygium on the cornea and cataract in the lens beginning in these regions.

Photokeratitis. A painful inflammation of the cornea of the eye.

Photoretinitis. Retinal injury due to viewing the sun or other high-intensity light source arising from photochemical damage due to short, visible wavelengths.

Pterygium. A fleshy growth that invades the cornea (the clear front window of the eye). It is an abnormal process in which the conjunctiva (a membrane that covers the white of the eye) grows into the cornea. A pterygium may be small or grow large enough to interfere with vision and commonly occurs on the inner (nasal) corner of the eye.

Solar retinitis. Retinal injury caused from staring at the sun—photochemical damage produced by short, visible (blue) light.

---

* *Note:* The opinions or assertions herein are those of the author and should not be construed as official policies of the U.S. Department of the Army or Department of Defense.

15.2   *INTRODUCTION*

Optical radiation hazards to the eye vary greatly with wavelength and also depend upon the ocular exposure duration. Depending upon the spectral region, the cornea, conjunctiva, lens, and/or retina may be at risk when exposed to intense light sources such as lasers and arc sources—and even the sun. Photochemical damage mechanisms dominate in the ultraviolet (UV) end of the spectrum, and thermal damage mechanisms dominate at longer wavelengths in the visible and infrared (IR) spectral regions. Natural aversion responses to very bright light sources limit ocular exposure, as when one glances at the sun, but artificial sources may be brighter than the solar disk (e.g., lasers) or have different spectral or geometrical characteristics that overcome natural avoidance mechanisms, and the eye can suffer injury. Guidelines for the safe exposure of the eye from conventional light sources as well as lasers have evolved in recent decades. Because of special concerns about laser hazards, safety standards and regulations for lasers exist worldwide. However, the guidelines for exposure to conventional light sources are more complex because of the likelihood of encountering more than one type of ocular hazard from the broader emission spectrum.

The adverse effects associated with viewing lasers or other bright light sources such as the sun, arc lamps, and welding arcs have been studied for decades.[1-4] During the last three decades, guidelines for limiting exposure to protect the eye have evolved.[4-12] The guidelines were fostered to a large extent by the growing use of lasers and the quickly recognized hazard posed by viewing laser sources. Injury thresholds for acute injury in experimental animals for corneal, lenticular, and retinal effects have been corroborated for the human eye from accident data. Exposure limits are based upon this knowledge.[3] These safe exposure criteria were based not only upon studies aimed at determining thresholds in different ocular structures, but also upon studies of injury mechanisms and how the different injury mechanisms scaled with wavelength, exposure duration, and the area of irradiated tissue. The exposure guidelines also had to deal with competing damage to different ocular structures such as the cornea, lens, and retina, and how these varied with wavelength.[3-11] Although laser guidelines could be simplified to consider only the type of damage dominating for a single wavelength and exposure duration, the guidelines for incoherent, broadband sources were necessarily more complex.[11] To understand the guidelines for safe exposure, it is first necessary to consider the principal mechanisms of injury and which ocular structures may be injured under different exposure conditions.

15.3   *INJURY MECHANISMS*

For most exposure durations ($t > 1$ s), optical radiation injury to ocular structures is dominated by either photochemically or thermally initiated events taking place during or immediately following the absorption of radiant energy. At shorter durations, nonlinear mechanisms may play a role.[12] Following the initial insult, biological repair responses may play a significant role in determining the final consequences of the event. While inflammatory, repair responses are intended to reduce the sequellae; in some instances, this response could result in events such as scarring, which could have an adverse impact upon biological function.[3]

Action Spectra

Photochemical and thermal effects scale differently with wavelength, exposure duration, and irradiated spot size. Indeed, experimental biological studies in humans and animals make use of these different scaling relationships to distinguish which mechanisms are playing a role in observed injury. Photochemical injury is highly wavelength dependent; thermal injury depends upon exposure duration.

Exposure Duration and Reciprocity

The Bunsen-Roscoe law of photochemistry describes the *reciprocity* of exposure rate and duration of exposure, which applies to any photochemical event. The product of the dose-rate (in watts per square centimeter) and the exposure duration (in seconds) is the exposure dose (in joules per square centimeter at the site of absorption) that may be used to express an injury threshold. Radiometrically, the irradiance E in watts per square centimeter, multiplied by the exposure duration t, is the radiant exposure H in joules per square centimeter, as shown in Eq. 1.

$$H = E \cdot t \tag{1}$$

This reciprocity helps to distinguish photochemical injury mechanisms from thermal injury (burns) where heat conduction requires a very intense exposure within seconds to cause photocoagulation; otherwise, surrounding tissue conducts the heat away from the absorption site (e.g., a retinal image). Biological repair mechanisms and biochemical changes over long periods and photon saturation for extremely short periods will lead to reciprocity failure.

Thermal injury is a *rate process* that is dependent upon the time-temperature history resulting from the volumic absorption of energy across the spectrum. The thermochemical reactions that produce coagulation of proteins and cell death require critical temperatures for detectable biological injury. The critical temperature for an injury gradually decreases with the lengthening of exposure duration. This approximate decrease in temperature varies over many orders of magnitude in time and deceases approximately as a function of the exposure duration t raised to the -0.25 power {i.e., $[f(t^{-0.25})]$}.

As with any photochemical reaction, the *action spectrum* should be known.[13] The action spectrum describes the relative effectiveness of different wavelengths in causing a photobiological effect. For most photobiological effects (whether beneficial or adverse), the full width at half-maximum is less than 100 nm and a long-wavelength cutoff exists where photon energy is insufficient to produce the effect. This is not at all characteristic of thermal effects, in which the effect occurs over a wide range of wavelengths where optical penetration and tissue absorption occur. For example, significant radiant energy can penetrate the ocular media and be absorbed in the retina in the spectral region between 400 and nearly 1400 nm, the absorbed energy can produce retinal thermal injury.

Although the action spectra for acute photochemical effects upon the cornea,[3,14–16] upon the lens,[3,16] and upon the retina[3,17] have been published, the action spectra of some other effects appear to be quite imprecise or even unknown.[14] The action spectrum for neuroendocrine effects mediated by the eye is still only approximately known. This points to the need to specify the spectrum of the light source of interest as well as the irradiance levels if one is to compare experimental results from different experimental studies.

Although both E and H may be defined over the entire optical spectrum, it is necessary to employ an action spectrum for photochemical effects. The International Commission on Illumination (CIE) photopic $V(\lambda)$, UV hazard $S(\lambda)$, and blue-light hazard $B(\lambda)$ curves of Fig. 1 are examples of action spectra that may be used to spectrally weight the incident light. With modern computer spreadsheet programs, one can readily spectrally weight a lamp's spectrum by a large variety of photochemical action spectra. These computations all take the following form:

$$E_{\text{eff}} = \sum E_{\text{eff}} \cdot A(\lambda) \cdot \Delta(\lambda) \tag{2}$$

where $A(\lambda)$ may be any action spectrum of interest. One can then compare different sources to determine the relative effectiveness of the same irradiance from several lamps for a given action spectrum.

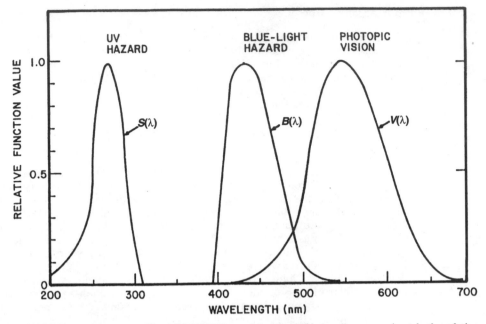

FIGURE 1   Action spectra. The ACGIH UV hazard function $S(\lambda)$ describes approximately the relative spectral risk for photokeratitis, and the blue-light hazard function $B(\lambda)$ describes the spectral risk for photoretinitis. For comparison, the CIE spectral sensitivity (standard observer) function $V(\lambda)$ for the human eye is also provided.

15.4   TYPES OF INJURY

The human eye is actually quite well adapted to protect itself against the potential hazards from optical radiation (UV, visible, and IR radiant energy) from most environmental exposures from sunlight. However, if ground reflections are unusually high, as when snow is on the ground, reflected UV radiation may produce "snow blindness." Another example is during a solar eclipse, if one stares at the solar disc for more than 1 to 2 min without eye protection, the result may be an "eclipse burn" of the retina.[3,17] Lasers may injure the eye, and under unusual situations other artificial light sources, such as quartz discharge lamps or arc lamps, may pose a potential ocular hazard. This is particularly true when the normal defense mechanisms of squinting, blinking, and aversion to bright light are overcome. Ophthalmic, clinical exposures may increase this risk if the pupil is dilated or the head is stabilized.

There are at least six separate types of hazards to the eye from lasers and other intense optical sources, and these are shown in Fig. 2 (along with the spectral range of responsible radiation):[3]

1. Ultraviolet photochemical injury to the cornea (photokeratitis); also known as "welder's flash" or "snow blindness" (180–400 nm)[3,15,16]
2. Ultraviolet photochemical injury to the crystalline lens (cataract) of the eye (295–325 nm, and perhaps to 400 nm)[16]
3. Blue-light photochemical injury to the retina of the eye (principally, 400–550 nm; unless aphakic, 310–550 nm)[3,4,17]
4. Thermal injury (photocoagulation) to the retina of the eye (400 to nearly 1400 nm) and thermoacoustic injury at very short laser exposure durations[3,4,12,18]

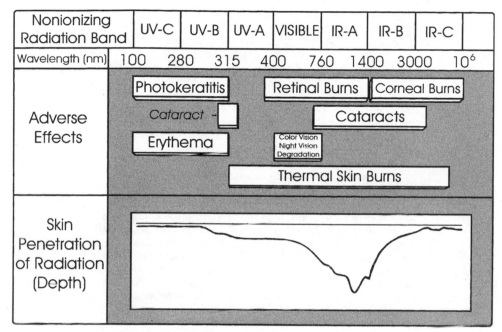

Nonionizing Radiation Band	UV-C	UV-B	UV-A	VISIBLE	IR-A	IR-B	IR-C	
Wavelength (nm)	100	280	315	400	760	1400	3000	10⁶

FIGURE 2   The separate types of hazards to the eye from lasers and other intense optical sources, along with the spectral range of responsible radiation.[3]

5. Near-IR thermal hazards to the lens (~800–3000 nm)[3,19]

6. Thermal injury (burns) of the cornea of the eye (~1300 nm to 1 mm)[3]

The potential hazards to the eye and skin are illustrated in Fig. 2, which shows how effects are generally dominant in specific CIE photobiological spectral bands, UV-A, -B, and -C and IR-A, -B, and -C.[20] Although these photobiological bands are useful shorthand notations, they do not define fine lines between no effect and an effect in accordance with changing wavelengths.

Photokeratitis

This painful but transient (1–2 d) photochemical effect normally occurs only over snow. The relative position of the light source and the degree of lid closure can greatly affect the proper calculation of this UV exposure dose. For assessing risk of photochemical injury, the spectral distribution of the light source is of great importance.

Cataract

Ultraviolet cataract can be produced from exposure to UV radiation (UVR) of wavelengths between 295 and 325 nm (and even to 400 nm). Action spectra can only be obtained from animal studies, which have shown that cortical and posterior subcapsular cataract can be produced by intense exposure delivered over a period of days. Human cortical cataract has been linked to chronic, lifelong UV-B radiation exposure. Although the animal studies and some epidemiological studies suggest that it is primarily UV-B radiation in sunlight, and not UV-A, that is most injurious to the lens, biochemical studies suggest that UV-A radiation may also contribute to accelerated aging of the lens. It should be noted that the wavelengths between 295 and 325 nm are also in the wavelength region where solar UVR increases significantly with ozone depletion. Because there is an earlier age of onset of cataract in equatorial zones, UVR exposure has frequently been one of the most appealing of a number of theories to explain this latitudinal dependence. The UVR guideline, established by the American Conference of Governmental Industrial Hygienists (ACGIH) and the International Commission on Non-Ionizing Radiation Protection (ICNIRP) is also intended to protect against cataract.

Despite the collection of animal and epidemiological evidence that exposure of the human eye to UVR plays an etiological role in the development of cataract, this role in cataractogenesis continues to be questioned by others.

Because the only direct pathway of UVR to the inferior germinative area of the lens is from the extreme temporal direction, it has been speculated that side exposure is particularly hazardous. For any greatly delayed health effect, such as cataract or retinal degeneration, it is critical to determine the actual dose distribution at critical locations. A factor of great practical importance is the actual UVR that reaches the germinative layers of any tissue structure. In the case of the lens, the germinative layer where lens fiber cell nuclei are located is of great importance. The DNA in these cells is normally well shielded by the parasol effect of the irises. However, Coroneo[21] has suggested that the focusing of very peripheral rays by the temporal edge of the cornea, those which do not even reach the retina, can enter the pupil and reach the equatorial region as shown in Fig. 3. He terms this effect, which can also produce a concentration of UVR at the nasal side of the limbus and lens, *ophthalmoheliosis*. He also noted the more frequent onset of cataract in the nasal quadrant of the lens and the formation of pterygium in the nasal region of the cornea. Figure 4 shows the percentage of cortical cataract that actually first appears in each quadrant of the lens from the data of Barbara Klein and her colleagues in their study of a population in the U.S. Midwest, in Beaver Dam, Wisconsin.[22] This relationship is highly consistent with the Coroneo hypothesis.

Pterygium and Droplet Keratopathies

The possible role of UVR in the etiology of a number of age-related ocular diseases has been the subject of many medical and scientific papers. However, there is still a debate as to the validity of these arguments. Although photokeratitis is unquestionably caused by UVR

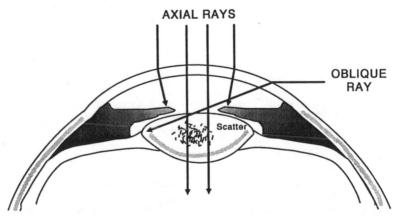

FIGURE 3   The Coroneo effect. Very oblique, temporal rays can be focused near the equator of the lens in the inferior nasal quadrant.

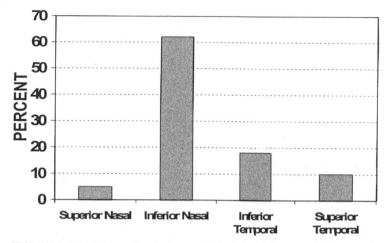

FIGURE 4   Distribution of cortical cataract by segment. The percentage of cortical cataract that actually first appears in each quadrant of the lens.
Source: From the data of Barbara Klein and her colleagues in their study of a population in the U.S. Midwest, in Beaver Dam, Wisconsin, *Am. J. Pub. Hlth.* **82**(12):1658–1662 (1992).

reflection from the snow,[3,23,24] pterygium and droplet keratopathies are less clearly related to UVR exposure.[24] Pterygium, a fatty growth over the conjunctiva that may extend over the cornea, is most common in ocean island residents (where both UVR and wind exposure is prevalent). Ultraviolet radiation is a likely etiological factor,[24,25] and the Coroneo effect may also play a role.[21,25]

To better answer these epidemiological questions, far better ocular dosimetry is required. Epidemiological studies can arrive at erroneous conclusions if assignments of exposure are seriously in error, and assumptions regarding relative exposures have been argued to be incorrect.[25] Before one can improve on current epidemiological studies of cataract (or determine the most effective UVR protective measures), it is necessary to characterize the actual solar UVR exposure to the eye.

Photoretinitis

The principal retinal hazard resulting from viewing bright continuous-wave (CW) light sources is photoretinitis [e.g., *solar retinitis* with an accompanying scotoma ("blind spot"), which results from staring at the sun]. Solar retinitis was once referred to as "eclipse blindness" and associated "retinal burn." At one time, solar retinitis was thought to be a thermal injury, but it has been since shown conclusively (1976) to result from a photochemical injury related to exposure of the retina to shorter wavelengths in the visible spectrum (i.e., violet and blue light).[3,17] For this reason, it has frequently been referred to as the "blue-light" hazard. The action spectrum for photoretinitis peaks at about 445 nm in the normal phakic eye; however, if the crystalline lens has been removed (as in cataract surgery), the action spectrum continues to increase for shorter wavelengths into the UV spectrum, until the cornea blocks shorter wavelengths, and the new peak shifts down to nearly 305 nm.[17]

As a consequence of the Bunsen-Roscoe law of photochemistry, blue-light retinal injury (photoretinitis) can result from viewing either an extremely bright light for a short time, or a less bright light for longer periods. The approximate threshold for a nearly monochromatic source at 440 nm is approximately 22 J/cm², hence a retinal irradiance of 2.2 W/cm² delivered in 10 s, or 0.022 W/cm² delivered in 1000 s will result in the same threshold retinal lesion. Eye movements will reduce the hazard, and this is particularly important for sources subtending an angle of less than 11 milliradians (mrad), because saccadic motion even during fixation is of the order of 11 mrad.[26]

Infrared Cataract

Infrared cataract in industrial workers appears only after lifetime exposure of 80 to 150 mW/cm². Although thermal cataracts were observed in glassblowers, steelworkers, and others in metal industries at the turn of the century, they are rarely seen today. These aforementioned irradiances are almost never exceeded in modern industry, where workers will be limited in exposure duration to brief periods above 10 mW/cm², or they will be (should be) wearing eye protectors. Good compliance in wearing eye protection occurs at higher irradiances if only because the worker desires comfort of the face.[3]

15.5 RETINAL IRRADIANCE CALCULATIONS

Retinal irradiance (exposure rate) is directly related to source radiance (brightness). It is *not* readily related to corneal irradiance.[3] Equation (3) gives the general relation, where E_r is the retinal irradiance in watts per square centimeter, L_s is the source radiance in watts per square centimeter per steradian (sr), f is the effective focal length of the eye in centimeters, d_e is the pupil diameter in centimeters, and τ is the transmittance of the ocular media:

$$E_r = \frac{(\pi L_s \cdot \tau \cdot d_e^2)}{(4 f^2)} \tag{3}$$

Equation 3 is derived by considering the equal angular subtense of the source and the retinal image at the eye's nodal point (Fig. 5). The detailed derivation of this equation is given elsewhere.[3] The transmittance τ of the ocular media in the visible spectrum for younger humans (and most animals) is as high as 0.9 (i.e., 90 percent).[3] If one uses the effective focal length f of the adult human eye (Gullstrand eye), where $f = 1.7$ cm, one has:

$$E_r = 0.27 L_s \cdot \tau \cdot d_e^2 \tag{4}$$

All of the preceding equations assume that the iris is pigmented, and the pupil acts as a true aperture. In albino individuals, the iris is not very effective, and some scattered light reaches the retina. Nevertheless, imaging of a light source still occurs, and Eq. 4 is still valid if the contribution of scattered light (which falls over the entire retina) is added.

15.6 EXAMPLES

As an example, a typical cool-white fluorescent lamp has an illumination of 200 lx, and the total irradiance is 0.7 mW/cm². By spectral weighting, the effective blue-light irradiance is found to be 0.15 mW/cm². From measurement of the solid angle subtended by the source at the measurement distance,[3,11] the blue-light radiance is 0.6 mW/(cm²·sr) and the radiance is 2.5 mW/(cm²·sr). When fluorescent lamps are viewed through a plastic diffuser, the luminance and blue-light radiance are reduced.

As another example, a 1000-watt tungsten halogen bulb has a greater radiance, but the percentage of blue light is far less. A typical blue-light radiance is 0.95 W/(cm²·sr) compared with a total radiance of 58 W/(cm²·sr). The luminance is 2600 candelas (cd)/cm²—a factor of

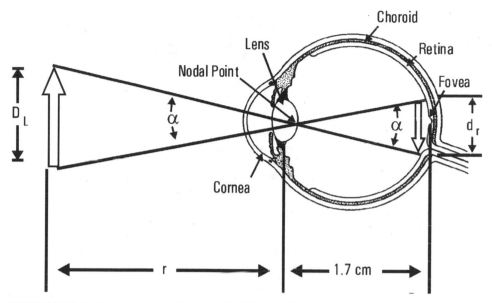

FIGURE 5 Retinal image size related to source size. The retinal image size can be calculated based upon the equal angular subtense of the source and the retinal image at the eye's nodal point (17 mm in front of the retina).

3000 times brighter than the cool-white fluorescent. However, because the retinal image is small, eye movements spread the exposure over a much larger area.

Once the source radiance, L or L_B, is known, the retinal irradiance E_r is calculated by Eq. 4. The preceding examples illustrate the importance of considering the size of a light source and the impact of eye movements in any calculation of retinal exposure dose. If one were exposed to a focal beam of light (e.g., from a laser or LED) that was brought to focus in the anterior chamber (the aqueous humor), the pupil plane, or the lens, the light beam would diverge past this focal point and could be incident upon the retina as a relatively large image. This type of retinal illumination is frequently referred to as *Maxwellian view* and does not occur in nature. The retinal irradiance calculation in this case would be determined by the depth of the focal spot in the eye; the closer to the retina, the smaller the retinal image and the greater the irradiance.

Because the iris alters its diameter (and pupil center) dynamically, one must be alert to vignetting. The pupil aperture also diminishes with age. Near-sighted, or myopic, individuals tend to have larger pupils, and far-sighted, or hyperopic, individuals tend toward smaller pupils. Accommodation also results in a decrease in pupil diameter.

15.7 *EXPOSURE LIMITS*

A number of national and international groups have recommended occupational or public exposure limits (ELs) for optical radiation (i.e., UV, light and IR radiant energy). Although most such groups have recommended ELs for UV and laser radiation, only one group has recommended for some time ELs for visible radiation (i.e., light). This one group is well known in the field of occupational health—the American Conference of Governmental Industrial Hygienists (ACGIH).[5,6] The ACGIH refers to its ELs as *threshold-limit values* (TLVs); these are issued yearly, so there is an opportunity for a yearly revision. The current ACGIH TLVs for light (400–760 nm) have been largely unchanged for the last decade; in 1997, with some revisions, the International Commission on Non-Ionizing Radiation Protection (ICNIRP) recommended these as international guidelines.[11] The ICNIRP guidelines are developed through collaboration with the World Health Organization (WHO) by jointly publishing criteria documents that provide the scientific database for the exposure limits.[4] The ACGIH TLVs and ICNIRP ELs are generally applied in product safety standards of the International Commission on Illumination (CIE), the International Electrotechnical Commission and consensus standards from the American National Standards Institute and other groups. They are based in large part on ocular injury data from animal studies and from data from human retinal injuries resulting from viewing the sun and welding arcs. All of the guidelines have an underlying assumption that outdoor environmental exposures to visible radiant energy is normally not hazardous to the eye except in very unusual environments such as snow fields, deserts, or out on the open water.

Applying the UV Limits

To apply the UV guideline, one must obtain the average spectrally weighted UV irradiance E_{eff} at the location of exposure. The spectrally weighted irradiance is

$$E_{\text{eff}} = \sum E_\lambda \cdot S(\lambda) \cdot \Delta\lambda \qquad (5)$$

where the summation covers the full spectral range of $S(\lambda)$. The maximum duration of exposure to stay within the limit t_{max} is determined by dividing the EL by the measured effective irradiance to obtain the duration in seconds, as noted in Eq. 6:

$$t_{\text{max}} = (3 \text{ mJ} \cdot \text{cm}^{-2})/(E_{\text{eff}}) \qquad (6)$$

In addition to the envelope action spectrum-based EL, there has always been one additional criterion to protect the lens and to limit the dose rate to both lens and the skin from very high irradiances. Initially, this was based only upon a consideration to conservatively protect against thermal effects. This was later thought essential not only to protect against thermal damage, but to also hedge against a possible unknown photochemical damage in the UV-A to the lens.

Sharply defined photobiological action spectra apply to the ultraviolet hazard $[S(\lambda)]$, the blue-light hazard function $[B(\lambda)]$, and the retinal thermal hazard function $[R(\lambda)]$. These are shown in Fig. 1.

Guidelines for the Visible

The ocular exposure limits for intense visible and IR radiation exposure of the eye (from incoherent radiation) are several, because they protect against either photochemical or thermal effects to the lens or retina.

The two primary hazards that must be assessed in evaluating an intense visible light source are (1) the photoretinitis (blue-light) hazard and (2) the retinal thermal hazard. Additionally, lenticular exposure in the near IR may be of concern. It is almost always true that light sources with a luminance less than 1 cd/cm^2 (10^4 cd/m^2) will not exceed the limits, and this is generally a maximal luminance for comfortable viewing. Although this luminance value is not considered a safety limit, it is frequently provided as a quick check to determine the need for further hazard assessment.[3,5]

The retinal thermal criteria, based upon the action spectrum $R(\lambda)$, applies to pulsed light sources and to intense sources. The longest viewing duration of potential concern is 10 s, because pupillary constriction and eye movements limit the added risk from greater exposure durations. The retinal thermal hazard EL is therefore not specified for longer durations.[12] The retinal thermal limit is:

$$\sum L_\lambda \cdot R(\lambda) \cdot \Delta\lambda \le 5/(\alpha \cdot t^{0.25}) \tag{7}$$

The blue-light photoretinitis hazard criteria were based upon the work of Ham et al.[6,11,17] The limit for time t is expressed as a $B(\lambda)$ spectrally weighted radiance:

$$\sum L_\lambda \cdot B(\lambda) \cdot t \cdot \Delta\lambda \le 100 \text{ J cm}^{-2} \text{ sr}^{-1} \tag{8}$$

Applying IR Limits

There are two criteria in the near-IR region. To protect the lens against IR cataract, the EL that is applicable to IR-A and IR-B radiant energy (i.e., 780–3000 nm) specifies a maximal irradiance for continued exposure of 10 mW/cm^2 (average) over any 1000-s period, but not to exceed an irradiance of $1.8t^{-0.75}$ W/cm^2 (for times of exposure less than 1000 s). This is based upon the fact that IR cataract in workers appears only after lifetime exposure of 80 to 150 mW/cm^2.

The second IR EL is to protect against retinal thermal injury from low-luminance IR illumination sources. This EL is for very special applications, where near-IR illuminators are used for night surveillance applications or IR LEDs are used for signalling. These illuminators have a very low visual stimulus and therefore would permit lengthy ocular exposure with dilated pupils. Although the retinal thermal limit [based upon the $R(\lambda)$ function] for intense, visible, broadband sources is not provided for times greater than 10 s because of pupillary constriction and the like, the retinal thermal hazard—for other than momentary viewing— will only realistically occur when the source can be comfortably viewed, and this is the intended application of this special-purpose EL. The IR illuminators are used for area illumination for nighttime security where it is desirable to limit light trespass to adjacent housing. If directly viewed, the typical illuminator source may be totally invisible, or it may appear as a deep cherry red source that can be comfortably viewed. The EL is proportional to $1/\alpha$ and is simply limited to:

$$\sum L_\lambda \cdot \Delta\lambda \le (0.6)/(\alpha) \text{ W/cm}^2 \tag{9}$$

It must be emphasized that this criterion is not applied to white-light sources, because bright light produces an aversion response.

15.8 DISCUSSION

Exceeding the Exposure Limits

When deriving the ELs for visible and infrared radiations, the ICNIRP generally includes a factor of 10 to reduce the 50 percent probability of retinal injury by a factor of 10. This is not a true "safety factor," because there is a statistical distribution of damage, and this factor was based upon several considerations. These included the difficulties in performing accurate measurements of source radiance or corneal irradiance, the measurement of the source angular subtense, as well as histological studies showing retinal changes occurring at the microscopic level at levels of approximately 2 below the ED-50 value.[3] In actual practice, this means that an exposure at two to three times the EL would not be expected to actually cause a physical retinal injury. At five times the EL, one would expect to find some injuries in a population of exposed subjects. The ELs are guidelines for controlling human exposure and should not be considered as fine lines between safe and hazardous exposure. By employing benefit-versus-risk considerations, it would be appropriate to have some relaxed guidelines; however, to date, no standards group has seen the need to do this.

Laser Hazards

The very high radiance (brightness) of a laser (MW and TW cm^{-2} sr^{-1}) is responsible for the laser's great value in material processing and laser surgery, but it also accounts for its significant hazard to the eye (Fig. 6). When compared with a xenon arc or the sun, even a small He-Ne alignment laser is typically 10 times brighter (Fig. 7). A collimated beam entering the relaxed human eye will experience an increased irradiance of about 10^5 (i.e., 1 W cm^{-2} at the cornea becomes 100 kW/cm^2 at the retina). Of course, the retinal image size is only about 10 to 20 μm, considerably smaller than the diameter of a human hair. So you may wonder: "So what if I have such a small lesion in my retina? I have millions of cone cells in my retina." The retinal injury is always larger because of heat flow and acoustic transients, and even a small disturbance of the retina can be significant. This is particularly important in the region of central vision, referred to by eye specialists as the *macula lutea* (yellow spot), or simply the *macula*. The central region of the macula, the fovea centralis, is responsible for your detailed 20/20 vision. Damage to this extremely small (about 150-μm diameter) central region can result in severe vision loss even though 98 percent of the retina is unscathed. The surrounding retina is useful for movement detection and other tasks but possesses limited visual acuity (after all, this is why your eye moves across a line of print, because your retinal area responsible for detailed vision has a very small angular subtense).

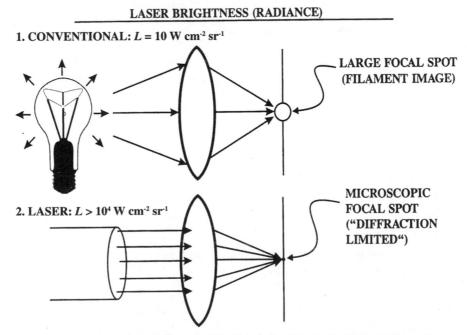

LASER BRIGHTNESS (RADIANCE)

1. CONVENTIONAL: $L = 10$ W cm^{-2} sr^{-1}

LARGE FOCAL SPOT (FILAMENT IMAGE)

2. LASER: $L > 10^4$ W cm^{-2} sr^{-1}

MICROSCOPIC FOCAL SPOT ("DIFFRACTION LIMITED")

FIGURE 6   The radiance of a light source determines the irradiance in the focal spot. Hence, the very high radiance of a laser permits one to focus laser radiation to a very small image of very high irradiance.

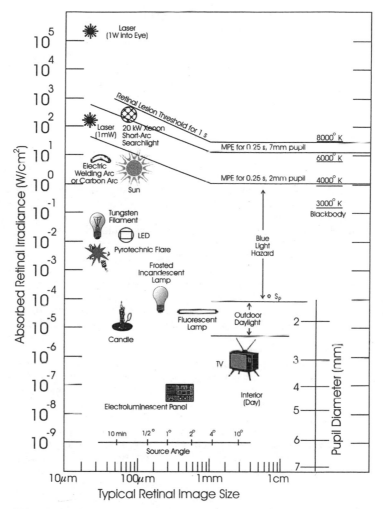

FIGURE 7 Relative retinal irradiances for staring directly at various light sources. Two retinal exposure risks are depicted: (1) retinal thermal injury, which is image-size dependent, and (2) photochemical injury, which depends on the degree of blue light in the source's spectrum. The horizontal scale indicates typical image sizes. Most intense sources are so small that eye movements and heat flow will spread the incident energy over a larger retinal area. The range of photoretinitis (the "blue-light hazard") is shown to be extending from the normal outdoor light levels and above. The xenon arc lamp is clearly the most dangerous of nonlaser hazards.

Outside the retinal hazard region (400–1400 nm), the cornea—and even the lens—can be damaged by laser beam exposure.

Laser Safety Standards. In the United States, the American National Standard, ANSI Z136.1-2000, *The Safe Use of Lasers,* is the national consensus standard for laser safety in the user environment. It evolved through several editions since 1973. Maximum permissible exposure (MPE) limits are provided as sliding scales, with wavelength and duration for all wavelengths from 180 nm to 1 mm and for exposure durations of 100 fs to 30 ks (8-h workday). Health and safety specialists want the simplest expression of the limits, but some more mathematically inclined scientists and engineers on standards committees argue for sophisticated formulas to express the limits (which belie the real level of biological uncertainty). These exposure limits (which are identical to those of ACGIH) formed the basis for the U.S. Federal Product Performance Standard (21 CFR 1040).[3] The latter standard regulates only laser manufacturers. On the international scene, the International Electrotechnical Commission Standard IEC 825-1.1 (1998) grew out of an amalgam of the ANSI standard for user control measures and exposure limits and the U.S. Federal product classification regulation. The ICNIRP, which has a special relationship with the WHO in developing criteria documents on laser radiation, now recommends exposure limits for laser radiation.[10] All of the aforementioned standards are basically in agreement.

Laser safety standards existing worldwide group all laser products into four general hazard classes and provide safe measures for each hazard class (classes 1–4).[4,5] The U.S. Federal Product Performance Standard (21 CFR 1040) requires all commercial laser products to have a label indicating the hazard class. Once one understands and recognizes the associated ocular hazards, the safety measures recommended in these standards are quite obvious (e.g., beam blocks, shields, baffles, eye protectors). The ocular hazards are generally of primary concern.

Laser Accidents.   A graduate student in a physical chemistry laboratory is aligning a Nd:YAG-pumped optical parametric oscillator (OPO) laser beam to direct it into a gas cell to study photodissociation parameters for a particular molecule. Leaning over a beam director, he glances down over an upward, secondary beam and approximately 80 μJ enters his left eye. The impact produces a microscopic hole in his retina, a small hemorrhage is produced over his central vision, and he sees only red in his left eye. Within an hour, he is rushed to an eye clinic where an ophthalmologist tells him he has only 20/400 vision. In another university, a physics graduate student attempts to realign the internal optics in a Q-switched Nd:YAG laser system—a procedure normally performed by a service representative that the student had witnessed several times before. A weak secondary beam reflected upward from a Brewster window enters the man's eye, producing a similar hemorrhagic retinal lesion with a severe loss of vision. Similar accidents occur each year and frequently do not receive publicity because of litigation or for administrative reasons.[3,26] Scientists and engineers who work with open-beam lasers really need to realize that almost all such lasers pose a very severe hazard to the eye if eye protection is not worn or if other safety measures are not observed![2–4]

A common element in most laser laboratory accidents is an attitude that "I know where the laser beams are; I do not place my eye near to a beam; safety goggles are uncomfortable; therefore, I do not need to wear the goggles." In virtually all accidents, eye protectors were available, but not worn. The probability that a small beam will intersect a 3- to 5-mm pupil of a person's eye is small to begin with, so injuries do not always happen when eye protectors are not worn. However, it is worthwhile to consider the following analogy. If these individuals were given an air rifle with 100 BBs to fire, were placed in a cubical room that measures 4 m on each side and is constructed of stainless-steel walls, and were told to fire all of the BBs in any directions they wished, how many would be willing to do this without heavy clothing and eye protectors?!! Yet, the probability of an eye injury is similar. In all fairness, there are laser goggles that are comfortable to wear. The common complaint that one cannot see the beam to align it are mere excuses that are readily solved with some ingenuity once you accept the hazard. For example, image converters and various fluorescent cards (such as are used to align the Nd:YAG 1064-nm beam) can be used for visible lasers as well.

Laser Eye Protectors.   In the ANSI Z136.1 standard, there is broad general guidance for the user to consider factors such as comfort and fit, filter damage threshold, and periodic inspection, as well as the critical specification of wavelength and optical density (OD). However, there has never been any detailed U.S. specifications for standard marking and laboratory proofing (testing) of filters. By contrast, the approach in Germany (with DIN standards) and more recently in Europe (with CEN standards) has been to minimize the decision making by the user and place heavy responsibility upon the eyewear manufacturer to design, test, and follow standardized marking codes to label the eye protector. Indeed, the manufacturer had been required to use third-party test houses at some considerable expense to have type eyewear tested. Each approach has merit, and a new standard in the United States is now being developed for testing for standardized marking that is understandable to the wearer. The most important test of a protective filter is to measure the OD under CW, Q-switched, and mode-locked pulse irradiation conditions to detect saturable absorption-reversible bleaching.

The marking of laser eye protection in an intelligible fashion to assure that the user will not misunderstand and select the wrong goggle has been a serious issue, and there appears to be no ideal solution. Several eye injuries appear to have been caused by a person choosing the wrong protector. This is particularly likely in an environment where multiple and different laser wavelengths are in use, as in a research laboratory or a dermatological laser setting. A marking of OD may be clearly intelligible to optical physicists, but this is seldom understandable to physicians and industrial workers. Probably the best assurance against misuse of eyewear has been the application of customized labeling by the user. For example, "Use only with the model 12A," or "Use only for alignment of mode-locked YAG," or "Use only for port-wine stain laser." Such cautions *supplement* the more technical marking. The terms *ruby, neodymium, carbon dioxide,* and similar labels in addition to the wavelength can reduce potential confusion with a multiwavelength laser. Marking of broadband eye protection, such as welding goggles, does not pose a problem to the wearer. The welding shade number (e.g., WG-8), even if misunderstood by the welder, poses little risk. The simple suggestion to select the filter based upon visual comfort prevents the welder from being injured; wearing too light a shade such that the EL for blue light is exceeded is virtually impossible because of the severe disability glare that would result. The same is not true for laser eye protectors: One could wear a goggle that protects against a visible wavelength but that doesn't protect against the invisible 1064-nm wavelength. Improved, standardized marking is certainly needed!

Lamp Safety Standards

Lamp safety is dealt with in ANSI/IESNA RP27.3-1996 [*Recommended Practice for Photobiological Safety for Lamps—Risk Group Classification & Labeling Safe Use of Lasers in Optical Fiber Communications Systems (OFCS)*], ANSI Z136.2 (1997), and also applies to LEDs used in this special application, because the special case of optically aided viewing with a microscope or eye loupe actually happens in realistic servicing conditions and the instrumentation to distinguish the narrow bandwidth of the laser from the wider, nominal 50-nm bandwidth of the LED is a needless burden. All current LEDs have a maximum radiance of about 2 W cm^{-2} sr^{-1} and can in no way be considered hazardous. Some observers of the IEC group accuse the group of having an ulterior motive of providing test houses and safety delegates with greater business!

15.9   REFERENCES

1. L. R. Solon et al., "Physiological Implications of Laser Beams," *Science* **134**:1506–1508 (1961).

2. D. H. Sliney and B. C. Freasier, "The Evaluation of Optical Radiation Hazards," *Applied Optics* **12**(1):1–24 (1973).

3. D. H. Sliney and M. L. Wolbarsht, *Safety with Lasers and Other Optical Sources.* Plenum Publishing, New York, 1980.

4. World Health Organization (WHO), *Environmental Health Criteria No. 23, Lasers and Optical Radiation,* joint publication of the United Nations Environmental Program, the International Radiation Protection Association, and the World Health Organization, Geneva, 1982.

5. American Conference of Governmental Industrial Hygienists (ACGIH), *TLVs, Threshold Limit Values and Biological Exposure Indices for 1999,* American Conference of Governmental Industrial Hygienists, Cincinnati, OH, 1999.

6. ———, *Documentation for the Threshold Limit Values,* 4th ed., American Conference of Governmental Industrial Hygienists, Cincinnati, OH, 1992.

7. A. S. Duchene et al. (eds.), *IRPA Guidelines on Protection Against Non-Ionizing Radiation,* MacMillan, New York, 1991.

8. D. H. Sliney, "Laser Effects on Vision and Ocular Exposure Limits," *Appl. Occup. Environ. Hyg.* **11**(4):313–319 (1996).

9. American National Standards Institute (ANSI), *Safe Use of Lasers,* ANSI Z136.1-2000, Laser Institute of America, Orlando, FL, 2000.

10. International Commission on Non-Ionizing Radiation Protection (ICNIRP), "Guidelines on Limits for Laser Radiation of Wavelengths between 180 nm and 1,000 μm," *Health Phys.* **71**(5):804–819 (1996); update (in press).

11. ———, "Guidelines on Limits of Exposure to Broad-Band Incoherent Optical Radiation (0.38 to 3 μm)," *Health Phys.* **73**(3):539–554 (1997).

12. W. P. Roach et al., "Proposed Maximum Permissible Exposure Limits for Ultrashort Laser Pulses," *Health Phys.* **77**:61–68 (1999).

13. T. P. Coohill, "Photobiological Action Spectra—What Do They Mean?" *Measurements of Optical Radiation Hazards,* ICNIRP, 27–39 (1998).

14. M. Waxler and V. Hitchens (eds.), *Optical Radiation and Visual Health,* CRC Press, Boca Raton, FL, 1986.

15. J. A. Zuclich, "Ultraviolet-Induced Photochemical Damage in Ocular Tissues," *Health Phys.* **56**(5):671–682 (1989).

16. D. G. Pitts et al., "Ocular Effects of Ultraviolet Radiation from 295 to 365 nm," *Invest. Ophthal. Vis. Sci.* **16**(10):932–939 (1977).

17. W. T. Ham, Jr., "The Photopathology and Nature of the Blue-Light and Near-UV Retinal Lesion Produced by Lasers and Other Optical Sources," in *Laser Applications in Medicine and Biology,* M. L. Wolbarsht (ed.), Plenum Publishing, New York, 1989.

18. W. T. Ham et al., "Evaluation of Retinal Exposures from Repetitively Pulsed and Scanning Lasers," *Health Phys.* **54**(3):337–344 (1988).

19. D. H. Sliney, "Physical Factors in Cataractogenesis—Ambient Ultraviolet Radiation and Temperature," *Invest. Ophthalmol. Vis. Sci.* **27**(5):781–789 (1986).

20. Commission Internationale de l'éclairage (International Commission on Illumination), *International Lighting Vocabulary,* CIE Publication No. 17.4, CIE, Geneva, Switzerland (1987).

21. M. T. Coroneo et al., "Peripheral Light Focussing by the Anterior Eye and the Ophthalmohelioses," *Ophthalmic Surg.* **22**:705–711 (1991).

22. B. E. K. Klein et al., "Prevalence of Age-Related Lens Opacities in a Population, the Beaver Dam Eye Study," *Ophthalmology* **99**(4):546–552 (1992).

23. D. H. Sliney, "Eye Protective Techniques for Bright Light," *Ophthalmology* **90**(8):937–944 (1983).

24. P. J. Dolin, "Assessment of the Epidemiological Evidence that Exposure to Solar Ultraviolet Radiation Causes Cataract," *Doc. Ophthalmol.* **88**:327–337 (1995).

25. D. H. Sliney, "UV Radiation Ocular Exposure Dosimetry," *Doc. Ophthalmol.* **88**:243–254 (1995).

26. D. H. Sliney, "Ocular Injuries from Laser Accidents," *SPIE Proceedings of Laser-Inflicted Eye Injuries: Epidemiology, Prevention, and Treatment,* San Jose, CA, 1996, pp. 25–33.

27. J. W. Ness et al., "Retinal Image Motion during Deliberate Fixation: Implications to Laser Safety for Long Duration Viewing," *Health Phys.* **72**(2):131–142 (2000).

CHAPTER 16
VISION PROBLEMS AT COMPUTERS

James E. Sheedy
University of California, Berkeley, School of Optometry
Berkeley, California

16.1 INTRODUCTION

Vision and eye problems are the most commonly reported symptoms among workers at computer monitors. The percentage of computer workers who experience visually related symptoms is generally reported to be from 70 to 75 percent. The current concern for, and popularization of, these problems is likely related to the rapid introduction of this new technology into the work place, the job restructuring associated with it, and a greater societal concern about discomfort associated with work.

The problems are largely symptoms of discomfort. There is little evidence of permanent physiological change or damage associated with extended work at a computer monitor. The vision and eye problems caused by working at a computer monitor have been collectively named "Computer Vision Syndrome (CVS)" by the American Optometric Association. The most common symptoms of CVS obtained from a survey of optometrists in order of frequency are: eyestrain, headaches, blurred vision, dry or irritated eyes, neck and/or backaches, photophobia, double vision, and afterimages.

The causes for these symptoms are a combination of individual visual problems and poor office ergonomics. Many individuals have marginal vision disorders that do not cause symptoms on less demanding visual tasks. On the other hand, there are numerous aspects of the computer monitor and the computer work environment that make it a more demanding visual task than others—therefore, more individuals are put beyond their threshold for experiencing symptoms.

16.2 WORK ENVIRONMENT

Lighting

Improper lighting is likely the largest environmental factor contributing to visual discomfort. The room lighting is a particular problem for computer workers because the horizontal gaze angle of computer work exposes the eyes to numerous glare sources such as overhead lights and windows. Other desk work is performed with a downward gaze angle and the glare sources are not in the field of view.

Bright (high luminance) light sources in the field of view create glare discomfort. The threshold luminance ratios and locations of visual stimuli that cause glare discomfort have been determined,[1] but the physiological basis for glare discomfort is not known. Since large luminance disparities in the field of view can cause glare discomfort, it is best to have a visual environment in which luminances are relatively equal. Primarily because of glare discomfort, the Illumination Engineering Society of North America (IESNA) established maximum luminance ratios that should not be exceeded.[2] The luminance ratio should not exceed 1:3 or 3:1 between the task and visual surroundings within 25 degrees, nor should the ratio exceed 1:10 or 10:1 between the task and more remote visual surroundings. Many objects in the field of view can cause luminance ratios in excess of those recommended by the IESNA. Table 1 shows common luminance levels of objects in an office environment. Relative to the display luminance, several objects can greatly exceed the IESNA recommended ratios. A major advantage of using light background displays compared to dark background displays is that it enables better conformity with office luminance levels.

Good lighting design can reduce discomfort glare. Light leaving the fixture can be directed so that it goes straight down and not into the eyes of the room occupants. This is most commonly accomplished with louvers in the fixture. A better solution is indirect lighting in which the light is bounced off the ceiling—resulting in a large low luminance source of light for the room. Proper treatment of light from windows is also important.

Screen Reflections

In most cases the most bothersome source of screen reflections is from the phosphor on the inner surface of the glass. The phosphor is the material that emits light when struck by the electron beam—it also passively reflects room light. The reflections coming from the phosphor are diffuse. Most computer monitors have a frosted glass surface; therefore, the light reflected from the glass is primarily diffuse with a specular component.

Since most of the screen reflections are diffuse, a portion of any light impinging on the screen, regardless of the direction from which it comes, is reflected into the eyes of the user and causes the screen to appear brighter. This means that black is no longer black, but a shade of gray. The diffuse reflections from the phosphor and from the glass reduce the contrast of the text presented on the screen. Instead of viewing black characters on a white background, gray characters are presented on a white background. Calculations[3] show that these reflections significantly reduce contrast—from 0.96 to 0.53 [contrast = $(L_t - L_b)/(L_t + L_b)$, where L_t and L_b are luminances of the task and background] under common conditions. Decreased contrast increases demand upon the visual system.

A common method of treating reflections is with an antireflection filter placed on the computer monitor. The primary purpose of the filter is to make the blacks blacker, thereby increasing the contrast. The luminance of light which is emitted by the screen (the desired

TABLE 1   Display and Glare Source Luminances[3]

Visual Object	Luminance (cd/m^2)
Dark background display	20–25
Light background display	80–120
Reference material with 750 lumens/M^2	200
Reference material with auxiliary light	400
Blue sky (window)	2500
Concrete in sun (window)	6000–12000
Fluorescent lights (poor design)	1000–5000
Auxiliary lamp (direct)	1500–10000

image) is decreased by the transmittance factor of the glass filter, whereas light that is reflected from the computer screen (undesired light) is decreased by the square of the filter transmittance, since it must pass through it twice. For a typical filter of 30 percent transmittance, the luminances of black and white are reduced respectively to nine percent and 30 percent of their values without the filter. This results in an increase in the contrast.

Some antireflection filters have a circular polarized element in them. This feature results in circular polarized light's being transmitted to the computer screen. The resulting reflection on the screen changes the rotation of the polarization of the light so that it is blocked from coming back out through the filter. Since only specular reflected light maintains its polarized properties after reflection, this polarizing feature provides added benefit for only the specular reflections. If significant specular reflections are present, it is beneficial to obtain an antireflection filter with circular polarization.

Monitor Characteristics

One obvious method of improving the visual environment for computer workers is to improve the legibility of the display at which they are working. The visual image on computer monitors is compromised in many ways compared to most paper tasks.

A major aspect of computer displays is that the images are composed of pixels. A greater pixel density will be able to display greater detail. Most computer screens have pixel densities in the range of from 80 to 110 dots/inch (dpi; 31–43 dots/cm). This pixel density can be compared to impact printers (60–100 dpi), laser printers (300 dpi), and advanced laser printers (400 to 1200 dpi). Computer monitors today have dpi specifications that are comparable to impact printers—greater pixel density would result in visually discernible differences, as for printers. Besides pixel density there are many other factors that affect display quality, such as pixel definition (measured by area under the modulation transfer function), font type, font size, letter spacing, line spacing, stroke width, contrast, color, gray scale, and refresh rate.[4] Several studies have shown that increased screen resolution characteristics increase reading speed and/or decrease visual symptoms.[3] Although technological advances have resulted in improved screen resolutions, there is still substantial room for improvement.

Workstation Arrangement

It is commonly stated that "the eyes lead the body." Since working at the computer is a visually intensive task, our body will do what is necessary to get the eyes in the most comfortable position—often at the expense of good posture and causing musculo-skeletal ailments such as sore neck and back.

The most common distance at which people view printed material is 40 cm from the eyes. This is the distance at which eye doctors routinely perform near visual testing and for which most bifocal or multifocal glasses are designed. Most commonly, the computer screen is farther from the eyes—from 50 to 70 cm. This distance is largely dictated by other workstation factors, such as desk space and having room for a keyboard. The letter sizes on the screen are commensurately larger compared to printed materials to enable this longer viewing distance.

The height of the computer screen is a very important aspect of the workstation arrangement. A person can adapt to the vertical location of their task by changing their gaze angle (the elevation of the eyes in the orbit) and/or by changing the extension/flexion of the neck. Typically, users will alter their head position rather than eye position to adjust to a different viewing height. This can cause awkward posture and result in neck and/or backache. The eyes work best with a depression of from 10 to 20 degrees[3]; therefore, this is the preferred viewing angle.

Whatever is viewed most often during daily work should be placed straight in front of the worker when seated at the desk. This applies to the computer and/or the reference documents—whatever is viewed most frequently. Although this seems self-evident, many computer workers situate their work so they are constantly looking off to one side.

16.3 VISION AND EYE CONDITIONS

Many individuals have marginal vision disorders that do not cause symptoms on less demanding visual work, but will cause symptoms when the individual performs a demanding visual task. Given individual variation in visual systems and work environments, individual assessment is required to solve any given person's symptoms. Following are the major categories of eye and vision disorders that cause symptoms among computer users.

Dry Eyes

Computer users commonly experience symptoms related to dry eyes. These symptoms include irritated eyes, dry eyes, excessive tearing, burning eyes, itching eyes, and red eyes. Contact lens wearers also often experience problems with their contact lenses while working at a computer display—related to dry eyes. In response to dry eye and ocular irritation, reflex tearing sometimes occurs and floods the eyes with tears. This is the same reflex that causes tearing when we cry or when a foreign particle gets into our eye.

Computer workers are at greater risk for experiencing dry eye because the blink rate is significantly decreased and, because of the higher gaze angle, the eyes are wide open with a large exposed ocular surface. Patel et al.[5] measured blink rate by direct observation on a group of 16 subjects. The mean blink rate during conversation was 18.4 blinks/min and during computer use it was 3.6—more than a five-fold decrease. Tsubota and Nakamori (1993) measured blink rates on 104 office workers.[6] The mean blink rates were 22 blinks/min under relaxed conditions, 10 while reading a book on the table, and 7 while viewing text on a computer screen. Although both book reading and computer work result in significantly decreased blink rates, a difference between them is that computer work usually requires a higher gaze angle, resulting in an increased rate of tear evaporation. Tsubota and Nakamori[6] measured a mean exposed ocular surface of 2.2 cm^2 while subjects were relaxed, 1.2 cm^2 while reading a book on the table, and 2.3 cm^2 while working at a computer. Since the primary route of tear elimination is through evaporation, and the amount of evaporation is a roughly linear function of ocular aperture area, the higher gaze angle when viewing a VDT screen results in faster tear loss. Even though reading a book and VDT work both reduce the blink rate, the VDT worker is more at risk because of the higher gaze angle.

Other factors may also contribute to dry eye. For example, the office air environment is often low in humidity and can contain contaminants. Working under an air vent also increases evaporation. The higher gaze angle results in a greater percentage of blinks that are incomplete, resulting in poor tear film and higher evaporation rates.

Dry eyes are common in the population, reported as being 15 percent of a clinical population. For example, postmenopausal women and arthritis sufferers are more prone to dry eyes. For many people with marginal dry eye problems, work at a computer will cause their dry eye problem to become clinically significant.

Refractive Error

Many patients simply need an accurate correction of refractive error (myopia, hyperopia, or astigmatism). The blur created by the refractive error makes it difficult to easily acquire visual information at the computer display. This reduces work efficiency and induces fatigue.

Patients with hyperopia must exert added accommodative effort to see clearly at near distances. Therefore, many hyperopic computer patients require refractive correction that they may not require for a less visually demanding job. Patients with from 2.00 to 3.50 D of myopia who habitually read without their glasses often have visual and musculoskeletal difficulties at the computer because, to see clearly without their glasses, they must work too closely to their computer screen. These patients often require a partial correction of their myopia for proper function at the computer display.

Some computer-using patients develop a late onset myopia of from 0.25 to 1.00 D. The preponderance of evidence supports the contention that near work places some people at greater risk for the development of myopia. However, there is no evidence that work at a computer monitor causes myopia more than other forms of extended near-point work. Environmentally determined myopia is not well understood; future study might identify work environment or individual risk visual factors.

Accommodation

Accommodation is the mechanism by which the eye changes its focus to look at near objects. Accommodation is accomplished by constriction of the ciliary muscle that surrounds the crystalline lens within the eye. The amplitude of accommodation (dioptric change in power) decreases with age. Presbyopia is the age-related condition that results when the amplitude of accommodation is no longer adequate to meet near visual needs. (See Chaps. 11 and 13.)

Many workers have reduced amplitude of accommodation for their age, or *accommodative infacility* (inability to change the level of accommodation quickly and accurately). These conditions result in blur at near working distances and/or discomfort. Extended near work also commonly results in *accommodative hysteresis*—i.e., the accommodative mechanism becomes "locked" into the near focus and it takes time (a few minutes to hours) to fully relax for distant focus. This is effectively a transient myopia that persists after extended near work. Accommodative disorders are diagnosed in approximately one third of the prepresbyopic segment of a clinical population. Accommodative disorders in the younger or prepresbyopic patient are usually treated with near prescription spectacles that enable the worker to relax accommodation. Vision training can also work for some patients.

Binocular Vision

Approximately 95 percent of people keep both eyes aligned on the object of regard. Those individuals who habitually do not keep their eyes aligned have *strabismus*. Even though most people can keep their eyes aligned when viewing an object, many individuals have difficulty maintaining this ocular alignment and experience symptoms such as fatigue, headaches, blur, double vision, and general ocular discomfort (see Chap. 12, "Binocular Vision Factors That Influence Optical Design").

Binocular fusion is the sensory process by which the images from each eye are combined to form a single percept. When the sensory feedback loop is opened (e.g., by blocking an eye), the eyes assume their *position of rest* with respect to one another. If the position of rest is outward or diverged, the patient has *exophoria*. If it is inward or converged, the condition is *esophoria*. Clinically, the phoria is measured by occluding one of the eyes and measuring the eye alignment while occluded.

If the patient has a phoria, as is the case for most people, then a constant neuromuscular effort is required to keep the eyes aligned. Whether a person experiences symptoms depends on the amount of the misalignment, the ability of the individual to overcome that misalignment, and the task demands. The symptoms associated with phoria can be eyestrain, double vision, headaches, eye irritation, and general fatigue.

Eye alignment at near viewing distances is more complex than at far viewing distances because of the ocular convergence required to view near objects and because of the interaction between the ocular convergence and accommodative mechanisms. Treatment of these conditions can include refractive or prismatic correction in spectacles, or vision training exercises.

Presbyopia

Presbyopia is the condition in which the normal age-related loss of accommodation results in an inability to comfortably maintain focus on near objects. This usually begins at about age 40. The usual treatment for presbyopia is to prescribe reading glasses or multifocal lenses that have a distance vision corrective power in the top of the lens and the near vision corrective power in the bottom of the lens. The most common lens designs for correcting presbyopia are bifocals and progressive addition lenses (PALs). As usually designed and fitted, these lenses work well for the most common every day visual tasks and provide clear vision at 40 cm with a downward gaze angle of about 25 degrees. The computer screen is typically farther away (from 50 to 70 cm) and higher (from 10 to 20 degrees of ocular depression). A presbyope who tries to wear their usual multifocal correction at the computer will either not see the screen clearly or will need to assume an awkward posture—resulting in neck and back strain. Many, if not most, presbyopic computer workers require a separate pair of spectacles for their computer work. Several newer lenses are designed specifically for people with occupations that require intermediate viewing distances. (See Chaps. 11 and 13.)

16.4 REFERENCES

1. S. K. Guth, "Prentice Memorial Lecture: The Science of Seeing—A Search for Criteria," *American Journal of Optometry and Physiological Optics* **58**:870–885 (1981).
2. "VDT Lighting. IES Recommended Practice for Lighting Offices Containing Computer Visual Display Terminals," Illumination Engineering Society of North America, New York, 1990.
3. J. E. Sheedy, "Vision at Computer Displays," Vision Analysis, Walnut Creek, CA, 1995.
4. ANSI/HFS 100. American National Standard for Human Factors Engineering of Visual Display Terminal Workstations, Human Factors Society, Santa Monica, CA, 1999.
5. S. Patel, R. Henderson, L. Bradley, B. Galloway, and L. Hunter, "Effect of Visual Display Unit Use on Blink Rate and Tear Stability," *Optometry and Vision Science* **68**(11):888–892 (1991).
6. K. Tsubota and K. Nakamori, "Dry Eyes and Video Display Terminals. Letter to Editor," *New England Journal of Medicine* **328**:524 (1993).

CHAPTER 17

HUMAN VISION AND ELECTRONIC IMAGING

Bernice E. Rogowitz
IBM T. J. Watson Research Center
Huwthorne, New York

Thrasyvoulos N. Pappas
Electrical and Computer Engineering, Northwestern University
Evanston, Illinois

Jan P. Allebach
Electronic Imaging Systems Laboratory
Purdue University, School of Electrical and Computer Engineering
West Lafayette, Indiana

17.1 INTRODUCTION

The field of *electronic imaging* has made incredible strides over the past decade as increased computational speed, bandwidth, and storage capacity have made it possible to perform image computations at interactive speeds. This means larger images, with more spatial, temporal, and chromatic resolution, can be captured, compressed, transmitted, stored, rendered, printed, and displayed. It also means that workstations and PCs can accommodate more complex image and data formats, more complex operations for analyzing and visualizing information, more advanced interfaces, and richer image environments, such as virtual reality. This, in turn, means that image technology can now be practically used in an expanding world of applications, including video, home photography, internet catalogues, digital libraries, art, and scientific data analysis. These advances in technology have been greatly influenced by research in human perception and cognition, and in turn, have stimulated new research into the vision, perception, and cognition of the human observer. Some important topics include spatial, temporal, and color vision, attentive and pre-attentive vision, pattern recognition, visual organization, object perception, language, and memory.

The study of the interaction between human vision and electronic imaging is one of the key growth areas in *imaging science*. Its scope ranges from printing and display technologies to image processing algorithms for image rendering and compression, to applications involving interpretation, analysis, visualization, search, design, and aesthetics. Different electronic imaging applications call on different human capabilities. At the bottom of the visual food chain are the visual phenomena mediated by the threshold sensitivity of low-level spatial, temporal, and color mechanisms. At the next level are perceptual effects, such as color constancy, suprathreshold pattern, and texture analysis. Moving up the food chain, we find cognitive effects, including memory, semantic categorization, and visual representation, and moving to the next level, we encounter aesthetic and emotional aspects of visual processing.

Overview of This Chapter

In this chapter, we review several key areas in the two-way interaction between human vision and technology. We show how technology advances in electronic imaging are increasingly driven by methods, models, and applications of vision science. We also show how advances in vision science are increasingly driven by the rapid advances in technologies designed for human interaction. An influential force in the development of this new field has been the Society for Imaging Science and Technology (IS&T)/Society of Photographic and Instrumentation Engineers (SPIE) Conference on Human Vision and Electronic Imaging, which had its origins in 1988 and since 1989 has grown annually as a forum for multidisciplinary research in this area.[1-14] This chapter has been strongly influenced by the body of research that has been presented at these conferences, and reflects their unique perspectives.

A decade ago, the field of *human vision and electronic imaging* focused on the threshold sensitivity of the human visual system (HVS) as it relates to display technology and still-image compression. Today, the scope of human vision and electronic imaging is much larger and keeps expanding with the electronic imaging field. We have organized this chapter to reflect this expanding scope. We begin with early vision approaches in Sec. 17.2, showing how the explicit consideration of human spatial, temporal, and color sensitivity has affected the development of algorithms for compression and rendering, as well as the development of image quality metrics. These approaches have been most successful in algorithmically evaluating the degree to which artifacts introduced by compression or rendering processes will be detected. Section 17.3 considers how image features are detected, processed, and perceived by the human visual system. This work has been influential in the design of novel gaze-dependent compression schemes, new visualization and user interface designs, and perceptually-based algorithms for image retrieval systems. Section 17.4 moves higher up the food chain to consider emotional and aesthetic evaluations. This work has influenced the development of virtual environments, high-definition TV, and tools for artistic appreciation and analysis. Section 17.5 concludes the chapter paper, and suggestions for further reading are given in Sec. 17.6.

17.2   *EARLY VISION APPROACHES: THE PERCEPTION OF IMAGING ARTIFACTS*

There are many approaches to characterizing image artifacts. Some approaches are based purely on physical measurement. These include measuring key image or system parameters to insure that they fall within established tolerances, or comparing an original with a rendered, displayed, or compressed version, on a pixel-by-pixel basis, using a metric such as mean square error. These approaches have the advantage of being objective and relatively easy to implement; on the other hand, it is difficult to generalize from these data. Another approach is to use human observers to judge perceived quality, either by employing a panel of trained experts or by running psychological experiments to measure the perception of image characteristics. These approaches have the advantage of considering the human observer explicitly; on the other hand, they are costly to run and the results may not generalize. A third approach is to develop metrics, based on experiments measuring human visual characteristics, that can stand in for the human observer as a means of estimating human judgments. These perceptual models are based on experiments involving the detection, recognition, and identification of carefully controlled experimental stimuli. Since these experiments are designed to reveal the behavior of fundamental mechanisms of human vision, their results are more likely to generalize. This section reviews several models and metrics based on early human vision which have been developed for evaluating the perception of image artifacts.

Early vision models are concerned with the processes mediating the threshold detection of spatial, temporal, and color stimuli. In the early days of television design, Schade[15] and his colleagues introduced the notion of characterizing human threshold sensitivity in terms of the response to spatial frequency patterns, thereby beginning a long tradition of work to model human spatial and temporal sensitivity using the techniques of linear systems analysis. The simplest model of human vision is a threshold contrast sensitivity function (CSF). A curve representing our sensitivity to spatial modulations in luminance contrast is called a spatial CSF. A curve representing our sensitivity to temporal modulations in luminance is a temporal CSF. These band-pass curves vary depending on many other parameters, including, for example, the luminance level, the size of the field, temporal modulation, and color. In early applications of visual properties to electronic imaging technologies, these simple threshold shapes were used. As the field has progressed, the operational model for early vision has become more sophisticated, incorporating interactions between spatial, temporal, and color vision, and making more sophisticated assumptions about the underlying processes. These include, for example, the representation of the visual system as a set of band-pass spatial-frequency filters (Campbell and Robson[16] and Graham and Nachmias[17]), the introduction of near-threshold contrast masking effects, and nonlinear processing.

One key area where these models of early vision have been applied is in the evaluation of image quality. This includes the psychophysical evaluation of image quality, perceptual metrics of image distortion, perceptual effects of spatial, temporal, and chromatic sampling, and the experimental comparison of compression, sampling, and halftoning algorithms. This work has progressed hand-in-hand with the development of vision-based algorithms for still image and video compression, image enhancement, restoration and reconstruction, image halftoning and rendering, and image and video quantization and display.

Image Quality and Compression

The basic premise of the work in perceptual image quality is that electronic imaging processes, such as compression and halftoning, introduce distortions. The more visible these distortions, the greater the impairment in image quality. The human vision model is used to evaluate the degree to which these impairments will be detected. Traditional metrics of image compression do not incorporate any models of human vision and are based on the mean squared error (i.e., the average squared difference between the original and compressed images). Furthermore, they typically fail to include a calibration step, or a model of the display device, thereby providing an inadequate model of the information presented to the eye.

The first perceptually based image quality metrics used the spatial contrast sensitivity function as a model of the human visual system.[18,19] The development of perceptual models based on multiple spatial-frequency channels greatly improved the objective evaluation of image quality. In these models, the visual system is treated as a set of spatial-frequency-tuned channels, or as Gabor filters with limited spatial extent distributed over the visual scene. The envelope of the responses of these channels is the contrast sensitivity function. These multiple channel models provide a more physiologically representative model of the visual system, and more easily model interactions observed in the detection of spatially varying stimuli. One important interaction is contrast masking, where the degree to which a target signal is detected depends on the spatial-frequency composition of the masking signal.

In 1992, Daly introduced the *visual differences predictor,*[6,20] a multiple-channel model for image quality that models the degree to which artifacts in the image will be detected, and thus, will impair perceived image quality. This is a spatial frequency model which incorporates spatial contrast masking and light adaptation. At about the same time, Lubin[21] proposed a similar metric that also accounts for sensitivity variations due to spatial frequency and masking. It also accounts for fixation depth and image eccentricity in the observer's visual field. The output of such metrics is either a map of detection probabilities or a point-by-point measure of the distance between the original and degraded image normalized by the human visual system (HVS) sensitivity to error at each spatial frequency and location. These detection probabilities or distances can be combined into a single number that represents the overall picture quality. While both models were developed for the evaluation of displays and high quality imaging systems, they have been adapted for a wide variety of applications.

Perceptually based image compression techniques were developed in parallel with perceptual models for image quality. This is not surprising since quality metrics and compression algorithms are closely related. The image quality metric is trying to characterize the human response to an image; an image compression algorithm is either trying to minimize some distortion metric for a given bit rate, or trying to minimize the bit rate for a given distortion. In both cases, a perceptually based distortion metric can be used.

In 1989, three important papers introduced the notion of "perceptually lossless" compression. (See Ramamoorthy and Jayant[2]; Daly[2]; and Watson.[2]) In this view, the criterion of importance in image compression was the degree to which an image could be compressed without the user's perceiving a difference between the compressed and original images. Safranek and Johnston[22] presented the Perceptual Subband Image Coder (PIC) which incorporated a perceptual model and achieved perceptually lossless compression at lower rates than state of the art perceptually lossy schemes. The Safranek-Johnston coder used an empirically derived perceptual masking model that was obtained for a given CRT display and viewing conditions. As with the quality metrics we discussed earlier, the model determines the HVS sensitivity to errors at each spatial frequency and location, which is called the *just noticeable distortion level* (JND). In subsequent years, perceptual models were used to improve the results of traditional approaches to image and video compression, such as those that are based on the discrete cosine transform (DCT). (See Peterson, Ahumada, and Watson[7]; Watson[7]; and Silverstein and Klein.[7])

Perceptually-based image quality metrics have also been extended to video. In 1996, Van den Branden Lambrecht and Verscheure[23] described a video quality metric that incorporates spatio-temporal contrast sensitivities as well as luminance and contrast masking adjustments. In 1998, Watson extended his DCT-based still image metric, proposing a video quality metric based on the DCT.[12,13] Since all the current video coding standards are based on the DCT, this metric is useful for optimizing and evaluating these coding schemes without significant additional computational overhead.

In recent years, significant energy has been devoted to comparing these methods, models, and results, and to fine-tuning the perceptual models. The Video Quality Experts Group (VQEG), for example, has conducted a cross-laboratory evaluation of perceptually based video compression metrics. They found that different perceptual metrics performed better on different MPEG-compressed image sequences, but that no one model provided a clear advantage, including the nonperceptual measure, peak signal-to-noise ratio (Corriveau et al.[14]). We believe that the advantages of the perceptual metrics will become apparent when this work is extended to explicitly compare fundamentally different compression schemes over different rates, image content, and channel distortions. An important approach for improving low-level image quality metrics is to provide a common set of psychophysical data to model. Modelfest (Carney et al.[14]) is a collaborative modeling effort where researchers have volunteered to collect detection threshold data on a wide range of visual stimuli, under carefully controlled conditions, in order to provide a basis for comparing the predictions of early vision models. This effort should lead to a converged model for early vision that can be used to develop image quality metrics and perceptually based compression and rendering schemes. (Comprehensive reviews of the use of perceptual criteria for the evaluation of image quality can be found in Refs. 24 and 25.)

Image Rendering, Halftoning, and Other Applications

Another area where simple models of low-level vision have proven to be quite successful is image halftoning and rendering. All halftoning and rendering algorithms make implicit use of the properties of the HVS; they would not work if it were not for the high spatial frequency cutoff of the human CSF. In the early 1980s, Allebach introduced the idea of halftoning based on explicit human visual models and models of the display device. However, the development of halftoning techniques that rely on such models came much later. In 1991, Sullivan et al.[26] used a CSF model to design halftoning patterns of minimum visibility and Pappas and Neuhoff[5] incorporated printer models in error diffusion. In 1992, methods that use explicit visual models to minimize perceptual error in image halftoning were independently proposed by Analoui and Allebach,[6] Mulligan and Ahumada,[6] and Pappas and Neuhoff.[6] Allebach et al. extended this model-based approach to color in 1993[7] and to video in 1994.[8] Combing the spatial and the chromatic properties of early vision, Mulligan[3] took advantage of the low spatial frequency sensitivity of chromatic channels to hide high spatial frequency halftoning artifacts.

Low-level perceptual models and techniques have also been applied to the problem of target detection in medical images (e.g., Eckstein et al.[11]) and to the problem of embedding digital watermarks in electronic images (Cox and Miller[11] and Podilchuk[11]).

Early Color Vision and Its Applications

The trichromacy of the human visual system has been studied for over a hundred years, but we are just in the infancy in applying knowledge of early color vision to electronic imaging systems. From an engineering perspective, the CIE 1931 model, which represents human color-matching as a linear combination of three color filters, has been the most influential model of human color vision. This work allows the determination of whether two patches will have matching colors, when viewed under a constant illuminant. The CIE 1931 color space, however, is not perceptually uniform. That is, equal differences in chromaticity do not correspond to equal perceived differences. To remedy this, various transformations of this space have been introduced. Although CIE L*a*b* and CIE L*u*v* are commonly used as perceptually uniform spaces, they still depart significantly from this objective. By adding a nonlinear gain control at the receptor level, Guth's ATD model[2,7] provided a more perceptually uniform space and could model a wide range of perceptual phenomena. Adding a spatial component to a simplified version of Guth's model, Granger[7] demonstrated how this approach could be used to significantly increase the perceived similarity of original and displayed images. The observation that every pixel in an image provides input both to mechanisms sensitive to color variations and to mechanisms sensitive to spatial variations has provided other important contributions. For example, Uriegas[8,10] used the multiplexing of color and spatial information by cortical neurons to develop a novel color image compression scheme.

Understanding how to represent color in electronic imaging systems is a very complicated problem, since different devices (e.g., printers, CRT or TFT/LCD displays, film) have different mechanisms for generating color, and produce different ranges, or gamuts, of colors. Several approaches have been explored for producing colors on one device that have the same appearance on another. Engineering research in this field goes under the name of *device independent color,* and has recently been energized by the need for accurate color rendering over the internet. (See Gille, Luszcz, and Larimer.[12])

Limitations of Early Vision Models for Electronic Imaging

The goal of the early vision models is to describe the phenomena of visual perception in terms of simple mechanisms operating at threshold. Pursuing this Ockham's razor approach has motivated the introduction of masking models and nonlinear summation models, and has allowed us to extend these simple models to describe the detection and recognition of higher-level patterns, such as textures and multiple-sinewave plaids. These models, however, eventually run out of steam in their ability to account for the perception of higher-level shapes and patterns. For example, the perception of a simple dot cannot be successfully modeled in terms of the response of a bank of linear spatial frequency filters.

In image quality, early vision approaches consider an image as a collection of picture elements. They measure and model the perceived fidelity of an image based on the degree to which the pixels, or some transformation of the pixels, have changed in their spatial, temporal, and color properties. But what if two images have identical perceived image fidelity, but one has a lot of blur and little blockiness while the other has little blur and a lot of blockiness? Are they equivalent? Also, since it is not always possible to conceal these artifacts, it is important to understand how their effects combine and interact in order to minimize their objectionable and annoying effects. This has led to the development of new suprathreshold scaling methods for evaluating the perceived quality of images with multiple suprathreshold artifacts (e.g., de Ridder and Majoor,[3] Roufs and Boschman[3]), and has led to the development of new dynamic techniques to study how the annoyance of an artifact depends on its temporal position in a video sequence (Pearson[12]).

In color vision, a key problem for electronic imaging is that simply capturing and reproducing the physical color of individual color pixels and patches in an image is not sufficient to describe the perceived colors in the image. Hunt,[27] for example, has shown that surrounding colors affect color appearance. Furthermore, as McCann[12] has demonstrated using spatially-complex stimuli, the perception of an image depends, not simply on the local luminance and color values, and not simply on nearby luminance and color values, but on a more global consideration of the image and its geometry.

17.3 HIGHER-LEVEL APPROACHES: THE ANALYSIS OF IMAGE FEATURES

Higher-level approaches in vision are dedicated to understanding the mechanisms for perceiving more complex features such as dots, plaids, textures, and faces. One approach is to build up from early spatial-frequency models by positing nonlinear mechanisms that are sensitive to two-dimensional spatial variations, such as t-junctions (e.g., Zetzsche and Barth[3]). Interesting new research by Webster and his colleagues[12] identifies higher-level perceptual mechanisms tuned to more complex visual relationships. Using the same types of visual adaptation techniques commonly used in early vision experiments to identify spatial, temporal, or color channel properties, they have demonstrated adaptation to complex visual stimuli, including, for example, the adaptation to complex facial distortions and complex color distributions.

Another interesting approach to understanding the fundamental building blocks of human vision is to consider the physical environment in which it has evolved, with the idea in mind that the statistics of the natural world must somehow have guided, or put constraints on, its development. This concept, originally introduced by Field,[2,28] has led to extensive measurements of the spatial and temporal characteristics of the world's scenes, and to an exploration of the statistics of the world's illuminants and reflective surfaces. One of the major findings in this field is that the spatial amplitude spectra of natural scenes falls off as $f^{-1.1}$. More recently, Field et al.,[10] have related this function to the sparseness of spatial-frequency mechanisms in human vision. This low-level description of the global frequency content in natural scenes has been tied to higher-level perception by Rogowitz and Voss.[3] They showed that when the slope of the fall-off in the amplitude spectrum is in a certain range (corresponding to a fractal dimension of from 1.2 to 1.4), observers see nameable shapes in images like clouds. This approach has also been applied to understanding how the statistics of illuminants and surfaces in the world constrains the spectral sensitivities of visual mechanisms (Maloney[4]), thereby influencing the mechanisms of color constancy (see also, Finlayson[8] and Brainard and Freeman[8]).

The attention literature provides another path for studying image features by asking which features in an image or scene naturally attract attention and structure perception. This approach is epitomized by Triesman's work on "preattentive" vision[29] and by Julesz' exploration of "textons."[2,30] Their work has had a very large impact in the field of electronic imaging by focusing attention on the immediacy with which certain features in the world, or in an image, are processed. Their work both identifies image characteristics that attract visual attention and guide visual search, and also provides a paradigm for studying the visual salience of image features. An important debate in this area is whether features are defined bottom up—that is, generated by successive organizations of low-level elements such as edges, as suggested by Marr[31]—or whether features are perceived immediately, driven by top-down processes, as argued by Stark.[32]

This section explores the application of feature perception and extraction in a wide range of electronic imaging applications, and examines how the demands of these new electronic tasks motivate research in perception. One important emerging area is the incorporation of perceptual attention or "importance" in image compression and coding. The idea here is that if we could identify those aspects of an image that attracted attention, we could encode this image using higher-resolution for the areas of importance and save bandwidth in the remaining areas. Another important area is image analysis, where knowing which features are perceptually salient could be used to index, manipulate, analyze, compare, or describe an image. An important emerging application in this area is Digital Libraries, where the goal is to find objects in a database that have certain features, or that are similar to a target object. Another relevant application area is visualization, where the goal is to develop methods for visually representing structures and features in the data.

Attention and Region of Interest

A new idea in electronic imaging is to analyze images to identify their "regions of interest," features that draw our attention and have a special saliency for interpretation. If we could algorithmically decompose an image into its salient features, we could develop compression schemes that would, for example, compress more heavily those regions that did not include perceptual features, and devote extra bandwidth to regions of interest. Several attempts have been made to algorithmically identify regions of interest. For example, Leray et al.[8] developed a neural network model that incorporates both the response of low-level visual mechanisms and an attentional mechanism that differentially encodes areas of interest in the visual image, and used this to develop a compression scheme.

Stelmach and Tam[8] measured the eye movements of people examining video sequences and concluded that there wasn't enough consistency across users to motivate developing a compression scheme for TV broadcast based on user eye movements. Geissler and Perry,[12] however, demonstrated that by guiding a user's eye movements to a succession of target locations, and only rendering the image at high resolution at these "foveated" regions, the entire image appeared to have full resolution. If the system knew in advance where the eye movements would be directed, it could adjust the resolution at those locations, on the fly, thus saving bandwidth. Finding those perceptually relevant features, however, is a very difficult problem. Yarbus[33] showed that there is not a single stereotypic way that an image is viewed; the way the user's saccades are placed on the picture depends on the task. For example, if the goal is to identify how many people there are in the picture, the set of eye movements differs from those where the goal is to examine what the people are wearing.

A most important voice in this discussion is that of Stark.[6,14] Noton and Stark's classic papers[32,34] set the modern stage for measuring the eye movement paths, or "scanpaths," of human vision. They used this methodology to explore the role of particular visual stimuli in driving attention (bottom up) versus the role of higher-level hypothesis-testing and scene checking goals (top down) in driving the pattern of visual activity. Stark and his colleagues have contributed both to our basic understanding of the processes that drive these scanpaths, and to integrating this knowledge into the development of better electronic imaging systems. Recent evidence for top-down processing in eye movements has been obtained using an eye tracking device that can be worn while an observer moves about in the world (Pelz et al.[14]). As observers perform a variety of everyday tasks, they perform what the authors call "planful" eye movements, eye movements that occur in the middle of a task, in anticipation of the subject's interaction with an object in an upcoming task.

Image Features, Similarity, and Digital Libraries Applications

In Digital Libraries applications, the goal is to organize and retrieve information from a database of images, videos, graphical objects, music, and sounds. In these applications, the better these objects are indexed and organized, the more easily they can be searched. A key problem, therefore, is to identify features and attributes of importance, and to provide methods for indexing and searching for objects based on these features. Understanding which features are meaningful, or what makes objects similar, however, is a difficult problem. Methods from signal processing and computer vision can be used to algorithmically segment images and detect features. Since these databases are being designed for humans to search and navigate, it is important to understand which features are salient and meaningful to human observers, how they are extracted from complex objects, and how object features are used to judge image and object similarity. Therefore, these algorithms often incorporate heuristics gleaned from perceptual experiments regarding human color, shape, and pattern perception. For example, the work by Petkovic and his colleagues[35] includes operators that extract information about color, shape, texture, and composition. More recently, knowledge about human perception and cognition has been incorporated more explicitly. For example, Frese et al.[11] developed criteria for image similarity based on a multiscale model of the human visual system, and used a psychophysical model to weight the parameters of their model. Another method for explicitly incorporating human perception has been to study how humans explicitly judge image similarity and use this knowledge to build better search algorithms. Rogowitz et al.,[12] for example, asked observers to judge the similarity of a large collection of images and used a multidimensional scaling technique to identify the dimensions along which natural objects were organized perceptually. Mojsilović et al.[13] asked observers to judge the similarity of textured designs, then built a texture retrieval system based on the perceptual results.

In this expanding area, research opportunities include the analysis of human feature perception, the operationalization of these behaviors into algorithms, and the incorporation of these algorithms into digital library systems. This includes the development of perceptual criteria for image retrieval, the creation of perceptual image similarity metrics, methods for specifying perceptual metadata for characterizing these objects, and perceptual cues for navigating through large multimedia databases. An important emerging opportunity lies in extending these methods to accommodate more complex multidimensional objects, such as 3-D objects and auditory patterns.

Visualization

One consequence of the expanding computer age is the creation of terabytes and terabytes of data, and an interest in taking advantage of these data for scientific, medical, and business purposes. These can be data from satellite sensors, medical diagnostic sensors, business applications, simulations, or experiments, and these data come in many different varieties and forms. The goal of visualization is to create visual representations of these data that make it easier for people to see patterns and relationships, identify trends, develop hypotheses, and gain understanding. To do so, the data are mapped onto visual (and sometimes auditory) dimensions, producing maps, bar charts, sonograms, statistical plots, etc. A key perceptual issue in this area is how to map data onto visual dimensions in a way that preserves the structure in the data without creating visual artifacts. Another key perceptual issue is how to map the data onto visual dimensions in a way that takes advantage of the natural feature extraction and pattern-identification capabilities of the human visual system. For example, how can color and texture be used to draw attention to a particular range of data values, or a departure from a model's predictions? (See Rogowitz and Treinish[36] for an introduction to these ideas.)

One important theme in this research is the return to Gestalt principles of organization for inspiration about the analysis of complex visual information. The Gestalt psychologists identified principles by which objects in the visual world are organized perceptually. For example, objects near each other (proximity) or similar to each other (similarity) appear to belong together. If these fundamental principles could be operationalized, or if systems could be built that take advantage of these basic rules of organization, it could be of great practical importance. Some recent papers in this area include Kubovy's[11] theoretical and experimental studies, experiments by Hon et al.[11] on the interpolation and segmentation of sampled contours, and an interactive data visualization system based on Gestalt principles of perceptual organization (Rogowitz et al.[10]).

User Interface Design

As more and more information becomes available, and computer workstations and web browsers support more interactivity and more color, we move into a new era in interface design. With so many choices available, we now have the luxury to ask how best to use color, geometry, and texture to represent information. A key vision paper in this area has been Boynton's paper[2] on "the eleven colors which are almost never confused." Instead of measuring color discriminability, Boynton asked people to sort colors, to get at a higher-level color representation scheme, and found that the responses clustered in eleven categories, each organized by a common color name. That is, although in side-by-side comparison, people can discriminate millions of colors, when the task is to categorize colors, the number of perceptually distinguishable color categories is very small. In a related experiment, Derefeldt and Swartling[8] tried to create as many distinct colors as possible for a digital map application; they were only able to identify thirty. More recently, Yendrikovskij[14] extracted thousands of images from the web and used a K-means clustering algorithm to group the colors of the pixels. He also found a small number of distinct color categories. This suggests that the colors of the natural world group into a small number of categories that correspond to a small set of nameable colors. This result is important for many applications that involve selecting colors to represent semantic entities, such as using colors to represent different types of tumors in a medical visualization. It is also important for image compression or digital libraries, suggesting that the colors of an image can be encoded using a minimal number of color categories. Several groups have begun developing user-interface design tools that incorporate perceptually based intelligence, guiding the designer in choices of colors, fonts, and lay-outs based on knowledge about color discriminability, color deficiency, color communication, and legibility. Some examples include Hedin and Derefeldt[3] and Rogowitz et al.[10]

17.4 VERY HIGH-LEVEL APPROACHES: THE REPRESENTATION OF AESTHETIC AND EMOTIONAL CHARACTERISTICS

The higher-level applications discussed above, such as visualization and digital libraries, push the envelope in electronic imaging technology. They drive developers to address the content of the data, not just the pixel representation. As the technology improves and becomes more pervasive, new application areas are drawn into the web, including applications that use visual media to convey artistic and emotional meaning. For these applications, it is not just the faithfulness of the representation, or its perceptual features, that are of concern, but also its naturalness, colorfulness, composition, and appeal. This may seem an enormous leap, until we realize how natural it is to think of the artistic and emotional aspects of photographic imaging. Interest in using electronic media to convey artistic and emotional goals may just reflect the maturation of the electronic imaging field.

Image Quality

Although the world "quality" itself connotes a certain subjectivity, the major focus of the research in image quality, discussed so far, has been aimed at developing objective, algorithmic methods for characterizing the judgments of human observers. The research in Japan on High-Definition TV has played a major role in shaping a more subjective approach to image quality. This approach is based on the observation that the larger format TV, because it stimulates a greater area of the visual field, produces images which appear not only more saturated, but also more lively, realistic, vivid, natural, and compelling (Kusaka[6]). In related work on the image quality of color images, DeRidder et al.[9] found that, in some cases, observers preferred somewhat more saturated images, even when they clearly perceived them to be somewhat less natural. The observers' responses were based on aesthetic characteristics, not judgments of image fidelity.

Virtual Reality and Presence

Virtual reality is a set of technologies whose goal is to immerse the user in a visual representation. Typically, the user will wear a head-tracking device to match the scene to the user's viewpoint and stereo glasses whose two interlaced views provide a stereoscopic view. A tracked hand device may also be included so that the user can interact with this virtual environment. Although these systems do not provide adequate visual resolution or adequate temporal resolution to produce a realistic visual experience, they do give the sense of "presence," a sense of being immersed in the environment. This new environment has motivated considerable perceptual research in a diverse set of areas, including the evaluation of depth cues, the visual contributions to the sense of presence, the role of auditory cues in the creation of a realistic virtual environment, and cues for navigation. This new environment also provides an experimental apparatus for studying the effects of large-field visual patterns and the interaction of vision and audition in navigation.

Color, Art, and Emotion

Since the beginning of experimental vision science, there have been active interactions between artists and psychologists. This is not surprising since both groups are interested in the visual impression produced by physical media. As artists use new media, and experiment with new visual forms, perceptual psychologists seek to understand how visual effects are created. For example, Chevreul made fundamental contributions to color contrast by studying why certain color dyes in artistic textiles didn't appear to have the right colors. The Impressionists and the Cubists read the color vision literature of their day, and in some cases, the artist and vision scientist were the same person. For example, Seurat studied the perception of color and created color halftone paintings. Artists have continued to inspire vision scientists to this day. For example, Papathomas,[14] struck by the impressive visual depth illusions in Hughes's wall sculptures, used this as an opportunity to deepen our knowledge of how the visual system constructs the sensation of depth. Koenderink[14] used 3-D sculptures to study how the human observer constructs an impression of 3-D shape from the set of 2-D views, where there is an infinite number of valid reconstructions. Tyler[13] brought psychophysical testing to the issue of portrait composition, and discovered that portrait painters systematically position one of the subject's eyes directly along the central median of the painting.

Another approach in this area is to understand the artistic process in order to emulate it algorithmically. The implementation, thus, is both a test of the model and a method for generating objects of an artistic process. For example, Burton[11] explored how young children express themselves visually and kinesthetically in drawings, and Brand[11] implemented his model of shape perception by teaching a robot to sculpt. Some of these goals are esoteric, but practical applications can be found. For example, Dalton[11] broadened the dialogue on Digital Libraries by exploring how algorithms could search for images that are the same, except for their artistic representation.

Also, as more and more artists, fabric designers, and graphic artists exploit the advances of electronic imaging systems, the more important it becomes to give them control over the aesthetic and emotional dimensions of electronic images. The MIT Media Lab has been a major player in exploring the visual, artistic, and emotional parameters of visual stimuli. In 1993,[7] Feldman and Bender[7] conducted an experiment where users judged the affective aspects (e.g., "energy," "expressiveness") of color pairs, demonstrating that even such seemingly vague and subjective dimensions could be studied using psychophysical techniques. They found that color energy depended on the color distance between pairs in a calibrated Munsell color space, where complementary hues, or strongly different luminances or saturations, produced greater emotional energy. These authors and their colleagues have gone on to develop emotion-based color palettes and software tools for artistic design.

17.5   CONCLUSIONS

The field of human vision and electronic imaging has developed over the last decade, evolving with technology. In the 1980s, the bandwidth available to workstations allowed only the display of green text, and in those days people interested in perceptual issues in electronic imaging studied display flicker, spatial sampling, and gray-scale/resolution trade-offs. As the technology has evolved, almost all desktop displays now provide color, and the bandwidth allows for high-resolution, 24-bit images. These displays, and the faster systems that drive them, allow the development of desktop image applications, desktop digital libraries applications, and desktop virtual reality applications. Application developers interested in art, visualization, data mining, and image analysis can now build interactive image solutions, integrating images and data from multiple sources worldwide. This rapid increase in technology enables new types of imaging solutions, solutions that allow the user to explore, manipulate, and be immersed in images; and these in turn pose new questions to researchers interested in perceptually based imaging systems.

Some of these new questions include, for example: How do we measure the image quality of virtual reality environments? How do we measure presence, and what are the factors (including interactions of multiple media) that contribute to it? How do we model the interactive visualization of multivariate data? How do we provide environments that allow people to search through a sea of images? How do we create tools that artists and designers can use? How do we use image technology to perform remote surgery? How do we build user interfaces that sense our mood? With each advance in technology, the systems we build grow increasingly richer, and the questions we ask about human observers become more complex. In particular, they increasingly require a deeper understanding of higher levels of human perception and cognition. Low-level models of retinal and striate cortex function are the foundation, but upon this base we need to consider other dimensions of the human experience: color perception, perceptual organization, language, memory, problem solving, aesthetic appreciation, and emotional response.

As electronic imaging becomes ever more prevalent in the home and the workplace, the two-way interaction between human vision and electronic imaging technology will continue to grow in scope and importance. Research in human vision and its interactions with the other senses will continue to shape the solutions that are developed to meet the requirements of the imaging industry, which in turn will continue to motivate our understanding of the complex process of vision that is the window to the world around us.

17.6 ADDITIONAL INFORMATION ON HUMAN VISION AND ELECTRONIC IMAGING

A number of archival journals and conferences have provided a venue for dissemination and discussion of research results in the field of human vision or in the field of electronic imaging. For human vision, these include the annual meeting of the Optical Society of America, the annual meeting of the Association for Research in Vision and Ophthalmology, and the European Conference on Visual Perception (ECVP). For electronic imaging, these include the IEEE Signal Processing Society's International Conference on Image Processing, the annual meeting of the Society for Information Display (SID), and several conferences sponsored or co-sponsored by the Society for Imaging Science and Technology (IS&T), especially the International Conference on Digital Printing Technologies (NIP), the Color Imaging Conference (CIC), co-sponsored with SID, the Conference on Picture and Image Processing (PICS), and several conferences that are part of the Symposium on Electronic Imaging cosponsored by the Society of Photographic and Instrumentation Engineers (SPIE) and IS&T.

The major forum for human vision *and* electronic imaging is the Conference on Human Vision and Electronic Imaging co-sponsored by the International Imaging Society (IS&T) and the Society for Photographic and Imaging Engineers (SPIE).

There are a number of excellent books on various aspects of human perception and electronic imaging, including *Human Color Vision* by Kaiser and Boynton,[37] *Foundations of Vision* by Wandell,[38] *Visual Perception* by Cornsweet,[39] *Spatial Vision* by De Valois and De Valois,[40] *The Senses,* edited by Barlow and Molon,[41] *The Artful Eye,* edited by Gregory, Harris, Rose, and Heard,[42] *How the Mind Works* by Pinker,[43] *Information Visualization: Optimizing Design For Human Perception,* edited by Colin Ware,[44] *Digital Image and Human Vision,* edited by Watson,[45] *Computational Models of Visual Processing,* edited by Landy and Movshon,[46] and *Early Vision and Beyond,* edited by Papathomas et al.[47]

Finally, we cite a number of review articles and book chapters.[24,25,48]

17.7 REFERENCES

1. G. W. Hughes, P. E. Mantey, and B. E. Rogowitz (eds.), *Image Processing, Analysis, Measurement, and Quality, Proc. SPIE,* vol. 901, Los Angeles, California, Jan. 13–15, 1988.

2. B. E. Rogowitz (ed.), *Human Vision, Visual Processing, and Digital Display, Proc. SPIE,* vol. 1077, Los Angeles, California, Jan. 18–20, 1989.

3. B. E. Rogowitz and J. P. Allebach (eds.), *Human Vision and Electronic Imaging: Models, Methods, and Applications, Proc. SPIE,* vol. 1249, Santa Clara, California, Feb. 12–14, 1990.

4. M. H. Brill (ed.), *Perceiving, Measuring, and Using Color, Proc. SPIE,* vol. 1250, Santa Clara, California, Feb. 15–16, 1990.

5. B. E. Rogowitz, M. H. Brill, and J. P. Allebach (eds.), *Human Vision, Visual Processing, and Digital Display II, Proc. SPIE,* vol. 1453, San Jose, California, Feb. 27–Mar. 1, 1991.

6. B. E. Rogowitz (ed.), *Human Vision, Visual Processing, and Digital Display III, Proc. SPIE,* vol. 1666, San Jose, California, Feb. 10–13, 1992.

7. J. P. Allebach and B. E. Rogowitz (eds.), *Human Vision, Visual Processing, and Digital Display IV, Proc. SPIE,* vol. 1913, San Jose, California, Feb. 1–4, 1993.

8. B. E. Rogowitz and J. P. Allebach (eds.), *Human Vision, Visual Processing, and Digital Display V, Proc. SPIE,* vol. 2179, San Jose, California, Feb. 8–10, 1994.

9. B. E. Rogowitz and J. P. Allebach (eds.), *Human Vision, Visual Processing, and Digital Display VI, Proc. SPIE,* vol. 2411, San Jose, California, Feb. 6–8, 1995.

10. B. E. Rogowitz and J. P. Allebach (eds.), *Human Vision and Electronic Imaging, Proc. SPIE,* vol. 2657, San Jose, California, Jan. 29–Feb. 1, 1996.

11. B. E. Rogowitz and T. N. Pappas (eds.), *Human Vision and Electronic Imaging II, Proc. SPIE,* vol. 3016, San Jose, California, Feb. 10–13, 1997.

12. B. E. Rogowitz and T. N. Pappas (eds.), *Human Vision and Electronic Imaging III, Proc. SPIE,* vol. 3299, San Jose, California, Jan. 26–29, 1998.

13. B. E. Rogowitz and T. N. Pappas (eds.), *Human Vision and Electronic Imaging IV, Proc. SPIE,* vol. 3644, San Jose, California, Jan. 25–28, 1999.

14. B. E. Rogowitz and T. N. Pappas (eds.), *Human Vision and Electronic Imaging V, Proc. SPIE,* vol. 3959, San Jose, California, Jan. 24–27, 2000.

15. O. H. Schade, "Optical and Photoelectric Analog of the Eye," *Journal of the Optical Society of America* **46:**721–739 (1956).

16. F. W. Campbell and J. G. Robson, "Application of Fourier Analysis to the Visibility of Gratings," *J. Physiology* **197:**551–566 (1968).

17. N. Graham and J. Nachmias, "Detection of Grating Patterns Containing Two Spatial Frequencies: A Comparison of Single-Channel and Multiple-Channels Models," *Vision Res.* **11:**252 259 (1971).

18. H. Snyder, "Image Quality and Observer Performance in Perception of Displayed Information," in *Perception of Displayed Information,* Plenum Press, New York; 1973.

19. P. G. J. Barten, "The SQRI Method: A New Method for the Evaluation of Visible Resolution on a Display," *Proc. Society for Information Display,* **28:**253–262 (1987).

20. S. Daly, "The Visible Differences Predictor: An Algorithm for the Assessment of Image Fidelity," in A. B. Watson (ed.), *Digital Images and Human Vision,* pp. 179–206, MIT Press, Cambridge, MA, 1993.

21. J. Lubin, "The Use of Psychophysical Data and Models in the Analysis of Display System Performance," in A. B. Watson (ed.), *Digital Images and Human Vision,* pp. 163–178, MIT Press, Cambridge, MA, 1993.

22. R. J. Safranek and J. D. Johnston, "A Perceptually Tuned Sub-Band Image Coder with Image Dependent Quantization and Post-Quantization Data Compression," in *Proc. ICASSP-89,* vol. 3, Glasgow, Scotland, pp. 1945–1948, May 1989.

23. C. J. Van den Branden Lambrecht and O. Verscheure, "Perceptual Quality Measure Using a Spatio-Temporal Model of the Human Visual System," in V. Bhaskaran, F. Sijstermans, and S. Panchanathan (eds.), *Digital Video Compression: Algorithms and Technologies, Proc. SPIE,* vol. 2668, San Jose, California, pp. 450–461, Jan./Feb. 1996.

24. M. P. Eckert and A. P. Bradley, "Perceptual Quality Metrics Applied to Still Image Compression," *Signal Processing* **70:**177–200 (1998).

25. T. N. Pappas and R. J. Safranek, "Perceptual Quality Assessment of Compressed Image and Video," in *Handbook of Image and Video Processing,* Academic Press, New York, 2000. To appear.

26. J. Sullivan, L. Ray, and R. Miller, "Design of Minimum Visual Modulation Halftone Patterns," *IEEE Trans. Sys., Man, Cyb.* **21:** 33–38 (Jan./Feb. 1991).

27. R. W. G. Hunt, *The Reproduction of Colour in Photography, Printing and Television.* Fountain Press, England, 1987.

28. D. J. Field, "Relations Between the Statistics of Natural Images and the Response Properties of Cortical Cells," *Journal of the Optical Society of America A,* **4:**2379–2394 (1987).

29. A. Treisman and G. Gelade, "A Feature Integration Theory of Attention," *Cognitive Psychology* **12:**97–136 (1980).

30. B. Julesz, "Textons, the Elements of Texture Perception and Their Interactions," *Nature* **290:**91–97 (1981).

31. D. Marr, *Vision.* W. H. Freeman Company, 1982.

32. D. Noton and L. Stark, "Eye Movements and Visual Perception," *Scientific American* **224**(6):34–43 (1971).

33. A. Yarbus, *Eye Movements and Vision.* Plenum Press, New York, 1967.

34. D. Noton and L. Stark, "Scanpaths in Saccadic Eye Movements While Viewing and Recognizing Patterns," *Vision Research* **11:**929–942 (1971).

35. M. Flickner, H. Sawhney, W. Niblack, J. Ashley, Q. Huang, B. Dom, M. Gorkani, J. Hafner, D. Lee, D. Petkovic, D. Steele, and P. Yanker, "Query by Image and Video Content; the QBIC System," *Computer* **28:**23–32 (Sept. 1995).

36. B. E. Rogowitz and L. Treinish, "Data Visualization: The End of the Rainbow," *IEEE Spectrum,* 52–59 (Dec. 1998).

37. P. K. Kaiser and R. M. Boynton, *Human Color Vision.* Optical Society of America, Washington, DC.

38. B. A. Wandell, *Foundations of Vision.* Sinauer, Sunderland, MA, 1995.

39. T. N. Cornsweet, *Visual Perception.* Academic Press, New York, 1970.

40. R. L. D. Valois and K. K. D. Valois, *Spatial Vision.* Oxford University Press, New York, 1990.

41. H. B. Barlow and J. D. Molon (eds.), *The Senses.* Cambridge University Press, Cambridge, 1982.

42. R. L. Gregory, J. Harris, D. Rose, and P. Heard (eds.), *The Artful Eye.* Oxford University Press, Oxford, 1995.

43. S. Pinker, *How the Mind Works.* Norton, New York, 1999.

44. C. Ware (ed.), *Information Visualization: Optimizing Design For Human Perception.* Morgan Kaufmann Publishers, 1999.

45. A. B. Watson (ed.), *Digital Image and Human Vision.* Cambridge, MA: The MIT Press, 1993.

46. M. S. Landy and J. A. Movshon (eds.), *Computational Models of Visual Processing.* MIT Press, Cambridge, MA, 1991.

47. T. V. Papathomas, C. Chubb, A. Gorea, and E. Kowler (eds.), *Early Vision and Beyond.* MIT Press, Cambridge, MA, 1995.

48. B. E. Rogowitz, "The Human Visual System: A Guide for the Display Technologist," in *Proceedings of the SID,* vol. 24/3, pp. 235–252, 1983.

CHAPTER 18
VISUAL FACTORS ASSOCIATED WITH HEAD-MOUNTED DISPLAYS*

Brian H. Tsou
Air Force Research Laboratory
Wright Patterson AFB, Ohio

Martin Shenker
Martin Shenker Optical Design, Inc.
White Plains, New York

18.1 GLOSSARY

Aniseikonia.   Unequal right and left retinal image sizes. (Also see Chaps. 11 and 12.)

C, F lines.   Hydrogen lines at the wavelengths of 656 and 486 nm, respectively.

D (diopter).   A unit of lens power; the reciprocal of focal distance in m. (Also see Chap. 11.)

Prism diopter.   A unit of angle; 100 times the tangent of the angle. (Also see Chaps. 11 and 12.)

18.2 INTRODUCTION

Some virtual reality applications require a head-mounted display (HMD). Proper implementation of a binocular and lightweight HMD is quite a challenge; many aspects of human-machine interaction must be considered when designing such complex visual interfaces. We learned that it is essential to couple the knowledge of visual psychophysics with sound human/optical engineering techniques when making HMDs. This chapter highlights some visual considerations necessary for the successful integration of a wide field-of-view HMD.

---

* *Acknowledgments:* The author (BT) thanks Tom Metzler, Mike Richey, Trang Bui, Luc Biberman, and Walter Hollis for the opportunities to be involved with Army HMD programs. We thank Jay Enoch, Neil Charman, and Vince Billock for their critical reviews of the manuscript. We also thank Sarah Tsou for her editorial support. Finally, we acknowledge the significant contributions by Jerry Gleason toward the advancements of HMD designs.

18.3 COMMON DESIGN CONSIDERATIONS AMONG ALL HMDs

A wide field-of-view HMD coupled with gimbaled image-intensifying (see Chap. 21 of Vol. 1 for details) or infrared (see Chap. 23 of Vol. 1 for details) sensor enables helicopter pilots to fly reconnaissance, attack, or search-and-rescue missions regardless of weather conditions. Before any HMD can be built to specifications derived from mission requirements, certain necessary display performance factors should be considered. These factors include functionality, comfort, usability, and safety. A partial list of subfactors that affect vision follows:

Functionality	Comfort	Usability	Safety
Field of view	Eye line-of-sight/focus	Eye motion box	Center-of-mass
Image quality	Alignment	Range of adjustment	Weight

Safety

An analytic approach to considering these factors has been proposed.[1] In general, factors can be weighted according to specific needs. At the same time, safety—especially in the context of a military operation—cannot be compromised.[2] The preliminary weight and center-of-mass safety requirements for helicopters[3] and jets[4] have been published.

It is extremely difficult to achieve the center-of-mass requirement. Hanging eye lenses in front of one's face necessitates tipping the center-of-mass forward. A recent survey disclosed that pilots routinely resort to using counterweights to balance the center-of-mass of popular night-vision goggles[5] in order to obtain stability and comfort, even at the expense of carrying more weight.

Light yet sturdy optics and mechanical materials have already made the HMD lighter than was previously possible. In the future, miniature flat-panel displays (see Chaps. 12, 13, and 27 of Vol. 1 and Chap. 14 of Vol. 2 for details) and microelectromechanical system (MEMS) technology[6] might further lessen the total head-borne weight. Previous experience[7] shows that both the field-of-view and exit pupil have a significant impact on HMD weight. Although we can calculate[1] the eye motion box (see Eq. 1) that defines the display exit pupil requirement, we still do not have a quantitative field-of-view model for predicting its optimum value.

Usability

In order to avoid any vignetting, the display exit pupil size should match the eye motion box. The eye motion box is the sum of lateral translation of the eye and the eye pupil projection for viewing the full field of view (FOV). We assume the radius of eye rotation is 10 mm.

Eye motion box in mm

$$= 2 \times 10 \times \sin (FOV/2) + \text{eye pupil diameter} \times \cos (FOV/2) + \text{helmet slippage} \qquad (1)$$

The range of pertinent optomechanical HMD adjustments for helicopter aviators[8] is listed as minimum–maximum in millimeters in the following table:

	Interpupillary distance	Eye to top of head	Eye to back of head
Female	53–69	107.7–141.7	152.5–190.0
Male	56–75	115.1–148.2	164.4–195.4

Comfort

Levy[9] shows that, in total darkness, the eye line-of-sight drifts, on average, about 5° down from the horizon. Figure 1 shows the geometric relationship of the visual axis to the horizon. Levy argues that the physiological position of rest of the eye orbit is downward pointing. Menozzi et al.[10] confirm that maximum comfort is in fact achieved when line of sight is pointed downward. The viewing comfort was determined by the perceived muscle exertion during gaze in a given direction for a given time. The median line of sight for the most comfortable gaze at 1 m is around 10° downward.

Our laboratory has also examined how normal ocular-motor behavior might improve the comfort of using any HMD. In a short-term wear study,[11] Gleason found that a range of eyepiece lens power of AN/AVS-6 Aviator Night Vision Imaging Systems (ANVIS) produced comparable visual acuity independent of luminance and contrast. The single best-overall eyepiece lens power produced visual acuity equal to, or better than, that of subject-adjusted eyepiece lens power, producing visual acuity within 2 percent of optimal. Infinity-focused eyepieces made visual acuity worse, reducing it by 10%. In his long-term wear (4 hours) study, −1.5 diopter (D) eyepiece lens power caused half of the subjects (n = 12) to complain of blurred or uncomfortable vision. These studies indicate that those users, who are optically corrected to a "most plus–best binocular visual acuity" endpoint (see Section 18.6, "Appendix"), achieve satisfactory comfort and near optimal binocular visual acuity for extended ANVIS viewing when an eyepiece lens power of approximately −0.75 D is added to the clinical refraction. This result, consistent with an earlier report by Home and Poole,[12] may extend to other binocular visual displays.

Alignment tolerances[13] are the maximum permissible angular deviation between the optical axes of a binocular device that displays separate targets to the two eyes (see Chapter 12, "Binoc-

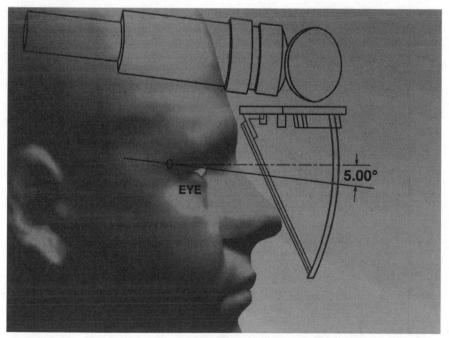

FIGURE 1   Geometric layout of the HMD showing the visual axis being 5° down from the horizon as discussed in the text. Also shown are the schematic depictions of miniature CRT, folding mirror, beam-splitter, and spherical combiner. Its prescription is given in Table 1.

ular Vision Factors That Influence Optical Design," of this volume for general issues). The ocular targets may be either identical or dichoptic (for stereoscopic presentation) but will be treated differently depending on whether there is a requirement for superimposition of the targets onto a direct view of the environment. If the direct view is either blocked out or too dark to see (e.g., flying at night), then it is classified as a nontransparent or closed HMD. A closed HMD resembles immersive virtual reality simulation where the real world is not needed.

Closed HMD.   Normal subjects have a substantial tolerance for angular deviation between the images in the two eyes for the three degrees of freedom of eye rotation: horizontal, vertical, and cyclorotational. These limits are especially of interest to optometrists and ophthalmologists. The optic axes of spectacle lenses must be aligned with the visual axes of the eyes of the wearer. Displacement of a spectacle lens from a centered position on the visual axis of the eye introduces a prismatic deviation that is entirely analogous to the misalignment of binocular devices. The American National Standards Institute (ANSI) has published ANSI 280.1–1987[14] and permits a maximum deviation of ⅔ prism diopter (23 arc minutes) horizontally and ⅓ prism diopter (11.5 arc minutes) vertically. This standard can be adopted for closed binocular HMD. These tolerance values are summarized as follows:

Recommended Tolerances for Closed HMD

	Tolerances (arc minutes)
Horizontal	±23
Vertical	±11.5
Cyclorotational	±12

Regarding the cyclorotational tolerances, both ophthalmic lenses and binocular displays can produce rotational deviation of the images about the line of fixation. ANSI spectacle standards do not exist for this alignment axis. Earlier research (dealing with distortion of stereoscopic spatial localization resulting from meridional aniseikonia at oblique axes) shows that there is considerable tolerance for cyclotorsional rotations of the images.[15] However, the effects of cyclorotational misalignment are complicated by the fact that these rotations may result in either an oculomotor (cyclofusional) or sensory (stereoscopic) response. If the eyes do not make complete compensatory cyclorotations, a declination error (rotational disparity) will exist. Declination errors result in an apparent inclination (rotation about a horizontal axis) of the display; that is, the display will appear with an inappropriate pitch orientation.

Sensitivity to declination (stereoscopic response to rotational disparity) is quite acute. Ogle[15] reports normal threshold values of ±6 arc minutes, which corresponds to an object inclination of 5° at 3 m distance. If this threshold value were adopted for the cyclotorsion tolerance measurement, it would be overly conservative, as it would deny any reflex cyclofusional capacity to compensate for misalignment errors. A 97 percent threshold[15] (±12 arc minutes) is suggested for adoption as the cyclofusional tolerance.

Transparent HMD.   Transparent HMD optically superimposes a second image upon the directly viewed image using a beam-combiner (see Section 18.6, "Appendix"). The difference in alignment angles between the direct and superimposed images equals the binocular disparity between target pairs.[16] Horizontal binocular disparity is the stimulus to binocular stereoscopic vision. Consequently, sufficient lateral misalignment will lead to the perception of a separation in depth between the direct and superimposed images.

The exact nature of the sensory experience arising from lateral misalignment depends on the magnitude of the relative disparity. Although the threshold disparity for stereopsis is generally placed at about 10 arc seconds for vertical rods,[17] it can approach 2 arc seconds under optimal conditions using >25 arc-minutes-long vertical line targets.[18] The normative value of threshold stereopsis in the standardized Howard-Dolman testing apparatus is 16 arc seconds. This defines 100 percent stereoacuity for clinical-legal requirements.

This extremely small stereothreshold (~2 arc seconds for optimal conditions) establishes an impractically small tolerance for the design of displays. However, this precision is only required if the relative depths of the direct and superimposed fields need to correspond and the direct and superimposed fields are dissimilar (nonfusible). An example would be the need to superimpose a fiduciary marker over a particular location in the direct field, which is important to military operations. If this is not a requirement, the binocular disparity between the direct and superimposed fields need not exceed the range for pyknostereopsis in order to permit sensory averaging of the fields. When two views are overlaid with small disparity separations (<20 arc minutes), they are depth averaged to form a single depth plane (called pyknostereopsis).[19,20] This sets an alignment tolerance of ±20 arc minutes (i.e., a maximum binocular disparity of 20 arc minutes). This limit is also within the disparity range required for maximum efficacy of oculomotor alignment of the eyes.[21] Because of the poor sensitivity of the human visual system to absolute disparity,[22] this amount of oculomotor misalignment will not influence the perception of absolute distances of objects in the display.

Vertical disparities do not result in local stereopsis, but they can have an effect on the overall stereoscopic orientation of objects in the binocular field. This response to vertical disparity is known as the *induced effect*. The sensitivity to vertical disparity for extended targets is similar to that of the horizontal disparity.[23] The other factor to consider when establishing vertical tolerances is the dimension of Panum's fusional area. Images falling within the Panum's area will fuse even if these images do not match exactly on the corresponding points between the two eyes. The vertical dimension of Panum's area can be as much as 2.5 times smaller than the horizontal dimension.[24]

The recommended cyclorotational tolerance is based on Ogle's[15] normal threshold value of ±6 arc minutes as discussed in the previous section. Because reflex cyclofusional would be restricted in transparent HMD by the direct field, no relaxation for compensatory eye movements has been included in this tolerance recommendation.

In formulating the final recommendations, the range of pyknostereopsis was reduced by half to introduce a margin for individual differences. Summary recommendations for tolerance in transparent HMD are provided as follows:

Recommended Tolerances for Transparent HMD

	Tolerances (arc minutes)
Horizontal	±10
Vertical	±4
Cyclorotational	±6

Functionality

A larger field of view equates with larger optics and, hence, a heavier piece of equipment. As early as 1962, optical engineers partially overlapped the individual telescopes to achieve a wider field of view without greatly increasing weight.[25] The impact of partial overlap on image quality, however, was not fully addressed until recently.[26,27] These psychophysical studies on binocular summation, rivalry, and stereopsis support a 40° binocular overlap in the middle of the field of view as both necessary and sufficient, with a divergent configuration of the two optical axes preferred over a convergent configuration (please note the visual axes are still parallel or convergent). Quantitative flight data (eye gaze) and pilots' subjective assessment are consistent with laboratory predictions.[28]

Still, how large should the total field of view be? Everyday experience tells us that it should be as large as our visual field. On the other hand, most studies consistently show there is no significant human performance gain as the field of view extends beyond the 40° to 60° range.[29] Restricting head movements or manipulating field-of-view aspect ratio[30] does not seem to affect the performance except for the smaller field of view (from 15° to 30°). These studies suggest that (1) we are basically foviated beings,[31,32,33] and (2) unless compensated by large target size, our peripheral sensitivity is very poor.[34] Nevertheless, quantitative models for the useful field-of-view[35,36] requirement are urgently needed so that HMD field of view is optimized to suit each application and scenario.

Image quality is perhaps the most important parameter for the acceptance of HMD since it is the aspect most noticeable by the users. Various image quality metrics have been devised (see Chap. 24 of Vol. 1 for reviews). The fundamental limitation on image quality in any display is either the loss of resolution, or lack thereof. Historically, the poor resolution of the miniature image source has hampered HMD progress; soon, the expected higher pixel density should significantly improve perceived image quality. For example, with high-definition TV's approximate 2000 pixels, a 50° field-of-view HMD will have 20 cy/deg resolution or a respectable 20/30. The loss of resolution in visual optics is primarily caused by the defocus and lateral chromatic smearing of an image point.[37] When the eye is centered in the exit pupil of a display, defocus is due to residual field curvature and astigmatism. As the eye is decentered (or off aperture) in the exit pupil, residual amounts of uncorrected spherical aberration and coma in the display cause additional astigmatism. Astigmatism is defined as the variation in focus as a function of orientation in the exit pupil. Recently, Mouroulis and Zhang[38] have shown that MTFa (the area under the modulation transfer function over a specified frequency range of from 5 to 24 cy/deg) gives good correlation with subjective sharpness judgment for all target orientations. There is another convenient approximation to use for image quality. Figure 2 shows plots of the variation in modulation transfer function (MTF) (see Chap. 32 of Vol. 2 for reviews) for a 3-mm aberration-free pupil out to a frequency of 1 arc minute per line pair (or 20/10) as its image is defocused from the best focus position. A 3-mm pupil has been chosen because the eye is generally considered to be diffraction-limited in image quality at near this value; we conclude that for "each tenth of diopter defocus 1 arc minute of resolution is lost" (see Fig. 2). We arrived at the same conclusion using Greivendamp's schematic eye.[39] Greivendamp et al. showed their model correlates quite well with Snellen visual acuity. It should be noted that, for small defocus errors (<0.5 D), a real eye can see slightly better than our calculation predicts.[40] This may be attributable to the residual spherical aberration in the real eye. While this spherical aberration lowers the MTF for the exact in-focus image, it increases the depth-of-focus over which the maximum resolution is visible. Regardless, our rule of thumb is quite useful for the purpose of design evaluation.

Similarly, we approximate the residual chromatic smear as the sum of the lateral color plus the off-aperture manifestation of the residual longitudinal color by using a pair of wedges with zero deviation at a middle wavelength but a specific chromatic spread between the C and F spectral lines. The loss of MTF for a 3-mm, otherwise aberration-free exit pupil is shown in Fig. 3. It can be seen that these defects have a very significant effect on the resolution capabilities at both the high and low spatial frequencies. Therefore, a rule of thumb of "⅔ arc minute of resolution is lost for each arc minute of chromatic spread between C and F lines" is applicable.

18.4 CHARACTERIZING HMD

Traditionally, we design and evaluate the image at the focal plane of photographic lenses with a plane, aberration-free wavefront, entering the entrance pupil of the objective. This is likely the arrangement in which the photographic lens will be utilized. The design of visual optics is usually performed in a similar manner with a perfect wavefront at the exit pupil of the display (which is defined as the entrance pupil of the eye in the design process). However, the use and the real performance of such a system is best defined, from a visual point of view, by ray trac-

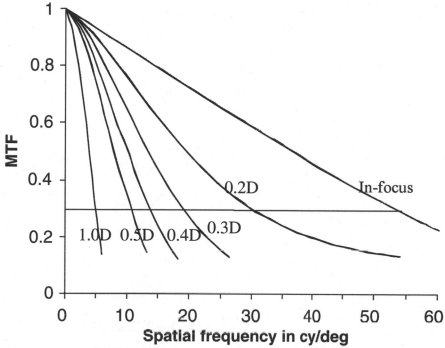

FIGURE 2 Loss of MTF in terms of resolution with defocus. The diopter values are the amount of defocus. At 30 percent modulation, the resolutions are about 1.1 arc minutes per line pair for in-focus, 2 for 0.2 D defocus, 3 for 0.3 D, 4.5 for 0.4 D, 6 for 0.5 D, and 12 for 1.0 D, respectively.

ing from a perfect object and evaluating the nature of the light exiting the system. There are at least two characteristics of this light that are of interest to the display designer.

The first of these is the direction of a ray that is defined as intersecting the center of the eye pupil that is viewing a particular point on the object. This datum is used for evaluation of the system magnification (i.e., mapping and distortion). If the system is supposed to be displaying an image projected to infinity (collimated display), then the apparent direction of a point in the field-of-view should not change as the eyes move relative to the display. Any deviation of this ray direction across the field-of-view will bring about an apparent change of the display distortion as well. This variation of the display distortion is often referred to as "dynamic distortion" or "swimming." Next, if all the rays over the small (from 3 to 4 mm) entrance pupil of the eye are in focus, a relatively well-corrected display will demonstrate essentially diffraction-limited performance. However, if the eye cannot accommodate for the defocus, then this can cause a significant loss of modulation in the image.

Thus, our approach to evaluating the display image quality is to assess the instantaneous image quality available to the observer as he/she looks at different field angles from a particular point within the eye motion box. We define a small eye pupil at various positions in the larger exit pupil of the display; then we compute the apparent sagittal and tangential foci for various points in the field-of-view as seen from this particular eye point. Both the exit pupils and the field-of-view are defined in terms of a clock face as shown in Fig. 4. The displacement of the small eye pupil is in the 12:00 direction within the larger exit pupil of the display. This radial direction is the same as the direction of the off-axis point in the field of view. We make the following further definitions:

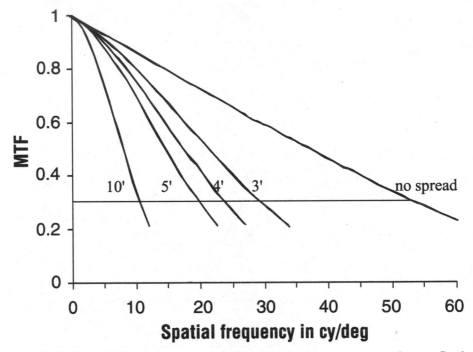

FIGURE 3 Loss of MTF with chromatic smear. The arc minute values represent smear between C and F lines. At 30 percent modulation, the resolutions are about 1.1 arc minutes per line pair for no chromatic spread, 2 for 3 arc minutes spread, 2.5 for 4′, 3.1 for 5′, and 6 for 10′, respectively.

6:00 is the direction in the field-of-view opposite to the direction in which the eye pupil is displaced or off-aperture within the large exit pupil of the display.

3:00 and 9:00 are directions in the field perpendicular to the direction in which the eye pupil is off-aperture in the large exit pupil of the display. These values are identical.

We simply trace a matrix of object points on the object and calculate the tangential and sagittal foci in diopters for each of these points. We are thus assuming that the aberration variation across the 3- to 4-mm eye pupil is small and that the only residual defect is defocus. The three possibilities are as follows:

If the tangential and sagittal dioptric values are equal and the defocus is such that the observer can accommodate for this apparent distance, then the resolution potential of the display at this point in the field reverts to the diffraction limited potential of the aperture of his/her eye.

If the tangential and sagittal values are unequal but he/she can accommodate to either of the foci, then the image quality loss is defined by a focus halfway between the two foci.

If the tangential and/or sagittal values are such that the observer cannot accommodate for the defocus, then the loss of image quality is defined by this total defocus value.

In this manner, a particular optical design can be evaluated for image quality. An indication that the design is not sufficiently corrected is when focus conditions at various points in

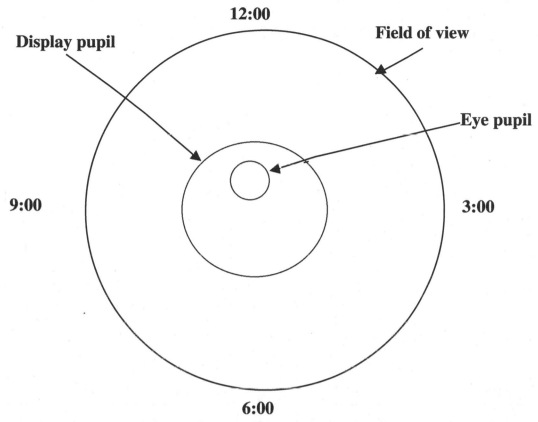

FIGURE 4 Spatial and directional relationships between display exit pupil and eye pupil with respect to field of view.

the field of view change rapidly for small displacements in the exit pupil. Our method for evaluating the potential display image quality is completely independent of their scale. It is valid for the 1.5-m focal length tilted mirrors used in the collimated displays for flight simulators, the 150- to 200-mm focal length refractive collimators used in heads-up displays, the 50- to 75-mm focal length eyepieces used in biocular magnifiers, and the 20-mm focal length erecting eyepiece systems used in HMD. However, for the HMD, the loss of image quality is likely to be the most severe. We will illustrate this method with a 50° field-of-view full-overlap binocular HMD, using only four elements to minimize weight. To conserve expense, all elements are simple spheres. The display has an overall focal length of 24.6 mm and is dimensioned to provide a 19-mm exit pupil with a horizontal field of view of 50° provided by a 23-mm image source format. Our electronic driver bandwidth can only adequately display 1000 video lines, resulting in a resolution of slightly better than 20/60 (10 cy/deg). This resolution is deemed minimal in order to support flying nap-of-the-earth (terrain following) or contour flight.[41] The prescription is given in Table 1.

Table 2 lists the optical defocus values exhibited by this display example. A negative value indicates that the light exiting the display is convergent and corresponds to images focused beyond infinity. Objective measurement of the display system confirms the defocus table. The display focus is then set to 0.87 D (see subsection entitled "Comfort"). Table 3 lists the visual defocus values according to the eye's ability to accommodate. Most observers cannot relax their accommodation beyond infinity (or greater than −0.87 D in our example) and thus suffer a loss in resolution corresponding to the full amount of defocus. The subjective evalua-

TABLE 1   Prescription for a 50° Field-of-View Display Optics

Surface	Radius (mm)	Thickness (mm)	Glass
Object	∞	0.519	Air
1	∞	2.54	BK7
2	20.559	5.715	Air
3	∞	8.89	SK16
4	−26.75	33.02	Air
5	119.907	2.54	SF58
6	30.384	11.43	LAF2
7	−48.552	7.62	Air
8	105.5	5.08	SK16
9	−105.5	95.325	Air
10	∞	−38.1	Mirror
11	98.298	76.2	Mirror
Stop	∞	0	
Image	∞		

tions agree with the predictions. Namely, while performance is satisfactory for the center of the field when the eye is positioned in the center of the display exit pupil, the performance falls off radically as the eye is decentered in the pupil. In addition, there is a significant falloff of performance across the field of view. We suspect the image quality would not be acceptable if the display resolution were much higher than the 20/60 used.[42]

TABLE 2   Optical Defocus Versus Pupil Displacements

Field direction	Field angle (deg)	Pupil displaced							
		0 mm		2 mm		4 mm		6 mm	
		12–6	3–9	12–6	3–9	12–6	3–9	12–6	3–9
12:00	25	−0.4*	+0.2†	−0.5	+0.1	−1.0	−0.1	−1.7	−0.5
12:00	20	−0.6	0	−0.6	−0.1	−1.2	−0.3	−2.0	−0.6
12:00	15	−0.4	−0.1	−0.6	−0.1	−1.2	−0.3	−2.0	−0.7
12:00	10	−0.2	0	−0.4	−0.1	−1.1	−0.4	−2.0	−0.7
12:00	5	−0.1	0	−0.3	−0.1	−1.1	−0.4	−2.1	−0.8
	0	0	0	−0.3	−0.1	−1.3	−0.4	−2.3	−0.9
6:00	5	−0.1	0	−0.5	−0.1	−1.5	−0.5	−2.6	−1.0
6:00	10	−0.2	0	−0.7	−0.2	−1.8	−0.6	−2.9	−1.1
6:00	15	−0.4	−0.1	−0.9	−0.2	−2.0	−0.6	−3.1	−1.1
6:00	20	−0.6	0	−1.1	−0.2	−2.1	−0.5	−2.9	−0.9
6:00	25	−0.4	+0.2	−0.7	+0.1	−1.5	−0.2	−1.7	−0.5
3:00	25	+0.2	−0.4	−0.2	−0.5	−1.0	−0.7	−1.8	−1.1
3:00	20	0	−0.6	−0.3	−0.7	−1.2	−1.0	−2.1	−1.4
3:00	15	−0.1	−0.4	−0.4	−0.5	−1.3	−0.9	−2.2	−1.3
3:00	10	0	−0.2	−0.4	−0.3	−1.3	−0.7	−2.3	−1.1
3:00	5	0	−0.1	−0.3	−0.2	−1.3	−0.5	−2.3	−1.0
	0	0	0	−0.3	−0.1	−1.3	−0.4	−2.3	−0.9

* Negative diopters indicate convergent rays or farther than optical infinity.
† Positive diopters indicate divergent rays or closer than optical infinity.

TABLE 3   Visual Defocus Versus Pupil Displacements

Field direction	Field angle (deg)	Pupil displaced							
		0 mm		2 mm		4 mm		6 mm	
		12–6	3–9	12–6	3–9	12–6	3–9	12–6	3–9
12:00	25	−0.3*	+0.3†	−0.3	+0.3	−0.9	0	−1.3	0
12:00	20	−0.3	+0.3	−0.3	+0.3	−0.2	0	−1.3	0
12:00	15	−0.3	+0.3	−0.2	+0.2	−0.2	0	−1.3	0
12:00	10	−0.1	+0.1	−0.2	+0.2	−0.1	0	−1.3	0
12:00	5	0	0	−0.1	+0.1	−0.1	0	−1.3	0
	0	0	0	−0.1	+0.1	−0.3	0	−2.3	−0.9
6:00	5	0	0	−0.2	+0.2	−1.0	0	−2.6	−1.0
6:00	10	−0.1	+0.1	−0.3	+0.3	−1.2	0	−2.9	−1.1
6:00	15	−0.3	+0.3	−0.7	0	−1.5	0	−3.1	−1.1
6:00	20	−0.3	+0.3	−0.9	0	−1.5	0	−2.9	−0.9
6:00	25	−0.3	+0.3	−0.4	+0.4	−1.3	0	−1.2	0
3:00	25	+0.3	−0.3	+0.2	−0.2	−0.2	0	−1.8	−1.1
3:00	20	+0.3	−0.3	+0.2	−0.2	−1.2	−1.0	−2.1	−1.4
3:00	15	+0.2	−0.2	+0.1	−0.1	−0.4	0	−2.2	−1.3
3:00	10	+0.1	−0.1	0	0	−0.6	0	−2.3	−1.1
3:00	5	0	0	−0.1	+0.1	−0.8	0	−2.3	−1.0
	0	0	0	−0.1	+0.1	−0.8	0	−2.3	−0.9

* Negative diopters indicate convergent rays or farther than where eye accommodates.
† Positive diopters indicate divergent rays or closer than where eye accommodates.

18.5   SUMMARY

We have described various visual issues dealing with a 50° binocular HMD breadboard. Both optical axes point downward by 5° and converge and intersect at a point about 1 m in front. Preliminary evaluation of the breadboard in the laboratory has not revealed any problems with respect to usability and comfort. However, the optics is undercorrected for both spherical aberration and astigmatism as shown by the defocus calculations. The astigmatism causes the variation of focus across the field of view when viewed from the center of the eye motion box. The spherical aberration causes the variation of focus when the eye pupil is decentered in the eye motion box. We found that using the defocus to predict display resolution is both valuable and intuitive in assessing HMD image quality during design and initial evaluation.

18.6   APPENDIX

There has been a concern[43] with respect to the transparent HMD and heads-up display that virtual images appear closer than the directly viewed image, inducing an undesired accommodation of the viewer. After investigating the issue, we did not find any support for misaccommodation. First, let us define the following terms:

Refractive error is defined by the eye's far point, which is the farthest distance to which the eye can focus. Eye focus is the dioptric distance at which the eye is focused, making objects at that distance optically conjugate with the eye's retina. Therefore, eye focus equals the dioptric sum of refractive error plus accommodation.

Accommodation is the eye's ability to focus closer than the far point by increasing its optical power through the action of the ciliary muscles. The hyperope's far point is beyond optical infinity, and therefore negative in value. Hyperopes must accommodate to focus to optical infinity. Myopia has a positive refractive error—that is, the eye is optically too strong to focus on distance objects. Do not confuse this with the clinical convention of associating myopia with the negative lens power of the correcting spectacles.

How is refractive error[44] determined? Optometers and Clinical Refraction Optometers (e.g., infrared,[45,46,47,48] laser,[49,50] and stigmatoscopy[51]) measure eye focus, not accommodation. Optometers cannot distinguish between nonaccommodating 1-D myopes, emmetropes accommodating 1 D, or 1-D hyperopes accommodating 2 D since all are focused 1 m away. Optometers approximate refractive error (i.e., far point) only when accommodation is minimized under cycloplegia. Cycloplegia is the pharmaceutically induced paralysis of accommodation.

Thus, clinical refraction does not determine refractive error in the strict sense. Clinical refraction determines the most-plus lens power resulting in best binocular visual acuity based on a subjective blur criterion (see Chap. 11, "Assessment of Refraction and Refractive Errors," of this volume for general issues). Therefore, its theoretical endpoint places the distal end of the eye's depth-of-field (see Chap. 24 of Vol. 1 for reviews) at the eye chart, and does not place eye focus at optical infinity. Since eye charts are usually located about 6 m (+0.17 D) away and the eye's depth of field is ±0.43 D for detecting blur of small sharp-edged high-contrast targets,[52] the clinical emmetrope is about +0.6 D myopic. This concept is supported by Morgan's[53] report that subjective tests tend to overcorrect by +0.75 D, and Bannon et al.'s[51] report that apparent emmetropes focused from +0.5 to +0.75 D closer than a fixated distant target. The presumed benefit of a myopic endpoint is that the eye's entire depth-of-field is maximized in real space when distant objects are viewed.

There are two potential causes for misaccommodation with transparent HMD: near-awareness of the virtual image and accommodative-trap by the transparent combiner. While proximal accommodation would be the result of the former, the latter may be related to the Mandelbaum's phenomenon. However, Jones[54] found that unless the exit pupil is abnormally constrained (<3 mm), proximal accommodation does not affect the accommodative accuracy for targets in virtual image displays. Therefore, those deleterious proximal effects in virtual image displays may be avoided by utilization of an exit pupil in binocular HMD sufficiently large to encompass a natural pupil (e.g., display exit pupil matches eye motion box).

Contrary to accepted opinion, Gleason[55] found that Mandelbaum's phenomenon is not caused by accommodative capture. Forty years ago, Mandelbaum[56] reported that distant scenery appears blurred when viewed through a window screen. The blur subsided with side-to-side head motion and returned after a delay when head motion ceased. These collective observations are called Mandelbaum's phenomenon. Mandelbaum hypothesized that the eye reflexively focused on the intervening screen instead of the desired distant scenery. Indeed, the assumed disruption of voluntary accommodation by intervening objects is now called the Mandelbaum effect.[57] Recently, Gleason investigated Mandelbaum's phenomenon using objective and continuous measurements of accommodation. In each trial, subjects looked directly at the distant E and were instructed to keep the E as clear as possible. After 7 s, a mesh screen was placed in the line of sight at the distance of either causing the greatest Mandelbaum phenomenon or at 40 cm in front. The 40-cm distance was near the subjects' dark focus. At 15 s, subjects were instructed to focus on the screen. At 23 s, subjects were instructed to refocus on the distant E. After the trial, the experimenter shook the screen side to side to verify that Mandelbaum's phenomenon was seen. Subjects were then questioned about what they saw during the trial. After each maximum Mandelbaum's phenomenon trial, subjects reported the distant letter blurred within one second of screen interjection; the letter remained blurred until the experimenter shook the screen at trial's end. These observations indicate that Mandelbaum's phenomenon occurred during the trial. Mandelbaum's phenomenon was weak or nonexistent in 40-cm trials.

A narrow range of screen distances (from 0.7 to 0.9 D) produced maximum Mandelbaum's phenomenon. These results nearly match Mandelbaum's report of from 0.6 to 1.1 D. The screen distance of maximum Mandelbaum's phenomenon was at 0.8 D. At this distance, the screen strands subtended 0.8 arc minute and the nominal screen gap width was 5.5 arc minutes. Dark focus averaged 1.7 D, matching Leibowitz and Owens'[58,59] reports of 1.7 and 1.5 D for 124 and 220 college-age observers, respectively.

Gleason found that eye focus was essentially equal for all viewing tasks requiring fixation of the distant letter. If Mandelbaum's phenomenon were caused by the screen's capture of accommodation, then accommodation would shift eye focus towards the screen. But the eye focus is essentially unchanged before and after the introduction of the intervening screen, indicating that the screen did not influence accommodation. Eye focus shifted to the screen only after the instruction to focus on the screen. Moreover, eye focus returned to its initial level after the instruction to refocus on the distant letter. Clearly, the intervening screen did not disrupt volitional accommodation, either. Even more interesting is the fact that accommodative accuracy was not diminished by the perceptual blur as eye focus shifted from the screen back to the distant letter. In fact, it turns out that clinical emmetropia promotes Mandelbaum's phenomenon. Since clinical emmetropes are actually myopic (+0.6 D nearsighted), when they look at the mesh screen (placed at +0.8 D), the distant scenery (0.0 D) is defocused by 0.6 D, but the screen is only defocused by 0.2 D. Thus, the more blurred distant scenery may be suppressed for the benefit of the screen. This imbalance of retinal blur increases with uncorrected clinical myopia and, consequently, less time is needed to initiate Mandelbaum's phenomenon. On the other hand, an uncorrected hyperope can actually focus to the distant scenery, making the nearby screen blurrier. Now, the screen is suppressed instead of the distant scenery, precluding Mandelbaum's phenomenon. In summary, all five subjects in Gleason's study saw Mandelbaum's phenomenon without a concomitant shift in accommodation, which eliminates the misaccommodation concern over transparent HMD.

18.7 REFERENCES

1. B. H. Tsou, "System Design Considerations for a Visually Coupled System," in S. R. Robinson (ed.) *The Infrared and Electro-Optics System Handbook: Vol. 8. Emerging Systems and Technologies.* SPIE Optical Engineering Press, Bellingham, WA, 1993, pp. 515–536.

2. D. A. Fulghum, "Navy Orders Contractors to Stop Work on Ejectable Night Vision Helmets," *Aviation Week and Space Technology* **133**(23):67–68 (Dec. 3, 1990).

3. B. J. McEntire and D. F. Shanahan, "Mass Requirements for Helicopter Aircrew Helmets," *Advisory Group for Aerospace Research and Development (AGARD) Conference Proceedings* **597**:20.1–20.6 (7–9 Nov. 1996).

4. C. E. Perry and J. R. Buhrman, "HMD Head and Neck Biomechanics," in J. E. Melzer and K. Moffitt (eds.), *Head Mounted Displays: Designing for the User,* McGraw-Hill, New York, 1997, pp. 147–172.

5. B. McLean, S. Shannon, J. McEntire, and S. Armstrong, "Counterweights Used with ANVIS," Report USAARL-96-30, US Army Aeromedical Research Laboratory, Fort Rucker, AL, 1990.

6. K. Y. Lau, "Microscanner Raster-Scanning Display: A Spyglass for the Future," *Optics and Photonics News* **10**(5):47–50, 84 (May 1999).

7. M. Shenker, "Optical Design Criteria for Binocular Helmet-Mounted Displays," *SPIE* **778**:70–78 (1987).

8. S. M. Donelson and C. C. Gordon, "1988 Anthropometric Survey of US Army Personnel: Pilot Summary Statistics," Report Natick-TR-91-040, US Army Natick Research, Development and Engineering Center, Natick, MA, 1991.

9. J. Levy, "Physiological Position of Rest and Phoria," *Am. J. Ophthal.* **68**(4):706–713 (1968).

10. M. Menozzi, A. Buol, H. Krueger, and Ch. Miege, "Direction of Gaze and Comfort: Discovering the Relation for the Ergonomic Optimization of Visual Tasks," *Ophthalmic and Physiological Optics* **14**:393–399 (1994).

11. G. A. Gleason and J. T. Riegler, "The Effects of Eyepiece Focus on Visual Acuity Through ANVIS Night Vision Goggles During Short- and Long-Term Wear," unpublished report performed for A. F.

Workunit 71842604, 1997. (Part of the report was first presented at 1996 Human Factors Society Annual Meeting.)

12. R. Home and J. Poole, "Measurement of the Preferred Binocular Dioptric Settings at a High and Low Light Level," *Opt. Acta* **24**(1):97–98 (1977).

13. R. Jones, "Binocular Fusion in Helmet Mounted Displays," unpublished report performed for A. F. Workunit 71842601, 1996. (Part of the report was first presented at 1992 Society for Information Display Annual Meeting.)

14. American National Standards Institute, 1430 Broadway, New York, NY 10018.

15. K. N. Ogle, *Researches in Binocular Vision,* Hafner Publishing Co., New York, 1964, pp. 283–290.

16. G. A. Fry, "Measurement of the Threshold of Stereopsis," *Optometric Weekly* **33**:1029–1033 (1942).

17. G. Sigmar, "Observations on Vernier Stereo Acuity with Special Consideration to Their Relationship," *Acta Ophthalmologica* **48**:979–998 (1970).

18. E. E. Anderson and F. Weymouth, "Visual Perception and the Retinal Mosaic," *Am. J. Physiol.* **64**:561–591 (1923).

19. R. A. Schumer, "Mechanisms in Human Stereopsis," Ph.D. dissertation, Stanford University, Palo Alto, CA, 1979.

20. C. W. Tyler, "Sensory Processing of Binocular Disparity," in Schor and Ciuffreda (eds.), *Vergence Eye Movements: Basic and Clinical Aspects,* Butterworths, Boston, 1983, pp. 199–295.

21. A. M. Norcia, E. E. Sutter, and C. W. Tyler, "Electrophysiological Evidence for the Existence of Coarse and Fine Disparity Mechanisms in Humans," *Vision Res.* **25**:1603 (1985).

22. C. J. Erkelens and H. Collewijn, "Motion Perception During Dichoptic Viewing of Moving Random Dot Stereograms," *Vision Res.* **25**:345–353 (1985).

23. K. N. Ogle, *Researches in Binocular Vision,* Hafner Publishing Co., New York, 1964, pp. 173–199.

24. C. M. Schor and C. W. Tyler, "Spatial Temporal Properties of Panum's Fusional Area," *Vision Res.* **21**:683–692 (1981).

25. M. Shenker, J. LaRussa, P. Yoder, and W. Scidmore, "Overlapping-Monoculars—An Ultrawide-Field Viewing System," *Appl. Opt.* **1**:399–402 (1962).

26. S. S. Grigsby and B. H. Tsou, "Grating and Flicker Sensitivity in the Near and Far Periphery: Naso-Temporal Asymmetries and Binocular Summation," *Vision Res.* **34**:2841–2848 (1994).

27. S. S. Grigsby and B. H. Tsou, "Visual Processing and Partial-Overlap Head-Mounted Displays," *J. Soc. for Information Display* **2/2**:69–73 (1994).

28. T. H. Bui, R. H. Vollmerhausen, and B. H. Tsou, "Overlap Binocular Field-of-View Flight Experiment," *1994 Society for Information Display International Symposium Digest of Technical Papers* **25**:306–308 (1994).

29. M. J. Wells, M. Venturino, and R. K. Osgood, "The Effect of Field-of-View Size on Performance at a Simple Simulated Air-to-Air Mission," *SPIE* **1116**:126–137 (1989).

30. B. H. Tsou and B. M. Rogers-Adams, "The Effect of Aspect Ratio on Helmet-Mounted Display Field-of-View," *Report of the 30th Meeting of Air Standardization Coordination Committee (ASCC) Working Party 61: Aerospace Medical and Life Support Systems Symposium: Aeromedical Aspects of Vision* **4**:136–146. Defense and Civil Institute of Environmental Medicine, Toronto, Canada, 1990.

31. A. T. Bayhill and D. Adler, "Most Naturally Occurring Human Saccades Have Magnitudes of 15 Degrees or Less," *Invest. Ophthal.* **14**:468–469 (1975).

32. G. H. Robinson, "Dynamics of the Eye and Head During Movement Between Displays: A Qualitative and Quantitative Guide for Designers," *Human Factors* **21**:343–352 (1979).

33. J. Wolfe, P. O'Neil, and S. C. Bennett, "Why Are There Eccentricity Effects in Visual Search? Visual and Attentional Hypotheses," *Percept. Psychophys.* **60**:140–156 (1998).

34. S. P. McKee and K. Nakayama, "The Detection of Motion in the Peripheral Visual Field," *Vision Res.* **24**(1):25–32 (1984).

35. K. K. Ball, B. L. Beard, D. L. Roenker, R. L. Miller, and D. S. Griggs, "Age and Visual Search: Expanding the Useful Field of View," *J. Opt. Soc. Am.* **A5**:2210–2219 (1988).

36. P. Havig and B. H. Tsou, "Is There a Useful Field of Visual Search?" [Abstract]. *Optical Society of America Annual Meeting,* Baltimore, MD, 4–9 October 1998.

37. L. N. Thibos and A. Bradley, "Modeling the Refractive and Neuro-Sensor Systems of the Eye," in P. Mouroulis (ed.), *Visual Instrumentation Handbook,* McGraw-Hill, New York, 1999, pp. 4.19–4.20.

38. P. Mouroulis and H. Zhang, "Visual Instrument Image Quality Metrics and the Effects of Coma and Astigmatism," *J. Opt. Soc. Am.* **A9**(1):34–42 (1992).

39. J. E. Greivenkamp, J. Schwiegerling, J. M. Miller, and M. D. Mellinger, "Visual Acuity Modeling Using Optical Raytracing of Schematic Eyes," *Am. J. Ophthal.* **120**(2):227–240 (1995).

40. W. N. Charman, and J. A. M. Jennings, "The Optical Quality of the Monochromatic Retinal Image As a Function of Focus," *Br. J. Physiol. Opt.* **31**:119–134 (1976).

41. D. Greene, "Night Vision Pilotage System Field-of-View (FOV)/Resolution Tradeoff Study Flight Experiment," Report NV-1-26, Center for Night Vision and Electro-Optics, Ft. Belvoir, VA, 1988.

42. G. E. Legge, K. T. Mullen, G. C. Woo, and F. W. Campbell, "Tolerance to Visual Defocus," *J. Opt. Soc. Am.* **A4**(5):851–863 (1987).

43. J. H. Iavecchia, H. P. Iavecchia, and S. N. Roscoe, "Eye Accommodation to Head-Up Virtual Images," *Human Factors,* **130**(6):687–702 (1988).

44. G. A. Gleason, and J. T. Riegler, "Do Clinical Refraction Result in Best Visual Acuity?", unpublished report performed for A. F. Workunit 71842604, 1997. (Part of the report was first presented at 1996 Association for Research in Vision and Ophthalmology Annual Meeting.)

45. T. N. Cornsweet and H. D. Crane, "Servo-Controlled Infrared Optometer," *J. Opt. Soc. Am.* **60**:548–554 (1970).

46. H. D. Crane and C. M. Steele, "Accurate Three Dimensional Eyetracker," *Appl. Opt.* **17**:691–705 (1978).

47. F. W. Campbell and J. G. Robson, "High-Speed Infra-Red Optometer," *J. Opt. Soc. Am.* **49**:268–272 (1959).

48. N. A. McBrien and M. Millodot, "Clinical Evaluation of the Canon Autoref R-1," *Am. J. Optom. Physiol. Opt.* **62**:786–792 (1985).

49. R. T. Hennessy and H. W. Leibowitz, "Laser Optometer Incorporating the Badal Principle," *Behav. Res., Meth. and Instru.* **4**(5):237–239 (1972).

50. A. Morrell and W. N. Charman, "A Bichromatic Laser Optometer," *Am. J. Optom. Physiol. Opt.* **64**:790–795 (1987).

51. R. E. Bannon, F. H. Cooley, H. M. Fisher, and R. T. Textor, "The Stigmatoscopy Method of Determining the Binocular Refractive Status," *Am. J. Optom. Arch. Am. Acad. Optom.* **27**:371 (1950). (Cited in I. M. Borish, *Clinical Refraction* third ed., Professional Press, Chicago, 1975, pp. 783–784.)

52. F. W. Campbell, "The Depth of Field of the Human Eye," *Opt. Acta* **4**:157–164 (1957).

53. M. W. Morgan, "The Meaning of Clinical Tests in the Light of Laboratory Measurements," *The Ind. Opt.* **2**:6 (1947). (Cited in I. M. Borish, *Clinical Refraction,* third ed., Professional Press, Chicago, 1975, p. 356.)

54. R. Jones, "Proximal Accommodation in Virtual Displays," *Society for Information Display International Symposium Digest of Technical Papers* **24**:195–198 (1998).

55. G. A. Gleason, and R. V. Kenyon, "Mandelbaum's Phenomenon Is Not Caused by Accommodative Capture," unpublished report performed for A. F. Workunit 71842604, 1998. (Part of the report was first presented at 1997 Association for Research in Vision and Ophthalmology Annual Meeting.)

56. J. Mandelbaum, "An Accommodative Phenomenon," *AMA Arch. Ophthal.* **63**:923–926 (1960).

57. D. A. Owens, "The Mandelbaum Effect: Evidence for an Accommodative Bias Toward Intermediate Viewing Distance," *J. Opt. Soc. Am.* **65**:646–652 (1975).

58. H. W. Leibowitz and D. A. Owens, "Anomalous Myopias and the Dark Focus of Accommodation," *Science* **189**:646–648 (1975).

59. H. W. Leibowitz and D. A. Owens, "New Evidence for the Intermediate Position of Relaxed Accommodation," *Documenta Ophthalmologica* **46**:133–147 (1978).

P · A · R · T · 3

X-RAY AND NEUTRON OPTICS

Carolyn A. MacDonald   Part Editor
Walter M. Gibson   Part Editor

University at Albany
State University of New York
Albany, New York

INTRODUCTION

CHAPTER 19
AN INTRODUCTION TO X-RAY AND NEUTRON OPTICS

Carolyn A. MacDonald
Walter M. Gibson
University at Albany, State University of New York
Albany, New York

19.1 A NEW CENTURY OF X-RAY AND NEUTRON OPTICS

One of Roentgen's earliest observations, shortly after his discovery of X rays in 1895,[1] was that while the rays were easily absorbed by some materials, they did not strongly refract. Standard optical lenses, which require weak absorption and strong refraction, were therefore not useful for manipulating the rays. New techniques were quickly developed. In 1914, van Laue was awarded the Nobel Prize for demonstrating the diffraction of X rays by crystals. By 1929, total reflection at grazing incidence had been used to deflect X rays.[2] A more detailed history of X-ray optics is presented in Volume II of this handbook,[3] but the available optics for X rays still can be classified by those three phenomena: refraction; diffraction; and total reflection. Surprisingly, given that the physics governing these optics has been well known for nearly a century, there has been a recent dramatic increase in the availability, variety, and performance of X-ray optics for a wide range of applications. This advance has been accompanied by a similar surge in the development of X-ray sources and detectors. The new abundance requires a fresh look at the problem of optimizing a wide range of X-ray and neutron applications. The development of X-ray technology has also advanced neutron science because a number of the optics and detectors are either applicable to neutrons or have inspired the development of neutron technology.

The purpose of this work is to provide an introduction to a variety of X-ray optics and contrast their use in a number of applications. Subparts 3.2 through 3.4 (Chaps. 21–30) of this volume are devoted to the discussion of the three classes of X-ray optics. Subpart 3.5 (Chaps. 31–35) elucidates source and detector properties which affect optics choices and system optimization and concludes with some example applications, including diffraction, fluorescence, and medical imaging. This list of optics and, particularly, applications cannot be exhaustive, but is meant to provide instructive examples of how applications requirements and source and detector properties affect optics choices. In Subpart 3.6 (Chap. 36), neutron optics are discussed and compared to X-ray optics. The appendix, following Chap. 37, includes tables of X-ray emission energies and X-ray optical properties of materials.

One question that arises immediately in a discussion of X-ray phenomena is precisely what spectral range is included in the term. Usage varies considerably by discipline, but for the purposes of this volume, the X-ray spectrum is taken to be roughly from 1 to 100 keV in photon energy (from 1.24 to 0.0124 nm wavelength). The range is extended down into the extreme ultraviolet (EUV) to include some microscopy and astronomical optics. A detailed study of astronomical optics, especially with respect to image transfer properties, and the metrology and effects of roughness and profile defects, is found in Vol. II of this handbook.[3] A review of recent progress in astronomical optics is made in Chap. 28 of this volume.

The reader is also referred to a number of useful textbooks on X-ray optics.[4–7]

19.2   X-RAY INTERACTION WITH MATTER

X rays are applicable to such a wide variety of fields because they are penetrating but interacting, have wavelengths on the order of atomic spacings, and have energies on the order of core electronic energy levels for atoms.

X-Ray Production

X rays are produced primarily by the acceleration of charged particles, the knockout of core electrons, or blackbody and characteristic emission from very hot sources such as laser generated plasmas or astronomical objects. The production of X rays by accelerated charges includes incoherent emission such as Bremsstrahlung radiation in tube sources and coherent emission by synchrotron undulators or free electron lasers. Highly coherent emission can also be created by pumping transitions between levels in ionic X-ray lasers.

The creation of X rays by the knockout of core electrons is the mechanism for the production of the characteristic lines in the spectra from conventional X-ray tube sources. The incoming electron knocks out a core electron, creating a vacancy, which is quickly filled by an electron dropping down from an outer shell. The energy difference between the outer shell energy level and the core energy level is emitted in the form of an X-ray photon. The photon energy is therefore characteristic of the element. K lines come from transitions from an outer shell to a K shell inner core vacancy. K levels have principal quantum number $n = 1$. Kα lines are due to transitions from an outer shell level with $n = 2$ to the $n = 1$ core level. Kβ lines come from outer shells with $n = 3$. L lines are due to transitions from outer shells with $n \geq 3$ to vacancies in the L core level, which has $n = 2$. M lines are from vacancies with $n = 3$. Lines can be doublet or triplet if there is more than one applicable level with the same principal quantum number n. Tables of characteristic lines are found in the appendix to this section.

X-ray sources are discussed at more length in Chaps. 31 through 33.

Refraction

In the X-ray regime, the real part of the index of refraction of a solid can be simply approximated by[8]

$$n_r = Re\left(\sqrt{\kappa}\right) = Re\left(\sqrt{\frac{\varepsilon}{\varepsilon_0}}\right) \cong \sqrt{1 - \frac{\omega_p^2}{\omega^2}} \cong 1 - \delta \qquad (1)$$

where n is the index of refraction, ε is the dielectric constant of the solid, ε_0 is the vacuum dielectric constant, κ is their ratio, ω is the photon frequency, and ω_p is the plasma frequency of the material. The plasma frequency, which typically corresponds to tens of eVs, is given by

$$\omega_p^2 = \frac{Ne^2}{m\varepsilon_0} \qquad (2)$$

where N is the electron density of the material, and e and m are the charge and mass of the electron. For X rays, the relevant electron density is the total density, including core electrons. Thus, changes in X-ray optical properties cannot be accomplished by changes in the electronic levels, in the manner in which optical properties of materials can be manipulated for visible light. The X-ray optical properties are determined by the density and atomic numbers of the elemental constituents. Tables of the properties of most of the elements are presented in the Appendix. Because the plasma frequency is very much less than the photon frequency, the index of refraction is slightly less than one.

Absorption and Scattering

The imaginary part of the index of refraction for X rays arises primarily from photoelectric absorption. This absorption is largest for low-energy X rays and has peaks at energies resonant with core ionization energies of the atom. A table of K edge absorption energies is given in the Appendix. X rays can also be deflected out of the beam by incoherent or coherent scattering from electrons. Coherent scattering from nearly free electrons (Thomson scattering) is responsible for the decrement to the real part of the index of refraction given in Eq. (1). Rayleigh scattering is coherent scattering from tightly bound electrons. The constructive interference of coherent scattering from arrays of atoms constitutes diffraction. Incoherent (Compton) scattering occurs when the photon exchanges energy and momentum with the electron. Compton scattering becomes increasingly important for high energy photons and thick transparent media.

19.3 OPTICS CHOICES

Refractive Optics

One consequence of Eq. (1) is that the index of refraction of all materials is very close to unity in the X-ray regime. Thus Snell's law implies that there is very little refraction at the interface,

$$(1)\ \sin\left(\frac{\pi}{2} - \theta_1\right) = n \sin\left(\frac{\pi}{2} - \theta_2\right) \tag{3}$$

where the first medium is vacuum and the second medium has index n, as shown in Fig. 1. In X-ray applications the angles θ are measured from the surface, not the normal to the surface. If n is very close to one, θ_1 is very close to θ_2. Thus, refractive optics are more difficult to achieve in the X-ray regime. The lens maker's equation,

$$f = \frac{\dfrac{R}{2}}{n - 1} \tag{4}$$

gives the relationship between the focal length of a lens, f, and the radius of curvature of the lens, R. The symmetric case is given. Because n is very slightly less than one, the focal length produced even by very small negative-radius (convex) lenses is rather long. This can be overcome by using large numbers of surfaces in a compound lens. Because the radii must still be small, the aperture of the lens cannot be large, and refractive optics are generally better suited to narrow synchrotron beams than to isotropic point sources. Refractive optics are discussed in Chap. 20.

Diffractive and Interference Optics

The coherent addition of radiation from multiple surfaces or apertures can only occur for a very narrow wavelength bandwidth. Thus, diffractive and interference optics such as crystals, gratings, multilayers, and zone plates, are all wavelength selective.

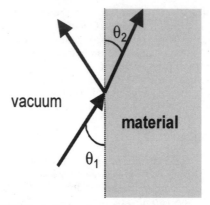

FIGURE 1 Refraction at a vacuum-to-material interface.

To achieve constructive interference, the spacing of a grating must be arranged so that the radiation from successive sources is in phase, as shown in Fig. 2. The circular apertures of zone plates, discussed in Chap. 23, are similar to the openings in a transmission grating. The superposition of radiation from the circular apertures results in focusing of the radiation to points on the axis.

Diffractive optics operate on the same principle—coherent superposition of many rays.[9] The most common diffractive optic is the crystal.

Arranging the beam angle to the plane, θ, and plane spacing, *d,* as shown in Fig. 3, so that the rays reflecting from successive planes are in phase (and ignoring refraction); yields the familiar Bragg's law,

$$n\lambda = 2d \sin \theta \qquad (5)$$

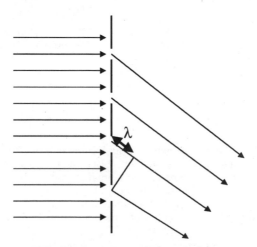

FIGURE 2 Schematic of constructive interference from a transmission grating. λ is the wavelength of the radiation. The extra path length indicated must be an integral multiple of λ. Most real gratings are used in reflection mode.

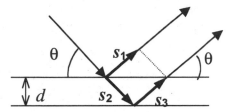

FIGURE 3 Bragg diffraction from planes with spacing d. The extra path length, $(s_2 + s_3) - s_1$, must be an integral multiple of λ.

where n is an integer and λ is the wavelength of the X ray. Bragg's law cannot be satisfied for wavelengths greater than twice the plane spacing, so crystal optics are limited to low wavelength, high energy, X rays. Multilayers, as shown in Fig. 4, are artificial crystals of alternating layers of materials. The spacing, Λ, which is the thickness of a layer pair, replaces d in Eq. (5), and can be much larger than crystalline plane spacings. Multilayers are therefore effective for lower-energy X rays. Diffractive and interference optics are discussed in Subpart 3.3, Chaps. 21 through 25. Diffraction is also discussed in Chap. 35.

Reflective Optics

Using Snell's law, Eq. (3), the angle inside a material is smaller than the incident angle in vacuum. For incident angles less than a critical angle, θ_c, no wave can be supported in the medium and the incident ray is totally externally reflected. The critical angle is given by

$$\sin\left(\frac{\pi}{2} - \theta_c\right) = n \times \sin\left(\frac{\pi}{2}\right) \tag{6}$$

Thus, using small angle approximations and Eq. (1),

$$\theta_c \cong \frac{\omega_p}{\omega} \tag{7}$$

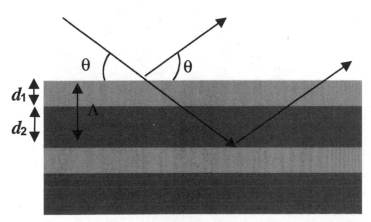

FIGURE 4 Bragg diffraction from multilayers with spacing Λ. d_1 is the thickness of the first layer, for example the silicon layer, and d_2 the thickness of the other, for example the Molybdenum in a Mo/Si multilayer.

Because the plasma frequency is very much less than the photon frequency, total external reflection only occurs for very small grazing incidence angles. This phenomenon is used extensively for single reflection mirrors for synchrotrons, X-ray microscopes, and X-ray telescopes. Multichannel plate optics are also single reflection mirrors. Since the solid angle that can be captured by a mirror is determined by θ_c, single reflection optics are most commonly used at lower energies, where θ_c is largest. Multiple reflections are used to transport X rays in glass capillary optics, and polycapillary arrays. Single and multiple total reflection optics are discussed in Subpart 3.4, Chaps. 26 through 30.

Simulations

One of the most significant tools for achieving rapid development of new X-ray optics has been the upsurge in modeling capability associated with the rise in available computing power. One commonly used program is SHADOW, which is available at http://www.xraylith.wisc.edu/shadow/shadow.html. Many other programs have been developed, particularly for more specialized analysis of particular optics, and are referenced in the following chapters.

Applications

X rays have long been essential tools for medical imaging and materials analysis. The application of X-ray optics to medical imaging, to the analysis of material structure by diffraction, and to the analysis of composition by microfluorescence is discussed in Chap. 35 in this section. Two somewhat newer applications are X-ray proximity lithography for the production of microelectronics devices, and X-ray microscopy. X-ray microscopy is particularly useful for the investigation of samples that are too opaque or have features that are too fine for light microscopy, but that are too delicate for electron microscopy. These two applications are also, more briefly, discussed. The best choice of optic for a particular application depends on the application requirements, but is also influenced by the properties of the available sources and detectors. A discussion of detector properties can be found in Chap. 34.

19.4   REFERENCES

1. W. C. Röntgen, "On a New Form of Radiation," *Nature* **53**:274–276 (January 23, 1896). (English translation from *Sitzungsberichte der Würzburger Physik-medic. Gessellschaft,* 1895.)

2. F. Jentzch, *Phys. Z.* **30**:1268 (1929).

3. James E. Harvey, "X-Ray Optics," chapter 11, in M. Bass (ed.), *Handbook of Optics,* second ed., vol. II, McGraw-Hill, New York, 1995.

4. A. G. Michette and C. J. Buckley (eds.), *X-Ray Science and Technology,* Institute of Physics Publishing, Philadelphia, 1993.

5. E. Spiller, *Soft X-Ray Optics,* SPIE, Bellingham, WA, 1994.

6. H. J. Queisser, *X-Ray Optics* (Topics in Applied Physics, vol. 22), Springer Verlag, New York, 1977.

7. A. I. Erko, V. V. Aristov, and B. Vidal, *Diffraction X-Ray Optics,* IOP Publishing, Philadelphia, 1996.

8. J. D. Jackson, *Classical Electrodynamics,* John Wiley & Sons, New York, 1962, p. 227.

9. Bernard D. Cullity, *Elements of X-Ray Diffraction,* Addison Wesley Longman, Reading, MA, 1978.

REFRACTIVE OPTICS

CHAPTER 20
REFRACTIVE X-RAY OPTICS

B. Lengeler, C. Schroer, J. Tümmler, and B. Benner
Zweites Physikalisches Institut, RWTH Aachen
Aachen, Germany
A. Snigirev and I. Snigireva
European Synchrotron Radiation Facility—ESRF
Grenoble, France

20.1 INTRODUCTION

Refractive lenses made out of glass are among the most widely used optical components for visible light, with a broad spectrum of applications in focusing and imaging. For a long time, refractive lenses for the hard X-ray range were considered unfeasible. The weak refraction of X rays in matter and their strong absorption seemed to be insurmountable obstacles. In 1996, it was shown that these obstacles can be circumvented and that linear lenses, shaped as a row of drilled holes in a bloc of aluminum or beryllium, can be manufactured with about 1 m focal length and a transmission of a few percent for a geometric aperture of 1 mm.[1–3] In the meantime the concept and the manufacture of the lenses have been improved considerably. The new lenses have a parabolic profile and rotational symmetry around the optical axis. Hence, they focus in two directions and are free of spherical aberration. They are genuine imaging devices, like glass lenses for visible light.[4–6]

20.2 CONCEPT AND MANUFACTURE OF PARABOLIC X-RAY LENSES

It has been shown in an extensive paper that most formulas known from ordinary optics, like those for the focal length, for the imaging properties and for the lateral and longitudinal resolution can be transferred to the X-ray regime.[6] The main difference is in the effect of absorption and surface roughness in the effective aperture. The most outstanding property of refractive X-ray lenses is the small value of their numerical aperture, NA, which is below 10^{-4}. The index of refraction n for X rays in matter reads[7]

$$n = 1 - \delta - i\beta \tag{1}$$

$$\delta = \frac{N_a}{2\pi} r_0 \lambda^2 \rho \frac{Z + f'}{A} \tag{2}$$

$$\beta = \lambda\mu/4\pi \tag{3}$$

Here, N_a is Avogadro's number, r_0 the classical electron radius, $\lambda = 2\pi/k_1 = 2\pi c/\omega$ the photon wavelength, $Z + f'$ the real part of the atomic scattering factor (in the forward direction) including the dispersion correction f', A the atomic mass, μ the linear coefficient of attenuation, and ρ the density of the lens material. δ is a very small number, of the order 10^{-6}. β is even smaller by three orders of magnitude. Since the real part of n is smaller than 1, a focussing X-ray lens must have concave shape (Fig. 1a). In order to obtain a focal length f in the range of one meter many single lenses have to be stacked behind each other to form a *compound refractive lens (CRL)* as shown in Fig. 1b. The focal length of such a CRL with parabolic profile $s^2 = 2Rw$ and N individual biconcave lenses is (Fig. 1)

$$f = R/2N\delta \tag{4}$$

as measured from the middle of the lens. R is the radius of curvature at the apex of the parabolas. A lens with thickness $2w_0 + d$ has an aperture $2R_0 = 2\sqrt{2Rw_0}$. An Al CRL for use at 15 keV needs $N = 42$ individual lenses in order to have a focal length of 1 m ($f = 0.986$ m) when $R = 200$ μm. For $2w_0 = 1$ mm, the CRL has a length of about 42 mm, which is short compared to the focal length. The thickness d between the apices of the parabolas is typically 10 μm (Fig. 1). The parabolic lenses, used up to now, have been manufactured at the University of Technology in Aachen.[6] They are made out of Al by a pressing technique similar to that used in coining metallic money pieces. The piece of Al which will become the lens is set in a metallic ring 12 mm in diameter and 1 mm thick. Two hard metal dies have convex parabolic protrusions which face each other and which press the lens profile simultaneously from both sides into the Al. For that purpose, the dies and the lens are guided in a centering ring. All high precision parts (lens ring, paraboloids, and centering ring) are manufactured with a precision of a μm by means of a computer controlled tooling machine. The individual lenses are stacked behind one another by means of two high-precision shafts, so that their optical axes are aligned with a precision of a μm. Figure 2 shows a sketch of an assembled CRL. It has the

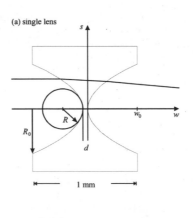

(a) single lens

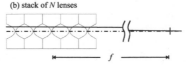

(b) stack of N lenses

FIGURE 1 Parabolic compound refractive lens (CRL). The individual lenses (*a*) are stacked to form a CRL (*b*).

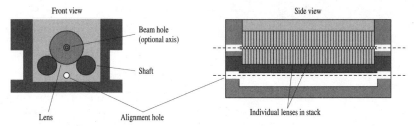

FIGURE 2   Compound refractive lens.

size of a match box. It is very easy to align the lens in the beam. A preadjustment is done by means of an alignment hole that allows one to set the lens axis parallel to the X-ray beam (Fig. 2). Since the focus stays on the optical axis, the whole alignment takes less than 15 min.

It turns out that CRLs are very robust and stable. They withstand the white beam of an undulator beam line at third generation synchroton radiation surces [like European Synchrotron Research Facility (ESRF) in Grenoble]. The 50 W of the beam increases the temperature of the CRL in still air to 150° C. No deterioration of the performance of the lens was observed.

20.3   *GENERATION OF A MICROFOCUS BY MEANS OF A CRL*

X rays are used extensively in analysis, like diffraction, small angle scattering, fluorescence, absorption, and reflectometry. There is a growing need to combine those techniques with high lateral resolution in the µm range. Refractive CRLs are excellently suited for that purpose. They are able to image the source onto the sample with a demagnification by a factor of typically 50. Undulators at third-generation synchrotron radiation sources have a size of typically $a_h = 700$ µm full width at half-maximum (FWHM) in the horizontal and $d_v = 35$ µm in the vertical. Figure 3 shows the image of the undulator source of beamline ID22 at the ESRF demagnified by an Al CRL ($E = 20.5$ keV, $N = 61$, $d = 25$ µm, $\mu = 8.37$/cm, $\delta = 1.29 \cdot 10^{-6}$). The intensity distribution behind the lens has been calculated in detail.[6] In the image plane at L_{20} behind the lens ($L_{20} = L_1 f/(L_1 - f)$), with L_1 being the source to lens distance), the intensity has a Gaussian profile with FWHM

$$B_v = \sqrt{(d_v L_{20}/L_1)^2 + (2\sqrt{2 \ln 2}\lambda L_{20}/\pi D_{eff})^2} \tag{5}$$

for the vertical direction and a similar expression B_h for the horizontal direction with d_h instead of d_v. Here D_{eff} is the effective aperture

$$D_{eff} = 2R_0 \sqrt{[1 - \exp{(-a_p)}]/a_p} \tag{6}$$

with

$$a_p = R_0^2(\mu NR + 2Nk_1^2 \, \delta^2\sigma^2)/2R^2 \tag{7}$$

B_v contains two contributions: The first one, $d_v L_{20}/L_1$, is the demagnification of the source according to geometrical optics. The second term, $0.75\lambda L_{20}/D_{eff}$, describes the influence of diffraction at the effective aperture of the lens. The parameter a_p takes into account the influ-

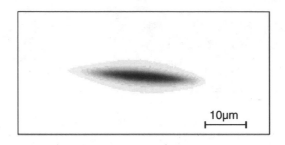

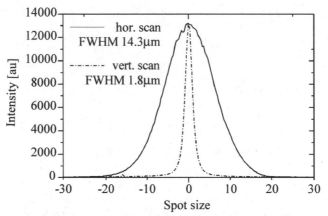

FIGURE 3 Image of the undulator source of beamline ID22 at the ESRF as imaged by an Al lens at 62.65 m from the source ($E = 20.5$ keV) and linear scans across the image in the horizontal and vertical directions.

ence of attenuation μ in the lens material and the influence of surface roughness σ in the lens surface. For Al lenses which have a surface roughness of $\sigma = 0.1$ μm rms, roughness makes a negligible contribution to the effective aperture of the lens. In the case of Fig. 3, it is $f = 1.271$ m, $L_1 = 62.65$ m, $L_{20} = 1.297$ m, $\mu NR = 10.211$, $2Nk_1^2\delta^2\sigma^2 = 0.022$, $a_p = 25.558$, $D_{\text{eff}} = 177.0$ μm.

Here, a comment is appropriate about the influence of roughness on the performance of X-ray mirrors as opposed to X-ray lenses. Roughness reduces the transmission and the reflectivity according to a Rayleigh damping factor $\exp(-2Q^2\sigma^2)$. For a mirror, the momentum transfer, Q, is $2k_1\theta$ with a typical value of 0.2° for the angle of reflection θ. For $\lambda = 1$ Å, it is $Q = 4.4 \cdot 10^{-2}$ Å^{-1}. For an Al CRL at 12.4 keV with $\delta = 3.54 \cdot 10^{-6}$, $N = 28$, and $f = 1$ m, the momentum transfer $\sqrt{2N} k_1\delta$ is $1.7 \cdot 10^{-4}$ Å^{-1}. The low value of Q for a CRL allows much larger values for the roughness σ. If a mirror for hard X rays needs a surface finish of a nm rms, a CRL needs a finish of a μm. This reduces drastically the requirements for manufacturing refractive X-ray lenses.

The transmission of a CRL is given by

$$T_p = \exp(-\mu Nd)[1 - \exp(-2a_p)]/2a_p \tag{8}$$

When surface roughness is neglected, a_p can be expressed in terms of the mass attenuation coefficient μ/ρ and the focal length f as

$$2a_p = \left(\frac{\mu}{\rho}\right)\frac{\pi R_0^2}{f}\frac{A}{N_a r_0 \lambda^2 (Z + f')} \tag{9}$$

For a given focal length f and a given wavelength λ, the transmission of the lens increases with decreasing μ/ρ. In other words, low Z materials, like Li, Be, B, C, and Al, are favorable. The photoabsorption decreases with the energy at about E^{-3}. However, at about 0.2 cm^2/g the decrease almost levels off when Compton scattering exceeds photoabsorption.[8] Compton-scattered photons do not contribute to the image formation. They generate a blur of the focal spot. In the end it is Compton scattering that limits the effective aperture and hence the resolution of CRL for X rays. The gain g of a CRL is defined as the ratio of the intensity in the focal spot and the intensity behind a pinhole of equal size. It is

$$g = T_p 4R_0^2/B_v B_h \tag{10}$$

The great advantage of beryllium as lens material as compared to aluminum becomes visible in the transmission and in the gain. A Be lens with $R = 0.2$ mm, $2R_0 = 0.895$ mm, $d = 20$ µm, $\sigma = 0.2$ µm rms, and $N = 52$ individual lenses has $f = 1.269$ m at 15 keV. In this case, $a_p = 1.483$, $D_{eff} = 646$ µm, and the transmission is 30 percent. The expected gain is larger than 10,000 compared to a value between 100 and 200 for a similar Al CRL.

Instead of focussing a beam, a CRL can also be used to make the beam highly parallel. The beam emitted by high-β undulators at ESRF has already a low divergence of 25 µrad and can be made even less divergent by means of a CRL, when the source is in the focal spot of the lens. The beam divergence is then

$$\Delta\vartheta_v = \sqrt{(d_v/L_1)^2 + (1.029\lambda/D_{eff})^2} \tag{11}$$

and a corresponding expression for the horizontal divergence $\Delta\vartheta_h$. A single Al lens with $2R_0 = 0.895$ mm, $R = 0.2$ mm, has at 15 keV a focal length $f = L_1 = 41.43$ m. This is the typical distance between the source and the first hutch at ESRF beamlines. In that case $D_{eff} = 0.708$ mm and $\Delta\vartheta_v = 0.85$ µrad, which is 29 times smaller than the divergence of undulator beams without lens. In the horizontal the reduction is less drastic, with $\Delta\vartheta_h = 17$ µrad. In both cases, it is the source size that determines the beam divergence behind the lens.

20.4 IMAGING OF AN OBJECT ILLUMINATED BY AN X-RAY SOURCE BY MEANS OF A CRL: HARD X-RAY MICROSCOPE

CRLs are not only useful optical devices in microanalysis, they can also be used as main components in a hard X-ray microscope. For that purpose, the object to be imaged is illuminated from behind by an undulator source. The lens is placed a distance L_1 behind the sample and the image is transferred by the lens onto a position sensitive detector a distance $L_{20} = L_1 f/(L_1 - f)$ behind the lens. As in any microscope, the object is located slightly outside of the focal length of the CRL in order to achieve a large magnification $f/(L_1 - f)$. Magnifications up to 50 can be achieved at the ESRF beamline ID22. The main advantage of a large magnification is in the relaxed requirements on the detector. Good detectors have a point spread function of 1 µm.[9] With a magnification above 20, details in the object below 0.1 µm can be resolved by the detector. The distribution of intensity in the image plane of the microscope at L_{20} has been calculated in detail.[6] The lateral resolution for incoherent illumination, defined as the distance between two object points that are separated in the image plane by the FWHM of the image of one point, is

$$d_t = 0.75\lambda/2\text{NA} \tag{12}$$

where the numerical aperture is

$$\text{NA} = D_{eff}/2L_1 \tag{13}$$

Figure 4 shows the image of a Fresnel zone plate (169 Au rings on a Si$_3$N$_4$ substrate, see Chap. 23) imaged by an Al CRL at 23.5 keV with $N = 62$, $f = 1.648$ m, $L_1 = 1.786$ m, and $L_{20} = 21.4$ m.

The theoretical resolving power was $d_t = 0.34\ \mu m$. The outermost zone of the object has a width of 0.3 μm and is clearly resolved in the image. For a Be CRL, a lateral resolution of 0.07 μm is feasible. As mentioned previously, Compton scattering ultimately limits the resolving power of the new microscope. As in normal optics, the longitudinal resolving power d_l of the microscope is

$$d_l = 2\lambda/\pi(NA)^2 \tag{14}$$

Since the NA is typically between 10^{-4} and 10^{-5}, d_l is in the 10 mm range. In other words, the hard X-ray microscope has no longitudinal resolution and the full 3-dimensional structure of a thick sample has to be obtained by tomographic techniques. The potential of the new microscope for investigating opaque objects is demonstrated in Fig. 5. A square Au grid with 15-μm period is imaged by an Al CRL at 23.5 keV. Attached to the lower part of the grid is a second linear array of Au lines with a period of 1 μm. This linear array is attached from behind and is therefore not visible in the scanning electron micrograph (Fig. 5b), whereas it is clearly visible in phase contrast in the X-ray microscope.

20.5   SUMMARY AND OUTLOOK

Compound refractive X-ray lenses with parabolic profile have been manufactured. Today, they are used up to about 55 keV. They have a geometric aperture of 1 mm and are best suited for undulator beams at synchrotron radiation storage rings. They are stable in the white beam of an undulator. Due to their parabolic profile they are free of spherical aberration and are genuine imaging devices. An object area of 300 μm in diameter can be imaged. The lateral resolution achieved up to now is 0.3 μm. A resolution below 0.1 μm can be expected. The transmission of aluminum CRL is of the order of a few percent, the gain is above 100. For Be CRL the transmission can be as high as 30 percent and the gain can be 10,000. The main fields of application of the new X-ray lenses are: (1) *microanalysis* with a beam in the μm range for diffraction, fluorescence, absorption, and reflection; (2) *imaging* of opaque objects in absorption and phase contrast, and (3) *coherent x-ray scattering*.

Although CRLs are best adapted to the small size of an undulator beam, this use is not the only one conceivable. The development of optics for the combination of a microfocus X-ray tube and a CRL is on the way.

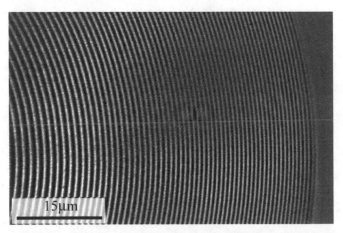

FIGURE 4   X-ray micrograph (23.5 keV) of a Fresnel-Zone plate (gold rings on Si_3Ni_4) with the outermost zone 0.3 μm in width.

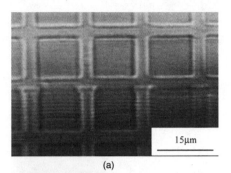

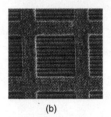

FIGURE 5 (*a*) X-ray micrograph of a gold mesh (15-μm period in both directions). The lower squares have attached to them from behind a second linear gold grid with 1-μm period. In contrast to the electron microscope [electron micrograph shown in (*b*)], the X-ray microscope is able to image the linear grid behind the bars of the square mesh.

20.6 REFERENCES

1. A. Snigirev, V. Kohn, I. Snigireva, and B. Lengeler, "A Compound Refractive Lens for Focusing High-Energy X-Rays," *Nature* **384**:49–53 (1996).

2. A. Snigirev, B. Filseth, P. Elleaume, T. Klocke, V. Kohn, B. Lengeler, I. Snigireva, A. Souvorov, and J. Tümmler, "Refractive Lenses for High Energy X-Ray Focusing," *SPIE* **3151**:164–170 (1998).

3. A. Snigirev, V. Kohn, I. Snigireva, A. Souvorov, and B. Lengeler, "Focusing High-Energy X Rays by Compound Refractive Lenses," *Applied Optics* **37**:653–662 (1998).

4. B. Lengeler, C. Schroer, M. Richwin, J. Tümmler, M. Drakopoulos, A. Snigirev, and I. Snigireva, "A Novel Microscope for Hard X-Rays Based on Parabolic Compound Refractive Lenses," *Appl. Phys. Lett.* **84**:3924–3926 (1999).

5. B. Lengeler, J. Tümmler, A. Snigirev, I. Snigireva, and C. Raven, "Transmission and Gain of Singly and Doubly Focusing Refractive X-Ray Lenses," *J. Appl. Phys.* **84**:5855–5861 (1998).

6. B. Lengeler, C. Schroer, J. Tümmler, B. Benner, M. Richwin, A. Snigirev, I. Snigireva, and M. Drakopoulos, "Imaging by Parabolic Refractive Lenses in the Hard X-Ray Range," *J. Synchrotron Rad.* **6**:1153–1167 (1999).

7. R. W. James, *The Optical Principles of the Diffraction of X rays*, Cornell University Press, Ithaca, NY, (1965).

8. B. L. Henke, E. M. Gullikson, and J. C. Davis, "X-Ray Interactions: Photoabsorption, Scattering, Transmission and Reflection," *Atomic Data and Nucl. Data Tables* **54**:181–342 (1993).

9. A. Koch, C. Raven, P. Spanne, and A. Snigirev, "X-ray Imaging with Submicrometer Resolution Employing Transparent Luminescent Screens," *J. Opt. Soc. Am.* **A15**:1940–1951 (1998).

DIFFRACTIVE AND INTERFERENCE OPTICS

CHAPTER 21

GRATINGS AND MONOCHROMATORS IN THE VUV AND SOFT X-RAY SPECTRAL REGION

Malcolm R. Howells
Advanced Light Source, Lawrence Berkeley
National Laboratory Berkeley, California

21.1 INTRODUCTION

Spectroscopy in the photon energy region from the visible to about 1 to 2 keV is generally done using reflection gratings. In the region above 40 eV, reasonable efficiency is only obtained at grazing angles and in this article we concentrate mainly on that case. Flat gratings were the first to be used and even today are still important. However, the advantages of spherical ones were recognized very early.[1] The first type of focusing grating to be analyzed theoretically was that formed by the intersection of a substrate surface with a set of parallel equispaced planes: the so-called "Rowland grating." The theory of the spherical case was established first,[1–3] and was described comprehensively in the 1945 paper of Beutler.[4] Treatments of toroidal[5] and ellipsoidal[6] gratings came later, and the field has been reviewed by Welford,[7] Samson,[8] Hunter,[9] and Namioka.[10]

The major developments in the last three decades have been in the use of nonuniformly spaced grooves. The application of holography to spectroscopic gratings was first reported by Rudolph and Schmahl[11,12] and by Labeyrie and Flamand.[13] Its unique opportunities for optical design were developed initially by Jobin-Yvon[14] and by Namioka and coworkers.[15,16] A different approach was followed by Harada[17] and others, who developed the capability to produce gratings with variable-line spacing through the use of a computer-controlled ruling engine. The application of this class of gratings to spectroscopy has been developed still more recently, principally by Hettrick.[18]

In this report we will give a treatment of grating theory up to fourth order in the optical path, which is applicable to any substrate shape and any groove pattern that can be produced by holography or by ruling straight grooves with (possibly) variable spacing. The equivalent information is available up to sixth order at the website of the Center for X-Ray Optics at the Lawrence Berkeley National Laboratory.[19]

21.2 DIFFRACTION PROPERTIES

Notation and Sign Convention

We adopt the notation of Fig. 1 in which α and β have opposite signs if they are on opposite sides of the normal.

Grating Equation

The grating equation may be written

$$m\lambda = d_0 \, (\sin \alpha + \sin \beta) \tag{1}$$

The angles α and β are both arbitrary, so it is possible to impose various conditions relating them. If this is done, then for each λ, there will be a unique α and β. The following conditions are used:

1. *On-blaze condition:*

$$\alpha + \beta = 2\theta_B \tag{2}$$

where θ_B is the blaze angle (the angle of the sawtooth). The grating equation is then

$$m\lambda = 2d_0 \sin \theta_B \cos (\beta + \theta_B) \tag{3}$$

2. *Fixed in and out directions:*

$$\alpha - \beta = 2\theta \tag{4}$$

where 2θ is the (constant) included angle. The grating equation is then

$$m\lambda = 2d_0 \cos \theta \sin (\theta + \beta) \tag{5}$$

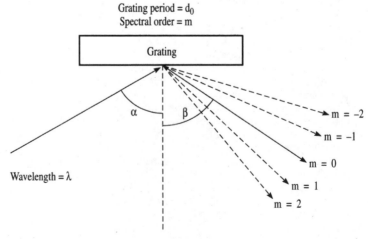

FIGURE 1   Grating equation notation.

In this case, the wavelength scan ends when α or β reaches 90°, which occurs at the horizon wavelength $\lambda_H = 2d_0 \cos^2 \theta$.

3. *Constant incidence angle:* Eq. (1) gives β directly.

4. *Constant focal distance (of a plane grating):*

$$\frac{\cos \beta}{\cos \alpha} = \text{a constant } c_{ff} \tag{6}$$

leading to a grating equation

$$1 - \left(\frac{m\lambda}{d_0} - \sin \beta\right)^2 = \frac{\cos^2 \beta}{c_{ff}^2} \tag{7}$$

Equations (3), (5), and (7) give β (and thence α) for any λ. Examples where the above α–β relationships may be used are as follows:

1. Kunz et al. plane-grating monochromator (PGM),[20] Hunter et al. double PGM,[21] collimated-light SX700.[22]

2. Toroidal-grating monochromators (TGMs),[23,24] spherical-grating monochromators (SGMs, also known as the *Dragon* system),[25] Seya-Namioka,[26,27] most aberration-reduced holographic SGMs,[28] and certain PGMs.[18,29,30] The variable-angle SGM[31] follows Eq. (4) approximately.

3. Spectrographs, *Grasshopper* monochromator.[32]

4. SX700 PGM[33] and variants.[22,34]

21.3 FOCUSING PROPERTIES[35]

Calculation of the Path Function *F*

Following normal practice, we provide an analysis of the imaging properties of gratings by means of the path function F.[16] For this purpose we use the notation of Fig. 2, in which the zeroth groove (of width d_0) passes through the grating pole O, while the nth groove passes through the variable point $P(\xi, w, l)$.

F is expressed as

$$F = \sum_{ijk} F_{ijk} w^i l^j$$

where

$$F_{ijk} = z^k C_{ijk}(\alpha, r) + z'^k C_{ijk}(\beta, r') + \frac{m\lambda}{d_0} f_{ijk} \tag{8}$$

and the f_{ijk} term, originating from the groove pattern, is given by one of the following expressions:

$$f_{ijk} = \begin{cases} 1 \text{ when } ijk = 100, 0 \text{ otherwise} & \text{Rowland} \\ \dfrac{d_0}{\lambda_0}\left[z_C^k C_{ijk}(\gamma, r_C) \pm z_D^k C_{ijk}(\delta, r_D)\right] & \text{holographic} \\ n_{ijk} & \text{varied line spacing} \end{cases} \tag{9}$$

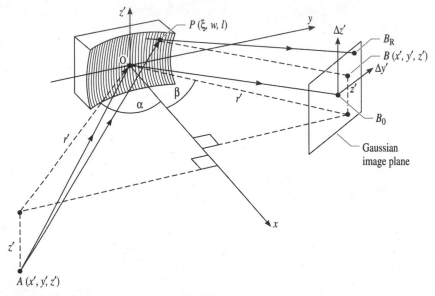

FIGURE 2   Focusing properties notation.

The holographic groove pattern in Eq. (9) is assumed to be made using two coherent point sources C and D with cylindrical polar coordinates (r_C, γ, z_C), (r_D, δ, z_D) relative to O. The lower (upper) sign refers to C and D, both real or both virtual (one real and one virtual), for which case the equiphase surfaces are confocal hyperboloids (ellipses) of revolution about CD. The grating with varied line spacing $d(w)$ is assumed to be ruled according to $d(w) = d_0(1 + v_1 w + v_2 w^2 + \ldots)$. We consider all the gratings to be ruled on the general surface $x = \sum_{ij} a_{ij} w^i l^j$ and the a_{ij} coefficients[36] are given for the important substrate shapes in Tables 1 and 2.

TABLE 1   Ellipsoidal Mirror a_{ij}'s[a,36]

$a_{20} = \dfrac{\cos\theta}{4}\left(\dfrac{1}{r} + \dfrac{1}{r'}\right)$	$a_{02} = \dfrac{a_{20}}{\cos^2\theta}$	$a_{22} = \dfrac{a_{20}(2A^2 + C)}{2\cos^2\theta}$
$a_{30} = a_{20}A$	$a_{12} = \dfrac{a_{20}A}{\cos^2\theta}$	$a_{04} = \dfrac{a_{20}C}{8\cos^2\theta}$
$a_{40} = \dfrac{a_{20}(4A^2 + C)}{4}$		

Other a_{ij}'s with $i + j \leq 4$ are zero.

[a] r, r' and θ are the object distance, image distance, and incidence angle to the normal, respectively, and

$$A = \frac{\sin\theta}{2}\left(\frac{1}{r} - \frac{1}{r'}\right), \quad C = A^2 + \frac{1}{rr'}.$$

The a_{ij}'s for spheres; circular, parabolic, or hyperbolic cylinders; paraboloids; and hyperboloids can also be obtained from Tables 1 and 2 by suitable choices of the input parameters r, r', and θ.

TABLE 2   Toroidal Mirror a_{ij}'s[a,36]

$a_{20} = \dfrac{1}{2R}$	$a_{02} = \dfrac{1}{2\rho}$	
$a_{30} = \dfrac{1}{8R^3}$	$a_{22} = \dfrac{1}{4\,\rho R^2}$	$a_{04} = \dfrac{1}{8R^3}$
Other a_{ij}'s with $i + j \leq 4$ are zero		

[a] R and ρ are the major and minor radii of the bicycle-tire toroid.

The coefficient F_{ijk} is related to the strength of the i, j, k aberration of the wavefront diffracted by the grating. The coefficients C_{ijk} and n_{ijk} are given in Tables 3 and 4 in which the following notation is used:

TABLE 3   Coefficients C_{ijk} of the Expansion of F[b,16]

$$C_{011} = -\frac{1}{r}$$
$$C_{020} = \frac{S}{2}$$

$$C_{022} = -\frac{S}{4r^2} - \frac{1}{2r^3}$$
$$C_{031} = \frac{S}{2r^2}$$
$$C_{100} = -\sin \alpha$$

$$C_{040} = \frac{4a_{02}^2 - S^2}{8r} - a_{04}\cos \alpha$$
$$C_{200} = \frac{T}{2}$$
$$C_{111} = -\frac{\sin \alpha}{r^2}$$

$$C_{120} = \frac{S \sin \alpha}{2r} - a_{12}\cos \alpha$$
$$C_{102} = \frac{\sin \alpha}{2r^2}$$
$$C_{202} = -\frac{T}{4r^2} + \frac{\sin^2 \alpha}{2r^3}$$

$$C_{300} = -a_{30}\cos \alpha + \frac{T \sin \alpha}{2r}$$
$$C_{211} = \frac{T}{2r^2} - \frac{\sin^2 \alpha}{r^3}$$

$$C_{220} = -a_{22}\cos \alpha + \frac{1}{4r}(4a_{20}a_{02} - TS - 2a_{12}\sin 2\alpha) + \frac{S \sin^2 \alpha}{2r^2}$$

$$C_{400} = -a_{40}\cos \alpha + \frac{1}{8r}(4a_{20}^2 - T^2 - 4a_{30}\sin 2\alpha) + \frac{T \sin^2 \alpha}{2r^2}$$

[b] The coefficients for which $i \leq 4$, $j \leq 4$, $k \leq 2$, $i + j + k \leq 4$, and $j + k =$ even are included in these tables.

TABLE 4   Coefficients n_{ijk} of the Expansion of F

$n_{ijk} = 0$ for $j, k \neq 0$

$$n_{100} = 1$$
$$n_{300} = \frac{v_1^2 - v_2}{3}$$

$$n_{200} = \frac{-v_1}{2}$$
$$n_{400} = \frac{-v_1^3 + 2v_1 v_2 - v_3}{4}$$

$$T = T(r, \alpha) = \frac{\cos^2 \alpha}{r} - 2a_{20}\cos \alpha, \quad S = S(r, \alpha) = \frac{1}{r} - 2a_{02}\cos \alpha \tag{10}$$

Determination of the Gaussian Image Point

By definition the principal ray AOB_0 arrives at the Gaussian image point $[B_0(r_0', \beta_0, z_0')$ in Fig. 2]. Its direction is given by Fermat's principle which implies $(\partial F/\partial w)_{w=0,\,l=0} = 0$, $(\partial F/\partial l)_{w=0,\,l=0} = 0$, from which

$$\frac{m\lambda}{d_0} = \sin\alpha + \sin\beta_0, \quad \frac{z}{r} + \frac{z_0'}{r_0'} = 0 \tag{11}$$

which are the grating equation and the law of magnification in the vertical direction. The tangential focal distance r_0' is obtained by setting the focusing term F_{200} equal to zero and is given by

$$T(r, \alpha) + T(r_0', \beta_0) = \begin{cases} 0 & \text{Rowland} \\[2mm] -\dfrac{m\lambda}{\lambda_0}\,[T(r_C, \gamma) \pm T(r_D, \delta)] & \text{holographic} \\[2mm] \dfrac{v_1 m\lambda}{d_0} & \text{varied line spacing} \end{cases} \tag{12}$$

Equations (11) and (12) determine the Gaussian image point B_0 and, in combination with the sagittal focusing condition ($F_{020} = 0$), describe the focusing properties of grating systems under the paraxial approximation. For a Rowland spherical grating the focusing condition [Eq. (12)] is

$$\left(\frac{\cos^2\alpha}{r} - \frac{\cos\alpha}{R}\right) + \left(\frac{\cos^2\beta}{r_0'} - \frac{\cos\beta}{R}\right) = 0 \tag{13}$$

which has the following important special cases:

1. A plane grating ($R = \infty$) implying $r_0' = -r\cos^2\alpha / \cos^2\beta = -r/c_{ff}^2$, so that the focal distance and magnification are fixed if c_{ff} is held constant.[37]
2. Object and image on the Rowland circle; $r = R\cos\alpha$, $r_0' = R\cos\beta$, and $M = 1$.
3. $\beta = 0$ (Wadsworth condition).

The tangential focal distances of TGMs and SGMs with or without moving slits are also determined by Eq. (13).

Calculation of Ray Aberrations

In an aberrated system, the outgoing ray will arrive at the Gaussian image plane at a point B_R displaced from the Gaussian image point B_0 by the ray aberrations $\Delta y'$ and $\Delta z'$ (Fig. 2). The latter are given by[38,39,40]

$$\Delta y' = \frac{r_0'}{\cos\beta_0}\frac{\partial F}{\partial w}, \quad \Delta z' = r_0'\frac{\partial F}{\partial l} \tag{14}$$

where F is to be evaluated for $A = (r, \alpha, z)$, $B = (r_0', \beta_0, z_0')$. By means of the series expansion of F, these equations allow the ray aberrations to be calculated separately for each aberration type, as follows:

$$\Delta y_{ijk}' = \frac{r_0'}{\cos\beta_0}\,F_{ijk}iw^{i-1}l^j, \quad \Delta z_{ijk}' = r_0' F_{ijk}w^i jl^{j-1} \tag{15}$$

Moreover, provided the aberrations are not too large, they are additive, so that they may either reinforce or cancel.

21.4 DISPERSION PROPERTIES

Angular Dispersion

$$\left(\frac{\partial \lambda}{\partial \beta}\right)_\alpha = \frac{d \cos \beta}{m} \tag{16}$$

Reciprocal Linear Dispersion

$$\left(\frac{\partial \lambda}{\partial (\Delta y')}\right)_\alpha = \frac{d \cos \beta}{mr'} \equiv \frac{10^{-3} d[\text{Å}] \cos \beta}{mr'[\text{m}]} \text{ Å/mm} \tag{17}$$

Magnification (M)

$$M(\lambda) = \frac{\cos \alpha}{\cos \beta} \frac{r'}{r} \tag{18}$$

Phase-Space Acceptance (ε)

$$\varepsilon = N\Delta\lambda_{S_1} = N\Delta\lambda_{S_2} \text{ (assuming } S_2 = MS_1) \tag{19}$$

where N is the number of participating grooves.

21.5 RESOLUTION PROPERTIES

The following are the main contributions to the width of the instrumental line spread function (an estimate of the total width is the vector sum):

1. *Entrance slit (width S₁):*

$$\Delta\lambda_{S_1} = \frac{S_1 d \cos \alpha}{mr} \tag{20}$$

2. *Exit slit (width S₂):*

$$\Delta\lambda_{S_2} = \frac{S_2 d \cos \beta}{mr'} \tag{21}$$

3. *Aberrations (of a perfectly made grating):*

$$\Delta\lambda_A = \frac{\Delta y' d \cos\beta}{mr'} = \frac{d}{m}\left(\frac{\partial F}{\partial w}\right)$$
(22)

4. *Slope error $\Delta\phi$ (of an imperfectly made grating):*

$$\Delta\lambda_{SE} = \frac{d(\cos\alpha + \cos\beta)\Delta\phi}{m}$$
(23)

Note that, provided the grating is large enough, diffraction at the entrance slit always guarantees a coherent illumination of enough grooves to achieve the slit–width limited resolution. In such cases, a diffraction contribution to the width need not be added to those listed.

21.6 EFFICIENCY

The most accurate way to calculate grating efficiencies is by the full electromagnetic theory for which code is available from Neviere.[41,42] However, approximate scalar-theory calculations are often useful and, in particular, provide a way to choose the groove depth (h) of a laminar grating. According to Bennett,[43] the best value of the groove-width-to-period ratio (r) is the one for which the area of the usefully illuminated groove bottom is equal to that of the top. The scalar theory efficiency of a laminar grating with $r = 0.5$ is given by Franks et al.[44] as the following:

$$E_0 = \frac{R}{4}\left[1 + 2(1-P)\cos\left(\frac{4\pi h \cos\alpha}{\lambda}\right) + (1-P)^2\right]$$

$$E_m = \begin{cases} R[1 - 2\cos Q^+ \cos(Q^- + \delta) + \cos^2 Q^+]/m^2\pi^2 & m = \text{odd} \\ R\cos^2 Q^+/m^2\pi^2 & m = \text{even,} \end{cases}$$
(24)

where

$$P = \frac{4h \tan\alpha}{d_0}, \quad Q^\pm = \frac{m\pi h}{d_0}(\tan\alpha \pm \tan\beta), \quad \delta = \frac{2\pi h}{\lambda}(\cos\alpha + \cos\beta)$$

and R is the reflectance at grazing angle $\sqrt{\alpha_g \beta_g}$, where $\alpha_g = \pi/2 - |\alpha|$ and $\beta_g = \pi/2 - |\beta|$.

21.7 REFERENCES

1. H. A. Rowland, "On Concave Gratings for Optical Purposes," *Phil. Mag.* **16**(5th ser.):197–210 (1883).
2. J. E. Mack, J. R. Stehn, and B. Edlen, "On the Concave Grating Spectrograph, Especially at Large Angles of Incidence," *J. Opt. Soc. Am.* **22**:245–264 (1932).
3. H. A. Rowland, "Preliminary Notice of the Results Accomplished in the Manufacture and Theory of Gratings for Optical Purposes," *Phil. Mag.* **13**(supp.) (5th ser.):469–474 (1882).
4. II. G. Beutler, "The Theory of the Concave Grating," *J. Opt. Soc. Am.* **35**:311–350 (1945).
5. H. Haber, "The Torus Grating," *J. Opt. Soc. Am.* **40**:153–165 (1950).
6. T. Namioka, "Theory of the Ellipsoidal Concave Grating: I," *J. Opt. Soc. Am.* **51**:4–12 (1961).

7. W. Welford, "Aberration Theory of Gratings and Grating Mountings," in E. Wolf (ed.), *Progress in Optics,* vol. 4, North-Holland, Amsterdam, pp. 243–282, 1965.

8. J. A. R. Samson, *Techniques of Vacuum Ultraviolet Spectroscopy,* John Wiley & Sons, New York, 1967.

9. W. R. Hunter, "Diffraction Gratings and Mountings for the Vacuum Ultraviolet Spectral Region," in *Spectrometric Techniques,* G. A. Vanasse (ed.), vol. IV, Academic Press, Orlando, pp. 63–180, 1985.

10. T. Namioka and K. Ito, "Modern Developments in VUV Spectroscopic Instrumentation," *Physica Scripta* **37**:673–681 (1988).

11. D. Rudolph and G. Schmahl, "Verfaren zur Herstellung von Röntgenlinsen und Beugungsgittern," *Umsch. Wiss. Tech.* **67**:225 (1967).

12. D. Rudolph and G. Schmahl, "Holographic Gratings," in *Progress in Optics,* E. Wolf (ed.), vol. 14, North-Holland, Amsterdam, pp. 196–244. 1977.

13. A. Laberie and J. Flamand, "Spectrographic Performance of Holographically Made Diffraction Grating," *Opt. Comm.* **1**:5–8 (1969).

14. G. Pieuchard and J. Flamand, "Concave Holographic Gratings for Spectrographic Applications," Final report on NASA contract number NASW-2146, GSFC 283-56,777, Jobin Yvon, Longjumeau, France, 1972.

15. T. Namioka, H. Noda, and M. Seya, "Possibility of Using the Holographic Concave Grating in Vacuum Monochromators," *Sci. Light* **22**:77–99 (1973).

16. H. Noda, T. Namioka, and M. Seya, "Geometrical Theory of the Grating," *J. Opt. Soc. Am.* **64**:1031–1036 (1974).

17. T. Harada and T. Kita, "Mechanically Ruled Aberration-Corrected Concave Gratings," *Appl. Opt.* **19**:3987–3993 (1980).

18. M. C. Hettrick, "Aberration of Varied Line-Space Grazing Incidence Gratings," *Appl. Opt.* **23**:3221–3235 (1984).

19. http://www-cxro.lbl.gov/, 1999

20. C. Kunz, R. Haensel, and B. Sonntag, "Grazing Incidence Vacuum Ultraviolet Monochromator with Fixed Exit Slit for Use with Distant Sources," *J. Opt. Soc. Am.* **58**:1415 (1968).

21. W. R. Hunter, R. T. Williams, J. C. Rife, J. P. Kirkland, and M. N. Kaber, "A Grating/Crystal Monochromator for the Spectral Range 5 ev to 5 keV," *Nucl. Instr. Meth.* **195**:141–154 (1982).

22. R. Follath and F. Senf, "New Plane-Grating Monochromators for Third Generation Synchrotron Radiation Light Sources," *Nucl. Instrum. Meth.* **A390**:388–394 (1997).

23. D. Lepere, "Monochromators with Single Axis Rotation and Holographic Gratings on Toroidal Blanks for the Vacuum Ultraviolet," *Nouvelle Revue Optique* **6**:173 (1975).

24. R. P. Madden and D. L. Ederer, "Stigmatic Grazing Incidence Monochromator for Synchrotrons (abstract only)," *J. Opt. Soc. Am.* **62**:722 (1972).

25. C. T. Chen, "Concept and Design Procedure for Cylindrical Element Monochromators for Synchrotron Radiation," *Nucl. Instr. Meth.* **A256**:595–604 (1987).

26. T. Namioka, "Construction of a Grating Spectrometer," *Sci. Light* **3**:15–24 (1954).

27. M. Seya, "A New Monting of Concave Grating Suitable for a Spectrometer," *Sci. Light* **2**:8–17 (1952).

28. T. Namioka, M. Seya, and H. Noda, "Design and Performance of Holographic Concave Gratings," *Jap. J. Appl. Phys.* **15**:1181–1197 (1976).

29. W. Eberhardt, G. Kalkoffen, and C. Kunz, "Grazing Incidence Monochromator FLIPPER," *Nucl. Inst. Meth.* **152**:81–4 (1978).

30. K. Miyake, P. R. Kato, and H. Yamashita, "A New Mounting of Soft X-Ray Monochromator for Synchrotron Orbital Radiation," *Sci. Light* **18**:39–56 (1969).

31. H. A. Padmore, "Optimization of Soft X-Ray Monochromators," *Rev. Sci. Instrum.* **60**:1608–1616 (1989).

32. F. C. Brown, R. Z. Bachrach, and N. Lien, "The SSRL Grazing Incidence Monochromator: Design Considerations and Operating Experience," *Nucl. Instrum. Meth.* **152**:73–80 (1978).

33. H. Petersen and H. Baumgartel, "BESSY SX/700: A Monochromator System Covering the Spectral Range 3 eV - 700 eV," *Nucl. Instrum. Meth.* **172**:191–193 (1980).

34. W. Jark, "Soft X-Ray Monochromator Configurations for the ELETTRA Undulators: A Stigmatic SX700," *Rev. Sci. Instrum.* **63**:1241–1246 (1992).

35. H. A. Padmore, M. R. Howells, and W. R. McKinney, "Grazing Incidence Monochromators for Third-Generation Synchrotron Radiation Light Sources," in J. A. R. Samson and D. L. Ederer (eds.), *Vacuum Ultraviolet Spectroscopy,* vol. 31, Academic Press, San Diego, pp. 21–54, 1998.

36. S. Y. Rah, S. C. Irick, and M. R. Howells, "New Schemes in the Adjustment of Bendable Elliptical Mirrors Using a Long-Trace Profiler," in P. Z. Takacs and T. W. Tonnessen (eds.), *Materials manufacturing and measurement for synchrotron-radiation mirrors,* Proc. SPIE, vol. 3152, SPIE, Bellingham, WA, 1997.

37. H. Petersen, "The Plane Grating and Elliptical Mirror: A New Optical Configuration for Monochromators," *Opt. Comm.* **40**:402–406 (1982).

38. M. Born and E. Wolf, *Principles of Optics,* Pergamon, Oxford, 1980.

39. T. Namioka and M. Koike, "Analytical Representation of Spot Diagrams and Its Application to the Design of Monochromators," *Nucl. Instrum. Meth.* **A319**:219–227 (1992).

40. W. T. Welford, *Aberrations of the Symmetrical Optical System,* Academic Press, London, 1974.

41. M. Neviere, P. Vincent, and D. Maystre, "X-Ray Efficiencies of Gratings," *Appl. Opt.* **17**:843–845 (1978). (Neviere can be reached at michel.neviere@fresnel.fr.)

42. R. Petit (ed.), *Electromagnetic Theory of Gratings (Topics in Current Physics, Vol. 22),* Springer Verlag, Berlin, 1980.

43. J. M. Bennett, "Laminar X-Ray Gratings," Ph.D. Thesis, London University, London, 1971.

44. A. Franks, K. Lindsay, J. M. Bennett, R. J. Speer, D. Turner, and D. J. Hunt, "The Theory, Manufacture, Structure and Performance of NPL X-Ray Gratings," *Phil. Trans. Roy. Soc.* **A277**:503–543 (1975).

CHAPTER 22

CRYSTAL MONOCHROMATORS AND BENT CRYSTALS

Peter Siddons
National Synchrotron Light Source
Brookhaven National Laboratory
Upton, New York

22.1 CRYSTAL MONOCHROMATORS

For X-ray energies higher than 2 keV or so, gratings become extremely inefficient, and it becomes necessary to utilize the periodicity naturally occuring in a crystal to provide the dispersion. Since the periodicity in a crystal is 3-dimensional, the normal single grating equation must be replaced by the three grating equations, one for each dimension, called the Laue equations,[1] as follows:

$$\mathbf{a}_1 \cdot (\mathbf{k}_{H_1 H_2 H_3} - \mathbf{k}_0) = H_1$$

$$\mathbf{a}_2 \cdot (\mathbf{k}_{H_1 H_2 H_3} - \mathbf{k}_0) = H_2 \qquad (1)$$

$$\mathbf{a}_3 \cdot (\mathbf{k}_{H_1 H_2 H_3} - \mathbf{k}_0) = H_3$$

where the $\mathbf{a}$'s are the repeat vectors in the three dimensions, the $\mathbf{k}$'s are the wave vectors for the incident and scattered beams, and the H's are integers denoting the diffraction order in the three dimensions. All of them must be simultaneously satisfied in order to have an interference maximum (commonly called a *Bragg reflection*). One can combine these equations into the well-known Bragg's law[2] for one component of the crystalline periodicity (usually referred to as a set of *Bragg planes*),

$$n\lambda = 2d \sin \theta \qquad (2)$$

where n is an integer indicating the order of diffraction from planes of spacing d, and θ is the angle between the incident beam and the Bragg planes. This equation is the basis for using crystals as X-ray monochromators. By choosing one such set of Bragg planes and setting the crystal so that the incident X rays fall on these planes, the wavelength of the light reflected depends on the angle of incidence of the light. Table 1 shows some commonly used crystals and the spacings of some of their Bragg planes. The most common arrangement is the symmetric Bragg case (Fig. 1a), in which the useful surface of the crystal is machined so that it is parallel to the Bragg planes in use. Under these conditions, the incident and reflected angles are equal. The literature on crystal diffraction differs from that on grating instruments in that the incidence angle for X rays is called the *glancing angle* for gratings. As the X-ray wavelength gets shorter and the angles get smaller, it can be difficult to obtain large enough crystals of good quality. In such cases it is possible to employ the Laue case (Fig. 1b), in which the

TABLE 1   Some Common Monochromator Crystals, Selected d-Spacings, and Reflection Widths at 1 Å (12.4 keV)

Crystal	Reflection	d-Spacing (nm)	Refl. width (microrad)	Energy resolution ($\Delta E/E$)
Silicon	(1 1 1)	0.31355	22.3	1.36×10^{-4}
	(2 2 0)	0.19201	15.8	5.37×10^{-5}
Germanium	(1 1 1)	0.32664	50.1	3.1×10^{-4}
	(2 2 0)	0.20002	37.4	1.37×10^{-4}
Diamond	(1 1 1)	0.20589	15.3	5.8×10^{-5}
	(4 0 0)	0.089153	5.2	7.4×10^{-6}
Graphite	(0 0 0 . 2)	0.3354	Sample-dependent	Sample-dependent

surface is cut perpendicular to the Bragg planes and the X rays are reflected through the bulk of the crystal plate. Of course, the wavelength should be short enough or the crystal thin enough so that the X rays are not absorbed by the monochromator. This is true, for example, in silicon crystals around 1 mm thick above an X-ray energy of around 30 keV ($\lambda = 0.04$ nm).

The detailed calculation of the response of a crystal to X rays depends on the degree of crystalline perfection of the material in use, as well as its chemical composition. Two main theoretical treatments are commonly used: the kinematical theory of diffraction and the dynamical theory. The kinematical theory assumes single-scattering of the X rays by the crystal, and is appropriate for crystals with a high concentration of defects. Such crystals are called *mosaic crystals,* following C. G. Darwin.[3] The dynamical theory, in contrast, explicitly treats the multiple scattering that arises in highly perfect crystals and is commonly used for the semiconductor monochromator materials that can be grown to a high degree of perfection. Of course, many crystals fall between these two idealized pictures and there exist approximations to both theories to account for some of their failures. We will not describe these theories in detail here, but will refer the reader to texts on the subject[4-6] and content ourselves with providing some of the key formulas that result from them.

Both theories attempt to describe the variation in reflectivity of a given crystal as a function of the incidence angle of the X-ray beam near a Bragg reflection. The neglect or inclusion of multiple scattering changes the result quite dramatically. In the kinematical case, the width of the reflectivity profile is inversely related to the size of the coherently diffracting volume, and for an infinite perfect crystal it is a delta function. The integrated reflectivity increases linearly with the illuminated crystal volume, and is assumed to be small. In the dynamical case, the X-ray beam diffracted by one part of the crystal is exactly oriented to be diffracted by the same Bragg planes, but in the opposite sense. Thus, there coexist two waves in the crystal, one with its wave vector along the incident beam direction and the other with its vector along the diffracted beam direction. These waves are coherent and can interfere. It is these interferences that give rise to all the interesting phenomena that arise from this theory. The integrated reflectivity in this case initially increases with volume, as in the kinematical case, but eventually saturates to a constant value that depends on which of the geometries in Fig. 1 is taken.

(a)   (b)

FIGURE 1   The two most usual X-ray diffraction geometries used for monochromator applications: (*a*) the symmetric Bragg case and (*b*) the symmetric Laue case.

For the kinematical theory, the reflectivity curve is approximated by[7]

$$r(\Delta) = \frac{a}{[1 + a + \sqrt{1 + 2a} \cdot \cot h(b\sqrt{1 + 2a})]} \tag{3}$$

where

$$a = \frac{w(\Delta)Q}{\mu}$$

$$b = \frac{\mu t}{\sin \theta_B}$$

$$Q = \left(\frac{e^2}{mc^2 V}\right)^2 \cdot \frac{|F_H|^2 \lambda^3}{\sin 2\theta_B}$$

in which e is the electronic charge and m its mass, c is the velocity of light, θ_B is the Bragg angle for the planes whose structure factor is F_H at wavelength λ. $w(\Delta)$ represents the angular distribution of the mosaic blocks, which depends on the details of the sample preparation and/or growth history. The reflection width will primarily reflect the width of this parameter $w(\Delta)$, but often the curve will be further broadened by extinction.

The equivalent equation for the dynamical theory is sensitive to the exact geometry under consideration and so will not be given here. The crystal becomes birefringent near the Bragg angle, and this causes some interesting interference effects that are worthy of study in their own right. Reference 8 is a review of the theory with a physical approach that is very readable. However, the main results for the purposes of this section are summarized in Fig. 2; the key points are the following:

1. In the Bragg case (Fig. 1a) the peak reflectivity reaches nearly unity over a small range of angles near the Bragg angle.

2. The range of angles over which this occurs is given by

$$\Omega = 2R\lambda^2 |C| \frac{\sqrt{|\gamma| F_h F_{\bar{h}}}}{\pi V \sin 2\theta} \tag{4}$$

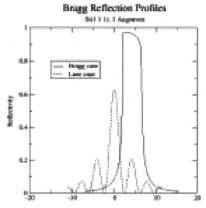

FIGURE 2   The perfect-crystal Bragg-case reflectivity profile for silicon with perpendicular-polarized X rays at 1 Å.

where R is the classical electron radius, 2.81794×10^{-15} m, λ is the X-ray wavelength, C is the polarization factor (1 for σ-polarized light and $\cos 2\theta$ for π-polarized light). γ is the asymmetry parameter (1 in the symmetric case), the Fs are the structure factors for reflections (hkl) and $(-h-k-l)$, V is the unit cell volume, and θ is the Bragg angle.

3. The effect of absorption in the Bragg case is to reduce the peak reflectivity and to make the reflectivity curve asymmetric.

4. In the Laue case, the reflectivity is oscillatory, with an average value of ½, and the effect of absorption (including increasing thickness) is to damp out the oscillations and reduce the peak reflectivity.

Even small distortions can greatly perturb the behavior of perfect crystals, and there is still no general theory that can handle arbitrary distortions in the dynamical regime. There are approximate treatments that can handle some special cases of interest.[4,5]

In general, the performance of a given monochromator system is determined by the range of Bragg angles experienced by the incident X-ray beam. This can be determined simply by the incident beam collimation or by the crystal reflectivity profile, or by a combination of both factors. From Bragg's law we have

$$\frac{\delta E}{E} = \frac{\delta \lambda}{\lambda} = \cot \theta \, \delta \theta + \delta \tau \tag{5}$$

where δE and E are the passband and center energy of the beam, and $\delta \tau$ is the intrinsic energy resolution of the crystal reflection as given in Table 1—essentially the dynamical reflection width as given previously, expressed in terms of energy. This implies that, even for a perfectly collimated incident beam, there is a limit to the resolution achievable, which depends on the monochromator material and the diffraction order chosen.

The classic Bragg reflection monochromator uses a single piece of crystal set to the correct angle to reflect the energy of interest. This is in fact a very inconvenient instrument, since its output beam direction changes with energy. For perfect-crystal devices, the reflectivity is sufficiently high that one can afford to use two of them in tandem, deviating in opposite senses in order to bring the useful beam back into the forward direction, independent of energy (Fig. 3). Such two-crystal monochromators have become the standard design, particularly for use at accelerator-based (synchrotron) radiation sources (as discussed in Chap. 32). Since these sources naturally generate an energy continuum, a monochromator is a common requirement for a beamline at such a facility. There is a wealth of literature arising from such applications.[9]

A particularly convenient form of this geometry was invented by Bonse and Hart,[10] in which the two reflecting surfaces are machined from a single-crystal block of material (in their case germanium, but more often silicon in recent years).

For the highest resolution requirements, it is possible to take two crystals that deviate in the same direction, the so-called ++ geometry. In this case the first crystal acts as a collimator, generating a fan of beams with each angular component having a particular energy. The second one selects which angular (and hence energy) component of this fan to transmit. In this arrangement the deviation is doubled over the single crystal device, and so the most successful arrangement for this so-called dispersive arrangement is to use two monolithic double-reflection devices like that in Fig. 2, with the ++ deviation taking place between the second and third reflections (Fig. 4).

22.2 BENT CRYSTALS

There are two reasons to consider applying a uniform curvature to a crystal plate for a monochromator system. One is to provide some kind of beam concentration or focussing, and the other is to improve the energy resolution in circumstances where a divergent incident

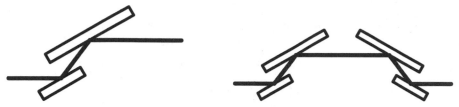

FIGURE 3 The +− double-crystal X-ray monochromator.

FIGURE 4 Two monolithic double reflectors arranged in a high-resolution configuration.

beam is unavoidable. The most common geometry is the Bragg-case one based on the Rowland circle principle, modified to account for the 3-dimensional nature of the crystal periodicity. This principle relies on the well-known property of a circle that an arc segment subtends a constant angle for any point on the circle. For a grating this is straightforwardly applied to a focussing spectrometer, but for crystals one has the added complication that the incidence angle for the local Bragg plane must also be constant. The result was shown by Johansson[11] (Fig. 5) to require the radius of curvature to be twice the radius of the crystal surface. This can be achieved by a combination of bending and machining the crystal. For applications in which the required optical aperture is small, the aberrations introduced by omitting the machining operation are small and quite acceptable.[12] Since this is a rather difficult operation, it is attractive to avoid it if possible.

Although Fig. 5 shows the symmetrical setting, it is possible to place the crystal anywhere on the circle and achieve a (de)magnification other than unity. In this case the surface must be cut at an angle to the Bragg planes to maintain the geometry. The Laue case can also be used in a similar arrangement, but the image becomes a virtual image (or source). For very short wavelengths (e.g., in a gamma-ray spectrometer) the Laue case can be preferable since the size of the crystal needed for a given optical aperture is much reduced. In both Laue and Bragg cases, there are changes in the reflection properties on bending. Depending on the source and its collimation geometry, and on the asymmetry angle of the crystal cut, the bending can improve or degrade the monochromator resolution. Each case must be considered in detail. Again, within the scope of this section we will content ourselves with some generalities:

1. If the angular aperture of the crystal as seen from the source location is large compared to the reflectivity profile of the crystal, then the Rowland circle geometry will improve things for the Bragg case and the symmetric Laue case.

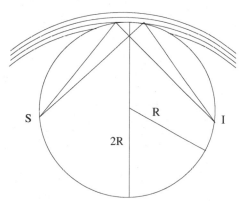

FIGURE 5 The Johansson bent/ground focussing monochromator.

FIGURE 6   The usual arrangement of a sagittally focussing monochromator, with the bent element as the second one of a two-crystal monochromator.

2. When the absorption length of the incident radiation becomes large compared to the extinction distance, then the X rays can travel deep into the bent crystal even in the Bragg case, and the deformation means that the incidence angle changes with depth, leading to a broadening of the bandpass.

3. If the bending radius becomes very small, such that the Bragg angle changes more than the perfect-crystal reflection width within the extinction depth, then the peak reflectivity will fall and the reflection width will increase, and consequently the resolution will deteriorate.

4. In the asymmetric Laue case, the reflectivity profile width for perfect crystals also depends on the curvature,[6] and so for strongly collimated beams such as synchrotron radiation sources, the bending may well degrade the resolution at the same time as it increases the intensity.

Arrangements of multiple consecutive curved crystals are unusual, but have found application in high-throughput synchrotron radiation (SR) monochromators, where the two-crystal device in Fig. 3 is modified by placing the first crystal on the Rowland circle and adjusting the convex curvature of the second to maximize its transmission of the convergent beam from the first.[13] There are also examples of combinations of curved Laue and curved Bragg reflectors.[14]

Another geometry in which the technique of bending a monochromator crystal can be used is in the so-called sagittal-focussing geometry.[15] This geometry, as its name indicates, has the bending radius perpendicular to that in the Rowland-circle geometry, and provides a focussing effect in the plane perpendicular to the diffraction plane. In the diffraction plane the crystal behaves essentially as though it were flat (Fig. 6). Anticlastic bending of the bent crystal can greatly reduce its efficiency, since this curvature is in the meridional plane so that parts of the crystal move out of the Bragg-reflection range. Attempts to counteract this by using stiffening ribs machined into the plate[16] or by making use of the elastic anisotropy of single crystals[17] have been made, with some success.

22.3   REFERENCES

1. M. von Laue, *Münchener Sitzungsberichte* 363 (1912), and *Ann. der Phys.* **41**:989 (1913).

2. W. H. Bragg and W. L. Bragg, *Proc. Roy. Soc. London* **88**:428 (1913), and **89**:246 (1913).

3. C. G. Darwin, *Phil. Mag.* **27**:315, 657 (1914), and **43**:800 (1922).

4. D. Taupin, *Bull. Soc. Franc. Miner. Crist.* **87**:469–511 (1964); S. Takagi, *Acta Cryst.* **12**:1311 (1962).

5. Penning and Polder, *Philips Res. Rep.* **16**:419 (1961).

6. R. Caciuffo, S. Melone, F. Rustichelli, and A. Boeuf, *Phys. Rep.* **152**:1 (1987); P. Suortti and W. C. Thomlinson, *Nuclear Instrum. and Meth.* **A269**:639–648 (1988).

7. A. K. Freund, A. Munkholm, and S. Brennan, *Proc. SPIE* **2856**:68–79 (1996).

8. B. W. Batterman and H. Cole, *Rev. Mod. Phys.* **36**:681–717 (1964). (For a fuller treatment, see W. H. Zachariasen, *Theory of X-Ray Diffraction in Crystals,* Dover, New York, 1967.)

9. See for example, the *Handbook of Synchrotron Radiation* or browse the *Proceedings of the International Conference on Synchrotron Radiation Instrumentation,* published in *Nucl. Instr. and Meth., Rev. Sci. Instrum.,* and *J. Synchr. Rad.* at various times.

10. U. Bonse and M. Hart, *Appl. Phys. Lett.* **7**:238–240 (1965).

11. T. Johansson, *Z. Phys.* **82**:507 (1933).

12. H. H. Johann, *Z. Phys.* **69**:185 (1931).

13. M. J. Van Der Hoek, W. Berne, and P. Van Zuylen, *Nucl. Instrum. and Meth.* **A246**:190–193 (1986).

14. U. Lienert, C. Schulze, V. Honkimäki, Th. Tschentschor, S. Garbe, O. Hignette, A. Horsewell, M. Lingham, H. F. Poulsen, W. B. Thomsen, and E. Ziegler, *J. Synchrot. Rad.* **5**:226–231 (1998).

15. L. Von Hamos, *J. Sci. Instr.* **15**:87 (1938).

16. C. J. Sparks, Jr., B. S. Borie, and J. B. Hastings, *Nucl. Instrum. and Meth.* **172**:237 (1980).

17. V. I. Kushnir, J. P. Quintana, and P. Georgopoulos, *Nucl. Instrum. and Meth.* **A328**:588 (1983).

CHAPTER 23
ZONE AND PHASE PLATES, BRAGG-FRESNEL OPTICS

Alan Michette
King's College London
Physics Department
Strand, London, United Kingdom

23.1 INTRODUCTION

The focusing properties of zone plates were first discussed in the latter part of the nineteenth century,[1-3] and their use as X-ray optical components was originally suggested by Baez.[4] In their most common form, they are circular diffraction gratings with radially increasing line densities; linear,[5] square,[6] elliptical,[7] and hyperbolic[7,8] zone plates (among others) have also been considered, but only the circular and, to a lesser extent, linear and elliptical forms have generally been used. The properties of zone plates may be described, without loss of generality, by considering the circular form (Fig. 1).

A linear diffraction grating (see Chaps. 21 and 22) with a constant period, d, working in the first order diffracts radiation through an angle, $\theta = \sin^{-1}\lambda/d$, $\approx \lambda/d$ in the small-angle approximation. A circular grating with constant period could therefore be used to form an axial-line focus of a point source, as shown in Fig. 2a; note that diffraction from opposite orders contributes either side of the axis. The distance from a radial point, r, on the grating to a point on the axis is $z = r/\tan \theta \approx rd/\lambda$. If the period is now made to decrease as the radius increases, then z can be made constant and the grating then acts as a lens, as shown in Fig. 2b, where positive

FIGURE 1   A circular zone plate.

(negative) diffraction orders are now defined as being on the opposite (same) side as the object. This is the basis of a zone plate, for which the focusing properties depend on the following:

- The relationship between d and r
- The number of zones*
- The zone heights and profiles

23.2 ZONE PLATE GEOMETRY

The required variation of the zone period with radius may be obtained by considering Fig. 2b. Here, radiation from an object point, A, is brought to a focus, via the zone plate, to the image point, B. To obtain constructive interference at B, the optical path difference between successive zone boundaries must be $\pm m\lambda/2$, where m is the diffraction order. Thus, for the first order and using the lengths defined in Fig. 2b,

$$a_n + b_n = z_a + z_b + \Delta + \frac{n\lambda}{2} \qquad (1)$$

where Δ is the optical path difference introduced by the central zone of radius, r_0. For a distant object $(a_n, z_a \to \infty)$ with $a_n - z_a \to 0$, writing $b_n = \sqrt{z_b^2 + r_n^2} = \sqrt{f_1^2 + r_n^2}$, where r_n is the radius of the nth zone and f_1 is the first-order focal length, squaring and simplifying,

$$r_n^2 = p\lambda f_1 + \left(\frac{p\lambda}{2}\right)^2 \qquad (2)$$

where $p = n + 2\Delta/\lambda$. For a finite object distance, Eq. (2) still holds with the addition of higher-order terms in λ and if the term in λ^2 is multiplied by $(M^3 + 1)/(M + 1)^3$, where M is the magnification. In most practical cases, terms in λ^2 and above are negligible and so, to a good approximation,

$$r_n^2 = n\lambda f_1 + 2\Delta f_1 = n\lambda f_1 + r_0^2 \qquad (3)$$

---

* For X-ray zone plates, the usual convention is that the area between successive boundaries is called a *zone*. Strictly speaking, and in keeping with the terminology used for diffraction gratings, this area should be called a *half-period zone*, but for consistency with other publications on X-ray zone plates, the term *zone* will be used here.

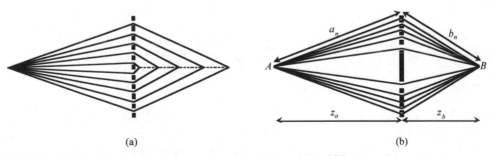

(a)   (b)

FIGURE 2 Diffraction by (*a*) a circular grating with constant period and (*b*) a zone plate.

because, for the central zone with $n = 0$, $r_0^2 = 2\Delta f_1$. Equation (3) describes the Fresnel zone plate; the special case with $r_0^2 = 0$ is the Fresnel-Soret zone plate, which is most commonly used and for which

$$r_n^2 = n\lambda f_1 = nr_1^2 \qquad (4)$$

The higher-order terms that have been ignored in deriving Eq. (4) are aberration terms; in particular, the term in λ^2 describes spherical aberration and only becomes comparable with the first term when $n \sim 4f_1/\lambda$, which is rarely the case for X-ray zone plates. Equation (4) shows that the focal length of a zone plate is inversely proportional to wavelength, and thus monochromatic radiation (with $\lambda/\Delta\lambda \sim N$, where N is the total number of zones[9]) must be used if chromatic aberration is to be avoided. The area of the nth zone is

$$\pi(r_n^2 - r_{n-1}^2) = \pi[n\lambda f_1 - (n-1)\lambda f_1] = \pi\lambda f_1 \qquad (5)$$

that is, it is constant so that each zone contributes equally to the amplitude at the focus, provided the zone plate is evenly illuminated. The width, d_n, of the nth zone is

$$d_n = r_n - r_{n-1} = \sqrt{n\lambda f_1} - \sqrt{(n-1)\lambda f_1} = \sqrt{n\lambda f_1}\left[1 - \left(1 - \frac{1}{n}\right)^{1/2}\right] \approx \frac{r_n}{2n} \qquad (6)$$

leading to an expression for the focal length

$$f_1 = \frac{r_n^2}{n\lambda} \approx \frac{D_n d_n}{\lambda} \qquad (7)$$

Because zone plates are diffractive optics, they do not have just one focus but an infinite number. The mth-order focus can be described by m zones acting in tandem; hence, the focal lengths are given by

$$f_m = \frac{f_1}{m}, \quad m = 0, \pm 1, \pm 2, \pm 3 \cdots \qquad (8)$$

positive values of m lead to real foci, negative values lead to virtual foci, and $m = 0$ corresponds to undiffracted (i.e., unfocused) radiation.

The sizes of the focal spots cannot be determined from the preceding analysis. However, when N is sufficiently large (theoretically, ≥ 100, but in practice much less than this), a zone plate acts as a thin lens, and the diffraction pattern at a focus approximates closely to an Airy pattern. For a lens of diameter, D, and focal length, f, the first zero of the Airy distribution, which defines the lateral resolution, δ, via the Rayleigh criterion, is at a radius $f \tan 1.22\lambda/D$. For a zone plate, using Eqs. (7) and (8), this gives, for the resolution in the mth order,

$$\delta_m = 1.22\frac{d_n}{m} \qquad (9)$$

where the small-angle approximation has been used. Equation (9) shows that, to achieve high resolution, the outermost zone width must be small, and that better resolutions may be obtained by using the higher-diffraction orders. This latter observation may seem to indicate that high-diffraction orders offer an advantage; however, the achievable diffraction efficiency negates this, as will be seen in the following.

The depth of focus, Δf_m, of a zone plate may also be determined using the thin-lens analogy. For a thin lens, the depth of focus is $\Delta f = \pm 2(f/D)^2\lambda$, which, for a zone plate, reduces to

$$\Delta f_m = \pm\frac{f}{2mN} \qquad (10)$$

The preceding results are determined primarily by the relative placement of the zone boundaries; the diffraction efficiencies into the various orders depend in part on the zone heights and profiles and in part on the accuracy of positioning these boundaries.

23.3 AMPLITUDE ZONE PLATES

If alternate zones are totally absorbing and totally transmitting (an amplitude zone plate), then 50 percent of the incident radiation is absorbed, leaving 50 percent to be diffracted into the various orders, of which half goes into the zeroth order.[10] If the zone boundaries are in the positions given by Eq. (4), then the even orders (except the zeroth) vanish, because the amplitudes from adjacent zones cancel. The only orders that contribute are $0, \pm 1, \pm 3, \ldots$, and, from symmetry, it is clear that the $+m$th and $-m$th diffraction efficiencies are equal. The peak amplitudes in each diffraction order are equal, but the focal spot areas decrease as m^2. Thus, if ε_m is the diffraction efficiency in the mth order,

$$0.25 = 2 \sum_{\substack{m=1 \\ m \text{ odd}}}^{\infty} \varepsilon_m = 2\varepsilon_1 \sum_{\substack{m=1 \\ m \text{ odd}}} \frac{1}{m^2} = 2\varepsilon_1 \frac{\pi^2}{8}$$

so that

$$\varepsilon_0 = 0.25; \qquad \varepsilon_m = \frac{1}{m^2 \pi^2} \quad m = \pm 1, \pm 3, \pm 5, \ldots; \qquad \varepsilon_m = 0 \quad m = \pm 2, \pm 4, \ldots \qquad (11)$$

The first diffraction order thus gives the highest focused intensity, but even so, it is only about 10 percent efficient. If the zone boundaries are displaced from their optimum positions, then intensity is put into the even orders, at the expense of the odd, to a maximum of $1/m^2\pi^2$, as shown in Fig. 3. If the clear zones are not totally transmitting but have amplitude transmission, A_1 (e.g., due to a supporting substrate) and the other zones have amplitude transmission, A_2, then the diffraction efficiencies are reduced by a factor $(A_2^2 - A_1^2)$.

The multiplicity of diffraction orders means that this type of zone plate normally has to be used in conjunction with an axial stop and a pinhole, the order-selecting aperture (OSA), as shown in Fig. 4, to prevent loss of contrast at the focus. The axial stop typically has a diameter of about 0.4 D, which has the effect of reducing the focused intensity and the width of the central maximum of the diffraction pattern, while putting more intensity into the outer lobes.

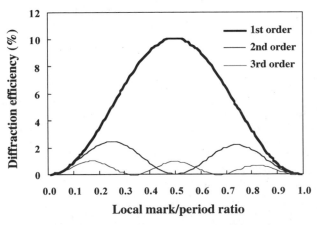

FIGURE 3 Diffraction efficiencies of an amplitude zone plate.

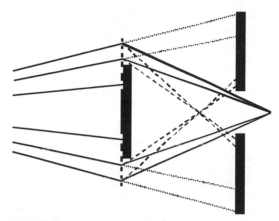

FIGURE 4 The use of an axial stop and aperture with a zone plate to remove the effects of the zero order (dotted lines) and higher orders (dashed lines; only the second order is shown for clarity) from the first-order focus. All negative orders are also removed.

In an alternative type of amplitude zone plate, the Gabor zone plate, the square wave transmittance amplitude is replaced by an approximately sinusoidal one

$$T(r) = \frac{1}{2}\left[1 + \sin\frac{\pi r^2}{\lambda f_1}\right]$$ (12)

in which case

$$\varepsilon_0 = 0.25; \qquad \varepsilon_{\pm 1} = \frac{1}{16}; \qquad \varepsilon_m = 0 \quad \text{all other } m$$ (13)

with the remaining ⅝ of the incoming intensity absorbed. The OSA is no longer needed, although the central stop is, but the first-order diffraction efficiency is less than that for an ordinary amplitude zone plate.

23.4 *PHASE ZONE PLATES*

In the absence of absorption, if alternate zones were made to change the phase of the radiation by π rad, the amplitude at a focus could be doubled so that the diffraction efficiency in the first order, for example, could be increased to 40 percent for rectangular zones. This is not possible for X rays because there is always some absorption, but a significant improvement in diffraction efficiency can be made if the zones are of the right thickness. The optimum thickness, which balances the gain due to the phase change with the loss due to absorption, is

$$t = \frac{\lambda\Delta\phi}{2\pi(1-n)}$$ (14)

Here, n is the real part of the complex refractive index, $\tilde{n} = n - ik$, and $\Delta\phi$ is the optimum phase shift given by the nontrivial solution of[11]

$$\eta\exp(-\eta\Delta\phi) = \sin\Delta\phi + \eta\cos\Delta\phi$$ (15)

where η is the ratio of the imaginary and real parts of the refractive index, $\eta = k/(1 - n)$. The diffraction efficiencies (for correctly placed zone boundaries) are now given by

$$\varepsilon_0 = 0.25\left[\sin^2 \Delta\phi + \left(2 \cos \Delta\phi + \frac{\sin \Delta\phi}{\eta}\right)^2 \right]$$

$$\varepsilon_1 = \frac{1}{\pi^2}\left(1 + \frac{1}{\eta^2}\right) \sin^2 \Delta\phi \qquad (16)$$

with $\varepsilon_m = \varepsilon_1/m^2$ for m odd and 0 for m even. The zeroth and first-order diffraction efficiencies are shown in Fig. 5 as functions of η, with the limits $\eta \to \infty$ and $\eta \to 0$ corresponding to amplitude and absorption-free phase zone plates, respectively. For energies of a few hundred electronvolts (eV), nickel is a good material ($\eta = 0.36$ at 350 eV, giving $\varepsilon_1 = 18$ percent), whereas for higher energies, materials such as gold or tungsten ($\eta = 0.24$ at 4 keV, giving $\varepsilon_1 = 22$ percent) are preferable.

Applying a similar analysis to a Gabor zone plate gives a corresponding increase in diffraction efficiency. Higher efficiencies could, in principle, be obtained by using zone profiles in which the phase shift varies continuously[12] in a similar fashion to a blazed linear transmission grating, but it has not yet been possible to make such zone plates.

23.5 MANUFACTURE OF ZONE PLATES

Because the spatial resolution of a zone plate is determined by the outermost (smallest) zone width [Eq. (9)], manufacturing techniques must be capable of delivering small linewidths over large areas (to give apertures as large as possible) with zone thicknesses of the correct values to give good diffraction efficiencies. The accuracy of the technique must be sufficient to allow zone boundaries to within about one-third of the outer zone width[13] and, to date, three main methods have been used to achieve this. Typically, zone plates with diameters up to about 200 μm and outer zone widths down to about 25 nm can now be made more or less routinely.

Electron-Beam Lithography

In this, an initial pattern is recorded in resist, followed by transfer of the pattern into a material that is suitable for a zone plate. One example of the electron-beam method is similar to the

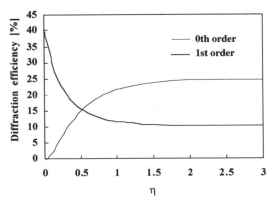

FIGURE 5 Diffraction efficiencies of phase zone plates.

technique employed in the manufacture of microcircuits,[14] using a polymer resist such as PMMA, followed by an etching step to reproduce the zone plate pattern in a suitable material. For X rays of a few hundred electronvolts of energy this could be nickel, requiring a thickness of approximately 100 to 200 nm for optimum efficiency. For higher-energy X rays, those of a few kilo-electronvolts, gold or tungsten would be more suitable, requiring thicknesses of approximately 500 nm to 1 μm. A variant of the electron-beam method uses vacuum-borne hydrocarbon vapor as the resist;[15] this is very accurate and allows the production of very small structures (<20 nm), but it is also very slow.

Interference Methods

These, in which the initial pattern is formed by the interference of two ultraviolet beams,[16] are also known as *holographic methods* because a zone plate is the hologram of a point. As the recording wavelength is much longer than the wavelength of use, the aberration terms neglected in the derivation of Eq. (4) must be corrected for by using an aplanatic lens system, which makes the technique unsuitable for making zone plates with different parameters. Large zone plates (with diameters approaching 1 cm) have been made, but these are not suitable for high-resolution imaging, because the aberration corrections become much more difficult.

The Sputter-and-Slice Method

In this method, alternating layers of two materials are deposited onto a rotating wire that is then sliced and thinned to the appropriate thickness.[17] This was put forward as a possible technique for making thick zone plates for high-energy (several kilo-electronvolts) X rays, although recent advances in electron-beam lithography also now allow this.[18]

23.6 BRAGG-FRESNEL OPTICS

As their name indicates, *Bragg-Fresnel optics,* which were first suggested by Aristov and his colleagues in 1988,[19] combine the Bragg reflection of crystals (Chap. 22) or multilayers (Chap. 24) with the Fresnel diffraction of gratings (Chap. 21) or zone plates. The discussion here will concentrate on the combination of zone plates with multilayers or crystals, but the generalization to grating geometries is obvious.

X-ray zone plates, as discussed earlier, normally work in transmission, but, like gratings, they can also be used in reflection. Then, because normal, or near-normal, incidence X-ray reflectivities are very small, to allow the (near) circular symmetry of a zone plate to be maintained, the in-phase addition of many reflections afforded by a periodic crystal or multilayer structure must be utilized. The way in which these Bragg-Fresnel lenses work has been analyzed in detail by Erko,[20] and so only a qualitative description is given here; the diffraction pattern at a focus can be determined in essentially the same way as that of an ordinary zone plate, and the intensity is determined by the diffraction efficiency combined with the Bragg reflectivity.

Figure 6 shows, schematically, how the optics work. Spherical waves from point sources at F_1 and F_2 cause an elliptical interference pattern with F_1 and F_2 at the foci. A slice across the diffraction pattern, perpendicular to the axis F_1F_2, gives the structure of a circular transmission zone plate that will focus radiation emitted at F_1 to F_2 (Fig. 6a). Moving F_1 to infinity gives a paraboloidal diffraction pattern and a standard Fresnel zone plate. Taking the slice at an angle to the axis (Fig. 6b) gives an elliptical zone plate that can form a reflected image of F_1 at

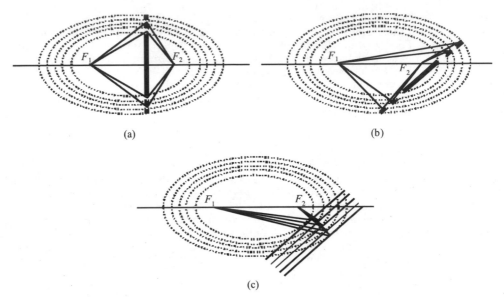

FIGURE 6 Construction of (*a*) a transmission zone plate, (*b*) a reflection zone plate, and (*c*) a Bragg-Fresnel lens.

F_2. If the reflecting surface is a crystal or multilayer, with period d equal to the distance between the peaks of the interference pattern (Fig. 6*c*), then the reflectivity is enhanced. At the same time, due to the requirement of satisfying the Bragg equation, the incident radiation is monochromatized with a bandpass, $\lambda/\Delta\lambda \sim N_L$, where N_L is the number of layer pairs. The monochromaticity requirement of the zone plate structure, $\lambda/\Delta\lambda \sim N_Z$, where N_Z is the number of zones, must also be satisfied.

Bragg-Fresnel optics have been made by masking the surfaces of multilayers (see Chap. 24) with absorbing zone plates[20] and by etching the zone plate pattern into a multilayer[21] or crystal.[22]

23.7 REFERENCES

1. J. L. Soret, "Concerning Diffraction by Circular Gratings," *Ann. Phys. Chem.* **156**:99–106 (1875).

2. Lord Rayleigh, "Wave Theory," in *Encyclopaedia Brittanica,* 9th ed., **24**:429–451 (1888).

3. R. W. Wood, "Phase-Reversed Zone Plates and Diffraction Telescopes," *Philos. Mag.* (5)**45**:511–522 (1898).

4. A. V. Baez, "Fresnel Zone Plates for Optical Image Formation Using Extreme Ultraviolet and Soft X-Ray Radiation," *J. Opt. Soc. Am.* **51**:405–412 (1961).

5. H. E. Hart et al., "Diffraction Characteristics of a Linear Zone Plate," *J. Opt. Soc. Am.* **18**:1018–1023 (1966).

6. L. C. Janicijevic, "Diffraction Characteristics of Square Zone Plates," *J. Opt.* (*Paris*) **13**:199–206 (1982).

7. C. Gomez-Reino et al., "Elliptical and Hyperbolic Zone Plates," *Appl. Opt.* **19**:1541–1545 (1980).

8. T. R. Welberry and R. P. Williams, "On Certain Non-Circular Zone Plates," *Opt. Acta* **23**:237–244 (1976).

9. A. G. Michette, *Optical Systems for Soft X-Rays,* Plenum, New York, 1986, pp. 176–177.

10. G. R. Morrison, "Diffractive X-Ray Optics," in *X-Ray Science and Technology,* A. G. Michette and C. J. Buckley (eds.), Institute of Physics Publishing, Bristol, 1993, pp. 305–306.

11. J. Kirz, "Phase Zone Plates for X-Rays and the Extreme UV," *J. Opt. Soc. Am.* **64**:301–309 (1974).

12. R. O. Tatchyn, "Optimum Zone Plate Theory and Design," in *X-Ray Microscopy, Springer Series in Optical Sciences,* G. Schmahl and D. Rudolph (eds.), Springer, Berlin, 1984, vol. 43, pp. 40–50.

13. A. G. Michette, *Optical Systems for Soft X-Rays,* Plenum, New York, 1986, pp. 217–222.

14. E. H. Anderson and D. P. Kern, "Nanofabrication of Zone Plates for X-Ray Microscopy," in *X-Ray Microscopy III, Springer Series in Optical Sciences,* A. G. Michette, G. R. Morrison, and C. J. Buckley (eds.), Springer, Berlin, 1992, vol. 67, pp. 75–78.

15. P. S. Charalambous and D. Morris, "The Status of Zone Plate Fabrication at King's College London," in *X-Ray Microscopy III, Springer Series in Optical Sciences,* A. G. Michette, G. R. Morrison, and C. J. Buckley (eds.), Springer, Berlin, 1992, vol. 67, 79–82.

16. G. Schmahl et al., "Zone Plates for X-Ray Microscopy," in *X-Ray Microscopy, Springer Series in Optical Sciences,* G. Schmahl and D. Rudolph (eds.), Springer, Berlin, 1984, vol. 43, pp. 63–74.

17. D. Rudolph et al., "Status of the Sputtered Sliced Zone Plates for X-Ray Microscopy," in *High Resolution Soft X-Ray Optics, Proc. SPIE,* E. Spiller (ed.), 1982, vol. 316, pp. 103–105.

18. B. Lai et al., "Advanced Zone Plate Microfocusing Optics," in *X-Ray Microfocusing: Applications and Techniques, Proc. SPIE,* I. McNulty (ed.), 1998, vol. 3449, pp. 133–136.

19. V. V. Aristov et al., "Principles of Bragg-Fresnel Multilayer Optics," *Rev. Phys. Appl.* **23**:1623–1630 (1988).

20. A. I. Erko, "Synthesized Multilayer Fresnel X-Ray Optics," *J. X-Ray Sci. Technol.* **2**:297–316 (1990).

21. A. I. Erko et al., "Fabrication and Tests of Multilayer Bragg-Fresnel X-Ray Lenses," *Microelectronic Engineering* **13**:335–338 (1991).

22. E. Aristova et al., "Phase Circular Bragg-Fresnel Lens Based on Germanium Single Crystal," in *X-Ray Microscopy IV,* V. V. Aristov and A. I. Erko (eds.), Bogorodskii Pechatnik, Chernogolovka, 1994, pp. 617–620.

CHAPTER 24
MULTILAYERS

Eberhard Spiller
Lawrence Livermore National Laboratories
Livermore, California

24.1 GLOSSARY

$\tilde{n} = 1 - \delta + ik$	refractive index
θ_i	grazing angle of propagation in layer i
ϕ_i	propagation angle measured from normal in layer i
q, q_s	momentum transfer perpendicular to and along surface
d	layer thickness
Λ	multilayer period, $\Lambda = d_1 + d_2$
f, Λ_s	spatial frequency and period on a surface or boundary
r_{nm}, t_{nm}	amplitude reflection and transmission coefficients
PSD	two-dimensional power spectral density
$a(f)$	roughness replication factor

24.2 INTRODUCTION

The reflectivity of all mirror materials is small beyond the critical grazing angle, and multilayer coatings are used to enhance this small reflectivity by adding the amplitudes reflected from many boundaries coherently. The multilayers for the extreme ultraviolet (EUV) and X-ray region can be seen as an extension of optical coatings toward shorter wavelengths or as artificial one-dimensional Bragg crystals with larger lattice spacings than natural crystals. In contrast to the visible region, no absorption-free materials are available for wavelengths $\lambda < 110$ nm. In addition, the refractive indices of all materials are very close to one, resulting in a small reflectance at each boundary and requiring a large number of boundaries to obtain substantial reflectivity. For absorption-free materials, a reflectivity close to 100 percent can always be obtained, independent of the reflectivity of an individual boundary, r_{12}, by making the number of boundaries, N, sufficiently large, $Nr_{12} \gg 1$. Absorption limits the number of periods that can contribute to the reflectivity in a multilayer to

$$N_{\max} = \frac{\sin^2 \theta}{2\pi k} \tag{1}$$

where k is the average absorption index of the coating materials. Multilayers became useful for $N_{max} \gg 1$, and very high reflectivities can be obtained if N_{max} is much larger than $N_{min} = 1/r_{12}$, the number required for a substantial reflectivity enhancement. In the "quarter-wave stack" of multilayer mirrors for the visible, all boundaries add their amplitudes in-phase to the total reflectivity, and each layer has the same optical thickness and extends between a node and an antinode of the standing wave that is generated by the incident and reflected waves. To mitigate the effect of absorption in the EUV and X-ray region, one first selects a *spacer* material with the lowest available absorption and combines it with a *reflector* layer with good contrast with the optical constants of the spacer. The optimum design minimizes absorption by reducing the thickness of the reflector layer and by attempting to position it close to the nodes of the standing wave field, whereas for the spacer layer, the thickness is increased and it is centered around the antinodes. The design accepts some dephasing of the contributions from adjacent boundaries and the optimum is a compromise between the effects of this dephasing and the reduction in absorption. Typically, between 0.25 and 0.45 of one period in an optimized coating is used by the reflector, the rest by the spacer layer.

The absorption index, k, is in the order of one for wavelengths around 100 nm and decreases to very small values at shorter wavelengths in the X-ray range. Materials that satisfy the condition $N_{max} > 1$ become available for $\lambda < 50$ nm; for $\lambda < 20$ nm and not too close to absorption edges, k decreases fast with decreasing wavelength as $k \propto \lambda^3$. The reflected amplitude, r, from a single boundary also decreases with wavelength, albeit at a slower rate as $r \propto \lambda^2$, and one can compensate for this decrease by increasing the number of boundaries or periods, N, in a multilayer stack ($N \propto 1/\lambda^2$). With this method, reflectivities close to 100 percent are theoretically possible at very short wavelengths in the hard X-ray range. A perfect crystal showing Bragg reflection can be seen as a multilayer that realizes this high reflectivity, with the atomic planes located at the nodes of the standing wave within the crystal.

The condition that all periods of a multilayer reflect in phase leads to the Bragg condition for the multilayer period Λ

$$\Lambda = \frac{m\lambda}{2\sin\theta} \approx \frac{m\lambda}{2\sin\theta_0\sqrt{1 - 2\delta/\sin^2\theta_0}} \tag{2}$$

where θ is the effective grazing angle of propagation within the multilayer material, and θ_0 the corresponding angle in vacuum. The refraction correction represented by the square root in Eq. (2) becomes large for small grazing angles, even for small values of δ. The path difference between the amplitudes from adjacent periods is $m\lambda$, and multilayers are most of the time used in order $m = 1$. The shortest period, Λ, for good multilayer performance is limited by the quality of the boundaries (roughness, σ, should be smaller than $\Lambda/10$), and this quality is in turn limited by the size of the atoms for noncrystalline materials. Practical values for the shortest period are around $\Lambda = 1.5$ nm, and the high reflectivities that are theoretically possible for very hard X rays can only be realized with multilayers of periods $\Lambda > 1.5$ nm that are used at small grazing angles. By introducing the momentum transfer at reflection as

$$q = \frac{4\pi}{\lambda}\tilde{n}\sin\theta \tag{3}$$

we can express the condition $\Lambda > 1.5$ nm also as $|q| < 4$ nm^{-1}. It is convenient to introduce the variable, q, in X-ray optics, because not to close to an absorption edge, the performance of optical components is determined mainly by q and not by the specific values of λ and θ.

Attempts to observe X-ray interference from thin films started around 1920 and Kiessig[1] observed and analyzed the X-ray interference structure from Ni films. Multilayer structures produced around 1940[2] lost their X-ray reflectivity due to diffusion in the structure, and this fact became a tool to measure small diffusion coefficients.[3-5] The usefulness of multilayers for normal incidence optics in the EUV was recognized in 1972.[6] The deposition processes to produce these structures were developed by many groups in the next 2 decades. Today, multilayer telescopes on the orbiting observatories *SOHO* and *TRACE* provide high-resolution EUV pictures of the solar corona (see *http://umbra.nascom.nasa.gov/eit/* and *http://vestige.lmsal.com/TRACE/*). Cameras with multilayer coated mirrors for the EUV are a main contender for the fabrication of the next generation of computer chips, and multi-layer mirrors are found at the beamlines of all synchrotron radiation facilities and in X-ray diffraction equipment as beam deflectors, collimators, filters, monochromators, polarizers, and imaging optics. (See Ref. 7 and Chaps. 21, 26–28, and 35 for more details.)

24.3 CALCULATION OF MULTILAYER PROPERTIES

The theoretical treatment of the propagation of X rays in layered structures does not differ from that for other wavelength regions as discussed by Dobrowolski in Chap. 42 of Vol. I of this handbook and in many textbooks and review articles. At the boundary of two materials or at an atomic plane, an incident wave is split into a transmitted and reflected part, and the total field amplitude in the structure is the superposition of all these waves. Figure 1 sketches two layers as part of a multilayer and the amplitudes of the forward a_n and backward b_n running waves in each layer. The amplitudes in each layer are coupled to those of the adjacent layers by linear equations that contain the amplitude transmission $t_{n,m}$ and reflection coefficients r_{nm}.

$$a_n = a_{n-1} t_{n-1,n} \, e^{i\varphi_n} + b_n e^{2i\varphi_n} r_{n,n-1}$$

$$b_n = a_n r_{n,n+1} + b_{n+1} \, e^{i\varphi_{n+1}} t_{n+1,n} \tag{4}$$

The phase delay, φ_n, due to propagation through layer n of thickness, d, and angle, ϕ_n, from normal is given by

$$\varphi_n = \frac{2\pi}{\lambda} \tilde{n} d_n \cos \phi_n \tag{5}$$

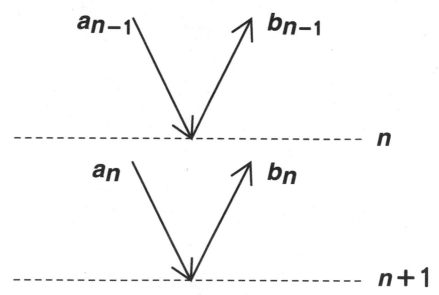

FIGURE 1 Two layers and their boundaries in a multilayer structure with the incoming wave amplitudes, a_n, and reflected amplitudes, b_n, in each layer.

and the transmitted and reflected amplitudes at each boundary are obtained from the Fresnel equations

$$r_{n,n+1} = \frac{q_n - q_{n+1}}{q_n + q_{n+1}} \tag{6}$$

$$t_{n,n+1} = \frac{2q_n}{q_n + q_{n+1}} \tag{7}$$

with the q values as defined in Eq. (3) for perpendicular-polarization and $q_i = (4\pi/\lambda\tilde{n}_i) \sin \theta$ for parallel-polarization.

Matrix methods[8,9] are convenient to calculate multilayer performance using high-level software packages that contain complex matrix manipulation. The transfer of the amplitudes over a boundary and through a film is described by[9]

$$\begin{pmatrix} a_i \\ b_i \end{pmatrix} = \frac{1}{t_{i-1,i}} \begin{pmatrix} e^{i\varphi_i} & r_{i-1,i}e^{-i\varphi_i} \\ r_{i-1,i}e^{i\varphi_i} & e^{-i\varphi_i} \end{pmatrix} \begin{pmatrix} a_{i+1} \\ b_{i+1} \end{pmatrix} \tag{8}$$

and the transfer between the incident medium ($i = 0$) and the substrate ($i = n + 1$) by

$$\begin{pmatrix} a_0 \\ b_0 \end{pmatrix} = \frac{\prod_{i=1}^{i=n+1} M_i}{\prod_{i=1}^{i=n+1} t_{i,i+1}} \begin{pmatrix} a_{n+1} \\ b_{n+1} \end{pmatrix} \tag{9}$$

with the matrices M_i as defined in Eq. (8). For incident radiation from the top ($b_{n+1} = 0$), we can calculate the reflected and transmitted amplitudes from the elements of the product matrix m_{ij} as

$$r = b_0/a_0 = m_{21}/m_{11}$$

$$t = a_{n+1}/a_0 = (\prod t_{i,i+1})/m_{11} \tag{10}$$

The reflected intensity is $R = rr^*$, and the transmitted intensity is $T = tt^* (q_{n-1}/q_0)$ with the q values of Eq. (3). In another convenient matrix formalism due to Abeles, each matrix contains only the parameters of a single film[8,10]; however, the transfer matrix method given earlier[9] is more convenient when one wants to include the imperfections of the boundaries.

The achievable boundary quality is always of great concern in the fabrication of X-ray multilayers. The first effect of boundary roughness or diffusion is the reduction of the reflected amplitude due to dephasing of the contributions from different depths within the boundary layer. If one describes the reflected amplitude as a function of depth by a Gaussian of width, σ, one obtains a Debye-Waller factor for the reduction of boundary reflectivity:

$$r/r_0 = \exp (-0.5 \, q_1 q_2 \sigma^2) \tag{11}$$

where r_0 is the amplitude reflectivity for a perfect boundary, and the q values are those of the films on both sides of the boundary.[7,11–14] Reducing the amplitude reflection coefficients in Eq. (8) by the Debye-Waller factor of Eq. (11) gives a good estimate of the influence of boundary roughness on multilayer performance. The roughness, σ, has been used by most authors as a fitting parameter that characterizes a coating. Usually, the Debye-Waller factor is given for the intensity ratio [absolute value of Eq. (11) squared] and with the vacuum q-values for both materials (refraction neglected). The phase shift in Eq. (11) due to the imaginary part of q produces a shift of the effective boundary toward the less absorbing material.

For a boundary without scattering (gradual transitition of the optical constants or roughness at very high spatial frequencies), the reduction in reflectivity is connected with an increase in transmission according to $t_{12}t_{21} + r_{12}^2 = 1$. Even if the reflectivity is reduced by a substantial factor, the change in transmission can in most cases be neglected, because reflectivities of single boundaries are typically in the 10^{-4} range and transmissions close to 1. Roughness that scatters a substantial amount of radiation away from the specular beam can decrease both the reflection and transmission coefficients, and the power spectral density (PSD) of the surface roughness over all relevant spatial frequencies has to be measured to quantify this scattering.

It is straightforward to translate Eqs. (4) through (11) into a computer program, and a personal computer typically gives a reflectivity curve of a 100-layer coating for 100 wavelengths or angles within a few seconds. Multilayer programs and the optical constants of all elements and of many compounds can be accessed on (http://www-cxro.lbl.gov) or downloaded[15] (http://cletus.phys.columbia.edu/~windt). The first site also has links to other relevant sites, and the Windt programs can also calculate nonspecular scattering.

24.4   *FABRICATION METHODS AND PERFORMANCE*

Multilayer X-ray mirrors have been produced by practically any deposition method. Thickness errors and boundary roughness are the most important parameters that have to be kept smaller than $\Lambda/10$ for good performance. Magnetron sputtering, pioneered by Barbee,[16] is most widely used: Sputtering systems are very stable, and one can obtain the required thickness control simply by timing. Thermal evaporation systems usually use an in situ soft X-ray reflectometer to control the thickness, and an additional ion gun for smoothing the boundaries.[17-20] The best designs of multilayers for high reflectivity deviate from the quarter-wave stack used for visible laser mirrors; to minimize absorption, the heavier, more absorbing material is made thinner than the lighter, more transparent material. The optimum design balances the advantage of reduced absorption with the penalty due to the phase error between the reflected amplitudes from adjacent boundaries. The best values for $\gamma = d_h/\Lambda$ are between 0.3 and 0.4.[7,21,22] The paper by Rosenbluth[22] gives a compilation of the best multilayer materials for each photon energy in the range from 100 to 2000 eV. Absorption is the main performance limit for multilayers of larger-period $\Lambda > 80$ nm used at longer wavelength near normal incidence. In this region, it is important to select the material with the smallest absorption as the first component of a structure (usually a light element at the long-wavelength side of an absorption edge), and the absorption of this material becomes the main limit for the multilayer performance. A list of materials of low absorption at selected wavelengths with their values for k and N_{max} is given in Table 1.

Optimum multilayers for longer wavelengths ($\lambda > 20$ nm) are quasi-periodic with decreasing γ from the bottom to the top of a multilayer stack. One always attempts to locate the stronger absorber close to the nodes of the standing wavefield within the coating to reduce absorption, and the absorption reduction is greater near the top of the stack where the standing wave between incident and reflected radiation has more contrast.[6]

High-Reflectivity Mirrors

A compilation of the reflectivities of multilayers obtained by different groups can be found at www-cxro.lbl.gov/multilayer/survey.html. The highest normal incidence reflectivities around 70 percent have been obtained with Mo/Be and Mo/Si around $\lambda = 11.3$ and 13 nm near the absorption edges of Be and Si, as shown in Fig. 2.[23] The peak reflectivity drops at longer wavelength due to the increased absorption. At shorter wavelengths, roughness becomes important and reduces the reflectivity of normal-incidence mirrors. One can, however, still obtain very high reflectivity at short wavelengths by keeping the multilayer period above 2 nm and using grazing angles of incidence to reach short wavelengths.

TABLE 1 Absorption index, k, and N_{max} for the largest q values (normal incidence for $\lambda > 3.15$ nm and $\sin\theta = \lambda/\pi$ for $\lambda < 3.14$ nm) of good spacer materials near their absorption edges and absorption of some materials at $\lambda = 0.154$ nm.

	λ_{edge} (nm)	k	N_{max}
Mg-L	25.1	7.5×10^{-3}	21
Al-L	17.1	4.2×10^{-3}	38
Si-L	12.3	1.6×10^{-3}	99
Be-K	11.1	1.0×10^{-3}	155
Y-M	8.0	3.5×10^{-3}	45
B-K	6.6	4.1×10^{-4}	390
C-K	4.37	1.9×10^{-4}	850
Ti-L	3.14	4.9×10^{-4}	327
N-K	3.1	4.4×10^{-5}	3580
Sc-L	3.19	2.9×10^{-4}	557
V-L	2.43	3.4×10^{-4}	280
O-K	2.33	2.2×10^{-5}	3980
Mg-K	0.99	6.6×10^{-6}	2395
Al-K	0.795	6.5×10^{-6}	1568
Si-K	0.674	4.2×10^{-6}	1744
SiC	0.674	6.2×10^{-6}	1182
TiN	3.15	4.9×10^{-4}	327
Mg$_2$Si	25.1	7.4×10^{-3}	21
Mg$_2$Si	0.99	6.8×10^{-6}	2324
Be	0.154	2.0×10^{-9}	189,000
B	0.154	5.7×10^{-9}	67,450
C	0.154	1.2×10^{-8}	32,970
Si	0.154	1.7×10^{-7}	5239
Ni	0.154	5.1×10^{-7}	750
W	0.154	3.9×10^{-6}	98

For hard X rays, where N_{max} is very much larger than the number needed to obtain good reflectivity, one can position multilayers with different periods on top of each other (depth-graded multilayers) to produce a coating with a large spectral or angular bandwidth. Such "supermirrors" are being proposed to extend the range of grazing incidence telescopes to higher energies up to 100 keV.[24,25] Supermirrors are common for cold neutrons where absorption-free materials are available.[26] The low absorption for hard X rays makes it also possible to produce coatings with very narrow bandwidths. The spectral width is determined by the effective number of periods contributing to the reflectivity $\lambda/\Delta\lambda = N_{eff}$. By reducing the reflectivity from a single period (e.g., by using a very small value of $\gamma = d_H/\Lambda$), radiation penetrates deeper into the stack allowing more layers to contribute, thus reducing the bandwidth.[7]

Multilayer Coated Optics

Multilayers for imaging optics usually require a grading of the multilayer period, Λ, according to Eq. (2), to adapt to the change in incidence angle over the optics. Many methods have been used to produce the proper grading of the multilayer period, among them shadow masks in front of the rotating optics, substrate tilt, speed control of the platter that moves the optics over the magnetrons, and computer-controlled shutters.[27–33] Figure 3 shows the control that has been achieved for a condensor mirror of an EUV camera[23,34]; the reproducibility of the profile from run to run is better than 0.1 percent. Multilayer mirrors with a linear thickness grading on parabolically bent Si wafers are commercially available (www.osmic.com) and are

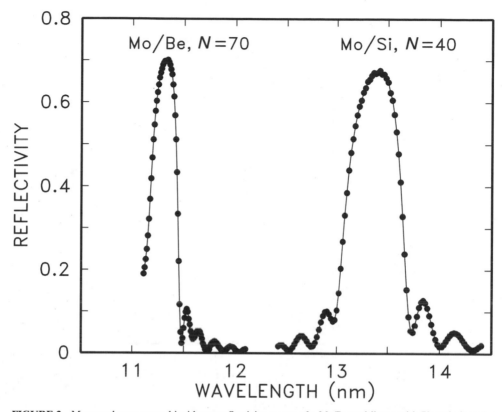

FIGURE 2 Measured near-normal incidence reflectivity curves of a Mo/Be multilayer with 70 periods and a Mo/Si multilayer with 40 periods in wavelengths regions of low absorption near the Be-K and Si-L edge.

being used as collimators in X-ray diffraction experiments at $\lambda = 0.154$ nm. At this wavelength, the reflectivity is over 75 percent, and the flux to the diffractometer can be increased by about an order of magnitude[35] (see Chap. 35).

Polarizers and Phase Retarders

The Brewster angle occurs close to 45° in the EUV and X-ray region for all materials, and all reflectors used near 45° are effective polarizers. Multilayer coatings do not change the ratio of the reflectivities for parallel and perpendicular polarization, but enhance the reflectivity for perpendicular-polarization to useful values.[36,37] The reflectivity for parallel-polarization at the Brewster angle is zero for absorption-free materials but increases with absorption. Therefore, the achievable degree of polarization is higher at shorter wavelengths where absorption is lower. Typical values for the reflectivity ratio, $R_\perp/R_{||}$, are 10 around $\lambda = 30$ nm and over 1000 around $\lambda = 5$ nm.

It is not possible to design effective 90° phase-retarding multilayer reflectors with high reflectivity for both polarizations. The narrower bandwidth of the reflectivity curve for parallel-polarization allows one to produce a phase delay at incidence angles or wavelengths that are within the high-reflectivity band for perpendicular- but not for parallel-polarization; however, because of the greatly reduced parallel-reflectivity of such a design, one cannot use them to transform linear polarized radiation into circular polarized radiation.[38] Multilayers used in

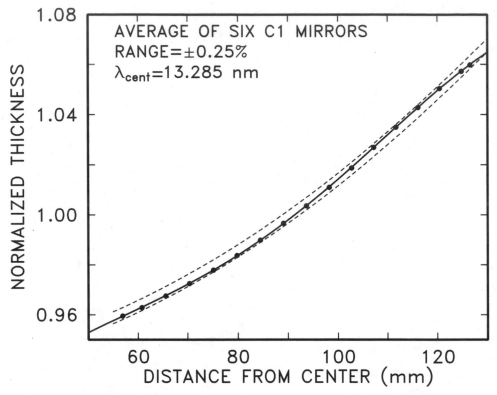

FIGURE 3 Grading of a multilayer achieved for the first condensor of an EUV camera. The dashed lines mark the ±0.25 percent tolerances from the ideal thickness.

transmission offer a better solution. Near the Brewster angle, parallel-polarized radiation is transmitted through a multilayer structure without being reflected by the internal boundaries and has a transmission determined mainly by the absorption in the multilayer stack. A high transmission for perpendicular-polarization is also obtained from a multilayer stack that is used off-resonance near a reflectivity minimum or transmission maximum. However, the transmitted radiation is delayed due to the internal reflection within the multilayer stack. Figure 4 gives an example of the calculated performance of such a design.[39–42] For a grazing angle of 50°, we obtain a phase retardation of 80° with a useful reflectivity for parallel polarization of $R_p = 15$ percent. One can tune such a phase retarder in wavelength by producing a graded coating or by using different incidence angles.[43] A 90° phase retarder is possible with high-quality multilayers that can obtain peak reflectivities above 70 percent. Boundary roughness reduces the phase retardation because it reduces the effective number of bounces within the structure. The maximum phase retardation measured at wavelengths $\lambda < 5$ nm are in the 10° range.[44]

Boundary Roughness

The quality of the boundary is the main limitation for the performance of a multilayer with short periods. Most good multilayer systems can be described by a simple growth model: Particles or atoms arrive randomly and can relax sideways to find locations of lower energy. The

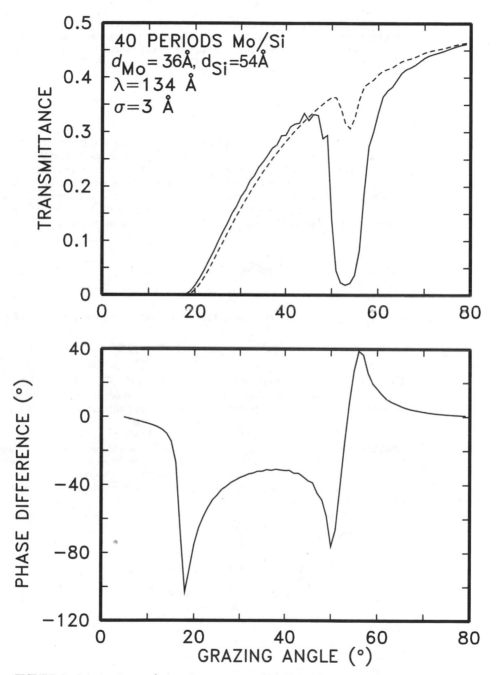

FIGURE 4 Calculated transmission of an unsupported Mo/Si multilayer for perpendicular-polarization (full curve) and parallel-polarization (dashed curve) and phase difference between the two polarizations. A phase shift of 80° is obtained for an angle of 50°. The value near 18° is of no use because there is practically no transmission.

random deposition produces a flat power spectrum for the surface roughness with most of the roughness at very high spatial frequencies, and the relaxation reduces roughness at the highest spatial frequencies. The two-dimensional power spectral density (see Church and Takacs in Vol. I, Chap. 7, and Harvey, Vol. II, Chap. 11 of this handbook) of a film on a perfectly smooth substrate is then given by[7,45-47]

$$\text{PSD}(q_s, d) = \Omega \, \frac{1 - \exp\left(-2l_r^{n-1} \, dq_s^n\right)}{2l_r^{n-1} \, q_s^n} \tag{12}$$

where Ω is the particle volume; l_r is a relaxation length; d is the total thickness of the coating; and $q_s = 2\pi f = 2\pi/\Lambda_s$, with f a spatial frequency or Λ, a spatial period on the surface. The parameter, n, characterizes the relaxation process; $n = 1$ indicates viscous flow; $n = 2$ is condensation and reevaporation; $n = 3$ is bulk diffusion; and $n = 4$ is surface diffusion. Equation (12) yields the flat PSD of the random deposition with roughness $\sigma \propto \sqrt{d}$ for small spatial frequencies and a power law with exponent, n, that is independent of thickness at high spatial frequencies. Roughness is replicated throughout a stack with a replication factor

$$a(f) = \exp - \left(l_r^{n-1} \, dq_s^n\right) \tag{13}$$

The PSD of a film on a substrate has a total PSD, PSD^{tot},

$$\text{PSD}^{tot} = \text{PSD}^{film} + a^2 \text{PSD}^{sub} \tag{14}$$

and the roughness that is contributed from the spatial frequency range from f_1 to f_2 is

$$\sigma^2 = 2\pi \int_{f_1}^{f_2} \text{PSD}(f) \, df \tag{15}$$

The intensity of the scattered light is proportional to the PSD, and consistent values for the PSD can be obtained by atomic force microscopy and from scatter measurements.[48-50] The growth parameters of some multilayer structures are given in Table 2.[51] Note that the roughness values

TABLE 2   Growth parameters of multilayer systems with multilayer period (Λ), number of periods (N), growth parameters (Ω, l_r, and n), total thickness (d), and roughness (σ) of the multilayer film calculated for a perfectly smooth substrate within the spatial frequency range of $f = 0.001$–0.25 nm. Parameters are the average value for both coating materials, for sputtered Mo/Si we give also the parameters for Mo and Si separately.

System	Λ (nm)	N	Ω (nm^3)	l_r (nm)	n	d (nm)	σ (nm)
Co/C	3.2	150	0.016	1.44	4	480	0.14
Co/C	2.9	144	0.016	1.71	4	423	0.12
Co/C	2.4	144	0.016	1.14	4	342	0.15
Co/C	3.2	85	0.016	1.71	4	274	0.19
Mo/Si	7.2	24	0.035	1.71	4	172	0.14
Ni/C[a]	5.0	30	0.035	1.71	4	150	0.14
W/B$_4$C[b]	1.22	350	0.01	1.22	4	427	0.10
W/B$_4$C[b]	1.78	255	0.018	1.22	3	454	0.12
Mo/Si[c]	6.8	40	0.035	1.36	4	280	0.19
Mo/Si[d]	6.8	40	0.055	1.20	2	280	0.12
Si in Mo/Si[c]	6.8	40	0.02	1.36	4	280	0.14
Mo in Mo/Si[c]	6.8	40	0.5	1.36	4	280	0.23

[a] Coating produced by dual-ion beam sputtering courtesy of J. Pedulla (NIST).
[b] Coating produced by dual-ion beam sputtering courtesy of Y. Platonov (Osmic).
[c] Data for Mo/Si produced by magnetron sputtering from D. Stearns et al.[45]
[d] Data for Mo/Si produced by ion beam sputtering from P. Kearney and D. Stearns (LLNL).

in the last column do not include contributions from spatial frequencies $f > 0.25$ nm^{-1} and from diffuse transition layers. These values have to be added in the form $\sigma_{tot}^2 = \sigma_1^2 + \sigma_2^2$ in Eq. (11) for the calculation of the reflectivity.

During the development phase of multilayer X-ray mirrors, it was fortuitous that practically perfect smooth substrates were available as Si wafers, float glass, and mirrors fabricated for laser gyros. It is, however, still a challenge to produce figured substrates that have both good figure and finish in the 0.1-nm range.[52] Multilayer mirrors can remove high spatial roughness with spatial periods of less than 50 nm, and graded coatings can be used to correct low-order figure errors; however, mirrors that do not meet specifications for high-resolution imaging usually have considerable roughness at spatial frequencies between these limits that cannot be modified by thin films.[51]

24.5 REFERENCES

1. H. Kiessig, "Interferenz von Röntgenstrahlen an dünnen Schichten," *Ann. der Physik 5. Folge* **10**:769–788 (1931).

2. J. DuMond and J. P. Joutz, "An X-Ray Method of Determining Rates of Diffusion in the Solid State," *J. Appl. Phys.* **11**:357–365 (1940).

3. H. E. Cook and J. E. Hilliard, "Effect of Gradient Energy on Diffusion in Gold-Silver Alloys," *J. Appl. Phys.* **40**:2191–2198 (1969).

4. J. B. Dinklage, "X-Ray Diffraction by Multilayered Thin-Film Structures and Their Diffusion," *J. Appl. Phys.* **38**:3781–3785 (1967).

5. A. L. Greer and F. Spaepen, "Diffusion," in *Synthetic Modulated Structures*, L. L. Chang and B. C. Giessen (eds.), Academic Press, Orlando, FL, 1985, pp. 419.

6. E. Spiller, "Low-Loss Reflection Coatings Using Absorbing Materials," *Appl. Phys. Lett.* **20**:365–367 (1972).

7. E. Spiller, *Soft X-Ray Optics*, SPIE Optical Engineering Press, Bellingham, WA, 1994.

8. F. Abelés, "Recherches sur la propagation des ondes électromagnétique inusoidales dans les milieux stratifies. Application aux couches minces," *Ann. de Physique* **5**:596–639 (1950).

9. O. S. Heavens, *Optical Properties of Thin Solid Films*, Dover, New York, 1966.

10. M. Born and E. Wolf, *Principles of Optics*, 5th ed. Pergamon Press, Oxford, 1975.

11. P. Croce, L. Névot, and B. Pardo, "Contribution a l'étude des couches mince par réflexion spéculaire de rayon X," *Nouv. Rev. d'Optique appliquée* **3**:37–50 (1972).

12. P. Croce and L. Névot, "Étude des couches mince et des surfaces par réflexion rasante, spéculaire ou diffuse, de rayon X," *J. De Physique Appliquée* **11**:113–125 (1976).

13. L. Névot and P. Croce, "Charactérisation des surfaces par réflection rasante de rayon X, Application à étude du polissage de quelque verres silicates," *Revue Phys. Appl.* **15**:761–779 (1980).

14. F. Stanglmeier, B. Lengeler, and W. Weber, "Determination of the Dispersive Correction $f(E)$ to the Atomic Form Factor from X-Ray Reflection," *Acta Cryst.* **A48**:626–639 (1992).

15. D. Windt, "IMD—Software for Modeling the Optical Properties of Multilayer Films," *Computers in Physics* **12**:360–370 (1998).

16. T. W. Barbee, Jr., "Multilayers for X-Ray Optics," *Opt. Eng.* **25**:893–915 (1986).

17. E. Spiller, A. Segmüller, J. Rife, and R.-P. Haelbich, "Controlled Fabrication of Multilayer Soft X-Ray Mirrors," *Appl. Phys. Lett.* **37**:1048–1050 (1980).

18. E. Spiller, "Enhancement of the Reflectivity of Multilayer X-Ray Mirrors by Ion Polishing," *Opt. Eng.* **29**:609–613 (1990).

19. E. J. Puik, M. J. van der Wiel, H. Zeijlemaker, and J. Verhoeven, "Ion Bombardment of X-Ray Multilayer Coatings: Comparison of Ion Etching and Ion Assisted Deposition," *Appl. Surface Science* **47**:251–260 (1991).

20. E. J. Puik, M. J. v. d. Wiel, H. Zeijlemaker, and J. Verhoeven, "Ion Etching of Thin W Layers: Enhanced Reflectivity of W-C Multilayer Coatings," *Appl. Surface Science* **47**:63–76 (1991).

21. A. V. Vinogradov and B. Y. Zel'dovich, "X-Ray and Far UV Multilayer Mirrors; Principles and Possibilities," *Appl. Optics* **16**:89–93 (1977).

22. A. E. Rosenbluth, "Computer Search for Layer Materials That Maximize the Reflectivity of X-Ray Multilayers," *Revue Phys. Appl.* **23**:1599–1621 (1988).

23. C. Montcalm, R. F. Grabner, R. M. Hudyma, M. A. Schmidt, E. Spiller, C. C. Walton, M. Wedowski, and J. A. Folta, "Multilayer Coated Optics for an Alpha-Class Extreme-Ultraviolet Ithography System," *Proc. SPIE* **3767**:1999.

24. P. Hoghoj, E. Ziegler, J. Susini, A. K. Freund, K. D. Joensen, P. Gorenstein, and J. L. Wood, "Focusing of Hard X-Rays with a W/Si Supermirror," *Nucl. Instrum. Meth. Phys. Res. B* **132**:528–533 (1997).

25. K. D. Joensen, P. Voutov, A. Szentgyorgyi, J. Roll, P. Gorenstein, P. Høghøj, and F. E. Christensen, "Design of Grazing-Incidence Multilayer Supermirrors for Hard-X-Ray Reflectors," *Appl. Opt.* **34**:7935–7944 (1995).

26. F. Mezei, "Multilayer Neutron Optical Devices," in *Physics, Fabrication, and Applications of Multilayered Structures,* P. Dhez and C. Weisbuch (eds.), Plenum Press, New York, 1988, pp. 311–333.

27. D. J. Nagel, J. V. Gilfrich, and J. T. W. Barbee, "Bragg Diffractors with Graded-Thickness Multilayers," *Nucl. Instrum. Methods* **195**:63–65 (1982).

28. D. G. Stearns, R. S. Rosen, and S. P. Vernon, "Multilayer Mirror Technology for Soft-X-Ray Projection Lithography," *Appl. Opt.* **32**:6952–6960 (1993).

29. S. P. Vernon, M. J. Carey, D. P. Gaines, and F. J. Weber, "Multilayer Coatings for the EUV Lithography Test Bed," presented at POS Proceedings on Extreme Ultraviolet Lithography, Monterey, CA, 1994.

30. E. Spiller and L. Golub, "Fabrication and Testing of Large Area Multilayer Coated X-Ray Optics," *Appl. Opt.* **28**:2969–2974 (1989).

31. E. Spiller, J. Wilczynski, L. Golub, and G. Nystrom, "The Normal Incidence Soft X-Ray, λ = 63.5 Å Telescope of 1991," *Proc. SPIE* **1546**:168–174 (1991).

32. D. L. Windt and W. K. Waskiewicz, "Multilayer Facilities Required for Extreme-Ultraviolet Lithography," *J. Vac. Sci. Technol. B.* **12**:3826–3832 (1994).

33. M. P. Bruijn, P. Chakraborty, H. W. v. Essen, J. Verhoeven, and M. J. v. d. Wiel, "Automatic Electron Beam Deposition of Multilayer Soft X-Ray Coatings with Laterally Graded *d* Spacing," *Opt. Eng.* **25**:916–921 (1986).

34. D. W. Sweeney, R. M. Hudyma, H. N. Chapman, and D. R. Shafer, "EUV Optical Design for a 100-nm CD Imaging System," *Proc. SPIE* **3331**:2–10 (1998).

35. M. Schuster and H. Göbel, "Parallel-Beam Coupling into Channel-Cut Monochromators Using Curved Graded Multilayers," *J. Phys. D: Appl. Phys.* **28**:A270–A275 (1995).

36. E. Spiller, "Multilayer Interference Coatings for the Vacuum Ultraviolet," in *Space Optics,* B. J. Thompson and R. R. Shannon (eds.), National Academy of Sciences, Washington, DC, 1974, pp. 581–597.

37. A. Khandar and P. Dhez, "Multilayer X Ray Polarizers," *Proc. SPIE* **563**:158–163 (1985).

38. E. Spiller, "The Design of Multilayer Coatings for Soft X Rays and Their Application for Imaging and Spectroscopy," in *New Techniques in X-ray and XUV Optics,* B. Y. K. a. B. E. Patchett, Ed. Chilton, U.K.: Rutherford Appleton Lab., 1982, pp. 50–69.

39. J. B. Kortright and J. H. Underwood, "Multilayer optical elements for generation and analysis of circularly polarized x rays," *Nucl. Instrum. Meth.,* vol. A291, pp. 272–277, 1990.

40. J. B. Kortright, H. Kimura, V. Nikitin, K. Mayama, M. Yamamoto, and M. Yanagihara, "Soft X-Ray (97-eV) Phase Retardation Using Transmission Multilayers," *Appl. Phys. Lett.* **60**:2963–2965 (1992).

41. J. B. Kortright, M. Rice, S. K. Kim, C. C. Walton, and C. Warwick, "Optics for Element-Resolved Soft X-Ray Magneto-Optical Studies," *J. of Magnetism and Magnetic Materials* **191**:79–89 (1999).

42. S. Di Fonzo, B. R. Muller, W. Jark, A. Gaupp, F. Schafers, and J. H. Underwood, "Multilayer Transmission Phase Shifters for the Carbon K Edge and the Water Window," *Rev. Sci. Instrum.* **66**:1513–1516 (1995).

43. J. B. Kortright, M. Rice, and K. D. Franck, "Tunable Multilayer EUV Soft-X-Ray Polarimeter," *Rev. Sci. Instr.* **66**:1567–1569 (1995).

44. F. Schäfers, H. C. Mertins, A. Gaupp, W. Gudat, M. Mertin, I. Packe, F. Schmolla, S. Di Fonzo, G. Soullié, W. Jark, W. Walker, X. Le Cann, and R. Nyh, "Soft-X-Ray Polarimeter with Multilayer Optics: Complete Analysis of the Polarization State of Light," *Appl. Opt.* **38**:4074–4088 (1999).

45. D. G. Stearns, D. P. Gaines, D. W. Sweeney, and E. M. Gullikson, "Nonspecular X-Ray Scattering in a Multilayer-Coated Imaging System," *J. Appl. Phys.* **84**:1003–1028 (1998).

46. E. Spiller, D. G. Stearns, and M. Krumrey, "Multilayer X-Ray Mirrors: Interfacial Roughness, Scattering, and Image Quality," *J. Appl. Phys.* **74**:107–118 (1993).

47. W. M. Tong and R. S. Williams, "Kinetics of Surface Growth: Phenomenology, Scaling, and Mechanisms of Smoothening and Roughening," *Annu. Rev. Phys. Chem.* **45**:401–438, 1994.

48. E. M. Gullikson, D. G. Stearns, D. P. Gaines, and J. H. Underwood, "Non-Specular Scattering from Multilayer Mirrors at Normal Incidence," *Proc. SPIE* **3113**:412–419 (1997).

49. E. M. Gullikson, "Scattering from Normal Incidence EUV Optics," *Proc. SPIE* **3331**:72–80 (1998).

50. V. Holý, U. Pietsch, and T. Baumbach, *High Resolution X-Ray Scattering from Thin Films and Multilayers,* vol. 149, Springer, Berlin, 1999.

51. E. Spiller, S. Baker, E. Parra, and C. Tarrio, "Smoothing of Mirror Substrates by Thin Film Deposition," *Proc. SPIE* **3767**:1999.

52. J. S. Taylor, G. E. Sommargren, D. W. Sweeney, and R. M. Hudyma, "The Fabrication and Testing of Optics for EUV Projection Lithography," *Proc. SPIE* **3331**:580–590 (1998).

CHAPTER 25
POLARIZING CRYSTAL OPTICS*

Qun Shen
Cornell High Energy Synchrotron Source and Department of
Materials Science and Engineering
Cornell University
Ithaca, New York

25.1 INTRODUCTION

Being able to produce and to analyze a general polarization of an electromagnetic wave has long benefited scientists and researchers in the field of visible light optics, as well as those engaged in studying the optical properties of condensed matter.[1-3] In the X-ray regime, however, such abilities have been very limited because of the weak interaction of X rays with materials, especially for production and analysis of circularly polarized X-ray beams. The situation has changed significantly in recent years. The growing interest in studying magnetic and anisotropic electronic materials by X-ray scattering and spectroscopic techniques has initiated many new developments in both the production and the analysis of specially polarized X rays. Routinely available, high-brightness synchrotron radiation sources can now provide naturally collimated X rays that can be easily manipulated by special X-ray optics to generate energy-tunable as well as polarization-tunable X-ray beams. The recent developments in X-ray phase retarders and multiple-Bragg-beam interference have allowed complete analyses of general elliptical polarization of X rays from special optics and special insertion devices. In this article, we will review these recent advances, especially in the area of the production and detection of circular polarization.

As for any electromagnetic wave,[1-3] a general X-ray beam defined by

$$\mathbf{E}(\mathbf{r},t) = (E_\sigma \hat{\boldsymbol{\sigma}} + E_\pi e^{i\varepsilon}\hat{\boldsymbol{\pi}})e^{i(\mathbf{k}\cdot\mathbf{r}-\omega t)} \tag{1}$$

can be linearly polarized if $\varepsilon = 0$ or $180°$, circularly polarized if $\varepsilon = \pm 90°$ and $E_\sigma = E_\pi$, and elliptically polarized for other values of ε, E_σ, and E_π. If $\varepsilon =$ random values, then the X-ray beam is unpolarized or has an unpolarized component. The polarization is in general characterized by three Stokes-Poincaré parameters (P_1, P_2, P_3):

* *Acknowledgments:* The author is grateful to B. W. Batterman, K. D. Finkelstein, S. Shastri, and D. Walko for many useful discussions and collaborations. This work is supported by the National Science Foundation through CHESS under Grant No. DMR-97-13424.

$$P_1 = \frac{I_{0°} - I_{90°}}{I_{0°} + I_{90°}} = \frac{E_\sigma^2 - E_\pi^2}{E_\sigma^2 + E_\pi^2}$$

$$P_2 = \frac{I_{45°} - I_{-45°}}{I_{45°} + I_{-45°}} = \frac{2E_\sigma E_\pi \cos \varepsilon}{E_\sigma^2 + E_\pi^2} \qquad (2)$$

$$P_3 = \frac{I_+ - I_-}{I_+ + I_-} = \frac{2E_\sigma E_\pi \sin \varepsilon}{E_\sigma^2 + E_\pi^2}$$

which represent the degrees of the 0° and 90° [σ (perpendicular) and π (parallel)] linear polarization, the ±45°-tilted linear polarization, and the left- and right-handed circular polarization, respectively.[1–3] The unpolarized portion in the beam is characterized by its fraction P_0 of the total intensity, given by $P_0 = 1 - (P_1^2 + P_2^2 + P_3^2)^{1/2}$. The unpolarized component is generally related to the incoherency in the X-ray beam, where the phase between the σ and the π wavefields is not well-defined, and can exist only in partially coherent radiation.[1]

25.2   *LINEAR POLARIZERS*

Today's high-brightness synchrotron sources provide natural, linearly polarized X rays on the orbital plane of the storage ring. The degree of linear polarization is usually better than from 90 to 95 percent. The actual degree of linear polarization is not important for most X-ray experiments, as long as it is known and unchanged during the course of an experiment. Certain types of synchrotron experiments, such as linear dichroism spectroscopy, magnetic scattering, and atomic and nuclear resonant scattering, do require or prefer a higher degree of linear polarization. One of the most demanding experiments is an optical activity measurement[4] in which the plane of polarization is defined and its rotation is measured to a high precision.

Using a 2θ = 90° Bragg reflection (Fig. 1a) from a perfect crystal such as silicon, the natural linear polarization from a synchrotron source can be further enhanced. In general, the linear polarization P_1' after a Bragg reflection in the vertical diffraction plane can be improved from the incoming polarization P_1 by a factor given by

$$P_1' = \frac{(1 + P_1) - (1 - P_1)p}{(1 + P_1) + (1 - P_1)p} \qquad (3)$$

where p is the polarization factor of the reflection: $p = \cos^2 2\theta$ for a kinematic and $p = |\cos 2\theta|$ for a dynamical Bragg reflection. Obviously, $P_1 \leq P_1' \leq 1$, depending on the actual 2θ angle used. The efficiency of a 90° Bragg reflection polarizer is very high, only limited by the effective reflectivity, which is usually better than 50 percent for a kinematic crystal and at least 90 percent for a perfect dynamical crystal.

Another type of linear polarizer for X rays is based on the principle of Borrmann transmission (Fig. 1b) in dynamical diffraction.[5] When a strong Bragg reflection [e.g., Si (220)], is excited in the Laue transmission mode, one of the σ-wave field branches has a smaller-than-usual absorption coefficient and is thus preferentially transmitted through the crystal in the forward direction. This anomalously transmitted beam is therefore highly σ-polarized, and the Laue-diffracting crystal (typically with μt ~ 10) can be used as a linear polarizer. This method has the advantage that the insertion of the polarizer does not alter the beam path, which is a common and desirable feature in visible light optics. However, because of its extremely narrow angular acceptance range (a fraction of the reflection angular width) and its intensity loss due to small but non-zero absorption, the Borrmann transmission method is seldom used in practice as a linear polarizer.

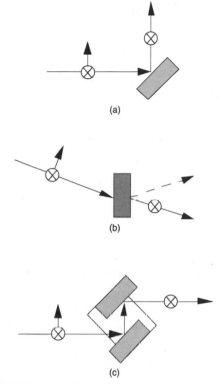

FIGURE 1 Possible linear polarizers for X rays: (*a*) 90° Bragg reflection, (*b*) Bormann transmission, and (*c*) multiple-bounce Bragg reflection with possible detuning.

The 90° Bragg reflection polarizer can be used in series to increase the purity of linear polarization even further. This can be easily achieved in practice using a channel-cut single crystal (Fig. 1*c*), in which the incident X-ray beam is diffracted multiple times between the two parallel crystal slabs. The highest degree of linear polarization (~99.99%) can be provided by a channel-cut crystal with a weak link between the two slabs.[6,7] By slightly detuning the two crystal slabs, the intensity ratio of the π-polarized to the σ-polarized X rays can be suppressed by another factor of from 10 to 100. It has another advantage that the requirement of 2θ = 90° can be further relaxed and the polarizer is effectively tunable in a wider energy range (e.g., ±20 percent) at 9 keV using Si (440).

Synchrotron experiments designed to study linear dichroism in materials sometimes require fast switching between two perpendicular linear polarization states in an incident X-ray beam. Although the switching can be done by a simple 90° rotation of a linear polarizer around the incident beam, a rapid switching can be achieved by a couple of methods. One is to use two simultaneously excited Bragg reflections whose scattering planes are perpendicular (or close to perpendicular) to each other. The other is to insert an X-ray half-wave phase retarder based on the effect of diffractive birefringence in dynamical diffraction from perfect crystals.[8] In the latter method, if one uses a transmission-type phase retarder, the incident beam direction and position are not affected by the polarization switching, which is a significant advantage.

25.3 *LINEAR POLARIZATION ANALYZERS*

In general, any linear polarizer can be used as an analyzer for linear polarizations P_1 and P_2. A simplest form of linear analyzer is to measure the 90° elastic scattering from an amorphous target[9] or a powder sample.[10] For better precisions, Bragg reflections with $2\theta = 90°$ from single crystals are usually used. Again, one can achieve very high precision by means of a multiply bounced Bragg reflection from a channel-cut perfect crystal. The requirement for $2\theta = 90°$ can be relaxed if one takes into account a scale factor involving the actual 2θ angle used in the measurement, as given in the following.

All these methods are based on measuring the intensity variation I_b as a function of the rotation angle χ of the scattering plane around the incident beam (Fig. 2) with respect to a reference polarization direction (e.g., the σ direction):

$$I_b(\chi) = \frac{1}{2}[(1+p) + (P_1 \cos 2\chi + P_2 \sin 2\chi)(1-p)] \tag{4}$$

where p is again the polarization factor of the scattering process as defined in Eq. (3). For $2\theta = 90°$, $p = 0$.

It is straightforward to show that for an arbitrary $2\theta \neq 0$, the incident linear polarizations P_1 and P_2 can be obtained by measuring two difference-over-sum ratios,

$$P_1(2\theta) = \frac{I_b(0°) - I_b(90°)}{I_b(0°) + I_b(90°)}$$

$$\tag{5}$$

$$P_2(2\theta) = \frac{I_b(45°) - I_b(-45°)}{I_b(45°) + I_b(-45°)}$$

and using a proper scale factor involving the actual 2θ,

$$\begin{bmatrix} P_1 \\ P_2 \end{bmatrix} = \frac{1+p}{1-p} \begin{bmatrix} P_1(2\theta) \\ P_2(2\theta) \end{bmatrix} \tag{6}$$

Equations (5) and (6) are useful when a single Bragg reflection is used for linear polarization analyses over a wide energy range.[11,12]

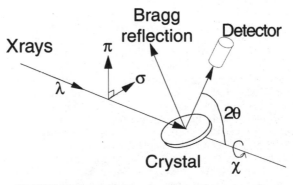

FIGURE 2 Bragg reflection from a crystal serves as a linear polarization analyzer.

25.4 PHASE PLATES FOR CIRCULAR POLARIZATION

In principle, circular polarization can be produced from linearly polarized radiation by one of the following three effects (Fig. 3): circular absorption dichroism, linear birefringence or double refraction, and reflection birefringence analogous to Fresnel rhomb. The last two effects only work with linearly polarized radiation, while the first effect can be used to convert both linear and unpolarized radiation.

Circular absorption dichroism is an effect of differential absorption between the left- and the right-handed circular polarizations. For X rays, the natural circular dichroism in materials is a very weak effect. To date the only X-ray circular dichroism that has been studied exten-

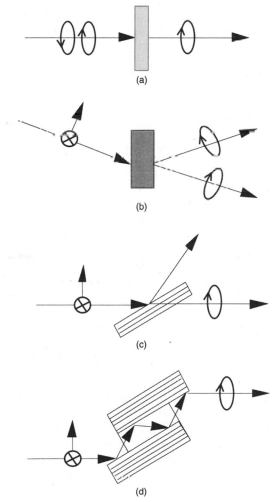

FIGURE 3   Circular phase plates for X rays based on (*a*) circular absorption dichroism, (*b*) linear birefringence in Laue geometry, (*c*) linear birefringence in Bragg transmission geometry, and (*d*) Fresnel rhomb in Bragg reflection geometry.

sively is the magnetic circular dichroism in magnetic materials. However, the largest difference experimentally observed in absorption coefficients between the left- and the right-handed circular polarization is on the order of 1 percent, which is too small for any practical use as an X-ray circular polarizer.[13]

Fresnel rhomb is based on the principle that the reflected waves from the surface of a material can have different phase shifts between two othogonal linear states, σ and π. The X-ray analog of this occurs at a Bragg reflection from a crystal and arises from the difference in the intrinsic angular widths (Darwin widths) of the σ and the π diffracted waves.[14,15] The width for the π polarization is smaller than that for the σ polarization. Because of a continous phase shift of 180° from one side of the reflection width to the other, a difference in the phase shift can be obtained between the σ and the π polarizations if the angular position is selected to be near the edge of the angular width. Experimentally, a multiply bounced Bragg reflection is needed to make a ±90° phase shift. The phase shift is independent of the crystal thickness, but this method requires a highly collimated incident beam, about $\frac{1}{10}$ of the reflection width, which is usually the limiting factor for its thoughput.[16]

The linear birefringence or double refraction effect relies on the difference in the magnitudes of the wavevectors of two orthogonal linear polarization states, σ and π, when a plane wave travels through a crystalline material. Because of this difference, a phase shift between the σ and the π traveling waves can be accumulated through the thickness of the birefringent material:[17]

$$\Delta = 2\pi(K_\sigma - K_\pi)t \qquad (7)$$

where t is the thickness, and K_σ and K_π are the magnitudes of the wavevectors inside the crystal for the σ and π wavefields, respectively. When the phase shift Δ reaches ±90°, circularly polarized radiation is generated, and such a device is usually termed a *quarter-wave phase plate* or a *quarter-wave phase retarder*.

For X rays, large birefringence effects exist near strong Bragg reflections in relatively perfect crystals. Depending on the diffraction geometry, one can have three types of transmission birefringence phase retarders: Laue transmission, Laue reflection, and Bragg transmission, as illustrated in Fig. 3b and c. The Laue reflection type[9,18] works at full excitation of a Bragg reflection, while the Laue and the Bragg transmission types[19–22] work at the tails of a Bragg reflection, which has the advantage of a relaxed angular acceptance. In the past few years, it has been demonstrated that the Bragg-transmission-type phase retarders are very practical X-ray circular polarizers. With good-quality diamond single crystals, such circular phase-retarders can tolerate a larger angular divergence and their throughputs can be as high as 0.25, with a degree of circular polarization in the range of from 95 to 99 percent. The handedness of the circular polarization can be switched easily by setting the diffracting crystal to either side of the Bragg reflection rocking curve. There have been some excellent review articles[20–22] in this area and the reader is referred to them for more details.

25.5   *CIRCULAR POLARIZATION ANALYZERS*

Circular polarization P_3 of an X-ray beam can be qualitatively detected by magnetic Compton scattering[18] and by magnetic circular dichroism. In general, these techniques are not suitable for quantitative polarization determination because these effects are relatively new and because of the uncertainties in the materials themselves and in the theories describing the effects.

Two methods have been developed in the past few years for quantitative measurements of circular polarization in the X-ray regime. One is to use a quarter-wave phase plate to turn the circular polarization into linear polarization which can then be measured using a linear polarization analyzer (Fig. 4a), as described in the previous sections. This method is entirely analo-

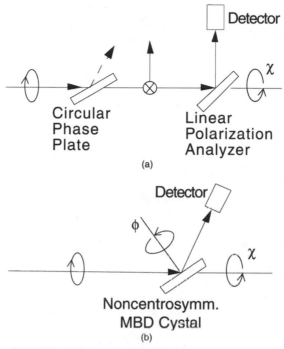

FIGURE 4 Two types of optics for analyzing circularly polarized X rays: (*a*) circular phase plate plus a linear polarization analyzer, and (*b*) multiple-beam Bragg diffraction from a noncentrosymmetric crystal.

gous to the similar techniques in visible optics and has been used by a couple of groups to characterize special insertion devices and quarter-wave phase plates.[20,23]

Multiple-beam Bragg diffraction (MBD) is the other way to measure the degree of circular polarization of X rays (Fig. 4*b*). This technique makes use of the phase shift δ between the σ and the π wave fields that arises from the phase-sensitive interference[24] and possible polarization mixing in an MBD process in a single crystal.[25] The phase shift δ is insensitive to crystal perfection and thickness as well as X-ray wavelength, since it is strictly determined by the crystal structure factors. Thus the MBD method has a broad applicable energy range and a good tolerance in the angular divergence of an X-ray beam.

Multiple-beam Bragg diffraction is also called simultaneous Bragg reflection, or *Umweganregung* (detour), as first noted by Renninger.[26] It occurs when two or more sets of atomic planes satisfy the Bragg's law simultaneously inside a crystal. A convenient way (Fig. 5) of realizing a multiple-beam reflection condition is exciting first one Bragg reflection, say $\mathbf{H}$, then rotating the diffracting crystal around $\mathbf{H}$ to bring another reciprocal node, say $\mathbf{L}$, onto the Ewald sphere.[27] The rotation around the $\mathbf{H}$ is represented by the azimuthal angle ϕ. It has been known[28–30] that the simultaneously excited Bragg waves interfere with each other and give rise to a diffracted intensity that is sensitive to the *structural phase triplet* $\delta - \arg(F_{\mathbf{L}}F_{\mathbf{H-L}}/F_{\mathbf{H}})$, where $F_{\mathbf{H}}$ is the structure factor of the main reflection $\mathbf{H}$ and $F_{\mathbf{L}}$ and $F_{\mathbf{H-L}}$ are the structure factors of the detoured reflections $\mathbf{L}$ and $\mathbf{H-L}$. Another effect that exists in MBD is polarization mixing due to the intrinsic double Thomson scattering process,[31,32] which causes a part of the π wave field amplitude to be scattered into the σ channel and vice versa. The combination of these two effects makes the MBD intensity sensitive to the phase difference between the σ and the π wave fields and thus to the circular polarization in the incident X-ray beam.

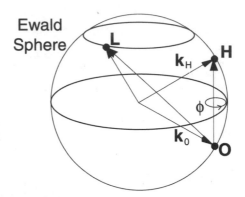

FIGURE 5   Ewald sphere construction showing a multiple-beam diffraction geometry.

For maximum sensitivity to circular polarizations, a non-centrosymmetric crystal structure can be chosen as the analyzer and an MBD reflection with $\delta = \pm 90°$ can be used. A complete determination of all Stokes-Poincaré parameters can be obtained either by measuring *three* independent MBD profiles[25,33] and solving for (P_1, P_2, P_3), or by separate measurements[11,12,34] of P_1 and P_2. This latter method provides a very convenient way for complete determination of Stokes-Poincaré parameters. Once the analyzer crystal is oriented and the proper multiple reflection is found, a complete determination of (P_1, P_2, P_3) requires four rocking curve measurements of the integrated 2-beam intensities, $I_b(0°)$, $I_b(90°)$, $I_b(-45°)$, and $I_b(45°)$, at $\chi = 0°$, $90°$, $-45°$, and $45°$, respectively, and two rocking curve measurements of the MBD intensities, $I(+\Delta\phi)$ and $I(-\Delta\phi)$, at each side of the multiple reflection peak. The (P_1, P_2, P_3) are then obtained by taking the three pairs of difference-over-sum ratios,[12] with a scale factor for P_3 involving the multiple-beam structure factors.

We would like to point out that for an MBD reflection to have maximum sensitivity to circular polarization, $\delta = \pm 90°$ is desired. This condition can be conveniently satisfied by choosing a noncentrosymmetric crystal such as GaAs. However, as we have previously shown,[32] because of the *dynamical* phase shift *within* the full excitation of a multiple reflection, an MBD in a centrosymmetric crystal such as Si or Ge can also be sensitive to circular polarization. The only drawback is that it requires an extremely high angular collimation. With the arrival of low-emittance undulator-type sources, one should keep in mind that the use of Si or Ge crystals for MBD circular polarimetry may be feasible for well-collimated X-ray beams.

25.6   *REFERENCES*

1. J. D. Jackson, *Classical Electrodynamics,* 2nd ed., John Wiley and Sons, New York, 1975.

2. D. J. Griffiths, *Introduction to Electrodynamics,* Prentice-Hall, Englewood Cliffs, New Jersey, 1975.

3. M. Born and E. Wolf, *Principles of Optics,* 6th ed., Pergamon, New York, 1983.

4. M. Hart and A. R. D. Rodrigues, "Optical Activity and the Faraday Effect at X-Ray Frequencies," *Phil. Mag.* **B43**:321 (1981).

5. B. W. Batterman and H. Cole, "Dynamical Diffraction of X Rays by Perfect Crystals," *Rev. Mod. Phys.* **36**:681 (1964).

6. M. Hart, "X-Ray Polarization Phenomena," *Phil. Mag.* **B38**:41 (1978).

7. D. P. Siddons, M. Hart, Y. Amemiya, and J. B. Hastings, "X-Ray Optical Activity and the Faraday Effect in Cobalt and Its Compounds," *Phys. Rev. Lett.* **64**:1967 (1990).

8. C. Giles, C. Malgange, J. Goulon, C. Vettier, F. de Bergivin, A. Freund, P. Elleaume, E. Dartyge, A. Fontaine, C. Giorgetti, and S. Pizzini, "X-Ray Phase Plate for Energy Dispersive and Monochromatic Experiments," *SPIE Proceedings* **2010**:136 (1993).

9. J. A. Golovchenko, B. M. Kincaid, R. A. Levesque, A. E. Meixner, and D. R. Kaplan, "Polarization Pendellosung and the Generation of Circularly Polarized X Rays with a Quarter-Wave Plate," *Phys. Rev. Lett.* **57**:202 (1986).

10. G. Materlik and P. Suortti, "Measurements of the Polarization of X Rays from a Synchrotron Source," *J. Appl. Cryst.* **17**:7 (1984).

11. Q. Shen and K. D. Finkelstein, "A Complete Characterization of X-Ray Polarization State by Combination of Single and Multiple Bragg Reflections," *Rev. Sci. Instrum.* **64**:3451 (1993).

12. Q. Shen, S. Shastri, and K. D. Finkelstein, "Stokes Polarimetry for X Rays Using Multiple-Beam Diffraction," *Rev. Sci. Instrum.* **66**:1610 (1995).

13. K. D. Finkelstein and Q. Shen, "Feasibility of Using Magnetic Circular Dichroism to Produce Circularly Polarized X Rays," unpublished.

14. O. Brummer, Ch. Eisenschmidt, and H. Hoche, "Polarization Phenomena of X Rays in the Bragg Case," *Acta Crystallogr.* **A40**:394 (1984).

15. B. W. Batterman, "X-Ray Phase Plate," *Phys. Rev.* **B45**:12677 (1992).

16. S. Shastri, K. D. Finkelstein, Q. Shen, B. W. Batterman, and D. Walko, "Undulator Test of a Bragg Reflection Elliptical Polarizer at ~7.1 keV," *Rev. Sci. Instrum.* **66**:1581 (1994).

17. V. Belyakov and V. Dmitrienko, "Polarization Phenomena in X-Ray Optics," *Sov. Phys. Uspekh.* **32**:697 (1989).

18. D. M. Mills, "Phase-Plate Performance for the Production of Circularly Polarized X Rays," *Nucl. Instrum. Meth.* **A266**:531 (1988).

19. K. Hirano, K. Izumi, T. Ishikawa, S. Annaka, and S. Kikuta, "An X-Ray Phase Plate Using Bragg Case Diffraction," *Jpn. J. Appl. Phys.* **30**:L407 (1991).

20. C. Giles, C. Malgrange, J. Goulon, F. de Bergivin, C. Vettier, A. Fontaine, E. Dartyge, S. Pizzini, F. Baudelet, and A. Freund, "Perfect Crystal and Mosaic Crystal Quarter-Wave Plates for Circular Magnetic X-Ray Dichroism Experiments," *Rev. Sci. Instrum.* **66**:1549 (1995).

21. K. Hirano, T. Ishikawa, and S. Kikuta, "Development and Application of X-Ray Phase Retarders," *Rev. Sci. Instrum.* **66**:1604 (1995).

22. Y. Hasegawa, Y. Ueji, K. Okitsu, J. M. Ablett, D. P. Siddons, and Y. Amemiya, "Tunable X-Ray Polarization Reflector with Perfect Crystals," *Acta Cryst.* **A55**:955–962 (1999).

23. T. Ishikawa, K. Hirano, and S. Kikuta, "Complete Determination of Polarization State in the Hard X-Ray Region," *J. Appl. Cryst.* **24**:982 (1991).

24. Q. Shen and K. D. Finkelstein, "Solving the Phase Problem with Multiple-Beam Diffraction and Elliptically Polarized X Rays," *Phys. Rev. Lett.* **45**:5075 (1990).

25. Q. Shen and K. D. Finkelstein, "Complete Determination of X-Ray Polarization Using Multiple-Beam Bragg Diffraction," *Phys. Rev.* **B45**:5075 (1992).

26. M. Renninger, "Umweganregung, eine bisher unbeachtete Wechselwirkungserscheinung bei Raumgitter-interferenzen," *Z. Phys.* **106**:141 (1937).

27. H. Cole, F. W. Chambers, and H. M. Dunn, "Simultaneous Diffraction: Indexing Umweganregung Peaks in Simple Cases," *Acta Crystallogr.* **15**:138 (1962).

28. M. Hart and A. R. Lang, *Phys. Rev. Lett.* **7**:120–121 (1961).

29. Q. Shen, "A New Approach to Multi-Beam X-Ray Diffraction Using Perturbation Theory of Scattering," *Acta Crystallogr.* **A42**:525 (1986).

30. Q. Shen, "Solving the Phase Problem Using Reference-Beam X-Ray Diffraction," *Phys. Rev. Lett.* **80**:3268–3271 (1998).

31. Q. Shen, "Polarization State Mixing in Multiple-Beam Diffraction and Its Application to Solving the Phase Problem," *SPIE Proceedings* **1550**:27 (1991).

32. Q. Shen, "Effects of a General X-Ray Polarization in Multiple-Beam Bragg Diffraction," *Acta Crystallogr.* **A49**:605 (1993).

33. K. Hirano, T. Mori, A. Iida, R. Colella, S. Sasaki, and Q. Shen, "Determination of the Stokes-Poincaré Parameters for a Synchrotron X-Ray Beam by Multiple Bragg Scattering," *Jpn. J. Appl. Phys.* **35**:5550–5552 (1996).

34. J. C. Lang and G. Srajer, "Bragg Transmission Phase Plates for the Production of Circularly Polarized X Rays," *Rev. Sci. Instrum.* **66**:1540–1542 (1995).

TOTAL REFLECTION OPTICS

CHAPTER 26
MIRRORS FOR SYNCHROTRON BEAMLINES

Andreas Freund

European Synchrotron Radiation Facility (ESRF)
Grenoble, France

26.1 SPECIFIC REQUIREMENTS FOR SYNCHROTRON X-RAY OPTICS

X-ray mirrors have been discussed in detail by Harvey in Chap. 11, Vol. II, of the *Handbook of Optics.*[1] In particular, the basic scattering processes of X rays by surfaces, the effects of surface figure and finish on mirror performance, and the influence of aberrations on the efficiency of various focusing schemes have been described. The applications mentioned were mainly in the field of astrophysics and often in the soft X-ray range. The present section will therefore be limited to facts that are specific to synchrotron radiation applications in the medium and hard X-ray range (i.e., for energies $E > 1$ keV). Recent more general reviews of synchrotron X-ray beam optics have been given by Hart,[2] Freund,[3] and Hart and Berman.[4] More detailed papers on the subject can be found in Vol. 34 of *Optical Engineering*[5] and in recent *Proceedings of SPIE Conferences,* for example, Vol. 3773.[6]

Specular reflection of X rays is based on the refractive index whose real part of the decrement, δ, is of the order of 10^{-5} for X rays, so that the so-called *critical angle* $\theta_c = (2\delta)^{1/2}$ below which total reflection is observed, is very small, of the order of 3 mrad for 20 keV radiation and a heavy metal such as platinum or palladium. The evanescent wave penetrates only a few nm and nearly 100 percent reflectivity is theoretically obtained for ideal surfaces. Mirrors are often the first optical element in a synchrotron beamline, positioned at a distance of typically from 20 to 30 m from the source. The beam height there is about 1 mm and then, depending on energy or glancing angle, the mirror length may exceed 1 m.

Mirrors on synchrotron X-ray beamlines have four main functions. First, they are efficient filters cutting off photons with energies above the critical energy corresponding to the critical angle given above. They thus eliminate higher order contamination of otherwise monochromatic beams provided by single crystal monochromators. Second, when used as the first optical element upstream of the monochromator, a mirror decreases the heat load carried by the absorbed high energy photons, typically by a factor of 2. Whereas these two functions are related to waste photon management, the two other functions affect beam divergence: mirror surfaces can be curved by machining or bending and then either collimate or focus the X rays. Classical schemes are

1. A toroidal mirror directly upstream of the monochromator and double-focusing the beam onto the sample position

2. A vertically collimating mirror upstream and a (re)focusing second mirror downstream of the monochromator to improve the energy resolution due to a decreased angular spread of the X-ray beam entering the monochromator

The beam conditioning by mirrors can be well predicted by calculations such as XOP[7] and by ray tracing codes such as SHADOW.[8]

Modern, so-called "third-generation" storage ring X-ray sources such as wigglers and undulators, are characterized by low emittance (i.e.), small source size and beam divergence, and high spectral brightness or brilliance. The source dimensions and opening angles are quite anisotropic, typically 1 mm wide × 0.02 mm high and 1 mrad horizontally and 0.02 mrad vertically. The stability of these parameters is about 10 percent of their absolute values which imposes severe tolerances on the optics. Along with the high brightness comes a very high heat load, up to several kW total power and 100 W mm^{-2} power density normal to the beam.[9] This means that mirrors in the white beam must be cooled, which implies thermal deformation that has to be minimized by appropriate cooling schemes and other techniques.

26.2 *MIRROR SUBSTRATE QUALITY*

A very common way to define the quality of a surface is the power spectral density function (PSD) that is a measure of the surface height variation as a function of the spatial frequency or the reciprocal of the corrugation length. The surface figure described by the slope error (SLE) or the shape error (SHE) corresponds to low frequencies, typically one inverse centimeter, and that described by the finish or microroughness (MR) corresponds to high frequencies, down to the inverse nanometer range. To avoid emittance and brilliance deterioration and excessive blur of the focus, the former should not exceed a fraction of the beam divergence, say, 1 μrad (rms value). The latter creates a halo around the beam or the focus by diffuse scattering described by a static Debye-Waller factor and thereby reduces the peak reflectivity. The MR should be of the same order as the X-ray wavelength [i.e., 0.1 nm (rms value)]. Because of the small glancing angle, only the meridional figure is of importance for X rays. The sagittal slope error can be much bigger. The basic X-ray scattering processes by real surfaces are quite well understood.[10–12] They can be incorporated in ray tracing codes.

The tolerances for mirror surfaces being very tight, the materials choice is much restricted. Machining and polishing capabilities are not the only criteria, but mechanical rigidity and dimensional stability are important as well. Radiation hardness, high thermal conductivity and low thermal expansion are crucial parameters for use in white beams; finally, cost must be considered too.[13,14] Although SiC shows slightly better performance, nowadays most mirrors for synchrotron hard X rays are made of single-crystal silicon that can be obtained as rods up to 1.3 m long and 0.15 m in diameter at a quite reasonable price. For monochromatic beams, also fused silica and Zerodur (a glass ceramic) are excellent substrate materials. Metals such as Glidcop, Invar, and stainless steel are in use too. Their major advantages are excellent brazing and machining capabilities (the latter by EDM) that are particularly helpful when integrating cooling loops or flexure hinge structures for bending purposes into the substrate itself.[15] The substrates made of Si or other materials are coated by thin layers of heavy metals such as Au, Pt, or Pd without degrading the surface.

The present state of the art of achievable surface (standard) quality is (rms values) as follows:

1. SLE: 2 μrad; SHE: 10 nm; MR: 0.2 nm for flat and spherical mirrors up to 1.2 m long;

2. SLE: 1 μrad; SHE: 5 nm; MR: 0.1 nm for flat mirrors up to 0.3 m long.[16]

Strong efforts are presently being made to further improve on these figures by developing new techniques.[17] Recently, an important step forward has been achieved using the ion beam machining technique and a deterministic, computer controlled process.[18] The surface topology of a 0.3 m long Si substrate of standard quality was recorded by visible-light and X-ray metrology. Based on this map the substrate was exposed to the ion beam figuring process. After a single iteration an almost 10-fold decrease of the shape and slope errors to 1.2 nm and 0.11 μrad were observed with X-ray metrology, without any increase of microroughness at the 0.1 nm level. The final aim is to reach both shape error and microroughness of about 0.1 nm.

The latter improvement is important for the conservation of the coherent flux that becomes increasingly important for applications such as speckle spectroscopy and phase tomography where until now the use of mirrors has been prohibited. A useful number for the shape error can be obtained from Maréchal's criterion[19] requiring exactly the above 1.2 nm shape error to get at least 80 percent of the intensity into the center of the image of a point source at a glancing angle of 3 mrad. Better surface quality will also be beneficial for microfocusing.

26.3 METROLOGY

Surface metrology is of crucial importance for the improvement of mirror performance: quality increase is only possible if it can be measured. Clearly the most appropriate metrology is based on evaluation with X rays, but mirror manufacturers need techniques that are independent of synchrotron beamlines. Several instruments based on visible light interferometry are commercially available. The long-trace profiler (LTP)[20,21] can measure slope errors with a precision of fractions of a μrad, and microroughness can be determined with sub-nm accuracy.[22] There are several other very useful optical devices such as the Shack-Hartmann analyzer that can even be used for in situ observation of the thermal deformation of a mirror under heat load[23] and for optimizing mirror alignment and focusing geometry on the beamline.

The ultimate technique is X-ray metrology with the same type of radiation for which the mirrors are intended. Two methods are applied for assessing the slope or shape error, which are still being developed:

1. Reconstruction of the surface shape after recording speckle patterns in a monochromatic beam (*coherent X-ray metrology*).[24,25]

2. Reconstruction of the surface height variation by scanning a broad-band X-ray pencil beam across the surface, with image processing of the scattered intensity recorded by a position-sensitive detector (*incoherent X-ray metrology*).[26]

The microroughness can be derived in a straightforward way from specular reflection profiles.[27,28] It is thus possible to obtain the whole PSD from a full X-ray examination. Experience has shown that in most cases the integrated MR parameter measured with visible light is up to 3 times smaller than the MR obtained by X-ray scattering, because the two methods cover a different range of the spatial frequencies.[29] This empirical fact must be taken into account in mirror specification. Also, atomic force microscopy (AFM) has been applied that shows relatively good agreement with X-ray results.

26.4 THE HEAT LOAD PROBLEM

The heat load on mirrors can represent a major problem even though the angle of incidence is small and the heat is spread out over a certain length. The main difficulty is to keep the mirror straight when it needs to be cooled. This can be achieved to some extent by efficient cooling through internal channels,[30] but for very high heat loads of up to 15 kW total power[31] it might be necessary to apply active-adaptive techniques that are complex and thus not very straightforward to utilize.[32] In many cases, however, side-cooling is still satisfactory. The mirror body has to be more than 10 cm thick and the heat is extracted by copper blocks that are attached along both sides close to the surface. Good thermal contact is ensured by a liquid (In 75 percent–Ga 25 percent) eutectic metal interface.

Efficient cooling is a whole chapter in itself and its detailed description is beyond the scope of this section. Regular conferences and workshops are devoted to this subject and interested readers are referred to the literature given here and to the publications cited therein. A related matter is mirror surface damage due to hydrocarbon cracking and carbon deposit on the highly polished and thus delicate surface. To avoid problems it is mandatory that mirrors are kept under ultrahigh vacuum.

26.5   FOCUSING WITH MIRRORS

This section would not be complete without mentioning the recent progress achieved in microfocusing X-ray beams. Whenever a wide bandpass, relative ease in alignment, and high focusing efficiency (gain factors) are required, mirrors are desirable. Many types of mirror benders have been developed and are frequently applied on beamlines.[14,33,34] Aberrations and other issues related to imaging have been extensively discussed.[35] The most widely used type of focusing mirror is a toroid that is in most cases obtained by elastic bending of a cylindrically shaped mirror. By applying different bending moments to the two ends of a beam, shapes other than circular can be obtained, such as elliptical and parabolic.

The capability to create aspherical shapes with a very high precision is of vital importance for the strong demagnification required for the microfocusing methods applied to microdiffraction or microfluorescence that are techniques of major industrial potential. Three following methods are presently in use and under further development to achieve ultimate performance:

1. Bending a very flat plate whose lateral profile is shaped to achieve the best matching to the ideal (elliptical) shape.[36,37]

2. Grooving an ellipse into a flat substrate by conventional machining followed by ion-beam figuring.[38]

3. Grooving an approximate elliptical shape into a flat substrate followed by differential deposition of a metal to correct residual shape errors.[39]

All three approaches seem very promising. Because the synchrotron source is very anisotropic, the most adapted double-focusing scheme is the well-known Kirkpatrick-Baez optics. Presently, submicron spot sizes and gain factors exceeding 10^4 are obtained, and in a few years the 100 nm limit should be reached.

26.6   REFERENCES

1. J. E. Harvey, "X-Ray Optics," chap. 11, in M. Bass, E. W. Van Stryland, D. R. Williams and W. L. Wolfe (eds.), *Handbook of Optics*, McGraw-Hill, New York, 1995, pp. 11.1–11.34.

2. M. Hart, "X-Ray Optical Beamline Design Principles," in A. Authier, S. Lagomarsino, and B. K. Tanner (eds.), *X-Ray and Neutron Dynamical Diffraction*, Plenum, New York, 1996, 73–90.

3. A. K. Freund, "Synchrotron X-Ray Beam Optics," in A. Furrer (ed.) *Complementarity Between Neutron and Synchrotron X-Ray Scattering*, World Scientific, Singapore, 1998, pp. 329–349.

4. M. Hart and L. E. Berman, "X-Ray Optics for Synchrotron Radiation; Perfect Crystals, Mirrors and Multilayers," *Acta Cryst.* **A54**:850–858 (1998).

5. A. Khounsary (ed.), "High Heat Flux Optical Engineering," *Opt. Eng.* **34**:312–452 (1995).

6. A. Khounsary, A. K. Freund, T. Ishikawa, G. Srajer, and J. Lang (eds.), "X-Ray Optics Design, Performance, and Applications," *Proceedings SPIE* **3773** (1999).

7. M. Sánchez del Río and R. J. Dejus, "XOP: A Multiplatform Graphical User Interface for Synchrotron Radiation Spectral and Optics Calculations," *Proceedings SPIE* **3152**:148–157 (1997).

8. M. Sánchez del Río and A. Marcelli, "Waviness Effects in Ray-Tracing of 'Real' Optical Surfaces," *Nucl. Instr. Meth.* **A319**:170–177 (1992).

9. K. J. Kim, "Optical and Power Characteristics of Synchrotron Radiation Sources," *Opt. Eng.* **34**:342–352 (1995).

10. L. Nevot and P. Croce, "Caractérisation des surfaces par réflexion rasante de rayons X. Application à l'étude du polissage de quelques verres silicates," *Rev. Phys. Appl.* **15**:761–779 (1980).

11. S. K. Sinha, E. B. Sirota, S. Garoff, and H. B. Stanley, "X-Ray and Neutron Scattering from Rough Surfaces," *Phys. Rev.* **B38**:2297–2311 (1988).

12. E. L. Church and P. Z. Takacs, "Specification of Surface Figure and Finish in Terms of System Performance," *Appl. Opt.* **32**:3344–3353 (1993).

13. M. R. Howells and R. A. Paquin, "Optical Substrate Materials for Synchrotron Radiation Beam Lines," in *Advanced Materials for Optica and Precision Structures, SPIE Critical Review* **CR67**:339–372 (1997).

14. J. Susini, "Design Parameters for Hard X-Ray Mirrors: The European Synchrotron Radiation Facility Case," *Opt. Eng.* **34**:361–376 (1995).

15. M. R. Howells, "Design Strategies for Monolithic Adjustable-Radius Metal Mirrors," *Opt. Eng.* **34**:410–417 (1995).

16. O. Hignette, private communication (1999).

17. Y. Furukawa, Y. Mori, and T. Kataoka (eds.), *Precision Science and Technology for Perfect Surfaces,* The Japanese Society for Precision Engineering, Tokyo, 1999.

18. O. Hignette, E. Chinchio, A. Rommeveaux, P. Villermet, H. Handschuh, and A. K. Freund, "Towards X-Ray Coherence Preservation of Optical Surfaces," *ESRF Highlights 1999,* European Synchrotron Radiation Facility, Grenoble, France, 2000, pp. 101–102.

19. M. Born and E. Wolf, *Principles of Optics,* Pergamon Press, Oxford, 1993, p. 469.

20. P. Z. Takacs, K. Furenlid, R. A. DeBiasse and E. L. Church, "Surface Topography Measurements Over the 1 Meter to 10 Micrometer Spatial Period Bandwidth," *Proceedings SPIE* **1164**:203–211 (1989).

21. P. Z. Takacs, "Nanometer Precision in Large Surface Profilometry," in Y. Furukawa, Y. Mori, and T. Kataoka (eds.), *Precision Science and Technology for Perfect Surfaces,* The Japanese Society for Precision Engineering, Tokyo, 1999, pp. 301–310.

22. O. Hignette and A. Rommeveaux, "Status of the Optics Metrology at the ESRF," *Proceedings SPIE* **2856**:314–323 (1996).

23. S. Qian, W. Jark, P. Z. Takacs, K. Randall, and W. Yun, "*In-Situ* Surface Profiler for High Heat Load Mirror Measurements," *Opt. Eng.* **34**:396–402 (1995).

24. I. Schelokov, O. Hignette, C. Raven, I. Snigireva, A. Snigirev, and A. Souvorov, "X-Ray Interferometry Technique for Mirror and Multilayer Characterization," *Proceedings SPIE* **2805**:282–292 (1996).

25. A. Rommeveaux and A. Souvorov, "Flat X-Ray Mirror as Optical Element for Coherent Synchrotron Radiation Conditioning," *Proceedings SPIE* **3773**:70–77 (1999).

26. O. Hignette, A. K. Freund, and E. Chinchio, "Incoherent X-Ray Mirror Surface Metrology," *Proceedings SPIE* **3152**:188–198 (1997).

27. V. E. Asadchikov, E. E. Andreev, A. V. Vinogradov, A. Yu, Karabkov, I. V. Kozhevnikov, Yu. S. Krivonosov, A. A. Postnov, and S. I. Sagitov, "Investigations of Microroughness of Superpolished Surfaces by the X-Ray Scattering Technique," *Surface Investigations* **14**:887–900 (1999).

28. T. Salditt, T. H. Metzger, J. Peisl, B. Reinker, M. Moske, and K. Samwer, "Determination of the Height-Height Correlation Function of Rough Surfaces from Diffuse X-Ray Scattering," *Europhys. Lett.* **32**:331–336 (1995).

29. O. Hignette, C. Morawe, and A. Rommeveaux, *ESRF* (1999), unpublished.

30. T. Tonnessen, A. Khounsary, W. Yun, and D. Shu, "Design and Fabrication of a 1.2 M Long Internally Cooled Silicon X-Ray Mirror," *Proceedings SPIE* **2855**:187–198 (1996).

31. A. Khounsary, "Thermal Management of Next-Generation Contact-Cooled Synchrotron X-Ray Mirrors," *Proceedings SPIE* **3773**:78–87 (1999).

32. J. Susini, R. Baker, M. Krumrey, W. Schwegle, and Å. Kvick, "Adaptive X-Ray Mirror Prototype: First Results," *Rev. Sci. Instr.* **66**:2048–2052 (1995).

33. L. Zhang, R. Hustache, O. Hignette, E. Ziegler, and A. K. Freund, "Design and Optimization of a Flexural Hinge-Based Bender for X-Ray Optics," *J. Synchrotron Rad.* **5**:804–807 (1998).

34. J. Susini, D. Pauschinger, R. Geyl, J. J. Fermé, and G. Vieux, "Hard X-Ray Mirror Fabrication Capabilities in Europe," *Opt. Eng.* **34**:388–395 (1995).

35. M. R. Howells, "Mirrors for Synchrotron Radiation Beamlines," Report LBL-34750, Lawrence Berkeley National Laboratory, Berkeley, CA, 1993.

36. P. J. Eng, M. Newville, M. L. Rivers, and S. Sutton, "Dynamically Figured Kirkpatrick-Baez X-Ray Micro-Focusing Optics," to be published.

37. O. Hignette, R. Hustache, G. Rostaing, and A. Rommeveaux, "Micro-Focusing of X-Rays by Elastically Bent Mirrors," in preparation.

38. A. Schindler, T. Haensel, D. Flamm, A. Nickel, H. J. Thomas, and F. Bigl, "Nanometer Precision (Reactive) Ion Beam Figuring of (Aspherical) Optical Surfaces," in Y. Furukawa, Y. Mori, and T. Kataoka (eds.), *Precision Science and Technology for Perfect Surfaces,* The Japanese Society for Precision Engineering, Tokyo, 1999, pp. 243–248.

39. G. E. Ice, J. S. Chung, J. Tischler, A. Lunt, and L. Assoufid, "Elliptical X-Ray Microprobe Mirrors by Differential Deposition," submitted to *Rev. Sci. Instr.* (1999).

CHAPTER 27
THE SCHWARZSCHILD OBJECTIVE

F. Cerrina
Department of Electrical and Computer Engineering
University of Wisconsin, Madison
Madison, Wisconsin

27.1 INTRODUCTION

The Schwarzschild objective is based on the use of two almost concentric spherical surfaces: a small convex mirror and a larger concave facing each other as shown in Fig. 1.

While this design is widely attributed to Schwarzschild, previous descriptions of the optical system had already been analyzed and published by Paul and Chretien. The design is simple and elegant, and well suited for optical systems with small field of view and high resolution. The Schwarzschild objective is an evolution of the Cassegrain telescope, where the primary and secondary are both nonspherical elements, providing good aberration correction and large field of view. Aspherical optics are, however, difficult and expensive to manufacture, and simpler designs are desirable. The Schwarzschild objective replaces the aspheres with spherical elements, as shown in Fig. 1.

The Schwarzschild objective has found its primary use in microscopy and astronomy at wavelengths where glass lenses are not suitable or where a truly achromatic optical is needed, like in microspectroscopy systems. The applications are thus mainly in the Infrared (FTIR microscopes) and the UV; recently, the Schwarzschild objective has been extended to the Extreme UV region ($\approx$13 nm) for microscopy[1,4–7] and for the development of advanced lithography.[3] Some of the first X-ray images of the sky were also acquired using Schwarzschild objectives.[4]

In the Schwarzschild objective each spherical surface forms an aberrated image, and it is a simple exercise in third-order expansion to show that the aberrations can be made to compensate each other. This is because the surfaces have curvatures of different sign, and hence each aberration term is of the opposite sign. All the third-order Seidel aberrations are exactly zero on axis, and only some fifth-order coma is present. However, the focal plane of the Schwarzschild objective is not flat; because of the spherical symmetry, the focal plane forms a spherical surface. Thus, the Schwarzschild objective has excellent on-axis imaging properties, and these extend to a reasonably large field. It is ideally suited for scanning systems, where the field of view is very narrow.

The excellent imaging properties of the Schwarzschild objective are demonstrated by an analysis of the objective's modulation transfer function (MTF). In Fig. 2 one can see the excellent imaging property of the Schwarzschild objective, essentially diffraction limited, as well as the effect of central obscuration (increase of side wings). As shown in Fig. 3 the curve extends to the diffraction limit and remains very close to the ideal (diffraction limited) MTF curve

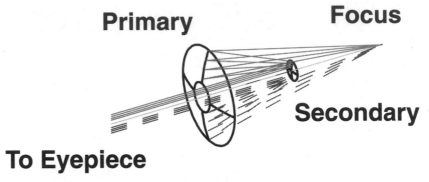

FIGURE 1   Layout of a Schwarzschild objective used as a microscope objective.

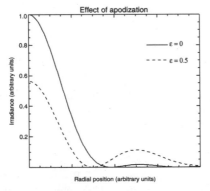

FIGURE 2   Image of a Schwarzschild objective used as a microscope objective at a wavelength of 13 nm. ε refers to the amount of obstruction.

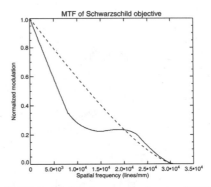

FIGURE 3   Modulation transfer function (MTF) of the previous Schwarzschild objective used as a microscope objective. Notice that in this figure the dotted line corresponds to $\varepsilon = 0$ (no obstruction).

even at wavelengths as short as 13 nm. However, in the midspatial frequency region, the MTF deviates considerably from the ideal. This behavior is typical of optical systems with central obstruction, and in general does not affect the ultimate resolution of the objective.

27.2   APPLICATIONS TO X-RAY DOMAIN

The high resolution of the microscope is achieved through the use of near normal-incidence optics; off-axis systems of comparable speed have unacceptable aberrations. This poses a big problem in the soft X-ray (EUV) region, where normal-incidence reflectivity is essentially negligible. The development of interference multilayer coatings for the soft X-ray region [now often called *Extreme UV* (EUV)] has in part solved this problem and made possible sophisticated optics.[4] (Also, see Chap. 24 of this handbook.) The combination of Schwarzschild objective and multilayer coatings, in its simplicity, is a very appealing design for the EUV region. Because the difference in optical path is so small, it is possible to achieve diffraction-limited performance even at wavelengths of 13 nm with good overall transmission.

At wavelengths longer than about $2d = 40$ Å (300 eV) it is possible to use interference filters to increase the reflectivity of surfaces in the soft X rays. These filters, often called "multilayers," are, in effect, synthetic Bragg crystals formed by alternating layers with large optical contrast.[4] From another point of view, they are a $\lambda/4$ stack in the very short X-ray region. The possibility of using near normal optical surfaces clearly simplifies the design and allows greater freedom to the optical designer. Furthermore, the relatively high reflectivity of the multilayers (up to 60 percent) allows the design of multiple surface optics.

Several microscopes have been built at synchrotron facilities, and are operated successfully.[5] Nonsynchrotron sources do not have enough brightness to deliver the flux required for

FIGURE 4   Mirrors forming a Schwarzschild objective during assembly.

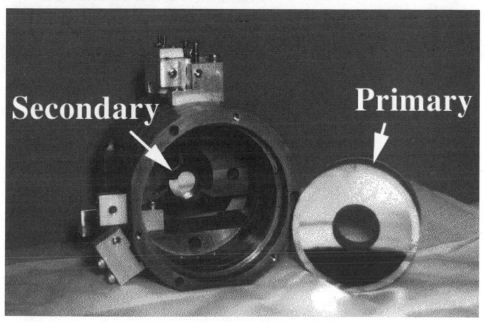

FIGURE 5   Schwarzschild objective in its mount with in situ adjustments.

practical experiments. The best resolution achieved with a Schwarzschild objective is of the order of 90 nm at the MAXIMUM microscope[2] at the Advanced Light Source. For a numerical aperture of 0.2 at the sample, the diffraction-limited resolution is approximately given by the Rayleigh criterion; at a wavelength of 13 nm, we have $\delta = \lambda/2 \, NA = 32$ nm. However, one is limited by the flux (i.e., by the finite brightness of the source) and by mounting and surface errors.[5] Diffraction limit operation has not yet been demonstrated.

High resolution is particularly easy to achieve if only a small imaging field is required as in the case of a scanning microscope. The requirement of UHV conditions for the surface physics experiments forced the design of in situ alignment systems, by including piezodriven actuators in the mirror holders. During the alignment procedure, the use of an at-wavelength knife edge test made it possible to reach the ultimate resolution of approximately 900 Å.[6] Excellent reflectivities were achieved by using Mo-Si multilayers for the region below 100 eV, and Ru-B$_4$C for the region around 135 eV.

An example of a Schwarzschild objective is shown in Fig. 4. The two mirrors are coated with a Mo-Si multilayer for 92 eV operation. Notice the groove on the large concave mirror for stress-free clamping. These mirrors are then mounted in the casing shown in Fig. 5, which includes piezoelectric adjustments suitable for ultrahigh vacuum operation. Finally, Fig. 6 shows some images acquired using this objective in a scanning X-ray microscope.[2,7]

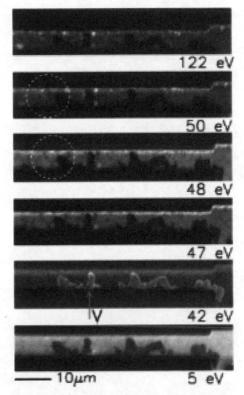

FIGURE 6 Images acquired from an Al-Cu interconnect wire in an integrated circuit after failure using the Schwarzschild objective as part of a scanning photoemission microscope.[7] The images correspond to different electron kinetic energy and show chemical contrast.

27.3 REFERENCES

1. I. Lovas, W. Santy, E. Spiller, R. Tibbetts, and J. Wilczynski, *SPIE High Resolution X-Ray Optics* **316**:90–97 (1981).

2. W. Ng, A. K. Ray-Chaudhuri, S. Liang, S. Singh, H. Solak, F. Cerrina, G. Margaritondo, L. Brillson, A. Franciosi, J. H. Underwood, J. B. Kortright, and R. C. C. Perera, *Synchr. Rad. News* **7**:25–29 (Mar/Apr 1994).

3. C. W. Gwyn, R. Stulen, D. Sweeney, and D. Attwood, *Journ. Vac. Sci. and Techn.* **16**:3142–3149 (1998).

4. D. Attwood, *Phys. Today* **45**:24 (1992).

5. F. Cerrina, *J. Electr. Spectr. and Rel. Phenom.* **76**:9–19 (1995).

6. A. K. Ray-Chaudhuri, W. Ng, S. Liang, and F. Cerrina, *Nucl. Instr. and Methods in Physics* **A347**:364–371 (1992).

7. H. H. Solak, G. F. Lorusso, S. Singh-Gasson, and F. Cerrina, *Appl. Phys. Lett.* **74**(1):22 (Jan. 1999).

CHAPTER 28
ASTRONOMICAL X-RAY OPTICS

Marshall K. Joy
Dept. of Space Sciences, NASA/Marshall Space Flight Center
Huntsville, Alabama

28.1 INTRODUCTION

Over the past two decades, grazing incidence optics have transformed observational X-ray astronomy into a major scientific discipline at the cutting edge of research in astrophysics and cosmology. This review summarizes the design principles of grazing incidence optics for astronomical applications, describes the capabilities of the current generation of X-ray telescopes, and explores several avenues of future development.

The first detection of a cosmic X-ray source outside of our solar system was made during a brief rocket flight less than 40 years ago (Giacconi et al., 1962). Numerous suborbital and Spacelab experiments were deployed over the next 15 years to follow up on this exciting discovery (Giacconi et al., 1981), and in 1978 the first general purpose X-ray telescope (the Einstein Observatory) was placed into orbit, followed by EXOSAT, the Roentgen satellite (ROSAT), the Advanced Satellite for Cosmology and Astrophysics (ASCA), and very recently, the Chandra X-ray Observatory and the X-ray Multimirror Mission (XMM). All of these observatories employ grazing incidence X-ray optics of the type investigated by Wolter (1952). The design of Wolter optics and the technologies for fabricating them are presented in Sec. 28.2 following. Kirkpatrick-Baez optics are described in Sec. 28.3, and the development of very high angular resolution X-ray optics is discussed in Sec. 28.4.

28.2 WOLTER X-RAY OPTICS

Optical Design and Angular Resolution

Wolter optics are formed by grazing-incidence reflections off of two concentric conic sections (a paraboloid and hyperboloid, or a paraboloid and ellipsoid). The most common case (Wolter type I) is conceptually similar to the familiar Cassegrain optical telescope: the incoming parallel beam of X-rays first reflects from the parabolic section and then from the hyperbolic section, forming an image at the focus (Fig. 1). To increase the collecting area, reflecting shells of different diameters are nested, with a common focal plane (Fig. 2).

A Wolter-I optic can be described by four quantities:

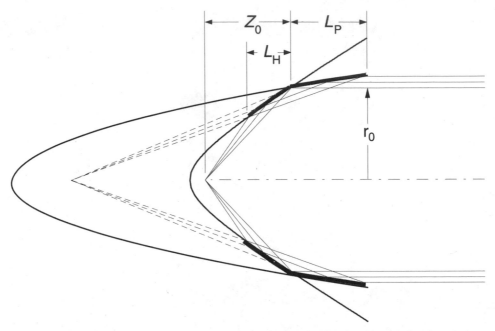

FIGURE 1   Geometry of a Wolter type I X-ray optic. Parallel light incident from the right is reflected at grazing incidence on the interior surfaces of the parabolic and hyperbolic sections; the image plane is at the focus of the hyperboloid.

1. The focal length, Z_0 (defined as the distance from the intersection of the paraboloid and hyperboloid to the focal point).

2. The mean grazing angle, α, which is defined in terms of the radius of the optic at the intersection plane, r_0:

$$\alpha \equiv \frac{1}{4} \arctan \frac{r_0}{Z_0} \tag{1}$$

3. The ratio of the grazing angles of the paraboloid and the hyperboloid, ξ, measured at the intersection point.

4. The length of the paraboloid, L_p.

   Wolter optics produce a curved image plane, and have aberrations that can cause the angular resolution of the optic to be significantly worsened for X-ray sources that are displaced from the optical axis of the telescope; for these reasons, designs are usually optimized using detailed ray-trace simulations. However, a good approximation to the optimum design can be readily obtained using the results of Van Speybroeck et al. (1972). The highest X-ray energy that the optic must transmit largely determines the mean grazing angle (see Sec. 28.3), which in turn constrains the focal ratio of the optic [Eq. (1)]. The grazing angles on the parabolic and hyperbolic sections are usually comparable, so $\xi \approx 1$. With the diameter and length of the optic as free parameters, the curves in Van Speybroeck et al. can be used to estimate the angular resolution and the collecting area for different designs. Very high resolution and good off-axis performance are possible with Wolter-I optics: an image with subarcsecond angular resolution from the Chandra X-ray Observatory is shown in Fig. 3, and Fig. 4 summarizes the image quality across the field of view of the detector.

FIGURE 2   An X-ray optics module for the XMM observatory. Fifty-eight electroformed nickel Wolter-I optics are nested to increase the effective X-ray collecting area. (*Photo courtesy of ESA.*)

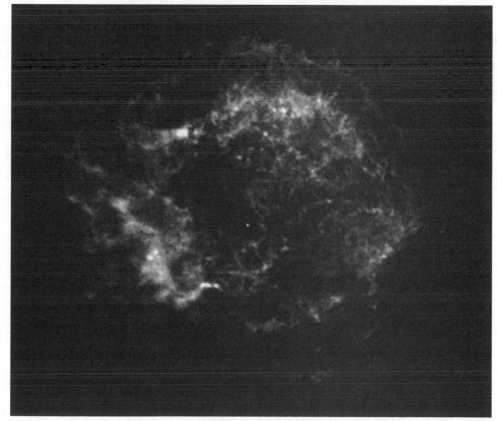

FIGURE 3   High resolution X-ray image of the Cassiopeia A (Cas A) supernova remnant. This 5000-s image was made with the Advanced CCD Imaging Spectrometer (ACIS) on the Chandra X-ray Observatory; the field of view is 6.3 arcmin. (*Photograph courtesy of the Chandra X-ray Observatory Center and NASA.*)

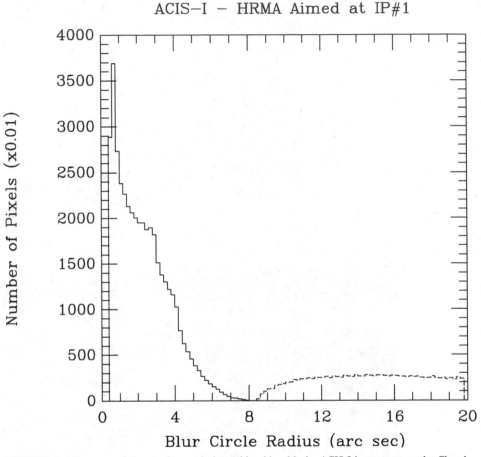

FIGURE 4 Histogram of the angular resolution achievable with the ACIS-I instrument on the Chandra X-ray Observatory. (*Chandra X-ray Observatory Guide, 2000.*)

Mirror Figure and Surface Roughness

Irregularities in the mirror surface will cause light to be scattered out of the core of the X-ray image, degrading the angular resolution of the telescope. If an incoming X ray strikes an area of the mirror surface that is displaced from the ideal height by an amount σ, the resulting optical path difference (OPD) is given by

$$\text{OPD} = 2\sigma \sin \alpha \tag{2}$$

and the corresponding phase difference is

$$\Delta = \frac{4\pi\sigma \sin \alpha}{\lambda} \tag{3}$$

where α is the grazing angle and λ is the X-ray wavelength. For a uniformly rough mirror surface with a Gaussian height distribution, rms values of σ and Δ can be used to calculate the scattered intensity relative to the total intensity (Aschenbach, 1985; Beckmann and Spizzichino, 1963):

$$\frac{I_s}{I_0} = 1 - e^{-\Delta_{rms}^2} \tag{4}$$

This result implies that high quality X-ray reflectors must have exceptionally smooth surfaces: in order for the scattered intensity I_S to be small, Δ_{rms} must be $\ll 1$. For a grazing angle of 0.5° and a wavelength of 1 Å, a surface with an rms microroughness of 4 Å will scatter ~20 percent of the incident intensity. A surface roughness of 8 Å will scatter more than 50 percent of the incident intensity at this X-ray wavelength and grazing angle.

To calculate the effect of surface imperfections on the X-ray point spread function in detail, it is necessary to measure the distribution of the deviations on all spatial scales of the optic, from the microroughness on submicron scales to the overall slope error of the full mirror. The surface deviations are characterized by the power spectral density, and the resulting X-ray scattering can be calculated using the methods described by Church (1979), Aschenbach (1985), O'Dell et al. (1993), and Hughes et al. (1993).

X-Ray Reflectivity

For most astronomical applications, X-ray telescopes are required to have high reflection efficiency across the operational energy band, and also to function at the highest possible energy for a given focal length and diameter. The overall reflection efficiency and the high energy response can both be increased by applying a high density coating to the polished reflecting surfaces of the optics, such as nickel, gold, platinum, or iridium [the latter element having the highest density and the best reflectivity (Elsner et al., 1991)]. As discussed above, the uniformity and smoothness of the reflective coating is critical to the imaging performance of the optic.

The reflection efficiency of an X-ray optic is strongly dependent on energy, due to the presence of atomic absorption edges and the rapid decrease in efficiency at high energies. The reflectivity can be readily calculated using the Fresnel reflection formulas expressed in terms of the complex dielectric constant of the reflecting surface, κ (Elsner et al., 1991). Dielectric constants can be easily derived from experimentally determined atomic scattering factors, f, using the relation

$$\kappa = 1 - \frac{r_e \rho N_A}{\pi A_W} \lambda^2 f \tag{5}$$

where r_e is the classical electron radius, N_A is Avogadro's number, and ρ and A_W are the density and molecular weight of the reflecting material. An extensive compilation of atomic scattering factors for elements 1 through 92 over the energy range from 50 eV to 30 keV is given by Henke et al. (1993) and in the appendix to this part, following Chap. 37.

Effective Collecting Area

The useful collecting area of a grazing incidence optic depends on several factors: the size and number of mirror shells, the projection of the grazing incidence mirrors onto the sky, blockage of the aperture by support structures, X-ray reflection efficiency, and vignetting of off-axis sources. The effective collecting area (A_e) is the convolution of all of these terms; it is highly energy dependent due to the mirror reflectivity, and falls to zero at the high energy cutoff of the optic. The on-axis effective area of several astronomical telescopes is shown in Fig. 5. The effective collecting area of the XMM observatory is notably larger by virtue of the large number of thin Wolter optics that are packed into each of the three optics modules. There is, however, con-

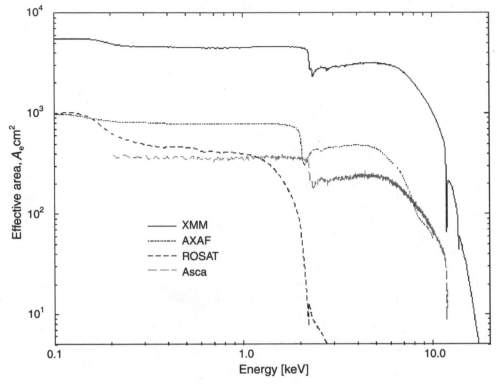

FIGURE 5   Effective on-axis collecting area of X-ray observatories as a function of energy.

siderable room for improvement: all of the current X-ray observatories have effective areas that are from 2 to 3 orders of magnitude lower than the premier ground-based optical telescopes.

Technologies for Fabricating Astronomical Optics

There are currently three primary methods for fabricating nested Wolter X-ray optics for astronomical applications. The approach that has produced the highest angular resolution involves the fabrication of Zerodur glass shells, a few cm thick, which are ground, figured, highly polished, and coated on the interior reflecting surface. Optics of this type have been built and flown on the Einstein observatory (Giacconi et al., 1979), ROSAT (Aschenbach, 1988, 1991), and the Chandra X-ray Observatory (*Chandra X-ray Observatory Guide,* 2000). Einstein and ROSAT each had ~5 arcsec angular resolution, while Chandra has subarcsecond on-axis resolution.

A different approach has been employed for the EXOSAT and XMM optics (Gondoin et al., 1998). The required parabolic and hyperbolic surfaces are machined into the outer surface of a mandrel, which is then highly polished and coated. A thin nickel shell is electroplated onto the mandrel; the shell and mandrel (which have different coefficients of thermal expansion) are then separated by cryogenic cooling. The XMM observatory has three optics modules, each containing 58 replicated Wolter-I shells (Fig. 2), which provide a high effective collecting area and 15 arcsec angular resolution (Stockman et al., 1999).

The X-ray optics fabricated for the BBXRT, ASCA, and ASTRO-E missions consist of a very large number of thin foil segments (Fig. 6) which are replicated off of a highly polished cylindrical mandrel; the foil segments are mounted so that they approximate the parabolic and hyperbolic figure of a Wolter-I optic. The resulting optics modules are compact, densely nested, and very lightweight, and have ~100 to 200 arcsec angular resolution.

28.3 KIRKPATRICK-BAEZ OPTICS

The first demonstration of grazing incidence X-ray imaging was performed by Kirkpatrick and Baez (1948) using mirrors with one-dimensional parabolic curvature (Fig. 7a). An optic of this type focuses in one plane only, so two mirrors are mounted orthogonally to each other to achieve 2-dimensional imaging (Fig. 7b); they can also be nested to increase the effective collecting area (Fig. 7c). Design studies of multi-element Kirkpatrick-Baez systems have been carried out by Van Speybroeck et al. (1971), Weisskopf (1973), and Kast (1975).

Kirkpatrick-Baez optics have not been widely used in astronomical applications because the required grazing angles are a factor of 2 larger than those of a Wolter design, for the same optic diameter and focal length. (This can be seen from Fig. 7b: an X-ray reflecting in the hor-

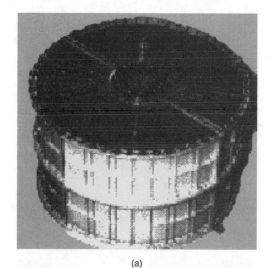

(a)

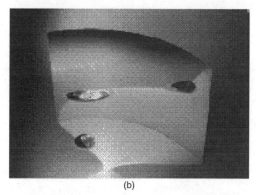

(b)

FIGURE 6   Replicated foil optics: (*a*) a full 2-reflection module, (*b*) a single replicated foil segment. (*Photo courtesy of NASA.*)

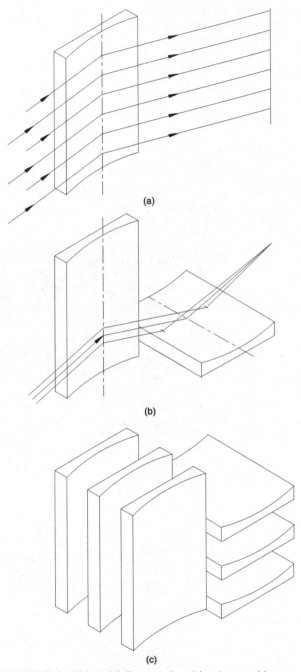

FIGURE 7 Kirkpatrick-Baez optics: (*a*) mirrors with one-dimensional parabolic curvature focus in one plane only, producing a line image, (*b*) two mirrors mounted orthogonally to each other can produce 2-dimensional imaging, (*c*) parabolic Kirkpatrick-Baez optics can be nested to increase the effective collecting area.

izontal plane receives the full angular displacement in a single reflection, while a Wolter optic achieves the same angular displacement via two half-angle reflections.) The larger grazing angles in the Kirkpatrick-Baez design significantly reduce the high-energy efficiency of the optic. However, optimal packing schemes and ease of manufacture may eventually permit Kirkpatrick-Baez designs to overcome this disadvantage; for example, mass-produced flat plates can be substituted for the curved optics in cases where the curvature of the mirrors is small (Fig. 8). In the small angle approximation the projected width of the flat plate is

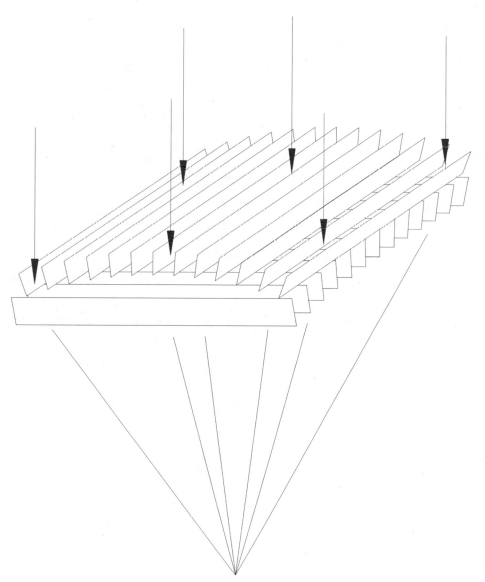

FIGURE 8 Flat plates can be substituted for the parabolic Kirkpatrick-Baez optics in cases where the curvature of the mirrors is small, and 2-dimensional imaging can be achieved by stacking two orthogonal sets of flat reflectors.

$$\delta = \frac{Lr_0}{2F} \tag{6}$$

where L is the length of the plate, r_0 is the distance of the plate from the optical axis, and F is the focal length. The limiting angular resolution due to the flat plate geometry, $\Delta\theta$, is

$$\Delta\theta(\text{arcsec}) \sim 10^5 \frac{Lr_0}{F^2} \tag{7}$$

Equation (7) indicates that the error introduced by the flat plate approximation can be made reasonably small: $\Delta\theta = 10$ arcsec for a mirror 0.1 meters in length and $r_0 = 0.1$ meters away from the axis of a 10 meter focal length optic. Analytic expressions for the number of flat plates, their spacing, and the resulting packing fraction are given by Weisskopf (1973).

28.4   *HIGH ANGULAR RESOLUTION X-RAY OPTICS*

It is difficult to construct grazing incidence X-ray optics that combine very high angular resolution and large effective area. The Chandra X-ray Observatory currently delivers subarcsecond angular resolution, a superlative achivement for an X-ray telescope, but still an order of magnitude below the resolution achieved by optical observatories, and more than three orders of magnitude below the diffraction limit. Improving conventional Wolter X-ray optics to produce milliarcsecond (or better) angular resolution is a formidable technical challenge. One alternative is to exploit an approach that is widely used in the radio and optical spectral regions: interferometry.

Several fundamental questions about the feasibility of X-ray interferometry are immediately apparent. First, interferometry requires that the phase errors be controlled to a fraction of a wavelength; is it possible to shape and control optics for X-ray wavelengths on the scale of Angstroms? Second, cosmic X-ray sources are notoriously faint (especially the cosmologically interesting sources at high redshift), and the effective collecting areas of X-ray telescopes are low; is it possible to build an X-ray interferometer that would collect a sufficient number of photons to construct a useful image in a finite amount of time?

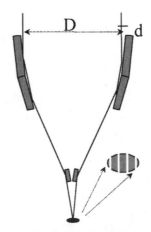

FIGURE 9 Schematic drawing of a grazing-incidence X-ray interferometer.

The scientific case for very high angular resolution X-ray imaging proves to be strong. Many of the most interesting objects in the universe (including active galactic nuclei, accretion disks, and the black holes that power them) are very small (<1 milliarcsec), very hot ($T > 10^8$ Kelvin), and radiate a large amount of thermal energy ($B_v \propto T^4$) primarily in the X-ray spectral region (Cash, 1999; Cash et al., 2000).

Figure 9 shows a schematic drawing of an X-ray interferometer, in which the beams from two X-ray telescopes are mixed to form fringes. Grazing-incidence mirrors are used for both the telescope optics and the beam combining optics.

How accurately must the grazing-incidence optics be figured and controlled in order to achieve an acceptable phase error at X-ray wavelengths? For a ½ degree grazing incidence optic, $\lambda = 2$ Å, and a $\lambda/10$ path difference, Eq. (2) indicates that the tolerance on surface deviations is large compared to the wavelength; $\sigma \approx 10$ Å. Thus, for typical grazing angles, the factor of sin α in Eq. (2) relaxes the tolerances on the mirror surface by ~2 orders of magnitude. Significantly, this brings the required X-ray optics and metrology solidly into a regime that

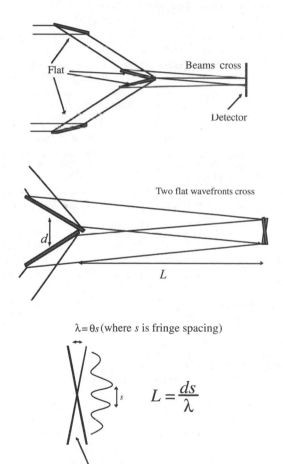

FIGURE 10   Flat-plate grazing-incidence optics and beam combiners can be used to create an X-ray interferometer; fringes are produced on the detector where the two wavefronts cross.

has been demonstrated in the laboratory, and is soon to be flight tested, for the optical Space Interferometry Mission (Space Interferometry Mission Information Document, 1996).

The practicality of an astronomical X-ray interferometer is beginning to be explored. One approach involving an array of grazing-incidence flat-plate collectors (Fig. 10) has been successfully demonstrated in the laboratory at extreme ultraviolet and X-ray wavelengths (Joy et al., 2000). Further details and other approaches are described by Shipley et al. (2000), Cash et al. (2000), and in the *Microarcsecond X-ray Imaging Mission Study Report* (2000).

28.5 REFERENCES

1. B. Aschenbach, "X-ray Telescopes," *Rep. Prog. Phys.* **48**:579–629 (1985).

2. B. Aschenbach, "Design, Construction, and Performance of the ROSAT High-Resolution X-Ray Mirror Assembly," *Applied Optics* **27**:1404 (1988).

3. B. Aschenbach, *Rev. Mod. Astron.* **4**:173 (1991).

4. P. Beckmann and A. Spizzichino, *The Scattering of Electromagnetic Waves from Rough Surfaces,* Pergamon, Oxford, (1963).

5. W. Cash, "High Resolution X-ray Imaging" in *Small Missions for Energetic Astrophysics,* S. P. Brumby (ed.), American Institute of Physics Press, 1999, pp. 44–57.

6. W. Cash, N. White, and M. Joy, "The Maxim Pathfinder Mission: X-ray Imaging at 100 Micro-Arcseconds," *Proc. SPIE* **4012**:(2000) (in press).

7. *Chandra X-ray Observatory Guide*, 2000, http://chandra.harvard.edu.

8. E. L. Church, "Role of Surface Topography in X-Ray Scattering," *Proc. SPIE* **184**:196 (1979).

9. R. F. Elsner, S. L. O'Dell, and M. C. Weisskopf, "Effective Area of the AXAF X-Ray Telescope: Dependence upon Dielectric Constants of Coating Materials," *Journal of X-Ray Science and Technology* **3**:35–44 (1991).

10. R. Giacconi, G. Branduardi, U. Briel, A. Epstein, D. Fabricant, E. Feigelson, S. S. Holt, R. H. Becker, E. A. Boldt, and P. J. Serlemitsos, "The Einstein HEAO-2 X-Ray Observatory," *Astrophysical Journal* **230**:540 (1979).

11. R. Giacconi, P. Gorenstein, S. S. Murray, E. Schreier, F. Seward, H. Tananbaum, W. H. Tucker, and L. Van Speybroeck, "The Einstein Observatory and Future X-ray telescopes" in *Telescopes for the 1980s,* Palo Alto: Annual Reviews, Inc., Palo Alto, CA, 1981, pp. 195–278.

12. B. Giacconi, H. Gursky, F. Paolini, and B. Rossi, "Evidence for X-Rays from Sources Outside the Solar System," *Physical Review Letters* **9**:435 (1962).

13. P. Gondoin, B. R. Aschenbach, M. W. Beijersbergen, R. Egger, F. A. Jansen, Y. Stockman, and J.-P. Tock, "Calibration of the First XMM Flight Mirror Module: I. Image quality," *Proc. SPIE* **3444**:278 (1998).

14. B. L. Henke, E. M. Gullikson, and J. C. Davis, "X-ray Interactions: Photoabsorption, Scattering, Transmission, and Reflection at E = 50-30000 eV, Z = 1-92," *Atomic Data and Nuclear Data Tables* **54**(2), 181–342 (1993). (This data, along with useful computational tools, is also available on the internet at http://www-cxro.lbl.gov/optical_constants/.)

15. J. P. Hughes, D. Schwartz, A. Szentgyorgyi, L. Van Speybroeck, and P. Zhao, "Surface Finish Quality of the Outer AXAF Mirror Pair Based on X-ray Measurements of the VETA-I," *Proc. SPIE* **1742**:152 (1993).

16. M. Joy, A. Shipley, W. Cash, J. Carter, D. Zissa, and M. Cuntz, "Experimental Results from a Grazing Incidence X-Ray Interferometer," *Proc. SPIE* **4012**:(2000) (in press).

17. J. W. Kast, "Scanning Kirkpatrick-Baez X-Ray Telescope to Maximize Effective Area and Eliminate Spurious Images," *Applied Optics* **14**:537 (1975).

18. P. Kirkpatrick and A. V. Baez, *J. Opt. Soc. Am.* **38**:776 (1948).

19. *Microarcsecond X-ray Imaging Mission Study Report*, Goddard Space Flight Center, Greenbelt, MD, 2000.

20. S. L. O'Dell, R. F. Elsner, J. J. Kolodziejeczak, M. C. Weisskopf, J. P. Hughes, and L. P. Van Speybroeck, "X-Ray Evidence for Particulate Contamination on the AXAF VETA-1 Mirrors," *Proc. SPIE* **1742**:171 (1993).

21. A. Shipley, W. Cash, and M. Joy, "Grazing Incidence Optics for X-ray Interferometry," *Proc. SPIE* **4012**:(2000) (in press).

22. *Space Interferometry Mission Information Document,* Jet Propulsion Laboratory, Pasadena, CA, 1996.

23. Y. Stockman, P. Barzin, H. Hansen, J. P. Tock, D. de Chambure, R. Laine, D. Kampf, M. Canali, G. Grisoni, P. Martin, and P. Radaelli, "XMM Flight Model Mirror Module Testing and Manufacturing," *Proc. SPIE* **3766**:(1999) (in press).

24. L. P. Van Speybroeck and R. C. Chase, "Design Parameters of Paraboloid-Hyperboloid Telescopes for X-ray Astronomy," *Applied Optics* **11**:440 (1972).

25. L. Van Speybroeck, R. C. Chase, and T. F. Zehnpfennig, *Applied Optics* **10**:945 (1971).

26. M. C. Weisskopf, *Applied Optics* **12**:1436 (1973).

27. H. Wolter, *Ann. Phys.* **10**:94 (1952).

CHAPTER 29

SINGLE CAPILLARIES

Donald H. Bilderback
Cornell High Energy Synchrotron Source
Cornell University
Ithaca, New York
dhb2@cornell.edu

Edward D. Franco
Advanced Research and Applications Corporation
Sunnyvale, California

29.1 BACKGROUND

Monocapillary X-ray optics can be used to increase the X-ray intensity onto the sample and control the divergence and profile of an X-ray beam. These optics collect and transport X rays of all energies up to a cutoff energy that is dependent on the capillary material and shape. The past decade has seen the rapid development of figured monocapillary optics that are designed to condense an X-ray beam or image and demagnify the X-ray source. These optics are being used as high-gain optical devices for both synchrotron radiation as well as X-ray tube sources.

29.2 DESIGN PARAMETERS

Monocapillary optics relies on total external reflection of the X rays from the internal surface of the tube to transport X rays. To keep the X rays from being absorbed in the wall of the capillary, the angle of incidence must be kept below the critical angle. Typical glass materials that have been used to fabricate capillary optics are borosilicate (Pyrex), lead-based, and silica glasses. The composition of these glasses is shown in Table 1. The critical angle, θ_c, is equal to $(2\delta)^{1/2}$, where δ is the refractive index decrement of the material at the energy of the X-ray photon. For borosilicate glass (Corning 7740) with a density of 2.23 g/cm^3, θ_c is approximately $(3.8 \times 10^{-2})/E$ radians, where E is the X-ray energy in kilo-electronvolts (keV). Lead-based glasses have a greater electron density and, consequently, can have a larger value of θ_c compared with borosilicate glasses. This advantage, however, can be offset by increased absorption and the production of interfering lead fluorescence radiation.

The simplest form of a monocapillary optic is a straight-glass tube, as shown in Fig. 1. This was first used in the 1920s when Jentzch and Nähring demonstrated that X-rays could be propagated through glass tubes with minimal losses.[1] These optics rely on the multiple reflection of

TABLE 1   Typical Glass Starting Materials

Glass code	Type	Density (g/cm³)	Softening point (°C)	Weight percent (%)							
				SiO_2	Na_2O	K_2O	CaO	MgO	PbO	B_2O_3	Al_2O_3
0080	Soda Lime	2.47	696	73.6	16	0.6	5.2	3.6	0	0	1.0
7050	Borosilicate	2.25	703	67.3	4.6	1.0	0	0.2	0	24.6	1.7
7740	Borosilicate	2.23	820	80.5	3.8	0.4	0	0	0	12.9	2.2
7900	96% Silica	2.18	1500	96.3	0.2	0.2	0	0	0	2.9	0.4
7910	99% Silica	2.18	1500	99.5	0	0	0	0	0	0	0
8870	High Lead	4.28	580	35	0	7.2	0	0	58	0	0

Source: From Corning Glass Works, Corning, NY.

X rays, at constant angle, from the inner glass surface to guide them down the length of the tube. They can be used to transport X rays from the anode of an X-ray tube to the sample surface, partially eliminating the $1/r^2$ losses that would otherwise occur. The intensity of the beam is proportional to the solid angle subtended by the capillary entrance or to $(d_i/L_1)^2$.[1] The ideal intensity gain over a pinhole at the exit is proportional to $[(L + L_1)/L_1]^2$, provided that the glancing angle at entrance of the capillary is less than or equal to θ_c. These properties have been used by several investigators to create small beams of X rays that ranged from 10 to 200 μm in diameter.[2–10]

Condensing capillary optics are figured so that the X-ray beam is compressed with each reflection, producing a gain in intensity. These optics produce a smaller, more intense beam at the expense of divergence. Stern et al.[11] were the first to quantitatively describe how X rays would propagate down a linearly tapered, or conical, optic. The design parameters for this optic are shown in Fig. 2 and are primarily driven by the following parameters:

- Divergence of the exit beam, standoff distance, and spot size
- Distance from the source, source dimensions, and source beam divergence

The acceptance angle, at a given X-ray energy, of the capillary is $2\theta_c$. Upon reflection, the angle of incidence increases by 2γ, and only those photons for which the angle of incidence is less than the critical angle will emerge from the capillary. The emerging beam has a divergence that is generally less than $2\theta_c$, and the sample must be placed close to the exit end of the capillary to preserve the small spot size. The beam is smallest at the tip and the sample should be positioned no further away than 20 to 100 times the exit diameter.[12]

The design of condensing capillary optics relies on the use of ray-tracing computer models that allow the capillary shape, figure errors, surface roughness, materials, and source parameters to be included.[13–15]

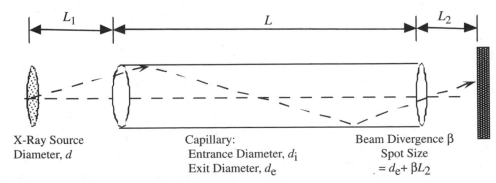

FIGURE 1   Design parameters of a straight capillary optic.

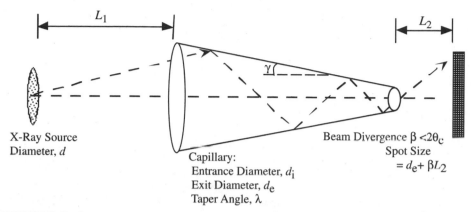

FIGURE 2 Design parameters for a condensing capillary optic.

Advanced condensing capillary designs are being investigated that include using elliptical and hybrid combinations of capillary shapes.[16–19] A simple extension is the combination of a straight-glass tube that is then rapidly tapered at the end.[18,19]

Imaging capillary optics produce an image of the X-ray source at a well-defined focus. A single hollow, tapered tube is used in a single-bounce geometry that resembles a parabolic or elliptical curve rotated about an axis of symmetry. For synchrotron applications, the source is typically located many meters away from the optic and the incident radiation has a low divergence. Figure 3 shows the design parameters for an elliptical imaging capillary. The two key parameters in the design of these optics are as follows:

1. The source demagnification that determines the spot size on the sample and the working distance
2. The divergence of the X-ray beam incident upon the sample

This form of imaging optics was first developed at the University of Melbourne in Australia, and has demonstrated the ability to produce a focused spot of 40 to 50 μm with monochromatic light, and a gain factor of 700 (the increase in flux through a small 5 μm × 5 μm aperture with the optic) with a convergence angle of approximately 6 mrad.[20,21] A typical imaging optic uses a large bore tube, with a 1-mm input diameter, an output diameter of 0.3 mm, and a length of 25 cm. The optic produces a demagnified image of the source in the image plane located 25 to 50 mm from the exit.[20–22] The real size and divergence of the X rays in the image plane depend on the size of the source, the shape of the glass tube,

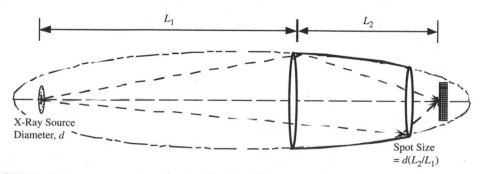

FIGURE 3 Design parameters for an imaging elliptical capillary.

and the figure errors introduced during the manufacture. Also, because of the spread in focal lengths of rays reflecting from different locations along the optic, there is no single image of the source. Instead, the single-bounce imaging optic acts like a series of extended demagnifying lenses. The rays closest to the exit tip have the strongest focusing and those closest to the input end have the weakest focusing (or smallest angular deflection).

29.3 FABRICATION

Capillary optics are typically fabricated by heating the glass tube and then pulling the glass out of the furnace in a controlled rate. The shape is varied by changing the rate at which the glass is fed into and pulled out of the furnace and by the effects of surface tension. A schematic of a typical puller is shown in Fig. 4.[22] This puller, located at the Cornell High Energy Synchrotron Source (CHESS), controls the furnace motion, glass tension, and the rate at which the glass enters and leaves the furnace. This puller has been used to fabricate condensing and imaging capillary optics that are 7 to 40 cm in length starting with tubing diameter up to 3 mm. The furnace limits the peak temperature to about 900°C, so that low-temperature glasses, such as leaded glass, borosilicate glass, soda lime glass, and the like can be pulled with the present equipment.

The fabrication of high-efficiency capillary optics requires three factors:

1. A diameter-versus-length profile that approaches an ideal taper
2. A concentric capillary bore
3. A glass that is smooth on an atomic scale

Other glass pullers have been fabricated for capillary drawing.[23,24]

A second method has been recently developed to fabricate capillary optics from metals. In this technique, the capillary optic is replicated from an ultrasmooth mandrel through sputtering, vacuum evaporation, or other coating techniques, followed by electroforming.[25] The mandrel is expendable and is formed by the precision etching of a glass or metal wire into the shape of the desired bore of the capillary optic. This technique has been successfully used to fabricate gold and copper paraboloidal imaging optics that produce a focused spot size of less than 10 µm in diameter with a collimated synchrotron radiation source. Advantages of this technique include wide latitude in the selection of materials compromising the optics, good control of the figure and straightness of the part, and high thermal conductivity.

Applications of Imaging Capillary Optics

The largest application for imaging optics is as a beam conditioner for crystallography experiments. Focusing capillaries have been used in application areas such as protein crystallography.[21,26] At CHESS, an imaging capillary optic has been used downstream of a 1.5 percent energy bandpass multilayer mirror on a bending magnet beamline to produce a 50-µm full width at half-maximum (FWHM) beam with a total X-ray flux of 6×10^{11} photons/s. Measurements that can make use of the wide bandpass and angular spread of the beam include small- and wide-angle scattering from partially ordered liquid systems such as phospholipids, proteins, synthetic, and natural fibers.[22]

Applications of Condensing Capillary Optics

Condensing capillary optics are used to produce small spots of X rays whose diameter ranges from 1 to about 25 µm with X-ray tubes and from 0.1 to 10 µm with synchrotron sources. A major application area is in determining the composition and structure of heterogeneous sam-

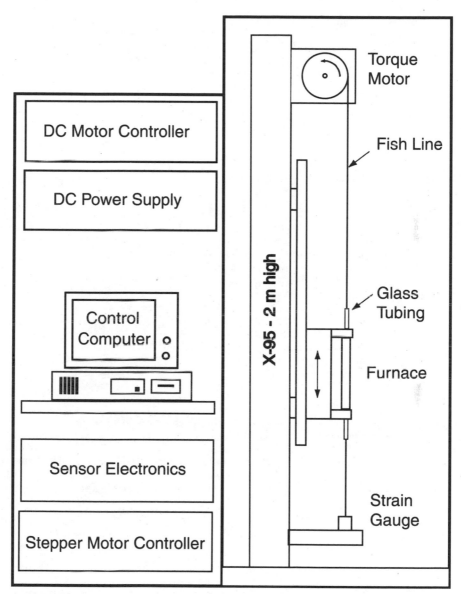

FIGURE 4 Schematic of the CHESS capillary puller. A hollow glass tube is placed in a vertical tube furnace. Fish line anchors the bottom of the glass to a strain gauge. The top is connected to a motor that can take up the drawn glass, which is kept under constant tension. The furnace is programmed to move based on the amount of glass extension to produce the appropriate glass shape (linear, parabolic, elliptical taper).

ples through X-ray fluorescence and diffraction. In this technique, capillaries are used to increase the flux density of the X-ray beam onto the sample. By scanning a sample through the beam and recording the resulting X-ray fluorescence spectra, one can obtain the spatial distribution of major, minor, and trace elements that are contained within the sample. The diffracted radiation can be used to identify the crystallographic structure of particular regions of the sample. Diffraction data has been obtained with X-ray beams from 0.8 to 0.05 μm in diameter.[17,27–29] Yamamoto developed an integrated spectrometer that he used for the simultaneous measurement of local strain and trace metal contaminants in integrated circuits.[17,28,30–33] York has developed and implemented multiple capillary-based microdiffractometers for defect analysis in magnetic disk read heads.[27] Capillary-based microbeams have been used to study the impact of crystallographic structure on the sensitivity of IC interconnects to the effects of electromigration.[28,30–35] Other laboratory-based microfluorescence instruments have been developed to produce X-ray microbeams that range from 200 to approximately 5 μm in diameter.[7–10,16,31–33,37]

29.4 CONCLUSION

Imaging and condensing monocapillary optics are still in the developmental stage. Usable, but not yet perfected, prototype optics are being produced and evaluated. Perhaps one of the most serious problems to be overcome in their production is to produce optics whose centerline is straight enough to not limit the calculated flux gains. One advantage these grazing incidence devices have over many other competing microbeam optics (e.g., zone plates, etc.) is that they work over a wide energy range without changing the position to which the beam focuses. The condensing monocapillary has the potential to produce very small beams, even at hard X-ray energies, because the beam size is produced by the exit hole diameter in the glass where hole sizes down to 10 nm diameter have been previously reported.[38] A potential drawback of the condensing optics for submicron beams is that the sample has to be placed near the tip, generally within 20 to 100 times the exit tip inside diameter. Nonetheless, condensing monocapillaries are an exciting way to produce some of the smallest possible diameter X-ray beams.

29.5 REFERENCES

1. F. Jentzch and E. Nähring, "Die Fortleitung von Licht—und Röntgenstrahlen durch Röhren," *Zeitschr. F. Techn. Phys.* **12**:185 (1931).

2. W. T. Vetterling and R. V. Pond, "Measurements on an X-Ray Light Pipe at 5.9 and 14.4 keV," *J. Opt. Soc. Am.* **66**(10):1048 (1976).

3. D. Mosher and S. Staphanakis, "X-ray Light Pipes," *Appl. Phys. Lett.* **29**:105 (1976).

4. P. S. Chung and R. H. Pantell, "Properties of X-Ray Guide Tubes," *Electron. Lett.* **13**:527 (1977).

5. H. Nakazawa, "X-Ray Guide Tube for Diffraction Experiments," *J. Appl. Crystallog.* **16**:239 (1983).

6. A. Rindby, "Application of Fiber Technique in the X-Ray Region," *Nucl. Instrum. Methods* **A249**:536 (1986).

7. P. Engström, S. Larsson, A. Rindby, and B. Stocklassa, "A 200 μm X-ray Microbeam Spectrometer," *Nucl. Inst. Meth. Phys. Res.* **B36**:222 (1989).

8. D. A. Carpenter, "An Improved Laboratory X-ray Source for Microfluorescence Analysis," *X-Ray Spectrometry* **18**:253 (1989).

9. H. Fukumoto, Y. Kobayashi, M. Kurahashi, and A. Kawase, "Development of a High Spatial Resolution X-Ray Fluorescence Spectrometer and its Application to Quantitative Analysis of Biological Systems," *Advances in X-Ray Analysis* **35**:1285 (1992).

10. S. Shimomura, and H. Nokazawa, "Scanning X-Ray Analytical Microscope Using X-Ray Guide Tube," *Advances in X-Ray Analysis* **35**:1289 (1992).

11. E. A. Stern, Z. Kalman, A. Lewis, and K. Lieberman, "Simple Method for Focussing X Rays Using Tapered Capillaries," *Appl. Optics* **27**(24):5135 (1988).

12. D. H. Bilderback, D. J. Theil, R. Pahl, and K. E. Brister, "X-ray Applications with Glass-Capillary Optics," *J. Synchrotron Rad.* **1**:37 (1994).

13. L. Vincze, K. Janssens, A. Rindby, and F. Adams, "Detailed Ray-Tracing Code for Capillary Optics," *X-Ray Spectrometry* **24**:27 (1995).

14. D. X. Balaic, and K. A. Nugent, "The X-ray Optics of Tapered Capillaries," *Appl. Optics* **34**:7263 (1995).

15. Thiel, D. J., "Ray-tracing analysis of capillary concentrators for macromolecular crystallography," *J. Synchrotron Rad.* **5**:820 (1998).

16. A. Attaelmanan, P. Voglis, A. Rindby, S. Larsson, and P. Engström, "Improved Capillary Optics Applied to Microbeam X-ray Fluorescence: Resolution and Sensitivity," *Rev. Sci. Instrum.* **66**(1):24 (1995).

17. N. Yamamoto, "A Micro-Fluorescent/Diffracted X-Ray Spectrometer with a Micro-X-Ray Beam Formed by a Fine Glass Capillary," *Rev. Sci. Instrum.* **69**(9):3051 (1996).

18. B. R. York, "Fabrication and Development of Tapered Capillary X-ray Optics for Microbeam Analysis," *Microbeam Analysis, Proc. 29th Annual MAS Meeting* (1995).

19. B. R. York, Q.-F. Xiao, and N. Gao, "Radiation Focussing Monocapillary with Constant Inner Dimension Region and Varying Inner Dimension Region," U.S. Patent 5,747,821 (1998).

20. D. X. Balaic, K. A. Nugent, Z. Barnea, R. Garrett, and S. W. Wilkins, "Focussing of X-Rays by Total Reflection from a Paraboloidally-Tapered Glass Capillary," *J. Synchrotron Rad.* **2**:296 (1995).

21. D. X. Balaic, Z. Barnea, K. A. Nugent, R. F. Garrett, J. N. Varghese, and S. W. Wilkins, "Protein Crystal Diffraction Patterns Using Capillary-Focused Synchrotron X-Ray Beam," *J. Synchrotron Rad.* **3**:289 (1996).

22. D. H. Bilderback and E. Fontes, "Glass Capillary Optics for Making X-Ray Beams of 0.1 to 50 Microns Diameter," in *Synchrotron Radiation Instrumentation: 10th US National Conference,* Ernest Fontes (ed.), American Institute of Physics, Melville, NY, 1997, pp. 147–155.

23. A. Rindby, "Production of Capillaries," in *Proceedings of the First International Developers Workshop on Glass Capillary Optics for X-ray Microbeam Applications,* Don Bilderback and Per Engstrom, (eds.), Cornell University, October 18–20, 1996, p. 157.

24. M. Haller, J. Heck, S. Kneip, A. Knöchel, F. Lechtenberg, and M. Radtke, "Stretching Tapered Capillaries from Bubbles," in *Proceedings of the First International Developers Workshop on Glass Capillary Optics for X-ray Microbeam Applications,* Don Bilderback and Per Engstrom, (eds.), Cornell University, October 18–20, 1996, p. 165.

25. G. Hirsch, "Tapered Capillary Optics," U.S. Patent 5,772,903 (1998).

26. J. N. Varghese, A. van Donkelaar, M. Hrmova, D. X. Baliac, and Z. Barnia, "Structure of an Exoglucanse Complexed with Conduritol B Epoxide from a 50-micron Crystal Using Monocapillary Optics," Abstract P11.04.017 of Int. Union of Crystallography, XVIII Congress, Glasgow, 329–330 (1999).

27. B. R. York, "Chemical Analysis Using X-Ray Microbeam Diffraction (XRMD) with Tapered Capillary Optics," *Microbeam Analysis, Proc. 28th Annual MAS Meeting,* p. 11 (1994).

28. N. Yamamoto, Y. Homma, S. Sakata, and Y. Hosokawa, "X-ray Spectrometer with a Submicron X-Ray Beam for ULSI Microanalysis," *Mat. Res. Soc. Symp. Proc.* **338**:209 (1994).

29. D. H. Bilderback, S. A. Hoffman, and D. J. Thiel, "Nanometer Spatial Resolution Achieved in Hard X-Ray Imaging and Laue Diffraction Experiments," *Science* **263**:201–203 (1994).

30. N. Yamamoto, and Y. Hosokawa, "Development of an Innovative 5 μmφ Focussed X-Ray Beam Energy Dispersive Spectrometer and Its Applications," *Jpn. J. Appl. Phys.* **27**(11):L2203 (1988).

31. N. Yamamoto, Y. Takano, Y. Hosokawa, and K. Yoshino, "Development of an Innovative 5 μmφ Focused X-Ray Beam Energy Dispersive Spectrometer for Analyzing Residual Stresses and Impurities in ULSIs," *Ext. Abstracts of the 20th* (*1988 International*) *Conference on Solid State Devices and Materials,* p. 463 (1988).

32. N. Yamamoto, and S. Sakata, "Applications of a Micro X-Ray Spectrometer and a Computer Simulator for Stress Analyses in Al Interconnects," *Jpn. J. Appl. Phys.* **28**(11):L2065 (1989).

33. N. Yamamoto, and S. Sakata, "Strain Analysis in Fine Al Interconnections by X-Ray Diffraction Spectrometry Using Micro X-Ray Beam," *Jpn. J. Appl. Phys.* **34**:L664 (1995).

34. P.-C. Wang, G. S. Cargill, I. C. Noyan, E. G. Liniger, C.-K. Hu, and K. Y. Lee, "X-ray Microdiffraction for VLSI," *Mat. Res. Soc. Symp. Proc.* **427**:408 (1997).

35. P.-C. Wang, G. S. Cargill, I. C. Noyan, E. G. Liniger, C.-K. Hu, and K. Y. Lee, "Thermal and Electromigration Strain Distributions in 10-mm-Wide Al Conductor Lines Measured by X-Ray Microdiffraction," *Mat. Res. Soc. Symp. Proc.* **473**:273 (1997).

36. P.-C. Wang, G. S. Cargill, I. C. Noyan, and C.-K. Hu, "Electromigration-Induced Stress in Aluminum Conductor Lines Measured by X-Ray Microdiffraction," *Appl. Phys. Letters* **72**:1296 (1998).

37. D. A. Carpenter, A. Gorin, and J. T. Shor, "Analysis of Heterogeneous Materials with X-Ray Microfluorescence and Microdiffraction," *Advances in X-Ray Analysis* **38**:557 (1995).

38. Sutter Instrument Company, 51 Digital Drive, Novato, CA 94949 [personal communication]. See also K. T. Brown, and D. G. Flaming, *Advanced Micropipette Techniques for Cell Physiology,* International Brain Research Organization Handbook Series: Methods in the Neurosciences, Vol. 9, John Wiley and Sons, New York, 1986.

CHAPTER 30
POLYCAPILLARY AND MULTICHANNEL PLATE X-RAY OPTICS*

Carolyn A. MacDonald and Walter M. Gibson
University at Albany, State University of New York
Albany, New York

30.1 INTRODUCTION

Polycapillary and multichannel plate optics are both arrays of a large number of small hollow glass tubes and both depend on total reflection at the glass surface to transport X rays. The critical angle for borosilicate glass is approximately

$$\theta_c \approx \frac{30}{E(keV)}\text{mrad} \tag{1}$$

which is 30 mrad (1.7°) for 1-keV photons and 1.5 mrad (0.086°) for 20-keV photons. The angles are somewhat larger for leaded glass. The resemblance between polycapillary and multichannel plate optics is largely superficial. Multichannel plate optics are thin sheets of glass perforated with holes, with length to diameter ratios less than 100. They are designed to depend on a single horizontal and vertical bounce to deflect the ray towards the desired direction and so, in operation, resemble elements of a series of nested mirrors. Polycapillary optics are collections of long hollow tubes, with length to diameter ratios of order 10,000. X rays are guided down these curved or tapered tubes by multiple reflections in a manner analogous to the way fiber optics guide light (see Chap. 29).

* *Acknowledgments:* The authors are grateful for useful discussions and data from a large number of collaborators, including Carmen Abreu, David Aloisi, Simon Bates, David Bittel, Cari, Dan Carter, Heather Chen, Greg Downing, Ning Gao, David Gibson, Mikhail Gubarev, Joseph Ho, Frank Hoffman, Huimin Hu, Huapeng Huang, Chris Jezewski, Kardinawarman, John Kimball, Ira Klotzko, David Kruger, K. Matney, David Mildner, Charles Mistretta, Johanna Mitchell, Robin Moresi, Scott Owens, Sushil Padiyar, Wally Peppler, Igor Ponomarev, Bimal Rath, Christine Russell, Francisca Sugiro, Suparmi, Christi Trufus-Feinberg, Johannes Ullrich, Hui Wang, Lei Wang, Brian York, Russel Youngman, and Qi Fan Xiao, and for grant support from the Department of Commerce, NASA, NIH, and the Breast Cancer Research Program.

30.2 *MULTICHANNEL PLATE OPTICS*

The basic principle of operation[1] of multichannel plate (MCP) optics is shown in Fig. 1. X rays are reflected from the source to the focal point, so long as the maximum angle at the outer edge of the optic is less than the critical angle. This limits the usable radius of the optic to $f\theta_c$, where f is the distance from the optic to the focal point. Similarly, several considerations limit the thickness for practical optics. Although a thicker optic intercepts a larger fraction of the incident beam, multiple reflections, which remove energy from the focal spot, become possible at thicknesses greater than the channel diameter divided by the incident angle. Additionally, the spread in the focal spot along the beam axis is roughly twice the optic thickness. The optic is optimized for a channel-diameter-to-thickness ratio equal to the critical angle.[1]

The schematic in Fig. 1 shows only the vertical reflections from the horizontal faces. A single horizontal reflection also occurs from the vertical faces. Square pores, as shown in Fig. 2,[2] are more efficient[3] than circular pores, shown in Fig. 3.[4] The optics are produced by assembling fibers with an inner core and outer clad glass and drawing the assembly down to the desired pore diameter. The assembly is then sliced into thin sheets and the inner core glass is etched away, leaving the hollow channels. The technology was developed for multichannel plate electron multipliers (see Chap. 35), hence the name of the optics. Because the optics are single bounce devices, the relatively rough surface produced by etching yields adequate reflectivity. Measured surface roughnesses are from 2.5 to 5 nm.[5]

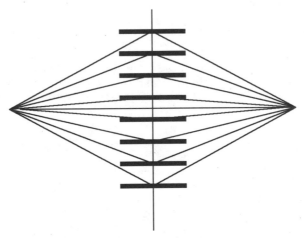

FIGURE 1 Focusing from a point source using a multichannel plate optic.

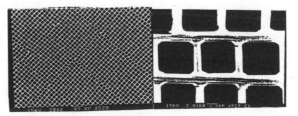

FIGURE 2 Square pore multichannel plate with 13.5-μm pore size, magnified at right. (*From Ref. 2.*)

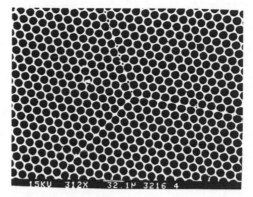

FIGURE 3   Circular pore microchannel plate. (*From Ref. 4.*)

To use the optic to focus a parallel beam, for example for astronomical applications, or to produce a quasiparallel beam from a point source, the plate must be curved, as shown in Fig. 4.[1,6] This can be achieved by thermally *slumping* the optic onto a spherical mandrel. Slumped multichannel plate optics closely resemble the structure of the eyes of lobsters, which do not have lenses, but convey light to the retina by arrays of reflective surfaces. Thus, multichannel plate optics are often termed *lobster-eye* optics. A curved microchannel plate has a relationship between input and output distances analogous to the thin lens equation,

$$\frac{1}{l_s} - \frac{1}{l_f} = \frac{2}{R_{\text{slump}}} \tag{2}$$

where l_s is the source-to-optic distance, l_f is the optic-to-focal-spot distance, and R_{slump} is the radius of the microchannel plate, as shown in Fig. 4.[3]

The results of a focusing experiment using low energy X rays are shown in Figs. 5 and 6. The *cross* pattern seen in Fig. 5 is typical of square channel MCP optics, and is predicted by the simulations. Used with a point source at focus, slumped MCP optics may be useful as collimators for X-ray lithography.[3,5,7]

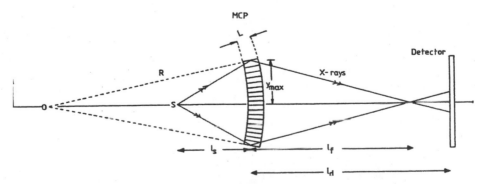

FIGURE 4   Creating a collimated beam with a slumped multichannel plate. (*From Ref. 6.*)

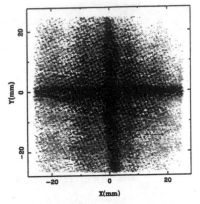

FIGURE 5 Output image from MCP with 11-μm square pores and thickness to diameter ratio of 40:1 illuminated with 1.74-keV Si K X rays. (*From Ref. 2.*)

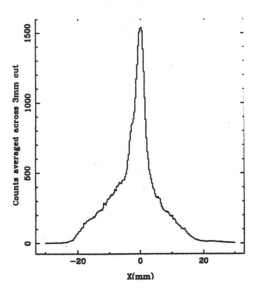

FIGURE 6 Scan of image with same MCP as in Fig. 5, taken with 0.28-keV C K X rays. (*From Ref. 2.*)

30.3 POLYCAPILLARY OPTICS

Introduction

Description. Systems involving the use of a large number of capillary channels for shaping X-ray beams were first suggested by M. A. Kumakhov and his collaborators in 1986.[8] The development and study of polycapillary optics and its applications in X-ray lithography,[9] X-ray astronomy,[10,11] diffraction analysis,[12,13] X-ray fluorescence,[14] and medicine[1–18] have been pursued since 1990.[19-21] Polycapillary optics are well-suited for broadband divergent radiation. Like multichannel plates, they differ from single-bore capillaries in that the focusing or collecting effects come from the overlap of the beams from hundreds of thousands of channels, rather than from the action within a single tube.

X rays can be transmitted down a curved fiber as long as the fiber is small enough and bent gently enough to keep the angles of incidence less than the critical angle. As shown in Fig. 7, the angle of incidence for the ray near one edge increases with tube diameter. The requirement that the incident angles remain less than the critical angle necessitates the use of tiny tube diameters. However, mechanical limitations prohibit the manufacture of capillary tubes with outer diameters smaller than about 300 μm. For this reason, polycapillary fibers, an example of which is shown in Fig. 8, are employed. Typical channel sizes are between 2 and 12 μm. Thousands of such fibers are strung through lithographically produced metal grids to produce a multifiber lens, such as is shown in Fig. 9. Alternatively, a larger-diameter polycapillary fiber can be shaped into a monolithic optic, as shown in Fig. 10.

Polycapillary Fiber Measurements and Numerical Simulations.

Detailed measurements of polycapillary fibers, including transmission, absorption, and exit divergence, have been performed as a function of length, bend radius, X-ray source position, X-ray source geometry, and X-ray energy.[22,23] Computer automated systems for measuring fiber quality have been developed. Repeatability of transmission measurements performed with such systems is within 1 percent.[24]

Extensive modeling programs describe the propagation of X rays along capillaries with complex geometries. These computer codes are based on Monte Carlo simulations of geometrical optics trajectories and provide essential information on performance, design, and potential applications of capillary optics. The geometric algorithm for the simulations is approximated in two dimensions by projecting the trajectory onto the local fiber cross section.[25] Reflectivities are computed from standard tables.[26] Simulation results are compared to experimental data taken on a nominally straight fiber in Fig. 11. The transmission drops off above 45 keV partly due to unintentional bending. However, the model cannot fit the entire spectra with bending alone. A bending radius smaller than 100 m would be necessary to fit the midrange energies, but underestimates the high energy transmission. Accurate simulation requires the roughness[27,28] and waviness[23,29] of the capillary walls to be taken into account.

Capillary surface oscillations with wavelengths shorter than the capillary length and longer than the wavelength of the roughness are called waviness. The detailed shape of waviness is unknown. Its average effect can be considered as a random tilt of the glass wall. It is assumed that these tilt angles, ϕ, are normally distributed in the range $(-\pi/2, \pi/2)$ with the mean value equal to zero. For high-quality optics, the standard deviation of this normal distribution, σ, is much smaller than the critical angle, θ_c. The model with waviness alone is shown in Fig. 12.

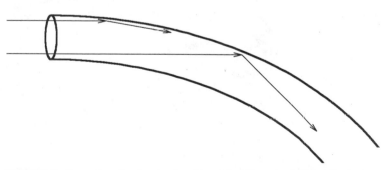

FIGURE 7   X-rays traveling in a bent capillary tube. The ray entering at the bottom (closest to the center of curvature) strikes at a large angle. (*Adapted from Ref. 15.*)

FIGURE 8 Cross-sectional scanning electron micrograph of a polycapillary fiber with 0.55-mm outer diameter and 50-µm diameter channels.

FIGURE 9 Multifiber collimating lens constructed from over a thousand individual polycapillary fibers strung though a metal grid. The lens is 10-cm long with an output of 20 mm × 20 mm. The fibers are parallel at the output end (shown) and at the input end point to a common focal spot.

FIGURE 10   Sketch of the interior channels of a monolithic polycap-illary optic. Monolithic optics can be focusing, as shown, or collimating, as in Fig. 9. (*Sketch from Ref. 12.*)

Both bending and waviness are required, as shown in Fig. 13. Very good agreement is found between simulation and experimental results for a wide range of geometries.

Measured polycapillary fiber transmissions are in excess of 60 percent at 15 keV and better than 50 percent from 1 to 45 keV for 10- to 12-cm-long fibers. At higher energies the fibers must be rigid; if they are thin and flexible, as for fiber A in Fig. 13, the fibers are difficult to hold straight in a groove and their transmissions are poor. However, deliberate bending of the fiber allows deflection of the X-ray beam. Transmission at 20 keV as a function of angular deflection is shown in Fig. 14.[30] Transmission greater than 50 percent is achieved for angular deflections greater than 25 times the critical angle.

For high-quality fibers, transmission varies slowly with fiber length, as shown in Fig. 15 for 8-keV photons. A standard technique for assessing capillary surface quality is to move the source off-axis. As the source is moved off-axis, the X-ray entrance angle and the number of reflections increase, while the transmission through the capillary rapidly decreases. A comparison of the measured and simulated transmission efficiency versus source position in the energy range from 10 to 80 keV is shown in Fig. 16. The data and the simulation are in extremely good agreement. The width of a source scan performed with a small spot source, as shown in Fig. 16, gives a measurement of the input acceptance spot size for the fiber or optic.

The low transmission of off-axis rays is responsible for the extremely high scatter rejection of capillary optics. The scatter transmission of two fibers, which is measured by moving the source off-axis to an angle much larger than the critical angle, is shown in Fig. 17.

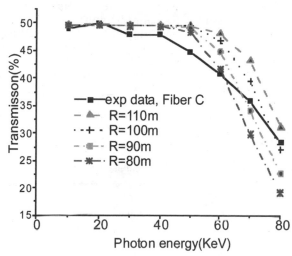

FIGURE 11   Transmission spectra of fiber C simulated with different bending curvature alone and compared with experimental data. (*From Ref. 29.*)

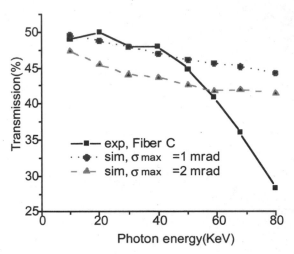

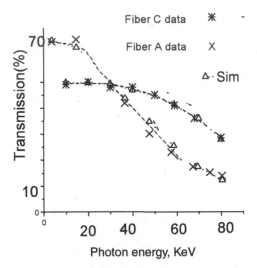

FIGURE 12 Simulations of transmission spectra with waviness only compared with the experimental data. Large values of waviness underestimate the low energy transmission. (*From Ref. 29.*)

FIGURE 13 Simulations of transmission spectra of fiber A and fiber C with their best-fit parameters compared with the experimental data. The fiber and fitting parameters are listed in Table 1.

Collimation

As shown in Fig. 18, the output from a multifiber polycapillary collimating optic has both global divergence, α, and local divergence, β. Even if the fibers are parallel ($\alpha = 0$), the output divergence, β, is not zero, but is determined by the critical angle and therefore the X-ray energy. The exit divergence from capillary optics is measured by rotating a high-quality crystal in the beam and measuring the angular width of a Bragg peak. Since the Darwin width and mosaicity of the crystal are typically much smaller than the exit divergence from the optic, the measurement yields the divergence directly. The result at 8 keV, for a 5-mm-diameter monolithic collimating optic, is a full width at half maximum of 3 mrad, which is approximately given by the critical angle for total reflection.[12] A large multifiber optic similar to that in Fig. 9 has a measured divergence width of 3.8 mrad at 8 keV over the full 3-cm-diameter field.[13] The small difference between the output divergence from a small optic and a large multifiber optic requires good alignment of the individual fibers.

The divergence of the beam, and therefore the angular resolution of the diffraction measurement when using polycapillary optics, does not depend on the source size, unlike the case for pinhole collimation. Thus, larger, higher-power sources may be used without adversely affecting the resolution of the measurement. The maximum useful X-ray source spot size is

TABLE 1 Parameters for Best-Fit Simulations

	Fiber description				Model	
Fiber no.	Outer diameter (mm)	Channel size (μm)	Area	Length (mm)	R (m)	σ (mrad)
A	0.5	12	65%	105	60	0.225
C	0.75	22	50%	136	225	0.2

Note: R is the bending radius and σ is the standard deviation of the waviness.

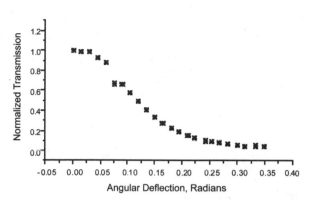

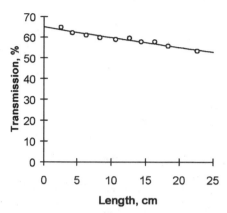

FIGURE 14 Measured (stars) and simulated (boxes) transmission at 8 keV as a function of deflection angle for uniform circular bending. (*From Ref. 30.*)

FIGURE 15 Measured transmission at 8 keV, as a function of length, for a borosilicate polycapillary fiber, with 17-μm channels. Solid line is an exponential fit with decay length = 120 cm. (*Adapted from Ref. 16.*)

limited by the acceptance area of the collimating lens, which is about 1 mm wide by 15 mm long for a multifiber lens, well matched for a typical rotating anode with a line focus viewed at a small takeoff angle.

The transmission of a 3-cm-square multifiber collimating lens developed for astronomical applications is shown as a function of photon energy in Fig. 19. The output of the lens is shown in Fig. 20.

Gain.   The *gain* obtained from any optic ultimately should be a figure of merit defined by a particular application. For a diffraction measurement which requires 2-D collimation (e.g., strain measurements), the intensity ratio delivered within a given output divergence angle is a reasonable figure.

The flux output from a collimating lens is given by

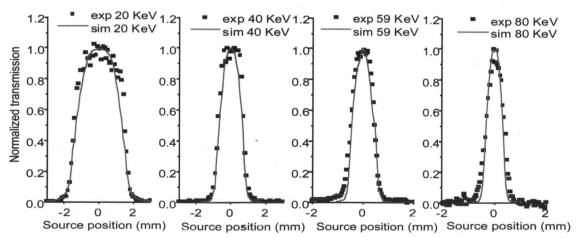

FIGURE 16 Simulated source scan curves compared with experimental data at four different photon energies. Parameters: $R = 125$ m, $\sigma_{max} = 0.35$ mrad, roughness height = 0.5 nm. (*From Ref. 29.*)

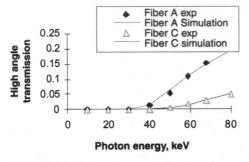

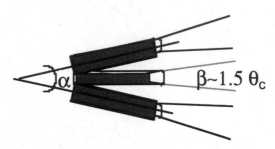

FIGURE 17   High-angle transmission through capillary walls as a function of photon energy. (*From Ref. 23.*)

FIGURE 18   The output divergence from a multifiber collimating optic is characterized by global (α) and local (β) divergence. (*From Ref. 21.*)

$$P_{\text{out}} = T \frac{\pi r^2}{4 \pi f^2} P = \frac{T}{4} \tan^2 \left(\frac{\Omega_0}{2} \right) P \tag{3}$$

where P is the source power, T is the transmission of the lens, r is the input radius, f is the source to lens distance, and Ω_0 is the capture angle, $\Omega_0 = 2 \tan^{-1}(r/f)$. The simple power gain, compared to a pinhole designed to have an output divergence of $2\theta_c$, similar to the optic, is thus

$$\text{output power gain} = \frac{\dfrac{T}{4} \tan^2 \left(\dfrac{\Omega_0}{2} \right)}{\dfrac{1}{4} \tan^2 \theta_c} \approx T \left(\frac{\Omega_0}{2\theta_c} \right)^2 \tag{4}$$

For a lens with 2-cm output area, 132-mm focal distance (source-to-lens distance), 7° capture angle, 25 percent transmission, and 8-keV operation, the predicted gain is 66. Kennedy et al.

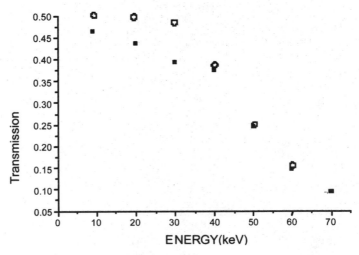

FIGURE 19   Simulated[10] (circles) and measured[31] (squares) transmission of a 3-cm-square multifiber collimating lens with 2-m focal length. (*Simulation from Ref. 10.*)

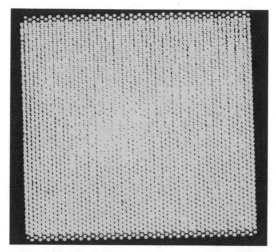

FIGURE 20   X-ray micrograph of multifiber collimator, exposed 5 min at 60 keV and 0.4 mA with a tungsten source and Polaroid film. (*From Ref. 38.*)

used this lens to measure diffraction gains.[32] An out-of-plane Ψ diffraction scan of a Si (100) test specimen across a <111> reflection gave a measured count rate a factor of 20 higher with the optic than with a 1.35-mm pinhole at 155 mm from the source. The resolution with the optic was 0.34° compared to 0.5° with the pinhole. Because the optic was retrofitted to the Philips X-Pert MRD diffractometer without optimization, the detector was 7 mm by 9 mm, smaller than the output beam of the collimating optic, resulting in a substantial reduction of gain. A measurement with a 2-cm-diameter detector, matched more optimally to the optic, would have yielded a gain of 60, compared to the gain of 66 calculated from Eq. (4). A measurement with a 3-cm-diameter lens, detector, and sample would yield a gain of ~100.

Measurements were performed by Matney et al. with a similar optic on a complex multilayer magnetic film with very thin interlayer thickness.[33] Intensity increases of more than 100 with signal-to-background improvement of more than an order of magnitude were obtained, as shown in Fig. 21. The parallel beam geometry also provides insensitivity to sample preparation, shape, position, and transparency as described in Chap. 35 of this volume.

Collimation for Small Samples.   For small beams, for example as required for protein crystallography, the monolithic lens construction, shown in Fig. 10, is employed. This results in a smaller optic, with smaller focal distance and smaller source acceptance spot. The 0.19° divergence from the optic is less than the ω crystal oscillations typically employed to increase the density of reflections captured in a single image in protein crystallography. The quality of the data with the optic was as good as for the standard measurement.[34] Using a low-power X-ray source which allows close access to the beam spot, giving a much larger collection angle, the X-ray intensity for a quasiparallel output beam 0.3 mm in diameter obtained with a 20-W X-ray source was comparable to that achievable with a 3.5-kW rotating anode source with pinhole collimation. Chicken egg white lysozyme data taken with a 20-W source and a monolithic collimating lens in 20 minutes per frame produced an R-factor (variance between the measured and model structure factors) of 5.1 percent and resolution to 1.6 Å, as good as equivalent rotating anode data taken with the same or longer exposure time. The data shown in Fig. 22 were taken in 10 minutes with the collimating optic and 20-W source.

Slightly Convergent Protein Crystallography.   Even slightly focusing the beam can result in a large intensity increase compared to collimation. Protein crystallography does not require

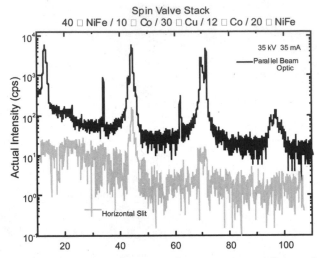

FIGURE 21 Spin Valve Stack. 40 Å NiFe / 10 Å Co/30 Å Cu/12 Å Co/20 Å NiFe. Dark line is measurements with optic; gray line is with horizontal slit and equal counting time. Note log intensity scale. Thin film peaks at 10, 30, 65, and 100° are not apparent without the optic. (*From Ref. 33.*)

highly parallel beams. The effects of slight convergence on full protein crystallographic data sets were studied.[34,35] A 4-mm diameter focusing lens with output focal length 150 mm was placed in a standard 40-kV, 35-mA rotating anode crystallography system. The lens gave a direct beam intensity increase of a factor of 20, and an average diffracted beam intensity increase of 21.6 for the spots in common for the two datasets. Data from the same sample of chicken egg white lysozyme were taken with and without the optic. Both datasets had a resolution of 1.6 Å. The dataset collected for 30 minutes per frame in the standard system without the optic had an *R*-factor of 6.4 percent. Data collected for 4 minutes per frame with the optic had an *R*-factor of 6.9 percent.

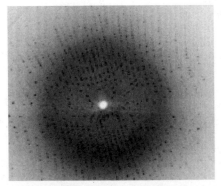

FIGURE 22 Lysozyme diffraction image taken with a 20-W microfocus source and collimating polycapillary optic in 10 minutes. (*From Ref. 34.*)

Energy Filtering

The critical angle for reflection from a surface at grazing incidence is inversely proportional to photon energy, as given by Eq. (1). This drop in reflectivity can be applied to the bent capillary channel of Fig. 1, which has a maximum incidence angle dependent on the bending radius and channel diameter. Above a certain photon energy, the critical angle for reflection will be smaller than the maximum incident angle, and X rays will no longer be efficiently guided down the channel, as shown in Fig. 11. Capillary optics can be used as a low-pass filter to remove high-energy bremsstrahlung photons above a given energy threshold. Thus high anode voltages can be used to increase the intensity of the characteristic lines without increasing the high-energy background. An example of the effect of a monolithic optic designed to pass 8-keV Cu K_α radiation is shown Fig. 23.[36] The lens reduces the Cu K_β 9-keV peak and suppresses the high-energy bremsstrahlung. For low-resolution diffraction applications, the energy filtration provided by the optic allowed the monochromator to be replaced with a simple filter.[12] The removal of the monochromator significantly reduces alignment complexity.

Focusing

Polycapillary lenses can be used to collect broadband divergent radiation and redirect it toward a focal spot. The first Kumakhov capillary lens was built to demonstrate focusing. It had channel sizes of 300 µm and was about 1 m long.[8,37] Smaller channel sizes allow for tighter bending and shorter lenses.

Intensity Gain. Focusing the beam increases the intensity on a small sample, compared to pinhole collimation. The intensity gain depends on the spot size produced by the optic, which is determined by the capillary channel size, c, output focal length, f, and X-ray critical angle, θ_c,

$$d_{spot} \approx \sqrt{c^2 + (1.3 \cdot f_{out} \cdot \theta_c)^2} \qquad (5)$$

The factor 1.3 is an experimentally determined parameter that arises from the fact that most of the beam has a divergence less than the maximum possible divergence of $2\theta_c$ produced by reflection. Because of the divergence, lenses with small focal lengths used at high energies have the smallest spot sizes. The critical angle, θ_c, at 20 keV is 1.5 mrad. A lens with $c = 3.4$ µm and $f_{out} = 9$ mm has a predicted spot size of 18 µm. An intensity distribution measurement, by scanning a small pinhole, gave a full width at half-maximum (FWHM) of 21 µm.[38]

For the case of a very small sample, the gain, relative to pinhole collimation, is given by

$$gain = \frac{P_{lens}}{P_{pinhole}} = \left(\frac{d^2 Lens}{f_{in}^2} \right) T_{Lens} \left(\frac{L_{source-sample}}{\sigma} \right)^2 \qquad (6)$$

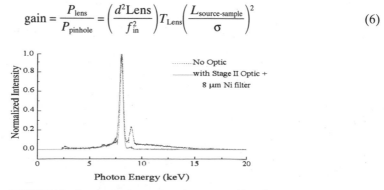

FIGURE 23 Spectrum of a copper tube source with and without a slightly focusing optic. The optic suppresses the high-energy bremsstrahlung. The nickel filter suppresses the K_β peak. (*From Ref. 36.*)

For the slightly convergent lens used to produce lysozyme diffraction data,[12] the computed gain for a 0.5-mm pinhole is 124. The measured intensity gain is 110. The intensity gain for the more highly convergent lens with the output focal spot of 21 μm is 2400 at a source-to-sample distance of 100 mm.

The intensity gain yields a real and dramatic ability to downsize the source, as evinced by the data of Fig. 22. The diffraction signal from high-z inorganics would be even stronger. The theory of convergent beam microdiffraction is discussed in Chap. 35 of this volume.

Microanalysis.   Focused X-ray beams provide a similar benefit for microfluorescence analysis.[14] In addition, a second optic could be used to collect the fluorescent radiation, as shown in Fig. 24.[39] This would provide the double benefit of enhanced signal intensity and 3-dimensional spatial resolution. The 3-dimensional resolution arises from the overlap of the cone of irradiation of the first lens and the cone of collection of the second, as shown in Fig. 24. Microfluorescence with polycapillary optics is discussed in Chap. 35.

Synchrotron Focusing.   Polycapillary optics can be used with both point source and synchrotron radiation. A 5-mm diameter focusing lens with a 17-mm focal length was measured using knife-edge scans in a synchrotron beam.[40] The measured spot size, transmission, and gain are shown in Table 2, along with calculated gains. Gains of two orders of magnitude were measured.

Radiation Damage

Use of polycapillary optics with intense sources such as synchrotrons raises the question of radiation damage. Because the X-ray optical properties of materials depend on total electron density,[41] the optical constants are insensitive to changes in electronic state. Color centers that form rapidly in glass during exposure to intense radiation are not indicative of a change in the X-ray transmission of a polycapillary optic. In addition to darkening, thin fibers exposed to intense beams undergo reversible deformation. However, rigid optics, if annealed in situ at 100° C, were shown to withstand in excess of 2 MJ/cm² of white beam bending magnet radiation without measurable change in performance at 8 keV.[42]

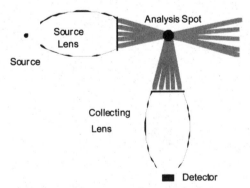

FIGURE 24   Sketch of microfluorescence experiment, showing that overlap of irradiation and collection volumes yields 3-dimensional spatial resolution. (*From ref. 21.*)

TABLE 2 Results for Monolithic Focusing Optic in a Synchrotron Beam

X-ray energy (keV)	Spot size (mm)	Transmission (%)	Measured gain 350 μm pinhole	Calculated gain 350 μm pinhole	Calculated gain 90 μm pinhole	Calculated gain 10 μm pinhole
6	0.09	36	78	81	645	911
8	0.08	49	96	110	933	1359
10	0.09	39	83	87	624	842
12	0.09	39	74	87	654	903
White	0.17	42	11	89	243	266

Source: From Ref. 40.

Discussion and Conclusions

Polycapillary X-ray optics are a powerful, relatively new control technology for X-ray beams. Using polycapillary optics to collimate the output from a point source provides in most cases much higher intensity than pinhole collimation, particularly if 2-dimensional collimation is required. Collimating optics collect from 6 to 15° (from 0.7 to 3 millisteradian) of a divergent beam. Output diameters range from 1 to 6 mm for a monolithic optic to ≥3 cm for a multifiber collimating optic. Output divergence varies inversely with the photon energy, and is around 0.17° (3 mrad) at 8 keV.

Focusing the beam yields even higher-intensity gains. Measured focused beam gains for sample sizes from 0.3 to 0.5 mm are around a factor of 100, and agree well with computations. Computed gains for smaller samples are even higher.

Polycapillary optics can also be used to perform low pass spectral filtering, which allows the use of increased source voltage. Further, the optics also remove the connection between source size and resolution, which allows the use of increased source current. Increasing the voltage and current of the source increases the useful intensity.

While not true imaging optics, polycapillary fibers can transmit an image in the same manner as a coherent fiber bundle. Polycapillary optics can be used to magnify and demagnify images and to remove the high-angle Compton scattering which can otherwise result in substantial image degradation. Medical applications are discussed in Chap. 35 of this volume.

30.4 REFERENCES

1. S. W. Wilkins, A. W. Stevenson, K. A. Nugent, H. Chapman, and S. Steenstrup, "On the Concentration, Focusing and Collimation of X-rays and Neutrons Using Microchannel Plates and Configurations of Holes," *Rev. Sci. Instrum.* **60**:1026–1036 (1989).

2. A. N. Brunton, G. W. Fraser, J. E. Lees, W. B. Feller, and P. L. White, "X-ray Focusing with 11 μm Square Porte Microchannel Plates," in *X-ray and Ultraviolet Sensors and Applications, SPIE Proc.* **2519**, 1995.

3. H. N. Chapman, K. A. Nugent, and S. W. Wilkins, "X-ray Focusing Using Square Channel-Capillary Arrays," *Rev. Sci. Instrum.* **62**:1542–1561 (1991).

4. G. W. Fraser, A. N. Brunton, J. E. Lees, and D. Lemberson, "Production of Quasi-parallel X-ray Beams Using Microchannel Plate X-ray Lenses," *NIM A*, **324**:404–407 (1993).

5. I. C. E. Turcu, A. N. Brunton, G. W. Fraser, and J. E. Lees, "Microchannel Plate (MCP) Focusing Optics for a Repetitive Laser-Plasma Source," in *Applications of Laser Plasma Radiation II, SPIE Proc.* **2523**, 1995.

6. G. W. Fraser, A. N. Brunton, J. E. Lees, J. F. Pearson, R. Willingale, D. L. Emberson, W. B. Feller, M. Stedman, and J. Haycocks, "Development of Microchannel Plate MCP X-ray Optics," in *Multilayer and Grazing Incidence X-ray/EUV Optics III, Proc. SPIE* **2011**:215–226 (1993).

7. W. M. Gibson and C. A. MacDonald, "Applications Requirements Affecting Optics Selection," Chap. 35, this volume.

8. V. A. Arkd'ev, A. I. Kolomitsev, M. A. Kumakhov, I. Yu. Ponomarev, I. A. Khodeev, Yu. P. Chertov, and I. M. Shakparonov, "Wide-Band X-ray Optics with a Large Angular Aperture," *Sov. Phys. Usp.* **32**(3):271 (March 1989).

9. I. Klotzko, Q. F. Xiao, D. M. Gibson, R. G. Downing, W. M. Gibson, A. Karnaukhov, and C. J. Jezewski, "Investigation of Glass Polycapillary Collimator for Use in Proximity Based X-ray Lithography," *SPIE Proc.* **2523**:175–182 (1995).

10. C. H. Russell, W. M. Gibson, M. V. Gubarev, F. A. Hofmann, M. K. Joy, C. A. MacDonald, Lei Wang, Qi-Fan Xiao, and R. Youngman, "Application of Polycapillary Optics for Hard X-ray Astronomy," in R. B. Hoover and A. B. C. Walker II (eds.), *Grazing Incidence and Multilayer X-ray Optical Systems, SPIE Proc.* **3113**:369–377 (1997).

11. C. H. Russell, M. Gubarev, J. Kolodziejczak, M. Joy, C. A. MacDonald, and W. M. Gibson, "Polycapillary X-ray Optics for X-ray Astronomy," in *Advances in X-ray Analysis, Proceedings of the 48th Denver X-ray Conference,* vol. 43, 1999.

12. S. M. Owens, F. A. Hoffman, C. A. MacDonald, and W. M. Gibson, "Microdiffraction Using Collimating and Convergent Beam Polycapillary Optics", *Advances in X-ray Analysis, Proceedings of the 46th Annual Denver X-ray Conference Proceedings,* vol. 41, Steamboat Springs, Colorado, August 4–8, 1997, pp. 314–318.

13. B. R. Kardiawarman, R. York, X.-W. Qian, Q.-F. Xiao, C. A. MacDonald, and W. M. Gibson, "Application of a Multifiber Collimating Lens to Thin Film Structure Analysis," in R. B. Hoover and M. B. Williams (eds.), *X-ray and Ultraviolet Sensors and Applications, SPIE Proc.* **2519**, pp. 197–206, July 1995.

14. N. Gao, I. Yu. Ponomarov, Q.-F. Xiao, W. M. Gibson, and D. A. Carpenter, *Appl. Phys. Lett.* **69**:1529 (1996).

15. C. C. Abreu and C. A. MacDonald, "Beam Collimation, Focusing, Filtering and Imaging with Polycapillary X-ray and Neutron Optics," *Physica Medica,* **XIII**(3):79–89 (1997).

16. C. C. Abreu, D. G. Kruger, C. A. MacDonald, C. A. Mistretta, W. W. Peppler, Q.-F. Xiao, "Measurements of Capillary X-ray Optics with Potential for Use in Mammographic Imaging," *Medical Physics* **22**(11-1):1793–1801 (November 1995).

17. W. M. Gibson, C. A. MacDonald, and M. S. Kumakhov, "The Kumakhov Lens: A New X-ray and Neutron Optics with Potential for Medical Applications," in S. K. Mun (ed.), *Technology Requirements for Biomedical Imaging, I.E.E.E. Press,* vol. 2580, 1991.

18. D. G. Kruger, C. C. Abreu, E. G. Hendee, A. Kocharian, W. W. Peppler, C. A. Mistretta, C. A. MacDonald, "Imaging Characteristics of X-ray Capillary Optics in Mammography," *Medical Physics* **23**(2):187–196 (February 1996).

19. C. A. MacDonald, C. C. Abreu, S. Budkov, H. Chen, X. Fu, W. M. Gibson, Kardiawarman, A. Karnaukhov, B. Kovantsev, I. Ponomarev, B. K. Rath, J. B. Ullrich, M. Vartanian, and X.-F. Xiao, "Quantitative Measurements of the Performance of Capillary X-ray Optics," in R. B. Hoover and A. Walker (eds.), *Multilayer and Grazing Incidence X-ray/EUV Optics II, SPIE Procs.* **2011**, 275–286, 1993.

20. W. M. Gibson and C. A. MacDonald, "Polycapillary Kumakhov Optics: A Status Report," in R. B. Hoover and M. Tate (eds.), *X-ray and UV Detectors, SPIE Proc.* **2278**, 156–167, 1994.

21. C. A. MacDonald, "Applications and Measurements of Polycapillary X-ray Optics," *Journal of X-ray Science and Technology* **6**:32–47 (1996).

22. J. B. Ullrich, V. Kovantsev, and C. A. MacDonald, "Measurements of Polycapillary X-ray Optics," *J. Appl. Phys.* **74**(10):5933–5939 (Nov. 15, 1993).

23. Wang Lei, B. K. Rath, W. M. Gibson, J. C. Kimball, and C. A. MacDonald, "Measurement and Analysis of Capillary Optic Performance for Hard X rays," *Jour. Appl. Phys.* **80**(7):3628–3638 (October 1, 1996).

24. B. Rath, R. Youngman, C. A. MacDonald, "An Automated Test System for Measuring Polycapillary X-Ray Optics," *Rev. Sci. Instru.* **65**:3393–3398 (Nov. 1994).

25. Q.-F. Xiao, Yu. Ponamarev, A. I. Kolomitsev, and J. C. Kimball, Numerical Simulations for Capillary-Based X-Ray Optics, *SPIE* **1736**:227 (1992).

26. B. L. Henke, E. M. Gullikson, and J. C. Davis, "Atomic Data," *Nucl. Data Tables,* **54**(2):181–342 (1993).

27. D. Bittel, and J. Kimball, "Theoretical Effects of Surface Roughness on Capillary Optics," *J. Appl. Phys.* **74**(2):877–883 (July 15, 1993).

28. J. E. Harvey, "X-Ray Optics," Chap. 11, in M. Bass (ed.), *Handbook of Optics,* 2nd ed., vol. II, McGraw-Hill, New York, 1995.

29. Hui Wang, Lei Wang, W. M. Gibson, and C. A. MacDonald, "Simulation Study of Polycapillary X-Ray Optics," in R. B. Hoover and A. B. C. Walker II (eds.), *X-Ray Optics, Instruments, and Missions, SPIE* **3444**:643–651 (July 1998).

30. Hui Wang, "Analysis and Measurement of Polycapillary X-ray Collimating Lenses for Medical Imaging," Ph.D. thesis, University at Albany, 1999.

31. Cari, C. A. MacDonald, W. M. Gibson, C. D. Alexander, M. K. Joy, C. H. Russell, and Z. W. Chen, "Characterization of a Long Focal Length Optic for High Energy X rays," to be published in C. A. MacDonald, A. M. Khounsary, *Advances in Laboratory-Based X-ray Sources and Optics, SPIE proc.* 4144, Bellingham, WA, 2000.

32. R. Kennedy, Q.-F. Xiao, T. Ryan, and B. R. York, *European Powder Diffraction Conference (EPDIC),* 1997.

33. K. M. Matney, M. Wormington, D. K. Bowen, and Q.-F. Xiao, Oral Presentation *Denver X-Ray Conf.,* Steamboat Springs, CO, Aug. 4–8, 1997.

34. F. A. Hofmann, W. M. Gibson, C. A. MacDonald, D. A. Carter, J.-X. Ho, and J. R. Ruble, "Polycapillary Optic-Source Combinations for Protein Crystallography," *Journal of Applied Crystallography,* submitted.

35. S. M. Owens, F. A. Hoffman, C. A. MacDonald, and W. M. Gibson, "Microdiffraction Using Collimating and Convergent Beam Polycapillary Optics", *Advances in X-ray Analysis, Proceedings of the 46th Annual Denver X-ray Conference Proceedings,* vol. 41, Steamboat Springs, Colorado, August 4–8, 1997, pp. 314–318.

36. S. M. Owens, J. B. Ullrich, I. Y. Ponomarev, D. C. Carter, R. C. Sisk, J.-X. Ho, and W. M. Gibson, "Polycapillary X-Ray Optics for Macromolecular Crystallography," in R. B. Hoover and F. P. Doty (eds.), *Hard X-Ray/Gamma-Ray and Neutron Optics, Sensors, and Applications, SPIE Proc.* **2859**, 200–209, 1996.

37. W. M. Gibson, and M. A. Kumakhov, in *Yearbook of Science & Technology,* McGraw-Hill, New York, 1993, pp. 488–490.

38. F. A. Hoffman, N. Gao, S. M. Owens, W. M. Gibson, C. A. MacDonald, and S. M. Lee, "Polycapillary Optics For In-Situ Process Diagnostics," in P. A. Rosenthal, W. M. Duncan, and J. A. Woollam (eds.), *In Situ Process Diagnostics and Intelligent Materials Processing, Materials Research Society Proceedings,* vol. 502, 1998, pp. 133–138.

39. W. M. Gibson, and M. A. Kumakhov, in *X-Ray Detector Physics and Applications, SPIE* **1736**:172 (1992).

40. F. A. Hofmann, C. A. Freinberg-Trufas, S. M. Owens, S. D. Padiyar, and C. A. MacDonald, "Focusing of Synchrotron Radiation with Polycapillary Optics," *Beam Interactions with Materials and Atoms: Nuclear Instruments and Methods* **133**:145–150 (1997).

41. See Chap. 19 of this volume.

42. B. K. Rath, W. M. Gibson, Lei Wang, B. E. Homan, and C. A. MacDonald, "Measurement and Analysis of Radiation Effects in Polycapillary X-ray Optics," *Journal of Applied Physics,* **83**(12):7424–7435 (June 15, 1998).

X-RAY OPTICS SELECTION: APPLICATION, SOURCE, AND DETECTOR REQUIREMENTS

SOURCES

CHAPTER 31
ELECTRON IMPACT SOURCES

Johannes Ullrich
Carolyn A. MacDonald
University at Albany, State University of New York
Albany, New York

31.1 INTRODUCTION

An X-ray measurement system includes a source, optics (at least pinhole optics or apertures), and a detector. To optimize the system, the requirements of the application and the properties of the available sources and detectors must be considered. Chapter 34 summarizes detector properties and Chap. 35 the requirements of a few sample applications. This chapter begins the discussion of source properties and how they affect optics optimization with the oldest and most common X-ray sources, electron beam impact sources.

31.2 SPECTRA FROM ELECTRON BEAM IMPACT X-RAY SOURCES

Electron beam impact X-ray sources generate X rays by accelerating electrons and directing them toward a target. Figure 1 shows a sketch of a typical X-ray source. Either the anode or the filament (cathode) is grounded. Many sources employ some electrostatic or magnetic optics to focus and steer the electron beam to impact only within a given area of the target. Other filament types, such as dispenser cathodes or field emission cathodes, can also be used and can be beneficial for low power, high brightness sources. Two basic processes convert the kinetic energy of the electrons into X rays: bremsstrahlung radiation and characteristic emission.

Bremsstrahlung Continuous Radiation

As electrons interact with the nuclei within the target material, they are decelerated, causing the kinetic energy to be converted into electromagnetic energy. X rays generated by this mechanism have a broad energy spectrum, with any energy up to the kinetic energy of the electron. The efficiency of conversion of electrical energy into continuum radiation is approximately

$$\eta = 1.1 \times 10^{-9} ZV \tag{1}$$

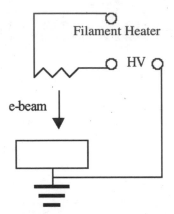

FIGURE 1 Typical electron impact source.

where Z is the atomic number of the anode material and V is the electron voltage, in volts.[1,2] For thin anodes the bremsstrahlung intensity is independent of the X-ray energy up to the maximum energy. For thick anodes, with multiple scattering of the electrons, the bremsstrahlung continuum spectrum has a characteristic triangular shape,

$$I(\nu) = C(\nu_{max} - \nu) \tag{2}$$

where ν is the photon frequency and ν_{max} is the maximum frequency, given by the incident electron energy.[3]

Characteristic Radiation

If the kinetic energy of the impacting electrons exceeds the binding energy of the electrons within the target material, the bound electrons can be removed. The resulting vacancy can be filled by a free electron or a bound electron from an upper shell with the emission of a photon. For inner shell electrons, the energy of the photon emitted by the electron is nearly equal to the binding energy of the electron. Inner shell vacancies (from s or p electrons) of common anode materials can release enough energy to generate X rays. Because these emitted X rays have an energy that depends on the core levels of the material, they are known as characteristic X rays. The lines are enumerated according to the vacancy core level, with K levels having principal quantum number $n = 1$, L levels $n = 2$, and M levels $n = 3$. α transitions have $\Delta n = 1$; β transitions $\Delta n = 2$. The energy of the core atomic level with quantum number n is given by

$$E = -E_0 \frac{(Z - \sigma)^2}{n^2} \tag{3}$$

where

$$E_0 = \frac{me^4}{8\varepsilon_0 h^2} = 13.5 \text{ eV} \tag{4}$$

Z is the atomic number of the anode material, and σ is a screening constant that describes the reduction in effective nuclear charge experienced by outer shell electrons, $\sigma \sim 1$ for K-level electrons and $\sigma \sim 7.5$ for L-level electrons. For copper, with $Z = 29$, the energy of a K shell electron is 8.98 keV. The $1/n^2$ dependence results in bunching of the upper levels. The separation of the Kβ line from the Kα line (8.9 and 8.0 keV for copper) is due to the energy difference between the $n = 3$ and $n = 2$ initial levels since both have $n = 1$ as the final level. This separation is much less than the separation of the Kβ and Lα lines (8.9 and 0.9 keV for copper), which is due to the energy difference between the $n = 2$ and $n = 1$ final levels since both are from an $n = 3$ initial level. Another consequence of Eq. (3) is that the square of the energy, or frequency, of a specific emission line is linearly proportional to the atomic number of the anode. This is Moseley's law. Thus, increasing the atomic number of the anode slowly increases the emission frequency of, for example, the Kα line. The Kα energy for Mo, with $Z = 42$, is 17 keV. Tables of characteristic lines are found in the appendix, following Chap. 37. Efficiencies for the production of characteristic X rays as a function of incident electron beam energy are tabulated in the *International Tables for X-Ray Crystallography*.[4] No characteristic emission is possible unless the impacting electron kinetic energy is larger than the binding energy of the core electron, E_i, given by Eq. (3). Above that energy the intensity, I, increases with voltage roughly as

$$I \propto (eV - E_i)^p \tag{5}$$

where V is the tube voltage.[2] The exponent, p, is a measured value approximately equal to 1.6, and is smaller at high voltages where the penetration of the electrons into the target (and thus the reabsorption of X rays) is higher.

A typical spectrum from a copper anode displaying both the characteristic lines and the continuum Bremsstrahlung radiation is shown in Fig. 2.[5]

31.3 EFFECT OF CURRENT ON SOURCE SIZE AND BRIGHTNESS

A basic limitation of any optic is Liouville's theorem,[6] which states that no optic can increase the brightness of a beam, and so limits the brightness delivered by an optic to that of the source. For electron impact sources, which emit X rays isotropically, source size and power determine the brightness of the source. The maximum power of an electron impact source is limited by the melting of the anode material. The power delivered to the anode by the electron beam is almost entirely converted into heat and must be dissipated by a cooling system. Only a few choices of cooling systems exist, as the electron beam must be operated in vacuum

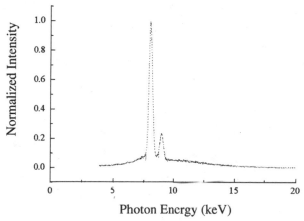

FIGURE 2 Spectrum from a copper tube. The width of the characteristic lines is determined by the resolution of the HPGe detector. (*From Ref. 5.*)

and the anode material has to be thick enough to separate the cooling medium from the vacuum. Most commonly, water jets are sprayed at the backside of the anode. As shown by Müller,[7] this causes the maximum possible power of an X-ray source to increase linearly with the diameter of the X-ray spot, not linearly with the area as one may expect. A plot showing the linear relationship of electrical energy to spot radius is shown in Fig. 3. This effect presents an opportunity for X-ray optics. Smaller, low power sources can be brighter and allow for the generation of brighter X-ray beams. These sources are more easily designed with short source-to-optic distances, which allow for larger solid angles to be collected by the optic. For the same solid angle to be collected from the source, optics with larger source-to-optic distances must have larger diameters. For sealed tube copper X-ray sources, the power of an X-ray source is about 1 kW/mm electron beam diameter, as shown in Fig. 3. For high power sources, the anode is often rotated at high speed under the electron beam, continually exposing the cooled anode to the beam. This increases the limiting power to 11 kW/mm for rotating anodes. Rotating anode systems typically are larger and have greater maintenance costs than fixed anode systems.

Effective Source Size

An important distinction has to be made when applying Liouville's theorem. The source size with respect to the optic is not the actual spot size of the X-ray source, but the area over which the optic can collect photons. Therefore, in order to use the advantage gained from a small, bright X-ray source, the optic has to be designed to only collect photons from that small area. On the other hand, large source sizes do not detract from beam brightness if the power density is maintained and the optic only collects from a small area. For sources with high peak brightness in a central region and lower brightness over an extended penumbral region, the optic has to be carefully optimized to collect X rays from the smallest possible region that will provide sufficient power for the selected application. As the optic will accept X rays within a specific angular range, increasing the source-to-optic distance increases the effective source size.

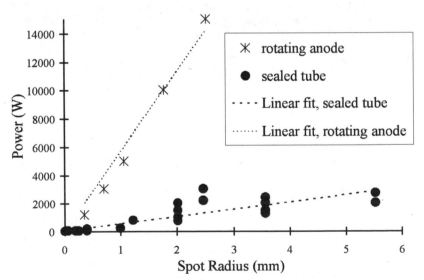

FIGURE 3 Electrical power versus source spot radius for a number of commercial sources. The linear fit for the sealed tube sources has a slope of 511 W/mm-radius, and for the rotating anode sources 5613 W/mm-radius. For spots with a noncircular shape, an equivalent radius to give the same area was taken.

Source Shape

Many sources have the electron beam focused to a line rather than a spot. The X rays are then collected at a small takeoff angle, resulting in a nearly square or circular apparent source, as shown in Fig. 4. Because the apparent size of the beam is reduced by a factor of the sine of the takeoff angle, the brightness is increased as

$$\text{brightness} \propto \frac{\text{power}}{\text{apparent area}} = \frac{\text{power}}{A \sin \alpha} \tag{6}$$

where A is the actual electron beam spot area and α is the takeoff angle. This can considerably improve the brightness for some applications. However, X rays are not produced at the surface of the anode, but at the depth of a few microns, depending on the electron stopping power of the anode material and the electron voltage. Absorption of the X rays traveling out of the anode material will eventually reduce the source brightness at small grazing angles. The transmitted power is reduced exponentially with the path length,

$$\text{brightness} \propto \frac{\text{power out}}{A \sin \alpha} = \frac{\text{power}}{A} \frac{e^{[(-d/\sin \alpha)/D]}}{\sin \alpha} \tag{7}$$

where d is the production depth and D the absorption length for the X rays. The factor on the right is plotted in Fig. 5 for the case that $d/D = 10$, which is approximately the case for a typical Cu Kα system. The brightness is usually maximized at from 6 to 12°. Excessively small takeoff angles are more affected by surface roughness, which may increase with use.

Optics with sufficient depth of field to collect from the entire length of the line source can take advantage of the increased power emitted by line-focus sources compared to a smaller square source. Optics designed for very short source-to-optic distances often do not have large depths of field.

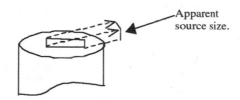

FIGURE 4   Apparent beam size from line focus source viewed at an angle.

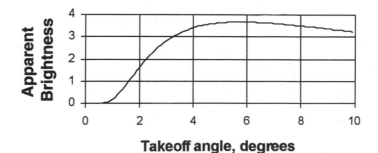

FIGURE 5   Source brightness as a function of takeoff angle for an X-ray production depth to absorption length ratio of 10.

31.4   EFFECT OF ANODE MATERIAL

As shown in Eq. (3), changing the anode material can be used to tune the energy of the characteristic emission to the desired value. According to Eq. (1), increasing the atomic number also increases the efficiency of continuum production. This may be a negative consideration for applications in which a high ratio of characteristic to continuum emission is desired, or a positive effect for applications such as imaging or white beam diffraction, where a large total X-ray power is desired. Equally important for high power density sources are the thermal properties of the anode, which prevent the anode from melting. Metals with good thermal conductivity and high melting temperatures, such as tungsten, are generally chosen.

31.5   EFFECT OF SOURCE VOLTAGE

According to Eqs. (1) and (5), both the characteristic and the continuum radiation intensities increase monotonically with tube voltage. The total bremsstrahlung continuum intensity for a given electrical power increases linearly with increasing voltage. For a fixed current rather than fixed power, the total bremsstrahlung intensity increases as the square of the tube voltage. The width of the output spectrum increases linearly with voltage, so that the bremsstrahlung intensity within a narrow spectral region increases roughly linearly with voltage. The characteristic intensity is low near the minimum energy for core shell ionization, but increases rapidly with increasing voltage. The intensity in a characteristic line becomes equal to the portion of bremsstrahlung intensity in a narrow range around the characteristic line, such as that passed by a monochromator crystal, at some voltage that depends on the monochromator bandwidth. The ratio of the characteristic intensity to the bremsstrahlung within a narrow bandpass increases monotonically with tube voltage. Using Eqs. (1) and (5), the ratio of the characteristic intensity to the total bremsstrahlung intensity (at all energies) peaks at a tube voltage corresponding to

$$eV = \frac{2}{2-p} E_i \qquad (8)$$

roughly five times the ionization energy, for example, about 45 kV for Cu K radiation. The ideal tube voltage may depend on whether the optics will pass undesirable high energy components, for example integral multiples of the characteristic energy. At a fixed current, increasing the tube voltage increases the power that must be dissipated in the anode, and so the comments on source size versus power above are relevant.

31.6   GENERAL OPTIMIZATION

Parallel Beam

As a result of Liouville's theorem, the X-ray source has to be smaller than the desired parallel beam diameter in order to achieve an intensity gain. The gain is produced by collecting over a solid angle from the source that is larger than the output divergence of the collimated beam. Therefore, it is particularly challenging to provide intense, small-cross-section, nearly parallel beams, as are required for some microdiffraction applications.

Focussed Beam

An optic can achieve a gain if it collects a diverging beam from a point source and refocuses it into a converging beam. The diameter of the optic is not limited by the application, but by manufacturing and integration constraints. For any optic, the minimum focal spot size will be smaller for a smaller optic-to-focus distance, as the output divergence of the optic is never zero; and therefore, the best spatial definition of the beam will occur for smaller focal distances.

31.7 EFFECT OF DIFFERENT OPTICS TYPES ON BRIGHTNESS

Imaging optics such as pinholes and mirrors are able to conserve brightness. Pinholes are the most basic optic and can be used to define unity gain. They are also useful to further condition the beam and eliminate scatter at the sample location. A straight, untapered, single capillary tube preserves the beam condition across its length if transport losses are ignored. Therefore, it is identical in performance to placing a pinhole at the input of the capillary. Straight single capillaries can be useful as improved pinholes or to provide more convenient geometries.[8] (Capillaries are discussed in Chap. 29.)

Mirrors are imaging optics and can preserve brightness. An ideal symmetric mirror produces a focal spot size equal to the source spot size. In order to demagnify, the output focal distance has to be reduced. Diffractive optics are a special case of mirrors with the ability to discriminate with respect to energy. Therefore, the same considerations as for regular mirrors are valid, with the exception that the Bragg condition must be met for the characteristic energy of the anode material. (Mirrors are discussed in Chaps. 26–28, and diffraction optics in Chaps. 21 and 22.)

Tapered single capillary tubes can provide a significantly enhanced flux. If shaped appropriately, they can serve as imaging micromirrors and fall under the mirror category. Polycapillary optics use many consecutive reflections to achieve a large overall deflection. The acceptance from the multiple capillary tubes can be overlapped so that the optic collects from a large solid angle with correspondingly large total power. However, polycapillary optics are not imaging and cannot conserve brightness. (Polycapillary optics are discussed in Chap. 30.)

31.8 CHOOSING A SOURCE/OPTIC COMBINATION

The requirements of the application for photon energy, energy bandwidth, beam size, beam divergence, and total power must be carefully analyzed. If total X-ray power is the only consideration, a large spot source with high Z anode material, high voltage, high current, small source-to-sample distance, and no optic is the most appropriate choice. If an optic is to be employed, it must be carefully optimized with the designed source parameters. Large spot sources with large source-to-output window distances are not well matched with optics designed to produce small bright beams. Detector properties, discussed in Chap. 34, also affect the system optimization.

31.9 REFERENCES

1. A. H. Compton and S. K. Allison, *X rays in Theory and Experiment,* second edition, D. Van Nostrand, New York, 1935.

2. A. G. Michette, "X-rays and Their Properties," in A. G. Michette and C. J. Buckley (eds.), *X-ray Science and Technology,* IOP Publishing, London, 1993.

3. A. W. Potts, "Electron Impact X-ray Sources," in A. G. Michette and C. J. Buckley (eds.), *X-ray Science and Technology,* IOP Publishing, London, 1993.

4. *International Tables for Crystallography, Volume C, Mathematical, Physical and Chemical Tables,* E. Prince and A. J. C. Wilson (eds.), Kluwer Academic Press/International Union of Crystallography, Dordrecht, The Netherlands, 1999.

5. J. B. Ullrich, W. M. Gibson, M. V. Gubarev, and C. A. MacDonald, "Potential for Concentration of Synchrotron Beams with Polycapillary Optics," *Nuclear Instruments and Methods in Physics Research* **A347**:401–406 (1994).

6. M. Born and E. Wolf, Principles of Optics, 6th ed., Cambridge University Press, Cambridge, 1999.

7. A. Müller, "On the Input Limit of an X-ray Tube with Circular Focus," *Proc. Roy. Soc.* **117**:30–42 (1927).

8. P. B. Hirsch and J. N. Keller, *Phys. Soc. (London)* **B64**:64–69 (1951).

CHAPTER 32

SYNCHROTRON RADIATION SOURCES

S. L. Hulbert and G. P. Williams

National Synchrotron Light Source
Brookhaven National Laboratory
Upton, New York

32.1 INTRODUCTION

Synchrotron radiation is a very bright, broadband, polarized, pulsed source of electromagnetic radiation extending from the infrared to the X-ray region. Brightness, defined as flux per unit area per unit solid angle, is normally a more important quantity than flux or intensity, particularly in throughput-limited applications, which include those in which monochromators are used.

It is well known from classical theory of electricity and magnetism that accelerating charges emit electromagnetic radiation. In the case of synchrotron radiation, relativistic electrons are accelerated in a circular orbit and emit electromagnetic radiation in a broad spectral range. The visible portion of this spectrum was first observed on April 24, 1947, at General Electric's Schenectady facility by Floyd Haber, a machinist working with the synchrotron team, although the first theoretical predictions were by Liénard[1] in the latter part of the 1800s. An excellent early history with references is presented by Blewett,[2] and a history covering the development of the utilization of synchrotron radiation is presented by Hartman.[3]

Synchrotron radiation covers the entire electromagnetic spectrum from the infrared region through the visible, ultraviolet, and into the X-ray region up to energies of many tens of kilovolts (kV). If the charged particles are of low mass, such as electrons, and if they are traveling relativistically, the emitted radiation is very intense and highly collimated, with opening angles, which depend inversely on the energy of the particle, on the order of 1 milliradian (mrad). In electron storage rings there are two distinct types of sources of synchrotron radiation: *dipole* (bending) *magnets* and *insertion devices*. Insertion devices are further classified either as *wigglers,* which act like a sequence of bending magnets with alternating polarities, or as *undulators,* which are also multiperiod alternating magnet systems, but in which the beam deflections are small, resulting in coherent interference of the emitted light.

In typical storage rings that are used as synchrotron radiation sources, several bunches of up to ~10^{12} electrons circulate in vacuum, guided by magnetic fields. The bunches are typically several tens of centimeters long, so that the light is pulsed, being on for a few tens to a few hundreds of picoseconds, and off for several tens to a few hundreds of nanoseconds depending on the particular machine and the radio-frequency cavity that restores the energy lost to synchrotron radiation. For example, the revolution time for a ring of circumference 30 m is 100 ns, so that each bunch of 10^{12} electrons is seen 10^7 times per second, giving a current of ~1 ampere (A).

The most important characteristic of accelerators built specifically as synchrotron radiation sources is that they have a magnetic focusing system that is designed to concentrate the electrons into bunches of very small cross section and to keep the electron transverse velocities small. The combination of high intensity with small opening angles and small source dimensions results in the very high brightness.

The first synchrotron radiation sources to be used were operated parasitically on existing high-energy physics or accelerator development programs. These were not optimized for brightness, and were usually accelerators rather than storage rings, meaning that the electron beams were constantly being injected, accelerated, and extracted. Owing to the successful use of these sources for scientific programs, a second generation of dedicated storage rings was built starting in the early 1980s. In the mid-1990s, a third generation of sources was built, this time based largely on insertion devices, especially undulators of various types. A fourth generation is also under development based on what is called *multiparticle coherent emission,* in which coherence along the path of the electrons, or longitudinal coherence, plays the major role. This is achieved by microbunching the electrons on a length scale that is comparable to or smaller than the scale of the wavelengths emitted. The emission is then proportional to the square of the number of electrons, N, which, if N is 10^{12}, can be a very large enhancement. These sources can reach the theoretical diffraction limit of source emittance (the product of solid angle and area).

32.2   THEORY OF SYNCHROTRON RADIATION EMISSION

General

The theory describing synchrotron radiation emission is based on classical electrodynamics and can be found in the works of Tomboulian and Hartman,[4] Schwinger,[5] Jackson,[6] Winick,[7] Hofmann,[8] Krinsky, Perlman, and Watson,[9] and Kim.[10] A quantum description, presented by Sokolov and Ternov,[11] is quantitavely equivalent.

Here we present a phenomenological description to highlight the general concepts involved. Electrons in circular motion radiate in a dipole pattern as shown schematically in Fig. 1a. As the electron energies increase and the particles start traveling at relativistic velocities, this dipole pattern appears different to an observer in the rest frame of the laboratory. Special relativity tells us that angles θ_t in a transmitting object are related to those in the receiving frame, θ_r, by

$$\tan \theta_r = \frac{\sin \theta_t}{\gamma (\cos \theta_t - \beta)} \tag{1}$$

with γ, the ratio of the mass of the electron to its rest mass, being given by $E/m_0 c^2$—E being the electron energy, m_0 the electron rest mass, and c the velocity of light. β is the ratio of electron velocity (v) to the velocity of light (c). Thus, for electrons at relativistic energies, $\beta \approx 1$ so the peak of the dipole emission pattern in the particle frame, $\theta_t = 90°$, transforms to $\theta_r \approx \tan \theta_r \approx \gamma^{-1}$ in the laboratory frame as shown in Fig. 1b. Thus, γ^{-1} is a typical opening angle of the radiation in the laboratory frame. For an electron viewed in passing by an observer, as shown in Fig. 2, the duration of the pulse produced by a particle under circular motion of radius ρ will be $\rho/\gamma c$ in the particle frame, or $\rho/\gamma c \times 1/\gamma^2$ in the laboratory frame owing to time dilation. The Fourier transform of this function will contain frequency components up to the reciprocal of this time interval. For a storage ring with a radius of 2 m and $\gamma = 1000$, corresponding to a stored electron beam energy of ~500 MeV, the time interval is 10^{-17} s, which corresponds to light of wavelength 30 Å.

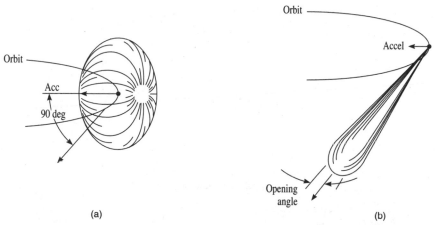

(a)　　　　　　　　　　　　　　　　　　(b)

FIGURE 1　Conceptual representation of the radiation pattern from a charged particle undergoing circular acceleration at (*a*) nonrelativistic and (*b*) relativistic velocities.

Bending Magnet Radiation

For an electron storage ring, the relationship between the electron beam energy E, bending radius ρ, and field B is

$$\rho = \frac{ecE}{B} = \frac{E\,(\mathrm{GeV})}{0.300B(\mathrm{T})}\ \mathrm{m} \tag{2}$$

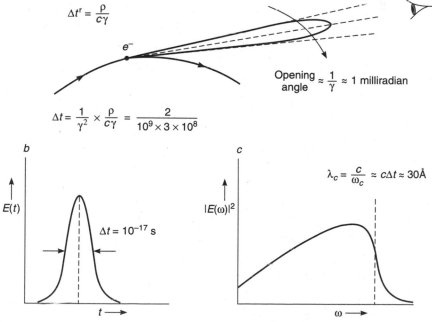

FIGURE 2　Illustration of the derivation of the spectrum emitted by a charged particle in a storage ring.

γ, the ratio of the mass of the electron to its rest mass is given by $\gamma = E/m_0 c^2 = E/0.511 \text{ MeV} = 1957E \text{ (GeV)}$, and λ_c, which is defined as the wavelength for which half the power is emitted above and half below, is

$$\lambda_c = \frac{4\pi\rho}{3\gamma^3} = 5.59 \text{ Å} \frac{\rho \text{ (m)}}{E^3 \text{ (GeV}^3\text{)}} = \frac{18.6 \text{ Å}}{[B \text{ (T)}][E^2 \text{ (GeV}^2\text{)}]} \tag{3}$$

The critical frequency and photon energy are

$$\omega_c = \frac{2\pi c}{\lambda_c} = \frac{3c\gamma^3}{2\rho}, \quad \varepsilon_c \text{ (eV)} = \hbar\omega_c \text{ (eV)} = 665.5 \, [E^2 \text{ (GeV}^2\text{)}][B \text{ (T)}] \tag{4}$$

The angular distribution of synchrotron radiation flux emitted by electrons moving through a bending magnet with a circular trajectory in the horizontal plane is given[9] by

$$\frac{d^2 F_{bm} (\omega)}{d\theta d\psi} = \frac{3\alpha}{4\pi^2} \gamma^2 \frac{\Delta\omega}{\omega} \frac{I}{e} \left(\frac{\omega}{\omega_c}\right)^2 (1 + \gamma^2\psi^2)^2 \left[K_{2/3}^2 (\xi) + \frac{\gamma^2\psi^2}{1 + \gamma^2\psi^2} K_{1/3}^2 (\xi) \right]$$

$$= 1.326 \times 10^{13} \text{ photons s}^{-1} \text{ mrad}^{-2} \cdot 0.1\% \text{ bandwidth}$$

$$\times [E^2 \text{ (GeV}^2\text{)}][I \text{ (A)}] (1 + \gamma^2\psi^2)^2 \left(\frac{\omega}{\omega_c}\right)^2 \left[K_{2/3}^2 (\xi) + \frac{\gamma^2\psi^2}{1 + \gamma^2\psi^2} K_{1/3}^2 (\xi) \right] \tag{5}$$

where θ is the observation angle in the horizontal plane, Ψ the observation angle in the vertical plane, α the fine-structure constant (1/137), ω the light frequency, I the beam current, and $\xi = (\omega/2\omega_c)(1 + \gamma^2\psi^2)^{3/2}$. The subscripted Ks are modified Bessel functions of the second kind. The $K_{2/3}$ term represents light linearly polarized parallel to the electron orbit plane, while the $K_{1/3}$ term represents light linearly polarized perpendicular to the orbit plane.

If one integrates over all vertical angles, then the total intensity is

$$\frac{dF_{bm}(\omega)}{d\theta} = \frac{\sqrt{3}}{2\pi} \alpha\gamma \frac{\Delta\omega}{\omega} \frac{I}{e} \frac{\omega}{\omega_c} \int_{\omega/\omega_c}^{\infty} K_{5/3} (y) dy$$

$$= 2.457 \times 10^{13} \text{ photons s}^{-1} \text{ mrad}^{-1} 0.1\% \text{ bandwidth} \times [E \text{ (GeV)}][I \text{ (A)}] \frac{\omega}{\omega_c} \int_{\omega/\omega_c}^{\infty} K_{5/3} (y) dy \tag{6}$$

The Bessel functions can be computed easily using algorithms of Kostroun[12]:

$$K_v(x) = h\left\{ \frac{e^{-x}}{2} + \sum_{r=1}^{\infty} e^{-x \cosh (rh)} \cosh (vrh) \right\} \tag{7}$$

and

$$\int_x^{\infty} K_v (\eta) d\eta = h\left\{ \frac{e^{-x}}{2} + \sum_{r=1}^{\infty} e^{-x \cosh (rh)} \frac{\cosh (vrh)}{\cosh (rh)} \right\} \tag{8}$$

for all x and for any fractional order v, where h is some suitable interval such as 0.5. In evaluating the series, the sum is terminated when the rth term is small, $<10^{-5}$ for example.

In Fig. 3 we plot the universal function $G_1(\omega/\omega_c) = (\omega/\omega_c) \int_{\omega/\omega_c}^{\infty} K_{5/3} (y) dy$ from Eq. (6), so that the photon energy dependence of the flux from a given ring can be calculated readily. It is found that the emission falls off exponentially as $e^{-\lambda_c/\lambda}$ for wavelengths shorter than λ_c, but only as $\lambda^{-1/3}$ at longer wavelengths.

The vertical angular distribution is more complicated. For a given ring and wavelength, there is a characteristic natural opening angle for the emitted light. The opening angle

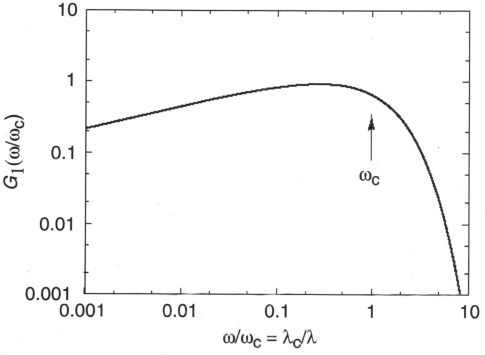

FIGURE 3 Universal synchrotron radiation output curve.

increases with increasing wavelength. If we define ψ as the vertical angle relative to the orbital plane, and if the vertical angular distribution of the emitted flux is assumed to be Gaussian in shape, then the root mean square (rms) divergence, σ_ψ is defined as $1/\sqrt{2\pi}$ times the ratio of Eqs. (5) and (6) evaluated at $\psi = 0$:

$$\sigma_\psi = \sqrt{\frac{2\pi}{3}} \frac{1}{\gamma} \left(\frac{\omega}{\omega_c} \right)^{-1} \frac{\int_{\omega/\omega_c}^{\infty} K_{5/3}(y)dy}{K_{2/3}^2 \left(\omega/2\omega_c \right)} \tag{9}$$

In reality, the distribution is not Gaussian, especially in view of the fact that the distribution for the vertically polarized component vanishes in the horizontal plane ($\psi = 0$). However, σ_ψ, defined by Eq. (9), is still a simple and useful measure of the angular divergence. The photon energy (ω) dependence of the electron-energy-independent quantity $\gamma\sigma_\psi$ is plotted in Fig. 4. At $\omega = \omega_c$, $\sigma_\psi = 0.647/\gamma$. The asymptotic values of σ_ψ can be obtained from the asymptotic values of the Bessel functions and are

$$\sigma_\psi \approx \frac{1.07}{\gamma} \left(\frac{\omega}{\omega_c} \right)^{-1/3} ; \quad \omega \ll \omega_c \tag{10}$$

and

$$\sigma_\psi \approx \frac{0.58}{\gamma} \left(\frac{\omega}{\omega_c} \right)^{-1/2} ; \quad \omega \gg \omega_c \tag{11}$$

In Fig. 5 we show examples of the normalized vertical angular distributions of both parallel and perpendicularly polarized synchrotron radiation for a selection of wavelengths.

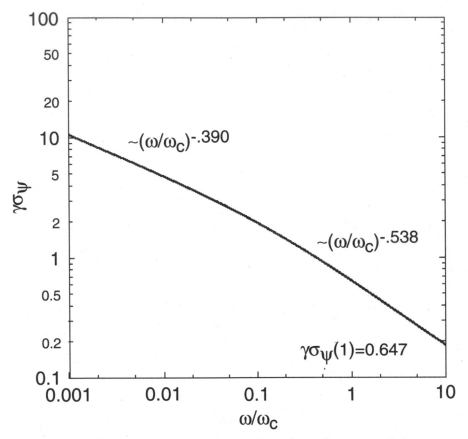

FIGURE 4 Plot of the normalized vertical opening angle $\gamma\sigma_\psi$ for bending magnet radiation.

Circular Polarization and Aperturing for Magnetic Circular Dichroism

Circularly polarized radiation is a valuable tool for the study of electronic, magnetic, and geometric structure of a wide variety of materials. The dichroic response in the soft X-ray spectral region (100–1500 eV) is especially important because in this energy range almost every element has a strong dipole transition from a sharp core level to its lowest unoccupied state.[13]

The production of bright sources of circularly polarized soft X rays is, therefore, a topic of keen interest, and is a problem that has seen a multitude of solutions, from special insertion devices (crossed undulators, helical undulators, elliptically polarized undulators/wigglers) to optical devices (multiple-bounce reflectors/multilayers and quarter-wave plates). However, standard bending magnet synchrotron radiation sources are good sources of elliptically polarized soft X rays when viewed from either above or below the orbital plane.

As discussed by Chen,[13] a practical solution involves acceptance of a finite vertical angular range, $\psi_{off} - \Delta\psi/2 < \psi < \psi_{off} + \Delta\psi/2$, centered about any vertical offset angle $\psi = \psi_{off}$ or, equivalently, about $\psi = -\psi_{off}$. This slice of bending magnet radiation exhibits a circular polarization[14]

$$P_c = -\frac{2A_h A_v}{(A_h^2 + A_v^2)} \tag{12}$$

where $A_h = K_{2/3}(\xi)$ and $A_v = \gamma\psi/(1 + \gamma^2\psi^2)^{1/2} K_{1/3}(\xi)$ are proportional to the square roots of the horizontally and vertically polarized components of bending magnet flux [Eq. (5)], that is, A_h

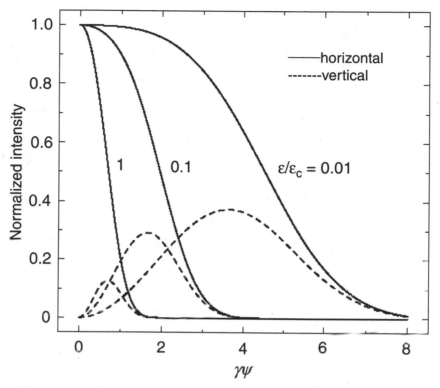

FIGURE 5 Normalized intensities of horizontal and vertical polarization components, as functions of the vertical observation angle for different photon energies.

and A_v are proportional to the horizontal and vertical components of the electric field, respectively. P_c depends on the vertical angle ψ, electron energy γ, and, through ξ, the emitted photon energy ω/ω_c. In Fig. 6 we plot values of P_c versus $\gamma\psi$ and ω/ω_c for $\gamma = 1565$ ($E = 0.8$ GeV) and $\rho = 1.91$ m ($h\nu_{crit} = 594$ eV).

Magnetic circular dichroism (MCD) measures the normalized difference of the absorption of right circular and left circular light. Assuming no systematic error, the signal-to-noise ratio in such a measurement defines a figure of merit

$$\text{MCD figure of merit} = (\text{average circular polarization}) \times (\text{flux fraction})^{1/2} \qquad (13)$$

where

$$\text{average circular polarization} = \frac{\int_{\psi_{off}-(\Delta\psi/2)}^{\psi_{off}+(\Delta\psi/2)} P_c(\psi) \dfrac{dF}{d\psi} d\psi}{\int_{\psi_{off}-(\Delta\psi/2)}^{\psi_{off}+(\Delta\psi/2)} \dfrac{dF}{d\psi} d\psi} \qquad (14)$$

and the fraction of the total (vertically integrated) flux emitted into the vertical slice $\psi = \psi_{off} \pm \Delta\psi/2$ is:

$$\text{flux fraction} = \frac{1}{dF_{bm}(\omega)/d\theta} \int_{\psi_{off}-\Delta\psi/2}^{\psi_{off}+\Delta\psi/2} \frac{d^2 F_{bm}(\omega)}{d\theta d\psi} d\psi \qquad (15)$$

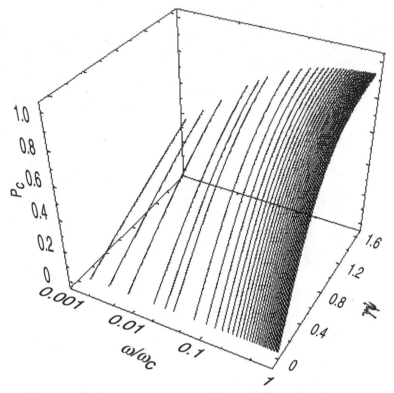

FIGURE 6 P_c versus $\gamma\psi$ versus ω/ω_c for $E = 0.8$ GeV, $\rho = 1.91$ m.

Here $d^2F_{bm}(\omega)/d\theta d\psi$ is the angular dependence of the bending magnetic flux from Eq. (5) and $dF_{bm}(\omega)/d\theta$ is the vertically integrated flux from Eq. (6). For an 0.8-GeV storage ring [e.g., the VUV ring at the National Synchrotron Light Source (NSLS), New York], the best choices of ψ and $\Delta\psi$ are 0.5 mrad and 0.66 mrad, respectively. This yields a flux fraction ~ 0.3, a circular polarization ~ 0.65, and a figure of merit ~ 0.35.

Bending Magnet Power

Integration of $(I/e)\hbar\omega(d^2F_{bm}(\omega)/d\theta d\psi)$ from Eq. (5) over all frequencies ω yields the angular distribution of power radiated by a bending magnet:

$$\frac{d^2P_{bm}}{d\theta d\psi} = \frac{I}{e}\int_0^\infty \hbar\omega \frac{d^2F_{bm}(\omega)}{d\theta d\psi} d\omega = \frac{I}{e}\frac{\alpha hc\gamma^5}{2\pi\rho}\frac{7}{16}F(\gamma\psi)$$

$$= 18.082 \text{ W/mrad}^2 \times \frac{[E^5 (\text{GeV}^5)][I(\text{A})]}{\rho(\text{m})}F(\gamma\psi) \qquad (16)$$

which is independent of the horizontal angle θ as required by symmetry, and the vertical angular dependence is contained in the factor

$$F(\gamma\psi) = \frac{1}{(1 + \gamma^2\psi^2)^{5/2}}\left[1 + \frac{5}{7}\frac{\gamma^2\psi^2}{(1 + \gamma^2\psi^2)}\right] \qquad (17)$$

The first term in $F(\gamma\psi)$ represents the component of the bending magnet radiation parallel to the orbital plane, whereas the second represents the perpendicular polarization component. $F(\gamma\psi)$ and its polarization components are plotted versus $\gamma\psi$ in Fig. 7. Note that the area under the F_{parallel} curve is approximately seven times greater than that for $F_{\text{perpendicular}}$.

Integrating Eq. (17) over the out-of-orbital-plane (vertical) angle ψ yields the total power radiated per unit in-orbital-plane (horizontal) angle θ:

$$\frac{dP_{\text{bm}}}{d\theta} = \frac{I}{e}\frac{hc\alpha\gamma^4}{3\pi\rho} = 14.08 \text{ W/mrad} \times \frac{[E^4\,(\text{GeV}^4)][I\,(\text{A})]}{\rho\,(\text{m})} \tag{18}$$

For example, a 1.0-GeV storage ring with 2-m radius bends generates 7.04 W mrad^{-1} A^{-1} of stored current. By contrast, a 2.5-GeV machine with 7-m radius bends generates 78.6 W mrad^{-1} A^{-1}, and a 7-GeV machine with 39-m radius bends generates 867 W mrad^{-1} A^{-1}.

Bending Magnet Brightness

Thus far, we have calculated the emitted flux in photons per second per milliradian^2 of solid angle. To calculate the brightness, we need to include the source size. In these calculations, we

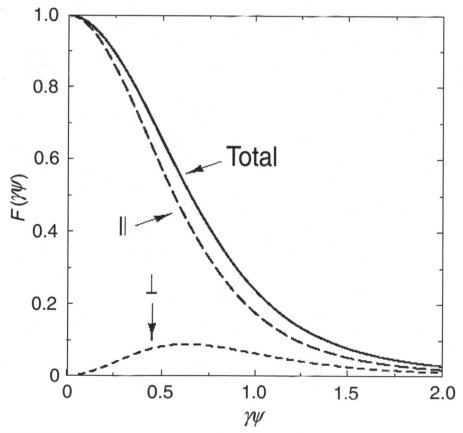

FIGURE 7   Vertical angle dependence of bending magnet power, $F(\gamma\psi)$ versus $\gamma\psi$.

calculate the central (or maximum) brightness, for which we use the natural opening angle to define both the horizontal and vertical angles. Using vertical angles larger than this will not increase the flux, as there is no emission. Using larger horizontal angles will increase the flux proportionately as all horizontal angles are filled with light, but owing to the curvature of the electron trajectory, the *average* brightness will actually be less. The brightness expression[15,16] is:

$$B_{\text{bm}} = \frac{[d^2 F_{\text{bm}}/(d\theta\,d\psi)]_{\psi=0}}{2\pi \sum_x \sum_y} \tag{19}$$

where

$$\sum_x = [\varepsilon_x \beta_x + \eta_x^2 \sigma_E^2 + \sigma_r^2]^{1/2}, \text{ and } \sum_y = \left[\varepsilon_y \beta_y + \sigma_r^2 + \frac{\varepsilon_y^2 + \varepsilon_y \gamma_y \sigma_r^2}{\sigma_\psi^2} \right] \tag{20}$$

ε_x and ε_y are the electron beam emittances in the horizontal and vertical directions, respectively, β_x and β_y are the electron beam beta functions in the horizontal and vertical planes, η_x is the dispersion function in the horizontal plane, and σ_E is the rms value of the relative energy spread. All the electron beam parameters are properties of a particular storage ring. The diffraction-limited source size is $\sigma_r = \lambda/4\pi\sigma_\psi$. The effective source sizes ($\sum_x$ and $\sum_y$) are photon energy dependent via the natural opening angle σ_ψ and the diffraction-limited source size σ_r.

Insertion Devices (Undulators and Wigglers)

General Insertion devices are periodic magnetic structures installed in straight sections of storage rings, as illustrated in Fig. 8, in which the vertical magnetic field varies approximately sinusoidally along the axis of the undulator. The resulting motion of the electrons is also approximately sinusoidal, but in the horizontal plane. One can understand the nature of the spectra that is emitted from these devices by again studying the electric field as a function of time, and this is shown in Fig. 9. This shows that the electric field and hence its Fourier transform, the spectrum, depend critically on the magnitude of the beam deflection in the device. At one extreme, when the magnetic fields are high, as in Fig. 9a, the deflection is large and the electric field is a series of pulses similar to those obtained from a dipole. Such a device is

Radiation from an insertion device

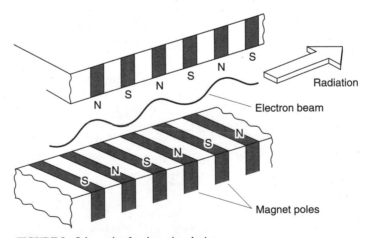

FIGURE 8 Schematic of an insertion device.

termed a *wiggler*. The Fourier transform for the wiggler is N times that from a single dipole. At the other extreme, as in Fig. 9b, the deflection of the electron beam is such that the electric field as a function of time is sinusoidal, and the Fourier transform is then a single peak with a width proportional to the inverse of the length of the wavetrain, L^*, according to $\lambda^2/\Delta\lambda = L^*$. L^* is obtained by dividing the real length of the device, L, by γ^2 because of relativistic effects. Thus, for a meter-long device emitting at a wavelength $\lambda = 10$ nm in a machine of energy 0.5 GeV ($\gamma \sim 1000$), we get $\lambda^2/\Delta\lambda = 10^{-6}$ m, and $\lambda/\Delta\lambda = 1000$. Interference occurs in an undulator because the electric field from one part of the electron path is added coherently to that from adjacent parts.

Formal Treatment   We assume that the motion of an electron in an insertion device is sinusoidal, and that we have a magnetic field in the vertical (y) direction varying periodically along the z direction, with

$$B_y = -B_0 \sin (2\pi z/\lambda_u), \quad 0 \le z \le N\lambda_u \tag{21}$$

where B_0 is the peak magnetic field, λ_u is the wiggler or undulator period length, and N the number of periods. By integrating the equation of motion, the electron transverse velocity $c\beta_x$ is found to be

$$\beta_x = \frac{K}{\gamma} \cos (2\pi z/\lambda_u) \tag{22}$$

where

$$K = eB_0\lambda_u/2\pi mc = [0.934\lambda_u(\text{cm})][B_0(\text{T})] \tag{23}$$

is a dimensionless parameter that is proportional to the deflection of the electron beam. The maximum slope of the electron trajectory is $\delta = K/\gamma$. In terms of δ, we define an undulator as a device in which $\delta \le \gamma^{-1}$, which corresponds to $K < 1$. When K is large, the device is called a *wiggler*. In most insertion devices, the field can be changed either electromagnetically or mechanically, and in some cases K can vary between the two extremes of undulator and wiggler operation.

Wigglers   For the wiggler, the flux distribution is given by $2N$ (where N is the number of magnetic periods) times the appropriate bending magnet formulas in Eqs. (5) and (6). However, ρ or B must be taken at the point in the path of the electron that is tangent to the direction of observation. For a horizontal angle θ,

$$\varepsilon_c(\theta) = \varepsilon_{c\,\text{max}} \sqrt{1 - (\theta/\delta)^2} \tag{24}$$

where

$$\varepsilon_{c\,\text{max}}(\text{keV}) = [0.665E^2 \,(\text{GeV}^2)][B_0\,(\text{T})] \tag{25}$$

from Eq. (4). Integration over θ, which is usually performed numerically, gives the wiggler flux.

The calculation of the brightness of wigglers needs to take into account the depth-of-field effects (i.e., the contribution to the apparent source size from different poles). The expression for the brightness of wigglers is:

$$B_w = \frac{d^2F_w}{d\theta \, d\psi} \sum_{\pm} \sum_{n=(-1/2)N}^{(1/2)N} \frac{1}{2\pi} \times \frac{\exp\left[-\frac{1}{2}\left(\frac{x_0^2}{\sigma_x^2 + z_{n\pm}^2\sigma_x'^2}\right)\right]}{\left[(\sigma_x^2 + z_{n\pm}^2\sigma_x'^2)\left(\frac{\varepsilon_y^2}{\sigma_\psi^2} + \sigma_y^2 + z_{n\pm}^2\sigma_y'^2\right)\right]} \tag{26}$$

where $z_{n\pm} = \lambda_w(n \pm 1/4)$, λ_w is the wiggler period, and σ_ψ is identical to Eq. (9), but evaluated, in the wiggler case, as the instantaneous radius at the tangent to the straight-ahead ($\theta = \psi = 0$)

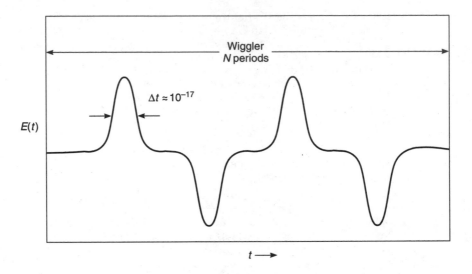

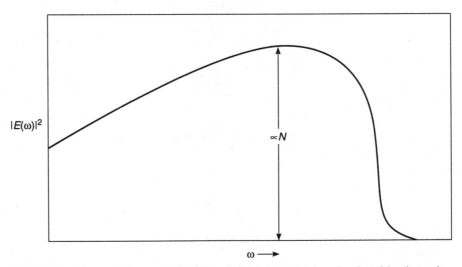

FIGURE 9a   Conceptual representation of the electric fields emitted as a function of time by an electron in a wiggler, with the corresponding spectra.

direction (i.e., minimum ρ, maximum ε_c), $\sigma_x = \sqrt{\varepsilon_x \beta_x}$ and $\sigma_y = \sqrt{\varepsilon_y \beta_y}$ are the rms transverse beam sizes, whereas $\sigma_x' = \sqrt{\varepsilon_x/\beta_x}$ and $\sigma_y' = \sqrt{\varepsilon_y/\beta_y}$ are the angular divergences of the electron beam in the horizontal and vertical directions, respectively. The exponential factor in Eq. (26) arises because wigglers have two source points that are separated by $2x_0$, where

$$x_0 = \frac{K}{\gamma} \frac{\lambda_w}{2\pi} \tag{27}$$

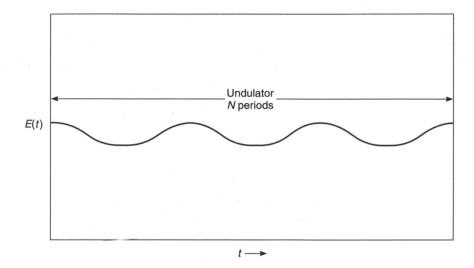

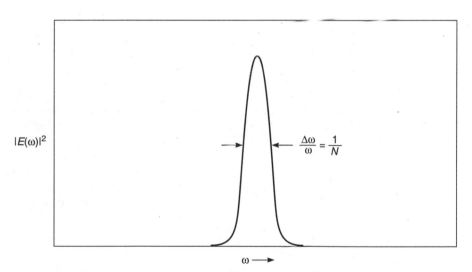

FIGURE 9b Conceptual representation of the electric fields emitted as a function of time by an electron in an undulator, with the corresponding spectra.

The summations in Eq. (26) must be performed for each photon energy because σ_ψ is photon energy dependent.

Undulators The interference that occurs in an undulator [i.e., when K is moderate ($K \leq 1$)], produces sharp peaks in the forward direction at a fundamental ($n = 1$) and all odd harmonics ($n = 3, 5, 7, \ldots$) as shown for a zero-emittance ($\varepsilon = 0$) electron beam in Fig. 10a (dotted line). In the $\varepsilon = 0$ case, the even harmonics ($n = 2, 4, 6, \ldots$) peak off-axis and do not appear in

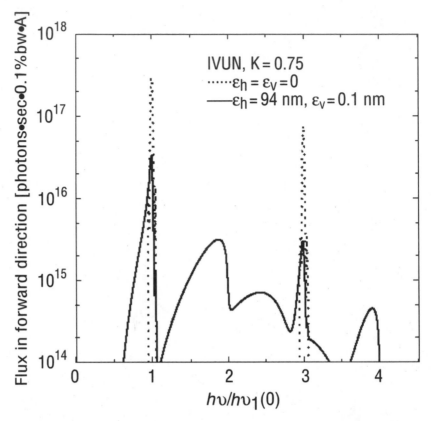

FIGURE 10a Spectral output and angular distribution of the emission from the NSLS In-Vacuum UNdulator (IVUN) for $K = 0.75$. Spectral output in the forward direction, with (solid line) and without (dotted line) the effect of electron beam emittance.

the forward direction. For real ($\varepsilon \neq 0$) electron beams, the spectral shape, angular distribution, and peak brightness are strongly dependent on the emittance and energy spread of the electron beam, as well as the period and magnitude of the insertion device field.

In general, the effect of electron beam emittance is to cause all harmonics to appear in the forward direction (solid line in Fig. 10a). The effect of angle integration on the spectrum in Fig. 10a is shown in Fig. 10b, a spectrum that is independent of electron beam emittance except for the presence of "noise" in the zero-emittance case. The effect of electron beam emittance on the angular distribution of the fundamental, second, and third harmonics of this device is shown in Fig. 10c, which also nicely demonstrates the dependence on harmonic number.

The peak wavelengths of the emitted radiation, λ_n, are given by

$$\lambda_n = \frac{\lambda_u}{2n\gamma^2}\left(1 + \frac{K^2}{2} + \gamma^2\theta^2\right) \tag{28}$$

where λ_u is the undulator period length. They soften as the square of the deviation angle θ away from the forward direction.

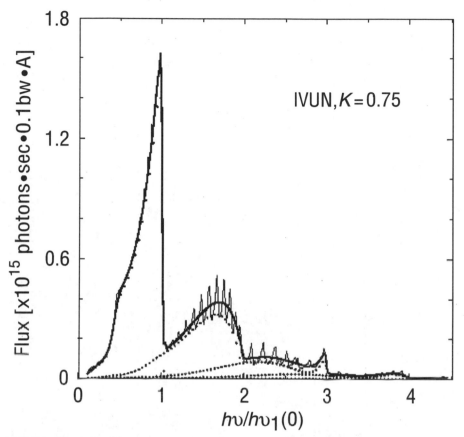

FIGURE 10b Spectral output and angular distribution of the emission from the NSLS In-Vacuum UNdulator (IVUN) for $K = 0.75$. Angle-integrated spectral output with (solid line) and without (faint solid line) the effect of electron beam emittance, and the decomposition into harmonics ($n = 1, 2, 3, 4$) (dotted lines).

Of main interest is the intense central cone of radiation. An approximate formula for flux integrated over the central cone is (for the odd harmonics)

$$F_u(K, \omega) = \pi \alpha N \frac{\Delta \omega}{\omega} \frac{I}{e} Q_n(K), \quad n = 1, 3, 5$$

$$= 1.431 \times 10^{14} \text{ photons s}^{-1} \ 0.1\% \text{ bandwidth} \times [I(A)][NQ_n(K)] \quad (29)$$

where

$$Q_n(K) = \left(1 + \frac{K^2}{2}\right) \frac{F_n(K)}{n}, \quad n = 1, 3, 5 \quad (30)$$

and

$$F_n(K) = \frac{K^2 n^2}{(1 + K^2/2)^2} \left\{ J_{(n-1)/2}\left[\frac{nK^2}{4\left(1 + \frac{1}{2}K^2\right)}\right] - J_{(n+1)/2}\left[\frac{nK^2}{4\left(1 + \frac{1}{2}K^2\right)}\right] \right\}^2 \quad (31)$$

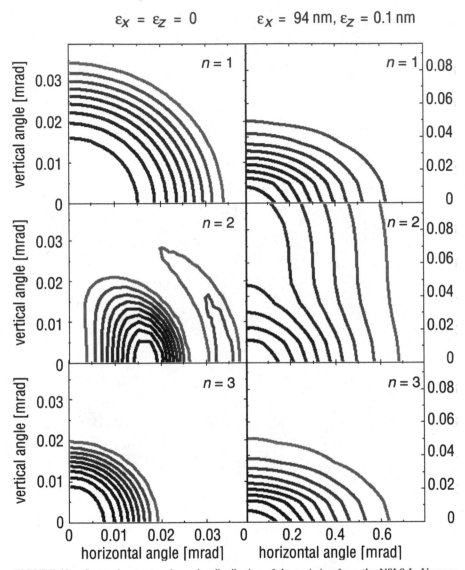

FIGURE 10c Spectral output and angular distribution of the emission from the NSLS In-Vacuum UNdulator (IVUN) for $K = 0.75$. Angular distribution of the first three harmonics ($n = 1, 2, 3$), with and without the effect of electron beam emittance. The emittance of the NSLS X-ray ring is 94 nm horizontal and 0.1 nm vertical.

To calculate the undulator flux angular distribution and spectral output into arbitrary solid angle, one can use freely available codes such as Urgent.[17] To include magnetic field errors (e.g., measured values), use Ur,[18] or SRW.[19]

The brightness of an undulator, B_u, is approximated by dividing the central cone flux by the effective angular divergence, Σ'_x (Σ'_y), and by the effective source size, Σ_x (Σ_y), in the horizontal (vertical) directions. These are given by convolution of the Gaussian distributions of the electron beam and the diffraction limited photon beam, in both space and angle:

$$\sum_{x'} = \sqrt{\sigma_{x'}^2 + \sigma_{r'}^2}, \quad \sum_{y'} = \sqrt{\sigma_{y'}^2 + \sigma_{r'}^2} \tag{32}$$

$$\sum_{x} = \sqrt{\sigma_x^2 + \sigma_r^2}, \quad \sum_{y} = \sqrt{\sigma_y^2 + \sigma_r^2} \tag{33}$$

Thus, B_u is given by

$$B_u = \frac{F_u}{(2\pi)^2 \sum_x \sum_y \sum_{x'}^{\delta} \sum_{y'}^{\delta}} \tag{34}$$

The diffraction-limited emittance of a photon beam is the minimum value in the inequality

$$\varepsilon = \sigma_r \sigma_{r'} \geq \frac{\lambdabar}{2} = \frac{\lambda}{4\pi} \tag{35}$$

where ε is the photon emittance and λ is the wavelength, in direct analogy to the Heisenberg uncertainty principle in nonrelativistic quantum mechanics. The space-versus-angle separation of this minimum emittance is both energy and harmonic dependent.[20] For the exact harmonic frequency in the forward direction, given by Eq. (28) with $\theta = 0$, there appears to be consensus that σ_r and $\sigma_{r'}$ are given by

$$\sigma_{r'} = \sqrt{\frac{\lambda}{2L}} \quad \text{and} \quad \sigma_r = \frac{\sqrt{2\lambda L}}{4\pi} \tag{36}$$

On the other hand, at the peak of the angle-integrated undulator spectrum, which lies a factor of $[1 - (1/nN)]$ below the exact harmonic energy, σ_r and $\sigma_{r'}$ are given by

$$\sigma_{r'} = \sqrt{\frac{\lambda}{L}} \quad \text{and} \quad \sigma_r = \frac{\sqrt{\lambda L}}{4\pi} \tag{37}$$

It is clear from Eqs. (32) and (33) that the choice of expression for σ_r and $\sigma_{r'}$ can have a non-negligible effect on the undulator brightness value, especially for small beam size and opening angle. Lacking a functional form for σ_r and $\sigma_{r'}$ as a function of photon energy, we shall use Eq. (37) in evaluating the expression for undulator peak spectral brightness from Eq. (34).

Insertion Device Power   The Schwinger[5] formula for the distribution of radiated power from an electron in a sinusoidal trajectory, which applies with reasonable approximation to undulators and, to a lesser extent, wigglers, reduces[21] to

$$\frac{d^2P}{d\theta d\psi} = P_{\text{total}} \frac{21\gamma^2}{16\pi K} G(K) f_K(\gamma\theta, \gamma\psi) \tag{38}$$

where the total (angle-integrated) radiated power is

$$P_{\text{total}} = \frac{N}{6} \frac{Z_0 I 2\pi ec}{\lambda_u} \gamma^2 K^2 = 633.0 \text{ W} \times [E^2 \text{ (GeV}^2)][B_0^2 \text{ (T}^2)][L \text{ (m)}] I \text{ [(A)]} \tag{39}$$

N is the number of undulator or wiggler periods, Z_0 is the vacuum impedance (377 Ω), I is the storage ring current, e is the electronic charge, c is the speed of light, $L = N\lambda_u$ is the length of the insertion device,

$$G(K) = \frac{K}{(1+K^2)^{7/2}}\left(K^6 + \frac{24}{7}K^4 + 4K^2 + \frac{16}{7}\right) \tag{40}$$

and

$$f_K(\gamma\theta, \gamma\psi) = \frac{16}{7\pi}\frac{K}{G(K)}\int_{-\pi}^{\pi}d\alpha\left(\frac{1}{D^3} - \frac{4(\gamma\theta - K\cos\alpha)^2}{D^5}\right)\sin^2\alpha \tag{41}$$

where

$$D = 1 + \gamma^2\psi^2 + (\gamma\theta - K\cos\alpha)^2 \tag{42}$$

The integral in the expression for f_K is best evaluated numerically.

For $K > 1$, which includes all wigglers and much of the useful range of undulators, an approximate formula for the angle dependence of the radiated power is

$$f_K(\gamma\theta, \gamma\psi) = \sqrt{1 - \left(\frac{\gamma\theta}{K}\right)^2}\, F(\gamma\psi) \tag{43}$$

where $F(\gamma\psi)$ is the bending magnet formula from Eq. (17). This form clearly indicates the strong weakening of insertion device power as θ increases, vanishing at $\theta = \pm K/\gamma$.

Because $f_K(0,0)$ is normalized to unity, the radiated power density in the forward direction (i.e., along the undulator axis) is

$$\frac{d^2P}{d\theta d\psi}(\theta = 0, \psi = 0) = P_{\text{total}}\frac{21\gamma^2}{16\pi K}G(K)$$

$$= 10.84 \text{ W/mrad}^2 \times [B_0 \text{ (T)}][E^4 \text{ (GeV}^4)][I \text{ (A)}][NG \text{ (K)}] \tag{44}$$

Polarization of Undulators and Wigglers   The polarization properties of the light that is emitted by wigglers is similar to that of dipoles. For both sources, the radiation is elliptically polarized when observed at some angle away from the orbital plane as given by Eq. (5). For radiation from planar undulators, however, the polarization is always linear. The polarization direction, which is in the horizontal plane when observed from that plane, rotates in a complicated way at other directions of observation. A comprehensive analysis of the polarization from undulators has been carried out by Kitamura.[22] The linear polarization of the undulator radiation is due to the symmetry of the electron trajectory within each period. The polarization can in fact be controlled by a deliberate breaking of this symmetry. Circularly polarized radiation can be produced by a helical undulator, in which the series of dipole magnets is arranged, each rotated by a fixed angle with respect to the previous one. For a variable polarization capability, one can use a pair of planar undulators oriented at right angles to each other. The amplitude of the radiation from these so-called crossed undulators is a linear superposition of two parts, one linearly polarized along the x direction and another linearly polarized along the y direction, x and y being orthogonal to the electron beam direction. By varying the relative phase of the two amplitudes by means of a variable-field magnet between the undulators, it is possible to modulate the polarization in an arbitrary way. The polarization can be linear and switched between two mutually perpendicular directions, or it can be switched between left and right circularly polarized. For this device to work, it is necessary to use a monochromator with a sufficiently small bandpass, so that the wave trains from the two undulators are stretched and overlap. Also, the angular divergence of the electron beam should be sufficiently small or the fluctuation in relative phase will limit the achievable degree of polarization. A planar undulator whose pole boundaries are tilted away from a right angle with respect to the axial direction can be used as a helical undulator if the electron trajectory lies a certain distance above or below the midplane of the device.

Transverse Spatial Coherence

As shown by Kim[23] and utilized in the earlier brightness formulas, in wave optics the phase-space area of a radiation beam is given by the ratio of flux (F_0) to brightness (B_0). A diffraction-limited photon beam (no electron size or angular divergence contribution) occupies the minimum possible phase-space area. From Eqs. (32) through (37), this area is

$$(2\pi\sigma_r\sigma_{r'})^2 = (2\pi\varepsilon)^2 = \left(\frac{\lambda}{2}\right)^2 \tag{45}$$

Thus, the phase space occupied by a single Gaussian mode radiation beam is $(\lambda/2)^2$, and such a beam is referred to as completely transversely coherent. It then follows that the transversely coherent flux of a radiation beam is

$$F_{\text{coherent}} = \left(\frac{\lambda}{2}\right)^2 B_0 \tag{46}$$

and the degree of transverse spatial coherence is

$$\frac{F_{\text{coherent}}}{F_0} = \left(\frac{\lambda}{2}\right)^2 \frac{B_0}{F_0} \tag{47}$$

Conversely, the number of Gaussian modes occupied by a beam is

$$\frac{F_0}{F_{\text{coherent}}} = \frac{F_0}{B_0(\lambda/2)^2} \tag{48}$$

Transverse spatial coherence is the quantity that determines the throughput of phase-sensitive devices such as Fresnel zone plates used for X-ray microscopy.

Fourth-Generation Sources

For completeness, we discuss fourth-generation sources at least conceptually. These sources are of even higher brightness than the devices discussed in the preceeding text and are based on multiparticle coherence that can be understood as follows. In Fig. 1 the electric field induced by one electron, and hence the intensity, is proportional to the charge on an electron. If N electrons are circulating together in a storage ring, the emission is simply proportional to N times the emission of a single electron. However, when the electrons circulating in the storage ring, or passing through an insertion device, are close together compared with the wavelength of the light being emitted,[24] the electric fields add coherently, so that the intensity scales like N^2. The electrons can be forced to microbunch when they are in the presence of the electric field of a superimposed light beam and a magnetic field. The degree of multiparticle enhancement depends on the degree to which the microbunching occurs. In these devices, a light beam either from a seed laser, or from back-reflection of light that is spontaneously emitted from the same electrons in a previous pass through the device, causes the electrons to microbunch. New, even brighter sources of VUV radiation are being planned based on these principles.[25]

32.3 CONCLUSION

We have attempted to compile the formulas needed to calculate the flux, brightness, polarization (linear and circular), and power produced by the three standard storage ring synchrotron radiation sources: bending magnets, wigglers, and undulators. Where necessary, these formulas have contained reference to the emittance (ε) of the electron beam, as well as to the electron beam size (σ) and its divergence (σ'). For all three types of sources, the source phase space area [i.e., the spatial and angular extent of the effective (real) source], is a convolution of its electron and photon components. For a more detailed description of these properties, see Ref. 26 and references therein.

32.4   REFERENCES

1. A. Liénard, *L'Eclairage Electrique* **16**:5 (1898).

2. J. P. Blewett, *Nuclear Instr. & Methods* **A266**:1 (1988).

3. P. C. Hartman, *Nuclear Instr. & Methods* **195**:1 (1982).

4. D. H. Tomboulian and P. L. Hartman, *Phys. Rev.* **102**:1423 (1956).

5. J. Schwinger, *Phys. Rev.* **75**:1912 (1949).

6. J. D. Jackson, *Classical Electrodynamics,* Wiley, New York, 1975.

7. H. Winick, *Synchrotron Radiation Research,* Chap. 2, Plenum Press, 1980.

8. A. Hofmann, *Physics Reports* **64**:253 (1980).

9. S. Krinsky, M. L. Perlman, and R. E. Watson, *Handbook of Synchrotron Radiation,* E. E. Koch (ed.), Chap. 2, North-Holland, 1983.

10. K. J. Kim, "Physics of Particle Accelerators," *AIP Proceedings* **184**:565 (1989).

11. A. A. Sokolov and I. M. Ternov, *Synchrotron Radiation,* Cambridge University Press, 1968.

12. V. O. Kostroun, *Nuclear Instruments and Methods* **172**:371 (1980).

13. C. T. Chen, *Review of Scientific Instruments* **63**:1229 (1992).

14. M. Born and E. Wolf, *Principles of Optics,* Pergamon Press, New York, 1964.

15. Lawrence Berkeley Laboratory Publication 643, Rev. 2, (1989).

16. S. L. Hulbert and J. M. Weber, *Nucl. Instruments and Methods* **A319**:25 (1992).

17. R. P. Walker and B. Diviacco, *Rev. of Scientific Instruments* **63**:392 (1992).

18. R. P. Dejus and A. Luccio, *Nucl. Instruments and Methods* **A347**:61 (1994).

19. http://www.esrf.fr/machine/support/ids/Public/Codes/SRW/srwindex.html.

20. M. R. Howells and B. M. Kincaid, LBL Report No. 34751 (1993).

21. K. J. Kim, *Nuclear Instruments and Methods* **A246**:67 (1986).

22. H. Kitamura, *Japanese Journal of Applied Physics* **19**:L185 (1980).

23. K. J. Kim, *Nuclear Instruments and Methods* **A246**:71 (1986). See also earlier work of A. M. Kondratenko and A. N. Skrinsky, *Opt. Spectroscopy* **42**:189 (1977) and F. Zernike, *Physica* **V**:785 (1938).

24. C. J. Hirschmugl, M. Sagurton, and G. P. Williams, *Phys. Rev.* **A44**:1316 (1991) and references therein.

25. See the World Wide Web Virtual Library: Free Electron Laser research and applications, at http://sbfel3.ucsb.edu/www/vl fel.html.

26. S. L. Hulbert and G. P. Williams, "Synchrotron Radiation Sources", in *Vacuum Ultraviolet Spectroscopy I, J. A. R. Samson and D. L. Ederer (eds.), Experimental Methods in the Physical Sciences,* vol. 31, p. 1, Academic Press, San Diego, 1998.

CHAPTER 33
NOVEL SOURCES

Alan Michette
King's College London
Physics Department
Strand, London,
United Kingdom

33.1 INTRODUCTION

The impetus behind the development of most novel sources, with the exception of X-ray lasers, is to bring something approaching the power of a synchrotron into a compact, essentially single-user, source. Of course, this cannot be done with the full versatility of a synchrotron, but such sources can be much cheaper and therefore more suited to a university, research, or industrial laboratory. Examples of novel sources include laser-generated or pinch plasmas, and typical output characteristics of these sources are summarized in Table 1. Other processes that could be used to provide compact X-ray sources include channeling radiation, transition radiation, and parametric radiation but, although these have been demonstrated, sources utilizing these phenomena have not yet been built.

X-ray lasers generally require the use of large facilities. They can be produced through the generation of plasmas by ultra-high-power, longer-wavelength lasers (e.g., NOVA, GEKKO, VULCAN) or by the use of an optical cavity in a synchrotron undulator, known as a *free-electron laser* (FEL). Typical output characteristics of X-ray lasers are summarized in Table 2, although it should be stressed that rapid advances are taking place and these parameters may therefore change in the very near future.

33.2 LASER-GENERATED PLASMAS

Laser plasmas are produced by focusing a pulsed laser beam (typically Nd:YAG at 1.064 μm, possibly frequency multiplied, or KrF excimer at 249 nm) onto a solid (tape[1]), liquid-droplet,[2] or gas-jet target.[3] The irradiance (focused intensity per unit area per unit time) required is typically from approximately 10^{17} to approximately 10^{19} W m^{-2}, which heats the target material to approximately 10^{6}K, thereby ionizing it to produce the plasma. The requirement for high irradiance means that high-beam energies, small focused-spot sizes, and short pulse lengths are needed. Typically, for repetitive systems,[1] pulse energies in the range of approximately 10 mJ to 1 J are used, with pulse lengths of several nanoseconds down to a few picoseconds and focal spot sizes of about 10 μm. For single-pulse systems, using lasers with much higher pulse energies, focal spot sizes can be much larger.

TABLE 1   Characteristics of Plasma Sources

Source	Nature	Duration	Spectral Emission	Source Size	Divergence
Laser-generated plasma (LGP)	Repetitive (<~kHz) or single shot	~fs–ns	<~2 keV, line or quasi-continuum	~10–100 µm	2π–4π steradians
Pinch plasma	Single shot	~ns	Line or quasi-continuum	>~100 µm	4π steradians

The spectral emission characteristics depend mainly on the target material, with the proviso that the irradiance must be high enough to produce the ionic state required to give a particular spectral feature. The use of a tape target, compared with those using liquid droplets or gas jets, allows a wider range of materials to be used, but the effects of increased debris emission must be alleviated by using a low-pressure buffer gas. For low-Z materials, it is common to nearly strip the atoms, and emission is then from H- and He-like ions. The spectrum largely consists of characteristic line emission, with small contributions of continuum radiation (bremsstrahlung or recombination radiation). A suitable material for many purposes is Mylar ($C_{10}H_8O_4$), in plentiful and cheap supply as the substrate of audio- and videotapes. This gives a spectrum with characteristic lines of H- and He-like carbon and oxygen, shifted compared with neutral hydrogen and helium into the soft X-ray region due to the nuclear charge, with some recombination radiation. The lines typically have a bandwidth of $\Delta\lambda/\lambda \sim 10^{-4}$.

For a repetitive source operated at about 100 Hz, the brightness (photons s^{-1} $mm^{-2}mrad^{-2}$ in 0.1 percent bandwidth) in a particular line [e.g., the H-like carbon Lyman-α line ($\lambda = 3.37$ nm)] can be comparable with that of a synchrotron like the National Synchrotron Light Source (NSLS), albeit only at this wavelength. Use of a higher-Z target (e.g., copper or iron) means that the ions are less fully stripped, and the emission from many closely spaced ionic energy levels merges into a quasi-continuum. The overall emission is higher than that of a low-Z target, but considerably less than that of a synchrotron, whereas the peak emission is much less than that of a low-Z material.

Choice of Optic

For applications requiring high spatial resolution, such as X-ray microscopy (Chap. 35, Sec. 35.3) or X-ray microprobes for studies of cellular radiation response, the optic of choice is a zone plate, which requires monochromatic radiation with high flux (for imaging microscopy) or brightness (for scanning microscopy). Thus, assuming efficient spectral conversion, a low-Z target should be used, with individual spectral lines selected by filters,[1] with the zone plate sufficiently far from the source that the demagnified source size is less than the outer zone width, so that the resolution is not compromised. For a zone plate diameter of 200 µm with an outer zone width of 50 nm, the focal length is approximately 3 mm at $\lambda = 3.37$ nm, and assuming a source size of 20 µm, a demagnification of about 300 is required for equality of the source-limited and zone plate–limited resolutions, necessitating a source distance of about 1 m. For applications such as X-ray lithography, needing high flux but with no strict monochromaticity requirements, the optic of choice would be a high-bandpass multilayer or a grazing incidence system.

TABLE 2   Characteristics of X-Ray Lasers

Source	Nature	Duration	Spectral Emission	Source Size	Divergence
LGP X-ray laser	Normally single shot	~fs–ns	Line (not tunable), <~0.5 keV	~0.1–1 µm	~mrad
FEL	Quasi-continuous	—	Line (tunable), <~0.5 keV	~10–100 µm	~µrad

33.3   PINCH PLASMAS

In pinch plasmas, an ionized gas is compressed by a magnetic field to form a hot dense plasma that can, just like an LGP, be a strong emitter of soft X rays. The various types of pinch plasma (e.g., z-pinch,[4] and θ-pinch[5]) differ primarily in the direction of the magnetic field. Compared with LGPs, pinch plasmas have higher overall conversion efficiency of input electrical power to X-ray emission, but the spatial repeatability is worse and the range of available materials is smaller. In addition, pinch plasmas tend to be single-shot systems. Thus, LGPs are more versatile, but a pinch plasma may be preferred for experiments requiring very high peak fluxes in single shots.

Choice of Optic

It is unlikely that pinch plasmas will be widely used in applications such as X-ray microscopy, which require the use of zone plates. The optic of choice, therefore, would tend to be a multilayer or a grazing incidence system.

33.4   CHANNELING RADIATION

Channeling radiation occurs when a charged particle moves along channels formed by crystal planes, and is deviated slightly by the periodic electric field in the crystal.[6] The spectrum of emitted energies depends on the characteristics of the particle and of the crystal, and the energies can be tuned by varying the particle energy or the incidence angle. For electrons or positrons of energy E_e passing through silicon, a rather broad X-ray emission spectrum is obtained, with peaks at a few tenths of E_e. The intensity scales rather rapidly with energy, meaning that it is of most interest in the hard X-ray region, where crystal optics are most suitable.

33.5   TRANSITION RADIATION

Transition radiation occurs when a relativistic charged particle crosses an interface between two materials of different dielectric properties.[7] The intensity of the radiation is roughly proportional to the particle energy, and the total irradiated energy depends on the squared difference of the plasma frequencies of the two materials. The angular distribution is peaked in the forward direction with an angular divergence of approximately $1/\gamma$, where γ is the usual relativistic factor, and the average number of radiated photons is about $\alpha\gamma$, where α is the fine structure constant. These properties mean that transition radiation can usually only be used to produce hard X rays, and so could be used with crystal optics.

33.6   PARAMETRIC RADIATION

Parametric radiation (PXR) is emitted when relativistic electrons pass through a single crystal, and is generated by the polarization of the lattice atoms by the electrons.[8] Parametric radiation can be considered to be due to the diffraction of virtual photons, and so the energy and emission angle of the emitted X rays must satisfy the Bragg law. Thus, two peaks, corresponding to positive and negative diffraction orders, are obtained, which in general are of different intensities. The absolute intensities are about three orders of magnitude smaller than in channeling radiation, but so are the linewidths, and thus the spectral densities are comparable. Similar comments concerning the choice of X-ray optics apply as made for transition radiation.

33.7 X-RAY LASERS

The most common form of X-ray laser is a particular type of LGP, in which the spontaneous emission from a collection of ions in population-inverted states is linearly amplified by the same ions. If the plasma is formed by focusing the laser beam to a line focus, with a cylindrical lens, then appreciable gain can be observed along the axis, provided that the target consists of a suitable material raised to suitable ionic levels.[9] Normally, the target will be irradiated simultaneously from several directions, by splitting the laser beam, to ensure that the plasma column forms evenly. The minimum X-ray wavelength that can be achieved is limited by three factors: (1) gain decreases rapidly, (2) pumping powers increase rapidly, and (3) cavity mirrors (even multilayers) become increasingly inefficient with decreasing wavelength. Because of this last point, most X-ray lasers are designed around single passes of the lasing plasma.

An alternative to this type of inversion X-ray laser is based on high-order parametric conversion of visible or UV laser radiation.[10]

Choice of Optic

Essentially the same comments apply as for LGPs, with the extra condition that the optic must be robust enough to withstand the extremely high pulse powers that can be achieved with current X-ray lasers.

33.8 FREE-ELECTRON LASERS

A free-electron laser[11] (FEL) produces laser radiation by passing a relativistic electron beam through an array of magnets that are arranged to produce an undulating field. In this sense, an FEL can be considered to be an undulator (see Chap. 32 on synchrotron radiation) in an optical cavity. The advantages of FELs over "conventional" X-ray lasers include the wavelength tunability (by varying the electron energy or the magnetic field strength) and the quasi-continuous time structure. A disadvantage, for some applications, is the somewhat larger source size.

Choice of Optic

To take advantage of the wavelength tunability, the most likely choice of optic is a grazing incidence system, as the focal length is independent of wavelength (unlike a zone plate) and, at angles below the critical angle, the reflectivity is roughly constant.

33.9 REFERENCES

1. C. E. Turcu, I. N. Ross, P. Trenda, C. W. Wharton, R. A. Meldrum, H. Daido, M. S. Schulz, P. Fluck, A. G. Michette, A. P. Juna, J. R. Maldonado, H. Shields, G. J. Tallents, L. Dwivedi, J. Krishnan, D. L. Stevens, T. L. Jenner, D. Batani, and H. Goodson, "Picosecond Excimer Laser-Plasma X-Ray Source for Microscopy, Biochemistry, and Lithography," *Applications of Laser Plasma Radiation, Proc. SPIE* **2015**:243–260 (1993).

2. H. M. Herz, L. Rymell, M. Berglund, and L. Malmqvist, "Debris-Free Liquid-Target Laser-Plasma X-Ray Sources for Microscopy and Lithography," in *X-Ray Microscopy and Spectromicroscopy,* Vols. 3–13, J. Thieme, G. Schmahl, D. Rudolph and E. Umbach (eds.), Springer, Berlin, 1998.

3. H. Fiedorowicz, A. Bartnik, H. Szczurek, H. Daido, N. Sakaya, V. Kmetik, Y. Kato, M. Suzuki, M. Matsumura, J. Tajima, T. Nakayama, and T. Wilhein, "Investigation of Soft X-Ray Emission from a Gas Puff Target Irradiated with a Nd:YAG Laser," *Opt. Commun.* **163**:103–114 (1999).

4. T. W. L. Sanford, R. C. Mock, R. B. Spielman, M. G. Haines, J. P. Chittenden, K. G. Whitney, J. P. Apruzese, D. L. Peterson, J. B. Greenly, D. B. Sinars, D. B. Reisman, and D. Mosher, "Wire Array Z-Pinch Insights for Enhanced X-Ray Production," *Phys. Plasmas* **6**:2030–2040 (1999).

5. M. H. Elghazaly, A. M. Abd Elbaky, A. H. Bassyouni, and H. Tuczek, "Spectroscopic Studies of Plasma Temperature in a Low-Pressure Hydrogen Plasma," *J. Quant. Spectrosc.* **61**:503–507 (1999).

6. M. A. Kumakhov and F. F. Komarov, *Radiation from Charged Particles in Solids,* American Institute of Physics, New York, 1989.

7. V. L. Ginzburg and V. N. Tsytovich, *Transition Radiation and Transition Scattering,* Adam Hilger, Bristol, 1990.

8. Y. Matsuda, T. Ikeda, H. Nitta, H. Minowa, and Y. H. Ohtsuki, "Numerical Calculation of Parametric X-Ray Radiation by Relativistic Electrons Channelled in a Si Crystal," *Nucl. Instrum. Meth. B* **115**:396–400 (1996).

9. R. C. Elton, *X-Ray Lasers,* Academic Press, San Diego, 1990.

10. S. Meyer, B. N. Chichkov, and B. Wellegehausen, "High-Order Parametric Amplifiers," *J. Opt. Soc. Am. B* **16**:1587–1591 (1999).

11. S. Doniach, "Studies of the Structure of Matter with Photons from an X-Ray Free-Electron Laser," *J. Synchrotron Rad.* **3**:260–267 (1996).

DETECTORS

CHAPTER 34
X-RAY DETECTORS

Walter M. Gibson
University at Albany, State University of New York
Albany, New York

34.1 INTRODUCTION

X rays can interact with gases, liquids, or solids to produce electronic, atomic, molecular, or collective excitations. Measurement of these excitations can be used to determine the X-ray intensity, energy, and position. A large variety of phenomena have been used for such detection. These include ionization of atoms and molecules, production of electrons and holes in semiconductors, excitation of both short-lived and metastable electronic states in liquids and solids, and excitation of phonons and Cooper pairs in superconductors. Selection of the detector or detectors to be used for a given application depends on the particular requirements and constraints inherent in that application, and will typically involve such factors as intensity, energy resolution, position resolution, size, convenience, and cost. To help with consideration of the kinds of trade-offs involved in selection of a particular detector, and, in the context of this handbook, the factors that influence the optimization of a particular optic-detector system, this chapter summarizes the detector choices in terms of their type and function. There are no comprehensive reviews specifically devoted to X-ray detectors. Most books cover detectors for several types of ionizing radiation.[1-4] A useful summary specifically devoted to X-ray detectors is given by Buckley,[5] and a summary of X-ray detectors for astronomy is given by Fraser.[6]

34.2 DETECTOR TYPE

Ionization

Ionization Chamber.   Ionization chambers contain a gas in a region of high electric field, usually produced with flat plates. X rays incident on a gas can excite bound electrons, resulting in their separation from the atom with energy equal to the difference between the absorbed X-ray energy and the binding energy of the electron. The emitted electron can, if it has sufficient energy, result in further ionization. The average X-ray (actually secondary electron, because these produce most of the ionization) energy loss to produce an electron-ion pair depends on the gas, but it is typically about 30 electronvolts (eV). This is considerably in excess of the energy needed to ionize the atom, with the difference going into dissociation and excitation of molecules. The electrons and positive ions can then be separated by an applied electric field. The resulting charge current in the external circuit is a measure of the number of electron-ion pairs produced. Because the electrons move much faster than the heavy positive ions, the short-term charge pulse is primarily a measure of the electron motion. Ionization chambers have frequently been used as spectrometers for energetic ions, such as alpha particles or fission fragments, because they have shorter range than X rays, permitting an electrode configuration that allows only electron motion to be measured.

The longer range of X rays precludes easy separation of electron and ion motion. Also, the statistical fluctuations are larger than in some other spectrometers, so ion chambers are infrequently used to measure X-ray energy distributions. However, ion chambers are the basis of frequently used radiation-monitoring devices, such as the pencil-type electrometers that are used as personal monitors. Another widespread application is to measure the X-ray beam intensity in synchrotron-based experiments. In this application, the ion chamber can be relatively thin (absorbing only a very small but proportional fraction of the incident X-rays), and the resulting current flow in the external circuit is used to measure the X-ray intensity. For example, an ion chamber before a sample and one after can be used to measure the X-ray absorption in the sample as in X-ray absorption fine structure (EXAFS)[7,8] and X-ray absorption near-edge structure (XANES)[9] experiments. The gas in an ion chamber can just be air at atmospheric pressure as in monitoring devices. Frequently, a rare gas such as helium or argon is used to suppress electron-ion recombination, often with added methane or other gas to increase the electron drift velocity. A general discussion of ionization chambers can be found in reviews of radiation detectors.[1-4]

Proportional Counter.   If the electric field gradient in an ionization chamber is increased high enough, even low-energy electrons released during the X-ray energy loss process can be accelerated enough to produce further ionization. Although, for charged particles or electrons, a parallel plate geometry is frequently used, for X-rays, thin wires are often used as the positive electrode so that the acceleration takes place in a relatively confined space, close to the anode. Over a relatively wide applied voltage range, the charge multiplication increases with applied field and is constant at a given applied field. This multiplication results in charge amplification that is fast, low noise, and because it takes place close to the anode, provides effective separation from the ion current. The multiplication factor can be large, typically in the range between 10^4 and 10^6. The multiplication gain can be stabilized and made less sensitive to the bias voltage as well as the particle energy or counting rate by addition of a small amount of a polyatomic gas to the rare gas. A typical gas mixture is 10 percent methane and 90 percent argon, known as "P-10 gas." Proportional counters have been extensively used as X-ray spectrometers. They are particularly useful for low-energy X rays because the intrinsic detector amplification reduces the effects of external amplifier noise, and because they can be used in a windowless mode. Measurements have been made for energies less than 1 kiloelectronvolt (keV).[10] At high X-ray energies, the sensitivity is limited by the reduced absorption of X rays in the gas, although the energy discrimination for even high-energy X rays that are absorbed remains good so long as the secondary electrons are stopped in the detector. The effective energy range is dependent on the gas pressure and type, the detector size and shape, and on the presence of applied magnetic or electric fields, which can cause the secondary electrons to adopt spiral paths. X rays with energy as high as 100 keV have been measured with special detector designs.[11]

One of the most useful applications of proportional counters for X rays is position determination.[12] If the charge current in an anode wire is collected from both ends of the wire, resistive separation in the wire can be used to determine the position along the wire at which the charge is collected. Position resolution of fractions of millimeters can be achieved this way. Furthermore, if a number of parallel anode wires are used, the relative charge collection on adjacent wires can be used to determine the position of the X-ray absorption between the wires. In this way, two-dimensional position determinations can be achieved. At the same time, energy discrimination can be obtained by combining the charge signal for a given event from both ends of the anode wire and from adjacent wires. This has been used for a number of applications, notably measurement of X-ray diffraction distributions and imaging of astrophysical X-ray sources.[13]

Geiger Counter. If the electric field in an X-ray proportional counter is increased beyond the proportional region (often by making the diameter of the anode wire smaller), the electron multiplication increases catastrophically, resulting in large-current pulses that are no longer proportional to the X-ray energy. This is called the *Geiger region*. In fact, the multiplication is large enough that it is usually necessary to include components in the gas mixture that will quench the resulting electrical discharge. One benefit of this mode of detection is the large amplification that makes it possible to detect X rays with simplified electronics. This is why the most common use of such detectors is in portable or remote radiation monitors.

Semiconductor Detector. In gases, X-ray absorption can produce electrons and positive ions that can be separated and measured by an applied electric field. In solids, the situation is somewhat different. In a conductor, the electrons that are excited by absorption of X rays are quickly dissipated in energy and lost among the abundant conduction electrons in the material, and the positive charge is almost immediately neutralized by conduction electrons. Furthermore, an electric field cannot be maintained in the material. In an insulator, it is easy to establish an electric field to sweep out excited electrons. However, the positive charge remains behind, trapped and immobile. The electric field due to this trapped charge quickly cancels the applied field. As a result, initial charge pulses from incident X rays decrease in size and intensity as a result of such polarization effects.[14] However, for semiconductors, the situation is much more favorable. By doping with an appropriate impurity or formation of a surface barrier, a rectifying *pn* junction can be formed that allows an electric field to be established in the solid. The thickness of the field region, called the *depletion layer* because it is free of mobile charge carriers, depends on the applied voltage and on the resistivity of the bulk semiconductor.

X rays absorbed within the depletion layer ionize bound electrons with energy equal to the difference between the absorbed X ray and the binding energy of the electron. These energetic electrons are free to move in the solid and produce further ionization by interaction with bound electrons. In this sense, the situation is similar to X-ray absorption in ionization chambers. Indeed, semiconductor detectors are sometimes referred to as *solid-state ion chambers*. However, there is an important difference. Instead of the ionization producing a heavy, slow-moving positive ion as in a gas, or an immobile positive charge as in an insulator, it produces a positively charged "hole." The hole is a vacancy in the electronic valence band of the solid, with low effective mass and high mobility, closer to that of free or conduction electrons than to positive ions. The negatively charged electrons and positively charged holes are swept apart by the electric field in the depletion layer, producing a current pulse in the external circuit. Contrary to the ion chamber, both the electron and hole motion contribute to the short-term charge pulse. The amplitude of the charge pulse is proportional to the energy of the absorbed X ray. The energy resolution depends on the statistics of the ionization process and on the electronic noise in the associated electronics, which depends in part on the capacitance of the semiconductor *pn* junction. Detailed reviews of semiconductor detectors are given in Refs. 1, 15, and 16.

To increase the depletion layer thickness and therefore the sensitive volume of the detector, very high-resistivity semiconductor material is required. For silicon, this is achieved by drifting lithium ions through the diode at an elevated temperature in the presence of an electric field.[17–19] The lithium compensates bound impurities in *p*-type silicon. Such detectors are called *lithium-drifted silicon,* or Si(Li), detectors. An alternative is to use very high-purity germanium (HpGe). Both Si(Li) and HpGe detectors are cooled (typically by liquid nitrogen) to reduce electronic noise due to thermally generated electrons and holes [and for Si(Li) to prevent out diffusion of the compensating lithium ions]. Typically, the energy resolution for such detectors is between 120 and 140 eV.

Wide bandgap semiconductors, such as CdZnTe or CdTe, have been used because of their stronger absorption and the possibility of room temperature operation. Such materials frequently exhibit trapping of minority carriers (typically holes) and therefore can show reduced resolution and sometimes polarization effects. They are of particular interest for use in arrays (typically by segmentation of one of the electrodes) for medical or astrophysical applications.[20,21]

Channel Electron Multipliers.   X-ray and visible photons can be detected by photocathodes and electron multipliers. Photoelectrons that are released from thin films or surfaces, accelerated, and electrostatically focused on another surface produce increasing numbers of secondary electrons as they move from surface to surface, until they are collected on an anode and measured as a charge pulse in an electronic circuit. The first such multipliers, similar to photomultiplier tubes, were introduced in the early 1960s. These were largely replaced with simpler and more stable semiconducting glass or ceramic tubes with a voltage applied between the input and output of the tube. These can be thought of as continuous-dynode electron multipliers. Modern channel electron multipliers (CEM) are made from lead glasses, which contain PbO, SiO_2, and several weight percent of alkali metal oxides. These are light, compact, and quite rugged and are used in X-ray pulse-counting detectors in applications ranging from electron microscopy to X-ray astronomy. The electron gain depends on the secondary electron yield of the surfaces employed, the applied voltage, and the channel diameter and length. If gas is present in the channel, ionization can take place, resulting in further electron multiplication and higher apparent gain. However, ionization can lead to feedback resulting from positive ions accelerating back up the channel, leading to long-term "ringing" of the signal due to electrons released by accelerated ions. Ion feedback can be suppressed if the channel is operated at very high vacuum. A more convenient solution is to curve the channel so that positive ions do not accelerate to high-enough velocity to produce secondary electrons before they strike the channel wall. Gain saturation results when the supply of charge to the channel wall lags the depletion rate by secondary electron emission. In practice, operational gains of from 10^4 to 10^6 are possible.[4]

Very high counting rates, and especially imaging of X rays, are possible by the use of microchannel plates (MCP), which are arrays of very fine bore CEMs, typically with channel diameters and channel separations of a few tens of micrometers.[22] Curving the channels in MCPs to suppress ion feedback is difficult and expensive, so two (or more) tilted channel MCPs are put together to form "chevron" multipliers in modern detectors. Energy resolution in CEM detectors is poor, and their efficiency decreases with increasing X-ray energy. Their simplicity, low cost, and, for MCPs, high spatial resolution make them very useful for many applications. Microchannel plates are also used as electron multipliers for low-noise amplification when used with scintillation detectors.

Scintillation

Incident X rays can excite electrons in atoms or molecules to excited electronic states, which then deexcite by returning to their ground state with accompanying photon emission. The excitation can be caused directly by the X ray or, more usually, by secondary energetic electrons excited by the X ray. The emitted photons (usually in the optical region of the spectrum) can be measured by film and also by photomultiplier tubes or by other electron multipliers such as MCPs. Various plastics, liquids, organic crystals, inorganic crystals, and gases can act as scintillators.[23] Polycrystalline solid light-emitting materials are sometime called *phosphors*. The deexcitation (and therefore, light emission) can be almost instantaneous or delayed. Both types are used for X-ray detectors.

Activated Phosphor.   The most-used scintillator detectors that also give the photon energy spectrum involve scintillating crystals with rapid response. These are often single crystals of sodium or cesium iodide with thallium added to activate the photo response and shift emission into the visible range. Because alkali halides are frequently hygroscopic, they must be sealed. Scintillators are almost never used for measurement of X rays with energy less than about 1 keV. The scintillating crystal is commonly mounted directly onto the face of the vacuum photomultiplier tube, which provides photoelectron conversion, and low-noise amplification of the electrons. The conversion efficiency, η, which is the fraction of incident energy that appears as scintillation, is typically about 5 percent for CsI(Tl) and 12 percent for NaI(Tl).[24] The quantum detection efficiency (QDE) is defined as the number of photoelectrons produced per incident photon and depends on absorption in the cladding, the thickness and absorption coefficient of the scintillator, and the nature of the photocathode of the photomultiplier tube.[5] The energy resolution of a well-matched NaI(Tl) system is $\Delta E/E \sim 1.7 E^{-1/2}$. Activated alkali halide scintillation detectors have working ranges of a few kiloelectronvolts to a few megaelectronvolts. They have good dynamic range with upper detection rates limited by the scintillator decay time, which is approximately 1 µs for NaI(Tl). Organic scintillators have faster (nanosecond) decay times but have lower efficiency and energy resolution. Scintillation detectors are convenient to use, can be large (several square centimeters), are readily available commercially, and are used for a large number of X-ray applications.

Scintillators are often used for X-ray imaging applications. In such cases, energy discrimination is often not important, and other phosphors, such as rare-earth oxides activated with terbium [e.g., $Gd_2O_2S(Tb)$, $La_2O_2(Tb)$, $Y_2O_2(Tb)$] or organic phosphors [e.g., tetraphenyl butadiene (TPB)] are used. The phosphor is deposited as a thin layer (1–10 µm) directly on a glass plate, which is viewed by a sensitive television camera, or directly on the face of a fused-glass fiber optic, which is tapered to allow the intensity distribution to match the size and shape of a charge-coupled (CCD) or charge-injected (CID) imaging camera.

Restimulable Phosphor.   In the last few years, a different type of scintillation detector has become widely used for X-ray imaging.[25,26] This makes use of phosphor materials in which the electron excitation is to a metastable state that can have long (minutes to hours) decay time. After exposure, the plate is scanned by a laser beam that stimulates the metastable excited states to deexcite, producing light that is detected by a photomultiplier tube. The time of the light emission gives the position of the X-ray exposure and the intensity of the light, the X-ray intensity. Such detectors have a number of useful features:

1. They produce a digital record that leads conveniently to computer processing, transmission, and storage.
2. They have a linear response (as opposed to film) with a large dynamic range.
3. The thickness and nature of the phosphor used can be varied to accommodate the need (e.g., low-energy X rays, high-energy X rays, neutrons, etc.), and the size of the plate can be adapted to existing systems. For example, such plates are frequently used to replace photographic film.
4. The photostimulable plates can be reused by "erasing" the stored image by exposure to intense white light for a few seconds. At the present time, a restimulable scanning plate system (including exposure plates, scanner, and eraser) can be less expensive and more flexible than other commercially available position-sensitive detectors, although some with automatic readout features are comparable or more expensive. Such systems, although faster than film-based systems, do not have the rapid readout needed for some applications.

Film

The oldest, and still the most widely used X-ray detector, is silver halide–based photographic film. Indeed, the mysterious exposure of light-protected film led to discovery of X rays by Roentgen in 1895, and to demonstration of their use for medical imaging, still the most widely used and important application. Metastable excitation of grains of silver halide crystals is induced by X rays or secondary electrons. The resulting latent image is developed by chemical reduction of the excited grains to produce the familiar photographic image. Sometimes a phosphor coating or an adjacent phosphor plate is used to increase the sensitivity to the penetrating X radiation, with the halide excitation being produced by the secondary light emission from the phosphor. An extensive discussion of the use of film for optical imaging is given in Vol. I of this handbook. Much of that discussion also applies to X-ray excitation. Detailed discussion of the use of film for X-ray detection and imaging is given in Refs. 1 and 5.

Cryogenic

Recent development of superconducting cryogenic X-ray detectors provides opportunity for important new applications, particularly in materials analysis and astrophysics. These make use of techniques to reach and maintain very low temperature (30–70 mK) for long periods of time.[27] They use the measurement of the temperature rise (microcalorimeter),[28–31] or tunnel junction current[32] in superconducting materials.

Microcalorimeter. Measurement of the temperature rise induced by absorption of an X-ray photon has been demonstrated by the use of semiconductor (typically Si or Ge) thermistors whose electrical conductivity is temperature dependent, and with bimetallic transition-edge sensors (TES). Of these, the TES detectors are the most studied and appear to have the highest potential for X-ray spectrometry. They operate by holding the temperature of the metallic absorber at the transition edge between the superconducting and normal state. At the transition edge, the conductivity is extremely temperature sensitive. The temperature rise that is induced by an absorbed X ray is detected by measuring the conductivity with a superconducting quantum (SQUID) detector. The transition-edge temperature is typically about 50 mK and is stabilized by thermoelectric feedback. Energy resolution as low as 2 eV and counting rates as high as 500 counts per second (cps) have been reported.[33,34] Both the energy resolution and the thermal recovery time (therefore the counting rate) are affected by the thermal capacitance of the absorber, which must be as low as possible. Absorber sizes of $300 \times 300 \times 50$ μm are typical for a very high-resolution detector. Detector arrays are under development to increase the effective area, the count rate capability, and to provide imaging for astrophysical applications.[26] Alternatively, polycapillary focusing optics[35] have been used to increase the effective area (to >7 mm^2).

TABLE 1 Typical Parameters That Affect Detector Choice, Including Counting Rate Limitations, Availability of Current Mode Operation, and Energy Resolution

| Detector | Intensity measurements | | Energy resolution |
	Pulse counting (cps)	Current mode	
Ionization			
Ionization chamber	1–10k	Yes	1–10 keV
Proportional counter	10 k		2–20 keV
Geiger counter	50 k		
Semiconductor	0–100k		
Si(Li)			120 eV
HpGe			140 eV
CdZnTe			250 eV
Pin diode		Yes	
Scintillation Detector	100–500 k		20–50%
Cryogenic			
Microcalorimeter detector	100–500 k		2–10 eV
STJ detector	2–20 k		20–50 eV

TABLE 2 Position-Sensitive X-Ray Detectors

Detector	Spatial resolution (μm)
Film	20–200
Ionization	
Multiwire proportional counter	100–300
Semiconductor	
Arrays	
Segmented electrode	100–500
Charge-coupled detector (CCD)	20–50
Charge injection detector (CID)	20–50
Active pixel	
CMOS	20–50
Amorphous silicon	50–100
Pulse height dispersive	100–300
Scintillation	
Segmented	500–2000
Video	50–100
Multichannel plate readout	20–100
Restimulable phosphor	50–200
Cryogenic	
Array	
Microcalorimeter	200–500
Tunnel junction	300–500
Dispersive	500–1000

Superconducting Tunneling Junction (STJ). Another type of cryogenic detector involves X-ray-induced breakup of superconducting Cooper Pairs, which leads to decreased tunneling current across a superconducting Josephson junction. Although such detectors must also be small and maintained at low temperature to keep the thermally induced current across the junction low, the dependence of the counting rate on the thermal capacitance is relaxed. Such detectors have demonstrated energy resolution <20 eV and counting rate >20,000 cps.[32] Again, arrays or optics have been used to increase the effective area.

34.3 SUMMARY

A list of properties of a number of single-pixel X-ray detectors is provided in Table 1 for comparative purposes. Spatial resolution for typical position-sensitive or imaging detectors is given in Table 2.

34.4 REFERENCES

1. W. J. Price, *Nuclear Radiation Detection, Second Edition,* McGraw-Hill, New York, 1964.
2. G. G. Knoll, *Radiation Detection and Measurement,* Wiley, New York, 1979.
3. W. R. Leo, *Techniques for Nuclear and Particle Physics Experiments,* Springer-Verlag, Heidelberg, 1987.
4. C. F. G. Delaney and E. C. Finch, *Radiation Detectors,* Oxford University Press, Oxford, 1992.
5. C. J. Buckley, in *X-Ray Science and Technology,* A. G. Michette and C. J. Buckley, (eds.), Institute of Physics Publishing, Bristol, 1993, pp. 207–252.

6. G. Fraser, *X-ray Detectors in Astronomy,* Cambridge University Press, Cambridge, 1989.

7. S. J. Gurman, *Mat. Sci.* **17:**1541 (1982).

8. E. A. Stearn and S. M. Heald, in *Handbook on Synchrotron Radiation,* Vol. 1b, E. E. Koch, (ed.), North Holland, Amsterdam, 1983, p. 955.

9. D. G. Stearns and M. B. Stearns, "Microscopic Methods in Metals", in *Topics in Current Physics,* U. Gonser, (ed.), 49, Springer, Berlin, 1983, p. 153.

10. S. C. A. Curran, A. L. Cockroft, and J. Angus, *Phil. Mag.* **40:**929 (1949).

11. B. D. Ramsey, J. J. Kolodziejczak, M. A. Fulton, J. A. Mir, and M. C. Weisskopf, *Proc. SPIE Proc.* **2280:**110–18 (1994).

12. G. Charpak, R. Bouclier, T. Bressani, J. Favier, and C. Zupancic, *Nucl. Instr. and Meth. in Phys. Res.* **62:**262–268 (1968).

13. B. D. Ramsey, R. A. Austin, and R. Decher, *Sp. Sci. Rev.* **69:**139–204 (1994).

14. R. Hofstader, *Nucleonics* **4:**2 (1949); *Proc. IRE* **38:**721 (1950).

15. G. L. Miller, W. M. Gibson, and P. F. Donovan, *Ann. Rev. Nucl. Sci.* **12:**189 (1962).

16. W. M. Gibson, G. L. Miller, and P. F. Donovan, in *Alpha, Beta, and Gamma Spectroscopy,* K. Siegbahn (ed.), North Holland, Amsterdam, 1964.

17. E. M. Pell, *J. Appl. Phys.* **31:**291 (1960).

18. J. H. Elliott, *Nucl. Instr. and Meth.* **12:**60 (1961).

19. J. L. Blankenship and C. J. Borkowski, *IRE Trans. on Nucl. Sci.* **NS-9:**213 (1963).

20. J. E. Gindley, T. A. Prince, N. Gehrels, J. Tueller, C. J. Hailey, B. D. Ramsey, M. C. Weisskopf, P. Ubertini, and G. K. Skinner, *SPIE Proc.* **2518:**202–210 (1995).

21. F. P. Doty, H. B. Barber, F. L. Augustine, J. F. Butler, B. A. Apotovsky, E. T. Young, and W. Hamilton, *Nucl. Instr. And Meth. In Phys. Res.* **A253:**356–360 (1994).

22. S. Dhawan, *IEEE Trans. Nucl. Sci.* **NS-28:**672 (1981).

23. R. B. Murray, in *Nuclear Instruments and Their Uses,* A. H. Snell (ed.), Wiley, New York, 1962.

24. R. L. Heath, R. Hofstader, and E. B. Hughes, *Nucl. Inst. and Methods* **162:**431 (1979).

25. H. Nanto, K. Murayama, T. Usuda, F. Endo, Y. Hirai, S. Tahiguchi, and N. Tagekuchi, *J. Appl. Phys.* **74:**1445–1447 (1993).

26. H. Nanto, Y. Hirai, F. Endo, and M. Ikeda, *SPIE Proc.* **2278:**108–117 (1994).

27. M. Altman, G. Angloher, M. Buhler, T. Hertrich, J. Hohne, M. Huber, J. Jochum, T. Nussle, S. Pfnur, J. Schnagl, S. Wanninger and F.v. Feilitsch, Proc. of 8th Int. Workshop on Low Temperature Detectors, (1999).

28. D. A. Wollman, K. D. Irwin, G. C. Hilton, L. L. Dulcie, D. E. Newbury, and J. M. Martinis, *J. Microscopy* 196–223 (1997).

29. D. McCammon, W. Cui, M. Juda, P. Plucinsky, J. Zhang, R. L. Kelley, S. S. Holt, G. M. Madejski, S. H. Moseley, and A. E. Symkowiak, *Nucl. Phys.* **A527:**821 (1991).

30. D. Lesyna, D. Di Marzio, S. Gottesman, and M. Kesselman, *J. Low Temp. Phys.* **93:**779 (1993).

31. E. Silver, M. LeGros, N. Madden, J. Beeman, and E. Haller, *X-Ray Spectrom.* **25:**115–122 (1996).

32. M. Frank, L. J. Hiller, J. B. Le Grand, C. A. Mears, S. E. Labov, M. A. Lindeman, H. Netel, and D. Chow, *Rev. Sci. Instrum.* **69:**25 (1998).

33. D. A. Wollman, S. W. Nam, D. E. Newbury, G. C. Hilton, K. D. Irwin, N. R. Bergren, D. Deiker, D. A. Rudman, and J. M. Martinis (submitted for publication August 1999).

34. D. A. Wollman, G. C. Hilton, K. D. Irwin, N. F. Bergren, D. A. Rudman, D. E. Newbury, and J. M. Martinis, 1999 NCSL Workshop and Symposium.

35. D. A. Wollman, C. Jezewski, G. C. Hilton, Q.-F. Xiao, L. L. Dulcie, and J. M. Martinis, *Microscopy and Microanalysis* **4:**172–173 (1998).

APPLICATIONS

CHAPTER 35
APPLICATIONS REQUIREMENTS AFFECTING OPTICS SELECTION

Carolyn A. MacDonald and Walter M. Gibson
University at Albany, State University of New York
Albany, New York

35.1 INTRODUCTION

X rays have a century-long history of medical and technological application. Optics have come to play an important role in many capacities. This chapter examines how various X-ray optics enable typical applications, and the factors that influence optics choices. It is not feasible in this chapter to discuss every possible utilization of X-ray optics. Rather, a representative selection has been included. Some applications, such as X-ray astronomy, have been thoroughly addressed in previous chapters.[1,2] There is, in addition, a class of applications for which the primary requirement is adequate intensity in an extremely narrow frequency bandwidth. These include the absorption spectroscopy tools, such as extended X-ray absorption fine structure (EXAFS) and X-ray absorption near-edge structure (XANES). These measurements are typically performed with multiple crystal monochromators[3] on a synchrotron beam line;[4] however, the discussions of intensity gain from X-ray optics in the following fluorescence, diffraction, and microscopy sections can be applied to these applications as well, which in some cases are then made feasible using laboratory sources.

35.2 COHERENCE AND FLUX REQUIREMENTS

X-ray optics are used over an extremely large range of X-ray fluxes, from the detection of fluorescence emission from part-per-billion impurities to beam shaping for third-generation synchrotrons. For a number of synchrotron and other high-flux applications, radiation hardness and thermal stability are important considerations. A large body of experience has been developed for high-heat-load synchrotron mirrors and crystals. (See, for example, Chap. 26.)[5,6] In addition, testing has been performed on a variety of other optics. Many optics such as zone plates, microscopy objectives, and glass capillary tubes are routinely used in synchrotron beam lines and are stable over acceptable flux ranges.

Another consideration with synchrotron sources is transverse coherence. Using classical geometric optics, the requirement in slit separation or beam radius, *a*, to observe diffraction effects is

$$a\Delta\theta << \lambda \qquad (1)$$

where λ is the wavelength of the radiation, and $\Delta\theta$ is the angle subtended by the source. For typical laboratory sources, the coherent beam radius is too small to have sufficient count rate for measurements. However, for synchrotron sources, $\Delta\theta$ is much smaller and the radius a is large enough to encompass optics such as zone plates, which require coherent superposition. On third-generation synchrotron beam lines, coherence effects such as speckle are routinely seen. Speckle patterns can be employed to measure time-dependent correlations in fluctuating samples.[7] When effects such as speckle are undesirable, coherence can be removed with multiple bounce or rough-surface optics.

35.3 MICROSCOPY

Diffraction effects limit the resolution of optical systems to within an order of magnitude of the wavelength of the light. Thus, in principle, X-ray microscopy systems could have resolution many orders of magnitude better than optical light systems. Electrons also have very small wavelength, and electron microscopes have extremely high resolution. However, electrons are charged particles and necessarily have low penetration lengths into materials. X-ray microscopy is capable of very high resolution imaging of relatively thick objects, including wet samples. However, diffraction-limited optics for X rays do not currently exist. In practice, X-ray microscopy covers a range of several orders of magnitude in wavelength and spot size.

The best resolutions are obtained by Schwarzschild objectives[8] or zone plates.[9] Because Schwarzschild objectives are used in normal incidence, they are essentially limited to the extreme ultraviolet (EUV) region. Near-diffraction-limited resolution of 90 nm at a wavelength of 13 nm has been demonstrated with a Schwarzschild objective. Synchrotron sources are required to provide adequate flux.

The resolution of zone plates is limited by the width of the outermost zone, typically around 20 to 40 nm. Zone plates are easiest to make for soft X rays, where the thickness required to absorb the beam is small. However, zone plates with high-aspect ratio have been demonstrated for hard X rays. Because the diameters of imaging zone plates are only a few hundred microns, and the efficiencies are no more than 10 percent for amplitude zone plates and 40 percent for phase zone plates, synchrotron sources are required.

Single capillary tubes can also have output spot sizes on the order of 100 nm or less, and thus can be used for scanning microscopy or microanalysis. Capillary tubes with outputs this small also require synchrotron sources.

Refractive optics are true imaging optics, with the potential for very small spots.

For larger spot sizes, a wider variety of sources and optics are applicable. Microscopes have been developed for laser-plasma sources with both Wolter[2,10] and bent-crystal optics[3,11] with resolutions of a few microns. These optics, and also capillary and polycapillary optics and nested mirrors, with or without multilayer coatings,[12] can produce spot sizes of a few tens of microns with laboratory e-beam sources. Optics designed to collect over large solid angles, such as bent crystals, graded multilayers, or polycapillary optics, will produce the highest intensities. Multilayers have very small Bragg angles for short-wavelength X-ray radiation. Bent crystals will yield monochromatic radiation; polycapillary optics can be used to produce higher-intensity, but broader-band radiation. Polycapillary optics have spot sizes no smaller than tens of microns, independent of source size. Mirrors and crystals will have smaller focal spot sizes for smaller sources and larger spot sizes for larger sources. Clearly, the optimal optic depends on the details of the measurement requirements and the sources available.

35.4 PROXIMITY LITHOGRAPHY

Just as there are inherent limits in the resolution that can be achieved with optical microscopes, the feature size that can be printed on a circuit using current technology will be limited by the resolution that can be achieved with ultraviolet light. X-ray lithography is an appealing alternative to produce smaller feature sizes. EUV projection lithography systems are also a current research and development area. Future EUV projection systems will produce small feature sizes and obviate some mask manufacturing issues by reducing the feature size from the mask to the substrate. In X-ray proximity lithography, the mask is placed close to the substrate and the mask-to-feature ratio is 1:1. X-ray lithography systems are expected to have higher depth of field and, therefore, greater applicability for complex multilayer chip processing than current UV systems. Because wafer throughput is a priority for a chip fabrication facility, high X-ray intensity is required. However, synchrotron sources are regarded as having too high an integration cost because of the difficulty of incorporating such sources in an existing fabrication line. Point sources of X rays, however, necessarily have global divergence. The effect of global divergence is shown in Fig. 1. To reduce the lateral shift (runout) and tilt errors to acceptable values, the divergence must be reduced[13] to less than (perhaps much less than) 20 mrad. For a point source without an optic and a field size of 22 mm, 20-mrad divergence would require a source to mask distance of more than 55 cm. Such large source distances would unacceptably reduce the intensity on the resist due to the $1/r^2$ divergence of the beam. A collimating optic is essential.

Challenges facing optic design are the need for uniformity in magnification, intensity, and divergence across the field as well as high intensity gain and inexpensive manufacture. An advantage of a point source/optic combination over a synchrotron source, aside from cost, is that some minimal local divergence is required to blur edge diffraction effects and prevent "ringing" of the image on the resist. Several optics designs have been considered, including a parabolic mirror[14] coated with a graded multilayer,[12] a slumped multichannel plate,[15] co-encentric nested conical or parabaloidal mirrors, a scanned aspherical mirror, and a flat mirror array.[13] Features of 200-nm width have been printed using laser plasma sources and scanned parabaloidal mirrors or polycapillary optics.[16] The parabaloidal mirror system had a more uniform field when scanned with a carefully designed mask, but lower gain than the polycapillary optic. It is expected that both the uniformity of the polycapillary optic and the gain of the scanned system will be increased.

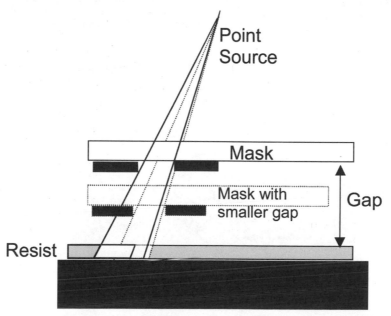

FIGURE 1 Effect of global divergence on lithographic feature (dark gray area on resist). Figure shows both feature tilt (edges of feature are not perpendicular to substrate) and the lateral shift produced by gap error (dotted lines).

35.5 *DIFFRACTION*

Introduction

In 1914, Max von Laue was awarded the Nobel prize in physics for the discovery of the diffraction of X rays by crystals. The following year, the William Braggs were awarded the prize for the analysis of crystal structure from X-ray diffraction data. For most of the last century, X-ray diffraction analysis has been used as the primary tool for structural analysis of materials. X-ray optics can be used to monochromate the beam, to increase the diffracted beam intensity, and to provide a parallel beam geometry from a divergent point source.

Kinematical Theory of Diffraction.   X-ray diffraction is the coherent superposition of Rayleigh scattering from planes of atoms in the crystal. The conditions necessary to assure that the superposition in phase occurs are illustrated in Fig. 2 for the Bragg and Ewald constructions.[17] In the Bragg construction, the condition that rays from sequential planes arrive in-phase requires that

$$\lambda = 2d \sin \theta \tag{2}$$

where d is the plane spacing, θ is the grazing angle between the incoming beam and the plane, and λ is the X-ray wavelength, and where the slight refraction that occurs at the sample surface has been neglected. The incoming and outgoing wavevectors, k_o and k_f, both have magnitude $2\pi/\lambda$ because the scattering is elastic. In the Ewald sphere construction, the identical requirement becomes

$$\mathbf{k_f} - \mathbf{k_o} = \mathbf{G} \tag{3}$$

where $\mathbf{G}$ is the vector of magnitude $2\pi/d$ in the direction perpendicular to the planes. The endpoints of the $\mathbf{G}$ vectors associated with the different sets of planes in the crystal are represented by the array of "reciprocal lattice" points in Fig. 2. Diffraction occurs when the sphere of possible final wavevectors with magnitude $2\pi/\lambda$, the Ewald sphere, intersects one of the reciprocal lattice points. Equation (3) can be shown to be equivalent to Eq. (2).

$$\mathbf{G} = \mathbf{k_f} - \mathbf{k_o}$$

$$G^2 = k_f^2 + k_o^2 - 2k_o k_f \cos 2\theta$$

$$\left(\frac{2\pi}{d}\right)^2 = 2\left(\frac{2\pi}{\lambda}\right)^2 - 2\left(\frac{2\pi}{\lambda}\right)^2 (1 - 2 \sin^2 \theta) \tag{4}$$

$$\lambda = 2d \sin \theta$$

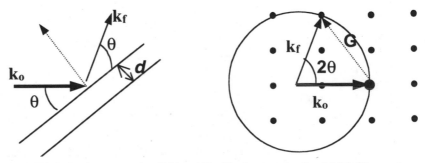

FIGURE 2   Bragg construction (left) and Ewald sphere construction (right). The constructions are shown for the same set of planes, with spacing d and reciprocal lattice vector $G = 2\pi/d$.

Formally, the diffraction amplitude is the three-dimensional spatial Fourier transform of the electron density in the material. Unfortunately, in measuring the X-ray intensity, the phase information is lost. This prevents a direct transformation to electron density from the diffraction pattern. To circumvent this problem, the intensity distribution that would occur from an assumed crystallographic structure can be computed and compared with the measured data. Alternatively, multiple-beam or standing-wave interference techniques can be used to infer the phase.[18]

Optimizing with Optics

The Ideal Plane Wave.   Clearly, analysis of the structure from the diffraction pattern would be simplest if the incoming beam were completely monochromatic and parallel. In practice, because all X-ray sources produce a distribution of angles and energies, selecting too narrow a range would leave one with a beam of zero intensity. Further, in examining Fig. 2, it is clear that there would be, at most, only a few accidental intersections of the reciprocal lattice with the Ewald sphere if a monochromatic parallel beam is used with a stationary perfect single crystal. In practice, a variety of techniques are used to force an intersection. The crystal can be rotated, thus rotating the reciprocal lattice points about the origin, as in a θ-2θ diffractometer, or oscillated, as is commonly done in protein crystallography. All possible rotations of the crystal can be presented by thoroughly grinding the crystal into a powder. Alternatively, in Laue diffractometry, a stationary single crystal is used with white radiation, giving a range of possible X-ray energies and thus Ewald sphere radii, and so a solid Ewald sphere.

Theoretical Limitations to Collimation.   As shown in Fig. 3, the output from a collimating optic has both global divergence, α, and local divergence, β. Even if the global divergence is made very small by employing an optic, the local divergence is usually not zero. For grazing incidence reflection optics, the local divergence is generally given by the critical angle for reflection and can be increased by profile errors in the optic. The degree of collimation achieved by the optic is limited not only by technology, but by the thermodynamic constraint that the beam brightness cannot be increased by any optic.[19] Liouville's theorem states that increasing the density of states in phase space is a violation of the second law of thermodynamics. This implies that the six-dimensional real space/momentum space volume occupied by the photons cannot be decreased. More simply, the angle-area product (i.e., the cross sec-

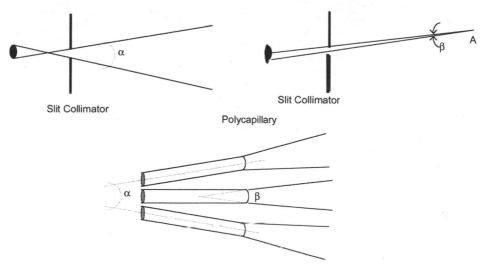

FIGURE 3   For either a slit collimator (top left and right) or an optic such as a polycapillary optic (bottom), the local divergence, β, seen at a point A is different than the angle subtended by the beam, α.

tional area of the beam multiplied by the divergence of the beam) cannot be decreased without losing photons. The beam can never be brighter than the source. An idealized point source of X rays has isotropic emission and occupies zero area. X rays from such a source could theoretically be perfectly collimated into any chosen cross section. However, a real source has a finite size and so limits the degree of possible collimation. This explains the need for small spot sources to produce bright, well-collimated beams.

Resolution and Count Rate.   The accuracy with which the planar d spacing of a sample can be determined can be calculated from Eq. (2) in terms of the uncertainty in measured angle θ and the spread of wavelengths λ

$$d = \frac{\lambda}{2 \sin \theta} \Rightarrow \Delta d = \frac{\Delta \lambda}{2 \sin \theta} - \frac{\lambda \cos \theta \Delta \theta}{2 \sin^2 \theta} \Rightarrow \frac{\Delta d}{d} = \frac{\Delta \lambda}{\lambda} - \frac{\Delta \theta}{\tan \theta} \tag{5}$$

Thus, high-resolution measurements require nearly parallel, extremely monochromatic beams, as can be created from multiple crystal monochromators using high-quality crystals. However, because any source has finite brightness, selecting a small angular range, $\Delta \theta$, and small wavelength range, $\Delta \lambda$, results in a decreased count rate. As X-ray measurements are usually count rate limited, the signal-to-noise ratio (SNR) is determined by Poisson statistics, $\Delta N \sim \sqrt{N}$, so that the SNR = $N/\Delta N \sim \sqrt{N}$. Thus, there is a trade-off between resolution, which is improved by limiting the beam, and SNR, which is improved by increasing the number of counts. Extremely high resolution measurements require high-brightness sources, such as undulator synchrotron lines, to have adequate SNR. It is generally a mistake to design a diffraction measurement to have much higher resolution than is actually required. A measurement of a sample to determine whether a suspected impurity is present, usually does not require high resolution, but may require high beam intensity to detect the impurity at low concentrations. Thus, a low-resolution monochromator, or even a simple filter, may be optimal. Conversely, an accurate measurement of strain in a large sample requires high resolution at the expense of beam intensity. For laboratory measurements, employing a more powerful source with higher beam current can increase the intensity and therefore the SNR. However, increased beam currents mean larger source sizes to avoid melting the anode, and as shown in Fig. 3, the global divergence of a beam defined by pinhole collimation depends on the source size. Thus, there is again a trade-off between intensity and resolution. However, the divergence of some optics, particularly multiple-bounce optics, does not depend on the source size. Thus, larger, higher-power sources may be used without adversely affecting the resolution of the measurement.

Peak-to-Background Ratio.   The detectability of a peak depends on both the SNR, which is based on the total number of counts in the peak, and also the peak-to-background ratio. Increasing the total beam intensity improves the counting statistics and the SNR, but does not necessarily affect the peak-to-background ratio. Alternatively, soller slits, arrays of flat metal plates arranged parallel to the beam direction, or other antiscatter optics, placed after the sample, serve to remove scattered radiation and reduce background, thus increasing the peak-to-background ratio without increasing the SNR.

Intensity Gain

A variety of X-ray optics choices exist for collimating or focusing X-ray beams. The best choice of the optic depends to a large extent on the geometry of the sample to be measured, the information desired from the measurement, and the source geometry and power. No one optic can provide the best resolution, highest intensity, easiest alignment, and shortest data acquisition time for all samples.

Five commonly used optics are nested cones (Chaps. 26 and 28), bent crystals (Chap. 22), multilayers (Chap. 24), single capillaries (Chap. 29), and polycapillary optics (Chap. 30). Bent crystals collect radiation from a point source and diffract it into a nearly monochromatic collimated beam. They collect from a solid angle of typically a few millisteradians and produce a beam with a bandwidth of a few electronvolts. Thus, bent crystals are an ideal choice for very high resolution measurements. However, for laboratory sources, the flux from such bent crystals is relatively low, because the collection angle is relatively small and because only a small fraction of the input beam will be within the energy bandwidth of the crystal. Typically, the alignment required for the bent-crystal geometries (Johansson, for example, which gives a wider bandwidth than Johann) is quite exacting and time consuming. Doubly bent crystals or two singly bent crystals are required for two-dimensional collimation or focusing.

Multilayer optics also work by diffraction, although in this case from the periodicity of the artificial compositional variation imposed in the multilayer. The beam is less monochromatic than for bent crystals. "Supermirrors," or "Goebel mirrors," are multilayers with graded or irregular spacing, and have wider energy bandwidths, larger output divergences, and smaller capture angles than multilayers.

Single capillaries are tubes which pipe radiation into a small output area. They have very short working distances but can have extremely small spot sizes and are the most practical choice for nanometer scale samples.

Polycapillary optics are arrays of hundreds of thousands of glass tubes that collect radiation and deflect it into a parallel or focused beam. Because the focal spot is produced by overlap, the spot sizes and working distances are larger than for single capillary tubes.

Curved mirrors utilize the same total external reflection employed by polycapillary optics. Because they are single-bounce optics, their maximum angular deflection is limited to the critical angle for their metallic coating, which can be about 10 mrad at 8 keV. The total capture angle is then determined by the length of the mirror. The output divergence is determined by the length of the optic and the source size. Because of the small grazing incidence angle, conical mirrors collect only a small annulus of the radiation from the source. One solution is to "nest" multiple optics. A series of even straight conical mirrors can be useful. At a source distance of 100 mm, a straight cone of 2.4-mm input radius, 3.4-mm output radius, and length 100 mm would have an output divergence of 7 mrad. This single cone would only collect an annulus of radiation from the source, with cone angle from 16.5 to 23.5 mrad. To fill in the annulus, smaller cones are "nested" inside the first. The divergence of the output beam would be a function of position within the beam, which can make interpretation of diffraction results more difficult. Nested parabolic mirrors are also used and have smaller output divergences, dependent on the source size.

Typical parameters for four optics types are listed in Tables 1 and 2. The relative X-ray intensity from each optic is

$$\tilde{P} \equiv \frac{P_{out}}{P_{source}} = \frac{\Omega}{4\pi}\eta \qquad (6)$$

TABLE 1   Typical Optics Parameters Needed to Compute Optic Output Power at 8 keV

	Capture angle			Efficiency (η)	
Optic	Degrees	Radians	Solid angle (Ω, milli-steradians)	R or T over input bandwidth (%)	Bandwidth
Polycapillary	15	0.26	50	40 (20% for focusing optic)	To 8 keV
Multilayer	2	0.04	1	30	100 eV
Bent crystal	4	0.07	4	10	10 eV
Nested mirrors	2	0.04	1	50	To 8 keV

Note: Parameters are approximate and may vary considerably for optics designed for different applications. Quoted reflectivities, transmissions, and capture angles assume the source size is small enough to be efficiently utilized by the optic and that the optic can be put close enough to the source.

TABLE 2   Typical Factors to Compute the Usable Fraction of the Output

Optic	Collimating optic			Focusing optic
	Output divergence (β, mrad)	Output beam radius (R, mm)	Need energy filter	Focused-beam spot size
Polycapillary	5	3	Filter	40–300 μm
Multilayer	0.2	0.5	No	Size of source
Bent crystal	0.02	0.5	No	Size of source
Nested mirror	3	3	Filter	Not applic.

where Ω is the capture angle of the optic, and η is the transmission or reflection efficiency of the optic over the whole input spectrum, and thus includes the fraction of the input spectrum, which is within the bandwidth of the optic. The resultant diffracted signal intensity, S, from a strongly diffracting sample of radius, s, with angular acceptance, $\Delta\theta$, assuming two-dimensional collimation, is

$$\tilde{S} = \frac{S_{\text{out}}}{P_{\text{source}}} = \tilde{P}\left(\frac{s}{R}\right)^2\left(\frac{\Delta\theta}{\beta}\right)^2 F = \frac{\Omega}{4\pi}\eta\left(\frac{s}{R}\right)^2\left(\frac{\Delta\theta}{\beta}\right)^2 F \tag{7}$$

where R is the radius and β the output divergence of the beam. The equation assumes that $s < R$, that the output field is of uniform intensity, and that $\Delta\theta < \alpha$. F is a factor representing the fraction of the optic output that is in the necessary energy bandwidth for the measurement, multiplied by the efficiency of any necessary monochromator or absorption filter.

A true comparison of two optics for a particular application requires careful analysis of the sample and measurement requirements, and adjustment of the source and optic design for the application. No global comparison of all optics for all applications is possible. For the purposes of comparing intensity gain alone, power and signal outputs for four collimating optics have been computed assuming 3- and 0.5-mm radii single-crystal samples. The results are given in Table 3.[20,21] Two different angular acceptance widths were assumed, $\Delta\theta = 0.25° = 4$ mrad (which is less than the sample oscillation often used in protein crystallography), and $\Delta\theta = 0.1$ mrad. Because very small samples would probably be used with focusing optics, the columns for $r = 3$ mm are the most appropriate comparison for the collimators. For the measurement with the high angular acceptance, the polycapillary optic gives the largest signal. For the high-resolution measurement with the small angular acceptance, the multilayer or bent-crystal optic give the highest signals.

Comparison of focusing optics requires a detailed analysis of the effect of the convergence angle on the diffracted signal intensity and so is very sample and measurement dependent. Decreasing the angle of convergence onto the sample improves the resolution and decreases the SNR, but also decreases the diffracted signal intensity relative to a large angle.[22] Conventional practice is to use a convergence angle that is less than the mosaicity of the sample. However, as discussed in the section entitled "Focused Beam Diffraction," which follows, convergent beams produce one-dimensional tangential streaking in a single-crystal diffraction pattern and so do not affect the radial resolution.[23] It is necessary that the beam cross section at the sample be larger than the sample to avoid the difficulty of correcting for intensity variations with sample angle. For some geometries, it may be possible to move the sample along the beam until the entire volume is illuminated.

A simple comparison of focused beam flux can be made using the result that

$$\tilde{F} \equiv \frac{P_{\text{focused on sample}}}{P_{\text{source}}} = \frac{\Omega}{4\pi}\eta\left(\frac{s}{f}\right)^2 F \tag{8}$$

where f is the radius of the focal spot from the optic, and F again accounts for an absorption filter, if any. These results are also presented in Table 3 for a 25-μm radius sample. It should

TABLE 3 Approximate Relative Signal Intensity for Various Optics and Samples

Optic	X-ray power output ($\tilde{P}$)	Angular acceptance ($\Delta\theta$)	Signal intensity ($\tilde{S}$)				Focused beam power ($\tilde{F}$) on 25-μm radius sample
		Sample radius (r)	0.15 mm	3 mm	0.15 mm	3 mm	
			4 mrad	4 mrad	0.1 mrad	0.1 mrad	
Polycapillary	1.6×10^{-3}		7.6×10^{-8}	3.0×10^{-5}	4.8×10^{-11}	1.9×10^{-8}	5×10^{-5}
Multilayer	6.4×10^{-7}		5.8×10^{-8}	6.4×10^{-7}	1.4×10^{-8}	1.6×10^{-7}	4×10^{-8}
Bent crystal	1.9×10^{-7}		1.7×10^{-8}	1.9×10^{-7}	1.7×10^{-8}	1.9×10^{-7}	1×10^{-8}
Nested mirrors	4.0×10^{-5}		3×10^{-9}	1.2×10^{-6}	3.3×10^{-12}	1.3×10^{-9}	NA

Note: For the multilayer and single crystal optics, a pair of optics for two-dimensional collimation was assumed, so the reflectivity is used twice. These optics are used without a filter or monochromator. The nested cone and polycapillary optic are assumed to be used with a nickel filter for $K\beta$ suppression, with transmission of 30 percent. The source is assumed to be a copper anode, with 6 percent of its radiation in the 8-keV copper $K\alpha_1$ characteristic line of width 3 eV, 3 percent in the $K\alpha_2$ doublet line, and 10 percent in the range required for the measurement.[20,21] The factor $\Delta\theta/\alpha$ was set equal to unity when the output divergence of the optic was less than the sample requirements. The quoted intensity for the "nested mirror" is highly dependent on the capture angle, which depends on the number of cones.

be taken into consideration in the comparison of intensities that the output of highly focused optics may be broader in angle than is usable.

Advantages of Parallel Beams

An important reason to use a collimating optic, aside from intensity gain, is that a parallel beam geometry provides insensitivity to sample preparation, shape, position, and transparency. An example of the insensitivity to sample position is shown in Fig. 4, which shows a silicon powder diffraction peak measured with and without a polycapillary optic for different sample positions.[24] Second, the symmetric beam profile and enhanced flux gives much improved particle statistics and measurement statistics. The constant peak width and resolu-

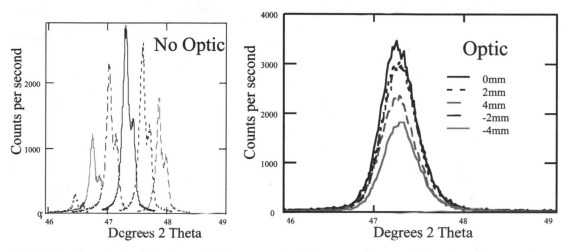

FIGURE 4 Silicon powder diffraction peak shift due to sample displacement with no optic (Bragg-Bretano geometry), left, and with a polycapillary collimating optic (parallel beam), right. Note that in addition to the peak shift, in the Bragg-Bretano case the peaks are asymetric due to the presence of the copper $K\alpha$ doublet. The resolution of the system in both cases was limited by the monochromator. (*From Ref. 24.*)

tion throughout the diffraction space facilitates very high precision residual stress and texture analysis and reciprocal space mapping. An example of the uniform peak shape in parallel beam geometry is shown in Fig. 5.[24] The peak shape is ideally suited to phase identification and full pattern analysis of phase content using wavelet transforms. The parallel beam geometry also greatly simplifies alignment.

Focused Beam Diffraction

For focused beam diffraction, the volume of reciprocal space that is accessed in a single measurement is greatly increased compared with parallel beam geometries. Additionally, the small irradiation spot on the sample reduces angular broadening due to sample size effects and allows the detector to be placed close enough to the sample that a small imaging detector will intercept diffracted beams at high 2-theta angles. The reduction in angular broadening due to the small source size is significant because the diffraction spot is not isotropically broadened by the convergence of the focused beam.[23]

Figure 6 displays a sketch of the diffraction condition for a single crystal with a monochromatic convergent beam. Diffraction conditions are satisfied for the two incident beam directions, $\mathbf{k}_0$ and $\mathbf{k}_1$, when they make the same angle with the reciprocal lattice vector, $\mathbf{G}$. Thus, changing from $\mathbf{k}_0$ to $\mathbf{k}_1$ rotates the diffraction triangle of $\mathbf{k}_0$, $\mathbf{G}$, and $\mathbf{k}_f$ about the vector $\mathbf{G}$ by an angle ϕ. This results in the diffracted beam, $\mathbf{k}_f$, moving to trace out a tangential line on the detector. The resultant streak, shown in Fig. 7, has length

$$x = 2z \tan\left(\frac{\phi}{2}\right) = 2\,\frac{D}{\cos(2\theta)}\tan\left(\frac{\phi}{2}\right) \tag{9}$$

The maximum value of ϕ is the convergence angle. There is no broadening in the radial direction. The straight-through beam shown as the vector $\mathbf{D}$ will vary within the range of incoming beam directions for different reciprocal lattice vectors. The effects of the one-dimensional streaking are shown in Fig. 8 for a single-crystal egg-white lysozyme diffraction pattern taken with a 2.1° focusing angle.[25] Comparisons of this measured pattern with a simulation based on the analysis of Eq. (9) give good results.[26] Serious overlap problems were not encountered except in low index directions, which are of less interest for structure determination. So long as the streaked diffraction spots are narrow compared with their separation, they can in principle be analyzed. Comprehensive analysis programs for convergent beam protein crystallography have been developed and the first stages reported. The initial convergent

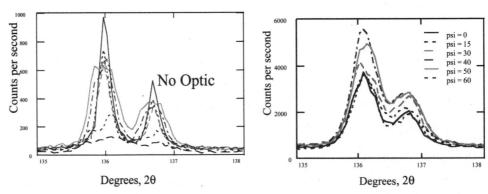

FIGURE 5   Strain measurements in a relaxed aluminum disk with no optic (Bragg-Bretano geometry), left, and with a polycapillary collimating optic (parallel beam), right. In the Bragg-Bretano geometry the peak intensity drops markedly with increasing psi angle. (*From Ref. 24.*)

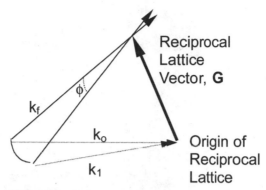

FIGURE 6 Ewald sphere description of focused beam diffraction on a single crystal.

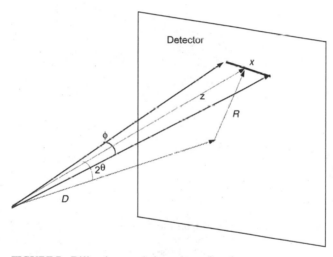

FIGURE 7   Diffraction streak due to beam focusing.

beam diffraction measurements have utilized polycapillary focusing optics and are therefore limited by their local divergence. However, other focusing optics can also in principle be used. A particularly attractive possibility is the use of doubly curved crystal diffraction optics, which, in addition to reduced local divergence, would produce a highly monoenergetic X-ray beam.

Protein diffraction patterns taken with smaller convergence can be analyzed with standard commercial software.[27] Patterns from inorganic crystals with smaller, less complex, unit cells will have lower diffraction spot density and, therefore, less potential overlap than for protein crystals. For polycrystalline samples, a small (≈20 μm) focused-beam spot on the sample increases the probability that a small number of grains will be simultaneously irradiated, simplifying the diffraction analysis, and relaxing the required angular resolution compared with normal powder diffraction analysis. The small spot size and high intensity of focused beams should make possible X-ray microdiffraction in laboratory, industrial, or field settings in ways that can currently only be realized at synchrotrons. These potential applications include: for thin films—scanning studies of thickness, composition, and crystallinity; for mineralogy, metallurgy, and pharmaceuticals—microphase analysis; for forensics and environmental science—field analysis. Indeed, a

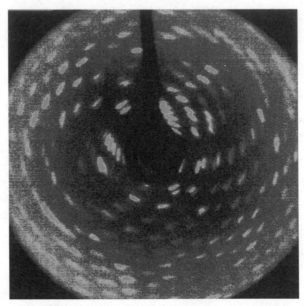

FIGURE 8   Lysozyme diffraction image taken in 5 min with a 1-m A tube source and a focusing polycapillary optic. (*From Ref. 26.*)

system for simultaneous microfluorescence and microdiffraction for planetary lander/rover applications utilizing a 1- to 2-W X-ray source has been proposed.

Protein Crystallography

One of the most dramatic and most important uses of X rays in recent years has been their application to elucidation of the detailed structure of biological macromolecules. This has not only led to discovery of the structure of DNA, RNA, and other essential biological building blocks, but also to development of protein structural analysis as a basis for drug design and the study of diseases. Nearly all synchrotron sources have dedicated macromolecular structure facilities where intense, monochromatic, collimated X-ray beams, together with modern position-sensitive detectors and powerful software analysis programs produce more new protein structures every week than were produced in a year, only a decade ago. Although beamline optics in the form of monochromators and mirrors of the type discussed in Chaps. 22 and 26 are important components of such systems, the central developments have been in sources, detectors, and software. It is generally accepted that final, high-resolution detailed structure determinations will nearly always be made with such facilities.

Even as hundreds, even thousands, of protein structures are determined, much more needs to be done. The most serious bottleneck in protein analysis is purification and growth of the protein crystals to be measured. Protein crystal growth is still a labor-intensive, arduous, complex, and uncertain process, sometimes taking weeks, months, or even years of effort to grow usable crystals of new proteins. This is such a sensitive and demanding process that NASA, in collaboration with many academic and industrial laboratories, has invested a great deal into study and application of microgravity as an enabling environment for protein crystal growth, and foresee this as an important activity on the planned Space Lab. The crucial need for laboratory-based measurements, which can provide rapid turnaround and fast feedback to the crystal growers, has resulted in active development of sources, optics, and detectors to increase the sensitivity and efficiency of laboratory analysis. These developments have basically followed two paths: high-power rotating anode generators, and high-brightness microfocus sources with closely coupled concentrating optics.

In the first of these, a number of manufacturers have developed water-cooled rotating anodes with line-focused beam spots that can be observed at small takeoff angles, allowing the electrical power in the electron bombardment beam to be increased from the 1 or 2 kW maximum for air-cooled tube sources to as much as 15 kW. Typically, such systems are operated at 3 to 6 kW. Such systems have continued to evolve with, for example, double-mirror optics similar to the Kirkpatrick-Baez optics described in Chaps. 26 and 28, and more recently with graded multilayer optics (Chap. 24) to give up to an order of magnitude intensity increase. Together with digital detectors (multiwire proportional counters, stimulated phosphor, and more recently, CCD), and integrated software analysis packages, such systems have become standard protein structural analysis systems. The principle concerns about such systems are their cost ($250K–400K), complexity, and size and maintenance requirements. The second, more recent, approach involves the use of a microfocus source with a grounded anode that allows close access of an optic that can collect over a large solid angle and produce either a parallel or weakly focusing output beam. Both elliptical single reflection[28] and polycapillary optics[27,29] have been used. Using a polycapillary focusing optic, the intensity through a 350-μm-diameter collimator was 16 times higher with a 40-W microfocus source than that obtained with a 2.8-kW rotating anode source with a graphite monochromator. The optic had a 20-μm-diameter focal spot with an input focal length of 3.5 mm and an output beam size of 450 μm diameter. An 8-μm-thick Ni filter was used with the optic to reduce the Cu $K\beta$ intensity to $\frac{1}{25}$ of the Cu $K\alpha$ line intensity. Analysis of a lysozyme protein crystal gave comparable quality diffraction for both the microfocus and the rotating anode system. The notable feature of this result was 16 times higher intensity with 76 times lower X-ray power.[29] A weakly focusing polycapillary optic was determined to produce even higher intensity—5 times that of a collimating optic, while giving similar statistical quality measures for full data sets with lysozyme crystals.[30] The use of weakly focused beams has also become commonplace with curved-crystal and multilayer optics used with rotating anode generators. The lower cost, smaller size, reduced complexity, and lower maintenance of the microfocus system along with its low power requirement make it attractive for laboratory or remote (Space Lab, etc.) use. An important limitation in the use of polycapillary optics for protein crystallography measurements arises from the local divergence inherent in such optics. For 8 keV Cu $K\alpha$ radiation, the local divergence is about 3 mrad. For protein crystals with unit cell dimensions <200 Å, such as lysozyme, this does not preclude structural determinations. However, for cell dimensions >200 Å, the diffraction spots are not completely separated. As discussed in the previous section, it may be possible to obtain even higher intensity on smaller samples by using strongly convergent beam diffraction.

35.6 FLUORESCENCE

Introduction

The use of secondary X rays that are emitted from solids bombarded by X rays, electrons, or positive ions to measure the composition of the sample is widely used as a nondestructive elemental analysis tool. Such secondary X rays are called *fluorescence X rays*. The "characteristic rays" emitted from a solid irradiated by X rays or electrons[31] were shown in 1913 by Moseley to have characteristic wavelengths (energies) corresponding to the atomic number of specific elements in the target.[32] Measurement of the wavelength of the characteristic X rays, as well as observation of a continuous background of wavelengths, was made possible by use of the single-crystal diffraction spectrometer first demonstrated by Bragg.[33] There was active development by a number of workers and by the late 1920s X-ray techniques were well developed. In 1923, Coster and von Hevesey[34] used X-ray fluorescence to discover the unknown element hafnium by measurement of its characteristic line in the radiation from a Norwegian mineral, and in 1932 Coster and von Hevesey published the classical text "Chemical Analysis by X-Ray and its Applications." Surprisingly, there was then almost no further activity until after World War II. In 1947 Friedman and Birks converted an X-ray diffractometer to an X-ray spectrometer for chemical analysis,[35] taking advantage of work on diffraction systems and detectors that had gone on in the previous decade. An X-ray fluorescence measurement in which the energy (or wavelength) spectrum is carried out by the use of X-ray diffraction spectrometry is called *wavelength-dispersive X-ray fluorescence* (WDXRF). There was then rapid progress with a number of companies developing commercial X-ray fluorescence (XRF) instruments. The early developments have been discussed in detail by Gilfrich.[36] During the 1960s, the development of semiconductor particle detectors that could measure the energy spectrum of emitted X rays with much higher energy resolution than possible with gas proportional counters or scintillators resulted in an explosion of applications of energy-dispersive X-ray fluorescence (EDXRF). Until recently, except for the flat or curved diffraction crystals used in WDXRF, X-ray optics have not played an important role in X-ray fluorescence measurements. This situation has changed markedly during the past decade. We will now review the current status of both WDXRF and EDXRF with emphasis on the role of X-ray optics without attempting to document the historical development.

Wavelength-Dispersive X-Ray Fluorescence (WDXRF)

There are thousands of XRF systems in scientific laboratories, industrial laboratories, and in manufacturing and process facilities worldwide. Although most of these are EDXRF systems, many use X-ray diffraction spectrometry to measure the intensity of selected characteristic X-rays. Overwhelmingly, the excitation mechanism of choice is energetic electrons, and many are built onto scanning electron microscopes (SEMs). Electron excitation is simple and can take advantage of electrostatic and magnetic electron optics to provide good spatial resolution and, in the case of the SEM, to give elemental composition maps of the sample with high resolution. In general, the only X-ray optics connected with these systems are the flat or curved analyzing crystals. Sometimes there is a single analyzing crystal that is scanned to give the wavelength spectrum (although multiple crystals, usually two or three, are used to cover different wavelength ranges). However, some systems are multichannel with different (usually curved) crystals placed at different azimuthal angles, each designed to simultaneously measure a specific wavelength corresponding a selected element or background wavelength. Sometimes a scanning crystal is included to give a less sensitive but more inclusive spectral distribution. Such systems have the benefit of high resolution and high sensitivity in cases where the needs are well defined. In general, WDXRF systems have not been designed to take advantage of recent developments in X-ray optics, although there are a number of possibilities and it is expected that such systems will be developed. One important role that X-ray optics can be used in such systems is shown in Fig. 9.

In this arrangement, a broad angular range of X-ray emission from the sample is converted into a quasi-parallel beam with a much smaller angular distribution. A variety of collimating optics could be used, for example, polycapillary (as shown), multilayer, nested cone, and so on. The benefit, represented as the gain in diffracted intensity, will depend on the optic used, the system design (for example, the diffracting crystal or multilayer film), and the X-ray energy. With a polycapillary collimator, 8-keV X-rays from the sample with divergence of up to approximately 12°, can be converted to a beam with approximately 0.2° divergence. With a diffraction width of 0.2° and a transmission efficiency for the optic of 50 percent, the gain in the diffracted beam intensity for flat crystal one-dimensional diffraction would be typically more than 30. Further discussion of the gains that can be obtained in X-ray diffraction measurements can be found in Sec. 35.5. Another potential benefit from the arrangement shown in Fig. 9 is confinement of the sampling area to a small spot defined by the collection proper-

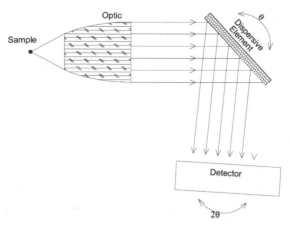

FIGURE 9 Schematic representation of collimating optic in WDXRF system.

ties of the optic as discussed in Chap. 30. Scanning of the sample will then give the spatial distribution of selected elements. This is also useful in so-called environmental, or high-pressure, SEMs where the position of the exciting electron beam is not so well defined.

Fine Structure in WDXRF Measurements. As noted previously, most of the WDXRF systems in use involve electron excitation either in SEM systems or in dedicated electron microprobe systems. Photon emission from electron excitation systems contains, in addition to the characteristic lines, a continuous background due to bremsstrahlung radiation resulting from slowing down of the electrons in the solid. Although this background does not seriously interfere with many measurements of elemental composition, it can limit the measurement sensitivity, and can preclude observation of very low intensity features. If the fluorescence X rays are excited by incident X rays, or energetic charged particles, the bremsstrahlung background can be avoided. It should be noted that a continuous background still is present when a broad X-ray spectrum is used as the exciting beam due to scattering of low-energy X rays. This can be largely avoided if monoenergetic X rays are used.[37]

A dramatic illustration of the value of a low background in XRF measurements is contained in recent studies in Japan of fine structure in fluorescence spectra.[38–41] Accompanying each characteristic X-ray fluorescence peak is an Auger excitation peak displaced typically approximately 1 keV in energy and lower in intensity by nearly 1000 times. This peak is not usually observable in the presence of bremsstrahlung background from electron excitation. By using X-ray excitation to get a low background and WDXRF to get high-energy resolution, Kawai and coworkers[38–41] measured the Auger excitation peaks from Silicon in elemental Si and SiO_2 and from Al. They showed that the observed structure corresponds to X-ray absorption fine structure (EXAFS) and X-ray absorption near-edge structure (XANES) that has been observed in high-resolution synchrotron studies. Very long measurement time was necessary to obtain sufficient statistical accuracy. This type of measurement could presumably be considerably enhanced by the use of collimating optics as shown in Fig. 9.

Energy-Dispersive X-Ray Fluorescence (EDXRF)

During the 1960s, semiconductor detectors were developed with dramatic impact on energy and later position measurement of energetic charged particles, electrons, and X rays.[42–43] Because of their high efficiency, high count rate capability, high resolution compared with gas counters, and their improved energy resolution compared with scintillation counters, these new detectors virtually revolutionized radiation detection and applications including X-ray fluorescence. Initially, the semiconductor junction detectors had a thin active area and, therefore, were not very efficient for X rays. However, by use of lithium compensation in the active area of the detector, it was possible to make very thick depletion layers[44] (junctions) and, therefore, to reach a detection efficiency of 100 percent for X rays. Later, high-purity germanium was used to make thick semiconductor junctions. Such large-volume semiconductor junctions need to be cooled to obtain the highest energy resolution (typically, 130–160 eV).

There are now thousands of XRF systems that use cooled semiconductor detectors, most of them mounted on SEMs. Most SEM-based XRF systems do not use any X-ray optics. The detectors can be made large enough (up to 1–2 cm^2) and can be placed close enough to the sample that they collect X-rays over a relatively large solid angle.

As pointed out previously, electron excitation produces a background of bremsstrahlung radiation that sets a limit on the signal-to-background ratio and, therefore, the minimum detection limit for impurities. This background can be avoided by using X rays or energetic charged particles as the excitation source. Consequently, it is common to see cooled lithium-drifted silicon Si(Li) or high-purity germanium (HpGe) detectors mounted on accelerator beamlines for materials analysis. The ion beam– (usually proton or helium ion) based technique is called *particle-induced X-ray emission* (PIXE). Again, these do not require optics because the detector can be relatively close to the sample. As with the electron-based systems, optics necessary for controlling or focusing the exciting beam are electrostatic or magnetic and will not be discussed here. When the exciting beam is composed of photons from a synchrotron or free-electron laser (FEL) source, the situation is virtually the same, with no optics required between the sample and the detector. Mirrors and monochromators used to control the exciting beam are discussed in Chaps. 22 and 26.

Monocapillary Micro-XRF (MXRF) Systems.   However, if the excitation is accomplished by X rays from a standard laboratory-based X-ray generator, X-ray optics have a very important role to play. In general, the need to obtain a high flux of exciting photons from a laboratory X-ray source requires that the sample be as close as possible to the source. Even then, if the sample is small or if only a small area is irradiated, practical geometrical considerations usually limit the X-ray flux. The solution has been to increase the total number of X rays from the source by increasing the source power, with water-cooled rotating anode X-ray generators becoming the laboratory-based X-ray generator of choice. (For a discussion of X-ray sources, see Chap. 31.) To reduce the geometrical $1/d^2$ reduction of X-ray intensity as the sample is displaced from the source (where d is the sample/source separation), capillaries (hollow tubes) have been used since the 1930s.[45] Although metal capillaries have been used,[46] glass is the overwhelming material of choice,[47–49] because of its easy formability and smooth surface. In most of the studies reported earlier, a straight capillary was placed between the X-ray source and the sample, and aligned to give the highest intensity on the sample, the capillary length (6–20 mm) being chosen to accommodate the source/sample spacing in a commercial instrument.

In 1988, Stern et al.[50] described the use of a linearly tapered or conical optic that could be used to produce a smaller, more intense but more divergent beam. This has stimulated a large number of studies of shaped monocapillaries. Many of these are designed for use with synchrotron beams for which they are especially well suited, but they have also been used with laboratory sources. A detailed discussion of monocapillary optics and their applications is given in Chap. 29.

In 1989, Carpenter[51] built a dedicated system with a very small and controlled source spot size, close-coupling between the capillary and source and variable distance to the sample chamber. The sample was scanned to obtain spatial distribution of observed elemental constituents. A straight 10-μm diameter, 119-mm-long capillary showed a gain of 180 compared to a 10-μm pinhole at the same distance and measurements were carried out with a much lower power X-ray source (12 W) than had been used before.

Polycapillary-Based MXRD.   As discussed in Chap. 30, a large number of capillaries can be combined to capture X rays over a large angle from a small, divergent source and focus them onto a small spot. This is particularly useful for microfocus X-ray fluorescence (MXRF) applications.[52–54]

Using the system developed by Carpenter, a systematic study was carried out by Gao[55] in which standard pinhole collimation, straight-capillary, tapered-capillary, and polycapillary focusing optics could be compared. A schematic representation of this system with a poly-capillary focusing optic is shown in Fig. 10. The X rays were generated by a focused electron beam, which could be positioned electronically to provide optimum alignment with whatever optical element was being used.[51]

The target material could be changed by rotating the anode as shown in Fig. 10. Measurement of the focal spot size produced by the polycapillary focusing optic was carried out by measuring the direct beam intensity while moving a knife edge across the focal spot. The result of such measurements for Cu $K\alpha$ and Mo $K\alpha$ X rays are shown in Fig. 11.

The intensity gain obtained from the polycapillary focusing optic depends on the size of the X-ray emission spot in the X-ray generator, on the X-ray energy, and on the input focal distance (distance between the source spot and the optic). This is because the effective collection angle for each of the transmitting channels is controlled by the critical angle for total external reflection (see Chap. 30). For the system shown, the flux density gain relative to the direct beam of the same size at 100 mm from the source, is shown in Fig. 12 for Cu $K\alpha$ and Mo $K\alpha$ X rays. The maximum gain is about 4400 at 8.0 keV and 2400 at 17.4 keV, respectively.

A secondary X-ray spectrum obtained by irradiating a standard NIST thin-film XRF standard sample, SRM1833, is shown in Fig. 13. Zirconium and aluminum filters were used before the optic to reduce the low-energy bremsstrahlung background from the source. The flux density of the beam at the focus was calculated to be 1.5×10^5 photons·s·μm^2 for Mo $K\alpha$ from the 12-W source operated at 40 kV. The minimum detection limits (MDLs) in picograms for 100-s measurement time were as follows: K, 4.1; Ti, 1.5; Fe, 0.57; Zn, 0.28; and Pb, 0.52. The MDL val-

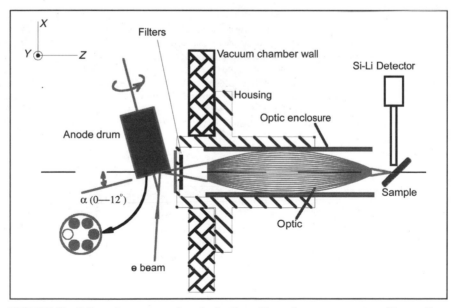

FIGURE 10   Schematic representation of microfocus X-ray fluorescence system. (*Source: From Ref. 52.*)

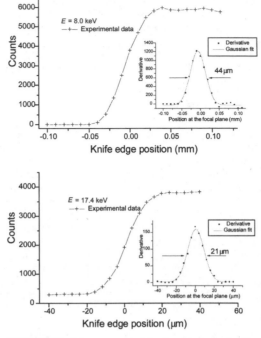

FIGURE 11 Measurement of the focal spot size for Cu and Mo X rays. (*From Ref. 55.*)

ues are comparable with those obtained by Engstrom et al.,[56] who used a 200-μm diameter straight monocapillary and an X-ray source with two orders of magnitude more power (1.7 kW) than the 12-W source used in the polycapillary measurements.

By scanning the sample across the focal spot of the polycapillary optic, the spatial distribution of elemental constituents was obtained for a ryolithic glass inclusion in a quartz phenocryst found in a layer of Paleozoic altered volcanic ash. This information is valuable in stratigraphic correlation studies.[57,58] The results are shown in Fig. 14. Also shown are the images obtained from Compton scattering (Comp) and Rayleigh scattering (Ray).

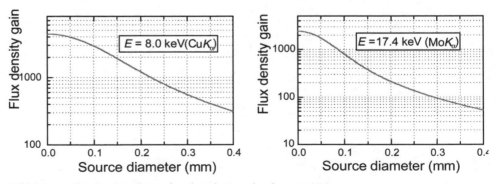

FIGURE 12 Flux density gain as a function of source size. (*From Ref 26.*)

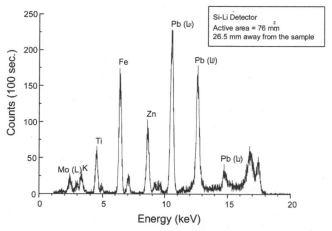

FIGURE 13   Spectrum of SRM 1833 standard XRF thin-film sample. (*From Ref. 52.*)

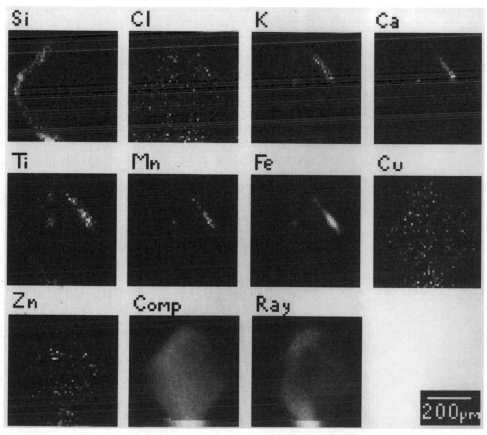

FIGURE 14   MXRF images of various elements in a geological sample that contains small volcanic glass inclusions (tens of microns in dimension) within a quartz phenocryst. The last two images are the Compton (energy-shifted) and Rayleigh (elastic) scattering intensity maps. (*From Ref. 55.*)

MXRF with Doubly Curved Crystal Diffraction.   Although X-ray-induced fluorescence has a significantly lower background than electron-induced fluorescence, there is still background arising from scattering of the continuous bremsstrahlung radiation in the sample. This can be reduced by filtering of high-energy bremsstrahlung in polycapillary focusing optics (see Chap. 30) and by use of filters to reduce the low-energy bremsstrahlung as done for the spectrum shown in Fig. 13. Recently, efficient collection and focusing of characteristic X rays by Bragg diffraction with doubly bent single crystals has been demonstrated by Chen and Wyttry.[59] The arrangement for this is shown in Fig. 15.[60]

Energy spectra taken with a thin hydrocarbon (acrylic) film with a polycapillary focusing optic, and with doubly curved crystal optics with a mica and with a silicon crystal are shown in Fig. 16.[60] The background reduction for the monoenergetic excitation is evident. Various order reflections are observed with the mica crystal. The angle subtended by the mica crystal is approximately $20° \times 5°$, giving an X-ray intensity only about a factor of three lower than that obtained with the polycapillary lens.

Ultra-High Resolution EDMXRF.   As discussed earlier, EDMXRF utilizing cooled semiconductor junction detectors is widely used in science and industry. The energy resolution of semiconductor detectors is typically 140 to 160 eV. During the past few years, very high resolution X-ray detectors based on superconducting transition-edge sensor (TES) microcalorimeters,[61] semiconductor thermistor microcalorimeters,[62–64] and superconducting tunnel junctions[65] have been developed. Although these detectors are still under active development, there have been demonstrated dramatic benefits for MXRF applications.

At the time of preparation of this review, TES microcalorimeter detectors have the best reported energy resolution (~2 eV at 1.5 keV).[66] A schematic representation of a TES microcalorimeter detector is shown in Fig. 17 and energy spectra for a titanium nitride thin film is shown in Fig. 18[66] and for a tungsten silicide thin film is shown in Fig. 19.[61] In each case, the spectrum in the same energy region from a silicon junction detector is also shown for comparison.

Because the absorbing element on microcalorimeter detectors must have a low thermal capacitance to achieve high resolution and short recovery time (for higher counting rates), and cannot operate closer than 5 mm to the sample (because of thermal and optical shielding), the sensitivity is low. However, by using a collecting and focusing optic between the sam-

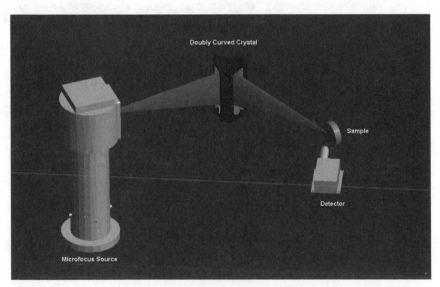

FIGURE 15   Doubly curved crystal MMEDXRF setup. (*Courtesy of XOS Inc. From Ref. 60.*)

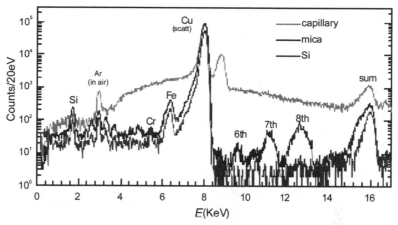

FIGURE 16   Energy spectrum obtained from a thin acrylic sample, measured with a mica doubly bent crystal, a silicon doubly bent crystal, and a polycapillary focusing optic. (*From Ref. 60.*)

ple and the detector, the effective area can be greatly increased.[67] A schematic of such an arrangement with a polycapillary focusing optic is shown in Fig. 20. With such a system, the effective area can be increased to approximately 7 mm², comparable with the area of high-resolution semiconductor detectors.

Figure 21 shows a logrithmic plot of the energy spectrum for an aluminum-gallium-arsenide sample.[61] This shows the bremsstrahlung background that is present when electrons (in this case, 5 keV) are used as the exciting beam. It is clear that X-ray excitation (especially with monochromatic X-rays) will be important in order to use such detectors to observe the low-intensity fine structure to get microstructure and microchemical information with high-resolution energy-dispersive detectors.

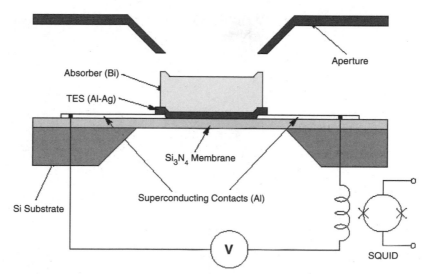

FIGURE 17   A schematic representation of a TES microcalorimeter detector. The operating temperature is ~50 mK. (*From Ref. 61.*)

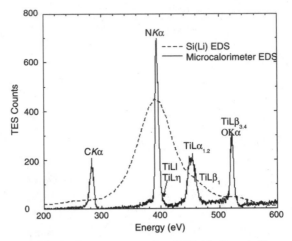

FIGURE 18 Energy spectrum for a TiN thin film on silicon. (*From Ref. 66.*)

35.7 MEDICAL APPLICATIONS

Introduction

The earliest announcements of the discovery of X rays included a radiograph of a human hand.[68] Within a few months, X radiation was being used to diagnose broken limbs on the other side of the Atlantic.[69] Within 3 years there had already been attempts to treat lupus by

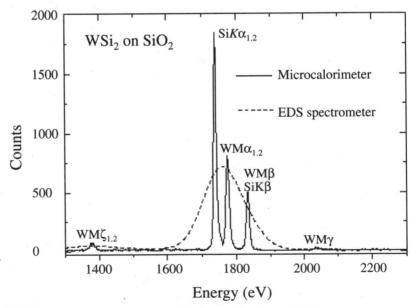

FIGURE 19 Energy spectrum for a Wsi$_2$ film on SiO$_2$. (*From Ref. 61.*)

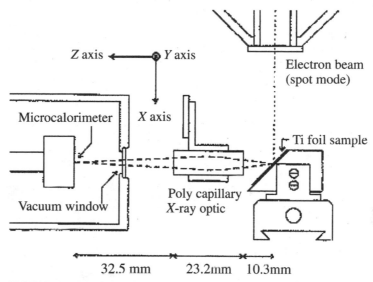

FIGURE 20 Schematic representation of focusing optic for microcalorimeter detector. (*From Ref. 67.*)

exposure to soft X rays, and so began the field of X-ray therapy.[70] One hundred years after its first use, despite the advent of ultrasound and magnetic resonance imaging (MRI), X-ray imaging is still the most common medical imaging modality. The relative simplicity of the required apparatus and the speed with which the image is obtained (and the resultant relatively low cost) make X-ray imaging an extremely important diagnostic tool. For example, despite decades of research on causes and cures, simple mammography has proven to be the most effective means of reducing breast cancer mortality.[71]

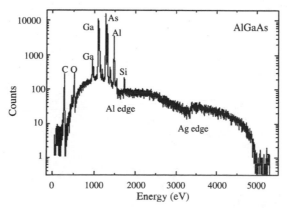

FIGURE 21 Energy spectrum from aluminum-gallium-arsenide sample measured with a TES microcalorimeter spectrometer. (*From Ref. 61.*)

Imaging

Theory.   The effectiveness of a medical image to be used in diagnosis depends on its contrast and resolution. For example, the detection of breast cancer from a mammogram relies upon the observation of either image contrast between the normal tissue and the only slightly denser cancerous lesion or observation of the existence of clusters of small, irregularly shaped microcalcifications. This requires both good contrast and good resolution. For mammography to be feasible for a screening modality, it must also be inexpensive, rapid, and require an extremely low radiation dose.

Contrast.   Three years after Roentgen's discovery of the X ray, his observations included a discussion of the proper hardness of the vacuum tube to obtain the best contrast:[72]

> With a very soft tube [one with a poor vacuum, which discharges at a low potential and produces low energy X rays] a picture is obtained in which the bones are only partially shown; by employing a harder tube the bones become very clear and visible in every detail; and with a very hard tube only very weak shadows of the bones are obtained. From what has been said, it follows that the choice of tubes must be regulated by the composition of the object to be radiographed.

Contrast is defined as

$$C = \ln \frac{E_{obj}}{E_{prim}} \approx \frac{\Delta E}{E_{prim}} \tag{10}$$

where E_{obj} is the X-ray exposure obtained behind the object of interest, and E_{prim} is the exposure obtained at an adjacent location.[73] For example, if the object of interest consists of a small cylinder of length L and absorption coefficient α', embedded in a rectangular block of absorption α and uniform thickness t, as shown in Fig. 22, then

$$E_{prim} = E_0 e^{-\alpha t}$$
$$E_{obj} = E_0 e^{-\alpha(t-L)} e^{-\alpha' L} = E_0 e^{-\alpha t} e^{(\alpha - \alpha')L} \tag{11}$$

where E_0 is the incident fluence. Thus, the contrast is proportional to the difference in absorption coefficients,

$$C = \ln \left(\frac{E_0 e^{-\alpha t} e^{\Delta \alpha L}}{E_0 e^{-\alpha t}} \right) = (\Delta \alpha) L \tag{12}$$

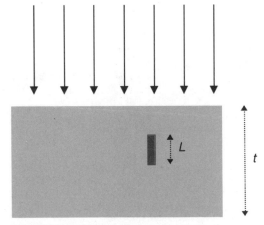

FIGURE 22   Model contrast-producing object.

Absorption coefficient differences, and therefore contrasts, are larger for soft than for hard X rays. However, the absorption coefficient itself is also larger for soft X rays, and consequently so is the absorbed patient dose, and also the Poisson quantum noise, which varies with the square root of the number of detected photons. There is thus a trade-off between contrast and absorption that determines the best photon energy for the application.

Scatter.   At diagnostic energies (20–120 keV), the primary interactions between X rays and tissue are due to the photoelectric effect and Compton scattering. The two processes have about equal probabilities in tissue at the lowest energies, with Compton scattering dominating at high energies. Both remove photons from the primary beam and are included in the coefficient α when calculating contrast. However, those photons that are Compton scattered in the direction of the detector contribute to a general background fog on the film. If the fluence of scattered photons is E_s, then the contrast becomes

$$C_s = \ln \frac{E_{obj} + E_s}{E_{prim} + E_s} \approx \frac{\Delta E}{E_{prim} + E_s} = \frac{C}{1 + (E_s/E_{prim})} \tag{13}$$

The scatter-to-primary ratio, E_s/E_{prim}, is typically from 40 to 90 percent for mammography, resulting in a contrast reduction of a factor of 1.4 to 1.9.[74] In current conventional radiography, a lead *grid* is placed after the patient to reduce the amount of scattered radiation incident on the film. The grid consists of lead strips about 20 μm across, separated by about 300 μm of carbon fiber resin. The thickness of the grid in the beam direction is about 5 to 12 times the strip spacing. A thicker grid, with a larger lead strip-height-to-spacing ratio, would more effectively stop high-angle scattered photons, but would also have a lower transmission for the primary beam. A conventional 5:1 grid transmits about 60 percent of the primary beam and only 20 percent of the scattered beam at 20 keV.

Resolution.   Diagnosis often requires the detection of the existence and even the shape of tiny objects, such as microcalcifications in mammography. A good film/screen mammography system has a resolution of 15 line pairs/mm (lp/mm), or about 33 μm. Other, higher-energy imaging typically uses lower-resolution systems. A computed radiography (CR) plate for chest radiography has a resolution of between 4 and 5 lp/mm.

Contrast Enhancement.   Contrast can be enhanced by removing the photons that have been Compton scattered out of the primary beam. These photons can be essentially eliminated from the image by inserting, between the patient and the detector, an optic with a low angular acceptance. For example, because polycapillary optics have an angular acceptance that is limited by the critical angle (1.5 mrad at 20 keV), scattered photons are not transported down optics channels that are aligned with the primary beam direction. These scattered photons are largely absorbed by the glass walls of the capillary optic. Measurements of absorption and transmission as a function of entrance angle on a 30-mm-diameter polycapillary prototype optic yielded full-scale scatter transmissions of less than 1 percent.[75] This was confirmed by scatter fraction measurements performed on a Lucite phantom with a smaller (3.6 mm) prototype optic.[75] Scatter transmission was 0.5 percent. This resulted in an average contrast enhancement for the Lucite phantom of a factor of 1.4 compared with a conventional antiscatter grid, with a maximum enhancement of more than a factor of two in regions of low contrast or high scatter. The image produced from the scan of the Lucite contrast phantom using the polycapillary optic is shown in Fig. 23, and using a conventional grid in Fig. 24. The measured contrast is shown in Fig. 25. Primary transmission for this optic was 45 percent. The primary transmission is expected to increase somewhat as more manufacturing experience is achieved.

Abdominal imaging and chest radiography are performed at higher photon energies than mammography. Typical abdominal images are produced using a tungsten target with a characteristic energy of 60 keV and maximum tube potential just slightly higher. Chest radiography is performed with a tungsten target and tube potential of 120 kV, producing a spectrum with a peak around 80 keV.

The potential performance of polycapillary optics at these energies has been calculated.[76] Performance estimates of three hypothetical lenses are given in Table 4. The ideal borosilicate

FIGURE 23 Image of contrast phantom made with scanned optic. (*From Ref. 75.*)

FIGURE 24 Image of contrast phantom produced with conventional grid. (*From Ref. 75.*)

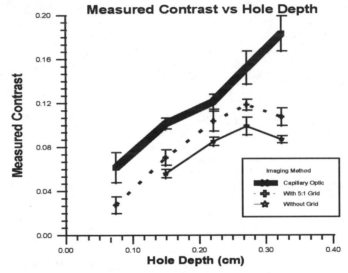

FIGURE 25 Contrast enhancement measured for a small scanned polycapillary optic using a phantom consisting of a Lucite block with drilled holes. (*From Ref. 8.*)

lens is scaled up from the center part of a measured polycapillary lens. Simulated transmission was used for the lead glass lens. The performance of the three lenses was compared with the performance of the conventional grids described in Table 5. The primary and scattered transmissions of the different grids in Table 5 were measured at 69 keV.[77]

The contrast after an optic depends on both the primary and scatter transmission of the optic. For example, if the scatter fraction before the capillary lens is F, then the scatter fraction after the lens, F_{optic}, is

$$F_{optic} = \frac{F \cdot T_{s\text{-}optic}}{F \cdot T_{s\text{-}optic} + (1-F) \cdot T_{p\text{-}optic}} = \frac{KF}{1 - F(1-K)} \tag{14}$$

where $T_{p\text{-}optic}$ is the primary transmission of the lens and K is the ratio

$$K = \frac{T_{s\text{-}optic}}{T_{p\text{-}optic}} \tag{15}$$

TABLE 4 Transmissions for Three Polycapillary Optics

Lens no.	Lens type	Length (mm)	Photon energy (keV)	Primary transmission ($T_{\text{p-optic}}$)	Scatter transmission ($T_{\text{s-optic}}$)
1	Ideal borosilicate optic; scaled up from the center part of the prototype lens	166	20	0.6	0.0
			45	0.55	0.002
			60	0.55	0.005
			70	0.45	0.03
2	Actual borosilicate prototype optic	166	20	0.33	0.0
3	Lead glass polycapillary optic	30	69	0.63	0.0076

Note: The results for the ideal borosilicate optic are measured results for the center part of a prototype lens. The results for lead glass lens are simulated with 120 kVp spectrum and a W target.

The contrast enhancement achieved by using the optic is

$$\frac{C_{\text{optic}}}{C_{\text{no-optic}}} = \frac{1 - F_{\text{optic}}}{1 - F} = \frac{1}{FK + (1 - F)} \tag{16}$$

The contrast enhancement alone is not the only relevant quality factor. The signal-to-noise ratio (SNR) is another important quality factor. According to the Rose model,

$$k^2 = C^2 \Phi A \tag{17}$$

where k is the SNR, C is the contrast, Φ is the photon flux, and A is the area of the target. To compare the SNR for different scatter rejection devices, relative SNR was used. Relative SNR is the SNR normalized by the SNR without any scatter rejection device,

$$k_{\text{r}}^2 = \frac{k^2}{k_0^2} = \frac{C^2}{C_0^2}(FT_{\text{s}} + (1 - F)T_{\text{p}}) \tag{18}$$

where k_{r} is the relative SNR. C and C_0 are the contrast with and without the scatter rejection device, respectively. T_{s} is the scatter transmission and T_{p} is the primary transmission of the device. The contrast enhancement and relative SNR were calculated for different scatter frac-

TABLE 5 Performance for Several Commercial Grids, Measured at a Scatter Fraction of 85 Percent[77]

Grid ratio	Primary transmission at 69 keV	Scatter transmission at 69 keV
6	0.73	0.27
8	0.64	0.13
12	0.61	0.071

tions and plotted in Fig. 26 and Fig. 27. The performance of the grids is assumed to be energy independent. For high scatter fractions, which occur at high energies where Compton scattering dominates, the optics could result in significantly enhanced SNRs.

Resolution Enhancement.   Magnification can improve image resolution, particularly if the image resolution is limited by the detector resolution, as is the case for some solid-state detectors and for computed radiography plates. Magnification is conventionally performed by increasing the air gap between the patient and the detector. This results in a loss of resolution due to the geometrical blurring from the finite source size, as shown in Fig. 28. However, magnification using X-ray optics, even with a large focal spot, would not be subject to a loss in resolution, as shown in Fig. 28 for a polycapillary optic. Although polycapillary optics are not true imaging optics in the full theoretical sense, images can be transported in the optics in the same manner as in a coherent optical fiber bundle. By changing the diameter of the monolithic bundle, images can be magnified. Measured modulation transfer functions (MTF) for CR plates using a magnifying polycapillary optic are compared with airgap magnification in Fig. 29. Computed radiography is a cheap, convenient digital medium, with a high dynamic range, readily available in a hospital environment. However, the resolution of CR is too poor for mammography. The modulation transfer function (MTF) of CR falls to 5 percent at a spatial frequency of 5 lp/mm. Using a tapered optic with a magnification factor of 1.8, the 5 percent MTF frequency was increased to 9.2 lp/mm.[75] The resolution was not degraded by the capillary structure, which was on a smaller scale (20-μm channel size) than the desired resolution. The optic yields an MTF increase at all spatial frequencies, which may be diagnostically more significant than the increase in limiting MTF. Magnifying capillary optics provide simultaneous contrast enhancement and resolution increase.

Demagnification and Digital Detectors.   In almost all radiological systems, detection of X rays is performed by the use of a *phosphor* screen, which converts absorbed X rays into visible-light photons. The visible light is then recorded on an analog medium, such as film, or detected by a digital detector. The phosphor screen is used because most detectors, including film, are not particularly sensitive to X rays. The use of a screen-film combination reduces the radiation requirement compared with direct-exposure film by a factor of 50 to 100.[78] This

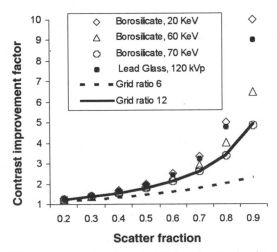

FIGURE 26   Contrast improvement versus scatter fraction for the two simulated polycapillary optics. The results were compared with the performance of two commercial grids. (*From Ref. 76.*)

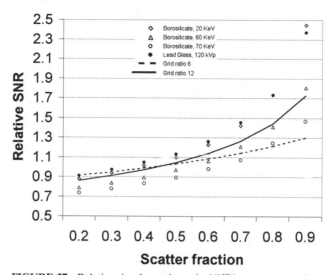

FIGURE 27   Relative signal-to-noise ratio (SNR) versus scatter fraction for two hypothetical optics. The results were compared with the performance of two commercial grids. (*From Ref. 76.*)

increase in sensitivity occurs at the expense of resolution. The MTF of a good film-screen combination drops to less than 0.1 at a frequency of 15 lp/mm.[78] The choice of phosphor screen thickness is a trade-off between detected quantum efficiency, which improves with increasing thickness, and resolution, which is degraded by light blur in a thick screen.[79] Efficient direct X-ray detection virtually eliminates the trade-off between spatial resolution and sensitivity because of the elimination of the phosphor screen. Direct X-ray detectors can pro-

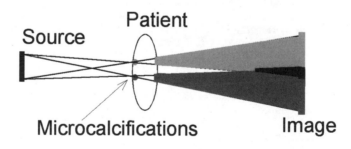

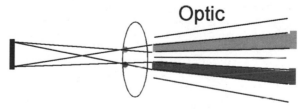

FIGURE 28   Airgap magnification (top) showing degradation of image due to finite source size. Capillary optic magnification (bottom) showing no increased blurring after the exit plane of the patient.

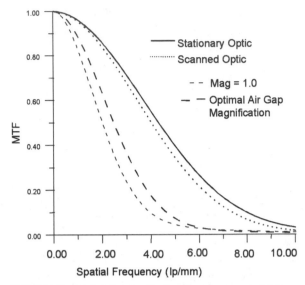

FIGURE 29 Modulation transfer function of a computed radiography plate with no optic, with airgap magnification, and with a polycapillary optic. (*From Ref. 75.*)

vide resolution of 20 lp/mm or better with nearly 100 percent detector quantum efficiency (DQE).

A technical difficulty with many direct X-ray detectors is that they are difficult to manufacture with large enough areas for medical imaging. With the use of X-ray optics, demagnification could be used to match the image size to a CCD, CID,[80] or other digital detector. A common digital imaging modality is the use of a fused fiber optic following a phosphor screen to demagnify the resulting visible-light image for recording with CCD technology. The use of capillary X-ray optics to demagnify the X-ray image in connection with direct X-ray-sensitive detectors may be more efficient and would avoid the loss of resolution in the optical conversion process.

In addition to the problem of resolution, conventional film/screen radiography suffers from limited dynamic range and from film granularity, which can reduce the sensitivity of detection of microcalcifications. Digital detection provides high dynamic range, improving contrast, and greatly increases the tolerance of the final image to under- or overexposure. Digital images can be enhanced and are amenable to computer-aided diagnosis. Spectral information can be included if it is available. Finally, digital images can be quickly transported for skilled consultation.

Advantages of Parallel Beams. Typical radiography uses divergent beams from point sources. The use of a uniform parallel beam would eliminate the variations in resolution from the different apparent source sizes viewed from different parts of the X-ray field and the image distortion arising from the different magnification of objects on the entry and exit side of the patient. In addition, scatter can be virtually eliminated by the large airgap made possible by a parallel beam. A parallel beam would also relax geometrical constraints that have affected system design. After collimation, changes in optic to patient distances would not affect magnification or resolution. Although X-ray beams from point sources can be made nearly parallel using large source-to-patient distances, only a small fraction of the output from the source is used. Large high-current sources are required. However, increasing the spot size of the source then adversely affects the resolution. Collecting X rays more efficiently by employing

X-ray optics reduces the required source power, or the time required to obtain an image for a given source power, thus reducing potential motion anomalies.

For a variety of small sample volume applications, such as specimen biopsy and small-volume computed tomography, the X-ray source is also utilized inefficiently (i.e., only a small fraction of the total solid angle of X-ray emission is collected). These applications can also benefit from a collimating optic, which collects from a larger solid angle and redirects the photons toward the sample.[81]

Monochromatization. Monochromatic beams could be produced from ordinary point sources by diffraction from a single crystal monochromator. Single-crystal silicon wafers, 20 cm in diameter, are routinely used in semiconductor manufacturing; such wafers with irrelevant electronic defects are available quite cheaply. However, diffraction from such a crystal is so selective in angle and energy that isotropic radiation from a point source, such as a conventional radiography tube, would not be reflected efficiently. This would result in insufficient intensity to produce an image quickly enough to prevent motion anomalies. For that reason, monochromatic imaging has been restricted to research sources such as synchrotrons and free-electron lasers. Use of a collimating optic could resolve the problem of delivering sufficient beam intensity.[82,83] Monochromatizing a conventional X-ray source with the combination of a collimating optic and a diffracting crystal would allow routine radiography to take advantage of the enhanced contrast produced by monochromatic radiation.[84] Dose reduction would result from the elimination, before exposure to the patient, of low-energy photons that would be heavily absorbed in the patient without contributing to contrast. The resulting parallel, tunable, monochromatic source would facilitate the use of a number of innovative imaging techniques such as refraction contrast,[85] K-edge tuning,[86] and dual-energy imaging.[87] Refraction contrast is observed for highly monochromatic beams when a second crystal is placed after the specimen. Gradients in the index of refraction deflect the beam at the edges of features. The deflected beams are not diffracted by the second crystal, resulting in greatly enhanced contrast in the vicinity of feature edges. K-edge tuning is the selection of a photon energy just above the absorption edge for an element that is preferentially concentrated in, for instance, cancerous tissue. Dual-energy imaging uses the tunability of the source to make a subtraction image taken at energies just above and below the critical edge.

Therapy

X-Ray Therapy. Conventional X-ray radiation therapy is currently performed with high-energy X-ray or gamma radiation. The patient is exposed to multiple parallel beams, produced with slit collimators, with an intersection at the tumor. High-energy photons are chosen to minimize the absorbed skin dose relative to the dose at the tumor, although energies as low as 100 keV are employed in the orthovoltage modality. The choice of high energies to reduce skin dose is necessary because in an unfocused beam the intensity is necessarily higher at the point of entry than at the tumor site. Use of focusing optics at energies less than 100 keV might substantially increase the tumor dose relative to the skin dose.

An X-ray optic designed to focus X rays with an absorption length in tissue of $L = 20$ mm (about 30 keV) can have an output radius $r_0 = 20$ mm, 50-mm output focal length, and focal spot radius of $r_f = 0.25$ mm. The optic could be focused on a tumor centered 25 mm under the surface. The radius of the beam at the surface is $r_s = 25/50 \cdot 20 = 10$ mm. The ratio of intensity at the focal spot to the intensity at the surface is then

$$\frac{I_f}{I_s} = \frac{\left(\dfrac{I_0}{\pi r_f^2} e^{-(25/20)}\right)}{\left(\dfrac{I_0}{\pi r_s^2}\right)} = 460 \tag{19}$$

ignoring Compton scattering. The ratio of the dose in a sphere of diameter $t = 1$ mm centered at the focal point compared with the surface dose in a 1-mm-thick layer is

$$\frac{\text{Dose}_f}{\text{Dose}_s} = \frac{\left(\dfrac{I_0 e^{-(25/20)}}{4/3\pi(t/2)^3} e^{-(t/20)}\right)}{\left(\dfrac{I_0}{\pi r_s^2} e^{-(t/20)}\right)} = 170 \tag{20}$$

A 12-kW X-ray source with a 0.3 percent electrical-to-X-ray conversion efficiency and an optic with a transmission of 20 percent will deliver 4.5 Gy of radiation to the central 1-mm diameter sphere in 3 min.

Neutron Therapy.   Neutron beams can also be focused.[88–91] A focused neutron beam could be used in boron neutron capture therapy (BNCT). Boron neutron capture therapy is based on the selective delivery of a boronated pharmaceutical to cancerous tissue followed by irradiation with thermal neutrons.[92] The reaction products (alpha and lithium particles) have a short range in tissue and a high relative biological effectiveness. Therefore, thermal neutrons are very effective at killing boronated cells while sparing healthy, unboronated tissue. Perrung[93] has described a BNCT system using capillary optics. The entrance of the lens accepts neutrons from the cooled moderator of a reactor and the exit of the lens directs a cold neutron beam into a hollow treatment tube. The treatment tube could be positioned against a surface of the patient or inserted to a depth to irradiate a volume. This approach limits the treatment to small volumes that are accessible by the tube without significant irradiation to other regions. An experimental study using a monolithic capillary optic has been performed to determine the radiation dose introduced by focusing a cold neutron beam through a tissue equivalent mass.[94] With the use of radiochromic images, it was found that neutron scattering within the material did not significantly alter the focal image. As a result, the deposition of dose peaks quickly at the depth of the focal point. This procedure could be useful in treating near-surface regions, such as ocular melanomas.

35.8   *REFERENCES*

1. James E. Harvey, "X-Ray Optics," *Handbook of Optics,* Vol. II.
2. Marshall Joy, "Astronomical X-Ray Optics," *Handbook of Optics,* Chap. 28, this volume.
3. Peter Siddons, "Crystal Monochromators," *Handbook of Optics,* Chap. 22, this volume.
4. S. L. Hulbert, "Synchrotron Radiation Sources," *Handbook of Optics,* Chap. 32, this volume.
5. Ali Khounsary (ed.), *Advances in Mirror Technology for Synchrotron X-ray and Laser Applications,* *SPIE* **3447**:(1998).
6. Ali Khounsary (ed.), *High Heat Flux Engineering I–IV,* *SPIE* **1739, 1997, 2855, 3153**:(1993–1997).
7. see various proceedings of the international conferences on synchrotion radiation instrumentalization, for example, *Nuclear Instruments and Methods B, Beam Interactions with Materials and Atoms* **133**:(1997).
8. F. Cerrina, "The Schwarzschild Objective," *Handbook of Optics,* Chap. 27, this volume.
9. A. Michette, "Zone and Phase Plates, Bragg-Fresnel Optics," *Handbook of Optics,* Chap. 23, this volume.
10. P. Trousses, P. Munsch, and J. J. Ferme, "Microfocusing Between 1 and 5 keV with Wolter Type Optic," in *X-ray Optics Design, Performance and Applications,* A. M. Khounsary, A. K. Freund, T. Ishikawa, G. Srajer, J. Lang (eds.), *SPIE* **3773**:60–69 (1999).
11. T. A. Pikuz, A. Y. Faenov, M. Fraenkel, A. Zigler, F. Flora, S. Bollanti, P. Di Lazzaro, T. Letardi, A. Grilli, L. Palladino, G. Tomassetti, A. Reale, L. Reale, A. Scafati, T. Limongi, and F. Bonfigli, "Large-Field High Resolution X-Ray Monochromatic Microscope, Based on Spherical Crystal and High

Repetition-Rate Laser-Produced Plasmas," in *EUV, X-ray and Neutron Optics and Sources,* C. A. MacDonald, K. A. Goldberg, J. R. Maldonado, H. H. Chen-Mayer, S. P. Vernon (eds.), *SPIE* **3767**:67–78 (1999).

12. E. Spiller, "Multilayers," *Handbook of Optics,* Chap. 24, this volume.

13. S. Lane, T. Barbee Jr., S. Mrowka, and J. R. Maldonalo "Review of X-ray Collimators for X-ray Proximity Lithography," in *EUV, X-ray and Neutron Optics and Sources,* C. A. MacDonald, K. A. Goldberg, J. R. Maldonado, H. H. Chen-Mayer, and S. P. Vernon (eds.), *SPIE* **3767**:172–182 (1999).

14. A. Freund, *Handbook of Optics,* Chap. 26, this volume.

15. C. A. MacDonald and W. M. Gibson "Polycapillary and Multichannel Plate Optics," *Handbook of Optics,* Chap. 30, this volume.

16. E. C. E. Turcu, R. Forber, R. Grygier, H. Rieger, M. Powers, S. Campeau, G. French, R. Foster, P. Mitchell, C. Gaeta, Z. Cheng, J. Burdett, D. Gibson, S. Lane, T. Barbee, S. Mrowka, and J. R. Maldonao, "High Power X-ray Point Source for Next Generation Lithography," in *EUV, X-ray and Neutron Optics and Sources,* C. A. MacDonald, K. A. Goldberg, J. R. Maldonado, H. H. Chen-Mayer, S. P. Vernon (eds.), *SPIE* **3767**:21–32 (1999).

17. Bernard Dennis Cullity, *Elements of X-Ray Diffraction,* Adison Wesley Longman, 1978.

18. Qun Shen, "Solving the Phase Problem Using Reference-Beam X-ray Diffraction," *Phys. Rev. Lett.* **80**(15):3268–3271 (1998).

19. D. L. Goodstein, *States of Matter,* Prentice-Hall, Upper Saddle River, NJ, 1975.

20. A. G. Michette and C. J. Buckley, *X-ray Science and Technology,* Institute of Physics Publishing, 1993.

21. L. T. Mirkin, *Handbook of X-ray Analysis of Polycrystalline Materials,* Consultants Bureau Enterprises, 1964.

22. U. W. Arndt, "Focusing Optics for Laboratory Sources in X-ray Crystallography," *J. Appl. Crystallography* **23**:161–168 (1990).

23. C. A. MacDonald, S. M. Owens, and W. M. Gibson, "Polycapillary X-Ray Optics for Microdiffraction," *Jour. of Appl. Crystallography* **32**:160–167 (1999).

24. S. Bates, European Powder Diffraction Conference 6, Budapest, 1998 (oral presentation).

25. S. M. Owens, F. A. Hoffman, C. A. MacDonald, and W. M. Gibson, "Microdiffraction Using Collimating and Convergent Beam Polycapillary Optics," in *Advances in X-ray Analysis, Vol. 41, Proceedings of the 46th Annual Denver X-ray Conference Proceedings,* Steamboat Springs, CO, August 4–8, 1997.

26. J. X. Ho, E. H. Snell, C. R. Sisk, J. R. Ruble, D. C. Carter, S. M. Owens, and W. M. Gibson, "Stationary Crystal Diffraction With a Monochromatic Convergent X-ray Beam Source and Application for Macromolecular Crystal Data Collection," *Acta Cryst.* **D54**:200–214 (1998).

27. F. A. Hofmann, W. M. Gibson, C. A. MacDonald, D. A. Carter, J. X. Ho, and J. R. Ruble, "Polycapillary Optic—Source Combinations for Protein Crystallography," *Jour. Appl. Crystallography* (submitted).

28. U. W. Arndt, P. Duncomb, J. V. P. Long, L. Pina, and A. Inneman, *Jour. Appl. Crystallography* **31**:733–741 (1998).

29. M. Gubarev, E. Ciszak, I. Ponomarev, W. Gibson, and M. Joy, *Jour. Appl. Crystallography* **33**:882–887 (2000).

30. H. Huang, F. A. Hoffman, C. A. MacDonald, W. M. Gibson, D. C. Carter, J. X. Ho, J. R. Ruble, and I. Ponomarev, "Low Power Protein Crystallography Using Polycapillary Optics, in C. A. MacDonald and A. M. Khounsary (eds.), *Advances in Laboratory-Based X-ray Sources and Optics SPIE* **4144**: (2000).

31. R. T. Beatty, *Proc. Roy. Soc.* **87A**:511 (1912).

32. H. G. J. Moseley, *Phil. Mag.* **26**:1024 (1913); **27**:703 (1914).

33. W. H. Bragg and W. L. Bragg, *Proc. Roy. Soc.* **88A**:428 (1913).

34. D. Coster and G. von Hevesey, *Nature* **111**:79, 182 (1923).

35. H. Friedman and L. S. Birks, *Rev. Sci. Instr.* **19**:323 (1948).

36. J. V. Gilfrich, "Advances in X-Ray Analysis," *Proc. of 44th Annual Denver X-Ray Analysis Conf.,* Vol. 39, Plenum Press, New York, 1997, pp. 29–39.

37. Z. W. Chen and D. B. Wittry, *J. Appl. Phys.* **84**:1064–1073 (1998).

38. J. Kawai, K. Hayashi, and Y. Awakura, *Jour. Phys. Soc. Japan* **66**:3337–3340 (1997).

39. K. Hayashi, J. Kawai, and Y. Awakura, *Spectrochimica Acta* **B52**:2169–2172 (1997).

40. J. Kawai, K. Hayashi, K. Okuda, and A. Nisawa, *Chem Lett* (*Japan*) 245–246 (1998).

41. J. Kawai, *Jour. Electron Spectr. and Related Phenomona* 101–103, 847–850 (1999).

42. G. L. Miller, W. M. Gibson, and P. F. Donovan, *Ann. Rev. Nucl. Sci.* **33**:380 (1962).

43. W. M. Gibson, G. L. Miller, and P. F. Donovan, in *Alpha, Beta, and Gamma Spectroscopy,* 2nd ed., K. Siegbahn (ed.), North Holland Publishing Co, Amsterdam, 1964.

44. E. M. Pell, *J. Appl. Phys.* **31**:291 (1960).

45. F. Jentzsch and E. Nahring, *Z. The. Phys* **12**:185 (1931); *Z. Tech. Phys.* **15**:151 (1934).

46. L. Marton, *Appl. Phys. Lett.* **9**:194 (1966).

47. W. T. Vetterling and R. V. Pound, *J. Opt. Soc. Am.* **66**:1048 (1976).

48. P. S. Chung and R. H. Pantell, *Electr. Lett.* **13**:527 (1977); *IEEE J. Quant. Electr.* **QE-14**:694 (1978).

49. A. Rindby, *Nucl. Instr. Meth.* **A249**:536 (1986).

50. E. A. Stern, Z. Kalman, A. Lewis, and K. Lieberman, *Appl. Optics* **27**:5135 (1988).

51. D. A. Carpenter, *X-Ray Spectrometry* **18**:253–257 (1989).

52. N. Gao, I. Yu. Ponomarev, Q. F. Xiao, W. M. Gibson, and D. A. Carpenter, *Appl. Phys. Lett.* **71**:3441–3443 (1997).

53. Y. Yan and X. Ding, *Nucl. Inst. and Methods in Phys. Res.* **B82**:121 (1993).

54. M. A. Kumakhov and F. F. Komarov, *Phys. Reports* **191**:289–350 (1990).

55. N. Gao, Ph.D. Thesis, University at Albany, SUNY, Albany, NY, 1998.

56. P. Engstrom, S. Larsson, A. Rindby, and B. Stocklassa, *Nucl. Instrum. and Meth.* **B36**:222 (1989).

57. J. W. Delano, S. E. Tice, C. E. Mitchell, and D. Goldman, *Geology* **22**:115 (1994).

58. B. Hanson, J. W. Delano, and D. J. Lindstrom, *American Mineralogist* **81**:1249 (1996).

59. Z. W. Chen and D. B. Wittry, *Jour. Appl. Phys.* **84**:1064–1073 (1998).

60. ZeWu Chen, R. Youngman, T. Bievenue, Qi-Fan Xiao, I. C. E. Turcu, R. K. Grygier, and S. Mrowka, "Polycapillary Collimator for Laser-Generated Plasma Source X-Ray Lithography," in C. A. MacDonald, K. A. Goldberg, J. R. Maldonado, H. H. Chen-Mayer, and S. P. Vernon (eds.), *EUV, X-Ray and Neutron Optics and Sources, SPIE* **3767:** 52–58 (1999).

61. D. A. Wollman, K. D. Irwin, G. C. Hilton, L. L. Dulcie, D. E. Newbury, and J. M. Martinis, *J. Microscopy* **188**:196–223 (1997).

62. D. McCammon, W. Cui, M. Juda, P. Plucinsky, J. Zhang, R. L. Kelley, S. S. Holt, G. M. Madejski, S. H. Moseley, and A. E. Szymkowiak, *Nucl. Phys.* **A527**:821 (1991).

63. L. Lesyna, D. Di Marzio, S. Gottesman, and M. Kesselman, *J. Low. Temp. Phys.* **93**:779 (1993).

64. E. Silver, M. LeGros, N. Madden, J. Beeman, and E. Haller, *X-Ray Spectrom.* **25**:112–115 (1996).

65. M. Frank, L. J. Hiller, J. B. le Grand, C. A. Mears, S. E. Labov, M. A. Lindeman, H. Netel, and D. Chow, *Rev. Sci Instrum.* **69**:25 (1998).

66. G. C. Hilton, D. A. Wollman, K. D. Irwin, L. L. Dulcie, N. F. Bergren, and J. M. Martinis, *IEEE Trans. on Appl. Superconductivity* **9**:3177–3181 (1999).

67. D. A. Wollman, C. Jezewiski, G. C. Hilton, Q. F. Xiao, K. D. Irwin, L. L. Dulcie, and J. M. Martinis, *Proc. Microscopy and Microanalysis* **3**:1075 (1997).

68. W. C. Röntgen, "On a New Form of Radiation," *Nature* **53**:274–276 (1896). English translation from Sitzungsberichte der Würzburger Physik-medic, Gesellschaft, 1895.

69. E. Trevert, *Something About X Rays for Everybody,* Bubier Publishing Co., Lynn, MA, 1896. Reprinted by Medical Physics Publishing, Madison, WI.

70. Experiments performed by Kummel in Hamburg, and referred to in C. L. Leonard, *American X Ray Journal* **iii**(5), abstracted in *Archives of the Roentgen Ray* **3**:80–88, 1899.

71. *Strategies for Managing the Breast Cancer Research Program: A Report to the U.S. Army Medical Research and Development Command,* 1993.

72. W. C. Röntgen, "Further Observations on the Properties of X Rays," *Archives of the Roentgen Ray,* **3**:80–88 (1899). English translation from *Annalen der Physik und Chemie.*

73. G. T. Barnes and G. D. Frey (eds.), *Screen Film Mammography,* Medical Physics Publishing, Madison, WI, 1991.

74. G. T. Barnes and I. A. Brezovich, "The Intensity of Scattered Radiation in Mammography," *Radiology* **126**:243–247 (1978).

75. D. G. Kruger, C. C. Abreu, E. G. Hendee, A. Kocharian, W. W. Peppler, C. A. Mistretta, and C. A. MacDonald, "Imaging Characteristics of X-Ray Capillary Optics in Digital Mammography," *Medical Physics* **23**(2):187–196 (1996).

76. Lei Wang, C. A. MacDonald, and W. W. Peppler, "Performance of Polycapillary Optics for Hard X-ray Imaging," *Medical Physics* (accepted).

77. C. E. Dick and J. W. Motz, "New Method for Experimental Evaluation of X-Ray Grids," *Medical Physics* **5**(2): (March/April 1978).

78. A. G. Haus, in *Screen Film Mammography,* G. T. Barnes and G. Donald Frey (eds.), Medical Physics Publishing, Madison, WI, 1991.

79. B. H. Hasegawa, *The Physics of Medical X-ray Imaging,* 2nd Ed., Medical Physics Publishing, Madison, WI, 1991.

80. R. Wentink, J. Carbone, D. Aloisi, W. M. Gibson, C. A. MacDonald, Q. E. Hanley, R. E. Fields, and M. B. Denton, "Charge Injection Device (CID) Technology: An Imaging Solution for Photon and Particle Imaging Applications," in *Proceedings of the SPIE,* vol. 2279.

81. S. M. Jorgensen, D. A. Reyes, C. A. MacDonald, E. L. Ritman, "Micro-CT Scanner with a Focusing Polycapillary X-Ray Optic," *Developments in X-Ray Tomography II,* in U. Bonse (ed.), *SPIE* **3772**: 158–166 (1999).

82. C. C. Abreu and C. A. MacDonald, "Beam Collimation, Focusing, Filtering and Imaging with Polycapillary X-ray and Neutron Optics," *Physica Medica,* **XIII**(3):79–89 (1997).

83. Francisca Sugiro, S. D. Padiyar, and C. A. MacDonald, "Characterization of Pre- and Post-Patient X-Ray Polycapillary Optics," to be published in *Advances in Laboratory-Based X-ray Sources and Optics, SPIE proc.* 4144.

84. M. J. Yaffe, R. Fahrig, "Optimization of Spectral Shape in Digital Mammography: Dependence on Anode Material, Breast Thickness, And Lesion Type," *Med Phys* 1994: **21**:1473–1481 (1994).

85. V. A. Somenkov, A. K. Tkalich, and S. S. Shilstein, "X-Ray Refraction Radiography of Biological Objects," *Zhurnal teknicheskoj fiziki* 1991: **11**:197 (1991).

86. F. E. Carroll, "Generation of Soft X-Rays by Using the Free Electron Laser as a Proposed Means of Diagnosing and Treating Breast Cancer," *Lasers in Surgery and Med.* **11**:72–78 (1991).

87. P. C. Johns, D. J. Drost, M. J. Yaffe, and A. Fenster, "Dual-Energy Mammography: Initial Experimental Results," *Medical Physics* **12**:297–304 (May/June 1985).

88. H. Chen, R. G. Downing, D. F. R. Mildner, W. M. Gibson, M. A. Kumakhov, I. Yu. Ponomarev, and M. V. Gubarev, "Guiding and Focusing Neutron Beams Using Capillary Optics," *Nature* **357**:391 (1992).

89. Q. F. Xiao, H. Chen, V. A. Sharov, D. F. R. Mildner, R. G. Downing, N. Gao, and D. M. Gibson, "Neutron Focusing Optic for Submillimeter Materials Analysis," *Rev. Sci. Instrum.* 1994: **65**:3399–3402 (1994).

90. D. F. R. Mildner, H. H. Chen-Mayer, and R. G. Downing, "Characteristic of a Polycapillary Neutron Focusing Lens," in *Proceed. of Intl. Symp. on Adv. in Neutron Optics and Related Research Facilities,* 19–21 March 1996, *J. Phys. Soc. Japan* (to be published).

91. David Mildner, "Neutron Optics," *Handbook of Optics,* Chap. 36, this volume.

92. R. F. Barth, A. H. Soloway, and R. G. Fairchild, "Boron Neutron Capture Therapy for Cancer," *Scientific American* **100**: (1990).

93. A. J. Peurrung, "Capillary Optics for Neutron Capture Therapy," *Medical Physics* **23**(4):487–494 (April 1996).

94. R. Mayer, J. Welsh, and H. Chen-Mayer, "Focused Neutron Beam Dose Deposition Profiles in Tissue Equivalent Materials: A Pilot Study of BNCT." Presented at the 5th International Conference on Neutron Techniques, Crete, Greece, 9–15 June 1996.

NEUTRON OPTICS

CHAPTER 36
NEUTRON OPTICS

David Mildner
Nuclear Methods Group
National Institute of Standards and Technology
Gaithersburg, MD 20899

36.1 INTRODUCTION

The neutron is a subatomic particle with zero charge, a mass of $m = 1.00897$ atomic mass units, a spin of ½, and a magnetic moment of $\mu_n = -1.9132$ nuclear magnetons. These four basic properties make thermal and cold neutrons not only such a rich and useful scientific tool for the investigation of condensed matter, but also a means for observing many beautiful optical phenomena with remarkable properties,[1] as well as for constructing numerous optical devices.[2] Because neutrons are quantum particles, a de Broglie wavelength, λ, is associated with a beam of thermal neutrons of energy, E, given by

$$E = (2\pi^2 \hbar^2/(m\lambda^2)) \tag{1}$$

that is comparable with interatomic distances within materials. This gives rise to the phenomenon of neutron *diffraction,* analogous to X-ray diffraction. Additionally, the mass and energy are such that the frequency of the radiation is comparable with the vibrational frequencies found in materials, which gives rise to neutron *inelastic scattering.* Hence, neutrons are ideally suited for the study of the atomic structure and dynamics in condensed matter. The magnetic moment of the neutron interacts with unpaired electrons in magnetic materials, giving rise to *magnetic diffraction* and *inelastic scattering.* Again, the wavelength and energy of thermal neutrons are ideal for the study of the magnetic structure and dynamics of spin systems. As a result of zero charge, the neutron has only a short-ranged interaction with the nucleus. This means that the amplitude of the interaction is small, so that neutrons penetrate into the bulk of most materials. In fact, the interaction probability of neutrons varies irregularly with the nuclear isotope, unlike X rays whose amplitudes increase monotonically with the atomic number. Finally, the neutron spin of ½ enables a neutron beam to exist in one of two polarized states. When the neutron scatters from a nucleus of nonzero spin, the strength of the interaction depends on the relative orientation of the neutron and the nuclear spin.

The propagation of neutron de Broglie waves in a potential field is analogous to the propagation of light waves in a medium with a continuously variable refractive index. The potential can be gravitational, magnetic, or nuclear. For example, slow neutrons have a parabolic path under the effect of gravity as in classical mechanics. There are magnetic interactions because the magnetic dipole moment of the neutron gives rise to a potential in a magnetic field that depends on whether the neutron spin is parallel or antiparallel to the magnetic field.

The magnetic field is a birefringent medium for an unpolarized neutron beam, with results analogous to the Stern-Gerlach experiment. Neutrons can be focused by refraction in an inhomogeneous magnetic field provided by a magnetic hexapole. Neutrons in bulk nonmagnetic materials interact with the atomic nuclei. Generally, the interaction gives rise to an isotropic spherical wave reradiation of the incident neutron beam. The potential of the neutron-nucleus interaction has an imaginary part that represents neutron absorption. The scattering of a beam of neutrons is described by a scattering length (b) that depends on a pointlike interaction potential. Generally, b is positive (but not always) and varies with different isotopes.

The similarity of the mathematical description for neutron wave propagation to photon propagation gives rise to phenomena that are analogous to those of classical optics. In fact, virtually all the well-known classical optical phenomena that are characteristic of light and X rays have also been demonstrated with neutrons. In geometric optics, there are not only the refraction and reflection of neutrons by materials, but also special properties in magnetic media. In wave optics, there is Bragg diffraction from crystalline materials, and so on, as for X rays, but there are also other phenomena completely analogous to classical optics. Various neutron optics experiments have demonstrated the quantum-mechanical phenomena on a macroscopic scale. For example, the perfect crystal neutron interferometer can measure the relative phase between two plane wave states, with the large separation of the beams enabling easy access for material phase-shifting devices and magnetic fields. These provide both quantitative verification of various fundamental quantum-mechanical principles applied to neutrons and accurate measurements of neutron scattering lengths. Finally, neutron optical devices have been developed to transport, collimate, focus, monochromate, filter, polarize, or otherwise manipulate neutron beams for applications in both basic and applied research, such as the study of the microscopic structure and dynamics of materials.

The optical phenomena arise from coherent elastic scattering of neutrons in condensed matter.[3-5] The neutron wave function, $\Psi(\mathbf{r})$, can be described by a one-body Schrödinger equation, upon which all neutron optics is based, viz.,

$$[-(\hbar^2/2m)\,\nabla + V(\mathbf{r})]\,\Psi(\mathbf{r}) = E\,\Psi(\mathbf{r}) \qquad (2)$$

in which E is the incident neutron energy. The optical potential $V(\mathbf{r})$ represents the effective interaction of the neutron with the medium. The scattering of a neutron by a single bound nucleus is based on the Born approximation and uses the Fermi pseudopotential, $V(\mathbf{r}) = (2\pi\hbar^2/m)b\delta(\mathbf{r})$, where $\mathbf{r}$ is the neutron position relative to the nucleus, to represent the effective interaction between the neutron and the nucleus. Coherent elastic scattering is not the only collision process than can occur. There are other scattering processes collectively referred to as *diffuse scattering*. In addition, the incident neutron might be absorbed by the nucleus. Both of these cause attenuation of the coherent wave, $\Psi(\mathbf{r})$, in the medium, so that the potential $V(\mathbf{r})$ and the bound scattering length, b, are in general complex.

The Schrödinger equation is a macroscopic equation that describes coherent elastic scattering and all neutron optical phenomena in terms of the interaction of the neutron wave with a potential barrier. The general solution in a medium of constant potential, V_0, can be expressed as a superposition of plane waves where the magnitudes of the wave vectors can be determined by the incident neutron energy,

$$E = (\hbar\mathbf{k})^2/2m = (\hbar\mathbf{k}')^2/2m + V_0 \qquad (3)$$

where $\mathbf{k}$ and $\mathbf{k}'$ are the incident and secondary wave vectors. The directions of the various wave vectors and their corresponding amplitudes are determined by requiring $\Psi(\mathbf{r})$ and $\nabla\Psi(\mathbf{r})$ to be continuous at the boundary between media. Some neutron optical phenomena are well described by the kinematic theory of diffraction. In geometric optics, neutron trajectories obey the same laws of reflection and refraction as in classical optics, though true mirror reflection only occurs for ultracold neutrons. Other phenomena require the dynamical diffraction theory that takes into account the interchange between the transmitted and reflected waves. Sears has shown[1] from the theory of dispersion that, with respect to neutron optics, all materials behave like a continuous macroscopic medium with a refractive index, n.

36.2 INDEX OF REFRACTION

All isotopes have a scattering length that characterizes the neutron-nucleus interaction (albeit some have a complex value).[6] Consequently, all materials have an index of refraction that depends on the scattering length of the isotopes that compose the material. The basic difference is that, as for X rays, the index of refraction for neutrons has a value of $n < 1$, whereas $n > 1$ for light. Moreover, the neutron wavelengths are about three orders of magnitude smaller than for visible light. This means much larger instruments are required in general to observe the optical phenomena with the same spatial resolution. In addition, the most intense thermal neutron sources are orders of magnitude weaker than conventional light and X-ray sources. This means that the phenomena are generally not as easy to observe, nor is the resolution as good. However, successful neutron optical devices have been developed and are now commonplace in instrumentation for scientific research studies, particularly those that use cold- or long-wavelength neutrons.

The index of refraction of a medium is defined by $n = \mathbf{k}'/\mathbf{k}$, and for slow neutrons is given by

$$n^2 = 1 - \lambda^2 \rho b / \pi \qquad (4)$$

where $\lambda = 2\pi/\mathbf{k}$ is the incident neutron wavelength, and ρ is the average atom density of the material. For thermal neutrons with wavelength $\lambda > 0.1$ nm, the refractive index for most materials differs from unity by about a part in 10^6. Consequently, the various optical features are similar to those in X-ray optics, and various analogous features have been demonstrated. A critical wavelength is defined by $\lambda_c = (\pi/\rho b)^{1/2}$. If b is real and positive, the index of refraction is real if $\lambda < \lambda_c$ and imaginary if $\lambda > \lambda_c$. The value of λ_c for materials is typically ≈ 140 nm. At the boundary between two media, A and B, the relative index of refraction is given by

$$n^2 = 1 - (\lambda^2/\pi)(\rho_A b_A - \rho_B b_B) \qquad (5)$$

where ρb is the scattering length density of a given medium.

Refraction and Mirror Reflection

The total external reflection of neutrons from material surfaces has been demonstrated, analogous to the total external reflection of X rays and the total internal reflection of light. Because the index of refraction is close to unity, total reflection is possible only at glancing incident angles, θ, on a material surface when $\cos \theta > n$. This is analogous to Snell's law in light (the continuity of the tangential component of the wavevector). From this we may define a critical angle, $\theta_c = \lambda \, (\rho b/\pi)^{1/2}$, for nonmagnetic materials. The measurement of the critical angle for various surfaces using monochromatic neutrons can be used for the determination of scattering lengths with an accuracy ≈ 0.2 percent. The reflectivity, R, is given by Fresnel's law,

$$R = |\{1 - [1 - (\theta_c/\theta)^2]^{1/2}\}/\{1 + [1 - (\theta_c/\theta)^2]^{1/2}\}|^2 \qquad (6)$$

where $\theta_c = \lambda/\lambda_c = \lambda(\rho b/\pi)^{1/2}$. If $b > 0$, $R = 1$ for $\theta < \theta_c$, the critical angle for total reflection. Because $(1 - n^2) \approx 10^{-5}$ for thermal neutrons, $\theta_c \approx 3$ mrad $\approx 0.2°$. Hence, the measurement of the reflectivity can be used to determine both the magnitude and sign of the bound coherent scattering length.

The mean optical potential of neutrons in most materials is comparable in magnitude with the gravitational potential energy corresponding to a height difference of order of 1 m. This is the principle of the gravity refractometer,[7] in which neutrons from a well-collimated initial horizontal beam fall under gravity in a long flight path onto the horizontal surface of a liquid. If the height of the fall exceeds a certain value, h_0, the neutrons will penetrate the mirror; otherwise, they will be totally reflected by the mirror and be detected in the counter. Consequently, h_0 is a measure of the quantity ρb, and hence of the scattering length, b. This experiment, illustrated in Fig. 1 demonstrates the unique particle-wave properties of neutron optics, with the motion obeying classical physics and the diffraction obeying quantum physics. The significant point is that the measurement is independent of the neutron wavelength.

Prisms and Lenses

The prism deflection of neutrons has been observed.[8] Because the refractive power, $(n-1)$, is negative for most materials, prism deflection is in the opposite direction compared with ordinary light optics. Deflection angles of a few seconds of arc can be measured with high accuracy, allowing the measurements of refractive indices to 1 part in 10^4. Because $(n-1)$ is proportional to λ^2, prism deflection of neutrons is highly dispersive. It has been used in a fused-quartz converging lens (which is concave in shape) and the focusing of 20-nm neutrons has been demonstrated.[9] More recently,[10] the focusing of shorter- (9–20 nm) wavelength neutrons has been demonstrated using a series of many closely spaced symmetric biconcave lenses, with gains ≈ 15 and focal lengths of 1 to 6 m, well matched to small-angle neutron scattering.

Neutron Polarization

It is the spin of ½ of the neutron that results in neutron optics having phenomena different from X-ray optics. The neutron spin, **s**, interacts with the nuclear spin, **I**, and the optical potential, $V(\mathbf{r})$, and the bound scattering length, b, are both spin-dependent for $I \neq 0$. The total spin, $\mathbf{J} = \mathbf{I} + \mathbf{s}$, is a constant of the motion with eigenvalues $J = I \pm \frac{1}{2}$. The index of refraction may be written

$$n^2 = 1 - \lambda^2 \left[\rho b/\pi \pm m/(2\pi^2\hbar^2)\, \mu_n B \right] \tag{7}$$

The sign of the magnetic scattering length depends on the neutron spin orientation relative to magnetic field, B, in the medium. The spins of neutrons may be oriented such that their projections are either parallel (+) or antiparallel (−) to the magnetic field. Because there are two

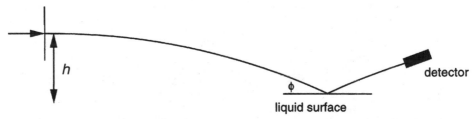

FIGURE 1   A schematic diagram of the neutron gravity spectrometer. A neutron of initial horizontal velocity, v, falls a vertical distance, h, and strikes the liquid surface at a grazing angle of $\phi = \sqrt{(2gh)}/v$ after traveling a horizontal distance $v\sqrt{(2h/g)}$. The critical height for total reflection is $h_0 = (2\pi\hbar^2/m^2g)\rho b$, independent of v or λ.

possible spin states, a beam with a fraction, ϕ_+, of its neutrons in the spin "up" state and a fraction, ϕ_-, in the "down" state has a magnitude of its polarization, **P**, in the field direction given by

$$|\mathbf{P}| = \phi_+ - \phi_- = 2\phi_+ - 1 = 1 - 2\phi_- \tag{8}$$

This polarization property of a neutron beam gives many extraspecial capabilities to neutron scattering and neutron optics.[11] For instance, magnetic mirrors have two critical glancing angles, and this property is used in polarizing mirror and polarizing guides. The reflection of neutrons from magnetized mirrors is widely used for obtaining polarized neutrons. From Eq. (7), if $\rho b < m\mu_n B/(2\pi\hbar^2)$, neutrons with spins antiparallel to the magnetization are not reflected, and only those neutrons with parallel spin orientation are in the reflected beam.

36.3 NEUTRON SCATTERING LENGTHS

The general theory of neutron scattering lengths and cross sections have been summarized by Sears.[6,12] The bound scattering length is in general complex, $b = b' - ib''$. For thermal neutrons, the scattering cross section is independent of the neutron wave vector, **k**, whereas the absorption cross section is inversely proportional to **k**, and therefore increases linearly with wavelength. For most nuclides, $V(\mathbf{r})$ is a strongly attractive potential and, therefore, the scattering length is positive. Indeed, there are only a few nuclides, such ^{1}H, ^{7}Li, ^{48}Ti, and ^{55}Mn, plus a few others, that have negative scattering lengths.

The scattering cross section is given by $\sigma_s = 4\pi<|b|^2>$, where the brackets denote a statistical average over neutron and nuclear spin states. The absorption cross section is given by $\sigma_{a\pm} = (4\pi/k)b''_+$. In practice, the imaginary part of the scattering length is of the same order of magnitude as the real part only when the absorption is large ($\sigma_a \approx 10^4\sigma_s$). Furthermore, the absorption is usually large for only one spin state of the compound nucleus. For nuclides such as ^{113}Cd, ^{155}Gd, or ^{157}Gd, the absorption is large only in the $J = I + \frac{1}{2}$ state, and $b''_- = 0$. For nuclides such as ^{3}He or ^{10}B, the absorption is large only in the $J = I - \frac{1}{2}$ state, $b''_+ = 0$. For a few nuclides such as ^{6}Li, the absorption is large for both $J = I \pm \frac{1}{2}$ states. Finally, the dependence of the scattering length on the particular isotope means that there is a further incoherence for a given element dependent on the abundance of the various constituent isotopes.

Neutron Attenuation

Finally, Lambert's law $\exp(-\mu L)$ for the transmission of radiation through matter also applies to the attenuation of a neutron beam in a given medium of length, L, though the absorption coefficient, μ, is usually written using a macroscopic cross section $\Sigma = \rho\sigma_t$, where $\sigma_t = \sigma_a + \sigma_s$ is the total cross section. Because the cross sections vary irregularly between elements in the periodic table, neutron radiography can give good contrast between neighboring elements, unlike X-ray radiography. In particular, the anomalously large scattering cross section for hydrogen makes neutron radiography important for the location of hydrogenous regions within materials.

Refractive Index Matching

The matching of the index of refraction of a material with the liquid in which it is immersed gives rise to a sharp minimum in the diffuse scattering.[13] This technique allows the measurement of the refractive index of a wide range of materials by matching the value of the neutron scattering density, ρb, of the material with a mixture of H_2O and D_2O, which can cover a wide range as shown in Fig. 2. This is analogous to the Christiansen filter in classical optics. This refractive index matching has become an important method for determining the macro-molecular structure of biological systems using neutron scattering by varying the contrast between different parts of the structure and the supporting medium. This is now the standard method for small-angle neutron scattering from biological macromolecules.[14] This technique of refractive index matching is also available for nuclear and magnetic components of the refractive index by matching one of the two spin states to the nonmagnetic material in order to highlight the other.

36.4 NEUTRON GUIDES

Total reflection gives rise to guide tubes[15] that are used to transport thermal neutron beams over relatively long distances with only small losses in intensity, and are analogous to light pipes in ordinary optics. A neutron entering the guide tube with an angle of incidence at the wall that is less than the critical angle for its particular wavelength is transported along the

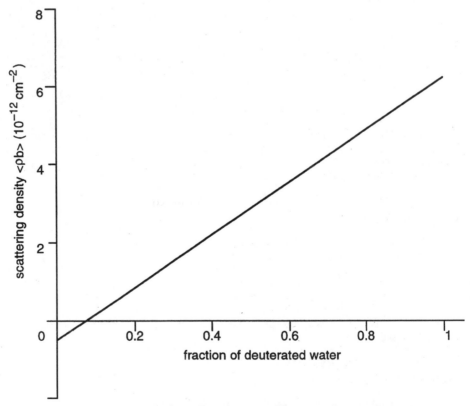

FIGURE 2 The scattering density of water as a function of the fraction that is deuterated. The scattering lengths of H and D are very different (-0.37×10^{-12} cm and 0.66×10^{-12} cm). The scattering densities of all bio-logical molecules lie between the limits of pure H_2O (-0.56×10^{-12} cm^{-2}) and pure D_2O (6.3×10^{-12} cm^{-2}) and can therefore be matched by some fraction of deuterated water.

hollow tube by multiple total reflection. The guides are usually highly polished nickel-plated glass tubes of rectangular cross section, though the greater scattering length for isotopically pure ^{58}Ni gives a 40 percent increase in transmission. Further increases are obtained by using glass coated with supermirror of nickel-titanium layers on top of nickel. The supermirror can extend the range of total reflection to a critical angle, θ_c, three or more times that of nickel. The effective collimation achieved by rectangular neutron guides corresponds in $2\theta_c$ in each orthogonal direction, and therefore depends on wavelength. If the guide has a gentle curvature (often a polygonal approximation to a curved guide), only those neutrons with wavelengths greater than some minimum are transported. Fast neutrons and gamma rays pass through the glass wall and are eliminated from the beam, so that guides can transport clean and highly collimated beams of thermal neutrons to regions of low-background far from the source.

36.5 ULTRACOLD NEUTRONS

The phenomenon of total reflection for all angles of incidence occurs when n is purely imaginary, that is when $\lambda > \lambda_c$. Typically, $\lambda_c \approx 140$ nm, which corresponds to a neutron energy of 4×10^{-8} eV or a temperature of 0.5 mK. It is therefore only for ultracold neutrons[16] that true mirror reflection can occur. Mirror reflections for ultracold neutrons can be used to confine neutrons in material cavities or bottles. Smooth, clean surfaces are needed for good storage, though this is limited by β decay of the free neutron (about 937 s). In practice, this time is shorter because of losses caused by inelastic scattering from surface impurities. Ultracold neutrons are useful for precision measurements of the free neutron lifetime, the neutron magnetic dipole moment and the search for its electrical dipole moment.

36.6 DIFFRACTION

The Bragg diffraction by single crystals, powders, amorphous materials, and liquids is analogous to that for electromagnetic waves, with formulas similar to those for X-ray diffraction except the magnitude is determined by the coherent scattering lengths. The diffraction pattern is simply a function of scattering angle and is proportional to the square of the Fourier transform of the structure of the diffracting sample. The small-angle neutron scattering from inhomogeneous materials also has expressions similar to those for X rays. The availability of the (partial) substitution of hydrogen by deuterium allows the ability to change the relative scattering powers for different features within the structure. These are standard techniques, and we now briefly review special measurements showing the wave nature of neutrons.

Fraunhofer Diffraction

Schull[17] has demonstrated Fraunhofer diffraction using a monochromatic beam of neutrons diffracted from a perfect single crystal of silicon that was analyzed by a similar crystal. The diffraction broadening of the transmitted beam was shown for several very narrow slit widths to agree with the classical formula $(\sin x/x)^{1/2}$ for diffraction from a single slit. Further experiments[18] with greater precision performed for single slit and double slit showed excellent agreement with wave diffraction theory. The diffraction of thermal neutron by ruled gratings at grazing incidence has also been demonstrated.[19]

Fresnel Diffraction

The Fresnel diffraction of an opaque straight edge[20] has been found to be in excellent agreement with theory. The focusing of slow neutrons by zone plates (phase gratings of variable spacings) has been demonstrated.[9,21] Conventional zone plates use constructive interference of diffracted waves through open zones, with absorption of radiation in dark zones. Neutron zone plates use phase reversing in alternate zones and can achieve greater focusing efficiency. Zone plates can be produced by photolithography on silicon wafer substrates from a computer-drawn pattern. The resolution is limited by the inherent chromatic aberration of the zone plate. Cylindrical zone plates have been used as a condensing lens.[22]

36.7   INTERFERENCE

There have been various demonstrations of the interference of coherent beams using thermal, cold, and ultracold neutrons.[23,24] Two beam interferometers allow the determination of the relative phases of two coherent wavefronts and the investigation of different influences, such as force fields and material media, on the phase of a wave. This is again analogous with classical optics. When neutrons are incident on perfect crystals under Bragg conditions, there is a dynamic interchange of flux between the incident beam and the Bragg diffracted beam inside the crystal medium. The radiation density traveling in either the Bragg or the forward direction is a periodic function of depth in the crystal. This results in extinction effects, with an associated extinction length of order 1 to 100 μm. The dynamical theory of diffraction leads to anomalous transmission effects, and there have been numerous experiments using perfect crystals to observe these special effects of neutron wave propagation in periodic media. For example, Schull[25] has observed the Pendellösung interference fringe structures in the diffraction from single-crystal silicon.

Interference by Wavefront Division

The superposition of two spatially separated parts of one coherent wavefront produces interference patterns from which the relative phases of the two parts can be deduced. The effective source for the original wavefront is a beam that is given spatial coherence by passage through a narrow slit from which a partially coherent cylindrical wave emerges. This beam illuminates a pair of closely spaced narrow slits, and the diffracted radiation gives rise to interference fringes in the plane of observation,[18] analogous to Young's slits in classical optics.

A similar experiment[26] has shown the interference of two spatially separated parts of one wavefront using the boundary between two domains of opposite magnetization in a crystal of Fe-3%Si. In traversing the ferromagnetic foil on either side of a domain wall in the sample, the spin of the neutron precesses in opposite directions. The two parts of the wavefunction acquire a relative phase shift that shows up in the interference pattern in the plane of observation. The expected pattern is the superposition of two Fresnel straight-edge patterns that for zero phase shift combine to give a uniform distribution. For π phase shift, destructive interference occurs in the center of the pattern, giving a deep minimum. The relative phase shift depends on the angle of precession of the neutron spin, which is proportional to the thickness of the region of the magnetic field traversed. The results of the experiment demonstrate that a phase shift of π (destructive interference) occurs when the spin is rotated by odd multiples of 2π, verifying the peculiar behavior of spin ½ particles under rotation, that the neutron is a spinor.

Interference by Amplitude Division

The coherent splitting of the amplitude of a wave by means of partial reflection from thin films gives rise to interference effects in visible light. For neutrons this partial reflection from the front and back surfaces of various thin films occurs at glancing angles close to critical reflection. The resultant interference phenomena give valuable information about the structure of the interfaces involved.[27] Various neutron optical devices based on thin-film interference have been developed, such as bandpass monochromators, supermirrors that are highly efficient neutron polarizers,[28] reflectivity and polarizing characteristics of Fe-Ge multilayers,[29] and devices involving multiple-beam interference effects, analogous to classical optics. Multiple-beam interference has been demonstrated[29] with ultracold neutrons in which thin films of copper and gold are used in the construction of a resonant structure analogous to a Fabry-Perot cavity.

Perfect Crystal Interferometers

Bonse and Hart[30] have pioneered the X-ray interferometer that uses Bragg reflection at crystal flats for splitting and recombining two interfering beams that are cut from a monolithic piece of large, highly perfect single-crystal silicon. This idea has been applied to neutron interferometry, and involves triple Laue transmission geometry optics.[31] This is analogous to the Mach-Zehnder interferometer of classical optics. The interferometer illustrated in Fig. 3 consists of three identical, perfect crystal slabs cut perpendicular to a set of strongly reflecting planes, with distances between slabs usually a few centimeters. There are of order 10^9 oscillations of the neutron wave function in each beam path, so that there are stringent requirements on the microphonic and thermal stability of the device. This requires vibrational isolation of the interferometer from the reactor hall, plus thermal isolation. The three crystal slabs must be aligned within the Darwin width of less than a second of arc for thermal neutrons in silicon. The precise definition of the wavelength is accomplished by the Bragg diffraction of the interferometer. Placing an object of variable thickness in one of the beams leads to interference contrast due to the spatially varying phase shift. The path difference must be

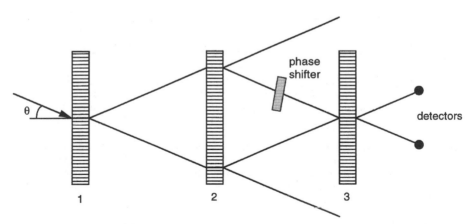

FIGURE 3 A schematic diagram of the triple Laue neutron interferometer cut from a monolithic perfect crystal of silicon in which the first slab splits the incident beam into two coherent parts, with the second slab acting as a mirror to bring the beams together at the third slab. Interference fringes are observed in the detectors by rotating the phase shifter in one of the paths.

small compared with the coherence length of the beam that depends on the degree of beam monochromatization. There have been many applications of neutron interferometry in fundamental physics, such as gravitationally induced quantum interference and nonstandard quantum mechanics.[23,24]

The analog of the classical Rayleigh interferometer is a two-crystal monolithic device cut from a large silicon crystal.[32] The incident beam is restricted by a narrow entrance slit and is Bragg-reflected by the first silicon crystal slab. Inside the crystal, the diffracted neutron beam fills the entire Bormann triangle, and the beam that leaves the back face of the crystal is broad, having a width that is dependent on the crystal thickness. There is a conjugate focal point on the exit face of the second crystal, from which neutrons along the complementary interfering rays leave the device in the incident and diffracted beam directions. If the central rays are blocked by a cadmium absorber, the parallelogram ray diagram is analogous to the three-crystal interferometer. Another variant uses Bragg reflection geometry rather than Laue transmission geometry. The incident beam reflected from the front surface of the first crystal is brought back to interfere with the beam originating from the back face of this crystal by using a second parallel Bragg reflecting crystal. The advantage is that the spatial definition of the outgoing interfering beam is as sharp as the beam incident on the interferometer.

Interferometric Measurement of Scattering Lengths

Perfect crystal neutron interferometry is a convenient and precise technique for measuring neutron-nuclear scattering lengths. The period of the intensity oscillations as a function of the orientation of a slab of material that is rotated about an axis perpendicular to the plane of the interferometer enables the determination of the refractive index and, hence, the scattering length of the scattering slab material. There is also a differential technique using a phase shifter in one of the beams. Neutron interferometry measures the difference in the average neutron-nuclear potential of the sample traversed by the two beams. Intensity oscillations are observed by varying the atom density, with more rapid interference oscillations for larger scattering lengths. The sensitivity of the technique allows a measurement of the hydrogen content in various transition metals at a level of about 0.05 at%.[33]

36.8 NEUTRON SOURCES

There are several types of thermal neutron beam sources with adequate fluxes for useful measurements.[34,35] Thermal nuclear reactors produce neutrons using the fission reaction, and are generally steady-state sources, though there are also pulsed reactors. However, pulsed sources are generally either electron accelerators producing neutrons through the bremsstrahlung, photoneutron reaction, or proton accelerators producing neutrons through spallation in heavy metal targets (though these may also be in a continuous mode). These sources produce fast neutrons that must be slowed down or moderated to thermal energies with velocities useful for neutron optics work. When in thermal equilibrium with the moderator, neutrons have an energy spectrum characteristic of the Maxwellian temperature of the moderator. In fact, it is best to have a cold moderator to produce large amounts of long-wavelength neutrons for optics work. Indeed, the greater availability of cold neutrons has increased the number of experiments involving neutron optics.

These beams must be transported from the moderator to the experimental hall, either through beam tubes, or better, through neutron guides using total reflection. This allows both thermal and subthermal beams to be brought to a region of much lower background. Because straight guides also transport fast neutrons and gamma rays from the source, either curved guides or neutron filters must be used. The various filters that are used are either Bragg cutoff filters (such as the liquid nitrogen–cooled polycrystalline beryllium filter), or single-crystal filters (such as the single-crystal sapphire filter that can be used at room temperature).

Though for some experiments the wide energy spectrum of the polychromatic beam after filtration is sometimes used, generally a monochromatic beam is used for neutron spectroscopy. The pulsed sources require special considerations because time-of-flight techniques are used to determine the neutron wavelength, and the neutron moderation time must be kept to a minimum. Though steady-state sources can also use time of flight using mechanical velocity selectors, monochromation produced by diffraction from a single crystal is generally preferred. The most appropriate beam for use depends on the particular measurement envisaged, though most neutron optics research has been performed on cold sources at reactors.

36.9 EXPERIMENTAL TECHNIQUES

Besides collimators, filters, crystal monochromators, and neutron choppers placed in the beam paths between the source, the sample, and the detector, various neutron optical devices[36] can also be inserted, such as neutron guides, beam benders, and polarizing mirrors.

Neutron Polarization

Because the neutron spin is ½, neutron beams can be polarized.[11] Beside polarizing monochromators such as $Co_{0.92}Fe_{0.08}$, magnetite Fe_3O_4, and Heusler alloy crystal Cu_2MnAl, polarized beams are obtained using total reflection from magnetized iron and cobalt mirrors. In addition, thin ($\approx$100 nm) evaporated films of magnetic alloys have been used. Two-dimensional multilayer structures have alternate layers of a (magnetized) magnetic material whose refractive index for the (−) spin neutrons is matched to the second (nonmagnetic) layer, for example, in Fe-Ge multilayers. Supermirrors can produce a broadband-polarized beam using evaporated multilayer films deposited to provide a lattice spacing gradient. The reflectivity profile extends in angle beyond the simple mirror critical glancing angle through thin-film interference.

Larmor Precession

The direction of the neutron spin precesses in a homogeneous magnetic field, B, with a Larmor frequency given by $\omega_L = 2|\mu_n B|/\hbar$. This phenomenon has given rise to the spin echo spectrometer,[37] in which the incoming and outgoing velocities of the neutron are measured, so that small changes in velocity that may take place on scattering can be inferred from the shift in phase in the Larmor precession angle. This is effectively a time-of-flight method in which polarized neutrons precess in the magnetic field to provide its own clock. Such a technique enables energy resolution in the neV region, orders of magnitude smaller than for conventional instruments.

Neutron Collimation

Neutrons may be collimated by defining the beam path with material of high cross section, such as cadmium or gadolinium. The transmission may be calculated using phase space acceptance diagrams indicating the position and angle in each of the two orthogonal transverse directions. The maximum angle transmitted relative to the beam centerline is given by the collimation angle. The transmission may be increased by the use of highly polished reflecting surfaces, such that transmission is uniform in position and angle up to θ_c, the critical angle, provided the device, now a guide, is sufficiently long. The principle of multiple-mirror reflection from smooth surfaces at small grazing angles enables the transport of high-intensity, slow-neutron beams to locations of low background for neutron scattering experiments and to provide facilities for multiple instruments. These neutron guides are common at research reactors, particularly those with cold sources. Originally coatings of nickel with a critical angle of 17.3 mrad nm^{-1} were used. Now ^{58}Ni guides with 20.4 mrad nm^{-1} are preferred with a 40 percent increase in transmission. Further increase in transmission is obtained with nickel-titanium supermirror coatings. Some guides are curved to take the beam away from the fast neutrons and gammas, and have good transmission for slow neutrons above a characteristic wavelength. Others are straight and require filters to obtain good transmission characteristics without the unwanted radiation.

Optical Filter

A new concept is the neutron optical filter,[38] in which the beam may be deflected by a small angle with a high transmission above a cutoff wavelength, without the spatial asymmetry of the curved guide. This is illustrated in Fig. 4. An initial long nickel guide is followed by a short length of supermirror guide, offset by a small angle, ξ, after which the nickel guide continues at a further small angle, ξ. The length of the intermediate supermirror guide is such that there is no direct line of sight through the entire system. The advantage of such a geometry is that it has the property that a parallel beam is transported unchanged, unlike the case of a curved guide. The supermirror critical angle must be at least twice that of nickel. A modification is to have the central section tapered and with a different critical angle. Consequently, this type of filter can be designed for particular experimental arrangements.

Converging Guides

The current density of a neutron beam may be increased by the use of a converging guide. The gain of the converging guide depends on three factors: (1) the convergence factor of the guide, (2) the ratio of the critical angles on the converging and straight guides, and (3) the

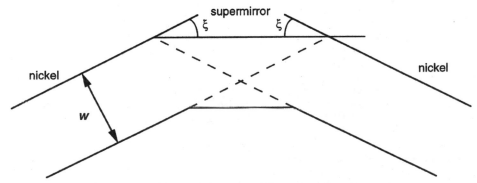

FIGURE 4   A schematic drawing of a neutron optical filter in which the initial long guide of nickel, of width *w*, is followed at an angle, ξ, by a short length of supermirror guide, and then another nickel guide at a further angle, ξ. If the angle ξ is made equal to the critical θ_c of nickel at the shortest wavelength of interest, then the supermirror must have a critical angle of at least $2\theta_c$.

neutron wavelength. The greatest gains are achieved when the critical angle of the converging guide is much greater than the divergence of the incoming beam defined by the critical angle of the straight guide. Even with supermirror coatings, intensity gains for the two-dimensional converging guide are limited to perhaps an order of magnitude on account of the increase in absolute angle that a neutron trajectory makes with the device axis upon successive reflections.[39]

Polycapillary Optics (See Chap. 30)

Capillary optics provides a better method for increasing the neutron current density by using many narrow guides. This allows the possibility of a greater curvature than for macroguides for a given wavelength because the transmission characteristics of a curved guide depend on the ratio of the transverse to longitudinal dimensions of the guide. These miniature versions of neutron guides can be used to transport, bend, and even focus thermal neutron beams. Glass polycapillary fibers (see Fig. 8 of Chap. 30) with thousands of hollow channels recently developed for transporting and focusing X rays can also be used for cold (≈ 0.4 nm) neutron beams. The advantage of the narrow (≈ 10 μm) channels is that the neutron beam can be bent more sharply than for the wide (≈ 50 mm) guides. The transmission properties of the capillaries depend on the internal diameter and bending radius of the channels, the glass composition, and the surface smoothness.

The focusing of neutrons comes at a price. Liouville's theorem requires that an increase in neutron current density is obtained with a necessary increase in beam divergence. This means that real-space focusing has limited applications in neutron scattering for which angular divergence is important, whereas neutron absorption techniques in analytical and materials research do not require a monochromatic neutron beam or a small angular divergence. Consequently, a high-intensity beam focused by a neutron capillary lens onto a small sample area can be useful for improving both the detection limits of individual elements and the spatial resolution of the measurement. Polycapillary fiber lenses (see Fig. 9 of Chap. 30) have been produced and tested for analytical applications.[40] The gain depends on the divergence and the energy spectrum of the incident beam. Within the focus of typical size of 0.5 mm, limited by the outer diameter of the fibers, the gain can be nearly two orders of magnitude. For the monolithic lens, a fused version of the polycapillary lens, the focus is less than 200 μm with a gain of about 10.

There are other modes of focusing available for neutrons. Focusing of neutrons in momentum space becomes important for diffraction and scattering applications. There is also time focusing for improving the resolution for time-of-flight measurements on instruments when the source, sample, analyzer (if any), and detector have extended areas.

36.10 NEUTRON DETECTION

Because neutrons are uncharged, they cannot be detected directly but need a converter using a nuclear reaction that results in the emission of a charge particle.[34,41] The three most common reactions used for the detection of slow neutrons all have cross sections that increase linearly with wavelength. The $^{10}B(n,\alpha)$ reaction has a thermal (velocity of 2200 m s^{-1}, or wavelength of $\lambda = 0.179$ nm) cross section of 3840 barns and a Q value of 2.79 MeV to the ground state (6 percent, or a Q of 2.31 MeV and the emission of a 478-keV gamma (94 percent). This reaction is used in BF$_3$ gas proportional counters or in B-loaded scintillators. The $^6Li(n,\alpha)$ reaction with a thermal cross section of 940 barns and a Q value of 4.78 MeV emits no gamma and is often the crucial ingredient of neutron scintillator detectors. The $^3He(n,p)$ reaction, with a thermal cross section of 5330 barns and a Q value of 0.764 MeV, is used in ^3He gas-proportional counters. These reactions are also used in the one- and two-dimensional detectors, and often form the basis of other detection methods, such as imaging detectors and neutron imaging plates. In general cadmium and gadolinium, or other rare earth elements, that have large thermal neutron cross sections are not considered good detector materials, because they result in the copious production of gammas.

36.24 REFERENCES

1. V. F. Sears, *Neutron Optics: An Introduction to the Theory of Neutron Optical Phenomena and Their Applications,* Oxford University Press, Oxford, 1989.

2. M. Utsuro (ed.), "Advance in Neutrons Optics and Related Research Facilities," *J. Phys. Soc. Japan* **65**(suppl A): (1996).

3. W. Marshall and S. W. Lovesey, *Theory of Thermal Neutron Scattering: The use of Neutrons for the Investigation of Condensed Matter,* Clarendon Press, Oxford, 1971.

4. G. L. Squires, *Thermal Neutron Scattering,* Cambridge University Press, Cambridge, 1978.

5. S. W. Lovesey, *Theory of Neutron Scattering from Condensed Matter,* Oxford University Press, Oxford, 1984.

6. V. F. Sears, "Neutron Scattering Lengths and Cross Sections," in *Methods of Experimental Physics: Neutron Scattering,* K. Sköld and D. L. Price (eds.), vol. 23A, Academic Press, San Diego, CA, 1986, pp. 521–550.

7. L. Koester, *Z. Phys.* **182**:326–328 (1965).

8. C. S. Schneider and C. G. Schull, *Phys. Rev.* **B3**:830–835 (1971).

9. R. Gähler, J. Kalus, and W. Mampe, *J. Phys.* **E13**:546–548 (1980).

10. M. R. Eskildsen, P. L. Gammel, E. D. Isaacs, C. Detlefs, K. Mortensen, and D. J. Bishop, *Nature* **391**:563–566 (1998).

11. W. G. Williams, *Polarized Neutrons,* Oxford University Press, Oxford, 1988.

12. V. F. Sears, *Neutron News* **3**(3):26–37 (1992).

13. L. Koester and H. Ungerer, *Z. Phys.* **219**:300–310 (1969).

14. H. B. Stuhrmann, "Molecular Biology," in *Methods of Experimental Physics: Neutron Scattering,* K. Sköld and D. L. Price (eds.), vol. 23C, Academic Press, San Diego, CA, 1986, pp. 367–403.

15. H. Maier-Leibnitz and T. Springer, *J. Nucl. Energy* **A/B17**:217–225 (1963).

16. A. Steyerl, "Neutron Physics," in *Springer Tracts in Modern Physics* Vol. 80, pp. 57–130, Springer-Verlag, Berlin, 1977.

17. C. G. Schull, *Phys. Rev.* **179**:752–754 (1969).

18. A. Zeilinger, R. Gähler, C. G. Schull, and W. Terimer, *AIP Conf. Proc.* **89**:93 (1981).

19. H. Kurz and H. Rauch, *Z. Phys.* **220**:419–426 (1969).

20. R. Gähler, A. G. Klein, and A. Zeilinger, *Phys. Rev.* **A23**:1611–1617 (1981).

21. P. D. Kearney, A. G. Klein, G. I. Opat, and R. A. Gähler, *Nature* **287**:313–314 (1980).

22. A. G. Klein, P. D. Kearney, G. I. Opat, and R. A. Gähler, *Phys. Lett.* **83A**:71–73 (1981).

23. S. A. Werner and A. G. Klein, "Neutron Optics," in *Methods of Experimental Physics: Neutron Scattering,* K. Sköld and D. L. Price (eds.), vol. 23A, Academic Press, San Diego, CA, 1986, pp. 259–337.

24. H. Rauch and S. A. Werner, *Neutron Interferometry: Lessons in Experimental Quantum Mechanics,* Oxford University Press, Oxford, 2000.

25. C. G. Schull, *Phys. Rev. Lett.* **21**:1585–1589 (1968).

26. A. G. Klein and G. I. Opat, *Phys. Rev. Lett.* **37**:238–240 (1976).

27. R. R. Highfield, R. K. Thomas, P. G. Cummins, D. P. Gregory, J. Mingins, J. B. Hayter, and O. Schärpf, *Thin Solids Films* **99**:165–172 (1983).

28. F. Mezei, *Commun. Phys.* **1**:81 (1976); F. Mezei and P. A. Dagleish, *Commun. Phys.* **4**:41 (1977).

29. K.-A. Steinhauser, A. Steyerl, H. Scheckhofer, and S. S. Malik, *Phys. Rev. Lett.* **44**:1306–1309 (1981).

30. U. Bonse and M. Hart, *Appl. Phys. Lett.* **6**:155–156 (1965).

31. H. Rauch, W. Treimer, and U. Bonse, *Phys. Lett.* **47A**:369–371 (1974).

32. A. Zeilinger, C. G. Schull, M. A. Horne, and G. Squires, in *Neutron Interferometry,* U. Bonse and H. Rauch (eds.), pp. 48–59, Oxford University Press, London, 1979.

33. H. Rauch, E. Seidl, A. Zeilinger, W. Bausspiess, and U. Bonse, *J. Appl. Phys.* **49**:2731–2734 (1978).

34. K. M. Beckurts and K. Wirtz, *Neutron Physics,* Springer-Verlag, 1964.

35. J. M. Carpenter and W. B. Yelon, "Neutron Sources," in *Methods of Experimental Physics: Neutron Scattering,* K. Sköld and D. L. Price (eds.), vol. 23A, Academic Press, San Diego, CA, 1986, pp. 99–196.

36. C. G. Windsor, "Experimental Techniques," in *Methods of Experimental Physics: Neutron Scattering,* K. Sköld and D. L. Price (eds.), vol. 23A, Academic Press, San Diego, CA, 1986, pp. 197–257.

37. F. Mezei, *Neutron Spin Echo,* Springer-Verlag, 1980.

38. J. B. Hayter, *SPIE Proc.* **1738**:2–7 (1992).

39. J. R. D. Copley and C. F. Majkrzak, *SPIE Proc.* **983**:93–104 (1988).

40. Q. F. Xiao, H. Chen, V. A. Sharov, D. F. R. Mildner, R. G. Downing, N. Gao, and D. M. Gibson, *Rev. Sci. Instrum.* **65**:3399–3402 (1994).

41. G. F. Knoll, *Radiation Detection and Measurement,* 2nd ed., chap. 14, John Wiley & Sons, 1989.

SUMMARY AND APPENDIX

CHAPTER 37

SUMMARY OF X-RAY AND NEUTRON OPTICS

Walter M. Gibson and Carolyn A. MacDonald
University at Albany, State University of New York
Albany, New York

Although discovered more than 100 years ago, X-ray science and technology has undergone a renaissance during the past 2 decades. This renewal of interest and activity has been led by the development of dedicated synchrotron sources of intense, coherent, polarized beams of photons in the ultraviolet to hard X-ray spectral range (30–0.025 nm, 0.040–50 keV). These sources have in turn led to development of a veritable alphabet soup of new tools for investigation of the structure and dynamics of atomic, molecular, and condensed systems (e.g., EXAFS, XANES, XPS, etc.) During the past decade, laboratory-based X-ray studies and applications have also advanced rapidly, with new developments in X-ray sources, detectors, and especially optics. These developments have allowed many of the synchrotron-developed tools to be applied in laboratory settings and increasingly in industrial analysis and process applications. In addition, the greatly increased efficiency with which X rays can be used is resulting in new in situ and remote applications of the unique properties of X rays. This X-ray and neutron optics part of the Handbook is intended to summarize the recent progress and the current state of these ongoing developments.

The next decade will almost certainly be as rich with new developments in X-ray optics as the past one has been. In addition to innovative new optics approaches and applications, we expect that numerous advances will be based on modifications and technological progress in existing optics. These should include improved fabrication and scaling of capillary optics, new applications in medicine and industry, increased effective area and energy range for astronomical optics, increased aperture and control for refractive optics, and increased specular and angular range for multilayer optics. Closer coupling and joint development of optics, sources, and detectors will also be important. Also important will be the development of many hybrid systems, which take advantage of the unique aspects of different optics. Some combinations that already appear to provide important benefits are polycapillary optics and bent crystal, tapered monocapillary, or refractive optics. There are undoubtedly many others, with combinations that will be surprising and exciting. A factor that will continue to drive optics and application development is the recent intense rapid development in simulation capability due to the increases in available microcomputing power.

Neutron optics and applications are also advancing. The crucial need for increased efficiency in neutron diffraction studies, especially of protein structure, has led to development of time-of-flight (TOF) Laue diffraction techniques and facilities. Comparable efficiency benefits are potentially possible from convergent beam neutron diffraction using the neutron focusing optics discussed in Chap. 36, in which two orders of magnitude intensity increase for thermal and cold neutrons have been reported. An important benefit from convergent beam diffraction is that it is particularly well suited to small samples, in the range available for protein studies. In fact, it should be possible to combine both TOF Laue and convergent beam diffraction to obtain remarkable gains in efficiency and, therefore, to enable revolutionary new advances in the use of neutron diffraction for a variety of applications.

We hope that this attempt to summarize the advancements and current status of X-ray and neutron optics will stimulate new and important developments.

APPENDIX
X-RAY PROPERTIES OF MATERIALS

E. M. Gullikson
Center for X-Ray Optics
Lawrence Berkeley National Laboratory
Berkeley, California

The primary interaction of low-energy X rays within matter, namely, photoabsorption and coherent scattering, have been described for photon energies outside the absorption threshold regions by using atomic scattering factors, $f = f_1 + if_2$. The atomic photoabsorption cross section, μ, may be readily obtained from the values of f_2 using the relation,

$$\mu = 2r_0\lambda f_2 \tag{1}$$

where r_0 is the classical electron radius, and λ is the wavelength. The transmission of X rays through a slab of thickness, d, is then given by,

$$T = \exp(-N\mu d) \tag{2}$$

where N is the number of atoms per unit volume in the slab. The index of refraction, n, for a material is calculated by,

$$n = 1 - N r_0\lambda^2(f_1 + if_2)/(2\pi) \tag{3}$$

The (semiempirical) atomic scattering factors are based upon photoabsorption measurements of elements in their elemental state. The basic assumption is that condensed matter may be modeled as a collection of noninteracting atoms. This assumption is, in general, a good one for energies sufficiently far from absorption thresholds. In the threshold regions, the specific chemical state is important and direct experimental measurements must be made. Note also that the Compton scattering cross section is not included. The Compton cross section may be significant for the light elements ($Z < 10$) at the higher energies considered here (10–30 keV).

The atomic scattering factors are plotted in Figs. 1 and 2 for every 10th element. Tables 1 through 3 are based on a compilation of the available experimental measurements and theoretical calculations. For many elements there is little or no published data, and, in such cases, it was necessary to rely on theoretical calculations and interpolations across Z. To improve the accuracy in the future, considerably more experimental measurements are needed.[1] More data and useful calculation engines are available at www-cxro.lbl.gov/optical_constants/.

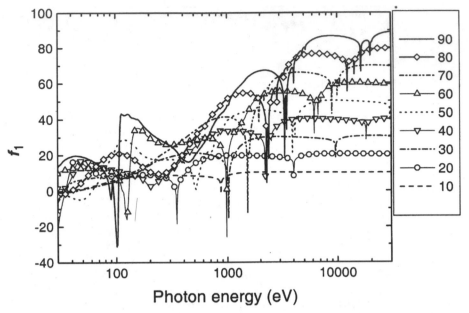

FIGURE 1 The atomic scattering factor f_1 as a function of photon energy from 30 to 30,000 eV for atomic numbers 10 (Ne), 20 (Ca), 30 (Zn), 40 (Zr), 50 (Sn), 60 (Nd), 70 (Yb), 80 (Hg), and 90 (Th).

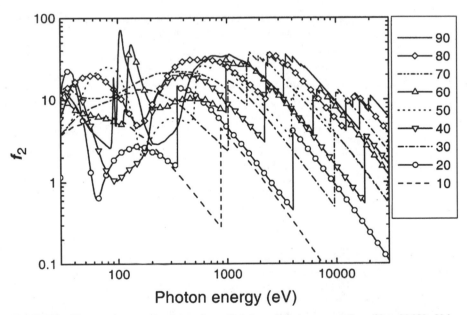

FIGURE 2 The atomic scattering factor f_2 as a function of photon energy from 30 to 30,000 eV for atomic numbers 10 (Ne), 20 (Ca), 30 (Zn), 40 (Zr), 50 (Sn), 60 (Nd), 70 (Yb), 80 (Hg), and 90 (Th).

A.1 ELECTRON BINDING ENERGIES, PRINCIPAL K- AND L-SHELL EMISSION LINES, AND AUGER ELECTRON ENERGIES

TABLE 1   Electron Binding Energies in Electronvolts (eV) for the Elements in Their Natural Forms*

Element	K1s	$L_1$2s	$L_2$2p$_{1/2}$	$L_3$2p$_{3/2}$	$M_1$3s	$M_2$3p$_{1/2}$	$M_3$3p$_{3/2}$	$M_4$3d$_{3/2}$	$M_5$3d$_{5/2}$	$N_1$4s	$N_2$4p$_{1/2}$	$N_3$4p$_{3/2}$
1 H	13.6											
2 He	24.6†											
3 Li	54.7†											
4 Be	111.5†											
5 B	188†											
6 C	284.2†											
7 N	409.9†	37.3†										
8 O	543.1†	41.6†										
9 F	696.7†											
10 Ne	870.2†	48.5†	21.7†	21.6†								
11 Na	1070.8‡	63.5‡	30.4‡	30.5†								
12 Mg	1303.0‡	88.6‡	49.6‡	49.2‡								
13 Al	1559.6	117.8†	72.9†	72.5†								
14 Si	1838.9	149.7†	99.8†	99.2†								
15 P	2145.5	189†	136†	135†								
16 S	2472	230.9†	163.6†	162.5†								
17 Cl	2833	270.2†	202†	200†								
18 Ar	3205.9†	326.3†	250.6†	248.4†	29.3†	15.9†	15.7†					
19 K	3608.4†	378.6†	297.3†	294.6†	34.8†	18.3†	18.3†					
20 Ca	4038.5†	438.4‡	349.7‡	346.2‡	44.3‡	25.4‡	25.4‡					
21 Sc	4492.8	498.0†	403.6†	398.7†	51.1†	28.3†	28.3†					
22 Ti	4966.4	560.9‡	461.2‡	453.8‡	58.7‡	32.6‡	32.6†					
23 V	5465.1	626.7‡	519.8‡	512.1‡	66.3‡	37.2‡	37.2‡					
24 Cr	5989.2	695.7‡	583.8‡	574.1‡	74.1‡	42.2‡	42.2‡					
25 Mn	6539.0	769.1‡	649.9‡	638.7‡	82.3‡	47.2‡	47.2‡					
26 Fe	7112.0	844.6‡	719.9‡	706.8‡	91.3‡	52.7‡	52.7‡					
27 Co	7708.9	925.1‡	793.3‡	778.1‡	101.0‡	58.9†	58.9‡					
28 Ni	8332.8	1008.6‡	870.0‡	852.7‡	110.8‡	68.0‡	66.2‡					
29 Cu	8978.9	1096.7‡	952.3‡	932.5‡	122.5‡	77.3‡	75.1‡					
30 Zn	9658.6	1196.2‡	1044.9†	1021.8†	139.8†	91.4‡	88.6‡	10.2†	10.1†			
31 Ga	10367.1	1299.0†	1143.2‡	1116.4‡	159.5‡	103.5‡	103.5‡	18.7‡	18.7‡			
32 Ge	11103.1	1414.6‡	1248.1†	1217.0†	180.1†	124.9†	120.8†	29.0†	29.0†			
33 As	11866.7	1527.0†	1359.1†	1323.6†	204.7†	146.2†	141.2†	41.7†	41.7†			
34 Se	12657.8	1652.0†	1474.3†	1433.9†	229.6†	166.5†	160.7†	55.5†	54.6†			
35 Br	13473.7	1782.0†	1596.0†	1549.9†	257†	189†	182†	70†	69†			
36 Kr	14325.6	1921.0	1730.9†	1678.4†	292.8†	222.2†	214.4	95.0†	93.8†	27.5†	14.1†	14.1†
37 Rb	15199.7	2065.1	1863.9	1804.4	326.7†	248.7†	239.1†	113.0†	112†	30.5†	16.3†	15.3†
38 Sr	16104.6	2216.3	2066.8	1939.6	358.7‡	280.3‡	270.0‡	136.0‡	134.2‡	38.9‡	20.3‡	20.3‡
39 Y	17038.4	2372.5	2155.5	2080.0	392.0†	310.6†	298.8†	157.7‡	155.8‡	43.8†	24.4†	23.1†
40 Zr	17997.6	2531.6	2306.7	2222.3	430.3‡	343.5‡	329.8‡	181.1‡	178.8‡	50.6‡	28.5‡	27.7‡
41 Nb	18985.6	2697.7	2464.7	2370.5	466.6‡	376.1‡	360.6‡	205.0‡	202.3‡	56.4‡	32.6‡	30.8‡
42 Mo	19999.5	2865.5	2625.1	2520.2	506.3‡	410.6‡	394.0‡	231.1‡	227.9‡	63.2‡	37.6‡	35.5‡
43 Tc	21044.0	3042.5	2793.2	2676.9	544†	445†	425†	257†	253†	68†	39†	39†
44 Ru	22117.2	3224.0	2966.9	2837.9	586.2‡	483.3‡	461.5‡	284.2‡	280.0‡	75.0‡	46.5‡	43.2‡
45 Rh	23219.9	3411.9	3146.1	3003.8	628.1‡	521.3‡	496.5‡	311.9‡	307.2‡	81.4‡	50.5‡	47.3‡
46 Pd	24350.3	3604.3	3330.3	3173.3	671.6‡	559.9‡	532.3‡	340.5†	335.2‡	87.6‡	55.7‡	50.9‡
47 Ag	25514.0	3805.8	3523.7	3351.1	719.0‡	603.8‡	573.0‡	374.0‡	368.0‡	97.0‡	63.7‡	58.3‡
48 Cd	26711.2	4018.0	3727.0	3537.5	772.0‡	652.6‡	618.4‡	411.9‡	405.2‡	109.8‡	63.9‡	63.9‡
49 In	27939.9	4237.5	3938.0	3730.1	827.2‡	703.2‡	665.3‡	451.4‡	443.9‡	122.7‡	73.5‡	73.5‡
50 Sn	29200.1	4464.7	4156.1	3928.8	884.7‡	756.5‡	714.6‡	493.2‡	484.9‡	137.1‡	83.6‡	83.6‡

TABLE 1 Electron Binding Energies in Electronvolts (eV) for the Elements in Their Natural Forms* (*Continued*)

Element	K1s	$L_1$2s	$L_2$2p$_{1/2}$	$L_3$2p$_{3/2}$	$M_1$3s	$M_2$3p$_{1/2}$	$M_3$3p$_{3/2}$	$M_4$3d$_{3/2}$	$M_5$3d$_{5/2}$	$N_1$4s	$N_2$4p$_{1/2}$	$N_3$4p$_{3/2}$
51 Sb	30491.2	4698.3	4380.4	4132.2	946[‡]	812.7[‡]	766.4[‡]	537.5[‡]	528.2[‡]	153.2[‡]	95.6[‡]	95.6[‡]
52 Te	31813.8	4939.2	4612.0	4341.4	1006[‡]	870.8[‡]	820.8[‡]	583.4[‡]	573.0[‡]	169.4[‡]	103.3[‡]	103.3[‡]
53 I	33169.4	5188.1	4852.1	4557.1	1072[†]	931[†]	875[†]	631[†]	620[†]	186[†]	123[†]	123[†]
54 Xe	34561.4	5452.8	5103.7	4782.2	1148.7[†]	1002.1[†]	940.6[†]	689.0[†]	676.4[†]	213.2[†]	146.7	145.5[†]
55 Cs	35984.6	5714.3	5359.4	5011.9	1211[†]	1071[†]	1003[†]	740.5[†]	726.6[†]	232.3[†]	172.4[†]	161.3[†]
56 Ba	37440.6	5988.8	5623.6	5247.0	1293[†]	1137[†]	1063[†]	795.7[†]	780.5[†]	253.5[†]	192178	.6[†]
57 La	38924.6	6266.3	5890.6	5482.7	1362[†]	1209[†]	1128[†]	853[†]	836[†]	247.7[†]	205.8	196.0[†]
58 Ce	40443.0	6548.8	6164.2	5723.4	1436[†]	1274[†]	1187[†]	902.4[†]	883.8[†]	291.0[†]	223.2	206.5[†]
59 Pr	41990.6	6834.8	6440.4	5964.3	1511.0	1337.4	1242.2	948.3[†]	928.8[†]	304.5	236.3	217.6
60 Nd	43568.9	7126.0	6721.5	6207.9	1575.3	1402.8	1297.4	1003.3[†]	980.4[†]	319.2[†]	243.3	224.6
61 Pm	45184.0	7427.9	7012.8	6459.3	—	1471.4	1356.9	1051.5	1026.9	—	242	242
62 Sm	46834.2	7736.8	7311.8	6716.2	1722.8	1540.7	1419.8	1110.9[†]	1083.4[†]	347.2[†]	265.6	247.4
63 Eu	48519.0	8052.0	7617.1	6976.9	1800.0	1613.9	1480.6	1158.6[†]	1127.5[†]	360	284	257
64 Gd	50239.1	8375.6	7930.3	7242.8	1880.8	1688.3	1544.0	1221.9[†]	1189.6[†]	378.6[†]	286	270.9
65 Tb	51995.7	8708.0	8251.6	7514.0	1967.5	1767.7	1611.3	1276.9[†]	1241.1[†]	396.0[†]	322.4[†]	284.1[†]
66 Dy	53788.5	9045.8	8580.6	7790.1	2046.8	1841.8	1675.6	1332.5	1292.6[†]	414.2[†]	333.5[†]	293.2[†]
67 Ho	55617.7	9394.2	8917.8	8071.1	2128.3	1922.8	1741.2	1391.5	1351.4	432.4[†]	343.5	308.2[†]
68 Er	57485.5	9751.3	9264.3	8357.9	2206.5	2005.8	1811.8	1453.3	1409.3	449.8[†]	366.2	320.2[†]
69 Tm	59398.6	10115.7	9616.9	8648.0	2306.8	2089.8	1884.5	1514.6	1467.7	470.9[†]	385.9[†]	332.6[†]
70 Yb	61332.3	10486.4	9978.2	8943.6	2398.1	2173.0	1949.8	1576.3	1527.8	480.5[†]	388.7[†]	339.7[†]
71 Lu	63313.8	10870.4	10348.6	9244.1	2491.2	2263.5	2023.6	1639.4	1588.5	506.8[†]	412.4[†]	359.2[†]
72 Hf	65350.8	11270.7	10739.4	9560.7	2600.9	2365.4	2107.6	1716.4	1661.7	538[†]	438.2[‡]	380.7[†]
73 Ta	67416.4	11681.5	11136.1	9881.1	2708.0	2468.7	2194.0	1793.2	1735.1	563.4[‡]	463.4[‡]	400.9[‡]
74 W	69525.0	12099.8	11544.0	10206.8	2819.6	2574.9	2281.0	1871.6	1809.2	594.1[‡]	490.4[‡]	423.6[‡]
75 Re	71676.4	12526.7	11958.7	10535.3	2931.7	2681.6	2367.3	1948.9	1882.9	625.4	518.7[‡]	446.8[‡]
76 Os	73870.8	12968.0	12385.0	10870.9	3048.5	2792.2	2457.2	2030.8	1960.1	658.2[‡]	549.1[‡]	470.7[‡]
77 Ir	76111.0	13418.5	12824.1	11215.2	3173.7	2908.7	2550.7	2116.1	2040.4	691.1[‡]	577.8[‡]	495.8[‡]
78 Pt	78394.8	13879.9	13272.6	11563.7	3296.0	3026.5	2645.4	2201.9	2121.6	725.4[‡]	609.1[‡]	519.4[‡]
79 Au	80724.9	14352.8	13733.6	11918.7	3424.9	3147.8	2743.0	2291.1	2205.7	762.1[‡]	642.7[‡]	546.3[‡]
80 Hg	83102.3	14839.3	14208.7	12283.9	3561.6	3278.5	2847.1	2384.9	2294.9	802.2[‡]	680.2[‡]	576.6[‡]
81 Tl	85530.4	15346.7	14697.9	12657.5	3704.1	3415.7	2956.6	2485.1	2389.3	846.2[‡]	720.5[‡]	609.5[‡]
82 Pb	88004.5	15860.8	15200.0	13035.2	3850.7	3554.2	3066.4	2585.6	2484.0	891.8[‡]	761.9[‡]	643.5[‡]
83 Bi	90525.9	16387.5	15711.1	13418.6	3999.1	3696.3	3176.9	2687.6	2579.6	939[‡]	805.2[‡]	678.8[‡]
84 Po	93105.0	16939.3	16244.3	13813.8	4149.4	3854.1	3301.9	2798.0	2683.0	995[†]	851[†]	705[†]
85 At	95729.9	17493	16784.7	14213.5	4317	4008	3426	2908.7	2786.7	1042[†]	886[†]	740[†]
86 Rn	98404	18049	17337.1	14619.4	4482	4159	3538	3021.5	2892.4	1097[†]	929[†]	768[†]
87 Fr	101137	18639	17906.5	15031.2	4652	4327	3663	3136.2	2999.9	1153[†]	980[†]	810[†]
88 Ra	103921.9	19236.7	18484.3	15444.4	4822.0	4489.5	3791.8	3248.4	3104.9	1208[†]	1057.6[†]	879.1[†]
89 Ac	106755.3	19840	19083.2	15871.0	5002	4656	3909	3370.2	3219.0	1269[†]	1080[†]	890[†]
90 Th	109650.9	20472.1	19693.2	16300.3	5182.3	4830.4	4046.1	3490.8	3332.0	1330[†]	1168[†]	966.4[†]
91 Pa	112601.4	21104.6	20313.7	16733.1	5366.9	5000.9	4173.8	3611.2	3441.8	1387[†]	1224[†]	1007[†]
92 U	115606.1	21757.4	20947.6	17166.3	5548.0	5182.2	4303.4	3727.6	3551.7	1439[†]	1271[†]	1043.0[†]
48 Cd	11.7[‡]	10.7[‡]										
49 In	17.7[‡]	16.9[‡]										
50 Sn	24.9[‡]	23.9[‡]										
51 Sb	33.3[‡]	32.1[‡]										
52 Te	41.9[‡]	40.4[‡]										
53 I	50[†]	50[†]										
54 Xe	69.5[†]	67.5[†]	—	—	23.3[†]	13.4[†]	12.1[†]					
55 Cs	79.8[†]	77.5[†]	—	—	22.7	14.2[†]	12.1[†]					
56 Ba	92.6[‡]	89.9[‡]	—	—	30.3[‡]	17.0[‡]	14.8[‡]					
57 La	105.3[†]	102.5[†]	—	—	34.3[†]	19.3[†]	16.8[†]					
58 Ce	109[†]	—	—	—	37.8	19.8[†]	17.0[†]					
59 Pr	115.1[†]	115.1[†]	—	—	37.4	22.3	22.3					
60 Nd	120.5[†]	120.5[†]	—	—	37.5	21.1	21.1					
61 Pm	120	120	—	—	—	—	—					

TABLE 1   Electron Binding Energies in Electronvolts (eV) for the Elements in Their Natural Forms* (*Continued*)

Element	K1s	$L_1 2s$	$L_2 2p_{1/2}$	$L_3 2p_{3/2}$	$M_1 3s$	$M_2 3p_{1/2}$	$M_3 3p_{3/2}$	$M_4 3d_{3/2}$	$M_5 3d_{5/2}$	$N_1 4s$	$N_2 4p_{1/2}$	$N_3 4p_{3/2}$
62 Sm	129	129	—	—	37.4	21.3	21.3					
63 Eu	133	127.7†	—	—	31.8	22.0	22.0					
64 Gd	140.5	142.6†	—	—	43.5†	20	20					
65 Tb	150.5†	150.5†	—	—	45.6†	28.7†	22.6†					
66 Dy	153.6†	153.6†	—	—	49.9†	29.5	23.1					
67 Ho	160†	160†	—	—	49.3†	30.8†	24.1†					
68 Er	167.6†	167.6†	—	—	50.6†	31.4†	24.7†					
69 Tm	175.5†	175.5†	—	—	54.7†	31.8†	25.0†					
70 Yb	191.2†	182.4†	—	—	52.0†	30.3†	24.1†					
71 Lu	206.1†	196.3‡	8.9†	7.5†	57.3†	33.6†	26.7†					
72 Hf	220.0‡	211.5‡	15.9‡	14.2‡	64.2‡	38†	29.9†					
73 Ta	237.9‡	226.4‡	23.5‡	21.6‡	69.7‡	42.2†	32.7†					
74 W	255.9‡	243.5‡	33.6‡	31.4†	75.6‡	45.3†	36.8†					
75 Re	273.9‡	260.5‡	42.9†	40.5‡	83‡	45.6†	34.6†					
76 Os	293.1‡	278.5‡	53.4‡	50.7‡	84‡	58†	44.5‡					
77 Ir	311.9‡	296.3‡	63.8‡	60.8‡	95.2†	63.0†	48.0‡					
78 Pt	331.6‡	314.6‡	74.5‡	71.2‡	101‡	65.3†	51.7‡					
79 Au	353.2‡	335.1‡	87.6‡	83.9‡	107.2†	74.2‡	57.2‡					
80 Hg	378.2‡	358.8‡	104.0‡	99.9‡	127‡	83.1‡	64.5‡	9.6‡	7.8‡			
81 Tl	405.7‡	385.0‡	122.2‡	117.8‡	136†	94.6‡	73.5‡	14.7‡	12.5‡			
82 Pb	434.3‡	412.2‡	141.7‡	136.9‡	147†	106.4‡	83.3‡	20.7‡	18.1‡			
83 Bi	464.0‡	440.1‡	162.3‡	157.0‡	159.3†	119.0‡	92.6‡	26.9‡	23.8†			
84 Po	500†	473†	184†	184†	177†	132†	104†	31†	31†			
85 At	533†	507†	210†	210†	195†	148†	115†	40†	40†			
86 Rn	567†	541†	238†	238†	214†	164†	127†	48†	48†	26		
87 Fr	603†	577†	268†	268†	234†	182†	140†	58†	58†	34	15	15
88 Ra	635.9†	602.7†	299†	299†	254†	200†	153†	68†	68†	44	19	19
89 Ac	675†	639†	319†	319†	272†	215†	167†	80†	80†	—	—	—
90 Th	712.1‡	675.2‡	342.4‡	333.1‡	290†	229†	182†	92.5‡	85.4‡	41.4‡	24.5‡	16.6‡
91 Pa	743†	708†	371†	360†	310†	232†	232†	94†	94†	—	—	—
92 U	778.3‡	736.2‡	388.2†	377.4‡	321†	257†	192†	102.8‡	94.2‡	43.9‡	26.8‡	16.8‡

* Electron binding energies for the elements in their natural forms, as compiled by G. P. Williams, Brookhaven National Laboratory, in Chap. 1, Ref. 1. The energies are given in electronvolts (eV) relative to the vacuum level for the rare gases and for H_2, N_2, O_2, F_2, and Cl_2; relative to the Fermi level for the metals; and relative to the top of the valence bands for semiconductors. Values are based largely on those given by J. A. Bearden and A. F. Barr, "Reevaluation of X-Ray Atomic Energy Levels," *Rev. Mod. Phys.* **39:**125 (1967).

† Values are from M. Cardona and L. Lay (eds.), *Photoemission in Solids I: General Principles,* Springer-Verlag, Berlin, 1978.

‡ Values are from J. C. Fuggle and N. Mårtensson, "Core-Level Binding Energies in Metals," *J. Electron. Spectrosc. Relat. Phenom.* **21:**275 (1980). For further updates, consult the website http://www.cxro.lbl.gov/optical_constants/.

TABLE 2 Photon Energies, in Electronvolts (eV), of Principal K- and L-Shell Emission Lines*

Element	$K\alpha_1$	$K\alpha_2$	$K\beta_1$	$L\alpha_1$	$L\alpha_2$	$L\beta_1$	$L\beta_2$	$L\gamma_1$
3 Li	54.3							
4 Be	108.5							
5 B	183.3							
6 C	277							
7 N	392.4							
8 O	524.9							
9 F	676.8							
10 Ne	848.6	848.6						
11 Na	1,040.98	1,040.98	1,071.1					
12 Mg	1,253.60	1,253.60	1,302.2					
13 Al	1,486.70	1,486.27	1,557.45					
14 Si	1,739.98	1,739.38	1,835.94					
15 P	2,013.7	2,012.7	2,139.1					
16 S	2,307.84	2,306.64	2,464.04					
17 Cl	2,622.39	2,620.78	2,815.6					
18 Ar	2,957.70	2,955.63	3,190.5					
19 K	3,313.8	3,311.1	3,589.6					
20 Ca	3,691.68	3,688.09	4,012.7	341.3	341.3	344.9		
21 Sc	4,090.6	4,086.1	4,460.5	395.4	395.4	399.6		
22 Ti	4,510.84	4,504.86	4,931.81	452.2	452.2	458.4		
23 V	4,952.20	4,944.64	5,427.29	511.3	511.3	519.2		
24 Cr	5,414.72	5,405.509	5,946.71	572.8	572.8	582.8		
25 Mn	5,898.75	5,887.65	6,490.45	637.4	637.4	648.8		
26 Fe	6,403.84	6,390.84	7,057.98	705.0	705.0	718.5		
27 Co	6,930.32	6,915.30	7,649.43	776.2	776.2	791.4		
28 Ni	7,478.15	7,460.89	8,264.66	851.5	851.5	868.8		
29 Cu	8,047.78	8,027.83	8,905.29	929.7	929.7	949.8		
30 Zn	8,638.86	8,615.78	9,572.0	1,011.7	1,011.7	1,034.7		
31 Ga	9,251.74	9,224.82	10,264.2	1,097.92	1,097.92	1,124.8		
32 Ge	9,886.42	9,855.32	10,982.1	1,188.00	1,188.00	1,218.5		
33 As	10,543.72	10,507.99	11,726.2	1,282.0	1,282.0	1,317.0		
34 Se	11,222.4	11,181.4	12,495.9	1,379.10	1,379.10	1,419.23		
35 Br	11,924.2	11,877.6	13,291.4	1,480.43	1,480.43	1,525.90		
36 Kr	12,649	12,598	14,112	1,586.0	1,586.0	1,636.6		
37 Rb	13,395.3	13,335.8	14,961.3	1,694.13	1,692.56	1,752.17		
38 Sr	14,165	14,097.9	15,835.7	1,806.56	1,804.74	1,871.72		
39 Y	14,958.4	14,882.9	16,737.8	1,922.56	1,920.47	1,995.84		
40 Zr	15,775.1	15,690.9	17,667.8	2,042.36	2,039.9	2,124.4	2,219.4	2,302.7
41 Nb	16,615.1	16,521.0	18,622.5	2,165.89	2,163.0	2,257.4	2,367.0	2,461.8
42 Mo	17,479.34	17,374.3	19,608.3	2,293.16	2,289.85	2,394.81	2,518.3	2,623.5
43 Tc	18,367.1	18,250.8	20,619	2,424.0	—	2,536.8	—	—
44 Ru	19,279.2	19,150.4	21,656.8	2,558.55	2,554.31	2,683.23	2,836.0	2,964.5
45 Rh	20,216.1	20,073.7	22,723.6	2,696.74	2,692.05	2,834.41	3,001.3	3,143.8
46 Pd	21,177.1	21,020.1	23,818.7	2,838.61	2,833.29	2,990.22	3,171.79	3,328.7
47 Ag	22,162.92	21,990.3	24,942.4	2,984.31	2,978.21	3,150.94	3,347.81	3,519.59
48 Cd	23,173.6	22,984.1	26,095.5	3,133.73	3,126.91	3,316.57	3,528.12	3,716.86
49 In	24,209.7	24,002.0	27,275.9	3,286.94	3,279.29	3,487.21	3,713.81	3,920.81
50 Sn	25,271.3	25,044.0	28,486.0	3,443.98	3,435.42	3,662.80	3,904.86	4,131.12
51 Sb	26,359.1	26,110.8	29,725.6	3,604.72	3,595.32	3,843.57	4,100.78	4,347.79
52 Te	27,472.3	27,201.7	30,995.7	3,769.33	3,758.8	4,029.58	4,301.7	4,570.9
53 I	28,612.0	28,317.2	32,294.7	3,937.65	3,926.04	4,220.72	4,507.5	4,800.9
54 Xe	29,779	29,458	33,624	4,109.9	—	—	—	—

TABLE 2 Photon Energies, in Electronvolts (eV), of Principal K- and L-Shell Emission Lines* (*Continued*)

Element	$K\alpha_1$	$K\alpha_2$	$K\beta_1$	$L\alpha_1$	$L\alpha_2$	$L\beta_1$	$L\beta_2$	$L\gamma_1$
55 Cs	30,972.8	30,625.1	34,986.9	4,286.5	4,272.2	4,619.8	4,935.9	5,280.4
56 Ba	32,193.6	31,817.1	36,378.2	4,466.26	4,450.90	4,827.53	5,156.5	5,531.1
57 La	33,441.8	33,034.1	37,801.0	4,650.97	4,634.23	5,042.1	5,383.5	5,788.5
58 Ce	34,719.7	34,278.9	39,257.3	4,840.2	4,823.0	5,262.2	5,613.4	6,052
59 Pr	36,026.3	35,550.2	40,748.2	5,033.7	5,013.5	5,488.9	5,850	6,322.1
60 Nd	37,361.0	36,847.4	42,271.3	5,230.4	5,207.7	5,721.6	6,089.4	6,602.1
61 Pm	38,724.7	38,171.2	43,826	5,432.5	5,407.8	5,961	6,339	6,892
62 Sm	40,118.1	39,522.4	45,413	5,636.1	5,609.0	6,205.1	6,586	7,178
63 Eu	41,542.2	40,901.9	47,037.9	5,845.7	5,816.6	6,456.4	6,843.2	7,480.3
64 Gd	42,996.2	42,308.9	48,697	6,057.2	6,025.0	6,713.2	7,102.8	7,785.8
65 Tb	44,481.6	43,744.1	50,382	6,272.8	6,238.0	6,978	7,366.7	8,102
66 Dy	45,998.4	45,207.8	52,119	6,495.2	6,457.7	7,247.7	7,635.7	8,418.8
67 Ho	47,546.7	46,699.7	53,877	6,719.8	6,679.5	7,525.3	7,911	8,747
68 Er	49,127.7	48,221.1	55,681	6,948.7	6,905.0	7,810.9	8,189.0	9,089
69 Tm	50,741.6	49,772.6	57,517	7,179.9	7,133.1	8,101	8,468	9,426
70 Yb	52,388.9	51,354.0	5,937	7,415.6	7,367.3	8,401.8	8,758.8	9,780.1
71 Lu	54,069.8	52,965.0	61,283	7,655.5	7,604.9	8,709.0	9,048.9	10,143.4
72 Hf	55,790.2	54,611.4	63,234	7,899.0	7,844.6	9,022.7	9,347.3	10,515.8
73 Ta	57,532	56,277	65,223	8,146.1	8,087.9	9,343.1	9,651.8	10,895.2
74 W	59,318.24	57,981.7	67,244.3	8,397.6	8,335.2	9,672.35	9,961.5	11,285.9
75 Re	61,140.3	59,717.9	69,310	8,652.5	8,586.2	10,010.0	10,275.2	11,685.4
76 Os	63,000.5	61,486.7	71,413	8,911.7	8,841.0	10,355.3	10,598.5	12,095.3
77 Ir	64,895.6	63,286.7	73,560.8	9,175.1	9,099.5	10,708.3	10,920.3	12,512.6
78 Pt	66,832	65,112	75,748	9,442.3	9,361.8	11,070.7	11,250.5	12,942.0
79 Au	68,803.7	66,989.5	77,984	9,713.3	9,628.0	11,442.3	11,584.7	13,381.7
80 Hg	70,819	68,895	80,253	9,988.8	9,897.6	11,822.6	11,924.1	13,830.1
81 Tl	72,871.5	70,831.9	82,576	10,268.5	10,172.8	12,213.3	12,271.5	14,291.5
82 Pb	74,969.4	72,804.2	84,936	10,551.5	10,449.5	12,613.7	12,622.6	14,764.4
83 Bi	77,107.9	74,814.8	87,343	10,838.8	10,730.91	13,023.5	12,979.9	15,247.7
84 Po	79,290	76,862	8,980	11,130.8	11,015.8	13,447	13,340.4	15,744
85 At	8,152	7,895	9,230	11,426.8	11,304.8	13,876	—	16,251
86 Rn	8,378	8,107	9,487	11,727.0	11,597.9	14,316	—	16,770
87 Fr	8,610	8,323	9,747	12,031.3	11,895.0	14,770	1,445	17,303
88 Ra	8,847	8,543	10,013	12,339.7	12,196.2	15,235.8	14,841.4	17,849
89 Ac	90,884	8,767	10,285	12,652.0	12,500.8	15,713	—	18,408
90 Th	93,350	89,953	105,609	12,968.7	12,809.6	16,202.2	15,623.7	18,982.5
91 Pa	95,868	92,287	108,427	13,290.7	13,122.2	16,702	16,024	19,568
92 U	98,439	94,665	111,300	13,614.7	13,438.8	17,220.0	16,428.3	20,167.1
93 Np	—	—	—	13,944.1	13,759.7	17,750.2	16,840.0	20,784.8
94 Pu	—	—	—	14,278.6	14,084.2	18,293.7	17,255.3	21,417.3
95 Am	—	—	—	14,617.2	14,411.9	18,852.0	17,676.5	22,065.2

* Photon energies in electronvolts (eV) of some characteristic emission lines of the elements of atomic number $3 \le Z \le 95$, as compiled by J. Kortright, "Characteristic X-Ray Energies," in *X-Ray Data Booklet* (Lawrence Berkeley National Laboratory Pub-490, Rev. 2, 1999). Values are largely based on those given by J. A. Bearden, "X-Ray Wavelengths," *Rev. Mod. Phys.* **39:**78 (1967), which should be consulted for a more complete listing. Updates may also be noted at the website www-cxro.lbl.gov.

TABLE 3   Curves Showing Auger Energies, in Electronvolts (eV), for Elements of Atomic Number $3 \leq Z \leq 92$

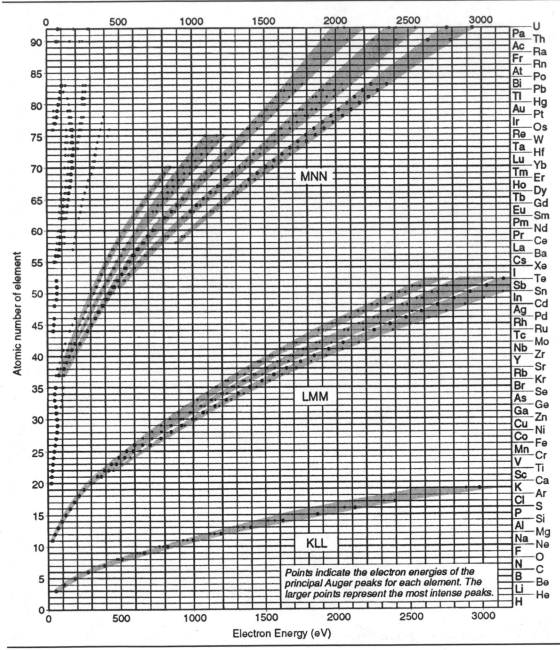

Only dominant energies are given, and only for principal Auger peaks. The literature should be consulted for detailed tabulations, and for shifted values in various common compounds.[2-4] (*Courtesy of Physical Electronics, Inc.*[2])

A.2 REFERENCE

1. B. L. Henke, E. M. Gullikson, and J. C. Davis, "X-Ray Interactions: Photoabsorption, Scattering, Transmission, and Reflection at $E = 50$–30000 eV, $Z = 1$–92, *Atomic Data and Nuclear Data Tables* **54**(2):181–342 (July 1993).

2. K. D. Childs, B. A. Carlson, L. A. Vanier, J. F. Moulder, D. F. Paul, W. F. Stickle, and D. G. Watson, *Handbook of Auger Electron Spectroscopy,* C. L. Hedberg (ed.), Physical Electronics, Eden Prairie, MN, 1995.

3. J. F. Moulder, W. F. Stickle, P. E. Sobol, and K. D. Bomben, *Handbook of X-Ray Photoelectron Spectroscopy,* Physical Electronics, Eden Prairie, MN, 1995.

4. D. Briggs, *Handbook of X-Ray and Ultraviolet Photoelectron Spectroscopy,* Heyden, London, 1977.

INDEX

ABOUT THE EDITORS

Michael Bass is Professor of Optics, Physics, and Electrical and Computer Engineering in the School of Optics/Center for Research and Education in Optics and Lasers at the University of Central Florida. He received his B.S. in physics from Carnegie-Mellon, and his M.S. and Ph.D. in physics from the University of Michigan.

Jay M. Enoch is Professor of the Graduate School and Dean Emeritus, School of Optometry, University of California at Berkeley. He also serves as a professor in the Department of Ophthalmology at the University of California at San Francisco. He received his B.S. in optics and optometry from Columbia University and his Ph.D. in physiological optics from Ohio State University.

Eric W. Van Stryland is Professor of Optics, Physics, and Electrical and Computer Engineering in the School of Optics/Center for Research and Education in Optics and Lasers at the University of Central Florida. He received his Ph.D. from the University of Arizona.

William L. Wolfe is a Professor Emeritus at the Optical Sciences Center at the University of Arizona. He received his B.S. in physics from Bucknell University, and his M.S. in physics and M.S.E. in electrical engineering from the University of Michigan.

The **Optical Society of America** is dedicated to advancing study, teaching, research, and engineering in optics.